SECOND EDITION

THE SCIENCE OF ZOOLOGY

PAUL B. WEISZ
Professor of Biology
Brown University

McGRAW-HILL BOOK COMPANY
New York St. Louis San Francisco Düsseldorf Johannesburg
Kuala Lumpur London Mexico Montreal New Delhi
Panama Rio de Janeiro Singapore Sydney Toronto

Library of Congress Cataloging in Publication Data

Weisz, Paul B.   1921–
  The science of zoology.

  Includes bibliographies.
  1. Zoology. I. Title.
QL47.2.W45 1973   591   72-4172
ISBN 0-07-069135-5

THE SCIENCE OF ZOOLOGY

  2 3 4 5 6 7 8 9 0 DODO 7 9 8 7 6 5 4 3

This book was set in Galaxy by York Graphic Services, Inc.
The editors were James R. Young, Jr., and Andrea Stryker-Rodda;
the designer was Nicholas Krenitsky; and the production
supervisor was Joe Campanella.
The printer and binder was R. R. Donnelley & Sons Company.

COLOR PHOTO CREDITS

1. Douglas P. Wilson.
2. R. H. Noailles.
3. *B*, M. C. Noailles.
4. M. C. Noailles.
5. *A*, Armed Forces Institute of Pathology.
6. General Biological Supply House, Inc.
7. General Biological Supply House, Inc.
9. *A*, *B*, R. H. Noailles. *C*, Douglas P. Wilson.
10. Douglas P. Wilson.
11. *A*, Dr. Boris Gueft. *B*, R. H. Noailles.
12. Eric V. Gravé.
13. Douglas P. Wilson.
14. Eric V. Gravé.
15. Eric V. Gravé.
16. Eric V. Gravé.
18. Douglas P. Wilson.
19. Douglas P. Wilson.
20. General Biological Supply House, Inc.
21. *B*, Douglas P. Wilson.
22. *A*, Douglas P. Wilson. *B*, General Biological Supply House, Inc.
23. Dr. Jonathan P. Green, Brown University.
24. *B*, Douglas P. Wilson. *C*, DS. Jonathan P. Green, Brown University.
25. R. H. Noailles.
26. General Biological Supply House, Inc.
27. *B*, Dr. Jonathan P. Green, Brown University.
29. R. H. Noailles.
30. *A*, Douglas P. Wilson, *B*, *D*, Ward's Natural Science Establishment, Inc. *C*, *E*, Dr. Jonathan P. Green, Brown University.

CHAPTER PHOTO CREDITS

2.3. R. H. Noailles.
2.4. *A*, American Museum of Natural History. *B*, R. H. Noailles.
2.6. Dr. Clifford Grobstein, University of California at Irvine.
2.8. General Biological Supply House, Inc.
2.10. American Museum of Natural History.
2.11. *A*, Ward's Natural Science Establishment, Inc. *B*, American Museum of Natural History, *C*, U.S. Department of Agriculture.

4.23. *A*, Eric V. Gravé. *B*, Dr. Elizabeth Leduc, Brown University.
4.24. *A*, *C*, Dr. Keith R. Porter, Harvard University. *B*, Drs. D. F. Poulson and C. W. Metz, from Fig. 1, *J. Morphol.*, vol. 63, p. 366, 1938.
4.26. *A*, Dr. Robert Brenner, Oregon Regional Primate Research Center. *B*, Dr. Keith R. Porter, Harvard University.
4.27. *A*, Dr. W. G. Whaley, University of Texas. *B*, Dr. Luis Biempica, Albert Einstein College of Medicine, N.Y.
4.28. Dr. Robert Brenner, Oregon Regional Primate Research Center.
4.29. *A*, Dr. Norman E. Williams, University of Iowa.
4.30. *B*, Dr. Dorothy R. Pitelka, University of California, Berkeley. *C*, Dr. Norman E. Williams, University of Iowa. *D*, Dr. Robert Brenner, Oregon Regional Primate Research Center.

6.1. *A*, Ward's Natural Science Establishment, Inc. *B*, Dr. Keith R. Porter, Harvard University.
6.6. *B*, American Museum of Natural History.
6.16. Dr. W. Beerman.
6.25. General Biological Supply House, Inc.
6.26. General Biological Supply House, Inc.
6.27. Dr. Herman B. Chase, Brown University.

7.1. Copyright Walt Disney Productions.
7.3. *B*, Ward's Natural Science Establishment, Inc. *C*, *D*, American Museum of Natural History.
7.4. *A*, Lynwood M. Chace.
7.6. Dr. G. W. Corner and Carnegie Institution of Washington.
7.16. *A*, General Biological Supply House, Inc. *C*, Eric V. Gravé.
7.17. *A*, General Biological Supply House, Inc. *B*, *C*, Dr. B. J. Serber.
7.18. *A*, General Biological Supply House, Inc. *B*, Ward's Natural Science Establishment, Inc.
7.19. *A*, Ward's Natural Science Establishment, Inc.
7.20. *A*, Dr. Mac E. Hadley, University of Arizona. *B*, R. H. Noailles. *C*, M. C. Noailles.
7.21. *A*, Ward's Natural Science Establishment, Inc. *B*, M. C. Noailles.

8.2. *A*, R. H. Noailles. *B*, Dr. William Montagna, Oregon Regional Primate Research Center.
8.4. *A*, Dr. William Montagna, Oregon Regional Primate Research Center.
8.5. Ward's Natural Science Establishment, Inc.
8.7. *B*, M. C. Noailles.
8.10. M. C. Noailles.
8.11. *A*, Ward's Natural Science Establishment, Inc.
8.13. *A*, Ward's Natural Science Establishment, Inc.
8.15. *B*, Ward's Natural Science Establishment, Inc. *D*, Dr. Jan Cammermeyer, National Institutes of Health, Bethesda, Md., and Zeitschrift für Anatomie und Entwicklungsgeschichte, vol. 124, pp. 543–561, 1965.
8.16. Inset, General Biological Supply House, Inc.
8.17. *C*, Ward's Natural Science Establishment, Inc.
8.18. *C*, *D*, Rhode Island Hospital Photographic Department.
8.20. Ward's Natural Science Establishment, Inc.
8.27. *A*, Ward's Natural Science Establishment, Inc. *B*, General Biological Supply House, Inc.
8.28. *B*, Dr. William Montagna, Oregon Regional Primate Research Center.
8.35. *B*, Ward's Natural Science Establishment, Inc.
8.38. *D*, Paul Popper, Ltd.
8.39. *D*, Ward's Natural Science Establishment, Inc.

9.2. *B*, Dr. B. J. Serber. *C*, General Biological Supply House, Inc.
9.3. *B*, Dr. Boris Gueft.
9.4. *C*, Rhode Island Hospital Photographic Department. *D*, Dr. Boris Gueft.
9.9. Dr. Robert Brenner, Oregon Regional Primate Research Center.
9.15. *A*, *D*, Dr. B. J. Serber. *B*, *C*, M. C. Noailles. *E*, General Biological Supply House, Inc.
9.23. *A*, Ward's Natural Science Establishment, from model by Dr. J. F. Mueller. *B*, Bell Telephone Laboratories, Inc.
9.31. *A*, General Biological Supply House, Inc.

10.2. *A*, *B*, Douglas P. Wilson. *C*, Ward's Natural Science Establishment, Inc.
10.7. *B*, R. H. Noailles.
10.11. *B*, *C*, General Biological Supply House, Inc.
10.19. *A*, J. Van Wormer. *B*, Jane Burton.
10.20. Gerhard Marcuse.
10.21. *A*, Ward's Natural Science Establishment, Inc.
10.22. *A*, Ward's Natural Science Establishment, Inc.

11.1. Dr. Richard J. Goss and M. W. Stagg, from J. Exp. Zool., vol. 137, p. 9, 1958.
11.2. Dr. Clifford Grobstein and 13th Growth Symposium, Princeton University Press, 1954.
11.3. *A*, *C*, Dr. Charles Thornton, Michigan State University. *B*, Dr. Richard J. Goss, Brown University.
11.5. *A*, Dr. Roberts Rugh and Burgess Publishing Company.

11.11. *D*, General Biological Supply House, Inc.
11.14. *A*, Dr. Roberts Rugh, from "Experimental Embryology," Harcourt, Brace and World, Inc.
11.15. Dr. Dietrich Bodenstein, from figs. 2 and 3, J. Exp. Zool., vol. 108, pp. 96, 97.
11.16. Drs. S. R. Detwiler and R. H. Van Dyke, from fig. 16, J. Exp. Zool., vol. 69, p. 157.
11.17. *B*, *C*, *D*, Douglas P. Wilson.
11.21. L. E. Perkins, Natural History Photographic Agency.
11.25. *B*, Dr. A. Gesell, from fig. 10, "The Embryology of Behavior," Harper and Row, Publishers, Inc.

12.1. *B*, *C*, R. H. Noailles.
12.13. Dr. W. Beerman.
12.20. Photo, U.S. Department of Agriculture.

13.5. New York Zoological Society.
13.6. New York Zoological Society.
13.7. *B*, U.S. Fish and Wildlife Service.
13.8. *A*, *B*, *C*, American Museum of Natural History. *D*, Chicago Natural History Museum.

15.1. American Museum of Natural History.
15.3. *A*, American Museum of Natural History. *B*, Chicago Natural History Museum.
15.5. *A*, *B*, Chicago Natural History Museum. *C*, American Museum of Natural History.
15.6. American Museum of Natural History.
15.8. American Museum of Natural History.
15.9. American Museum of Natural History.
15.10. *A*, Peabody Museum, Yale University. *B*, American Museum of Natural History.
15.11. *A*, American Museum of Natural History. *B*, Chicago Natural History Museum.
15.13. American Museum of Natural History.
15.14. Chicago Natural History Museum.
15.15. Chicago Natural History Museum.
15.16. American Museum of Natural History.
15.17. American Museum of Natural History.
15.18. American Museum of Natural History.
15.20. American Museum of Natural History.
15.21. *A*, Paul Popper, Ltd. *B*, Lynwood M. Chace.
15.22. *A*, W. Suschitzky. *B*, American Museum of Natural History.
15.25. *A*, adapted from painting by Peter Bianchi, copyright National Geographic Society, 1961. *B*, American Museum of Natural History.
15.26. American Museum of Natural History.
15.27. American Museum of Natural History.
15.28. *A*, American Museum of Natural History. *B*, Chicago Natural History Museum.
15.31. American Museum of Natural History.

16.1. Ron Church, photo Researchers, Inc.
16.2. Ringling Bros. and Barnum and Bailey Circus photo.
16.3. Ringling Bros. and Barnum and Bailey Circus photo.
16.4. Wisconsin Regional Primate Research Center.
16.6. Leonard L. Rue, from the National Audubon Society.
16.9. Stan Menscher, McGraw-Hill, Inc.
16.12. *A*, American Museum of Natural History. *B*, Ward's Natural Science Establishment, Inc. *C*, U.S. Fish and Wildlife Service.
16.13. U.S. Department of Agriculture.
16.14. *A*, *B*, *C*, U.S. Department of Agriculture. *D*, Buffalo Museum of Science.
16.15. *D*, Dr. Phyllis J. Dolhinow, University of California, Berkeley.
16.17. Irven DeVore.
16.18. U.S. Department of Agriculture.
16.19. Dr. N. Tinbergen.
16.20. Jeanne White, from the National Audubon Society.
16.21. U.S. Fish and Wildlife Service.
16.22. Dr. N. Tinbergen.

17.2. New York Zoological Society.
17.3. American Museum of Natural History.
17.4. *A*, Standard Oil Company, N.J. *B*, Jeanne White, from the National Audubon Society.
17.15. *A*, S. Dalton, Natural History Photographic Agency. *B*, U.S. Department of Agriculture. *C*, Paul Popper, Ltd.
17.16. *A*, New York Zoological Society. *B*, Paul Popper, Ltd.
17.17. Dr. R. W. G. Wyckoff, from "Electron Microscopy," Interscience Publishers, Inc., 1949.
17.18. *A*, Leonard L. Rue. *B*, Armed Forces Institute of Pathology.

18.3. Douglas P. Wilson.
18.4. American Museum of Natural History.
18.5. *A*, U.S. Fish and Wildlife Service. *B*, *C*, R. H. Noailles.
18.6. *A*, U.S. Department of Agriculture. *B*, *C*, copyright Walt Disney Productions.
18.7. *A*, U.S. Forest Service. *B*, South African Tourist Corporation. *C*, British Overseas Airways Corporation.
18.8. *A*, U.S. Forest Service. *B*, National Park Service.
18.9. *A*, U.S. Forest Service. *B*, copyright Walt Disney Productions. *C*, Lynwood M. Chace.
18.10. *A*, U.S. Forest Service. *B*, Leonard L. Rue. *C*, W. Suschitzky.
18.11. *A*, U.S. Fish and Wildlife Service. *B*, W. Suschitzky. *C*, copyright Walt Disney Productions.
18.12. *A*, Paul Popper, Ltd. *B*, Australian News and Information Bureau.
18.16. U.S. Department of Agriculture.
18.22. Kit and Max Hunn, John H. Gerard, Maurice E. Landre, from the National Audubon Society.
18.23. World Health Organization, United Nations.

19.7. *B*, Natural History Photographic Agency.
19.10. *A*, R. H. Noailles.
19.16. Eric V. Gravé.
19.18. *B*, Dr. Roman Vishniac.
19.24. *B*, Dr. Maria A. Rudzinska and J. Gerontol., vol. 16, p. 213, 1961.

20.2. *A*, M. Woodbridge Williams. *B*, *C*, American Museum of Natural History.
20.17. *B*, M. Woodbridge Williams.
20.18. *C*, M. C. Noailles.
20.19. *B*, Douglas P. Wilson.
20.21. Ward's Natural Science Establishment, Inc.
20.22. *A*, American Museum of Natural History.
20.23. *A*, Douglas P. Wilson.
20.25. Douglas P. Wilson.
20.28. Douglas P. Wilson.

21.3. *A*, General Biological Supply House, Inc.
21.7. *A*, Ward's Natural Science Establishment, Inc. *B*, Douglas P. Wilson.
21.9. Photo, Ward's Natural Science Establishment, Inc.
21.11. *A*, Ward's Natural Science Establishment, Inc.
21.12. *A*, *B*, *C*, General Biological Supply House, Inc.
21.13. *A*, Douglas P. Wilson.
21.17. *A*, M. C. Noailles. *B*, American Museum of Natural History.
21.22. *A*, Dr. W. F. Mai, Cornell University.
21.25. *A*, Ward's Natural Science Establishment, Inc. *B*, United Nations photo.

22.4. *E*, Douglas P. Wilson.
22.9. *A*, Douglas P. Wilson. *B*, R. H. Noailles.
22.10. *A*, General Biological Supply House, Inc.
22.12. *A*, General Biological Supply House, Inc.

23.4. *A*, American Museum of Natural History.
23.6. *C*, Douglas P. Wilson.
23.8. R. H. Noailles.

CONTENTS

x

Inasmuch as knowledge, like life, always changes, zoology courses too are subject to continuous change. Indeed, while the basic function of the introductory courses remains what it has always been, the thematic classroom setting for performing this function has changed particularly rapidly in the last few years. Thus, from any scientific or pedagogic standpoint it is now indefensible (and in fact hardly possible) to ignore the new fields of study in which fundamental and exciting advances continue to be made—biochemistry, genetics, and the molecular basis of life in general; or to ignore the even newer fields that currently also happen to have great social significance—animal behavior, ecology and the environmental crisis, and the work on populations and races.

Quite as emphatically, however, well-established zoological knowledge cannot be sacrificed in the teaching schedule either. A wholly ''topical'' course, without strong underpinnings of morphology, physiology, embryology, taxonomy, and phylogeny, can hardly amount to more than a misguided overreaction to some sophomoric notion of ''relevance.'' On the contrary, the really relevant point still is the ancient realization that new knowledge becomes meaningful only on the basis of already established understanding—just as old knowledge acquires ever fresh significance in the light of new insights.

Accordingly, this edition attempts to bring into balance the areas of study that are or by now have become standard and those that today are in the forefront of research and general interest. Some topics—particularly the basic chemical ones—have been condensed and streamlined to standardized form, and the space thereby gained has been employed to present new subject matter: a phylogenetic reassessment of body cavities and metameric divisions; an account of the social and racial status of man; the zoology of individual and social behavior; the crisis aspects of ecology; and several more. Most other sections of the book have been rewritten, moreover, so that hardly any paragraph now remains unchanged.

Nevertheless, the basic organization of the book still is the same as before. Unit 1 again represents an examination of zoological principles, and Unit 2, a systematic study of animal groups. In the first unit the six Parts of the previous edition are retained, although, as noted, the chapter contents in most cases are altered greatly. Part 1 again provides a scientific, zoological, and chemical introduction. Part 2, now rewritten extensively, deals with the organization and operation of animal cells—their chemical and physical structure and their metabolic and self-perpetuative functions. Part 3 focuses on the whole animal—body forms, morphology and physiology of organ systems, and interrelations between taxonomy, embryology, and adult anatomy.

The historical continuity of animals is the subject of Part 4, which concentrates on reproduction, development, and heredity, and of Part 5, which covers evolution, phylogeny, and paleontology. These portions contain accounts on the origins of life, of animals, and of the main animal groups, and they also include a new section on the history of man and the zoology and genetics of race. Part 6, on animal associations, begins with a new chapter on behavior and social groupings and continues with chapters on species, ecosystems, and the large-scale environment. A discussion of the ecological crisis concludes this part and the whole unit. As in the previous edition, therefore, the first unit again forms a detailed outline of the common animal heritage, level by level from molecule to biosphere.

Unit 2 now contains 12 chapters arranged in three parts, the first on noncoelomate types, the second on protostomial coelomates, and the third on deuterostomial coelomates. Phylum characteristics again are described as far down as the ordinal ranks, and the connecting theme of evolution and embryology has been preserved throughout the unit. Even so, this portion of the book likewise has undergone a thorough recasting and rewriting.

All the approximately 500 line diagrams have been executed anew for this edition, and the use of color screens now enhances both the clarity and the interest of the line art. The book also contains well over 400 black-and-white photographs and 16 pages of full-color photos. The review questions and lists of collateral readings at the end of each chapter are revised and updated, and nearly every reading entry now also includes a brief description of the topic covered. A detailed two-part glossary and an index are at the end of the book, as before, and the Instructor's Manual has been recast to conform with the new text.

Like the first, this edition should be adaptable to widely different course structures and should permit flexible use in the classroom. The hoped-for goal in any event is to eradicate rote memorizing of empty ''facts'' as *the* aim of academic work, and to promote instead a real comprehension of zoological knowledge; far too often is education equated simply with ''information transfer,'' and not, as it should be, with conceptual synthesis and *understanding*.

PAUL B. WEISZ

Despite their numerous and often obvious differences, animals of all kinds have far more in common than might at first be suspected. For example, all are alike in that they occupy particular parts of the environment; in that they have particular past histories; in that they have bodies that reflect both their present places in the environment and their historical past; and, above all, in that the events which make them *living* creatures are very similar in all. In effect, it is only in their more superficial traits that animals differ to greater or lesser degree; in their most fundamental respects all animals are, and indeed must be, very much alike.

All the common features represent an animal heritage, a set of life-maintaining properties inherited from early, nonanimal, ancestors. The first unit of this book, roughly half of the whole volume, is devoted to a study of this common heritage of animals. The objective is to examine the structural parts and functional processes that underlie all life generally and animal life particularly. The second unit then deals with the specific animals as such, the actual embodiments of the ancestral heritage.

UNIT 1
THE HERITAGE
OF ANIMALS

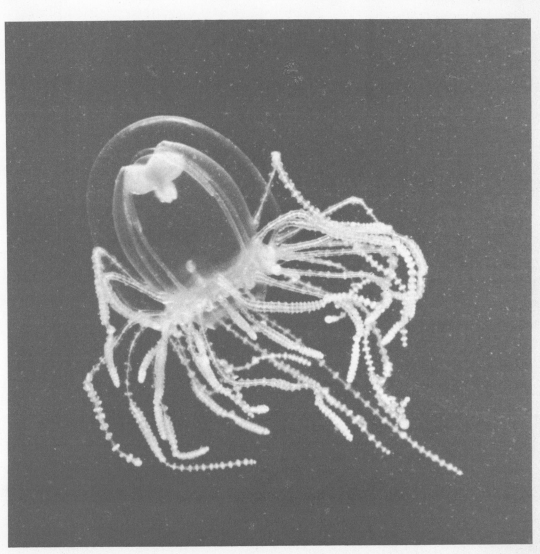

The investigation of animal life is the concern of the science of zoology. As a science, zoology is interrelated closely with all other natural sciences, chemistry in particular. Therefore, a study of zoology appropriately can begin with three questions. First, what is a science? Second, what do we mean by "living" and by "animal?" And third, what areas of chemistry and other natural sciences are important in understanding animal life?

Answers to these questions are outlined in the three chapters of this first part.

PART 1
ANIMAL STUDIES

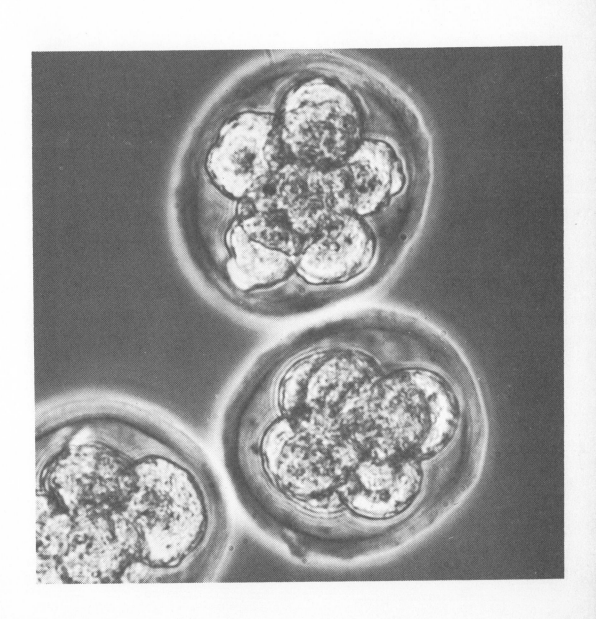

1

SCIENCE AND ZOOLOGY

It is sometimes useful to distinguish two forms of science, basic research, or *pure* science, and technology, or *applied* science. Basic research promotes our understanding of how the universe and its parts operate, and technology puts the results of basic research to practical uses. However, every scientific activity actually has both pure and applied aspects. Every pure scientist depends on equipment produced by technology, and every technologist depends on the ideas and insights generated by basic research. The relative emphasis between pure and applied work can vary greatly, yet both kinds are always required in any science.

Methods of Science

Everything that is science is ultimately based on some *scientific method*. Taken singly, most of the steps of such a method involve commonplace procedures carried out daily by every person. Taken together, they amount to the most powerful tool man has devised for learning about nature and making natural processes serve human purposes.

Observation, Problem, Hypothesis

Science generally begins with *observation*, the usual first step of scientific inquiry. This step immediately limits the scientific domain; something that cannot be observed, directly or indirectly, cannot be investigated by science. Furthermore, for reasons that will become clear presently, it is necessary that an observation be *repeatable*, actually or potentially. One-time events are outside science (the one-time origin of the universe possibly excepted).

Correct observation is a most difficult art, acquired only after long experience and many errors. Everyone observes, with eyes, ears, touch, and all other senses, but few observe correctly. The problem here is largely unsuspected bias. People forever see what they *want* to see or what they think they *ought* to see. It is extremely hard to rid oneself of such unconscious prejudice and to see just what is actually there, no more and no less. Past experience, ''common knowledge,'' and often teachers can be subtle obstacles to correct observation, and even experienced scientists may not always avoid them. That is why a scientific observation is not taken at face value until several scientists have repeated it independently and have reported matching data. That is also a major reason why one-time, unrepeatable events generally cannot be investigated scientifically.

After an observation has been made, a second usual step of scientific procedure is to define a *problem;* one asks a question about the observation. How does so and so come about? What is it that makes such and such happen in this or that fashion? Question asking again distinguishes the scientist from the layman; everyone makes observations, but not everyone has the curiosity to go further. Indeed, few become aware that a particular observation actually might pose a problem. For thousands of years, even curious people simply took it for granted that a detached, unsupported object falls to the ground. It took genius to ask "How come?" and not many problems have ever turned out to be more profound.

Thus, scientists take nothing for granted and they ask questions, even at the risk of irritating others. Question askers are notorious for getting themselves into trouble, and so it has always been with scientists. But they have to continue to ask questions if they are to remain scientists, and society has to expect annoying questions if it wishes to have science.

Like good observing, good questioning is a high art. To be valuable scientifically, a question must be *relevant* as well as *testable*. Often it is difficult or impossible to tell in advance whether a question is relevant or irrelevant, testable or untestable. If a man collapses on the street and others want to help him, it may or may not be irrelevant to ask when the man had his last meal. Without experience one cannot decide on the relevance of this question, and a wrong procedure might be followed. As to testability, it is clear that proper testing techniques must be available. But this cannot always be guaranteed. For example, Einstein achieved fame for showing that it is impossible to tell whether or not the earth moves through an "ether," an assumption held for many decades. All questions about an ether therefore became untestable. Einstein reformulated such questions and came up with relativity, an idea that posed fully testable problems.

In general, science does best with "how" or "what" questions. "Why" questions are more troublesome. Some of them can be rephrased to ask "How?" or "What?" But others, such as "Why does the universe exist?," fall into the untestable category. These are outside the domain of science.

Once a proper question has been asked, the third step of scientific methodology usually involves the seemingly quite unscientific procedure of guessing what the answer to the question might conceivably be. Scientists refer to this as postulating a *hypothesis*. Hypothesizing distinguishes the scientist still further from the layman. For while many people observe and ask questions, most stop there. Some do wonder about likely answers, and scientists are among these.

Since a particular question usually has thousands of possible answers but in most cases only a single right one, chances are excellent that a random guess will be wrong. The scientist will not know if his guess was or was not correct until he has completed the fourth step of scientific inquiry, *experimentation*. It is the function of experiments to test the validity of scientific guesses. If experiments show that a first guess was wrong, the scientist then must formulate a new or modified hypothesis and perform new experiments. Clearly, guessing and guess testing could go on for years and a right answer might not be found. This happens.

But here artistry, genius, and knowledge of the field usually provide shortcuts. There are good guesses and bad ones, and the experienced scientist is generally able to decide at the outset that, of a multitude of possible answers, so and so many are unlikely answers. This is also the place where hunches, intuitions, and lucky accidents aid science enormously. The ideal situation the scientist will strive for is to reduce his problem to just two distinct alternative possibilities. Experimental tests should then answer one of these with a clear "yes," the other with a clear "no." It is exceedingly difficult to streamline problems in this way, and with many it cannot be done. Very often the answer obtained is "maybe." But if a clear yes or no does emerge, the result well might be a milestone in science.

Experiment, Theory

With the next general step in a scientific inquiry, *experimentation*, science and nonscience part company completely. Most people observe, ask questions, and also guess at answers. But the layman then stops: "My answer is so logical, so reasonable, and it sounds so 'right,' that it must be correct." The listener considers the argument, finds that it is indeed logical and reasonable, and is convinced. He goes out and in his turn converts others. Before long, the whole world rejoices that it has the answer.

Now the small, killjoy voice of the scientist is heard in the background: "Where is the evidence?" Under such conditions in history, it has often been easier and more convenient to ignore the scientist than to change emotionally fixed public opinions. But disregarding the scientist does not alter the fact that answers without evidence are at best wishful thinking, at worst fanatical illusions. Experiments *can* provide the necessary evidence.

Experimenting is the hardest part of the scientific process. There are few rules, and each experiment must be tackled in its own particular way. The general nature of an experiment can be illustrated by the following example. Suppose that a chemical substance *X* has spilled accidentally into a culture dish full of certain disease-causing bacteria, and you observe that this chemical kills all the bacteria in the dish. Problem: can drug *X* be used to protect human beings against these disease-causing bacteria? Hypothesis: yes. Experiment: you find a patient with that bacterial disease and inject some of the drug into him. One possible result is that the patient gets well fairly quickly, in which case you would consider your hypothesis confirmed. Another possible result is that the patient remains ill or dies, and you would then conclude that your drug is worthless or dangerous.

However, in this example the so-called experiment was not really an experiment at all. First, no allowance was made for the possibility that different people might react differently to the same drug. Obviously, one would have to test the drug on many patients. Besides, one would make preliminary tests on mice or guinea pigs or monkeys. Second, the quantity of drug to be used was not determined. Clearly, a full range of dosages would have to be tested. Third, and most important, no account was taken of the possibility that your patient might have recovered (or died) even without your injecting the drug. What is needed here is *experimental control;* for every group of patients treated with the drug solution, a precisely equal group must be treated with a plain solution that does not contain the drug. Then, by comparing the control and the experimental groups, one can determine to what extent the results are actually attributable to the drug.

Every experiment thus requires at least two parallel sets of tests identical in all respects except one. One set is the control series that provides a standard of reference for assessing the results of the experimental series. In drug experiments on people, up to 100,000 to 200,000 tests, half of them controls, half of them experimentals, must sometimes be performed. Such a drug-testing program is laborious, expensive, and time-consuming, but the design of the experiment is nevertheless extremely simple. There are few steps to be gone through, and it is fairly clear what these steps must be. By contrast, experiments in many cases do not take more than an hour or two, whereas thinking up appropriate, foolproof plans for the tests can take several years.

And despite a most ingenious design and a most careful execution, the result still may not be a clear yes or no. In a drug-testing experiment, for example, it is virtually certain that some patients in the experimental group will not recover and that some of the untreated control patients will get better. The actual results might be something like 70 percent recovery in the experimentals and perhaps 20 percent recovery in the controls. In other words, 30 percent of the experimentals do not recover despite treatment, and 20 percent of the controls get well even without treatment. The drug therefore is effective in only 70 minus 20, or 50 percent, of the cases.

Such a result can be a major medical accomplishment, for having the drug is obviously better than not having it. Scientifically, however, one is confronted with an equivocal "maybe" result. It will probably lead to research based on the new observation that some people respond to the drug and some do not, and to the new problem of why and what can be done about it.

The result of any experiment represents *evidence;* the original guess about the answer to a problem is confirmed as correct or is invalidated. If invalidated, a new hypothesis and new experiments must be thought up. This process must be repeated until a hypothesis is hit upon that can be supported with confirmatory experimental evidence.

As with legal evidence, scientific evidence can be strong and convincing, or merely suggestive, or poor. In any case nothing has been "proved." Depending on the strength of the evidence, one merely obtains a basis for regarding the original hypothesis with a certain degree of confidence. Our new drug, for example, may be just what we claim it to be when we use it in this country. In another part of the world it might not work at all or it might work better. All we can confidently say is that our evidence is based on so and so many local experiments, and that we have shown the drug to have an effectiveness of 50 percent. Experimental results are never better or broader than the experiments themselves.

This is where many who have been properly scientific up to this point become unscientific. Their claims exceed the evidence; they mistake their partial answer for the whole answer; they contend that they have "proof" for a "fact," while all they actually have is some evidence for a hypothesis. There is always room for more and better evidence, or for new contradictory evidence, or indeed for better hypotheses.

Experimental evidence is the basis for a fifth general step in scientific procedure, the formulation of a *theory*. In our drug example, a simple theory would be the statement that "against such and such a bacterial disease, drug *X* is effective in 50 percent of the cases." To be sure, this statement cannot be regarded as a particularly significant or far-reaching theory. Nevertheless, it implies, for example, that drug *X* will be 50 percent effective anywhere in the world, under any conditions, and can be used also for animals other than man. Direct evidence for these extended implications does not exist. But inasmuch as drug *X* is already known to work within certain limits, the theory expresses the belief, or *probability*, that it will also work within certain wider limits.

To that extent every good theory has *predictive* value; it forecasts certain results. In contrast to nonscientific predictions, scientific ones always have a substantial body of evidence to back them up. Moreover, a scientific forecast does not say that something will certainly happen, but says only that something is likely to happen with a stated degree of probability.

A few theories have proved to be so universally valid and to have such a high degree of probability that they are spoken of as *natural laws*. For example, no exception has ever been found to the observation that an apple disconnected from a tree and not otherwise supported will fall to the ground. A law of gravitation is based on such observations. Yet even laws do not pronounce certainties. For all practical purposes it well might be irrational to assume that some day an apple will rise from a tree, but there simply is no evidence that can absolutely guarantee the future. Evidence can be used only to estimate probabilities.

Most theories have rather brief life spans. For example, if our drug *X* should be found to perform not with 50 percent but with 80 percent efficiency in chickens, then our original theory becomes untenable and obsolete. The exception to the theory now becomes a new observation, the start of a new cycle of scientific investigation. New research might show, for example, that chickens contain a substance in their blood that enhances the action of the drug substantially. This finding might lead to isolation, identification, and mass production of the booster substance, hence to world-wide improvement in curing the bacterial disease. And we would also have a new theory of drug action, based on the new evidence.

Thus, science is never finished. One theory predicts and holds up well for a time, exceptions are found, and a new, more inclusive theory takes over—for a while. Science is steady progression, not sudden revolution. Clearly, knowledge of scientific methodology does not by itself make a good scientist, any more than knowledge of English grammar alone makes a Shakespeare. At the same time, the demands of scientific inquiry should make it evident that scientists cannot be the cold, inhuman precision machines they are so often and so erroneously pictured to be. Scientists are essentially artists who require a sensitivity of eye and of mind as great as that of any master painter, and an imagination and keen inventiveness as powerful as that of any master poet.

Limitations of Science

Observing, problem-posing, hypothesizing, experimenting, and theorizing—these are the most common procedural steps in scientific investigations. To determine what science means in wider contexts, we must examine what scientific methodology implies and, more especially, what it does not imply.

Aims and Values

First, scientific investigation defines the domain of science. Anything that is amenable to scientific investigation, now or in the future, is or will be within the domain of science; anything that is not amenable to such investigation is not within the scientific domain.

An awareness of these limits can help us avoid many inappropriate controversies. For example, does the idea of God lend itself to scientific scrutiny? Suppose we wish to test the hypothesis that God is universal and exists everywhere and in everything. Being untested as yet, this hypothesis could be right or wrong. An experiment about God then would require experimental control, that is, two situations, one with God and one without, but otherwise identical.

If our hypothesis is correct, God would indeed exist everywhere. Hence He would be present in every test we could possibly make, and we would never be able to devise a situation in which God is not present. Yet we need such a situation for a controlled experiment. But if our hypothesis is wrong, He would not exist and would therefore be absent from any test we could possibly make. We would then never be able to devise a situation in which God *is* present. Yet we would need such a situation for a controlled experiment.

Right or wrong our hypothesis is untestable, since we cannot run a controlled experiment. Therefore, we cannot carry out a scientific investigation. The point is that the concept of God falls outside the domain of science, and science cannot legitimately say anything about Him. It should be carefully noted that this is a far cry from saying "Science disproves God," or "Scientists must be godless; their method demands it." Nothing of the sort. Science specifically leaves anyone perfectly free to believe in any god whatsoever or in none. Many first-rate scientists are priests; many others are agnostics. Science commits us to nothing more and to nothing less than adherence to the ground rules of proper scientific inquiry.

It may be noted that such adherence is a matter of faith, just as belief in God or confidence in the telephone directory is a matter of faith. Whatever other faiths they may or may not hold, all scientists certainly have strong faith in scientific methodology. So do those laymen who feel that having electric lights and not having bubonic plague are good things.

A second consequence of scientific methodology is that it defines the aim and purpose of science. The objective of science is to make and to use theories. Many believe that the objective of science is to discover "truth," to find out "facts." We must be very careful here about the meaning of words. The word "truth" is popularly used in two senses. It may indicate a temporary correctness, as in saying, "It is true that my hair is brown." Or it may indicate an absolute, eternal correctness, as in saying, "In plane geometry, the sum of the angles of a triangle is 180°."

From the earlier discussion of the nature of scientific investigation, it should be clear that science cannot deal with truth of the absolute variety. Something absolute is finished, known completely once and for all, and nothing further needs to be found out. Science can only supply evidence for theories, and "theory" is simply another word for relative truth. Because the word "truth" is ambiguous if not laboriously qualified, scientists try

not to use it at all. The words "fact" and "proof" have a similar drawback. Both can indicate either something absolute or something relative. If absolute they are not science; if relative, we actually deal with evidence. Thus, science is content to find evidence for theories, and it does not deal with truths, proofs, or facts.

A third important implication of scientific methodology is that it does not make value judgments or moral decisions. Very often, of course, we do place valuations on scientific results, but such assessments are human valuations and different people frequently assess the same results quite differently. Scientific results by themselves do not contain any built-in values, and nowhere in scientific inquiry is there a value-revealing step.

Thus the science that produces weapons for healing and creating and weapons for destroying and killing cannot of itself determine if such tools are good or bad. The decision in each case rests on the moral opinions of humanity, those of scientists included. Similarly, beauty, love, evil, happiness, virtue, justice, liberty, financial worth—all these are human values about which science as such is silent and noncommittal. For the same reason, it also would be folly to strive for a strictly "scientific" way of life or to expect strictly "scientific" government. To be sure, the role of science well might be enlarged in areas of personal and public life where science can make a legitimate contribution. But a civilization that adhered exclusively to the rules of scientific methods could never tell, for example, whether it is right or wrong to commit murder or whether it is good or bad to love one's neighbor. Science cannot and does not give such answers. This circumstance does not mean, however, that science does away with morals. The implication merely is that science cannot determine if one ought to have moral standards, or what particular set of moral standards one ought to live by.

Scientific Philosophy

A fourth and most important consequence of scientific methodology is that it determines the philosophical foundation on which scientific pursuits must be based.

Since the domain of science is the whole material universe, science must inquire into the nature of the forces that govern the universe and all happenings in it. What makes given events in the universe take place? What determines which event out of many possible ones will occur? And what controls or guides the course of any event to a particular conclusion?

We already know the framework within which the scientific answers to such questions must be given. If certain answers can be verified wholly or even partly through experimental analysis, they will be valuable scientifically. But answers that cannot be so verified will be without value in science, even though they well may be valuable in other human concerns.

Vitalism versus Mechanism   In the course of history two major types of answer have been proposed about the governing forces of the universe. They are incorporated in two systems of philosophy called *vitalism* and *mechanism.*

Vitalism is a doctrine of the supernatural. It holds, essentially, that the universe, and particularly its living components, are controlled by supernatural powers. Such powers have been variously called gods, spirits, or simply ''vital forces.'' Their influence is held to guide the behavior of atoms, planets, stars, living things, and indeed of all components of the universe. Most religious philosophies are vitalistic ones.

Whatever value a vitalistic philosophy might have elsewhere, it cannot have value in science because the supernatural is by definition beyond reach of the natural. Inasmuch as scientific inquiry deals with the natural world, it cannot be used to investigate the supernatural. As already noted, for example, science cannot prove or disprove anything about God. Any other vitalistic concept is similarly untestable by experiment and is therefore unusable as a *scientific* philosophy of nature.

A philosophy that *is* usable in science is the idea of mechanism. According to this view the universe is governed by a set of natural laws, the laws of physics and chemistry man has discovered by experimental analysis. The mechanistic philosophy holds that if all physical and chemical events in the universe can be accounted for, no other events will remain. Therefore, life too must be a result of physcial and chemical processes *only,* and the course of life must be determined automatically by the physical and chemical occurrences in living matter.

These differences between vitalism and mechanism clearly point up a conceptual conflict between religion and science. However, this conflict is not necessarily irreconcilable. To bridge the conceptual gap, one might ask how the natural laws of the universe came into being to begin with. A possible answer is that they were created by God. On this view, it could be argued that the universe ran vitalistically up to the time that natural laws were created and ran mechanistically thereafter. The mechanist would then have to admit the existence of a supernatural Creator at the beginning of time (even though he has no *scientific* basis for either affirming or denying this; mechanism cannot, by definition, tell anything about a time at which natural laws might not have been in operation). Correspondingly, the vitalist would have to admit that, so long as the natural laws continue to operate without change, supernatural control would not be demonstrable.

Thus it is not necessarily illogical to accept both scientific and religious philosophies at the same time. However, it is decidedly illogical to try to use religious ideas as explanations of scientific problems or scientific ideas as explanations of religious problems. Correct science does demand that supernatural concepts be kept out of those natural events that can be investigated scientifically. However much a vitalist he might be in his nonscientific thinking, man in his scientific thinking must be a mechanist. And if he is not he ceases to be scientific.

Many people, some scientists included, actually find it exceedingly difficult to keep vitalism out of science. Biological events, undoubtedly the most complex of all known events in the universe, have in the past been particularly subject to attempts at vitalistic interpretation. How, it has been asked, can the beauty of a flower ever be understood simply as a series of physical and chemical events? How can an egg, transforming itself into a baby, be nothing more than a ''mechanism'' like a clock? And how can a man who thinks and who experiences visions of God conceivably be regarded as nothing more than a piece of ''machinery''? Mechanism *must* be inadequate as an explanation of life, it has been argued, and only something supernatural superimposed on the machine, some vital force, is likely to account for the fire of life.

In such replacements of mechanistic with mystical thought, the connotations of words often play a supporting role. For example, the words ''mechanism'' and ''machine'' usually bring to mind images of crude iron engines or clockworks. Such analogies tend to reinforce the suspicion of vitalists that those who regard living things as mere machinery must be simple-minded indeed. Consider, however, that the machines of today also include electronic computers that can learn, translate languages, compose music, play chess, make decisions, and improve their performance of such activities as they gather experience. In addition, theoretical knowledge now available would permit us to build a machine that could heal itself when injured and that could feed, sense, reproduce, and even evolve. Thus the term "mechanism" is not at all limited to crude, stupidly "mechanical" engines. And there is certainly nothing inherently simple-minded or reprehensible in the idea that living things are exquisitely complex chemical mechanisms, some of which even have the capacity to think and to have visions of God.

On the contrary, if it could be shown that such a mechanistic view is at all justified, it would represent an enormous advance in our understanding of nature. In all the centuries of recorded history, vitalism in its various forms has hardly progressed beyond its original assertion that living things are animated by supernatural forces. Just how such forces are supposed to do the animating has not been explained, nor have programs of inquiry been offered to find explanations. Such inquiries actually are ruled out by definition, since natural man can never hope to fathom the supernatural. In the face of this closed door, mechanism provides the only way out for the curious. But is it justifiable to regard living things as pure mechanisms, even complicated chemical ones?

A mechanistic interpretation of life turns out to be entirely justifiable, and interjection of touches of vitalism is entirely unjustifiable. Science today *can* account for many living properties in purely mechanistic terms. Moreover, biologists are well on their way to being able to create a truly living entity "in the test tube," solely by means of physical and chemical procedures obeying known natural laws. We shall discuss some of the requirements for such laboratory creation in the course of this book. Evidently, vitalistic "aids" to explain the mechanistic universe are unnecessary and, because they do not foster inquiry, also unjustifiable.

It may be noted in this connection that, historically, vitalism has tended to fill the gaps left by incomplete scientific knowledge. Early man was a complete vitalist, who for want of better knowledge regarded even inanimate objects as "animated" by supernatural spirits. As scientific insight

later increased, progressively more of the universe was reclaimed from the domain of the supernatural. Thus it happened repeatedly that events originally thought to be supernatural, many living events included, were later shown to be explicable naturally. And those today who may still be prompted to fill gaps in scientific knowledge with vitalism must be prepared to have red faces tomorrow. We conclude that a mechanistic view of nature is one philosophic attitude required in science. A second can now be considered.

Teleology versus Causalism   Even a casual observer must be impressed by the apparent nonrandomness of natural events. Every part of nature seems to follow a plan, and there is a definite directedness to any given process. In living processes, for example, developing eggs behave *as if* they knew exactly what the plan of the adult is to be. A chicken soon produces two wings and two legs, *as if* it knew that these appendages were to be part of the adult. All known natural processes, living or otherwise, similarly start at given beginnings and proceed to particular endpoints. This observation poses a philosophical problem: how is a starting condition directed toward a specific terminal condition; how does a starting point appear to "know" what the endpoint is to be?

Such questions have to do with a detailed aspect of the more general problem of the controlling forces of the universe. We should expect, therefore, that two sets of answers would be available, one vitalistic and the other mechanistic. This is the case. According to vitalistic doctrines, natural events *appear* to be planned because they actually *are* planned. A supernatural "divine plan" is held to fix the fate of every part of the universe, and all events in nature, past, present, and future, are programmed in this plan. All nature is therefore directed toward a preordained goal, the fulfillment of the divine plan. As a consequence, nothing happens by chance but everything happens on purpose.

Being a vitalistic, experimentally untestable concept, the notion of purpose in natural events has no place in science. Does the universe exist for a purpose? Does man live for a purpose? We cannot hope for an answer from science, for science is not designed to tackle such questions. Moreover, if we already hold certain beliefs in these areas, we cannot expect science either to prove or to disprove them for us.

Yet many arguments have been attempted to show purpose from science. For example, it has been maintained by some that the whole purpose of the evolution of living things was to produce man—the predetermined goal from the very beginning. This conceit implies not only that man is the finest product of creation, but also that nothing could ever come after man, for he is supposed to be the last word in living magnificence. As a matter of record, man is sorely plagued by an army of parasites that cannot live anywhere except inside people. And it is clear that you cannot have a man-requiring parasite before you have a man.

Many human parasites did evolve after man. Thus, the purpose argument at best would show that the whole purpose of evolution was to produce living organisms causing influenza, diphtheria, gonorrhea, and syphilis. Even the most ardent purpose arguer would probably not care to maintain such a conclusion. If one is so inclined, he is of course perfectly free to believe that man is the pinnacle of it all. But one cannot maintain that such beliefs are justified by evidence from science. The essential point is that a statement that such and such is the purpose or goal of any material thing or event is to state a belief, not evidence obtained by scientific inquiry. Nowhere in such inquiries is there a purpose-revealing step.

The form of argumentation that has recourse to purposes and supernatural planning generally is called *teleology*. In one system of teleology, the preordained plan exists outside natural objects, in an external Deity, for example. In another system, the plan resides within objects themselves. According to this view, a starting condition of an event proceeds toward a particular end condition because the starting object has built into it supernatural foreknowledge of the end condition. For example, an egg develops toward the goal of the adult because the egg is endowed with information about the precise nature of the adult state. Clearly, this and all other forms of teleology "explain" an end state by simply asserting it to be already mapped out at the beginning. And in thereby putting the future in the past, the effect before the cause, teleology negates time.

The scientifically useful alternative to teleology is *causalism,* a form of thought based on mechanistic philosophy. Causalism denies foreknowledge of terminal states, preordination, purposes, goals, and fixed fates. It holds instead that natural events take place stepwise, each one conditioned by and dependent on earlier ones. Events occur only as previous events *permit* them to occur, not as preordained goals or purposes make them occur. End states are consequences, not foregone conclusions, of beginning states. A headless earthworm regenerates a new head because conditions in the headless worm are such that only a head—*one* head—can develop. It becomes the task of the zoologist to find out what these conditions are and to see if, by changing the conditions, two heads or another tail can be produced. Because scientists actually *can* obtain different end states by changing the conditions of initial states, the idea of predetermined goals loses all validity in scientific thought.

Care must therefore be taken in scientific endeavors not to fall unwittingly into the teleological trap. Consider often-heard statements such as "The *purpose* of the heart is to pump blood"; "The ancestors of birds evolved wings *so that* they could fly"; "Eggs have yolk *in order to* provide food for development." The last statement, for example, implies that eggs can "foresee" that food will be required in development and they therefore store up some. In effect, eggs are given human mentality. The teleologist is always anthropocentric, that is, he implies that the natural events he discusses are governed by minds like his. In making zoological statements, some of the teleological implications can be avoided by replacing every "purpose" with "function," every "so that" or "in order to" with "and."

Clearly then, science in its present state of development must operate within carefully specified, self-imposed limits. The basic philosophic attitude must be mechanistic and causalistic, and we note that the results obtained through science are inherently without truth, without value, and without purpose. But it is precisely because science is limited in this fashion that it advances. Truth is as subjective as ever, values change with time and place, and purposes basically express little more than man's desire to make the universe behave according to his own very primitive understanding. It therefore has proved difficult to build a knowledge of nature on the shifting foundations of values and truths or on the dogma of purpose. What little of nature we really know and are likely to know in the foreseeable future stands on the bedrock of science.

The Voice of Science

Fundamentally science is a *language,* a system of communication comparable to the systems of religion, art, politics, English, or French. Like the latter, science enables man to travel in new countries of the mind, and to understand and be understood in such countries. Like other languages, moreover, science has its grammar—the methods of scientific inquiry; its authors and its literature—the scientists and their written work; and its various dialects or forms of expression—physics, chemistry, and zoology, for example.

Indeed, science is one of the few truly universal languages, understood all over the globe. Art, religion, and politics are also universal. But each of these has several forms, with the result that Baptists and Hindus, for example, have little in common either religiously, artistically, or politically. By contrast, science has the same single form everywhere, and Baptists and Hindus do speak the same scientific language.

None of these systems of communication is "truer" or "righter" than any other. They are only *different* systems, each serving its function in its own domain. Many an idea is an idiom of a specific language and is best expressed in that language. For example, one cannot discuss morality in the language of science, thermodynamics in the language of religion, or artistic beauty in the language of politics. To the extent that each system of communication has its own idioms, there is no overlap or interchangeability among the systems.

But many ideas can be expressed equally well in several languages. The English "water," the Latin "*aqua,*" and the scientific "H_2O" are entirely equivalent, and no one of these is truer or more correct than the others. They are merely different. Similarly, in one language man was created by God, in another he is a result of chance reactions among chemicals and of evolution. Again, neither the scientific nor the religious interpretation is the truer. The theologian might argue that everything was made by God, including scientists who think that man is the result of chance chemical reactions. The scientist will then argue back that chance chemical reactions created men with brains, including those theological brains that can conceive of a god who made everything. The impasse is permanent, and within their own systems

of communication the scientist and the theologian are equally right. Many, of course, assume without warrant that it is the compelling duty of science to prove or disprove religious beliefs and of religion, to prove or to disprove scientific theories.

The point is that there is no single "correct" formulation of any idea that spans various languages. There are only different formulations, and in given circumstances one or the other may be more useful, more satisfying, or more effective. Clearly, anyone who is adept in more than one language will be able to travel that much more widely, and he will be able to feel at ease in the company of more than one set of ideas.

We are, it appears, forever committed to multiple standards, according to the different systems of communication we use. But to be multilingual in his interpretation of the world has been the unique heritage of man from the beginning. Though different proportions of the various languages are usually mixed in the outlook of different individuals, science, religion, art, politics, spoken language, all these and many more are always needed to make a full life.

In the language of science as a whole, one important dialect is *biology,* the domain of *living things.* Biology includes *zoology* as a major subdialect, the domain of the animal world. Man probably was a zoologist very early during his history. His own body in health and illness; the phenomena of birth, growth, and death; and the other animals that gave him food and clothing undoubtedly were matters of serious concern to even the first of his kind. The motives were sheer necessity and the requirements of survival. These same motives still prompt the same zoological studies today; animal husbandry, fisheries, veterinary science, wildlife management, medicine, and various fields allied to these are among the important branches of modern applied zoology. In addition, zoology today also is strongly experimental, pure research being done extensively all over the world. Some of this research promotes zoological technology; all of it increases our understanding of how animals are constructed and how they operate.

Over the decades the frontiers of zoological investigation have been extended to smaller and smaller realms. Some 100 to 150 years ago, when modern zoology began, the chief interest was the whole animal, how it lived, where it could be found, and how it was related to other whole living things. Such studies have been carried on ever since. In addition, techniques have gradually become available for the investigation of progressively smaller parts of the whole, their structures, their functions, and their interrelations. As a result, the frontiers of zoology have been extended down to the chemical level during the last few decades. And while research with larger living units continues as before, the newest zoology attempts to interpret living operations in terms of the chemicals out of which animals are constructed. Zoology today therefore attempts to show how chemicals are put together to form, on the one hand, something like a rock or a piece of metal and, on the other, something like a cat or a dog or a man.

This book is an outline of how successful the attempt has been thus far.

Review Questions

1. What are the aims and the limitations of science? Review fully. In what sense is science a language, and how does it differ from other languages?

2. Review the procedures commonly employed in scientific inquiries and discuss the nature of each such procedure. Define controlled experiment.

3. How would you show by controlled experiment whether or not (*a*) temperature affects the rate of growth of living things, (*b*) houseflies can perceive differently colored objects, (*c*) earthworms use up some of the soil they grow in?

4. Suppose it were found in the experiment described in question 3*a* that, at an environmental temperature of 28°C, fertilized frog eggs develop into tadpoles roughly twice as fast as at 18°C. What kinds of theories could such evidence suggest?

5. Which of the ideas you have previously held about science should you now, after studying this chapter, regard as popular misconceptions?

6. Can you think of observations or problems that have so far not been investigated scientifically? Try to determine in each case whether or not such investigation is inherently possible.

7. Describe the philosophic foundations of science. Define mechanism and causalism and contrast these systems of thought with those of vitalism and teleology. Can conceptual conflicts between science and religion be reconciled?

8. Consider the legal questions "Do you swear to tell the truth, the whole truth, and nothing but the truth?" and "Is it not a fact that on the night of . . . ?" If questions of this sort were to be used in a strictly scientific context, how should they properly be formulated?

9. Zoology is one of the so-called *natural sciences,* all of which deal with the composition, properties, and behavior of matter in the universe. What other sciences are customarily regarded as belonging to this category, and what distinguishes them from one another and from zoology?

10. Do natural sciences differ procedurally from (*a*) social sciences, (*b*) mathematics? Explain.

Collateral Readings

Bronowski, J.: "Science and Human Values," Harper Torchbooks 505, Harper & Row, New York, 1959. A well-known paperback, containing a stimulating discussion of the role of science in modern society.

Butterfield, H.: The Scientific Revolution, *Sci. American,* Sept., 1960. A popularly written historical survey of the growth of science since the time of the Renaissance.

Conant, J. B.: "Modern Science and Modern Man," Columbia, New York, 1952.

————: "Science and Common Sense," Yale, New Haven, Conn., 1951.

————: "On Understanding Science," Yale, New Haven, Conn., 1947.

    In these three works a noted educator discusses the scientific method, its application in research, and the role of science in society.

Mausner, B., and J. Mausner: A Study of the Anti-scientific Attitude, *Sci. American*, Feb., 1955. The issue of water fluoridation is analyzed.

Russell, Bertrand: "The Scientific Outlook," Norton, New York, 1931. A famous philosopher writes most penetratingly on the philosophic and logical foundations of science and on the relation of science to religion.

Terman, L. M.: Are Scientists Different? *Sci. American*, Jan., 1955. A psychologist examines the traits generally characteristic of scientists and compares them with those of nonscientists.

Wilson, E. Bright, Jr.: "An Introduction to Scientific Research," McGraw-Hill, New York, 1952. A good discussion of the nature of the scientific method and its application in scientific investigations.

2

LIFE
AND ANIMAL

An animal is a particular kind of *organism,* an individual creature. Animals share with all other kinds of organisms the property of being or having been alive. What does "being alive" actually signify, and what is an "organism," animal or otherwise?

Surely the most obvious difference between something living and something nonliving is that the first *does* certain things the second does not do. The essence of "living" evidently lies in characteristic activities, or processes, or *functions*.

"Nonliving" could mean either "dead" or "inanimate," terms that are not equivalent. If a chicken does not perform its living functions it is dead, but then it is still readily distinguishable from an inanimate object such as a stone. Chickens, either living or dead, are organisms; stones are not. All organisms are put together in such a way that the functions of life are or once were actually possible. Accordingly, the essence of "organism" lies in particular building materials and building patterns, or *structures*.

A "living organism," therefore, is what it is by virtue of its functions and structures; the functions endow it with the property of life, and the structures permit the life-sustaining functions to be executed. What are these functions and structures?

The Nature of Life

A main activity of organisms is *nutrition,* a process that provides the raw materials for maintaining life. All living matter depends unceasingly on such raw materials, for the very act of living continuously uses up two basic commodities, energy and matter. In this respect a living organism is like a mechanical engine or indeed like any other action-performing system in the universe. Energy is needed to power the system, to make the parts operate, to keep activity going—in short, to maintain function. And matter is needed to replace parts, to repair breakdowns, to continue the system intact and *able* to function—in short, to maintain structure. Therefore, by its very nature as an action-performing unit, a "living" organism can remain alive only if it continuously expends energy and matter. Both must be replenished from the outside through nutrition.

The external raw materials used in this function are *nutrients*. One general class of nutrients includes water, salts, and other materials obtainable directly from the physical environment of the earth. Another class comprises *foods*, available in the biological environment—microorganisms, plants, and animals, living or dead.

Nutrients are chemicals, and as such they contain chemical energy. All living matter is maintained on the chemical energy obtained from nutrients. The nutrients become decomposed inside an organism, through a series of energy-yielding chemical reactions; and the energy generated by these reactions sustains living activities. In this respect living systems are in principle quite similar to man-made machines. In a gasoline or steam engine, for example, fuel is decomposed by burning, and this process releases energy that drives the motor. In the living "motor" nutrients likewise function as fuels; indeed, foods and engine fuels belong to the same families of chemical substances. Moreover, foods too are decomposed by a form of burning, and the energy so generated then drives the living "machine."

In living organisms, the process of obtaining energy through decomposition of foods is called *respiration*. In most cases respiration requires environmental oxygen as an accessory ingredient, and the collection of oxygen by an organism is a phase of *breathing*. Respiration is a second major activity of living matter; it is the basic power-generating process that maintains *all* living functions—including nutrition and even respiration itself. Continued nutrition and respiration depend on energy made available through previous nutrition and respiration.

The second main role of nutrients is to serve as construction materials. The whole structure of the living organism must be built from and kept intact with nutrient "bricks." In effect, the chemical stuff of living matter is fundamentally the same as that of nutrients. This consideration leads to an interesting inference. If nutrients and living matter are basically equivalent and if nutrients also are respiratory fuels, it follows that living matter should be able to use *itself* as fuel. This is indeed the case; all living matter is inherently self-decomposing and self-consuming. The living "motor" cannot tell the difference between external fuel and internal structural parts, because both are fundamentally the same. Organisms therefore are unstable structurally, but they counteract this instability with the aid of nutrients. New structural parts are manufactured continually out of nutrients, and the new parts replace those that burn away. New parts also make up for those that are occasionally lost through injury and disease and those that rub off, evaporate, or dissipate in other ways.

Evidently, living matter is never the same from moment to moment. As wear and tear and reconstruction occur side by side, the substance of living matter always "turns over": the structural pattern remains the same, but almost every bit of the building material is replaced sooner or later. In a man, complete turnover of this sort is estimated to take about seven years. Moreover, if new building materials accumulate faster than old ones wear away, the living organism will *grow*. Growth is a characteristic outcome of nutrient use in the construction of living matter. The processes by which nutrients are fashioned into new structural parts can be referred to collectively as *synthesis* activities. They represent a third basic function of all living things. Like other functions of life synthesis requires energy, and respiration must provide it.

The three functions of nutrition, respiration, and synthesis together constitute a broad living activity known as *metabolism*. Taken as a whole, metabolism is roughly equivalent to the operation of the living machinery. Metabolism also permits this machinery to continue in operation; a system that nourishes, respires, and synthesizes is capable of undertaking more nutrition, more respiration, and more synthesis (Fig. 2.1).

However, actual continuation of metabolism requires *control*. In this respect living matter is again like an engine. Continuous engine operation demands that the different parts of the engine act in harmony and become adjusted or readjusted in response to internal or external events that might change engine performance. In the same manner, continuation of metabolism in living matter depends on harmonized activity. Metabolizing by itself is not equivalent to "living," but controlled metabolizing in a general sense is.

The necessary control is provided by *self-perpetuation*, a broadly inclusive set of processes. Self-perpetuation ensures that the metabolizing machinery continues to run indefinitely, and despite internal and external happenings that might otherwise alter or stop its operation (Fig. 2.2).

The most direct regulation of metabolism is brought about by the self-perpetuative function of *steady-state control*. This function permits a living organism to receive *information* from within itself and from the external environment, and to act on this information in a (usually) self-preserving manner. The information is partly inherited, partly received in the form of *stimuli;* and the ensuing actions are *responses*. For example, a common stimulus is a decrease in internal nutrient supplies,

Figure 2.1 Metabolism. The main processes of metabolism and their interrelations.

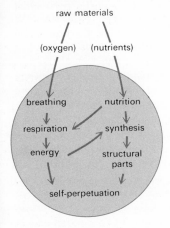

raw materials

(oxygen) (nutrients)

breathing / nutrition

respiration / synthesis

energy / structural parts

self-perpetuation

Figure 2.2 Self-perpetuation. The main processes of self-perpetuation and their interrelations.

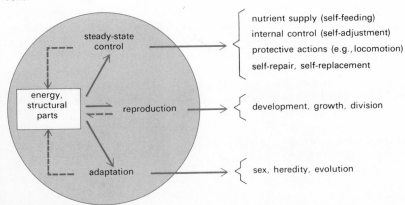

steady-state control → nutrient supply (self-feeding) / internal control (self-adjustment) / protective actions (e.g., locomotion) / self-repair, self-replacement

energy, structural parts

reproduction → development, growth, division

adaptation → sex, heredity, evolution

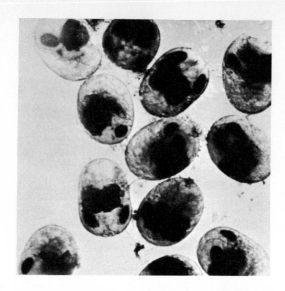

Figure 2.3 Reproduction: growth in size and number. These prawn embryos contribute to growth in numbers of individuals; and growth in size eventually will enable each to become adult and then in turn to contribute to growth in numbers. Reproduction thus can maintain the living succession indefinitely.

and the typical response is procurement of more nutrients from the environment. Or if the stimulus is produced by a situation of external danger, the response could be a protective activity such as movement away from danger. By thus making possible appropriate responses to given stimuli, the internal controls preserve *adequate operating conditions* in living matter; they make adjustments that result in a steady, living state in which an organism can remain intact and functioning.

But the span of existence of an organism is invariably limited. Death is a built-in attribute of living matter because the parts of the organism that maintain steady states are themselves subject to breakdown or destruction. When some of its controls become inoperative, the organism suffers *disease.* Diseases can be regarded generally as failures of steady-state controls, or as temporary *unsteady* states. Other, still intact controls can often initiate self-repair, but sooner or later control breakdown is so extensive that repairs are no longer possible. The organism then is in an irreversibly unsteady state and it must die.

However, before it dies it can bring into play a second major self-perpetuative function, *reproduction.* With the help of energy and raw materials the living organism can enlarge, and this growth in size prepares the way for later growth in numbers. Reproduction in a sense anticipates and compensates for unavoidable individual death. By means of reproduction successive generations are produced and in this manner life can be carried on indefinitely (Fig. 2.3).

Reproduction implies a still poorly understood capacity of *rejuvenation.* The material out of which the offspring is made—part of the parent—is really just as old as the rest of the parent. Yet the one lives and the other dies. Evidently there is a profound distinction between "old" and "aged." Reproduction also implies the capacity of *development,* for the offspring is almost always not only smaller than the parent but also less nearly complete in form and function.

As generation succeeds generation through reproduction, long-term environmental changes are bound to have major effects on the living succession. In the course of thousands and millions of years, for example, climates can change profoundly; ice ages may come and go; mountains, oceans, vast tracts of land can appear and disappear. Moreover, living organisms themselves in time alter the nature of a locality in major ways. Consequently, two related organisms many generations apart could find themselves in greatly different environments. And whereas the steady-state controls of the ancestor might have coped effectively with the early environment, such controls could be overpowered rapidly by the new environment. In the course of many generations, therefore, organisms must change *with* the environment if they are to continue in existence. They actually do change, through *adaptation.* As will become apparent later, this self-perpetuative function itself consists of three subfunctions, *sex, heredity,* and *evolution* (Fig. 2.4 and see Fig. 2.2).

Self-perpetuation as a whole therefore comprises three basic kinds of activities. First, steady-state controls maintain appropriate operating conditions in each organism. Second, reproduction ensures a continuing succession of individual organisms. And third, adaptation molds and alters the members of this succession in relation to slowly changing environments. Self-perpetuation thus adds the dimension of time to metabolism; regardless of how the environment might change in time, self-perpetuation virtually guarantees the continuation of metabolism. Metabolism in turn makes possible uninterrupted self-perpetuation, and the system so able to metabolize and to perpetuate itself can persist indefinitely; it becomes a "living" system.

The fundamental meaning of "living" can now be defined: *any structure that metabolizes and perpetuates itself is alive.* Moreover, the metabolic functions of nutrition, respiration, and synthesis

A

B

Figure 2.4 Adaptation: change with the environment. The upper figure is a drawing of a placoderm, a type of fish long extinct but very common some 300 million years ago. Fishes of this group were the ancestors of modern fish, of which one, a speckled brook trout, is shown in the lower figure. Most of the differences between ancestor and descendant, here as in other cases, appear to be a result of adaptation to changes in the physical and biological environment.

make possible, and are themselves made possible by, the self-perpetuative functions of steady-state control, reproduction, and adaptation.

A first implication of this definition is that, by their very nature, living systems collectively are highly permanent kinds of matter, perhaps the most permanent in the universe. They are certainly the most enduring on earth. Every inanimate or dead object on earth sooner or later decomposes and crumbles to dust under the impact of the environment. But every living object metabolizes and perpetuates itself and consequently can avoid such a fate. Oceans, mountains, even whole continents have come and gone several times during the last 3 billion years, but living matter has continued indestructibly during that time and, indeed, has become progressively more abundant.

A second implication is that any structure that does not satisfy the above definition in every particular is either inanimate or dead. Life must cease if even one of the functions of metabolism or self-perpetuation ceases. This criterion of life offers an instructive contrast to the operation of modern machines, many of which perform functions that also occur in living organisms. As noted, for exam-

ple, a machine can take on "nourishment" in the form of fuel and raw materials. The fuel can be "respired" to provide operating energy, and, with this energy the raw materials can then be "synthesized" to nuts, bolts, and other structural components out of which such a machine might be built. Evidently, machines can carry out activities fully equivalent to those of metabolism. Like living systems, moreover, many automated machines have ingenious steady-state controls built into them. For example, such controls could make a machine automatically self-feeding and self-adjusting.

But no machine is as yet self-protecting, self-repairing, or self-healing to any major extent, and no machine certainly is capable of growing. Furthermore, whereas living matter can reproduce before death, machines cannot. It is in this capacity of reproduction that living systems differ most critically from inanimate ones. However, the theoretical knowledge of how to build a self-reproducing machine now exists. A device of this kind would metabolize, maintain steady states, and eventually "die" but, before that, would produce "offspring." It would be almost living. If it had the additional capacity of adaptation, it would be fully living. Here too the theoretical know-how is already available. On paper, machines have been designed that could carry out "sexual" processes of a sort, that could pass on hereditary characteristics to their "progeny," and that could "evolve" and change their properties in the course of many "generations." If such machines should actually be built some day then the essential distinction between "living" and "machine" will have disappeared.

This consideration points up a third implication of the definition above: the property of life basically does not depend on a particular substance. *Any substance of whatever composition will be "living"* provided that it metabolizes and perpetuates itself. It happens that only one type of such a substance is now known. We call it "living matter," or often also *protoplasm,* and it exists in the form of organisms. But if some day we should be able to build a fully metabolizing and self-perpetuating system out of nuts, bolts, and wires, then it too will have to be regarded as being truly alive. Similarly, if some day out in space we should encounter a metabolizing and self-perpetuating being made up in a hitherto completely unknown way, it too will have to be considered living. It will not be "life as we know it," or life based on the earthly variety

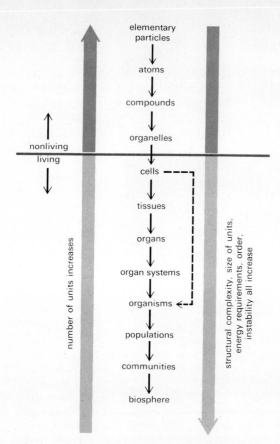

elementary
particles

↓

atoms

↓

compounds

↓

organelles

nonliving
——————————
living

cells

↓

tissues

↓

organs

↓

organ systems

↓

organisms

↓

populations

↓

communities

↓

biosphere

number of units increases

structural complexity, size of units,
energy requirements, order,
instability all increase

Figure 2.5 Hierarchy of levels in the organization of matter. The biosphere represents the sum total of all living things on earth.

of protoplasm, but in any case it will be truly living if it metabolizes and perpetuates itself.

It can be concluded that an object is defined as living or nonliving on the basis of its functional properties, not its structural ones. On the other hand, structural properties do determine whether an object is an "organism" or not. Linguistically as well as biologically, the root of "organism" is *organization:* a characteristic *structural order*.

The Nature of Organisms

Levels of Organization

The smallest structural units of all forms of matter, living matter included, are *subatomic particles*— mainly electrons, protons, and neutrons. The next larger units are *atoms,* each of which consists of subatomic particles. Atoms in turn form still more complex combinations called *chemical compounds;* and the latter are variously joined as even more elaborate *complexes of compounds*.

These units can be regarded as representing successively higher *levels of organization* of matter. They form a pyramid, or hierarchy, in which any given level contains all lower levels as components and is itself a component of all higher levels. For example, atoms contain subatomic particles as components, and atoms are themselves components of chemical compounds (Fig. 2.5).

All structural levels up to and including those of complexes of compounds are encountered in both the nonliving and the living world. For example, two familiar chemical compounds found in living as well as nonliving matter are water and table salt. Examples of complexes of compounds in the nonliving world are rocks, which are composed of several types of compounds (water and table salt among them). Mere differences in physical bulk do not necessarily indicate differences in level of organization. Thus, a pebble and a mountain range have the same organizational level, that of complexes of compounds. In living matter, complexes of compounds often occur as microscopic and submicroscopic bodies called *organelles*. But even in their most elaborate and complicated forms, neither organelles nor complexes of compounds of any kind can qualify as living units. The level of life is reached only with the next higher structural level, that of *cells*.

A cell is a specific combination of organelles, a usually microscopic bit of matter organized just complexly enough to contain all the necessary apparatus for the performance of metabolism and self-perpetuation. A cell in effect represents the least elaborate known structure that can be fully alive. It follows that a living organism must consist of at least one cell. Indeed, *unicellular* organisms probably constitute the majority of living creatures on earth. All other organisms are *multicellular,* each composed of up to many trillions of joined cells.

Several distinct levels of organization can be distinguished within multicellular organisms. The simplest multicellular types contain comparatively few cells. If all such cells are more or less alike, the organism often is referred to as a cellular *colony*. If two or more groups of different cells are present, each group usually forms a *tissue*. Structurally more complex organisms contain several tissues, and some of these tissues are joined further as one or more units called *organs*. The most complex organisms consist of many tissues and organs, and groups of organs are united even further as one or more *organ systems*. Thus, living

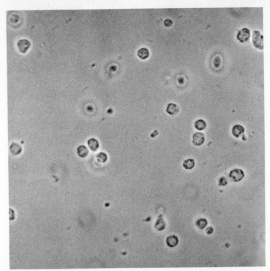

Figure 2.6 Life and levels of organization. These loose tissue cells of a mouse embryo, cultured in a nutrient solution, originally were part of a compact tissue. Disaggregation destroyed the tissue level of organization but did not destroy the cellular level; the individual cells shown here remain alive.

organisms exhibit at least five levels of structural complexity: the single-celled form, the colonial form, the organism with tissues, the type with organs, and the type with organ systems.

Several still higher levels of life can be distinguished beyond the organism level. A few individual organisms of one kind together sometimes make up a *family*. Groups of families of one kind often form a *society*. Groups of families, societies, or simply large numbers of organisms of a particular kind make up a geographically localized *population*. All populations of the same kind together form a *species*. Several different species are represented in a local *community*. And the sum of all local communities represents the whole living world.

This hierarchial organization of matter permits formulation of a structural definition of life and nonlife. Up to and including the level of the complex of compounds, matter is nonliving. At all higher levels matter can be living, provided that at each such level metabolic and self-perpetuative functions are carried out. For example, if a society is to be living it must metabolize and perpetuate itself on its own level as well as on every lower level down to that of the cell.

Moreover, as life is organized by levels so is death. Structural death occurs when one level is

disrupted or decomposed to the next lower. For example, if a tissue is disjoined as separate cells, the tissue ceases to exist. However, the lower levels need not necessarily be affected. Thus the separated cells of a tissue often carry on individually; or if a family is disrupted, the member organisms still can survive on their own (Fig. 2.6). By contrast, death at a given level often does entail death at higher levels as well. If many or all of its tissues are destroyed, the whole organ will be destroyed; or if many or all of its families are dismembered, a society may cease to exist. In general, the situation is comparable to a pyramid of cards. Removal of a top card need not affect the rest of the pyramid, but removal of a bottom card usually topples the whole structure. Evidently, neither life nor death is a singular state but is organized and structured by levels.

This hierarchy of levels provides a rough outline of the past history of matter. The universe as a whole is now believed to have begun in the form of subatomic particles. These then became joined as atoms and formed galaxies, stars, and planets. The atoms of planets later gave rise to chemical compounds and complexes of compounds. On earth, some of the complexes of compounds eventually produced living matter in the form of cells, and unicellular types were the ancestors of multicellular types. Among the latter, colonial types arose first, forms with organ systems last (see also Chap. 14). Considered historically, therefore, matter appears to have become organized progressively, level by level, and the presently existing hierarchy is the direct result.

Each level of organization includes fewer units than any lower level. There are fewer communities than species, fewer cells than organelles; and there is only one living world, but there are uncountable numbers of subatomic particles. Also, each level is structurally more complex than lower ones; a particular level combines the complexities of all lower levels and has an additional complexity of its own. For example, social complexity results from the characteristics of each member organism *as well as* from numerous special characteristics that arise out of the ways in which the members are organized as a society.

In any hierarchy of levels, moreover, a jump from one level to the next higher often can be achieved only at the expense of energy. It takes energy to build atoms to chemical compounds, and it takes energy to create cells out of chemical complexes. Similarly, energy is needed to produce tissues out of cells or societies out of separate families. And once a higher living level has been created, energy expenditures usually must continue thereafter to maintain that level. For example, if the energy supply to the cell, the organ, or the organism is stopped, death and decomposition soon follow and reversion to lower levels occurs. Similarly, maintenance of a family or a society requires work over and above that needed to maintain the organization of subordinate units.

This requirement is an expression of the *second law of thermodynamics*, about which more will be said in the next chapter: if left to itself, any system tends toward a state of greatest stability. "Randomness," "disorder," and "probability" are equivalent to this meaning of stability. When a system includes high levels of organization, it also exhibits a high degree of order and it is correspondingly nonrandom and improbable. The second law states that such a system is unstable and that if we leave it to itself, it will eventually become disordered, more random, more probable, and therefore more stable. Living systems are the most ordered, unstable, and improbable systems known. If they are to avoid the fate predicted by the second law, a price must be paid. That price is external energy—energy to maintain the order despite the continuous tendency toward disorder.

But this energy price is well worth paying, for each higher level exhibits new and useful properties over and above those found at lower levels. For example, a cell exhibits the property of life in addition to the various properties of the organelles that compose it; a multicellular organism with organs such as eyes and brains can see and distinguish objects, whereas cells taken singly at best can only sense the presence or absence of light. In general, the basic new property attained at each higher level is united, integrated function: disunited structure means independent function and, by extension, *competition*; united structure means joint function and, by extension, *cooperation*. If atoms, for example, remain structurally independent they can be in functional competition for other suitable atoms with which they might combine. Once they do unite as a compound, however, they have lost structural independence and thereafter must function jointly, as a single "cooperative" unit. Similarly, cells can remain independent structurally and can compete for space and raw materials. But if they form a multicellular unit, they surrender their independence and become a cooperative, integrated system.

This generalization applies at every other organizational level as well. The results on the human level are very familiar. Men can be independent and competing, or they can give up a measure of independence, form families and societies, and begin cooperating. The sociological laws that govern the organizational groupings of men are based on and are reflections of the more fundamental laws that govern the organization of all matter, from atoms to the whole living world.

Specialization

The fundamental advantage of cooperation is *operational efficiency;* the cooperating whole is more efficient in performing the functions of life than its lower-level components separately and competitively. For example, separate cells must expend more energy and materials for survival than if that same number of cells were integrated as a tissue. Similarly for all other organizational levels.

One underlying reason for this difference is that, in the integrated unit, duplication of effort can be avoided. In a set of separate cells, for example, every cell is exposed to the environment on all sides and therefore must expend energy and materials on all sides to cope with the environment. However, if the same cells are grouped together as a compact tissue only the outermost cells are in direct contact with the environment, and inner cells then need not channel their resources into protective activities.

Also, in addition to avoiding duplication of effort, cooperative groupings make possible continuity of effort. The general principle can be illustrated by contrasting unicellular and multicellular organisms, for example. A unicellular form must necessarily carry out all survival functions in its one cell. In many instances, however, the performance of even one of these functions requires most or all of the capacities of the cell. Thus the *entire* cell surface often must serve as gateway for entering nutrients and departing wastes. And *all* parts of the cell may have to participate directly in locomotion or in feeding. Very often, therefore, two such functions cannot be performed at the same time, for one function can preclude the simultaneous performance of the other. Mutual exclusion of some functions by others is a common occurrence in all unicellular organisms (Fig. 2.7).

In multicellular types, by contrast, continuity of a given effort becomes possible through *division of labor* among the cells. The total task of survival

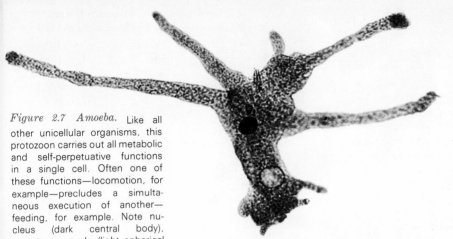

Figure 2.7 Amoeba. Like all other unicellular organisms, this protozoon carries out all metabolic and self-perpetuative functions in a single cell. Often one of these functions—locomotion, for example—precludes a simultaneous execution of another—feeding, for example. Note nucleus (dark central body), excretory vacuole (light spherical body), and the pseudopodia, fingerlike extensions that serve in locomotion and feeding (approx. X5,000). (Carolina Biological Supply Company.)

degree and the range of sensitivity can become enormously greater in multicellular forms, in which the cells can become specialized. Similarly for all other functions.

The fundamental advantage of higher organizational levels generally and of multicellularity specifically now can be understood. First, a multicellular structure permits division of labor, through which duplication of effort can be avoided and several efforts can continue simultaneously and uninterruptedly. Second, division of labor leads to specialization, which permits any given effort to become highly effective. The overall result is an enormous saving of energy and materials, hence cheaper operation, and an enormous gain in efficiency. This is one reason why living history has produced multicellular organisms, equipped successively with tissues, organs, and organ systems, rather than only bigger and better unicellular organisms.

To be sure, loss of functional versatility in a specialized cell is never total. A cell cannot be so completely specialized that it performs just a single function. Certain irreducible "housekeeping" functions must be carried out by every living cell of a multicellular organism. Each such cell must absorb nutrients, must respire and synthesize, and must maintain steady states relative to its immediate environment. These metabolic and self-perpetuative functions cannot be specialized. Performed continuously and simultaneously in every cell, they are the bedrock of cellular survival. Specialization only affects additional functions, and the fewer of such additional functions a cell performs the more specialized it is. Conversely, a cell cannot be totally unspecialized and so versatile functionally that it could survive under any or all conditions. All cells, even the most independent, still depend on, for example, specific kinds of environments that are suitable for life. Cells therefore are only more or less highly specialized; and within limits the relative degree of functional versatility is an inverse measure of the relative degree of specialization (Fig. 2.8).

Most multicellular organisms actually consist of cells that exhibit widely different degrees of specialization. But would it not be most efficient if a multicellular organism consisted exclusively of highly specialized cells? Probably not, because certain functions need not be performed continuously. For example, it would be quite wasteful to maintain a permanent set of specialized scar-tissue cells—the organism might not sustain an injury for

can be divided up into several subtasks, and each can become the continuous responsibility of particular cells only. Some cells might function in feeding, continuously so, and others in locomotion, again continuously. Frequently such division of labor is so pronounced that many or most cells are permanently limited in functional capacity; they can perform only certain jobs and no others. For example, mature nerve cells can conduct nerve impulses only and are unable to reproduce or move. Mature muscle cells can move by contracting but they normally do not reproduce. Indeed, the majority of the cells of a multicellular organism generally is restricted in some ways in functional versatility. Such cells exhibit greater or lesser degrees of *specialization*.

An individual *specialized* cell therefore cannot perform all the functions necessary for survival. This is why, when some cells are removed from a whole organism, as in injury, for example, such cells usually die; the specialized cell has lost independence. The whole task of survival can be carried out only by the entire multicellular system with its many *differently* specialized cells.

Specialization makes possible not only division of labor but also increased effectiveness of labor. For example, all unicellular organisms are sensitive to environmental stimuli, but the degree of this relatively unspecialized sensitivity is quite modest. By contrast, many multicellular organisms have highly specialized sensory cells that are exceedingly sensitive and respond to even very weak stimuli. Moreover, several kinds of sensory cells are often present, some specialized specifically for light stimuli, others for sound stimuli, still others for mechanical stimuli, and so on. Thus both the

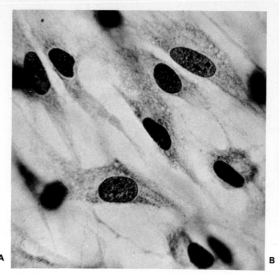
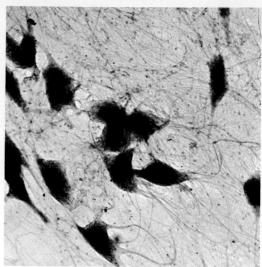

A   B

Figure 2.8   Cell specialization. *A,* a few relatively unspecialized cells (*fibroblasts*) from a rat. Such cells can transform to certain other cell types (pigment cells, fat-storing cells, membrane-forming cells, and others), none of them very specialized themselves. *B,* a few highly specialized nerve cells. Such cells function in transmission of nerve impulses and in controlling the activities of a whole animal, including exceedingly complex neural operations such as thinking.

months. Similarly, it is a decided advantage that reproductive structures often become fully developed only during seasons when reproduction actually occurs. The actual cell composition of the multicellular system thus permits the greatest possible economy of energy and materials: the most vital and continuously required functions are performed by permanently specialized cells; and less critical functions, and those required intermittently or only under unusual circumstances, are carried out by initially more versatile, less highly specialized cells.

The specializations at one level determine the specializations at higher levels. If the cells composing a tissue are specialized as muscle cells, then the whole tissue corespondingly will be specialized for contraction and movement. If the organs of an organ system include teeth, stomach, and intestine, then, since the organs perform nutritional functions, the whole organ system will be specialized for nutrition.

Therefore, every organism as a whole is specialized in accordance with the specializations of its subordinate parts. As a result an organism can live only in a very *particular* environment and can pursue only a very *particular* way of life. A fish must lead an aquatic existence, an earthworm cannot do without soil, and man too is specialized in his own way. He requires a terrestrial environ-

ment of certain properties and a social environment of variously specialized human beings, and he must live in a community that contains appropriate food organisms. In effect, the specializations of his body allow him to pursue no other but a characteristically human mode of life. And by being specialized, every organism in effect is a dependent, necessarily cooperating unit of a higher living level: the popuation, the whole species, the community of several species. These higher-level units are specialized in their turn, according to the particular specializations of their individual members.

Thus, although all organisms are alike in their general characteristics, they differ in their detailed characteristics because of specialization. Functionally, all organisms pursue life identically through metabolism and self-perpetuation; structurally all organisms are composed of cells. But in each organism both the functions and the structures are in some respects specialized, and these specializations differ in different cases. Evidently there are many ways of making a living; and it is precisely because living matter is able to specialize that the problems of life can have different solutions in different cases.

What are the particular specializations that make some organisms animals?

Table 1. *A few of the phyla of animals and some of their representative members*

PHYLUM NAME	GENERAL NAME AND CHIEF CHARACTERISTIC	REPRESENTATIVE ANIMALS
Protozoa	protozoans, predominantly one-celled types	amoebas, paramecia, malarial parasites
Porifera	sponges, or animals with water pores	chalk, glass, and horny sponges
Cnidaria	coelenterates, or animals with digestive sacs	hydras, jellyfishes, corals, sea anemones, sea fans
Platyhelminthes	flatworms	planarians, flukes, tape-worms
Aschelminthes	sac worms	rotifers, roundworms, hair worms
Mollusca	mollusks, or soft-bodied animals	snails, clams, squids, octopuses
Annelida	segmented worms	clam worms, earthworms, leeches
Arthropoda	arthropods, or jointed-legged animals	crustacea: copepods, barnacles, water fleas, shrimps, crabs, lobsters, crayfishes insects, scorpions, spiders, ticks, mites, centipedes, millipedes
Echinodermata	echinoderms, or spiny-skinned animals	starfishes, sea urchins, sea cucumbers, brittle stars, sea lilies
Chordata	chordates, or notochord-possessing animals	tunicates, amphioxus vertebrates: jawless fishes (lampreys), bony fishes (herring, tuna, etc.), cartilage fishes (sharks), amphibia (salamanders, newts, toads, frogs), reptiles (turtles, lizards, snakes, alligators), birds, mammals

The Nature of Animals

Altogether some two dozen major animal groups can be distinguished; ten fairly familiar ones are listed in Table 1, which can provide a preliminary background for discussion. Each such group is a *phylum*, with a technical and a general name. Every phylum is distinguished from others by specific identifying criteria, a subject dealt with in detail in Unit 2. It is often also useful to categorize animals more broadly as *vertebrates* and *invertebrates*, the latter group encompassing all animals without a vertebral column.

Most of the organisms traditionally called "animals" have two basic traits in common, one structural and one functional. Structurally, the animal body usually exhibits a *multicellular organization;* in the vast majority of cases it contains organs and organ systems. Functionally, the typical mode of animal nutrition is *heterotrophism:* food cannot be manufactured inside the organism (as by photosynthesis in plants, for example) but must be acquired in some way from organisms that already exist, living or dead.

Taken separately, neither of these two traits is unique to animals or necessarily characteristic of all animals. Much depends on how the concept "animal" is defined, a problem discussed separately in a later context (Chap. 14). Here it need be noted only that certain organisms exhibit both of the traits together, and that such organisms constitute the majority of what are commonly called "animals." In this majority, the two traits form the very essence of animal nature; they can be shown to condition virtually every other aspect of animal organization and way of life.

Most important, the combined traits of elaborate body structure and heterotrophism virtually necessitate the further trait of *motility,* or capability of active movement. Many nonanimal organisms also are heterotrophic (most bacteria and all fungi, for example), yet most of these are incapable of self-generated movement. However, the body of such organisms is not structured elaborately; in many cases it is unicellular and microscopic. As a result, the correspondingly small amounts of food required usually can be obtained at the spot where the organism happens to be located, or where wind, water currents, and other means of passive dispersal might have carried the organism by chance. But movement by passive means would be inadequate for animals. Being complexly structured and on the whole therefore more or less bulky, animals require correspondingly larger quantities of food; and random passive dispersion in most cases would not carry animals to adequate amounts or kinds of food.

In the majority of animals the motion-producing equipment is a *muscular system,* composed of contractile organs, muscles, which move body parts such as tentacles, legs, wings, or creeping surfaces. In many cases motion also is made possible by *cilia* or *flagella,* tiny hairlike projections from the surfaces of certain cells, and by *pseudopodia,* flowing fingerlike extensions from amoeboid

cells (see Fig. 2.7). Regardless of the particular means of motion, animals employ their motor capacity in two major ways to obtain food. In the more common case an animal carries out some form of *locomotion;* it propels its whole body toward the location of a likely food source. In the second case, the animal remains stationary, or *sessile,* and lets the food source move toward it. All sessile animals are aquatic. They employ their motion-producing equipment to create water currents that carry food organisms to them or to trap food organisms that happen to pass close by (Color Fig. 1).

Once the capacity of motion is given, it can serve not only in the search for food but also in other vital activities. As is well known, for example, locomotion aids animals in protection against danger, both biological danger occasioned by other organisms and physical danger from harmful environmental situations. Animals respond primarily by moving, either to safer areas nearby or to more suitable territory farther away. Locomotion also plays an important role in mate selection and in reproduction generally. Yet the most frequent locomotor effort made by animals probably is the search for food.

Before an animal can effectively move toward food or vice versa, the animal must be able to recognize that, at a certain spot in the environment, an object is located that is or appears to be a usable food. Moreover, after food identification and localization have been accomplished, it is essential to control the motion response—to set and adjust course and speed, and to determine when movement is to begin and to terminate. What is needed here is a complete guidance apparatus, and most animals actually have one in the form of a *nervous system.* In it, sense organs of various kinds permit recognition of environmental detail, impulses in nerves provide control over muscles, and a brain or brainlike organ coordinates a given set of recognitions with an appropriate set of motions.

Indeed, nervous systems must regulate more than muscular activity as such; for if muscles are to function at all, any structure or process that contributes to the operation of muscles requires regulation as well. It happens that virtually all internal components of animals play at least some role in maintaining the fitness of muscles. Correspondingly, nervous systems coordinate almost all internal operations of the body, and these systems thereby become major controllers of steady states.

In most animals steady states are regulated also by chemical controls. Chemical coordination usually involves components of other organ systems (for example, blood, kidneys, gills), and some groups, notably arthropods and vertebrates, also have chemical regulators in the form of hormone-producing *endocrine systems*. All chemical coordinators generally operate in conjunction with the neural ones, and both basically serve in making the actions of an animal dovetail sensibly with given environmental situations. One of the major actions here is motion. Thus, inasmuch as internal coordination is necessary for motion and motion in turn is necessary for heterotrophism, nervous and chemical steady-state controllers can be regarded as direct corollary requirements to the mode of animal nutrition.

Motility relates significantly to the architecture of animals. Clearly, motion will be most efficient if the external medium offers the least possible resistance. Unlike a tree, therefore, which is constructed in a branching shape for maximum exposure to light, air, and soil, an animal is built as compactly as possible for minimum surface exposure. Most motile animals also tend to be *bilaterally symmetric* and *elongated* in the direction of motion, a shape that aids in reducing resistance to movement. Moreover, since one end of an elongated animal necessarily enters new environments first, that end will serve best as the place for the chief sense organs and nerve centers and for the food-catching apparatus. The leading part of the body thus becomes a *head*. At the same time, elimination products of all kinds are best released at the hind end, where they do not impede forward progression. A general structure of this sort actually is standard and nearly universal among moving animals (Fig. 2.9).

By contrast, sessile animals, and also many of the slow and sluggish types, face their environment more or less equally from all sides, like plants, and their architecture reflects this. They are or tend to be *radially symmetric*, and a distinct head is usually absent (for example, corals, starfish, and see Color Fig. 1). Also, sense organs and other components of nervous systems tend to be greatly reduced (an observation that underscores the primary movement-control function of such systems).

Once animal and food source are near each other, the animal must make actual use of food. Here again the condition of heterotrophism imposes characteristically animal traits. Two basic "strategies" of handling foods are encountered. In one, the animal eats food in bulk lots. Bulk feeding is a form of heterotrophic nutrition called *holotrophism*, and the eating process itself is *ingestion*. However, most food generally is not immediately usable in the bulk form in which it is eaten. Consequently, *digestion* of food to manageable smaller components becomes a necessary further activity. And since ingested food normally contains various indigestible or otherwise unusable materials, these must eventually be eliminated, or *egested*. Taken together, ingestion, digestion, and egestion constitute *alimentation*, and all structures associated with this function form an *alimentary system*.

The second basic method of food handling is *parasitism*, another form of heterotrophic nutrition. Here the food source does not enter the animal

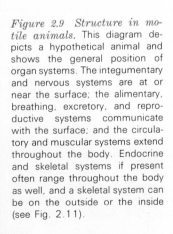

Figure 2.9 Structure in motile animals. This diagram depicts a hypothetical animal and shows the general position of organ systems. The integumentary and nervous systems are at or near the surface; the alimentary, breathing, excretory, and reproductive systems communicate with the surface; and the circulatory and muscular systems extend throughout the body. Endocrine and skeletal systems if present often range throughout the body as well, and a skeletal system can be on the outside or the inside (see Fig. 2.11).

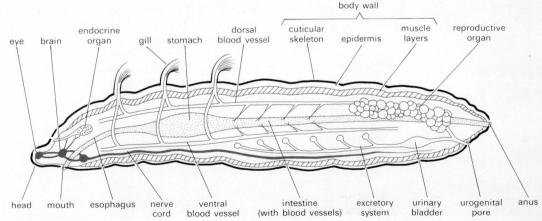

but the animal enters the food source—another living organism, or *host*. If an infected host dies, the parasite must find another living host or it will die itself. Parasites obtain food by siphoning usable nutrient chemicals directly from the host's body. Parasitic animals are believed to have evolved from nonparasitic, bulk-feeding ancestors, and many still have complete alimentary systems. But in many parasites such systems are reduced, and in some they are no longer present at all. Food molecules then enter the parasite directly through its skin.

Because the supply of appropriate food is limited, animals must compete for it more or less openly. It is the pressure of this competition that has made some animals parasites; and among the nonparasitic, free-living types, some have become predators whereas others are prey. Associated with such ways of life are many familiar specializations in offensive or defensive modes of *behavior* and in eating habits. Thus, *herbivores* normally eat only

plant foods; *carnivores* subsist on other animals; and *omnivores* can eat both animal and plant foods, living or dead.

Plant food is easily come by, and indeed more herbivorous animal types exist than any other. Also, since plants do not put up a fight before being eaten, herbivores generally tend to be more adept in defense than in offense. A plant diet presents its own special problems. Large amounts of cellulose make plant tissues tough and difficult to tear. Animals therefore must scrape or grind or chew plant foods or suck the juices out of them. Correspondingly, rasping, crushing, grinding, and sucking structures are particularly common ingestive devices of herbivores. Such animals also tend to have comparatively long digestive tracts, which offer more surface and more time for digestion of plant foods. A pound of fresh plant material consists largely of water and cellulose and of correspondingly less usable food. Accordingly, herbivores generally eat more, and more often, than other animal types.

Carnivores are specialized to overcome not only herbivores but also smaller carnivores and omnivores. Speed, strength, and varied prey-killing equipment in mouth or body appendage are familiar adaptations to a carnivorous way of life. Virtually none of the carnivores kills wantonly, but only when hungry or threatened; it is to the carnivore's advantage to live in a thriving population of herbivores. Animal tissue is softer than that of plants and tears fairly easily. Correspondingly, sharply pointed ingestive aids adapted for tearing are predominant among carnivores. Absence of cellulose in animal foods also tends to reduce chewing time and makes for easier digestion and relatively shorter alimentary tracts (Fig. 2.10).

Omnivores subsist on whatever nourishment they can find or catch. Many omnivores scavenge among the leftovers of carnivore kills. Others live on minute plant and animal debris in soil or water. The food-trapping and other alimentary structures of omnivores usually combine herbivorous and carnivorous features, as might be expected.

Numerous important characteristics of animals result from the bulkiness and compactness of the body. Since animal cells have comparatively little inherent rigidity, a bulky collection of such cells is likely to sag to a formless mass under the influence of gravity. Animals therefore require antigravity supports, and they have them in the form of muscular and particularly *skeletal systems*. That

Figure 2.10 Herbivores and carnivores. A, molar of elephant, showing broad, flat surface of the grinding teeth of such herbivores. *B,* skull of lion, showing fangs and pointed tearing teeth of such carnivores.

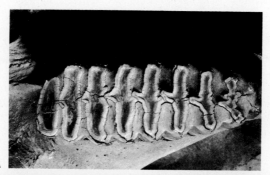

A

B

muscles function not only in motion but also in support is well illustrated in animals such as earthworms, which lack skeletons. The same muscles that move such animals also contribute to holding them together and maintaining their shapes. Moreover, even an animal with a skeleton would become a formless mass if muscles did not maintain its shape. Conversely, that skeletons function not only in support but also in locomotion is also clear. A large, heavy animal could neither hold its

Figure 2.11 Skeletal types. A, the calcareous endoskeleton of man. B, the silica endoskeleton of a glass sponge. The ropy tuft anchors the animal to the sea floor. C, the horny chitin exoskeleton of a stag beetle.

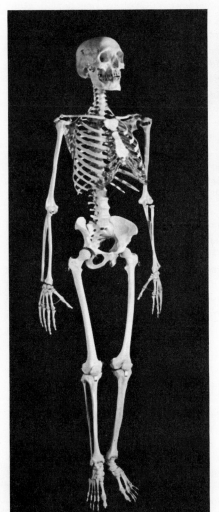

B

A

C

shape nor propel itself by muscles alone, without rigid supports.

Animal skeletons are either calcium-containing, *calcareous* supports or silicon-containing, *silicaceous* supports or variously composed *horny* supports. The skeletons are organized as *exoskeletons* or *endoskeletons.* In an exoskeleton, exemplified by the shell of a clam or snail, the supporting material is on the outside of the animal and envelops the body partly or wholly. In endoskeletons, as in the cartilage- and bone-possessing vertebrates, the supports are internal and soft tissues are draped over them. With increasing body size, an exterior skeletal envelope rapidly becomes inadequate for support of deep-lying tissues. Interior supports, however, can buttress all parts of even a large animal. It is not an accident, therefore, that the largest animals are the vertebrates, and that animals with exoskeletons or without skeletons of any kind usually are comparatively small (Fig. 2.11).

The bulky construction of animals also creates problems of internal logistics. For example, after food is eaten and digested, the usable nutrients must be distributed to all internal parts of the animal body. If the distance between the alimentary system and the farthest body parts is appreciable, as in the majority of animals, then some sort of internal transport system becomes essential. The *circulatory systems* of animals serve this requirement. In such networks of vessels, the transport vehicle of food is blood, and one or more muscular pumping organs, or hearts, maintain a circulation of blood throughout the body (Fig. 2.12). Blood is not pigmented in all animals. Where it is, the pigments function specially in transport of oxygen, not food, a circumstance that points up another problem of internal logistics.

Because of the compact construction of an animal, most of the cells are not in immediate contact with the external environment. Yet all cells require environmental oxygen for respiration, and every cell also must release waste substances to the environment. In the majority of animals blood in the circulatory system again serves as the main traffic vehicle between the environment and the interior of the body. In animals such as earthworms and frogs, exchange of gases between blood and the external environment can occur through the whole skin, which is thin and permeable. In most animals, however, the skin, or *integumentary system,* is elaborated more complexly and is relatively impermeable. Animals so covered can exchange materials with the environment only at restricted areas, where surface thinness and permeability are preserved and where the blood supply is particularly abundant. For protection, such thin and sensitive areas are frequently tucked well into the body. These areas represent parts of *breathing systems* and *excretory systems*. Gills and lungs are the main types of oxygen collectors, but these organs also contribute importantly to waste excretion. Serving primarily in excretion are kidneys and other, functionally equivalent types of organs (see Fig. 2.12).

Bulk and complex organization affect yet another aspect of animal nature, the process of reproduction and the pattern of the life cycle. Animal *reproductive systems* manufacture reproductive cells, sperms and eggs. After such cells fuse pairwise in a sexual process, the resulting fertilized egg, or *zygote,* still is a single cell. But the adult animal is complexly multicellular. The zygote therefore must divide repeatedly and give rise to numerous cells, and these must be fashioned to form structurally and functionally distinct body parts. Moreover, such parts must come to interrelate in highly specific ways, both in location and operation. In short, a specialized, lengthy course of *development* must take place. Animal development actually does occur in a unique manner; it includes two distinctive major stages, the *embryo* and the *larva.* The embryonic phase starts with the zygote and usually terminates in a process of *hatching.* The following larval phase then continues up to *metamorphosis,* or transformation to the adult. Both the embryo and the larva generally are rather unlike the adult in structure or function; both developmental phases can be regarded as specifically animal devices that provide the neces-

sary time, and the means, for the production of a complexly structured adult out of a single cell (Color Fig. 2).

All major aspects of animal nature thus can be considered consequences of the basic conditions of heterotrophism and great structural complexity. As noted, these two conditions at once necessitate the presence of nervous, muscular, skeletal, alimentary, circulatory, excretory, breathing, and integumentary systems. And if to these we add a reproductive system and in some cases also an endocrine system, we have a complete list of all the architectural ingredients that compose an animal. Implied also is a good deal about how an animal moves, behaves, feeds, develops, copes with its environment—in short, pursues life. Moreover, we know in broad outline how the structural ingredients of the animal must be put together to form a functioning whole. As suggested in Fig. 2.9, some of the organ systems must be in surface positions in whole or in part (integumentary, nervous); others can lie deep but must at least communicate with the surface (alimentary, breathing, excretory, and also reproductive); and still others must range over and through the whole body (skeletal, circulatory, nervous, endocrine).

In the line with this preliminary sketch, the basic anatomy of a motile, elongated animal can be visualized as a complex tube that has a triple-layered construction. The outermost layer of the tube is the body wall, which includes mainly the integumentary system and parts of the nervous system. The innermost layer, which encloses the open channel through the tube, is represented chiefly by the alimentary system. And the bulky middle layer contains all other organs and systems. Such a triple-layered picture of animal architecture actually is more than a rough analogy, for at an early stage of development most animal embryos do consist of just three layers, one inside the other and each originally not more than one cell thick. From the outside inward, these *primary germ layers* are the *ectoderm,* the *mesoderm,* and the *endoderm.* Later they each proliferate greatly and give rise to the triple-layered adult "tube" (Fig. 2.13).

In the resulting adult, all the structural parts are necessary and just sufficient to maintain the life-sustaining functions of metabolism and self-perpetuation. Certain organs and systems often are

Figure 2.12 Surface structure and animal complexity. In complexly structured animals, as symbolized here, the surface layers generally are thick and impermeable and the breathing and excretory surfaces usually are parts of specialized, interiorized organ systems. Internal transport of materials is accomplished by distinct circulatory systems.

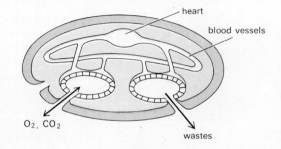

heart

blood vessels

O_2, CO_2

wastes

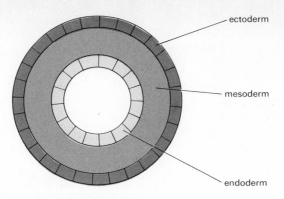

Figure 2.13 Primary germ layers. These embryonic layers will give rise to the triple-layered "tube" of the adult body (as in Fig. 2.9). Ectoderm generally forms epidermis and nervous system, endoderm forms the alimentary system, and mesoderm develops into muscular, circulatory, and reproductive systems. The other systems arise from different germ layers in different animal groups.

ectoderm

mesoderm

endoderm

concerned more directly with some functions than with others, but all systems nevertheless play at least some role in all functions. Moreover, all functions contribute importantly to the maintenance of all systems. For example, although the alimentary system serves primarily in nutrition, it also serves indirectly in all other functions of metabolism and in all processes of self-perpetuation, including even evolution: without a properly nourished reproductive system there can be no reproduction, hence also no heredity and no evolution. Similarly, it is easy to verify that every body part must provide functional support for every other part, and that only through such interdependent activities can all of metabolism and self-perpetuation take place.

Underlying these considerations is the firm recognition that all animals are composed of cells and that these in turn consist of atoms and chemicals. Accordingly, the chemical foundations of life must now be examined before the larger-scale aspects of animal life can be appreciated.

Review Questions

1. What is metabolism? Self-perpetuation? What are the main component functions of each of these, and what specific roles do these functions play in the maintenance of life?

2. What are the fundamental differences between inanimate, dead, and living systems? Discuss carefully and fully. Define living.

3. Review the hierarchy of levels in the organization of matter, and discuss how living matter is characterized in terms of levels.

4. Review the relation of levels of organization to energy, to complexity, to competition and cooperation, and to operational efficiency.

5. Define cell (*a*) structurally and (*b*) functionally. Define organelle, tissue, organ, organ system, organism.

6. In terms of cellular specializations, how does a cell of a single-celled organism differ from a cell of a multicellular organism? Cite examples of specialization on the tissue, organ, organism, and species levels of organization.

7. What are heterotrophism, holotrophism, parasitism, alimentation, hatching, metamorphosis, primary germ layers, integuments, hormones?

8. What basic traits identify animals generally? In what architectural respects are moving animals usually different from sessile ones?

9. Show in some detail how the basic traits of animals determine and influence the characteristics of all other aspects of animal nature. As far as you can, contrast such typically animal attributes with those generally considered to be characteristic of plants. Do you think that animals can be defined adequately as motile, heterotrophic organisms, and plants, contrastingly, as sessile, photosynthetic organisms? (Defer a final answer until after you have studied Chap. 14, the section on animal origins.)

10. Name the organ systems characteristic of animals and describe the fundamental functions of each. Show how each system contributes specifically to metabolism and self-perpetuation. In man, which familiar organs belong to each of the organ systems? What are some of the specific functions of such organs?

Collateral Readings

Grobstein, C.: "The Strategy of Life," Freeman, San Francisco, 1965. A thoughtful and stimulating paperback containing a section on levels of organization as well as discussions of other general phenomena of life.

Kemeny, J. G.: Man Viewed as a Machine, *Sci. American,* Apr., 1955. The article shows that computing machines can be built which can learn and even reproduce, like living systems.

Minsky, M. L.: Artificial Intelligence, *Sci. American,* Sept., 1966. On the "thinking" processes a computer is capable of.

Nagel, E.: Self-regulation, *Sci. American,* Sept., 1952. A good account on this component of control operations.

Penrose, L. S.: Self-reproducing Machines, *Sci. American,* July 1959. Models of self-duplicating mechanical systems are described, paralleling the action of self-duplicating biological systems.

Schrödinger, E.: "What is Life?" Cambridge University Press, Cambridge, 1944. A stimulating essay by a noted physicist, discussing some of the basic characteristics of living materials.

Wald, G.: Innovation in Biology, *Sci. American,* Sept., 1958. The characteristics of living systems are interpreted in terms of our present knowledge in the physical sciences.

3

CHEMICAL FOUNDATIONS

Regardless of what animal we examine, we ultimately find it to consist entirely of chemicals. And regardless of what particular function of an animal we examine, the function is ultimately always based on the properties and interactions of the chemicals present. Moreover, we now know that before there were living creatures of any kind on earth, there were only chemicals; organisms originated out of chemicals. Basically, therefore, the story of life, animal life included, is largely a story of chemicals.

Chemical Substances

Atoms

The material universe consists of just over one hundred fundamental kinds of substances called chemical *elements*. Iron, carbon, gold, oxygen, and aluminum are familiar examples. Some others, most of them present also in living matter, are listed in Table 2. The smallest whole units of an element are called *atoms*. Thus, a sample of the element gold consists of a collection of gold atoms.

Each element has a chemical symbol, often the first or the first two letters of its English or Latin name. For example, the symbol for hydrogen is H, that for carbon is C, and that for silicon is Si (see also Table 2). To represent a single atom of an element, one simply writes the appropriate symbol. Thus the letter H stands for one atom of hydrogen. If more than one atom is to be indicated, the appropriate number is put before the atomic symbol. For example, five separate hydrogen atoms are written as 5 H.

An atom is made up of *subatomic particles*. Two kinds of these, *neutrons* and *protons*, occur in varying numbers in the center of an atom, where they form an *atomic nucleus*. Neutrons are electri-

Table 2. *Some common chemical elements*

ELEMENT	SYMBOL	COMMON VALENCES	COMMON OXIDATION STATES IN COMPOUNDS
hydrogen	H	1	+1
sodium	Na	1	+1
potassium	K	1	+1
chlorine	Cl	1	−1
iodine	I	1	−1
calcium	Ca	2	+2
magnesium	Mg	2	+2
sulfur	S	2	−2, +6
oxygen	O	2	−2
copper	Cu	1, 2	+1, +2
iron	Fe	2, 3	+2, +3
carbon	C	2, 4	+4, +2, −2, −4
silicon	Si	4	+4
aluminum	Al	3	+3
nitrogen	N	3, 5	−3, +3, +5
phosphorus	P	3, 5	−3, +5

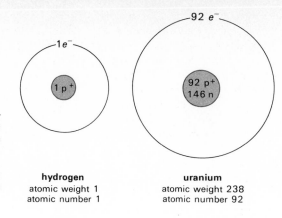

Figure 3.1 Atomic structure: hydrogen and uranium. The atomic nucleus of hydrogen contains a single proton (p^+); that of uranium, 92 protons and 146 neutrons (n). The number of electrons orbiting around the nucleus equals the number of protons.

cally neutral, protons carry one unit of positive charge. The atomic nucleus as a whole therefore is electrically positive. Orbiting around the nucleus are certain numbers of a third type of particle, *electrons.* Each of these carries one unit of negative electric charge, and an atom normally contains exactly as many electrons as there are protons in its nucleus. The electrons are maintained in orbit because they are attracted by the positively charged nucleus. Since the total positive charge equals the total negative charge, an atom as a whole is electrically neutral (Fig. 3.1).

Just as planetary orbits lie at various distances from the sun, so electron orbits are spaced out from the atomic nucleus at a number of fixed distances. The orbital paths at these distances

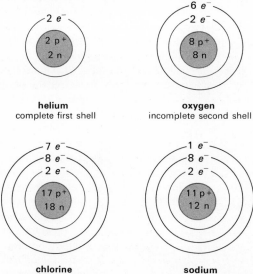

Figure 3.2 Atomic electron shells. Helium has a complete (first) shell of two electrons. In oxygen the first shell is complete but the second shell is incomplete by two electrons. Chlorine has complete first and second shells and a nearly complete third shell. And sodium has a nearly empty third shell.

form so-called electron *shells,* one outside the other. The first shell, closest to the atomic nucleus, can hold a maximum of two electrons; the second shell, a maximum of eight. Known maximums also characterize all other shells. Electrons usually fill these shells from the innermost outward. Thus, depending on the number of electrons in an atom, the outermost shell will be either complete and filled to capacity (*closed* shell) or incomplete to greater or lesser degree (*open* shell). Hydrogen, for example, has an incomplete outer shell since the single electron present does not fill the first shell to capacity. But helium, with two electrons, does fill this shell completely. In an oxygen atom similarly, two of the total of eight electrons fill the first shell and the remaining six occupy the second. Since this second shell can hold as many as eight electrons, oxygen has an incomplete outer shell (Fig. 3.2).

An atom is most stable when its electron shells are complete. Helium, for example, with a complete first shell, is stable and therefore also quite inert chemically; normally it does not react with other atoms. Similarly stable and chemically inactive are neon, with two complete shells, argon with three, krypton with four, xenon with five, and radon with six. These elements are the *inert gases.* In all other elements the outermost electron shells of the atoms are incomplete, and such atoms are more or less unstable. They reveal this comparative instability by their *chemical reactivity;* when two or more atoms come into contact, their incomplete outer shells can make them undergo chemical reactions. More specifically, atoms can become attached to one another through electric bonding forces, or *chemical bonds,* and the resulting atomic combinations are *compounds.* In effect, the chemical properties of atoms are determined by their outermost electron shells.

Each compound has a chemical name and a formula, both reflecting the kinds and numbers of atoms present. For example, table salt is technically the compound "sodium chloride," the name indicating the presence of sodium and chlorine. The formula NaCl also shows the quantitative ratio of these components: one sodium atom is linked to one chlorine atom. Water is technically the compound "hydrogen oxide," and the formula H_2O indicates the presence of two hydrogen atoms for every one of oxygen. The number of like atoms in a compound is customarily shown as a subscript. For example, iron oxide, Fe_2O_3, contains two iron atoms for every three oxygen atoms. A more complex compound is calcium phosphate, $Ca_3(PO_4)_2$. The formula here is a shorthand notation for the following combination of atoms: three calcium atoms are bonded to two subcombinations, each of which consists of one phosphorus and four oxygen atoms. Thus, thirteen atoms together form one unit of the compound calcium phosphate. If more than one unit of a compound is to be written in symbols, the appropriate number is put before the formula. For example, H_2O stands for one unit of the compound water and $5\ H_2O$ stands for five such units.

Different kinds of atoms from bonds and compounds in different ways.

Ions

Every atom has a tendency to complete its outer electron shell and thereby to become as stable as possible. How can an originally incomplete electron shell become complete? Consider an atom of chlorine. Of its 17 orbital electrons, 2 form a complete first shell, 8 a complete second shell, and the remaining 7 an incomplete third shell (see Fig. 3.2). Like the second shell, the third similarly can hold a maximum of 8 electrons. Evidently, the chlorine atom is just one electron short of having a complete outer shell. If the atom could in some way *gain* one more electron, it would satisfy its strong tendency for electronic completeness and greatest stability.

Consider now an atom of sodium. Of its 11 electrons, 2 form a complete first shell, 8 a complete second shell, and the remaining 1 a highly incomplete third shell (see Fig. 3.2). If this atom were to *lose* the single electron in the third shell, its second shell would then become the outermost shell. Since this second shell is already complete, the atom would have satisfied its tendency for completeness and would be stable. Thus, chlorine is unstable because it has one electron too few, and sodium, because it has one electron too many. Both atoms then could become stable simultaneously if an electron were transferred from sodium to chlorine. Such *electron transfers* actually can occur, and they represent one major class of chemical reactions (Fig. 3.3).

Since electrons carry negative charges, their transfer has important electric consequences. In the sodium-chlorine reaction, for example, neither atom is electrically neutral after the transfer. Sodium has lost a negative charge and therefore has become electropositive; and chlorine has

gained a negative charge, hence has become electronegative. Atoms or groups of atoms that have lost or gained electrons are called *ions;* electron transfer produces an *ionization* of the participating atoms. One of the ions formed is always electropositive, the other electronegative. And since positively and negatively charged particles attract each other, a positive and a negative ion similarly will exert mutual attraction. It is this attraction that forms a chemical bond between the ions, and it is this bond that unites the ions as a compound. Atoms that tend to lose electrons and thereby become positively charged ions are otherwise known as *metals;* and atoms that gain electrons and become negatively charged ions are *nonmetals*. Sodium is a metal; chlorine, a nonmetal.

What determines whether an atom is a metallic electron loser or a nonmetallic electron gainer? For example, could chlorine not become stable by losing its seven outer electrons instead of gaining an additional one? No, since it is exceedingly difficult to dislodge as many as seven electrons from an atom. The seven negatively charged electrons are attracted very strongly to the positively charged nucleus, and they cannot be removed readily in one batch. Indeed, the nucleus exerts a sufficiently strong attracting force to capture and hold on to an additional electron from another atom.

Conversely, sodium cannot become stable by gaining seven more electrons instead of losing the one in its outer shell: the attracting force of a nucleus that normally holds only a single electron in the outer shell is not strong enough to capture seven additional electrons. Thus, sodium—and other kinds of atoms with highly incomplete outer shells—normally acts as a metallic electron *donor;* and chlorine—as also other kinds of atoms with nearly complete outer shells—usually acts as a nonmetallic electron *acceptor*. Certain kinds of atoms—those with roughly half-filled outer shells—can act either as electron donors or electron acceptors depending on what other types of atoms they react with.

Compounds formed by electron transfers and mutual attraction of the resulting ions are *ionic compounds*. They represent a major category of chemical substances, and they are abundant in living matter. The chemical bonds in an ionic compound are *ionic*, or *electrovalent*, *bonds*. The number of such bonds in a compound generally equals the number of electrons transferred. For example, the ionic compound sodium chloride

Figure 3.3 Electron transfer and ionic compounds. In a reaction between one atom of sodium and one of chlorine, the single electron in the third shell of sodium is transferred to the third shell of chlorine. Sodium thereby acquires a complete outer (second) shell of eight electrons, and chlorine, a complete outer (third) shell of eight electrons. In this form sodium and chlorine are ions, and both together represent the ionic compound sodium chloride.

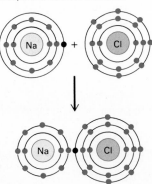

contains a single bond; the attracting force between the sodium and chloride ion results from transfer of a single electron.

Because sodium and chlorine can form one ionic bond, they are each said to have an electrovalence of 1. The numbers and signs of the electric charges on ions indicate so-called *oxidation states*. For example, since the sodium ion carries one unit of positive charge, it is said to have an oxidation state of $+1$. Similarly, the chloride ion with one unit of negative charge has an oxidation state of -1. Atoms that transfer two, three, or more electrons between them form compounds containing a correspondingly larger number of electrovalent bonds. For example, magnesium can lose two electrons by transfer to other atoms: $Mg \rightarrow Mg^{++} + 2\ e^{-}$. Magnesium thus has an electrovalence of 2, and the magnesium ion has an oxidation state of $+2$. The two electrons could be transferred to two fluorine atoms, each of which could gain one: $2e^{-} + 2\ F \rightarrow 2\ F^{-}$. Fluorine has an electrovalence of 1, and the fluoride ion, an oxidation state of -1. Two ionic bonds would then maintain the compound magnesium fluoride, $Mg^{++}F_2^{-}$.

In general, metal ions have positive oxidation states and nonmetal ions, negative ones (see Table 2). Since oxidation states indicate the numbers and signs of charges, whole neutral atoms have oxidation states of zero. Also, the arithmetic sum of oxidation states of all the ions in an ionic compound adds up to zero, since the compound as a whole is electrically neutral.

Molecules

Atoms can become stable not only by transferring but also by *sharing* electrons. For example, suppose that chlorine atoms are prevented from gaining electrons from other kinds of atoms because such other atoms are not present. A chlorine atom then can complete its outer shell of seven electrons by interacting with another chlorine atom. We already know that a chlorine atom can attract one additional electron quite strongly. If two such atoms come into contact, each therefore will attract an electron of the other with equal force. But each atom holds on to its own electrons, and an electronic "tug of war" will take place that neither atom can "win"; and without a decision, the mutual tugging will continue indefinitely. As a result the atoms will remain linked together, just as equally matched opponents of a real tug of war

remain linked by the mutual pull they exert on each other. In this case electrons are not transferred and ions are not formed. Instead, as each atom pulls on an electron of the other, a pair of electrons is shared: each atom has its own seven outer electrons plus one that it attracts from the other. Both atoms then behave as if they actually contained eight outer electrons each, and this suffices to establish their stability (Fig. 3.4).

Hydrogen atoms too can share electrons:

$$H\cdot + H\cdot \longrightarrow H\colon H$$

The two electrons then "belong" to both atoms equally, and each atom has a sphere of attracting influence over the two electrons it requires for a complete orbital shell. In many cases more than one pair of electrons can be shared. For example, an oxygen atom with six outer electrons can share two pairs with another oxygen atom:

$$\ddot{O}\colon + \colon\!\ddot{O} \longrightarrow \ddot{O}\colon\colon\ddot{O}$$

Each oxygen atom now has a sphere of attracting influence over eight electrons, the required number for a complete outer shell. Oxygen can also share two pairs of electrons with, for example, hydrogen:

$$\cdot H + \cdot H + \cdot\ddot{O}\cdot \longrightarrow H\colon\ddot{O}\colon H$$

Here each of the three participating atoms attracts just enough electrons for a complete outer shell—oxygen a shell of eight electrons and each hydrogen a shell of two.

Atoms such as chlorine or hydrogen either transfer or share electrons, depending on what other kinds of atoms are available for reaction. Another group of atoms (including sodium, magnesium, and many other metals) reacts almost exclusively by electron transfer. And in a third group (represented by oxygen, nitrogen, carbon, and others), electrons are almost always shared. Carbon, for example, can share its four outer electrons with four hydrogen atoms:

$$\cdot\overset{\cdot}{C}\cdot + 4H\cdot \longrightarrow H\colon\overset{\overset{\textstyle H}{\cdot}}{\underset{\underset{\textstyle H}{\cdot}}{C}}\colon H$$

Or, carbon can react with two oxygen atoms:

$$\colon\! C\cdot + 2\ \ddot{O}\colon \longrightarrow \ddot{O}\colon\colon C\colon\colon\ddot{O}$$

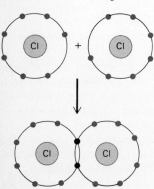

Figure 3.4 Electron sharing and molecular compounds. Two chlorine atoms with their seven outer electrons are shown at top. If the two atoms share one pair of electrons as at bottom, each atom acquires a complete outer shell of eight electrons. The result is a molecule of chlorine, Cl_2.

As the electron distributions here indicate, each of the participating atoms again attracts just the number of electrons required for a complete outer shell.

Electron-sharing represents a second major way in which compounds are formed. Compounds of this type are *molecular compounds,* or simply *molecules*. The chemical bonds in molecules are *covalent bonds,* each represented by a shared electron pair. The number of such bonds an atom can form indicates its *covalence*. Thus, hydrogen has a covalence of 1, oxygen of 2, nitrogen usually of 3, and carbon usually of 4. The atoms of a molecule also have oxidation states, according to how strongly they attract shared electron pairs. In

$$H : \overset{\overset{H}{\cdot\cdot}}{\underset{H}{C}} : H$$

for example, the four shared pairs are attracted more strongly to the carbon atom than to the hydrogen atoms; carbon here is said to have an oxidation state of -4, and each H atom, an oxidation state of $+1$. Conversely, in

$$\ddot{O} :: C :: \ddot{O}$$

it is the oxygen atoms that attract electron pairs more strongly, hence each oxygen atom is assigned an oxidation state of -2, whereas the carbon atom has an oxidation state of $+4$. Since a molecule is electrically neutral, the arithmetic sum of all the oxidation states of the participating atoms is zero.

In present contexts it usually will not be necessary to distinguish between electrovalent and covalent bonds, but the position of bonds will need to be shown fairly often. A compound then can be symbolized by means of a *structural* formula, in which the location of bonds (either electrovalent or covalent) is indicated by short dashes between appropriate atoms. For example,

NaCl	or	Na—Cl	sodium chloride
Cl_2	or	Cl—Cl	chlorine molecule
H_2O	or	H—O—H	water molecule
O_2	or	O=O	oxygen molecule
N_2	or	N≡N	nitrogen molecule
CH_4	or	$H-\overset{\overset{H}{|}}{\underset{\underset{H}{|}}{C}}-H$	methane molecule
CO_2	or	O=C=O	carbon dioxide molecule

Multiple dashes signify double or triple bonds.

The chemical properties of a compound are determined by the *arrangement,* the *numbers,* and the *types* of atoms present. Two molecules might contain the same set of atoms, but if these are arranged differently the molecules will have different properties. For example, the molecules

$$H-\overset{\overset{H}{|}}{\underset{\underset{H}{|}}{C}}-\overset{\overset{H}{|}}{\underset{\underset{H}{|}}{C}}-\overset{\overset{H}{|}}{\underset{\underset{H}{|}}{C}}-\overset{\overset{H}{|}}{\underset{\underset{H}{|}}{C}}-H \quad \text{and} \quad H-\overset{\overset{H}{|}}{\underset{\underset{H}{|}}{C}}-\overset{\overset{H}{|}}{\underset{\underset{\overset{|}{\underset{|}{H}-\overset{|}{\underset{|}{C}}-H}}{|}}{C}}-\overset{\overset{H}{|}}{\underset{\underset{H}{|}}{C}}-H$$

contain identical atoms, and both molecules can be symbolized as C_4H_{10}. But since their atoms are bonded in different patterns, they actually are different molecules with different properties. Variations in the bonding patterns of otherwise similar molecules are particularly significant in the chemistry of living matter.

Dissociation

Virtually all chemicals of zoological interest occur in a water medium. Ionic compounds in water generally exist as free ions, physically separate from one another; water molecules are interposed between the ions of the compound. When molecular compounds dissolve in water, one of two events can occur. Some types of molecular compounds simply continue to exist as whole, intact molecules, with water interposed between individual ones. Ordinary sugar is a good example. By contrast, other kinds of molecules do not remain intact in water; they break up, or *dissociate,* to free ions. For example,

$$\underset{\substack{\text{acetic} \\ \text{acid}}}{CH_3COOH} \xrightarrow{H_2O} \underset{\substack{\text{acetate} \\ \text{ion}}}{CH_3COO^-} + \underset{\substack{\text{hydrogen} \\ \text{ion}}}{H^+}$$

$$\underset{\substack{\text{ammonium} \\ \text{hydroxide}}}{NH_4OH} \xrightarrow{H_2O} \underset{\substack{\text{ammonium} \\ \text{ion}}}{NH_4^+} + \underset{\substack{\text{hydroxyl} \\ \text{ion}}}{OH^-}$$

Inasmuch as dissociation produces equal amounts of positive and negative electric charges, solutions containing dissociated compounds remain electrically neutral. However, the presence of free ions permits passage of electric currents through such solutions. Dissociable compounds are therefore called *electrolytes,* and undissociable ones are *nonelectrolytes*.

Note that, in the first reaction above, acetic

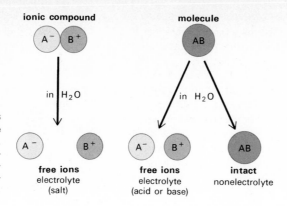

Figure 3.5 Dissociation. Salts in water exist in the form of the ions of which they are composed. Molecules in water either dissociate as free ions if they are electrolytes or remain as intact molecules if they are nonelectrolytes.

acid dissociates in such a way that hydrogen ions are formed. This is what actually makes acetic acid an acid; any compound that dissociates and yields hydrogen ions (H⁺) is an *acid*. Similarly, any compound that dissociates and yields hydroxyl ions (OH⁻), as in the second reaction above, is a *base*, or *alkali*.

Acids and bases are molecular compounds in pure form; they dissociate in water and give rise to ions as a result. If ions derived from a dissociated acid interact chemically with ions from a dissociated base, a *salt* is produced. Thus, because they are composed of ions, salts are ionic compounds. For example, sodium chloride (NaCl) is a salt formed by interaction of hydrochloric acid (HCl) and the base sodium hydroxide (NaOH):

$$HCl \xrightarrow{\text{dissociation}} H^+ + Cl^-$$
$$NaOH \longrightarrow Na^+ + OH^-$$

$$HCl + NaOH \longrightarrow NaCl + H_2O$$

Every electrolyte is an acid or a base or a salt (Fig. 3.5).

An electrolyte is said to be "strong" or "weak" according to the *extent* to which it is dissociated.

With some exceptions unimportant in zoological contexts, salts dissociate completely and *all* their chemical units exist in the form of ions. Salts therefore are very strong electrolytes. By contrast, acids and bases are not always dissociated fully, different ones dissociating to different degrees. For example, the molecular compound hydrochloric acid (HCl) is a strong acid, for in water virtually all the molecules of the compound dissociate to ions. But the molecular compound acetic acid (CH_3COOH) is a weak acid, for in water only a few ion pairs are formed; most of the molecules remain whole and intact, and only relatively few break up as free acetate and hydrogen ions. Strong and weak bases can be distinguished similarly. The weakest of all electrolytes in effect are the non-electrolytes, which do not dissociate at all.

It is often important to determine the acid or alkaline strength of a solution, or the degree to which the compounds in the solution are dissociated. This can be done with an electrical apparatus that measures the relative number of H⁺ ions (actually H⁺ bound to water in the form of H_3^+O) and OH⁻ ions in the solution; for the more of these ions found, the more the acids and bases present are dissociated. The result is expressed as a number, called the *pH* of the solution. In living material, the numbers that indicate pH usually range from 0 to 14. A pH of 7 indicates chemical *neutrality;* the solution is neither acid nor basic because the number of H⁺ ions equals the number of OH⁻ ions. Below 7, a solution is the more acid the lower its pH; it contains more H⁺ ions than OH⁻ ions. Conversely, a solution is the more alkaline the more its pH exceeds 7 (Fig. 3.6).

Animal matter, which contains a mixture of variously dissociated acids, bases, and salts, has a pH usually very near neutrality. For example, the pH of human blood generally is 7.3. Distinctly higher or lower pH levels do occur, however, but usually in restricted regions only. For example, stomach cavities are characteristically quite acid. Living animal material does not tolerate significant variations in its normal acid-base balance, and its pH must remain within fairly narrow limits. If these limits are greatly exceeded, major chemical and physical disturbances result that can be highly damaging or even lethal.

Many normal processes in animals yield small amounts of excess acids or bases, but these hardly affect the pH because living matter is *buffered;* it is protected to some extent against pH change. For example, one of the usual constituents of

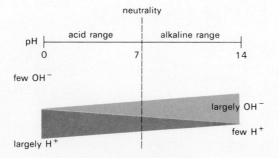

Figure 3.6 The pH scale.

animal material is the bicarbonate ion, HCO_3^-. If an acid is now added, this acid will yield an excess of hydrogen ions (H^+) that ordinarily might be damaging. However, when H^+ and HCO_3^- are present together they cannot remain free ions to any great extent, for they react together and form carbonic acid, H_2CO_3. This compound happens to be a weak acid that cannot dissociate very much:

1. $HCl \longrightarrow H^+ + Cl^-$

2. $HCO_3^- \overset{\longrightarrow}{\longleftarrow} H_2CO_3$

─────────────────────────────

3. $HCl + HCO_3^- \overset{\longrightarrow}{\longleftarrow} H_2CO_3 + Cl^-$

Dissociation of H_2CO_3 to free ions therefore will take place to only a minor extent (short arrow to left in reaction 2, above), but the reverse process that joins H^+ and HCO_3^- as whole H_2CO_3 molecules will occur abundantly (long arrow to right in reaction 2, above). As a result, the H^+ ions derived from the added acid are "taken out of circulation" by combination with HCO_3^-, and the pH therefore will not change appreciably despite the addition of the acid (summary reaction 3, above).

The bicarbonate ions in effect function as buffers against pH change. Phosphate ions have a similar buffering effect against added acids, and several positively charged ions in living matter protect against added bases. To be sure, if living material or any buffered system is flooded with large quantities of additional acids or bases, then pH protection will be inadequate.

Chemical Changes

Reactions

Any process in which at least one bond between two atoms is formed or broken can be regarded as a chemical reaction. Such processes generally occur as a result of *collisions* among molecules or ions; direct contact is necessary if compounds are to be close enough together for bond breaking or bond formation.

Inasmuch as reactions always imply changes in chemical bonding, the invariable outcome of any reaction is a *rearrangement* of atoms and bonding patterns. For example:

Or generally,

$A \longrightarrow B$

If reactions take place between two or more types of compounds, then the atomic rearrangements are usually accompanied by three kinds of broader results. First, two or more molecules or ions can come together and form a single larger unit; a *synthesis* can occur. For example:

$$HCO_3^- + H^+ \longrightarrow H_2CO_3$$
bicarbonate   hydrogen   carbonic
ion   ion   acid

Or generally,

$A + B \longrightarrow AB$

Reactions of this general type include those processes of synthesis that, as noted in the preceding chapter, represent a major metabolic activity of living matter.

Second, a single molecule or ion can become fragmented to two or more smaller units; a chemical *decomposition* can take place, the reverse of synthesis. For example:

$$H_2CO_3 \longrightarrow H^+ + HCO_3^-$$

Or generally,

$AB \longrightarrow A + B$

Reactions of this general type include the decomposition reactions of respiration, another metabolic activity referred to in Chap. 2.

Third, synthesis and decomposition can occur simultaneously; the decomposition products of one set of compounds can become the building blocks for the synthesis of another set. In such a process, one or more of the atoms or ions of one compound trade places with one or more of the atoms or ions of another compound. This is an *exchange* reaction. For example:

$$H{-}Cl + Na{-}OH \longrightarrow H{-}OH + Na{-}Cl$$
hydrochloric   sodium   water   sodium
acid   hydroxide   chloride

Or generally,

$$AB + CD \longrightarrow AD + BC$$

Note that each of these reaction types actually does include a rearrangement of atoms and bonding patterns. In every case one or more bonds are broken and one or more new ones are formed. However, the total numbers and types of atoms are exactly the same before and after the reaction; atoms are neither gained nor lost. In symbolic representations of reactions, therefore, it is important to make sure that equations balance; the total numbers and types of atoms to the left of the reaction arrow must exactly equal those on the right.

Apart from belonging to one of the reaction types just described, any chemical process has two fundamental characteristics. The first is *direction*. In the examples above the directions of the reactions have been indicated by arrows, but this has been possible only because of knowledge gained from earlier experience. If such experience were lacking, it would not necessarily be clear whether a given reaction would proceed to the right or to the left. What determines the actual direction? The second characteristic is reaction *rate*, a measure of how fast and for how long a reaction proceeds. Reactions result from collisions of chemical units, as noted, and the generalization can be made that reaction rate is directly proportional to collision rate. Any factor or condition that promotes collisions will also promote rates. What are these rate determinants?

The direction of a reaction is governed by the *energetics* of the reacting chemicals; and the rate is determined by *temperatures*, by *pressures* and *concentrations*, and by *catalysts*.

Figure 3.7 Thermodynamics and stability. A stone on a hillside can be regarded as a closed thermodynamic system. If the stone is left to itself a unidirectional process can occur; the stone can fall to the valley and thereby assume a more stable state.

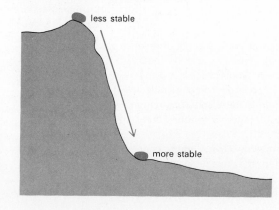

less stable

more stable

Energetics

Like all other kinds of changes in the universe, chemical changes obey and are governed by certain laws of *thermodynamics*. These most fundamental rules of nature deal with the *energy* relations of a system and its surroundings. The general concept of energy is roughly equivalent in meaning to *work potential*, or the capacity to do work. A "system" can be regarded as any set of materials on which attention is focused at the moment, such as a group of chemicals undergoing reactions. The surroundings then are all other parts of the universe, the total environment in which the system exists.

A *first law* of thermodynamics, also called the law of *conservation of energy*, states that, in any process, the sum of all energy changes must be zero. Expressed differently, energy can be neither created nor destroyed. Thus if a chemical system gains energy, that amount of energy must be lost by the environment of the system. Or if a chemical system loses energy, that amount must be gained by the environment. The first law therefore implies that energy can only be redistributed, changed in form, or both.

A *second law* of thermodynamics, already referred to in Chap. 2, states that if left to itself any system tends toward a state of greatest stability. A stone near the top of a hillside represents a less stable (and also a more ordered and less probable) system than a stone at the bottom of the slope. The second law indicates that, whereas such a stone might roll downhill to a more stable state, by itself it can never roll uphill. To be sure, a stone can be *brought* uphill if an environmental agent expends energy and pushes the stone up; but in that case the system is no longer "left to itself." Evidently, natural processes tend to be *unidirectional*, even though they are inherently reversible. Correspondingly, all chemical processes are theoretically reversible, but in actuality they proceed either in one direction or the other, according to which one of these leads to a more stable state (Fig. 3.7).

The actual stability of chemical systems depends mainly on two factors, both involving the energy of compounds. One factor is *enthalpy*, symbolized by the letter H; it denotes the *total energy content* of a chemical system. As already pointed out, a compound is held together by chemical bonds that result from mutual electric attraction between atoms or ions. These bonding forces represent *chemical energy*, or *bond energy*. The greater the attraction between two atoms or ions, the greater is the bond energy. Two bonded atoms or ions will become disunited only if some external force pushes them apart and thereby breaks the bond. Such forcible separation requires work, or energy, and the amount of energy needed clearly must be at least great enough to overcome the attraction between the two atoms or ions. In other words, the energy required to break a bond equals, and defines, the bond energy. Accordingly, the enthalpy, or total chemical energy, of a compound can be defined as the energy needed to break all the bonds in the compound.

If now two or more compounds are close together, such a chemical system will be most stable if and when its total chemical energy is at a *minimum*. The second law of thermodynamics shows that, for any system, a state of minimum available energy is the most likely; and it is because of this greatest likelihood that minimum energy states are the most stable. Chemical reactions therefore will tend to proceed in such a way that, at the end, the total energy content of all participants will be least. For example, assume that in the generalized reaction

$$A + B \longrightarrow C + D$$

the total energy of all the bonds in *A* and *B* together is greater than the total energy of all the bonds in *C* and *D* together. In other words, more potential energy is available in the bonds of the starting materials than is needed to form the bonds of the endproducts. If all other conditions are suitable, such a reaction can occur readily because *C + D*, containing less total energy, is more stable than *A + B*. The energy difference will be lost from the reaction system to the environment. The energy differential, or *total energy change*, is customarily symbolized as ΔH. And since ΔH passes from the reaction system to the environment, it is given a negative sign:

$$A + B \longrightarrow C + D - \Delta H$$

Such a liberation of energy can become evident, for example, through a spontaneous rise of temperature in the reaction mixture. Reactions that yield energy as above are said to be *exergonic* (or *exothermic*, if the energy is liberated in the form of heat). Decomposition reactions, including respiratory decompositions in living matter, tend to be of this type (Fig. 3.8).

Conversely, assume that the total energy of *A + B* is less than that of *C + D*. In this case *A + B* is more stable, and the second law stipulates that no process can proceed by itself from a more stable to a less stable state. However, the reaction could be *made* to occur if energy were supplied *from* the environment *to* the reaction system, in an amount at least equal to ΔH. For example, the starting materials could be heated over a flame. Then, so long as external energy continued to be supplied, the reaction could proceed:

$$A + B \longrightarrow C + D + \Delta H$$

Reactions that *require* external energy, as above, are said to be *endergonic* (or *endothermic*, if the energy is required in the form of heat). Synthesis reactions generally, including those in living matter, are of this type. Evidently, chemical (or any other) synthesis tends to be expensive in terms of energy.

The second factor that determines the relative stability of a chemical system is its *entropy*, or *energy distribution*, symbolized as *S*. A system is most stable if *S* is at a maximum, with energy distributed as uniformly or randomly as possible. It can be shown that, of all possible energy distributions in a system, the most random or unordered is the most likely; and it is because of this greatest likelihood that maximum entropy states are the most stable.

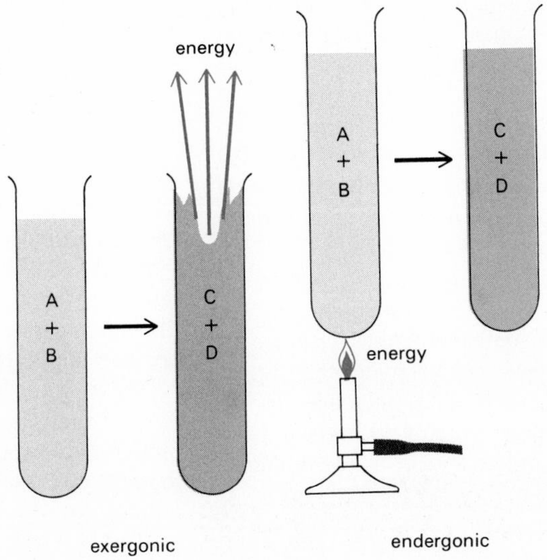

Figure 3.8 Energy and reactions. Left, exergonic reaction. The mixture *A + B* contains more total energy than *C + D*, and the energy difference escapes the reacting system. *Right,* endergonic reaction. The mixture *A + B* contains less energy than *C + D*. Hence if a reaction is to take place the energy difference must be supplied to the reacting system from an external source.

energy

exergonic

endergonic

For example, consider again the reaction $A + B \rightarrow C + D$, and assume that the total energy content is the same before and after the reaction ($\Delta H = 0$). Assume, however, that A is energy-rich and B energy-poor, and that C and D each contain intermediate amounts of energy. The energy distribution in $A + B$ therefore is more uneven or less random than in $C + D$, and $C + D$ will be more stable. The reaction to the left thus can occur, and it will increase the entropy of the system. If the entropy differential between $A + B$ and $C + D$ is symbolized as ΔS, the reaction can be written as

$$A + B \longrightarrow C + D + \Delta S$$

Consider now the reaction

$$A + B + C \longrightarrow D$$

and again assume that ΔH is zero. On the left side the total energy is distributed among three particles, but on the right all the energy is concentrated in one. With everything else equal, the $A + B + C$ state thus has a more scattered, more random energy distribution, and this state will be more stable and have a greater entropy than the D state. The reaction therefore will not be possible if the system is left to itself. However, energy supplied from the environment conceivably could lower the entropy of $A + B + C$ sufficiently to make the reaction to D possible. Symbolically,

$$A + B + C \longrightarrow D - \Delta S$$

Entropy is temperature-dependent. For example, if equal amounts of water under otherwise equivalent conditions exist in the form of ice, liquid water, and steam, then ice will have the lowest and steam the highest entropy; water molecules are most scattered and disordered in the form of steam. In general, entropy tends to increase with rising temperature. For this reason entropy changes usually are symbolized as a mathematical product of ΔS and T, where T is the absolute temperature at which a reaction occurs.

The overall stability change during a reaction therefore is a function of two variables, ΔH and $T\Delta S$. By conceptual definition these two are of opposite algebraic sign, since stability changes as enthalpy increases and entropy decreases (or as enthalpy decreases and entropy increases). This circumstance finds expression in the formulation

$$\Delta F = \Delta H - T\Delta S$$

an important equation that represents a symbolic statement of the second law. The term ΔF denotes overall change in system stability and is generally referred to as the *free-energy change*.

This change is a measure of the theoretical amount of *usable* work that can be obtained from a reacting system. For example, if a piece of wood is burned in a fireplace, the original energy of the wood less the energy of all the ashes and other combustion products represents ΔH, the total energy change. This total is not usable as such to perform work—to heat a room, for example; for in the process of burning, the compact piece of wood has scattered as ashes and escaping gases. This dispersal represents an increase in entropy, or loss of some of the potential energy of the original wood. The amount of this lost energy is equivalent to the energy that would have to be spent to reassemble the dispersed combustion products back in one place. This quantity, equal to $T\Delta S$, therefore is not available to heat the room, and the net amount that does remain available is $\Delta H - T\Delta S$, or ΔF.

Changes in free energy, enthalpy, and entropy are expressed as *heat equivalents*. The unit of heat is a *calorie* (cal), the amount of heat (or equivalent quantities of other energy forms) required to raise the temperature of one gram (g) of water by one degree Celsius (1°C). A commonly used zoological unit is a kilocalorie (kcal), or "dietary" calorie, equal to 1000 cal. (For example, when an average slice of bread is burned completely, either in a furnace or in the body, the energy liberated is roughly 100 kcal.)

A negative ΔF is largely characteristic of *decomposition* reactions; a positive ΔF, of *synthesis* reactions. Under suitable conditions some of the energy liberated in a decomposition reaction can be made to perform useful work. For example, it could drive an engine, as in the decomposition of fuels by burning, or it could simply provide useful heat as such. Or, indeed, it can make possible energy-requiring synthesis reactions. In living systems, actually, decompositions are intimately *coupled* with syntheses; as already pointed out in Chap. 2, energy obtained from respiratory decomposition supports metabolic synthesis.

The energetics of a chemical system thus determine one basic attribute of reactions, their direction. The second attribute, rate, is specified by factors that influence the collision frequency among chemical units.

Pressure, Temperature, Concentration

The effect of pressure on reacting systems varies greatly, depending on the gaseous, liquid, or solid nature of the participating chemicals. Reaction rates among gases are the higher the greater the environmental pressure; increasing pressure forces gases into progressively smaller volumes, and the possibility of direct contact among gas ions or molecules is thereby enhanced. Pressure has far less effect on liquids, which are virtually incompressible, and on solids, which may or may not be compressible.

More significant than pressure is the environmental temperature. Reactions depend on contact among chemical units, and the usual agency that brings chemical units into contact is *heat* motion. All atoms, ions, and molecules, regardless of whether they are in a gas, a liquid, or a solid, vibrate uninterruptedly in random back-and-forth movements, and in gases and liquids they also move from place to place over relatively greater distances. We feel these movements as heat and

we measure them as temperature. At a high temperature chemical units are in violent motion. Conversely, at $-273°C$, the theoretical absolute zero of temperature, heat is by definition entirely absent and all chemical units are stationary. But every known natural or experimentally produced material always contains at least some heat, and the chemical units undergo more or less intense motions. Such thermal movements produce collisions among the chemical units. At any given temperature chemical units therefore have a characteristic collision rate, and the collision rate in turn is a major determinant of the reaction rate among the units.

Ordinary room temperature often provides enough heat to agitate chemical units adequately for chemical reaction. But for many reactions room heat is insufficient. For example, a mixture of fat and water reacts so slowly at room temperature that the result is quite unnoticeable. However, molecular collisions, hence the speed of reaction, can be increased by heating the mixture, for example.

The external energy needed to produce significant reaction rates is called *activation energy*. Most materials in the physical world, as also in the living world, actually require activation energies far greater than those provided by ordinary temperatures. That is why, for example, the oxygen molecules in air and the molecules of wood or coal do not interact spontaneously, despite the numerous collisions between these substances. As is well known, combustible fuels must be heated substantially, or activated effectively, before they will react with aerial oxygen and thus burn. To start any reaction, therefore, enough activation energy must first be supplied from the outside. Interaction of the chemical units then can begin.

Activation energies interrelate with free-energy changes, as can be illustrated by considering a stone lying near the edge of a cliff. If the stone fell over the edge, its descent to the valley would yield energy that could be used to perform work. But some external force first must move the stone over the cliff edge, and this force is equivalent to the activation energy. Assume that, in Fig. 3.9, point *A* represents the degree of stability of a chemical system before a reaction has started, and position *C*, the degree of stability at the end of the reaction. The net stability change, or free-energy change ΔF, therefore can be represented as the difference in the levels of *A* and *C*. The

Figure 3.9 Activation energy. If *A* represents a less stable chemical system than *C*, an exergonic reaction from *A* to *C* can occur when activation energy (E_{act}) is first applied to *A*. E_{act} brings the reacting system over an energy barrier (symbolized by *B*), and as *A* then becomes *C*, energy is released. This energy release "pays back" the E_{act} expended earlier, and it also yields additional energy, ΔF, the "free-energy change." The net energy gain of the reaction thus is ΔF, an amount of energy that varies in proportion with the stability difference between *A* and *C*. Conversely, in an endergonic reaction from *C* to *A*, *C* must not only be activated (E_{act}) but must also be supplied with external energy in the amount of ΔF. The only energy gain then is E_{act}, hence the net energy expenditure is ΔF.

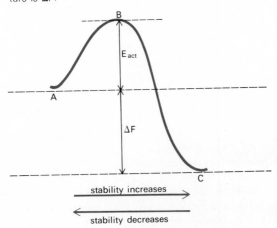

lower position of C signifies that, as with a falling stone, stability is greater at the end of the reaction. But note that the stability curve from A to C passes through point B, located at a higher level than either A or C. This means that the stability of the system first decreases from A to B before it increases from B to C. The amount of this decrease from A to B represents the activation energy, E_{act}, which must be supplied from the outside to bring the reaction system "over the cliff." Thereafter, just as a stone falls on its own, the reaction proceeds by itself; and in the process it not only "pays back" the energy expended for activation but also yields net free energy usable for work, in the amount of ΔF.

In the reverse reaction from C to A, adequate activation requires external energy in an amount equal to ΔF plus E_{act}. Here only E_{act} is paid back when the reaction later "falls" from B to A. The net energy expenditure therefore is ΔF, far greater than in the reaction from A to C; more work must be done to move a stone up from a valley onto a cliff than in the opposite direction.

After having been activated, a reaction will proceed the faster the more heat is supplied to it. It has been found that every temperature rise of 10°C increases the speed of reactions approximately two to three times (or, stated in technical language, the *temperature coefficient* Q_{10} is 2 to 3 for chemical reactions in general). The implications for reactions in living systems are important. Most animals lack internal temperature controls, and the temperatures in their bodies largely match those of the external environment. As a result, the life-maintaining chemical processes in a housefly, for example, will occur two to three times as fast on a day that is 10°C warmer than on another. By contrast, temperature controls are present in birds and mammals, and these animals maintain constant internal temperatures. In these "warm-blooded" types, therefore, the rates of chemical reactions do not fluctuate directly with changes in environmental temperature.

Apart from pressure and temperature, reaction rates also depend on the amount per volume, or *concentration*, of the reacting compounds present; for the greater the concentration of the starting materials, the more frequently collisions can occur and the faster the reaction will proceed. Other factors being equal, therefore, the rate of a reaction is proportional to the concentrations of the participating compounds. This generalization is

sometimes referred to as the *law of mass action*.

Assume that the reversible reaction

1 glycerin + 3 fatty acids $\rightleftharpoons$ 1 fat + 3 water

is in *equilibrium;* concentrations are constant and all other conditions are such that the reaction to the right occurs at the same rate as the reaction to the left. A *net* change will not occur so long as the equilibrium conditions are maintained. However, the equilibrium will be disturbed when some fat or water is removed from the system. The glycerin and fatty acids then in effect will have a relatively higher concentration, and according to the principle of mass action the reaction to the right now will take place at a greater rate than the reaction to the left. If we keep on removing the endproducts as fast as they form, the reaction to the right will proceed to completion and the yield of fat and water will become maximal. Reactions often proceed to completion and maximum yield if one of the endproducts is an escaping gas ($\uparrow$) or a relatively insoluble precipitate ($\downarrow$); either circumstance is equivalent to removing one of the participating substances. For example,

$$H_2CO_3 \longrightarrow H_2O + CO_2\uparrow$$
$$Ca^{++} + CO_3^{=} \longrightarrow CaCO_3\downarrow$$

Another way of upsetting the equilibrium in the glycerin-fat reaction above would be to *add* glycerin, fatty acids, or both to the reaction system. The concentrations of these substances then would always remain high, and according to the principle of mass action the reaction would always be "driven" to the right. More and more fat and water would form as more and more glycerin and fatty acids were added.

This rule of mass action also applies to the energy changes of exergonic and endergonic reactions. Consider the reversible reaction

$$A \rightleftharpoons B - \Delta H$$

Proceeding to the right, the reaction is exergonic. Energy here is as much an endproduct as B, and if energy escapes as fast as it is produced the reaction to the right will not reach equilibrium but will proceed to completion. The reverse reaction is endergonic. Here energy must be added continuously if all of B is to be converted to A. If the supply of energy is stopped prematurely, the reac-

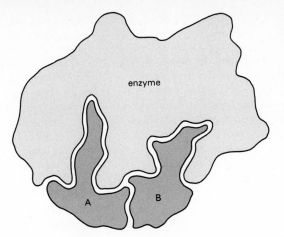

Figure 3.10 Enzyme action.
The surfaces of molecules *A* and
B fit into the surface of the enzyme. Reaction between *A* and *B*
can thereby be speeded up, for
contact between *A* and *B* now does
not depend on chance collision.

tion to the left will cease and the reverse reaction
to the right will take place and proceed to completion.

Catalysis

Most chemical reactions of zoological interest require fairly high activation energies—so high, indeed, that environments far hotter than room temperature would be needed. As noted, for example,
a fat-water mixture at room temperature (or even
at body temperature) remains virtually unchanged
for days. In the body, however, fat and water react
appreciably within an hour or so, yet body temperature remains comparatively low. In living matter
generally, sufficiently high rates of molecular collisions are produced without additional heat
through *catalysis,* or acceleration of reactions by
means of *catalysts* instead of heat.

Catalysts of various kinds are well known and
widely used by chemists dealing with nonliving
processes. The special catalysts that occur in living
matter are called *enzymes.* These substances are
proteins, compounds about which more will be
said in the next chapter. Virtually every one of the
thousands of chemical reactions in living matter
is speeded up enormously by a particular enzyme
protein. Without enzymes the reactions could not
occur fast enough at ordinary temperatures to
sustain life. Enzymes evidently represent a supplement to thermal motion, a means by which reactions requiring high temperatures in test tubes can
occur at low temperatures in living organisms. The
effect of enzymes is to lower the activation-energy

requirements. These catalysts thereby promote
appreciable reaction rates at lower temperatures
than would be possible otherwise.

An enzyme produces such an effect by combining temporarily with the reacting compounds. Mutual contact of these compounds then is no longer
a matter of chance collision but a matter of cer-
tainty, hence reactions are faster. The protein nature of enzymes is essential to this accelerating
action. Protein molecules are huge, and an almost
unlimited number of different kinds of proteins
exists. The surface shapes of such molecules appear to be the key to their effectiveness. Consider
the reaction

fat + water $\longrightarrow$ fatty acids + glycerin

A fat molecule has a particular surface geometry,
and so does a water molecule. Enzymatic acceleration of this reaction now occurs if the surfaces
of both fat and water happen to fit closely against
the surface of a particular protein molecule. When
the reacting molecules then become attached to
this suitably shaped enzyme surface, they will be
so close to each other that they interact chemically
(Fig. 3.10).

The enzyme itself remains almost passive here.
It merely provides a uniquely structured "plat-
form," or *template,* on which certain molecules
can become trapped. Such trapping brings reacting molecules into contact far faster than chance
collisions at that temperature. Reactions are therefore accelerated. When they are held by the enzyme, fat and water interact and become fatty acid
and glycerin, and these endproducts then disengage from the enzyme surface. The enzyme itself
remains unaffected. It reappears unchanged at the
end of the reaction, free to combine with a new
set of starting compounds. Being passive plat-
forms, enzymes speed up reaction in *either* direction; the reaction to the right, above, is accelerated by the same enzyme that speeds up the
reverse reaction. Thus, like heat, enzymes influence only the rates of reactions, not the directions.

In describing enzyme-accelerated reactions, it
is customary to speak of reacting molecules such
as fat and water as the *substrates.* When substrate
molecules are attached to an enzyme, the whole
is an *enzyme-substrate complex.* Formation of
such complexes can be thought of as "lock-
and-key" processes. Only particularly shaped keys
fit into particular kinds of keyholes. Just so, only

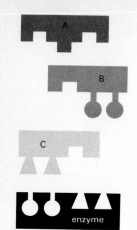

Figure 3.11 Enzyme specificity. Reactants *A* and *B* fit partially into the surface of the enzyme, but reactant *C* does not. Hence the enzyme can speed up the reactions involving *A* and *B* but not those involving *C*.

molecules of certain shapes will establish a close fit with a given type of enzyme protein (Fig. 3.11).

Because enzyme proteins are huge molecules compared to most substrates, the whole surface of an enzyme often is not required to promote a reaction; in many cases it has been shown that only one or at most a few limited surface regions, called *active sites*, are involved. Hence even if other parts of enzyme molecules become altered chemically or physically, the enzymes still can be effective if their active sites remain intact.

Until recently enzymes generally were believed to be rigid molecular structures, templates of fixed shapes, and enzyme activity was thought to depend on this permanence of the geometry. Newer research indicates, however, that enzyme molecules actually are flexible in a physical sense and that the structure of a particular substrate can *induce* the enzyme to bend or mold itself over the substrate. Such an "induced fit" effect is consistent with the observation that only active sites, not necessarily whole enzyme molecules, need to retain permanent configurations. Moreover, induced fits also account in part for the long-established observation that many enzymes operate adequately only in the presence of certain *cofactors*. These are of various kinds and include, for example, metal ions such as Mg^{++}. Cofactors appear to be agents that, by virtue of their own properties, aid in molding an enzyme (or its substrates) to the shape required for a proper enzyme-substrate fit (Fig. 3.12).

Differences in the surface configuration of different protein types result in *enzyme specificity:* a particular type of enzyme normally accelerates only one particular type of reaction. For example,

the enzyme in the fatty acid–glycerin reaction above, called *lipase,* is specific and catalyzes only that particular reaction. Living animal matter actually contains almost as many different kinds of enzymes as different varieties of reactions. This specificity of enzymes is an important corollary of the more general phenomenon of *protein specificity,* about which more will be said in the next chapter. Because of protein specificity, some proteins are enzymes to begin with and some are not. If a protein happens to have surface regions that can fit with the surfaces of some other compounds, then the protein could function as an enzyme for such compounds.

The effectiveness of enzymes varies greatly with changes of temperature, pH, and environmental conditions in general, just as all proteins are affected by such changes. More specifically, as either temperature or pH rises, the effectiveness of enzymes increases up to a certain optimum but decreases thereafter. Most enzymes operate optimally in a temperature range of about 25 to 40°C (human body temperature is 37°C) and in a pH range of about 6.0 to 7.5. At lower temperatures reactions of all kinds decrease in rate, as noted earlier, and at higher temperatures the thermal motion of the atoms in an enzyme becomes intense enough to disrupt the physical structure of the enzyme protein. Similarly, pH changes to either side of the optimum range produce structural changes that result in enzyme inactivation. These are basic reasons why excessive temperature or pH changes can be lethal to living matter.

Enzymes usually are named according to the kinds of compounds they affect. Thus, enzymes that accelerate reactions of so-called "carbohydrate" compounds are *carbohydrases.* Similarly, *proteinases* and *lipases* are enzymes that catalyze certain reactions of, respectively, proteins and fatty substances (lipids). The suffix *-ase* always identifies a name as one of an enzyme (although all enzyme names need not have this suffix). In writing an enzymatic reaction symbolically, the enzyme conventionally is indicated just above the reaction arrow. For example,

$$\text{fatty acids} + \text{glycerin} \xrightarrow{\text{lipase}} \text{fat} + \text{water}$$

Most of the discussion in this chapter applies to chemicals generally. The specific chemicals of animals now can be considered, in relation to the cells in which they occur.

Figure 3.12 Cofactors. The enzyme diagramed here is assumed to have an active site at each end. A cofactor can aid in producing a fit between the substrate and the active sites of the enzyme. Cofactors include various mineral ions and a variety of coenzymes. These organic substances are specific, each operating in conjunction with a particular enzyme (often called "apoenzyme" in this context).

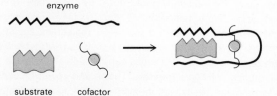

Review Questions

1. Define element, atom, compound, ion, molecule, chemical energy, chemical bond.

2. What is an electrovalent bond? How is such a bond formed? Explain in terms of atomic structure. What is a covalent bond? Describe a way in which such a bond can be formed. Again explain in terms of atomic structure.

3. Consider the following reaction:

$$Ca(OH)_2 + 2\ HCl \longrightarrow CaCl_2 + 2\ H_2O$$

   a. Identify the different atoms by name.
   b. Rewrite the equation to show bonds in each compound.
   c. Is the equation balanced?
   d. Is this an exchange, synthesis, or decomposition reaction?

4. Define dissociation, electrolyte, acid, base, salt. Is H_2SO_4 an acid, a base, or a salt? How does sodium sulfate (Na_2SO_4) dissociate? The magnesium ion is Mg^{++} and the nitrate ion is NO_3^-; write the formulas for magnesium hydroxide, nitric acid, and magnesium nitrate.

5. What does the pH of a solution indicate? What would you expect the pH of a solution of NaCl to be? Of HCl? Of NaOH?

6. What is a chemical reaction and what kind of event produces it? What general types of reaction are known? What factors determine the directions and rates of reactions?

7. State the first and second laws of thermodynamics and describe their general implications. Describe also their specific implications for chemical processes. Why are most natural processes unidirectional in practice?

8. Define enthalpy, entropy, free-energy change. How does each of these concepts apply to a reacting system? What does the ΔF of a reaction indicate? How do energy changes differ for exergonic and endergonic reactions? Define calorie.

9. How do pressure and concentration affect a reaction? Review the effect of environmental heat on reactions. What is activation energy? How is this energy related to free-energy changes?

10. What is a catalyst? What is an enzyme and how does it work? What is an active site of an enzyme? A cofactor? How does enzyme activity affect the activation energy of a reaction?

11. Show how enzyme activity varies with temperature and pH. What are the usual optimal conditions for enzyme activity? Review the general characteristics of enzymes.

12. Why is a carbohydrase ineffective in accelerating the reaction glycerin + fatty acids $\rightarrow$ fat + water? What kind of enzyme does such a reaction actually require?

Collateral Readings

Baker, J. J. W., and G. E. Allen: "Matter, Energy, and Life," Addison-Wesley, Reading, Mass., 1965. This paperback contains accounts on reaction energetics, catalysis, reaction rates and equilibria, and thermodynamics. Recommended.

Frieden, E.: The Enzyme-Substrate Complex, *Sci. American,* Aug. 1959. A popularly written article on enzymatic reactions.

Grunwald, E., and R. H. Johnsen: "Atoms, Molecules, and Chemical Change," Prentice-Hall, Englewood Cliffs, N.J., 1964. Introductory, directly pertinent to the topics of this chapter.

King, E. L.: "How Chemical Reactions Occur," Benjamin, New York, 1963. A paperback especially relevant to the contents of this chapter. Highly recommended.

Lessing, L.: The Life-saving Promise of Enzymes, *Fortune,* Mar., 1969. An excellent popular review of recent advances in our knowledge and applications of enzyme action.

Phillips, D. C.: The Three-dimensional Structure of an Enzyme Molecule, *Sci. American,* Nov., 1966. The makeup and manner of action of an antibacterial enzyme are examined.

Sienko, M. J., and R. A. Plane: "Chemistry," 3d ed., McGraw-Hill, New York, 1966. A recommended basic text, containing good accounts on the energetics and nature of reactions.

Speakman, J. C.: "Molecules," McGraw-Hill, New York, 1966. This paperback contains detailed but not too difficult accounts on atomic and molecular structure, written from both a chemical and a biochemical point of view. Some parts are clearly relevant to the topics of this chapter.

White, E. H.: "Chemical Background for the Biological Sciences," Prentice-Hall, Englewood Cliffs, N.J., 1964. A chapter on chemical reactions is included in this paperback.

The first chapter of this part examines the structure, or *organization*, of cells, and the two following chapters analyze cellular function, or *operation*. The discussion of structure deals with the chemical compounds that compose a cell; the physical properties that these compounds confer on a cell; and the ways in which the compounds are joined as higher levels of organization inside a cell. With respect to function, it is shown how, by virtue of its particular organization, a cell is able to metabolize and perpetuate itself, and how it actually performs these activities. In short, to the extent of current understanding, this examination outlines how cells actually can be alive.

PART 2
ANIMAL CELLS

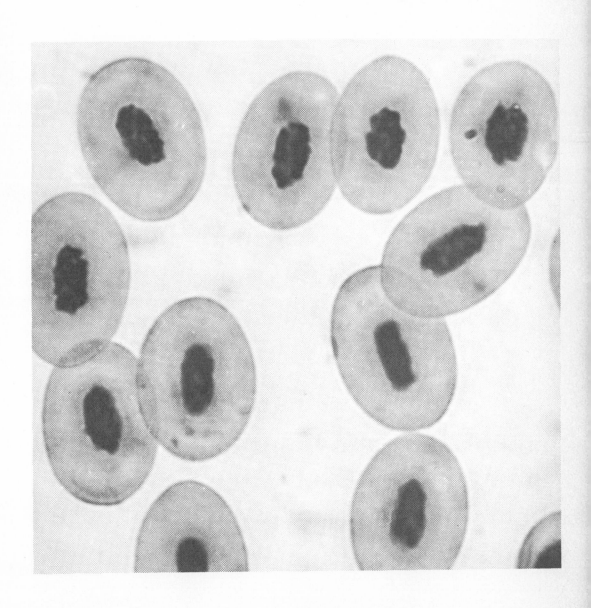

4

CELLULAR ORGANIZATION

No living cell is ever exactly like any other or exactly the same from moment to moment; living matter is not a static, passive material. New substances enter continuously, wastes and manufactured products leave continuously, and substances in the cell interior are continuously transformed chemically and redistributed physically. As a result, the internal components of a cell always are in an agitated state. To the casual human observer a sponge or a bone might appear to be a rather placid, inactive structure; but if the contents of sponge or bone cells could be seen, they all would be noted to be in unceasing motion, interacting and changing. Consequently, the parts of an animal change continuously, and so indeed does every animal as a whole.

However, despite such changes between and within cells, certain very basic features nevertheless remain the same. Representing the universal heritage passed on by the very first cells on earth, these basic characteristics are in part *chemical,* in part *physical,* and in part *biological.*

Chemical Organization

Elements and Compounds

Four of the most widely distributed chemical elements on earth—oxygen, carbon, hydrogen, and nitrogen—make up approximately 95 percent of the weight of cellular animal matter (Table 3). Some thirty other elements contribute the remaining 5 percent of the weight. All these elements occur in the ocean; animals have originated in water, and their cells still reflect the composition of the sea.

Most of the elements are joined together as compounds. One major class of such compounds comprises water, minerals, metallic and nonmetallic materials, and in general those substances that

Table 3. **The relative abundance of chemical elements in animal matter**

ELEMENT	SYMBOL	WEIGHT, PER CENT
oxygen	O	62
carbon	C	20
hydrogen	H	10
nitrogen	N	3
calcium	Ca	2.50
phosphorus	P	1.14
chlorine	Cl	0.16
sulfur	S	0.14
potassium	K	0.11
sodium	Na	0.10
magnesium	Mg	0.07
iodine	I	0.014
iron	Fe	0.010
		99.244
trace elements		0.756
		100.00

make up the bulk of the physical, nonliving universe. A good many of such substances occur also in animal matter. This class represents the so-called *inorganic compounds*. Directly or indirectly, all inorganic compounds in cells are of mineral origin and ultimately are derived as finished nutrients from the external physical environment.

The most abundant cellular mineral is *water,* present in amounts ranging from 5 to 90 or more percent. For example, the cellular water content of tooth enamel is about 5 to 10 percent; of bone, about 25 percent if without marrow and about 40 percent if with marrow; of muscle, 75 percent; of brain or milk, 80 to 90 percent; and of jellyfish, 90 to 95 or more percent. As a general average, cellular matter is about 65 to 75 percent water.

The other inorganic components of cells are *mineral solids*. Such substances are present in amounts averaging about 1 to 5 percent. A considerable fraction of the minerals often forms hard deposits, such as crystals inside cells or secreted precipitates on the outside of cells. Silicon- or calcium-containing deposits are common. For example, certain protozoa are protected externally with layers of glasslike silica; the hard part of bone is largely calcium phosphate, secreted in layers around bone-forming cells; clamshells consist of secreted calcium carbonate.

Other cellular minerals are in solution, either as free ions or combined with other compounds. The most abundant positively charged inorganic ions are H^+, hydrogen ions; Ca^{++}, calcium ions; Na^+, sodium ions; K^+, potassium ions; and Mg^{++}, magnesium ions. Abundant negatively charged mineral constituents include OH^-, hydroxyl ions; $CO_3^=$, carbonate ions; HCO_3^-, bicarbonate ions; $PO_4^{\equiv}$, phosphate ions; Cl^-, chloride ions; and $SO_4^=$, sulfate ions. In general, the kinds of minerals found in cells occur also in the ocean and in rocks. This is not a coincidence, for rocks are dissolved by water, water finds its way to the ocean and to soil, and animals ultimately draw their mineral supplies from these sources.

The second large group of cellular chemicals comprises the *organic compounds*. These are so named because they occur almost exclusively in living or once-living matter. Organic compounds often are exceedingly complex, and they in particular are responsible for the ''living'' properties of cells. Moreover, organic compounds represent the *foods* required by animals; inorganic compounds too serve as nutrients, but they are not foods. Chemically, organic substances are compounds of *carbon,* or, more specifically, compounds in which the main chemical bonds join two or more carbon atoms or carbon and hydrogen atoms. Thus, carbon dioxide (CO_2), carbonate ions ($CO_3^=$), and materials derived from them are not organic, since the carbon here is bonded to oxygen. But methane, CH_4, with its carbon—hydrogen bonds, is an organic compound. Most of the organic compounds of zoological interest contain not only carbon—hydrogen but also numerous carbon—carbon bonds.

In this respect carbon is a rather unusual element. The atoms of most other elements link to atoms of like kind too, but the number of atoms so bondable is usually quite limited. For example, a hydrogen atom can join with one other hydrogen atom at most ($H—H$, H_2); sulfur can form molecules of eight atoms, S_8. But a carbon atom is far more versatile. It can form as many as four (covalent) bonds with other atoms, including other carbon atoms. Long *chains* of carbon atoms can form in this way:

Such chains represent parts of organic molecules in which various other atoms are attached to the carbons. Further, carbon atoms can be joined in ringlike fashion, as in *benzene,* for example:

Many other types of configuration exist. Thus, carbon chains can be branched, rings and chains can become joined to one another, and any of these carbon structures can also be three-dimensional. Carbon combinations therefore can be exceedingly complex and varied. Organic substances actually display more complexity and more variety than all other chemicals put together.

Cells contain hundreds of different classes of organic compounds. Of these, four classes in particular are found in all cells and form the organic basis of living animal matter. The four are

1. Carbohydrates
2. Lipids
3. Proteins
4. Nucleotides

Like mineral compounds, some of these organic substances contribute to the formation of hard parts. For example, the various kinds of horny materials present in many skeletons and also in claws and hoofs are predominantly organic. More generally, however, organic materials are dissolved or suspended in cellular water. An animal such as man contains about 15 percent protein, about 15 percent fat (lipids), and other organic components to the extent of about 1 percent. Evidently, the inorganic matter (mainly water) far outweighs the organic, and this is true of animals generally.

Carbohydrates

These compounds are so named because they consist of carbon, hydrogen, and oxygen, the last two in a 2 : 1 ratio, as in water. The general atomic composition usually corresponds to the formula $C_x(H_2O)_y$, where x and y are whole numbers.

If x and y are low numbers, from 3 to about 7, then the formula describes the composition of the most common carbohydrates, the simple sugars, or *monosaccharides*. In these, the carbon atoms form a chain to which H and O atoms are attached. Several classes of monosaccharides are distinguished on the basis of the numbers of car-

bons present: C_3 sugars are *trioses*; C_4 sugars, *tetroses*; C_5 sugars, *pentoses*; C_6 sugars, *hexoses*; and C_7 sugars, *heptoses*. The suffix *-ose* always identifies a sugar.

According to the pattern in which H and O atoms are attached to the carbon chains, two series of sugars can be distinguished, the *aldose* and the *ketose* sugars. As shown in Fig. 4.1, their main difference is that a $-\overset{|}{C}=O$ group is terminal

in aldoses and subterminal in ketoses. Among individual sugars referred to in several later contexts are certain pentoses, particularly the aldose *ribose*, and certain hexoses, notably the aldose *glucose* (possibly the most common of all sugars) and the ketose *fructose* (Fig. 4.2).

Two or more similar or identical monosaccharides can become joined together end to end as chainlike larger molecules. If two monosaccharides are joined in this way, a double sugar, or *disaccharide*, is formed. For example, combination of two glucose units yields the disaccharide *maltose*, malt sugar; combination of glucose and fructose yields *sucrose*, the cane or beet sugar used familiarly as a sweetening agent; and combination of glucose and the hexose sugar galactose yields *lactose*, milk sugar. All three of these disaccharides have the formula $C_{12}H_{22}O_{11}$, and their formation is described by the same equation:

$$2\ C_6H_{12}O_6 \longrightarrow H_2O + C_{12}H_{22}O_{11}$$

If many more than two monosaccharide units are joined together, multiple sugars, or *polysaccharides*, are the result. The general chemical process through which large molecules are built up from smaller units of like type is known as *polymerization*. Polysaccharides are said to be *polymers* of simple sugars. A polymer consisting of some hundreds or thousands of glucose units forms *glycogen*, an animal polysaccharide of considerable importance. Another polysaccharide is *cellulose*, a polymer of up to 2,000 glucose units, rare among animals but very common in plants.

In a good many polysaccharides the joined sugar units form straight, unbranched chains. The union between adjacent sugar units in such cases is said to be a 1,4 link: the first carbon of one unit links to the fourth carbon of the adjacent unit. Cellulose is a good example of polysaccharides that contain 1,4 links exclusively. In glycogen and other polysaccharides, however, so-called 1,6

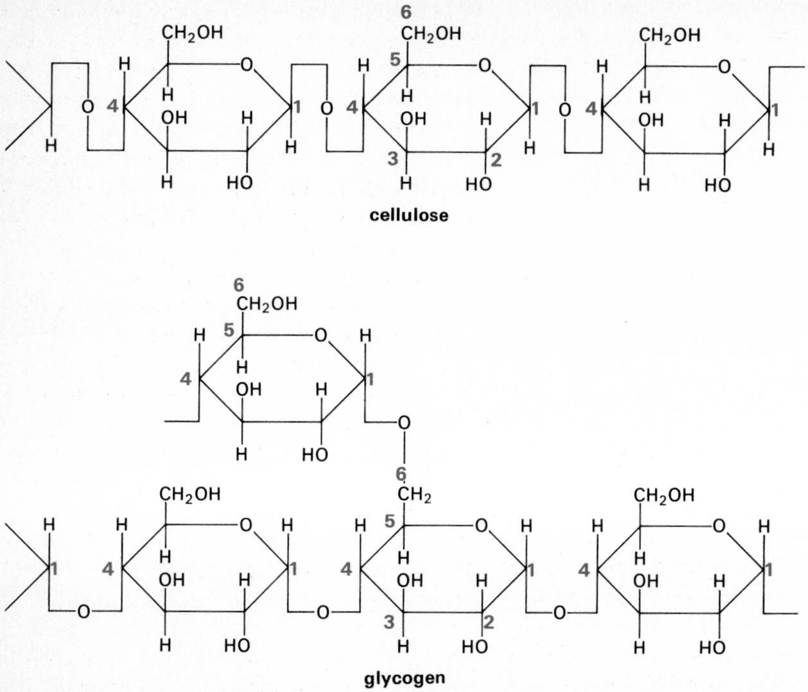

cellulose

glycogen

Figure 4.3 Polysaccharides. The two shown here are composed entirely of glucose units. In cellulose two adjacent glucose units are joined by a 1,4 link that forms a straight-chain compound. In glycogen the straight-chain portions have 1,4 links and branch-chain portions are joined by 1,6 links (carbon 6 of one glucose unit joins carbon 1 of another glucose unit).

are energy-rich molecules suitable as fuels in respiration. Carbohydrates therefore are important animal foods. Glycogen specifically represents the chief form in which carbohydrates are stored in animals, and glucose is the chief form in which carbohydrates are transported from cell to cell and also over greater distances by blood.

Lipids

Fats and their derivatives are known collectively as lipids. The chief lipids are *fatty acids*. Like the sugars, these acids are composed of C, H, and O, the carbon atoms being arranged as chains of various lengths. At one end such a chain carries a *carboxyl* group, —COOH, which confers acid properties on a fatty acid; the H of the carboxyl group can dissociate: $—COOH \rightarrow —COO^- + H^+$

The simplest fatty acid is *formic acid* (HCOOH). It occurs occasionally in sweat and urine and also plays a protective role in some ants, where it can be squirted out as an irritant spray against potential enemies. A series of increasingly more complex fatty acids is formed by successive addition of —CH_2— groups to HCOOH. Addition of one such group produces *acetic acid* (CH_3COOH), the active ingredient of vinegar. Beyond acetic acid, fatty acids have the general formula $CH_3(CH_2)_nCOOH$, where *n* is an integer other than zero (Fig. 4.4).

In most animal fatty acids *n* is an even number; cells synthesize fatty acids from acetic acid (2-carbon, or even-numbered) building units. Common fatty acids in most animal matter include, for example, *palmitic acid* and *stearic acid* (see Fig. 4.4). Fatty acids like these are said to be *saturated;* all available bonds of the carbon chains are filled with hydrogen atoms. By contrast, *unsaturated* fatty acids have one or more double bonds in the carbon chains. For example, the widely occurring *oleic acid*, $CH_3(CH_2)_7CH{=}CH(CH_2)_7COOH$ (or $C_{18}H_{34}O_2$), contains one double bond. Fatty acids having more than one double bond are said to be *polyunsaturated.*

In a fat molecule, three fatty acid units are joined to one glycerin unit (Fig. 4.5). The properties of a fat are determined by the chain lengths and the degrees of saturation of the fatty acids present. Fats containing fatty acids that are short-chained, unsaturated, or both tend to be volatile or oily liquids. For example, oleic acid is oily, and fats containing oleic acid tend to be oily too. By contrast, fats with long-chained and saturated fatty acids tend to be hard tallow. This is the case in,

linkages occur as well: not only carbon 4 but also carbon 6 of a given sugar unit links to carbon 1 of adjacent units. Such polysaccharide molecules are variously forked and branched as a result (Fig. 4.3).

As a group, carbohydrates function in animal cells in two general capacities: they are structural building blocks of the cellular substance, and they

Figure 4.4 Fatty acids. The (incomplete) series is arranged according to increasing molecular complexity. Carbon positions are identified by Greek letters, starting at the carbon next to the carboxyl group.

structure	composition	name
H—COOH α	CH_2O_2	formic
CH_3—COOH β $\quad$ α	$C_2H_4O_2$	acetic
CH_3—CH_2—COOH γ $\quad$ β $\quad$ α	$C_3H_6O_2$	propionic
CH_3—CH_2—CH_2—COOH	$C_4H_8O_2$	butyric
$CH_3(CH_2)_4COOH$	$C_6H_{12}O_2$	caproic
$CH_3(CH_2)_5COOH$	$C_8H_{14}O_2$	caprylic
$CH_3(CH_2)_{14}COOH$	$C_{16}H_{32}O_2$	palmitic
$CH_3(CH_2)_{16}COOH$	$C_{18}H_{36}O_2$	stearic

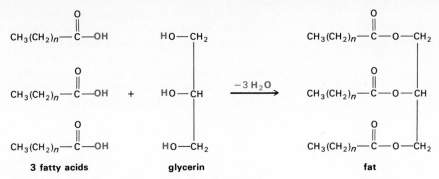

Figure 4.5 Fat formation. Three molecules of fatty acid combine with one molecule of glycerin, resulting in three water molecules and one fat molecule.

boxyl group, and —R an atomic grouping that can vary in composition considerably. For example, the simplest amino acid is *glycine,* where R = H; if R = CH_3, the amino acid is called *alanine.* Many other amino acids are characterized by comparatively more complex —R groups. Cells typically contain 23 different types of amino acid (Fig. 4.6).

Hundreds and even thousands of amino acid units can be joined together in a single protein molecule. Whenever molecules attain exceedingly large sizes, they are referred to as *macromolecules.* Proteins very often are macromolecules, and some of them are among the largest chemical structures known. As such, proteins are associated most intimately with the phenomenon we call ''life.''

Adjacent amino acids in a protein are united in such a way that the amino group of one acid links to the carboxyl group of its neighbor; the bond is formed by removal of one molecule of water (Fig. 4.7). The resulting grouping —NH—CO— represents a *peptide bond,* and two amino acids so joined form a *dipeptide.* If many amino acids are polymerized by means of peptide bonds, the whole chainlike complex is a polypeptide.

Chemically, polypeptides can vary in practically unlimited fashion:

1. They can contain any or all of the 23 different naturally occurring *types* of amino acids.

2. They can contain almost any *number* of each of these amino acid types.

3. The specific *sequence* in which given numbers and types of amino acids are joined as a chain can vary almost without restriction.

Thus, amino acid units can be envisaged as an ''alphabet'' of 23 ''letters,'' and an astronomically large number of different polypeptide ''sentences''

for example, tristearin, a common animal fat that contains three stearic acids per molecule.

Fats and fatty acids are the most abundant food-storage compounds in most animals. Also, like carbohydrates, fats play significant roles as structural components of cells. For example, they are present in cellular membranes, where they probably contribute to controlling the traffic of materials into and out of cells. Moreover, fats and fatty acids are even richer sources of respiratory energy than carbohydrates.

Proteins

These compounds are polymers of molecular units called *amino acids.* The general structure of an amino acid is represented by the formula

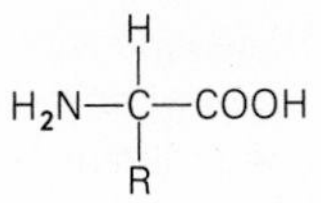

$$H_2N-\overset{\overset{\displaystyle H}{|}}{\underset{\underset{\displaystyle R}{|}}{C}}-COOH$$

where —NH_2 is an amino group, —COOH a car-

Figure 4.6 Amino acids. The structure of eight representative ones is shown.

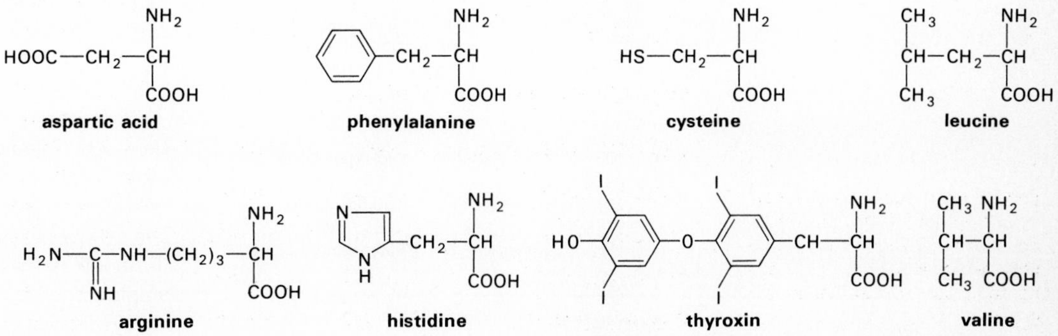

aspartic acid

phenylalanine

cysteine

leucine

arginine

histidine

thyroxin

valine

59

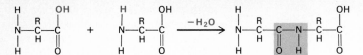

Figure 4.7 Amino acid bonding. Two amino acids combine with loss of water. The result is a peptide bond, as indicated in the tinted area at right.

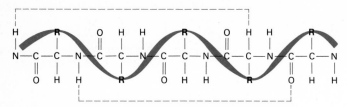

Figure 4.8 The spiral structure of protein. A line connecting the R— fractions of consecutive amino acid units in a polypeptide chain marks out a spiral called an α-helix. Such a helix is held together by hydrogen bonds between the H of the —NH of one amino acid unit and the O of a —CO— in another amino acid three units distant (colored broken lines).

can be constructed from this alphabet. Correspondingly, the possible number of chemically different proteins likewise is astronomical. Indeed no two animals or organisms of any kind have exactly the same types of proteins. A polypeptide chain with its particular sequence of amino acid units and peptide bonds represents the so-called *primary* structure of a protein.

Such a chain has the physical form of a twisted ribbon. If a line were drawn through all the —R portions of the consecutive amino acids present, the line would mark out a spiral (Fig. 4.8). Such spirals in some cases are "right-handed" (*α-helix*), in others "left-handed" (*β-helix*). In either configuration the backbone of the spiral is a bonded sequence of

$$-N-C-C-N-C-C-N-$$

Figure 4.9 Bonds in protein. Separate polypeptide chains or different segments of a single chain (vertical lines) can be held together by bonds such as shown. In a hydrogen bond, an H atom is held in common by two side groups on the polypeptides (top and bottom). Disulfide bridges are formed by S-containing amino acid units, mainly cysteine. Ionic links between charged side groups on polypeptide chains hold together by electric attraction.

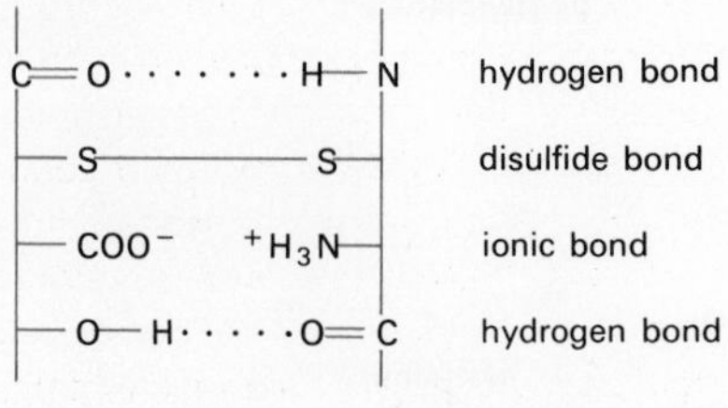

C=O · · · · · · H—N	hydrogen bond
—S————S—	disulfide bond
—COO⁻ ⁺H₃N—	ionic bond
—O—H · · · · O=C	hydrogen bond

atoms, each —N—C— portion representing the skeleton of one amino acid unit. Projecting out from this ribbon are the H—, O=, and R— groups of the amino acid unit. In most proteins the polypeptide spirals have fairly similar geometric properties: there are on the average 3.7 amino acid units per turn of a spiral. Accordingly, a chain of some 18 units forms a helix with five complete turns.

This spiral configuration is held together by *hydrogen bonds*. Such a bond is formed here when a H atom is shared between the nitrogen of one amino acid unit and an oxygen of a nearby unit. More precisely, the H of the —NH of one unit is bonded to the O of —C=O three amino acid units away (see Fig. 4.8). All —NH and —C=O groups of a polypeptide chain are hydrogen-bonded in this fashion, each such bond linking amino acids three units apart. The resulting spiral configuration is relatively stable, and it represents the *secondary structure* of a protein.

If long coils of this sort remain extended and threadlike, the protein molecule is said to be *fibrous*. In many cases, however, the coils are looped and twisted and folded back on themselves, in an almost infinite variety of ways. Protein molecules then are *globular*, balled together somewhat like entangled twine. Such loops and bends give a protein a *tertiary structure*. Where present, a tertiary configuration is held together chiefly by three kinds of bonds (Fig. 4.9). One again is the hydrogen bond, which in this case links together more or less distant portions of a polypeptide chain. Another is an *ionic bond*, formed when a carboxyl group and an amino group of two distant terminal amino acid units come to lie near each other. A third type of link is the *disulfide bond*, which arises between sulfur-containing amino acids. Sulfur occurs most often in the form of —SH groups, and the most common —SH-containing amino acid is *cysteine*. If then two distant cysteine units in a folded polypeptide come to lie close to each other, their —SH groups can link together and form a "disulfide bridge," —S—S—.

Some proteins consist of not one but several separate polypeptide chains bonded to one another, often in the form of a bundle. Proteins of this type are said to have a *quaternary structure*. The polypeptides here are held together largely by the bond types already referred to above. For example, one of the hormones of the pancreas, *insulin*, is a protein made up of two parallel poly-

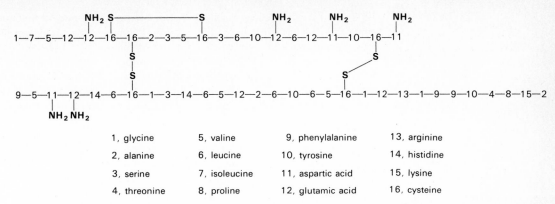

1, glycine   5, valine   9, phenylalanine   13, arginine

2, alanine   6, leucine   10, tyrosine   14, histidine

3, serine   7, isoleucine   11, aspartic acid   15, lysine

4, threonine   8, proline   12, glutamic acid   16, cysteine

Figure 4.10 Insulin: a protein consisting of 51 amino acid units. One chain consists of 21 units, another of 30 units, and the two are held together by sulfur-containing cysteine units.

peptide chains, and these are held together by two disulfide bonds (Fig. 4.10). Similarly, in a hemoglobin molecule of vertebrate blood the protein component consists of four linked polypeptide chains. Differences in the properties of proteins thus can arise from differences in as many as four aspects of structure. Therefore, even if two proteins have identical primary structure, they will have different properties if their secondary, tertiary, or quaternary structures are not the same.

Of all the bonds that maintain a globular protein configuration, hydrogen bonds are the weakest. Such bonds are disrupted readily by physical or chemical changes in the environment of a protein— for example, by excessive heat, pressure, pH, electricity, heavy metals, and other agents that create suboptimal conditions. When some or many of its hydrogen bonds are broken, a protein cannot retain its globular configuration and its quaternary, tertiary, or even secondary structure will be lost. An originally highly folded molecule then will become a straight, fibrous one. Also, in the original globular state the protein might have been soluble in water, for ball-like particles disperse readily; but long fibrous molecules pile together like a log jam and generally become insoluble in water.

Such changes in protein configuration are called *denaturation.* If the disruptive effect is mild and of brief duration, denaturation can be temporary and the protein later can revert to its original *native* state. But if the environmental change is drastic and lasting, then denaturation becomes permanent and irreversible and the protein will be *coagulated.* For example, the protein of egg white, albumen, is globular (and water-soluble) in the raw native state, but becomes fibrous (and water-insoluble) in the cooked, coagulated state (like

boiled egg white). Denaturation usually destroys the biological properties a protein has in the native state—a major reason why excessive heat or other excessive environmental changes kill cells.

Proteins differ considerably from carbohydrates or fats in their nearly unlimited structural variations. Even a highly complex carbohydrate, for example, is structurally the same whether we obtain it from mushrooms or mangoes, from mice or from men. A given lipid, similarly, is the same lipid regardless of where we find it. Not so for proteins, however; these compounds vary so much that, as noted, no two organisms contain precisely the same types. Even twin animals have slightly different proteins, and the differences are the greater the less closely two animals are related. Proteins evidently have a high degree of *specificity:* the proteins of an animal are unique for that animal.

Protein specificity has well-known consequences. For example, transfer of protein from one organism into the cells of another amounts to an introduction of foreign bodies, and disease can follow. Thus the proteins of plant pollen can produce allergic reactions in man. Or when blood of one animal is mixed with incompatible blood of another, the recipient can suffer protein shock that can be lethal. Partly because bacterial proteins differ from those of other organisms, infectious bacteria produce many diseases. And when body parts of one animal are grafted to another animal, the transplants normally do not heal in and are rejected sooner or later because the two sets of proteins differ. Special medication can sometimes retard or forestall graft rejection in man.

To some extent, normally far less than carbohydrates or fats, proteins are used in cells as foodstuffs. But proteins serve mainly in two far more important cellular roles. First, they represent vital construction materials out of which much of the basic framework of cells is built. Carbohydrates, fats, minerals, and other cellular components are organized around such *structural proteins* that form the molecular "scaffolding" of living material. Good examples here are *myosin*, the characteristic protein of muscle cells; *keratin*, the characteristic protein of hair and skin cells in mammals; and *collagen*, the fiber-forming protein produced by cells in bone, cartilage, tendons, and many other animal tissues.

Second, many proteins serve as reaction-catalyzing enzymes. Life depends on enzymatic acceleration of reactions, and "living" therefore means dependence on enzymatic *functional proteins*. Enzymes share the chemical and physical properties of proteins generally. As proteins are specific, so enzymes are specific; and if denaturation of enzyme proteins alters or destroys the native configuration, then specific enzymatic capacities will be lost (unless the active sites remain intact and accessible).

For both structural and functional reasons, therefore, cellular life would probably not be possible without molecular agents such as proteins. But even with proteins and all the other cellular compounds already described, a cell could not yet be alive; the chemicals discussed thus far only endow a cell with the *potential* of having a structure (proteins and other constituents), the *potential* of

performing functions (enzymes), and the *potential* of accumulating usable foods (carbohydrates and fats). The cell has not yet been equipped chemically to make these potentials actual: *how* to use the foods, *what* actual structure to develop, and *which* functions to carry out. These all-important capacities emerge from the organic compounds considered next.

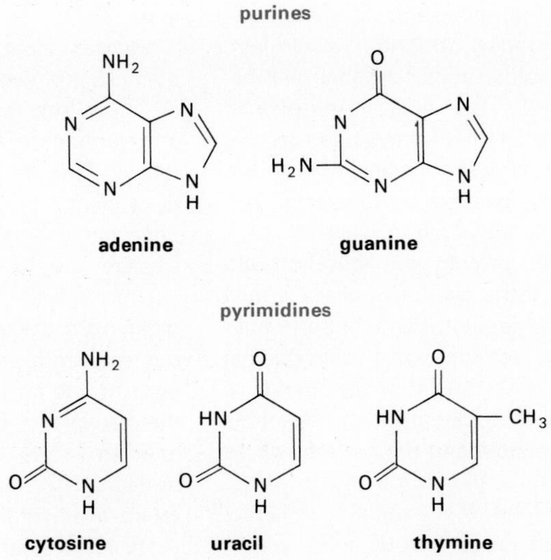

Nucleotides

A nucleotide is a molecular complex of three units: a *phosphate group*, a *pentose sugar*, and a *nitrogen base*. Phosphate groups are derivatives of phosphoric acid (H_3PO_4), an inorganic mineral substance. If this formula is rewritten as $H-O-H_2PO_3$, then the $-O-H_2PO_3$ part represents the phosphate group of present concern. And if the $-H_2PO_3$ portion of the group is represented simply as P, the whole phosphate group can be symbolized as $-O-P$.

The pentose sugar in a nucleotide is one of two kinds, *ribose* or *deoxyribose* (see Fig. 4.2). Therefore, according to the kind of sugar present, two types of nucleotide can be distinguished: *ribose nucleotides* (or *ribotides*) and *deoxyribose nucleotides* (or *deoxyribotides*).

The nitrogen base of a nucleotide is one of a series of ring compounds that contain nitrogen as well as carbon. A single ring occurs in *pyrimidines*, and a double ring characterizes *purines*. Pyrimidines include three variants of significance, *thymine, cytosine,* and *uracil*. Among purines are two important types, *adenine* and *guanine* (Fig. 4.11).

The nitrogen base in a ribose nucleotide usually is either uracil or cytosine or adenine or guanine. Similarly, a deoxyribose nucleotide typically contains either thymine or cytosine or adenine or guanine. Thus, nucleotides occur in the following two series:

RIBOTIDES	DEOXYRIBOTIDES
adenine—ribose—O—P	adenine—deoxyribose—O—P
guanine—ribose—O—P	guanine—deoxyribose—O—P
cytosine—ribose—O—P	cytosine—deoxyribose—O—P
uracil—ribose—O—P	thymine—deoxyribose—O—P

Uracil occurs only in the ribose series, thymine only in the deoxyribose series; adenine, guanine, and cytosine occur in both. Each of these nucleo-

Figure 4.11 Nitrogen bases.

tides has at least three different but equivalent names. For example, adenine—ribose—O—P can be called *adenine ribotide* or *adenylic acid* or *adenosine monophosphate* (AMP for short). The other nucleotides have similarly constructed designations (for example, guanylic acid or guanosine monophosphate or GMP; cytidylic acid or cytidine monophosphate or CMP; uridylic acid or uridine monophosphate or UMP; thymidylic acid or thymidine monophosphate or TMP).

Nucleotides are building blocks of larger molecules that serve three crucial functions in cells: some are *energy carriers;* others are *coenzymes;* and still others form *genetic systems.*

Energy Carriers A nucleotide can link up in serial fashion with one or two additional phosphate groups. For example, if to adenosine monophosphate (AMP) is added one more phosphate, then *adenosine diphosphate,* or ADP, is formed; and if a third phosphate is added to ADP, the result is *adenosine triphosphate,* or ATP:

adenine—ribose—O—P   AMP
adenine—ribose—O—P—O∼P   ADP
adenine—ribose—O—P—O∼P—O∼P   ATP

The wavy symbol in the added —O∼P groups indicates the presence of a so-called *high-energy bond.* The significance of such bonds will become clearer in Chap. 5. Here it need be noted only that the formation of bonds between certain atoms, notably the —O∼P— bonds, requires particularly large amounts of energy. They also release correspondingly large amounts of energy when they are broken. Thus, to convert AMP to ADP and ADP to ATP requires not only additional phosphate groups but also considerable energy inputs. The energy is derived in cells from respiratory fuels, and we shall find that the primary function of respiration actually is to create the high-energy bonds of ATP. This compound is the significant energy-rich endproduct of respiration. In effect ATP is an *energy carrier,* the most widespread of such carriers in cells. Derivatives of some of the other nucleotides play a more limited energy-carrying role. For example, by addition of phosphate groups and energy, UMP can become UDP and UTP.

Coenzymes A coenzyme is a carrier molecule that functions in conjunction with a particular enzyme. It happens often in a metabolic process that a group of atoms is removed from one compound and transferred to another. In such cases a specific enzyme accelerates the decomposition reaction that brings about the removal, but a specific coenzyme must also be present to carry out the transfer. The coenzyme temporarily joins with, or accepts, the removed group of atoms and later transfers it to another compound (Fig. 4.12).

Figure 4.12 Coenzymes. *A,* the general function of these substances: if *AX* is to be converted to *BX,* an enzyme catalyzes the transformation of *A* to *B,* while a coenzyme transfers *X* from *A* to *B. B,* many coenzymes are derived from vitamins and nucleotides. 1, from the B vitamin riboflavin cells manufacture the coenzyme FAD. 2, the B vitamin niacin is the cellular source for the formation of nicotinamide, a compound used in the synthesis of coenzyme NAD. 3, the B vitamin pantothenic acid is a component in the construction of coenzyme A in cells.

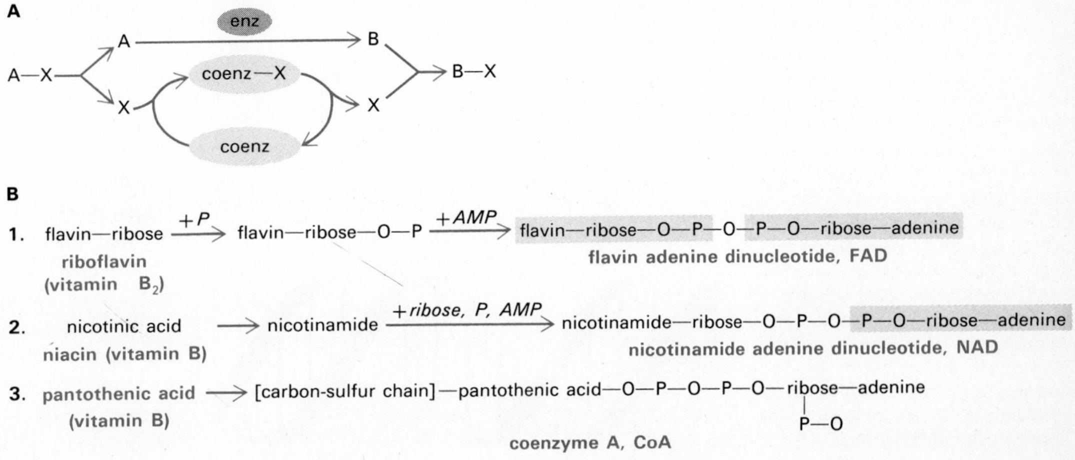

A

A—X ⟨ A —[enz]→ B ⟩ B—X
     X —coenz—X— X
        coenz

B

1. flavin—ribose $\xrightarrow{+P}$ flavin—ribose—O—P $\xrightarrow{+AMP}$ flavin—ribose—O—P—O—P—O—ribose—adenine
    riboflavin   flavin adenine dinucleotide, FAD
    (vitamin B$_2$)

2.   nicotinic acid $\longrightarrow$ nicotinamide $\xrightarrow{+ribose, P, AMP}$ nicotinamide—ribose—O—P—O—P—O—ribose—adenine
    niacin (vitamin B)   nicotinamide adenine dinucleotide, NAD

3.   pantothenic acid $\longrightarrow$ [carbon-sulfur chain]—pantothenic acid—O—P—O—P—O—ribose—adenine
    (vitamin B)   P—O

                                        coenzyme A, CoA

The majority of coenzymes happen to be chemical derivatives of nucleotides. More specifically, in many coenzymes the nitrogen base of a nucleotide is replaced by another chemical unit, usually a derivative of a particular vitamin. For example, one of the B vitamins in animal diets is *riboflavin* (B_2). This compound consists of a ribose portion and, attached to it, a complex *flavin* portion. In cells a phosphate group becomes linked to riboflavin, and the result is the nucleotidelike complex flavin—ribose—O—P. This complex then can become joined to the nucleotide AMP, resulting in a double nucleotide known as *flavin adenine dinucleotide*, FAD (see Fig. 4.12). FAD is a cellular coenzyme that serves as carriers in many processes in which hydrogen is transferred from one compound to another (see also Chap. 5). Another hydrogen-carrying coenzyme, called NAD, is constructed from adenine-containing nucleotides and from nicotinamide, a derivative of the B vitamin nicotinic acid (niacin). Nucleotides and still another B vitamin, pantothenic acid, contribute to the structure of *coenzyme A*, or CoA, a compound that carries not hydrogen but another specific group of atoms. In the next chapter we shall encounter not only the coenzymes mentioned here but also some others that are not nucleotide derivatives.

Genetic Systems If any single group of chemicals could qualify as the "secret" of life, that group would unquestionably have to be the *nucleic acids*. (But since we can actually make such an identification today, it is really no longer possible to speak of any "secret.") Nucleic acids are *polynucleotides*, extended chains of up to thousands of joined nucleotide units.

Such chains are of two types, according to whether the nucleotides composing them belong to the ribose series or the deoxyribose series. A chain consisting of ribotides is a *ribose nucleic acid*, or RNA for short; and a chain of deoxyribotides is *deoxyribose nucleic acid*, DNA for short. In either type, the sugar component of one nucleotide unit bonds to the phosphate component of the next. Thus, the sugar (*S*) and phosphate (*P*) components form an extended molecular thread from which nitrogen bases (*N*) project as side chains:

$$-P-S-P-S-P-S-P-S-P-$$
$$\quad\ N\quad\ N\quad\ N\quad\ N$$

In the case of RNA, the particular types, numbers, and sequences of the four possible kinds of nitrogen bases can vary almost infinitely. A short segment of a long RNA molecule might, for example, contain nitrogen bases in a sequence such as

$$-P-R-P-R-P-R-P-R-P-R-$$
$$\quad\ A\quad\ U\quad\ G\quad\ G\quad\ C$$

where *R* stands for ribose and *A, U, G, C,* for adenine, uracil, guanine, and cytosine, respectively. In effect, RNA molecules differ as the sequences of their nitrogen bases differ; and that, actually, is the key to their importance. For the four possible nitrogen bases can be regarded as a four-letter "alphabet" out of which, just as with amino acids in proteins, any number of "words" and "sentences" can be constructed. As will become apparent later, the protein "sentences" precisely correspond to, and indeed are determined by, the RNA "sentences."

The original source of the sentences is not RNA itself, but DNA. This type of nucleic acid is a long *double* chain of nucleotides; two parallel single chains are held together by hydrogen bonds between pairs of nitrogen bases (Fig. 4.13). By virtue of their particular structure these bases can be paired only in four different ways: adenine with thymine or the reverse, and guanine with cytosine or the reverse. But apparently there is no limit to the number of times each of these combinations can occur in a long double chain. Nor do there appear to be restrictions as to their sequence. Thus $A \cdot T$, $T \cdot A$, $G \cdot C$, and $C \cdot G$ can be

Figure 4.13 DNA structure. A, the Watson-Crick model (*P*, phosphate; *D*, deoxyribose; *A, T, G, C,* purines and pyrimidines. A *P—D—A* unit represents one of the nucleotides.) In the —*P—D—P—D*— double chain, four kinds of purine-pyrimidine pairs are possible: $A \cdot T$, $T \cdot A$, $G \cdot C$, and $C \cdot G$. Each of the four can occur many times, and the sequence of the pairs can vary in unlimited fashion. B, the spiral structure of DNA. The two spirals symbolize the —*P—D—P—D*— chains, and the connections between the spirals represent the purine-pyrimidine pairs.

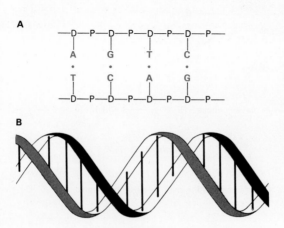

regarded as an original four-symbol alphabet, and sequences of any length can be constructed by using these symbols as often as desired and in any order. The possible number of different DNA's therefore is virtually unlimited. A final structural characteristic of DNA is that the double chain of the molecule is spiraled, not straight—a structure referred to as the *Watson-Crick model,* after the investigators who proposed it on the basis of X-ray diffraction studies of DNA (see Fig. 4.13).

Functionally, DNA exhibits three properties that make it the universal key to life. First, as will be shown in Chap. 6, DNA provides the cell with instructions on how to manufacture specific proteins. And through this control of protein manufacture, DNA ultimately controls the entire structural and functional makeup of every cell. Second, cellular DNA has the property of being self-duplicating; DNA is a *reproducing* molecule. That a chemical should be able to multiply itself under certain conditions may perhaps be astounding, but this capacity is nevertheless a known, unique property of DNA. Reproduction of DNA is at the root of all reproduction, and in a fundamental sense even the reproduction of a whole animal is, after all, a reproduction of "chemicals." Third, under certain conditions DNA can undergo *mutations,* or become slightly but permanently altered in its nitrogen-base content. When such changes occur the structural and functional traits of a cell can become changed correspondingly. Through changes in its cells a whole animal and its progeny then can become changed in the course of successive generations—a process equivalent to *evolution.*

In short, DNA is the material that forms *genes.* These cellular components have long been recognized as the carriers of heredity, but now they are known to be the ultimate controllers of all metabolism and all self-perpetuation. DNA molecules, or genes, thus are the basis of life. Chapter 6 will

show how the structure of the genetic nucleic acids actually permits these molecules to function as they do.

Other Constituents

Carbohydrates, lipids, proteins, and various nucleotide derivatives form the organic bulk of living matter. However, hundreds of other kinds of organic substances exist in cells. Although such substances often occur in only very small quantities, they are nevertheless of extreme importance in the maintenance of life. Some of these compounds are not related chemically to the four main categories above. Others are derivatives of one of the four groups, and still others are combinations of two or more of the basic four.

Few of these organic constituents occur universally in cells. More usually they are special components of particular cell types only, serving special functions. For example, *glucosamine* is a glucose derivative in which the —OH group on carbon 2 is replaced by the amino group —NH_2. A polymer of glucosamine forms the polysaccharidelike compound *chitin,* a hard, horny substance encountered widely in the exoskeleton of invertebrate animals such as arthropods. Another important glucose derivative is *ascorbic acid,* or vitamin C (Fig. 4.14).

Special fat derivatives include various compounds in which not three but only two fatty acid units are attached to glycerin. The third carbon of glycerin holds some other atomic grouping. Among substances of this type are *lecithin* and *cephalin,* found in small amounts in most accumulations of animal fat. Also related are compounds in which fatty acids are joined not to glycerin but to some other carrier molecule. Most *waxes* are of this type (beeswax, earwax, and waterproofing secretions on the surfaces of many animals; see Fig. 4.14).

Figure 4.14 Carbohydrate and lipid derivatives. Chitin and ascorbic acid are glucose derivatives. Lecithin resembles a fat (*R* stands for fatty acids), except that one of the fatty acids is replaced by another unit (a phosphate-choline combination). In a wax, one or more fatty acids are joined to an alcohol molecule (which substitutes for glycerin).

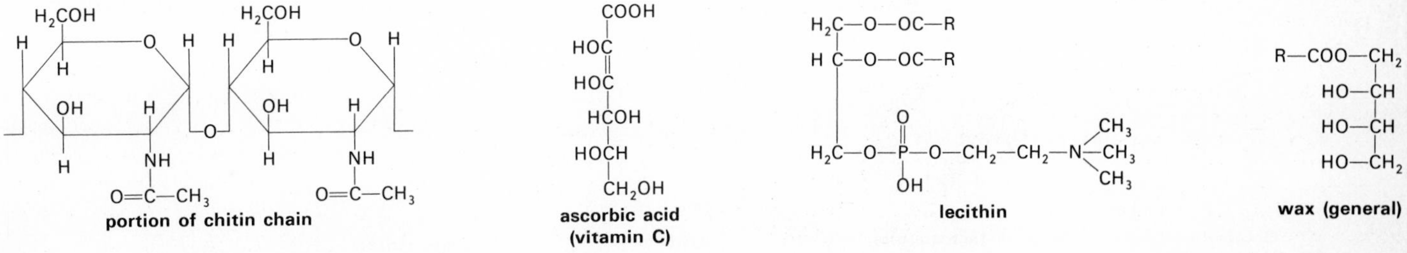

portion of chitin chain

ascorbic acid (vitamin C)

lecithin

wax (general)

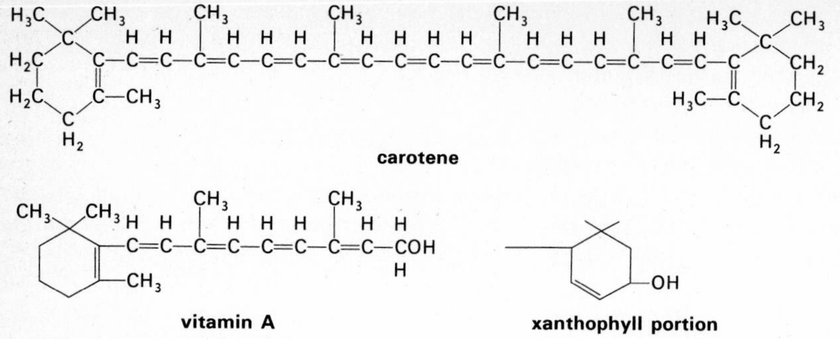

Figure 4.15 Lipid derivatives. When a molecule of carotene splits halfway along the carbon chain, two molecules of vitamin A are formed. Xanthophyll pigments resemble the carotenes in structure, except for differences in the terminal rings (only one end of a xanthophyll molecule is sketched here).

Among other lipid derivatives are the *carotenes*, fatty acid–like carbon chains that carry carbon rings at each end (Fig. 4.15). These compounds are pigments responsible for the yellow and cream colors of, for example, animal fat, egg yolk, milk, butter, and cheese. Vitamin A is a derivative of carotene. Closely related to the carotenes are the *xanthophyll* pigments, which produce the greens and reds on the body surfaces and in the interior of lobsters, crayfish, and other crustaceans, as well as the yellows in the feathers of canaries and some other birds.

Also related to lipids are *steroids,* complex ring structures that form the molecular framework of *cholesterol,* of vitamin D, and of the sex and certain adrenal hormones of vertebrates (Fig. 4.16). Still more distantly related to lipids are the *tetrapyrrols,* a group of pigmented compounds responsible for the red, blue, yellow, and other colors of, for example, the shells of robin and other bird eggs and of mammalian feces and urine. Some of the tetrapyrrols are complex ringlike molecules that contain a single atom of a metal in the center of such rings. Two important tetrapyrrols of this type are *cytochrome* and *heme,* red pigments in which the metal atom is iron (Fig. 4.17). Cytochrome is a hydrogen-transporting compound required in cell respiration (see Chap. 5); and heme is the oxygen-transporting component of hemoglobin in the blood of many animals. (The green pigment of photosynthetic organisms, *chlorophyll,* similarly is a ringlike tetrapyrrol, the central metal atom here being magnesium.)

Amino acids too form many special derivatives. For example, the amino acid tyrosine gives rise to the important *melanin* pigments. These are responsible for all yellow-brown, brown, and particularly black colors, both on animal body surfaces and in interior tissues (as also in the "inks" squirted out by excited squids and octopuses). Specialized pigment cells produce melanin, which accumulates in granules inside such cells. Among other tyrosine derivatives are various red and purple

Figure 4.16 Lipid derivatives: steroids. The four basic fused rings of a steroid are well seen in cholesterol. Note the great similarity between the male sex hormone testosterone, the female sex hormone estradiol, and the pregnancy hormone progesterone. The last in particular is similar also to cortisone, one of the hormones of the adrenal cortex.

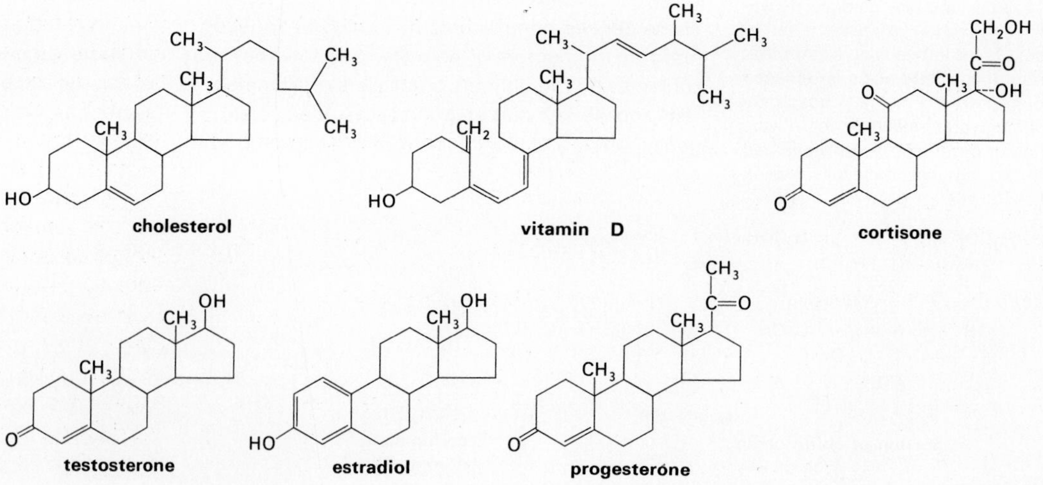

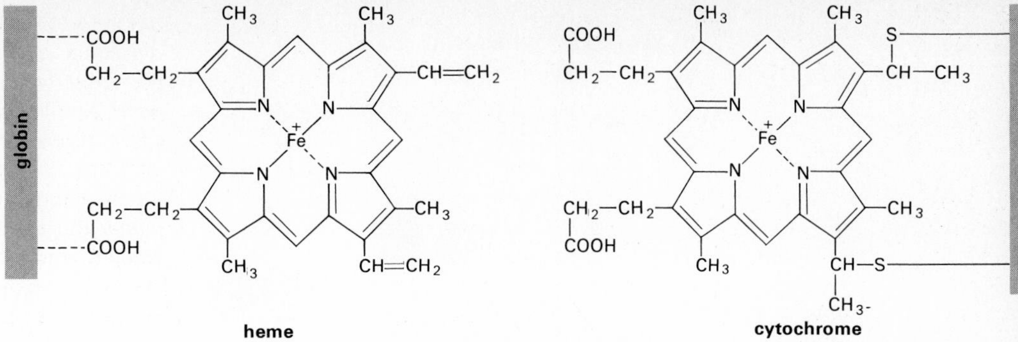

Figure 4.17 Lipid derivatives: tetrapyrrols. The ringlike tetrapyrrols heme and cytochrome are each bonded to a protein in cells. Hemoglobin transports oxygen in vertebrate blood, and cytochrome is a hydrogen carrier in cellular respiration.

pigments found among certain mollusks, as well as two vertebrate hormones, *thyroxin* and *adrenaline* (Fig. 4.18).

Thus, even without a lengthy listing of the enormous number of different cellular compounds, a general conclusion can be discerned: almost the whole vast array of organic substances in cells is related to or derived from only a half dozen or so fundamental types of compounds; and among these the main types are sugars, fatty acids, amino acids, and nucleotides. Nature apparently builds with but a limited number of fundamental construction units, yet the possible combinations and variations among them are virtually unlimited.

It should be kept in mind also that the diverse chemical components of a cell are not "just there," randomly and passively dissolved or suspended in water like the ingredients of a soup. Instead, the components interact and form a highly organized *living* system. This system con-

sists of a mixture of particular macromolecular and micromolecular substances, some organic and some inorganic. And as certain of them do and others do not dissolve in the water medium of a cell, the whole cell acquires a well-defined physical organization as described in the following section.

Physical Organization

Colloids

Any system composed of particles that are dispersed in another medium can be classified as belonging to one of three categories, depending on the size of the particles. If the particles are small enough to dissolve in the medium, the system is a true *solution*. If the particles are large—the size of soil grains, for example—they soon settle by gravity to the bottom of a container. Such a system is a coarse *suspension*. But, if the particles are of intermediate size, from about 1/1,000,000 to 1/10,000 mm in diameter, they neither form a solution nor settle out. Such a system is a *colloid*.

Any system composed of two kinds of components is a colloid if one of the components consists of particles of appropriate size. Eight general types of colloidal systems exist: a gas dispersed in either a solid or a liquid; a liquid in a liquid, a solid, or a gas; and a solid in a liquid, a solid, or a gas. The most common types are *sols*, in which solid colloidal particles are dispersed in liquids (for example, colloidal $Mg(OH)_2$ in water, as in milk of magnesia); *gels*, in which liquid col-

Figure 4.18 Amino acid derivatives. The amino acid tyrosine gives rise to the hormones thyroxin and adrenaline and also to the black pigment melanin (not shown), which is formed from dihydroxyphenyl alanine. Other amino acids have their own sets of chemical derivatives.

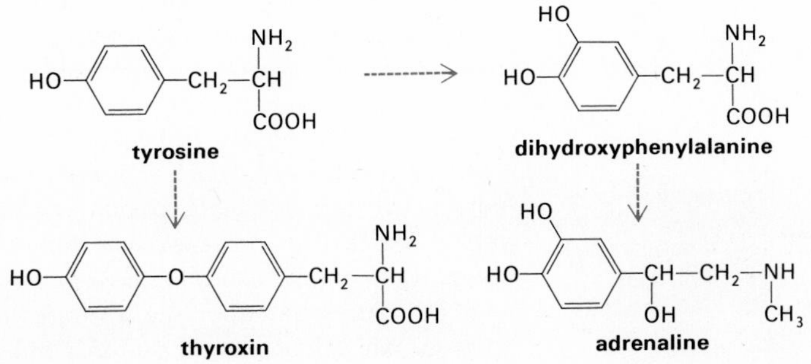

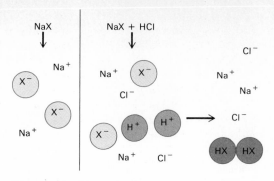

Figure 4.19 Dispersion in colloids. *Left,* if a material dissociates in water and if one of the ion types formed is of colloidal size (X^-), then the repulsion of like charges can maintain dispersion of the colloid. *Right,* if particles of opposite charge (H^+) are added to dispersed and charge-carrying colloidal particles (X^-), then the colloid can become neutralized electrically by formation of HX and settle out.

loidal particles are dispersed in solids (for example, colloidal water in protein, as in stiff gelatin or Jell-O); *emulsions,* in which colloidal liquids are dispersed in liquids (for example, colloidal fat in water, as in milk); and *aerosols,* in which either solids or liquids are dispersed in a gas (for example, colloidal ash in air, as in cigarette smoke, or colloidal water in air, as in fog).

The cell substance is partly a true solution, partly a colloidal system. Water is the medium in which many materials are dissolved, and it also is the *liquid phase* in which many insoluble mate-

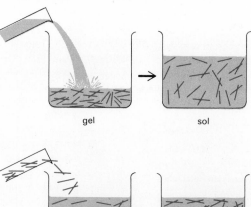

Figure 4.20 Phase reversals. A gel can be transformed to a sol either by addition of more liquid (top) or by withdrawal of solid particles. A sol can be transformed to a gel either by addition of more solid particles (bottom) or by withdrawal of liquid.

rials of colloidal size are dispersed. This colloidal *dispersed phase* includes, for example, macromolecular solids such as proteins and nucleic acids, and liquids such as oily fats. What prevents these dispersed colloidal particles from settling out? As noted in Chap. 3, the molecules of a liquid are in continuous thermal motion, the more so the higher the temperature. If dispersed particles are present, they are buffeted and bombarded constantly by the moving molecules of the liquid. Very large particles are unaffected by these weak forces, and they settle to the bottom of a container. But smaller bodies of colloidal size will be pushed in all directions. Such movements override the effect of gravity and the particles will remain suspended. This random movement of small particles, called *Brownian motion,* is readily demonstrable under the microscope.

Brownian movement aids in keeping colloidal particles from settling out, but they cannot remain suspended by this force alone. They stay dispersed mainly because most types of colloids are ionized and therefore carry *electric charges.* One member of an ion pair usually is noncolloidal and fully dissolved, and the other member is colloidal. Thus all colloidal particles of a particular type carry like charges, and the particles are kept apart by their electric repulsion. If the charges are neutralized by addition of oppositely charged substances that reduce the degree of ionization, then the colloid particles often do settle out (Fig. 4.19).

The relative quantities of the two components of a colloid often determine whether the system will form a *sol* or a *gel.* For example, if a little gelatin is added to a large amount of water, a sol will be formed. If more gelatin is added or, alternatively, if water is withdrawn, the gelatin particles will be brought closer together until they ultimately come into contact. They will then interlock as a spongelike meshwork of solid material, and in the spaces of the mesh discontinuous droplets of water will be dispersed. The original sol becomes transformed in this manner to a quasi-solid, pliable jelly, or gel. Conversely, addition of water or removal of solid particles can transform a gel to a sol (Fig. 4.20). In living materials *sol-gel transformations* of this sort occur normally and repeatedly as the concentrations of particles in a given region increase or decrease with time. In some cases a region remains more or less permanently in a gel state, like skin, or in a sol state, like blood plasma (which becomes a gel on clotting). Gels maintain definite forms and shapes despite high water contents. For example, a jellyfish is predominantly a gel, with a water content of 95 or more percent. Similarly, brain tissue is largely a gel, with a water content of about 80 percent.

Sol-gel reversals can be brought about also by temperature changes. At higher temperatures the thermal motion of the colloidal particles in a gel is more intense, and the gelled meshwork is disrupted (as in the liquefaction of Jell-O by heating). Conversely, lower temperatures promote conversion of sols to gels. Many other physical and chemical influences—low or high pH or pressure, for example—likewise affect sol-gel conditions.

All colloids *age*. The particles in a young, freshly formed colloidal system are enveloped by layers of water that are adsorbed to the particle surface by electric attraction. It is largely because of these forces that water in a gel does not "run out" through the gel meshes. With time, however, the binding capacity of the particles decreases and some of the water does run out. The colloid "sets," contracting and gelating progressively; examples are exudation of water from long-standing milk curd, custard, or mustard.

In colloids, and also in true solutions, *migratory movements* occur as a direct result of the thermal motion of the particles. If ions, molecules, or colloidal particles are distributed unevenly, more collisions take place in more concentrated regions. For example, if a particle in the circle in Fig. 4.21 is displaced by thermal motion or by Brownian bombardment *toward* a region of higher concentration, it will soon be stopped by collision with other particles. But if it is displaced *away* from a high concentration, its movement will not be interrupted so soon, since neighboring particles are farther apart. On an average, therefore, a greater number of particles is displaced to more dilute regions than to more concentrated ones. In time, particles throughout the system will become distributed evenly. This equalization resulting from migration of particles is called *diffusion*.

Diffusion plays an important role in living cells, for it often happens that particles inside a cell are distributed unevenly. Diffusion then will tend to equalize the distribution, one means by which materials in cells can be transported from place to place.

Membranes

At the boundary between a colloidal system and a different medium (air, water, solid, surfaces, or another colloid of different type), the molecules usually are subjected to complex physical forces that act on the boundary from both sides. The result is that the molecules there pack together tightly and become oriented in parallel or in layers and form a *membrane*. Good examples are the "skins" on puddings, custards, and boiled milk. Complex molecular skins also arise in living colloids. This property well may be the fundamental reason why living material does not occur in large undivided masses, but is organized in discrete units such as cells and smaller bodies inside cells, all separated from one another by membranes. The structure of these living membranes is more complex than that of simple nonliving ones, yet a simple colloidal structure appears to be the basis on which greater complexity is superimposed. If a living membrane is punctured, a new *surface precipitation membrane* develops over the opening within seconds, before appreciable amounts of material can flow through—a property displayed also by the boundary membranes of nonliving colloids.

Representing the gateways through which the chemical traffic to either side must pass, living membranes have differing *permeability* to different substances. Most membranes are completely permeable to water molecules, which can pass through freely in either direction. As for other materials, organic or inorganic, there is no rule by which their passage potential can be determined beforehand. In general, three classes of materials can be distinguished: those that can pass through a membrane in either direction; those that can pass in one direction but not in the other; and those that cannot penetrate at all. These categories vary considerably for different kinds of membranes.

In the past, traffic through living membranes

Figure 4.21 Diffusion. In the initial state at left, particles are distributed unevenly. A given particle (for example, the circled one) will therefore have more freedom to move in the direction of lower concentrations. This eventually leads to an even distribution of particles, as in the end state at right.

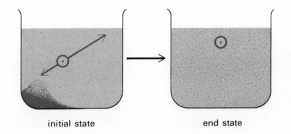

initial state   end state

has been compared with traffic through nonliving ones like cellophane. Such nonliving membranes let water or small ions through, but not proteins, for example. Particle penetration here can be explained by diffusion. Particles would strike the barrier, most of them would bounce off, but some would pass through *pores* in the membrane. If the concentration were greater on one side of the membrane than on the other, more particles on an average would migrate to the dilute side, thus equalizing concentrations.

However, a hypothesis of diffusion through pores generally is inadequate for living membranes. If such membranes were indeed inert films with holes like cellophane, then it should not be possible to poison them—passive inert structures could not be affected by poisons. But the functioning of cellular membranes actually is impaired or stopped by poisons, indicating that such membranes are not simply passive films. Moreover, if living membranes really contained small holes, then only the size of a particle should determine whether or not it could pass through. Yet particle size often is of little importance. For example, under certain conditions large protein molecules can pass through a given membrane readily, whereas very small molecules sometimes cannot. Also, sugars such as glucose, fructose, and galactose all have the same molecular size ($C_6H_{12}O_6$), yet they pass through living membranes at substantially different rates (fructose passes most readily and galactose least readily).

In short, membranes are highly *selective*. Although ordinary diffusion appears to play some role in most cases, membranes often act as if they "knew" which substances to transmit and which to reject. Moreover, energy-consuming work is often done by a living membrane in transmitting materials, and in a few cases complex chemical reactions are known to take place in the process (see Chap. 5). Cellular membranes therefore must be regarded as dynamic structures in which entering or leaving particles are actively "handed" across from one side to the other. The precise means by which cells accomplish such *active transport* across their boundaries is understood as yet for only a few types of materials.

Osmosis

When a membrane separates one solution or colloid from another kind of medium, it often happens that some of the particles on either side cannot go through the membrane. Suppose that one side has a very low concentration of such particles (side *A* in Fig. 4.22), and the other side (*B*) a very high concentration. What events will occur in such a system?

1. In the beginning, relatively more water molecules are in contact with the membrane *X* on the *A* side than on the *B* side, since fewer of the solid particles occupy membrane space on the *A* surface than on the *B* surface.

2. Therefore, more water molecules on an average are transmitted through the membrane from *A* to *B* than from *B* to *A*.

3. As a result, the water content decreases in *A* and increases in *B*. Particles in *A* become crowded in a smaller and smaller volume, and more and more of them therefore take up membrane space on the *A* surface. On the *B* side, the increasing water content permits the particles to spread into progressively larger volumes; hence the particle concentration along the *B* surface of the membrane will fall.

4. A stage will be reached at which the number of particles along the *A* surface equals that along the *B* surface. From then on the number of water molecules transmitted from *A* to *B* will equal the number transmitted from *B* to *A*. Thereafter no further *net* shift of water will occur.

This movement of water is called *osmosis*. The amount of osmosis depends on the *concentration differential*, the relative *numbers* of particles in *A* and *B*. If the difference in particle number is great enough—for example, if *A* contains only pure water but *B* contains a large number of particles—then *A* might dehydrate completely and collapse, while *B* might burst and thus also collapse. The

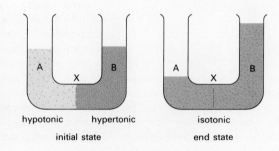

Figure 4.22 Osmosis. In the initial state *A* is less concentrated than *B*, and water therefore will move from *A* to *B*. This eventually leads to an isotonic end state, where concentrations in *A* and *B* are equal. From this point on further *net* migration of water does not occur (just as much water moves from *A* to *B* as from *B* to *A*). The broken line (*X*) represents a semipermeable membrane.

hypotonic   hypertonic

initial state

isotonic

end state

side that loses water in osmosis is said to be *hypotonic,* and the side that gains water, *hypertonic.* When neither side gains or loses the two sides are said to be *isotonic.*

The net effect of osmosis is to pull water *into* the region of higher concentration, from the hypotonic to the hypertonic side. The process will continue until the two sides are isotonic. And note that osmosis will occur whenever certain particles cannot pass through a membrane. Then nothing moves across except *water,* as well as those particles that *can* diffuse through the membrane. In living cells osmosis serves as one of the means by which water is distributed and redistributed across membranes.

In its physical organization, therefore, the substance of animal cells is a mixed colloidal system that has variously permeable boundary membranes, that undergoes localized sol-gel transformations, and that is kept agitated continuously by thermal and Brownian motion, by diffusion displacements, and by osmotic forces. As a result, the living material is subjected unceasingly to physical changes as profound as the chemical ones. Indeed, physical changes initiate chemical ones and vice versa. From any small-scale point of view, cellular matter clearly is never the same from moment to moment.

Superimposed on and ultimately resulting from its chemical and physical organization, the cell substance also displays a highly distinctive biological organization.

Biological Organization

The generalization that all animals, and living organisms generally, consist entirely of cells and cell products is known as the *cell theory.* Formulated in 1838 by the German biologists Schleiden and Schwann, this theory rapidly became one of the cornerstones of modern zoology and, with minor qualifications, it still has that status today.

Nucleus and Cytoplasm

Examination of living or killed animal cells under various kinds of microscopes shows that cell diameters vary considerably, from about $0.2\ \mu$ to as much as several millimeters and more [1 micron $(\mu) = 1/1,000$ mm]. However, the order of size of the vast majority of cells is remarkably uniform, a diameter of 0.5 to 15 μ being fairly characteristic generally. Too small a size presumably would not provide enough space for the necessary parts, and too large a size would increase the maintenance problem and at the same time reduce operating efficiency. For as a cell increases in size its surface enlarges as the *square* of its radius, and the available surface area determines how much nutrient uptake and waste elimination are possible. However, cell volume increases with the *cube* of the radius, and the volume determines how much mass a cell must keep alive. Hence if a cell kept on enlarging, its mass eventually would outrun the food-procuring capacity of its surface and cell growth then would cease. The actual sizes of cells appear to be governed in part by these surface-volume relations.

The two fundamental subdivisions of animal cells are the *nucleus* and the substance surrounding the nucleus, called the *cytoplasm.* The nucleus is bounded by a *nuclear membrane,* the cytoplasm by a *cell membrane,* also called *plasma membrane* (Fig. 4.23).

Most cells contain a single nucleus each, but there are many exceptions. For example, when mammalian red blood cells become mature, they lose their nuclei. Conversely, cells of many tissues are *binucleate* or *multinucleate,* with two or more nuclei each. In some cases the individuality of cells becomes obscured altogether. In certain tissues, for example, the membranes of adjacent cells dissolve at particular stages of development and the results are fused, continuous living masses with numerous nuclei dispersed through them. Such a structure is a *syncytium.*

A nucleus typically consists of three kinds of components: the more or less gel-like nuclear sap, or *nucleoplasm,* in which are suspended *chromosomes* and one or more *nucleoli* (Fig. 4.24).

The chromosomes are the main nuclear organelles. Indeed, a nucleus as a whole can be regarded primarily as a protective housing for these threadlike bodies. Chromosomes consist largely of nucleic acids and proteins, joined here as complexes called *nucleoproteins.* DNA is the main nucleic acid, but RNA is present also. Functionally, chromosomes are the carriers of the genes which, as noted previously, are the ultimate controllers of cellular processes.

Chromosomes are conspicuous only during cell reproduction, when they become thickly coated with additional nucleoprotein. At other times such

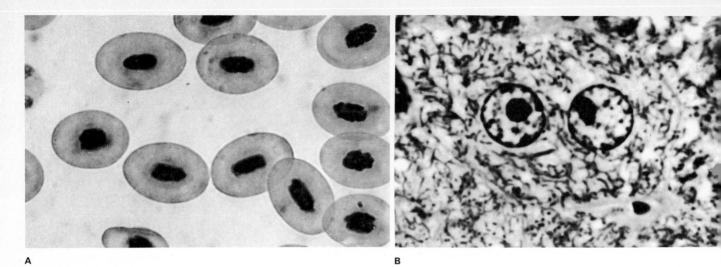

A   B

Figure 4.23 General cell structure. A, blood cells of a frog. Note the darkly stained cell nuclei, the cell membranes, and the cytoplasms between nuclei and cell membranes (approx. (×2,000). *B,* a binucleate cell from a mouse liver. Binucleates are fairly common in liver and various other organs.

coats are absent, and chromosomes then are very fine filaments. The exact number of chromosomes in each cell nucleus is an important species-specific trait. For example, cells of human beings contain 46 chromosomes each. The cells of every other type of animal have their own characteristic chromosome number.

A nucleolus ("little nucleus") is a spherical body composed largely of protein and RNA. Produced by chromosomes, nucleoli participate in the process of protein synthesis. Each cell is characterized by a fixed number of nucleoli. The whole nucleus is separated from the surrounding cytoplasm by the nuclear membrane, composed like most other living membranes mainly of proteins and lipids. It governs the vital traffic of materials between cytoplasm and nucleus. Examination with the electron microscope shows that the nuclear membrane actually is a double layer pierced by tiny pores (see Fig. 4.24).

By virtue of its genes the nucleus is the control center of cellular activities. The cytoplasm is the executive center in which the directives of the nucleus are carried out. But although the nucleus primarily controls, it also executes many directives of the cytoplasm; and although the cytoplasm primarily executes, it also influences many nuclear processes. A reciprocal interdependence thus links nucleus and cytoplasm, and experiment has repeatedly shown that the one cannot long survive without the other. For example, amoebas can be cut into halves so that one half includes the nucleus. Such a nucleated half then carries on in

every respect like a normal amoeba, but the nor nucleated half invariably dies eventually. Yet jus as survival of the cytoplasm depends on the nu cleus, so survival of the nucleus depends on th cytoplasm; a naked isolated nucleus soon dies Cellular life therefore must be viewed against background of cyclic interactions between nucleu and cytoplasm.

Cytoplasm consists of a semifluid *ground sub stance,* which is in a sol or a gel state at differer times and in different cellular regions, and in whic are suspended large numbers of several kinds c organelles. The following are widespread among many or all animal cell types (Fig. 4.25):

Endoplasmic Reticulum Composed largely o lipids and proteins, the endoplasmic reticulum i a network of exceedingly fine double membrane that traverse all regions of the cytoplasm and ar continuous with the cell and nuclear membrane (Fig. 4.26 and see Fig. 4.25). In some cases th endoplasmic reticulum also passes from one ce to adjacent ones. Attached to this membrane sys tem are many of the other cytoplasmic organelles The endoplasmic reticulum appears to functior broadly in two ways. First, it can serve as a net work of traffic pathways and conducting channels Materials in transit from one cell region to another or from the cell to the outside and vice versa, car migrate along the narrow spaces between th membranes. Second, by interconnecting othe organelles, the endoplasmic reticulum forms ar ultrastructural framework that keeps the nonfluic

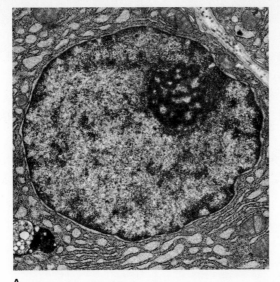

A

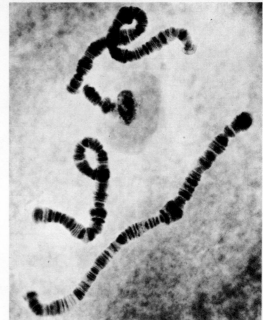

B

Figure 4.24 The cell nucleus. *A*, electron micrograph of a whole nucleus, covering most of the photo. Inside the nucleus the large dark patch is a nucleolus and the dark speckle elsewhere is the gene-containing chromosomal material (which in stained preparation has a filamentous appearance only during cell division; ×40,000). *B*, a stained preparation of insect chromosomes (×3,000). Note the characteristic crossbands, found in all chromosomes studied. See also Chap. 12. *C*, the nuclear membrane. In this electron-micrographic close-up of a cell in the spinal cord of a bat, the nuclear substance is at bottom of photo, cytoplasmic substance at top. Note that the nuclear membrane is a double layer with pores (×40,000).

Figure 4.25 Cell structure. No single cell contains all the organelles shown in this composite drawing. For clarity and simplicity most components are shown in only part of the cell. For example, the endoplasmic reticulum traverses all parts of the cytoplasm.

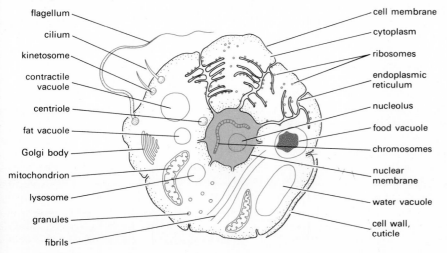

flagellum
cilium
kinetosome
contractile vacuole
centriole
fat vacuole
Golgi body
mitochondrion
lysosome
granules
fibrils

cell membrane
cytoplasm
ribosomes
endoplasmic reticulum
nucleolus
food vacuole
chromosomes
nuclear membrane
water vacuole
cell wall, cuticle

C

components of cytoplasm in certain relative positions. The membranes can form, dissolve, and re-form rapidly, often in conjunction with the frequent shifts of position and streaming movements of the other cell contents. Thus although the cell as a whole is readily deformable and the interior contents flow and intermix unceasingly, the endoplasmic reticulum nevertheless keeps other organelles properly stationed and distributed in relation to one another. As a result little is really "loose" in a cell, and an orderly organization is maintained.

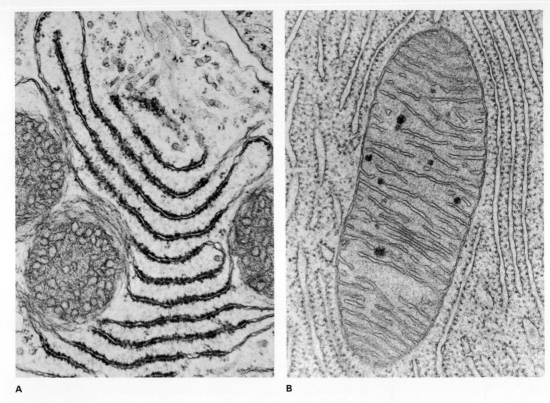

A   **B**

Figure 4.26 Endoplasmic reticula and mitochondria. *A,* electron micrograph of a cell portion of a monkey adrenal gland. Note the endoplasmic reticulum, a continuous double membrane passing throughout the cytoplasm. Note also the (dark) granular ribosomes that adhere to the membranes. The large round structures are mitochondria (×70,000). *B,* electron-micrographic section through a mitochondrion of a mammalian pancreatic cell (×70,000). Note the double-layered exterior boundary and the many infoldings of the inner membrane (mitochondrial cristae). The dark spots in the interior of the mitochondrion are calcium-rich granules. Outside the mitochondrion are parts of the endoplasmic reticulum (double membranes with clear channels inside) and numerous ribosomes (small dark granules).

Mitochondria These organelles too are constructed predominantly out of lipids and proteins. Small amounts of DNA and RNA are known to be present as well. The main functional components are respiratory enzymes and coenzymes; mitochondria are the chief chemical "factories" in which cellular respiration is carried out. Under the light microscope mitochondria appear as short rods or thin filaments averaging 0.5 to 2 μ in length. The electron microscope shows that the surface of a mitochondrion consists of two fine membranes (see Fig. 4.26). The inner one has folds that project into the interior of the mitochondrion. These folds (*mitochondrial cristae*) bear numerous tiny stalked microgranules, probably the specific locations where respiratory reactions take place.

Ribosomes These are tiny granules visible under the electron microscope (see Fig. 4.26). Many of them usually are attached to the endoplasmic reticulum; others lie free in the cytoplasm. Ribosomes contain RNA (hence the *ribo-* portion of their name) and the enzymes for protein manufacture. The granules are the "factories" where protein synthesis is carried out, a function probably performed by groups of them (*polyribosomes*) acting in concert.

Golgi Bodies Under the electron microscope these bodies are seen as stacks of thin, platelike layers (Fig. 4.27). The organelles function in the synthesis of cellular secretion products, and Golgi bodies are particularly conspicuous in actively secreting cells. For example, whenever gland cells manufacture their characteristic secretions, the Golgi bodies of such cells become very prominent.

Lysosomes Known to occur in various types of animal cells, these organelles are tiny membrane-bounded sacs or vesicles (see Fig. 4.27). They contain digestive enzymes that are released into the cytoplasm when the vesicles burst open. The enzymes probably participate in various normal decomposition processes in a cell—chemical breakdown of nutrients prior to their utilization, breakdown of cellular organelles prior to their reconstruction or remodeling, breakdown of foreign particles, and the like. It is known also that simultaneous disruption of many or all lysosomes leads to rapid dissolution and death of a cell. Lysosomes therefore might play a role in processes of tissue maintenance, in which old cells are replaced by new ones formed through cell reproduction.

Centrioles and Kinetosomes A centriole is a small granule located near (and in some cases inside) the nucleus of animal cells (see Fig. 4.25). As will be shown later, these granules play a spe-

cific role in cell reproduction. Another type of granule, called *basal granule* or *kinetosome*, is found in cells that have surface flagella or cilia (see below). Kinetosomes anchor and control the motion of such surface structures. Centrioles and kinetosomes appear to have arisen from a common evolutionary source, a single type of granule that served simultaneously as a centriole and a kinetosome. Some protozoan cells today still contain such granules with joint dual functions. In most other cases the originally single granule evidently evolved into two separate ones, one retaining the centriole function only, the other, the kinetosome function only.

The electron microscope reveals centrioles and kinetosomes to have a common, complex structure (Fig. 4.28). Each such granule is made up of a ringlike array of nine sets of parallel tubules, with a fine layer of boundary material around the ring. On one side of the granule the tubules continue into the cytoplasm as a spreading bundle of *microtubules;* and on the other side in a kinetosome the tubules form the basal parts of a flagellum or cilium (see also below).

Figure 4.27   Golgi bodies and lysosomes. A, electron micrograph of cytoplasm with two Golgi bodies, each composed of a stack of parallel double membranes (×50,000). *B,* portion of a rat liver cell, with dark membrane-covered lysosomes in lower half of photo. In the upper half are portions of a mitochondrion and the endoplasmic reticulum (×48,500).

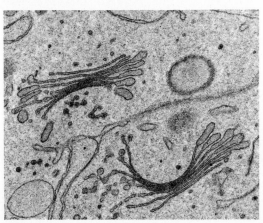

A

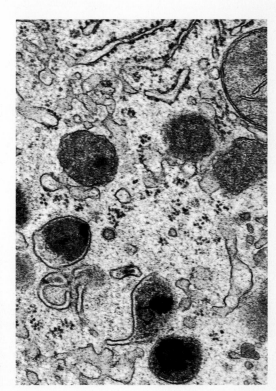

B

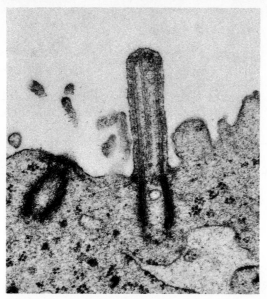

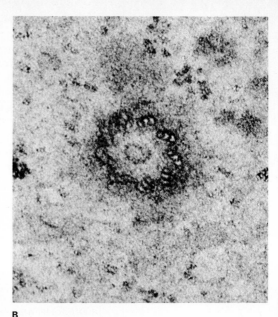

A   **B**

Figure 4.28 Kinetosomes and centrioles. A, electron micrographic section through a kinetosome (of a cell in a monkey oviduct), with a cilium growing out from this basal granule. Note that the ciliary fibrils (tubules) are continuous with those of the basal granule. Also note the small vesicle inside the kinetosome (×40,000). *B,* electron-micrographic cross section through a centriole (of a cell in a monkey oviduct). Note the central core material and the surrounding circular array of nine (triplet) tubules (which correspond to a ring of nine tubules in a kinetosome) (×50,000).

Additional Structures Apart from structures just described, cytoplasm generally contains additional *granules* and fluid-filled droplets called *vacuoles.* Such organelles perform a large variety of functions. For example, they transport nutrients from the cell surface to the interior (*food vacuoles*) or finished products in the opposite direction (*secretion granules*); they are places of storage (*glycogen granules, fat vacuoles, water vacuoles, pigment granules*); or they carry waste materials to points of elimination (*excretory vacuoles*).

In addition to all these, cytoplasm often contains a variety of long, thin protein *fibrils,* such as contractile *myofibrils* or conducting *neurofibrils.* Various other inclusions, unique to given cell types and serving unique functions, can be present as well. In general, every function a cell performs involves a particular structure in which the machinery for that function is housed.

Cytoplasm as a whole is normally in motion. Irregular eddying or streaming occurs often, and at other times the substance of a cell is subjected to cyclic currents known as *cyclosis.* The organelles, nucleus included, are swept along passively in these streams. The specific cause of such motions is unknown, but there is little doubt that they are a consequence of the unceasing chemical and physical changes taking place in a cell. Whatever the specific causes might be, the apparently random movements might give the impression that nothing is fixed in a cell and that cytoplasm is simply a collection of loose bodies suspended in ''soup.'' As already pointed out, however, such an impression is erroneous, for a cell does have an orderly, though deformable, interior organization.

Cell Surfaces

Composed predominantly of protein and lipid substances, the cell membrane plays a critical role in *all* cell functions, since directly or indirectly every such function depends on *absorption* of materials from the exterior, *excretion* of materials from the interior, or both.

Unlike plant cells, most animal cells and also many protozoan cells lack exterior walls. In many cases the surfaces of such naked cells are fairly smooth, but in certain compact animal tissues numerous fingerlike extensions project from one cell and interlock with similar extensions from neighboring cells. Occasionally one such cell nips off a protrusion of an adjacent cell—one of the means by which material can be transferred from cell to cell. Also, a cell sometimes develops a deepening surface depression that eventually nips off on the inside as a fluid vacuole. Through this type of fluid engulfment, or *pinocytosis*, a cell can transfer liquid droplets to its interior (Fig. 4.29).

All such transfers are specialized forms of the more general phenomenon of *amoeboid motion*, in which temporary fingerlike extensions, or *pseudopodia*, are again formed at any point on the cell surface (see Fig. 2.7). Some protozoa move and feed by means of pseudopods. In feeding, pseudopods engulf a bit of food by flowing around it and forming a food vacuole that comes to lie in the cell interior. Many kinds of nonprotozoan cells similarly are capable of amoeboid movement. For example, many types of eggs engulf sperm cells in fertilization, and several categories of blood cells engulf foreign bodies, bacteria, and other potentially harmful materials. When one cell "swallows" particles in amoeboid fashion, the process is often called *phagocytosis*.

Animal cells exposed directly to the external environment usually are not naked but are enveloped partially or wholly by wall-like *cuticles* or *pellicles*. For example, a protective horny coat of secreted *chitin* is found on the skin cells of insects and related groups. Numerous other animals secrete horny protein coats on their outer cells; and the skin cells of most vertebrates secrete coats of *keratin*, a protein that also covers surface structures such as feathers and hair. In many other cases surface cells secrete slimy protective films of mucus or *shells* of lime, glass (*silica*), or various organic materials.

Many cell types have the capacity of locomotion. Those that do not move in amoeboid fashion usually are equipped with specialized locomotor organelles on the cell surface. Among these are *flagella,* long, slender, threadlike projections from cells (Fig. 4.30). The base of a flagellum is anchored in the cell cytoplasm on a motion-controlling kinetosome granule. In some cases a threadlike fibril (*rhizoplast*) connects the kinetosome with the centriole in the cell interior. Numerous types of cells have shorter variants of flagella called *cilia*. These usually are present in large numbers and cover all or major portions of a cell like tiny bristles. Each cilium has its own separate

Figure 4.29 Cell surfaces. A, many cells without surface secretions develop fingerlike projections like those shown here. If two such cells are in contact, their projections usually interlock. *B,* cellular engulfments. Top, pinocytosis, or cellular engulfment of fluid droplets. Bottom, phagocytosis, or cellular engulfment of solid particles.

A

B

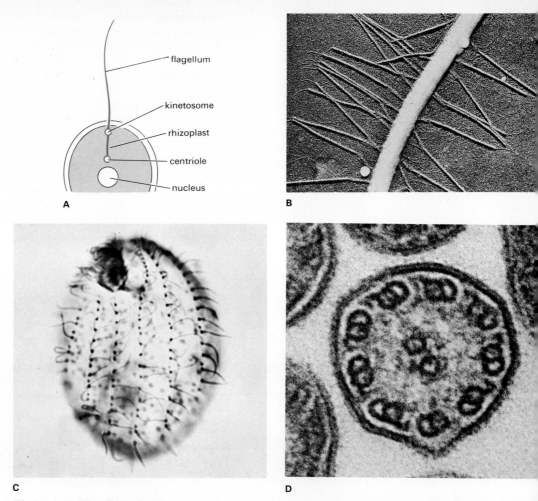

Figure 4.30 Flagella and cilia. A; the insertion of a flagellum in a cell. The flagellum drawn here is of the whiplash type—its surface is smooth. *B,* electron micrograph of portion of a tinsel-type flagellum, so named because of the fine lateral filaments along its surface (×35,000). *C,* the ciliate protozoan *Tetrahymena* stained to show the rows of cilia on the body surface (×500). *D,* electron-micrographic cross section through a cilium (from a cell in a monkey oviduct), showing the central two and the nine peripheral fibrils inside. The peripheral fibrils are continuous with the tubules of the kinetosome (and/or the centriole) (×100,000).

kinetosome at its base. Internally all flagella and cilia have the same structure. The electron microscope reveals a flagellum or cilium to be a bundle of eleven exceedingly fine fibrils, two of them central and nine arranged in a ring around the central two. The nine in the ring are continuations of the microtubules of the kinetosome. How motion is actually produced is as yet understood only poorly.

Flagella are the locomotor structures in numerous protozoa where the whiplike beat usually pulls the cell behind it. By contrast, in most animal sperms the flagellum is at the posterior end and its beat pushes the cell forward. Flagella or cilia also occur in many embryos, larvae, and small adult animals, which have flagellate or ciliated skins that serve in locomotion, in creating food-bearing water current, or both. Moreover, most animals have flagellate or ciliated cells in the interior of the body. For example, such cells occur in the lining tissues of breathing, alimentary, and reproductive channels; in sense organs (ears, for example, and also eyes, where greatly modified flagellate cells form the rod and cone cells of retinas); in the excretory organs of many invertebrates, where so-called *flame cells* maintain a flow of water-laden body fluids (see Chap. 9); and generally in any location where air, water, or solid materials must be moved over a surface or through a duct.

In any cell, evidently, certain of the structural components of the nucleus, the cytoplasm, or the surface serve directly as the machinery for particular cell functions. Thus, respiration and protein synthesis are distinct functions performed in distinct cytoplasmic structures. But many cell functions cannot be localized so neatly. For example,

cell reproduction or amoeboid movement requires the cooperative activity of many or all of the structural components of a cell. Functions of this kind cannot be referred to any particular part of a cell but are performed by the cell as a whole.

Also, whereas many organelles are bulky enough to be visible under the microscope, many more are not visible; individual molecules in a cell "function" no less than larger bodies. Be it a single dissolved molecule or a whole group of large suspended organelles, each cellular structure performs a function, and as the structures differ among cells so do the functions.

The actual functioning of animal cells is the subject of the next two chapters.

Review Questions

1. What are inorganic compounds? Organic compounds? What main classes of each occur in living matter, and in what relative amounts? Which of these substances are electrolytes and which are nonelectrolytes?

2. Review the chemical composition and molecular structure of carbohydrates. What are pentoses, hexoses, aldoses, and ketoses? Give examples of each.

3. What are monosaccharides, disaccharides, and polysaccharides? Give examples of each. Distinguish between 1,4 and 1,6 linkages in polysaccharides, and give specific examples of compounds having such links.

4. Review the composition and structure of fatty acids. Distinguish between saturated and unsaturated kinds, and give examples. How is a fat formed? What general functions do carbohydrates and fats have in living matter?

5. Review the structure of amino acids, and show how these compounds differ among themselves. Write out the formation of a peptide bond. What is a polypeptide, and in what ways can polypeptides differ in composition?

6. What is the primary, secondary, tertiary, and quaternary structure of a protein, and what kinds of bonds produce such structures? What is an α-helix? What are globular and fibrous proteins?

7. What is protein specificity? How is a coagulated protein different from a native or a denatured protein? Review the general role of proteins in cells.

8. Distinguish between nitrogen bases, nucleotides, ribotides, and deoxyribotides. Give examples of each. What are adenosine phosphates? Review the structure and function of coenzymes, and give the names of some of them.

9. What is the chemical composition and molecular structure of nucleic acids? In chemical terms, what are DNA and RNA? What different kinds of nucleotides occur in nucleic acids?

10. What makes nucleic acids specific, and what are the general functions of such acids? How are they related to genes?

11. What main classes of pigments occur in animal cells? How do these substances differ chemically? What are some of their functions? What derivatives of carbohydrates and fats are common in animals?

12. What is a colloidal system? How does such a system differ from a solution? Review the properties of colloidal systems.

13. Define diffusion and show how and under what conditions this process will occur. What is the significance of diffusion in cells?

14. How and where do colloidal membranes form? What are the characteristics of such membranes? What roles do they play in cellular processes?

15. Define osmosis. Show how and under what conditions this process will occur. Distinguish carefully between osmosis and diffusion. What are isotonicity, hypertonicity, and hypotonicity?

16. What are the structural subdivisions of cells? What are the main components of each of these subdivisions, where are they found, and what functions do they carry out?

17. List cytoplasmic inclusions encountered in all cell types and inclusions found only in certain cell types. What is cyclosis?

18. What structures are found on the surfaces of various cell types? Which of these structures are primarily protective? What do they protect against? What are the functions of other surface organelles? Describe the structure of such organelles.

19. Define "cell" (*a*) structurally and (*b*) functionally. Define syncytium, nucleolus, chromosome, kinetosome, endoplasmic reticulum.

20. Construct a list or table that relates particular cell structures to particular cell functions.

Collateral Readings

Allen, R. D.: Amoeboid Movement, *Sci. American*, Feb., 1962. A hypothesis as to how this form of cellular motion might be brought about.

Allison, A.: Lysosomes and Disease, *Sci. American*, Nov., 1967. On the role of these organelles in normal and abnormal cellular processes.

Baker, J. J. W., and G. E. Allen: "Matter, Energy, and Life," Addison-Wesley, Reading, Mass., 1965. This paperback contains a good general discussion of all classes of cellular chemicals. Recommended for further reading.

Brachet, J.: The Living Cell, *Sci. American*, Sept., 1961. A very good general account on our modern knowledge of cell structure.

deDuve, C.: The Lysosome, *Sci. American*, May, 1963. The structure and function of this cytoplasmic organelle are discussed in a nontechnical style.

Dippell, R. V.: Ultrastructure of Cells in Relation to Function, in "This Is Life," Holt, New York, 1962. An exceedingly good and well-illustrated article on the electron-microscopic fine structure of various cellular organelles. Highly recommended.

Green, D. E.: The Mitochondrion, *Sci. American*, Jan., 1964. An instructive description of the structure and function of this organelle.

Hokin, L. E., and M. R. Hokin: The Chemistry of Cell Membranes, *Sci. American*, Oct., 1965. The article describes the role of the chemical components of membranes and the transport of materials through the membranes.

Loewy, A. G., and P. Siekevitz: "Cell Structure and Function," Holt, New York, 1963. Part 2 of this paperback is highly recommended for further background on the topics dealt with in this chapter.

Neutra, M., and C. P. Leblond: The Golgi Apparatus, *Sci. American*, Feb., 1969. The functioning of of this organelle is examined by means of radioautography.

Nomura, M.: Ribosomes, *Sci. American*, Oct., 1969. Structure and function is studied by disassembling and then reassembling these organelles.

Racker, E.: The Membrane of the Mitochondrion, *Sci. American,* Feb., 1968. On the structure and respiratory function of the mitochondrial cristae.

Robertson, J. D.: The Membrane of the Living Cell, *Sci. American,* Apr., 1962. A good account on the exterior cell membrane and its continuations in the cellular interior.

Satir, P.: Cilia, *Sci. American,* Feb., 1961. A thorough description of the electron-microscopic fine structure of these surface organelles.

Solomon, A. K.: Pores in the Cell Membrane, *Sci. American,* Dec., 1960. The structure of cellular membranes is discussed in relation to the traffic of materials through them.

Swanson, C. P.: ''The Cell,'' 2d ed., Prentice-Hall, Englewood Cliffs, N.J., 1964. This paperback is recommended particularly for further background reading on the structure and function of cells.

5

CELL OPERATIONS: NUTRITION, RESPIRATION

As pointed out in Chapter 2, the function of metabolism includes the subfunction of *nutrition*, which supplies raw materials; the subfunction of *respiration*, which derives usable energy from some of the raw materials; and the subfunction of *synthesis*, which, with the aid of the remaining raw materials and with respiratory energy, produces new living matter.

This chapter considers the first two of these processes on the cellular level.

Raw Materials: Cell Nutrition

The nutrition of an individual cell depends on prior nutrition of the whole animal. By some form of alimentation the animal must have procured necessary nutrients, and these must have been distributed inside the animal body to all cells. *Cellular* nutrition, and cell metabolism in general, begins at that point. Note again that the term "food" is customarily restricted to *organic* nutrients only; the terms "nutrient" and "metabolite" apply to *all* external materials a cell requires for its survival.

If cellular intake by phagocytosis and pinocytosis is disregarded, cell nutrition is mainly a process of *absorption*. In almost all cases the nutrients are individual ions or molecules, and these are absorbed through the surface of a cell in at least three ways: water is absorbed in part by *osmosis*; compounds dissolved in water are absorbed in part by *diffusion*; and dissolved compounds also are absorbed by energy-consuming *active transport*, very often against prevailing osmotic or diffusion gradients. As noted earlier, the precise mechanisms of active transport are still poorly understood. But in the case of at least one important metabolite, glucose, some of the details of absorption are known.

Except to the extent that purely physical diffusion occurs, cells apparently do not take up glucose as such. Instead, glucose absorption often depends on *phosphorylation*, or chemical addition of a phosphate (—O—P) group at or in the cell surface. The result is phosphorylated glucose, or glucose-phosphate. The phosphorylation reaction requires energy, and ATP supplies both the energy and the phosphate; as the terminal third phosphate group of ATP is split off and added to glucose, its high-energy bond makes energy available for the reaction (Fig. 5.1).

In vertebrates, moreover, glucose uptake into cells also is facilitated by the pancreatic hormone *insulin*. This circumstance explains why insulin deficiency, as in diabetes, leads to impairment of cell function; without the hormone, cells cannot absorb sufficient amounts of glucose. The sugar then accumulates uselessly in blood (and is eventually excreted in urine).

Once nutrients have been absorbed into a cell, their distribution inside the cell is achieved mainly

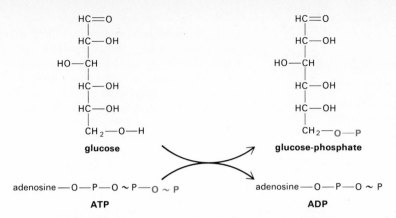

glucose → **glucose-phosphate**

$$adenosine—O—P—O \sim P—O \sim P \qquad\qquad adenosine—O—P—O \sim P$$

ATP **ADP**

Figure 5.1 The phosphorylation reaction. Glucose reacts with ATP, resulting in phosphorylated glucose and ADP. Note that ATP serves both as energy and phosphate donor. The —O—H at one end of the glucose molecule is replaced by —O~P of ATP (color), and in the process the energy of the high-energy bond is used up; a low-energy bond is left in glucose-phosphate (P stands for —PO_3H_2 here and in following illustrations).

Figure 5.2 Cellular nutrition. The direction of osmosis and diffusion depends on concentration differences of the indicated materials between the outside (left) and inside (right) of a cell. Phosphorylation and active transport can occur in either direction regardless of concentration differences.

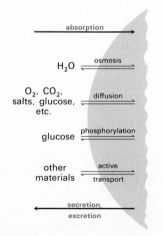

by diffusion and cyclosis. Through these processes every part of a cell comes to have access to all the nutrients the cell may have acquired only at limited regions of its surface. In the course of such internal distribution, some of the organic nutrients reach the mitochondria, the organelles in which respiration then occurs.

Apart from nutrient molecules, the only other raw material an animal cell requires for respiration is oxygen. Many animals are so constructed that most or all of their cells are in direct or nearly direct contact with the external environment. Oxygen intake here takes place in correspondingly direct fashion, by diffusion into individual cells. Specialized breathing systems occur only in the most complexly constructed animals, and in these cases oxygen diffuses to each cell from the body fluids, circulating blood in particular (see Chap. 9).

As cells then continuously use up oxygen, the concentration of the gas inside a cell always tends to be lower than in the surroundings. Consequently, the pressure gradient, or *tension gradient,* points into the cell and more oxygen diffuses in. At the same time, respiration continuously liberates CO_2, and the concentration of this byproduct therefore tends to be greater in a cell than outside it. Diffusion then accomplishes the removal of CO_2 from the cell (Fig. 5.2).

So supplied with nutrients and oxygen, a cell can respire.

Respiration: The Pattern

Respiration can be defined as a conversion of the chemical energy of organic molecules to energy usable in living cells. As noted, the organelles in a cell where most respiration occurs are the mitochondria; and the process itself consists of a series of exergonic, energy-yielding decomposition reactions: respiration is a form of burning, or combustion, in which chemical bonds in nutrient molecules are broken and the bond energy becomes available for metabolic work.

If respiration is equivalent to combustion, why does it not produce the high temperatures of a fire? For two reasons. First, a fire is an *uncontrolled* combustion, in the sense that all the bonds in a fuel molecule may be broken simultaneously. A maximum amount of energy then can be released all at once. Such sudden, explosive release generates the high temperatures of a fire. But respiration is a *controlled* combustion; energy is obtained from one or a few bonds at a time. If a nutrient fuel is respired completely, the total energy yield is the same as if it were burned in a furnace, but in respiration the energy is removed bit by bit, bond by bond. Temperatures therefore stay low.

Second, the energy produced in a fire is dissipated largely as heat and to some extent as light. But in respiration only some of the available energy escapes as heat and virtually none as light. Instead, much of it is "packaged" directly as new *chemical* energy. This energy-packaging process, already referred to in Chap. 4, is represented by the reaction that converts adenosine diphosphate (ADP) to adenosine triphosphate (ATP):

ADP + phosphate + energy ⟶ ATP

Energy from a fuel molecule becomes incorporated in ADP and phosphate, and ATP is thereby produced. Thus, fuel energy creates *new* chemical bonds in the form of ATP, and it is in this form that metabolic energy is used in cells. Since chemical bonds are not "hot," temperatures stay low during respiration.

The actual fuels in cells are organic compounds that contain bond energies—in effect, *any* organic constituent of cells: carbohydrates, fats, proteins, nucleotides, their various derivatives, vitamins, other special compounds, and indeed all the innu-

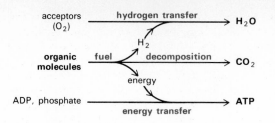

Figure 5.3 The three-phase pattern of respiration: breakdown of fuel, hydrogen transfer from fuel, and energy transfer from fuel.

merable substances that together make up a cell. Like a fire, respiration is no respecter of materials. However, under normal conditions a cell receives a steady enough supply of external foods to make *them* the primary fuels rather than the structural parts of a cell. Nevertheless, the formed parts of a cell, including even those that make up the decomposing apparatus itself, *are* decomposed gradually. A cell therefore can remain intact and functioning only by continuous construction of new living components. Destructive energy metabolism (*catabolism*) and constructive synthesis metabolism (*anabolism*) normally are in balance, and foods serve as both fuel for the one and building materials for the other. The components of a cell in effect are continuously "turned over," existing parts being replaced by new ones, and the living substance consequently is never quite the same from instant to instant.

Just how is energy released from a fuel molecule? Since respiration involves energy-yielding decomposition reactions, the endproducts will contain less energy and therefore will be more stable than the starting fuels. As noted in Chap. 3, reactions tend to proceed in such a way that the least energetic, most stable states are attained. It is this basic thermodynamic circumstance that ultimately "drives" all respiratory processes. The energy liberated in respiration comes from the carbon bonds in food molecules, and different carbon bonds have different stabilities. The least stable, hence most energy-rich, carbon bonds generally occur in *hydrocarbon* groups, atomic groupings that contain only carbon and hydrogen. These exist in organic molecules in forms such as CH_4, $—CH_3$, $—CH_2—$, $=CH—$. On the other hand, the most stable carbon combination is CO_2, a so-called *anhydride,* or hydrogen-free grouping. In general, therefore, usable respiratory energy will result from conversions of hydrocarbons to anhydrides, or from the replacement of H atoms bonded to carbon by O atoms.

Removal of hydrogen, or *dehydrogenation,* actually is the important energy-yielding event in respiration. Through dehydrogenation a hydrocarbon-containing starting fuel is decomposed to CO_2, and energy and hydrogen are released in the process. The energy then is used in converting ADP to ATP. The hydrogen does not remain free either but combines with a *hydrogen acceptor.* Of several substances serving as hydrogen acceptors in cells, the most essential usually is oxygen. That is why this gas must be supplied to all cells as a respiratory raw material. Hydrogen combines with oxygen, and water then becomes one of the endproducts of respiration.

Respiration as a whole thus consists of three interrelated processes (Fig. 5.3). First, a nutrient fuel molecule is decomposed by dehydrogenation. Smaller fuel fragments generally form, which can be decomposed in turn until the original food molecule has been degraded completely to CO_2. This phase of respiration can be termed *fuel decomposition.* Second, as a result of fuel decomposition hydrogen is freed. It becomes attached to an appropriate acceptor, and if the acceptor is oxygen water forms. This phase of respiration can be referred to as *hydrogen transfer.* Third, as a further result of fuel breakdown, energy is released. Some of this energy escapes as heat, but much of it is harvested by the ADP/ATP system. ATP then represents the main product of respiration. This phase can be called *energy transfer.*

Decomposition: Oxidation

Dehydrogenations are instances of so-called *oxidation-reduction,* or *redox,* reactions. Every such reaction can be considered to consist of two subreactions, one an oxidation, the other a reduction.

An oxidation fundamentally is any process that makes an atom more electropositive and thereby endows it with a higher oxidation state (Fig. 5.4). In electron-transfer reactions, for example, an atom that loses electrons becomes oxidized. Thus, a neutral sodium atom is oxidized when it becomes an ion; the ion is more electropositive and has a higher oxidation state ($+1$) than the neutral atom. Oxidation need not involve an outright loss of electrons, however. If electrons merely shift farther away from an atom, resulting in decreased attraction between the electrons and the atomic nucleus, such changes represent oxidations, too. For example, as pointed out in Chap. 3, the electron pairs that are shared between C and H of methane (CH_4) are attracted strongly by the C nucleus and actually lie much nearer to that nucleus than to the H atoms. In CO_2, by contrast, it is the oxygen atoms that attract the shared electron pairs more strongly than the C atom. Hence if CH_4 is converted to CO_2, the carbon atom becomes oxidized; its electrons shift farther away, the atom becomes more electropositive as a result, and its oxidation state increases (from -4 to $+4$ in this case; see Fig. 5.4). This shift is why removal of hydrogen from carbon compounds, as in respiration, actually represents an oxidation; dehydrogenation makes the carbon atoms more electropositive.

Conversely, a reduction is a process that makes an atom more electronegative and thus lowers its oxidation state. Reductions can occur through outright gains of electrons or through shifts of electrons closer to atomic nuclei (see Fig. 5.4). Oxidations and reductions always occur together, because every redox process requires both an electron donor (which becomes oxidized) and an electron acceptor (which becomes reduced). Thus if a redox process involves hydrogen, as in respiration, a hydrogen-donor compound becomes dehydrogenated and thereby oxidized, and a hydrogen-acceptor compound must be present to bond with the removed hydrogen and thereby become reduced.

In the context of respiration the important point is that every redox change is accompanied by energy changes. Since electrons are energy carriers, oxidation leads to an energy *loss* in the atom being oxidized, inasmuch as electrons are shifted away or removed. Conversely, reduction results in an energy *gain* in the atom being reduced, inasmuch as electrons are gained. In line with the second law of thermodynamics, a lower energy content signifies greater stability, and a higher content, lesser stability. Accordingly, oxidation tends to

increase the stability of a chemical unit; reduction, to decrease it. A complete redox change therefore includes an energy transfer: energy is released by a unit that is being oxidized and made more stable, and some or all of this energy is accepted by another unit that is reduced and made less stable. The units that are reduced usually cannot become quite as unstable as those that are oxidized; as in respiration, for example, the endproducts are more stable than the starting materials. A redox change then leads to a net release of some energy to the environment: of the energy resulting from an oxidation, a portion is transferred to the accompanying reduction and the rest becomes free and potentially available for other uses. It is this energy remainder that is the direct source of all the respiratory energy later harvested as ATP. Moreover, it is because energy is released in this fashion that respiratory redox reactions continue to take place unidirectionally rather than reach equilibrium.

Consider, for example, the complete redox decomposition of methane:

$$CH_4 + 2\ O_2 \longrightarrow CO_2 + 2\ H_2O + energy$$

Here the endproducts CO_2 and water are more stable and collectively poorer in energy than the starting materials methane and oxygen; the energy difference appears free in the environment. To symbolize the simultaneous oxidation and reduction separately, we can write:

$CH_4 + O_2 \longrightarrow CO_2 + 2\ H_2$	oxidation of carbon
$2\ H_2 + O_2 \longrightarrow 2\ H_2O$	reduction of oxygen
$CH_4 + 2\ O_2 \longrightarrow CO_2 + 2\ H_2O$	net redox change

Methane is the hydrogen donor, and the carbon of methane is oxidized to CO_2. At the same time oxygen is the hydrogen acceptor, and oxygen is reduced to H_2O. For convenience we can refer to the whole process simply as an "oxidation." This actually is common practice in discussions of respiration reactions, which are often called "biological oxidations." Notwithstanding the incomplete name, however, every oxidation implies and is accompanied by a reduction.

As noted earlier, a respiratory decomposition does not take place in a single large step but in several small ones; a *series* of consecutive redox reactions occurs, in which hydrogen is removed from fuel bit by bit, in which energy is released bit by bit, and in which the oxidation level of

Figure 5.4 Oxidation and reduction. The properties of these two types of reactions are summarized.

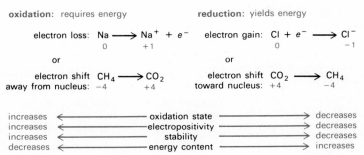

oxidation: requires energy   reduction: yields energy

electron loss: Na $\longrightarrow$ Na$^+$ + e^- electron gain: Cl + e^- $\longrightarrow$ Cl$^-$

or   or

electron shift $CH_4 \longrightarrow CO_2$ electron shift $CO_2 \longrightarrow CH_4$
away from nucleus: $-4 \qquad +4$ toward nucleus: $+4 \qquad -4$

increases $\longleftarrow$	oxidation state	$\longrightarrow$ decreases
increases $\longleftarrow$	electropositivity	$\longrightarrow$ decreases
increases $\longleftarrow$	stability	$\longrightarrow$ decreases
decreases $\longleftarrow$	energy content	$\longrightarrow$ increases

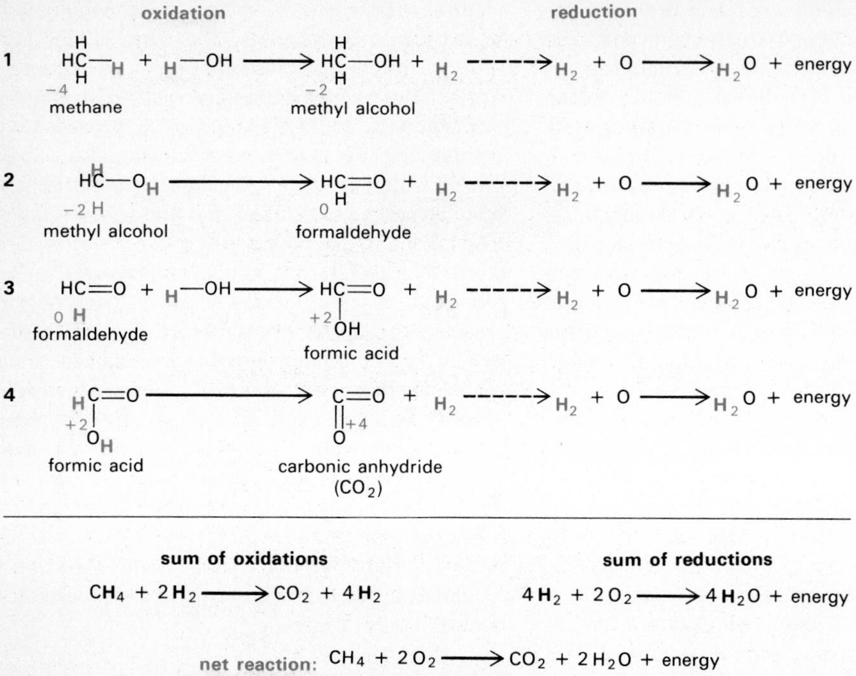

1 $HC-H + H-OH \longrightarrow HC-OH + H_2 \dashrightarrow H_2 + O \longrightarrow H_2O + energy$
methane → methyl alcohol (oxidation state $-4 \to -2$)

2 $HC-OH \longrightarrow HC=O + H_2 \dashrightarrow H_2 + O \longrightarrow H_2O + energy$
methyl alcohol → formaldehyde ($-2 \to 0$)

3 $HC=O + H-OH \longrightarrow HC=O\,(OH) + H_2 \dashrightarrow H_2 + O \longrightarrow H_2O + energy$
formaldehyde → formic acid ($0 \to +2$)

4 $HC(=O)OH \longrightarrow C=O\,(O) + H_2 \dashrightarrow H_2 + O \longrightarrow H_2O + energy$
formic acid → carbonic anhydride (CO_2) ($+2 \to +4$)

sum of oxidations	sum of reductions
$CH_4 + 2H_2 \longrightarrow CO_2 + 4H_2$	$4H_2 + 2O_2 \longrightarrow 4H_2O + energy$

net reaction: $CH_4 + 2O_2 \longrightarrow CO_2 + 2H_2O + energy$

Figure 5.5 *Stepwise decomposition of methane.* In each of the four steps an oxidizing half-reaction liberates H_2. The hydrogen then combines with oxygen in a reducing half-reaction, yielding water and energy. The whole reaction sequence is exergonic; it liberates more energy than it requires, and the endproducts are more stable than the starting materials (net reaction at bottom).

carbon atoms increases in steps. For example, the stepwise oxidation of methane is outlined in Fig. 5.5. Four consecutive redox reactions take place, in which methane is transformed successively to methyl alcohol, formaldehyde, formic acid, and carbon dioxide. Each of the four oxidative sub-

reactions yields H_2, and in the accompanying reducing subreactions the hydrogen pairs are accepted by oxygen and form water. Each of these four redox processes also yields energy. If the four redox reactions then are added together, we end up with the same final equation and the same net energy yield as in a single-step oxidation of methane.

Stepwise oxidation occurs generally in the respiration of all foods. Indeed, although methane itself is a poison, not a food, the oxidation steps for actual foods nevertheless are the same as those outlined for methane: *the carbon groupings of food molecules are transformed successively from hydrocarbon to alcohol to aldehyde to acid to anhydride.* Moreover, the oxidation states of carbon here change successively from -4 to -2 to 0 to $+2$ to $+4$ (Fig. 5.6).

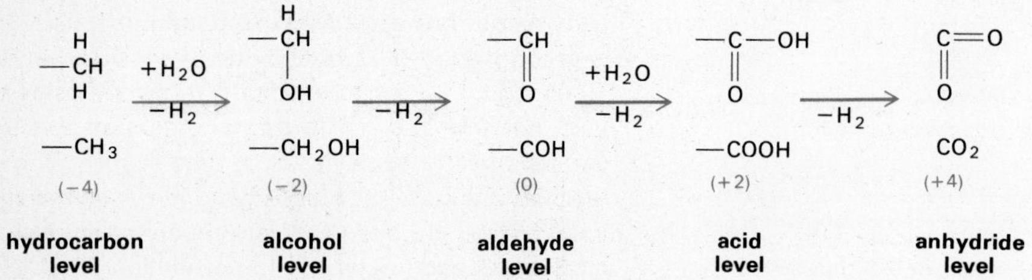

Energy Transfer

An appreciable fraction of the energy released in biological oxidations dissipates to the environment largely in the form of heat, and respiration actually is the most important source of internal heat in all animals (see also Chap. 6). However, a valuable fraction of the energy from respiratory oxidations does not become "free"; instead, as noted earlier, respiration includes a transfer of energy from chemical bonds in fuel to the chemical bonds of ATP. But if fuel energy already exists in the form of chemical bonds, what is the point of respiration if it only creates other chemical bonds?

Some bonds contain more energy than others, and one can distinguish between *high-energy bonds* and *low-energy bonds*. To create the first kind, a relatively large amount of energy must be expended. A correspondingly large amount then becomes available when such a bond is broken. However, most bonds in organic molecules are of the *low*-energy type. Thus the carbon-carbon, carbon-hydrogen, or carbon-oxygen links in fuel compounds are low-energy bonds. If one of these is broken only a comparatively small amount of energy becomes available. Yet much more concentrated packets of energy are required in metabolic work. An energy-*concentrating* process is therefore needed, one that pools the many low-energy packets of a fuel molecule and forms a smaller number of high-energy packets.

Respiration does just that. Through fuel oxidation, the bond energies of fuel become concen-

Figure 5.6 *Foods and oxidation levels.* Carbon groups in food molecules (symbolized in two ways, top and bottom rows) are at one of the oxidation levels shown, and during respiration these groups are transformed to the successively more oxidized levels indicated. At each step the oxidative event is hydrogen removal, and the resulting level is more stable than the preceding one. Combination of the hydrogen with oxygen in reducing half-reactions (not shown) then yields a net energy gain, as in Fig. 5.3.

hydrocarbon level	alcohol level	aldehyde level	acid level	anhydride level
$-CH_3$ (-4)	$+H_2O \xrightarrow{-H_2}$ $-CH_2OH$ (-2)	$\xrightarrow{-H_2}$ $-COH$ (0)	$+H_2O \xrightarrow{-H_2}$ $-COOH$ $(+2)$	$\xrightarrow{-H_2}$ CO_2 $(+4)$

trated in one or more high-energy bonds. Atomic groupings containing these bonds then are transferred from fuel to ADP, and ATP results. Respiration therefore accomplishes more than the mere making of new bonds out of old ones; it makes high-energy bonds out of low-energy bonds. The creation of high-energy from low-energy bonds is achieved essentially by *internal reorganizations* of a fuel molecule. In the course of stepwise oxidation, the bond energies of a fuel become redistributed in such a way that one of the bonds comes to hold a great deal of energy, whereas others hold less than before. In effect, a high-energy bond is created at the expense of several low-energy bonds.

The chief atomic groupings in cells in which high-energy bonds can be created are phosphate groups. If we symbolize such a group as —O—P, as on earlier occasions, then the bond that joins O and P is a *phosphate bond*. A bond of this type can be either of the low-energy or the high-energy variety; its properties are such that it can join —O and —P weakly or strongly. In the latter case the phosphate bond can be considered to contain *extra energy* and to represent a high-energy bond. To indicate such a bond, the symbol ∼ is used. Thus, a low-energy phosphate bond is symbolized as —O—P, and a high-energy phosphate bond, as —O∼P.

The energy-harvesting phase of respiration now can be symbolized as follows:

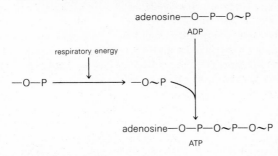

This scheme clearly implies that, before —O∼P groups can be transferred to ADP to form ATP, such groups must first be created. They are indeed created in respiration. Two general events take

place: (1) in a preparatory process a fuel molecule is phosphorylated; a phosphate group (—O—P) is attached to it; (2) this phosphorylated fuel is oxidized by dehydrogenation. The process yields energy, but inasmuch as an —O—P group is now part of the fuel molecule, a significant part of this energy is not released as heat. Instead it becomes redistributed in the fuel molecule and —O—P is transformed to —O∼P. The latter is then transferred to ADP (Fig. 5.7).

The ultimate source of the phosphate groups required in these (and other) phosphorylations is the external environment, which supplies inorganic phosphates as mineral nutrients. However, the immediate phosphate source in cells usually is ATP itself. As noted earlier, for example, glucose uptake by a cell requires an accompanying phosphorylation, and ATP is the phosphorylating agent (see Fig. 5.1). In this and all similar instances, the terminal high-energy phosphate of ATP is split off (and ATP thus becomes ADP), but an ordinary low-energy phosphate is added to the fuel molecule.

$$ADP—\overset{O∼P}{\underset{fuel}{\huge\curlyvee}}\overset{ADP}{\underset{fuel—O—P}{}}$$

ATP here supplies more than enough energy to produce a phosphorylated compound, and any energy excess dissipates unavoidably as heat. Evidently, some of the endproduct of respiration, ATP, is needed at the starting point for preliminary phosphorylations; prior respiration is a necessary condition for further respiration.

The amount of ATP that can be formed in respiration varies with the nature of the fuel, or more specifically, with the number of possible oxidation steps in the fuel. In a fatty acid, for example, the carbon in the —COOH group is at the acid level, the other carbons are at the hydrocarbon level; a sugar such as glucose contains one carbon at the aldehyde level, five others at the alcohol level (Fig. 5.8). Oxidations of individual carbon groups here will proceed stepwise from any given starting level until the anhydride stage is attained. Correspondingly, the energy yields will differ according to how many oxidation steps still are possible for each carbon group.

Moreover, every oxidation step does not necessarily yield enough energy for ATP formation. Con-

Figure 5.7 High-energy bonds. The general pattern of their creation by phosphorylation and oxidation, and the transfer of such bonds to ADP.

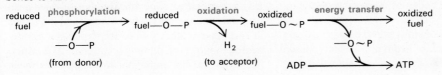

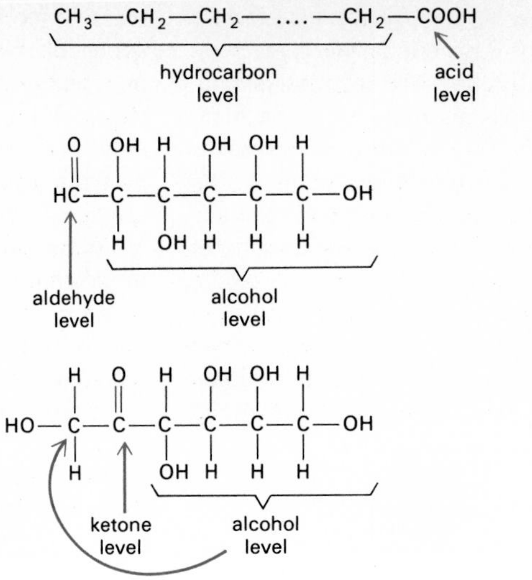

$$CH_3 - CH_2 - CH_2 - .... - CH_2 - COOH$$

hydrocarbon level   acid level

HC—C—C—C—C—C—OH

aldehyde level   alcohol level

HO—C—C—C—C—C—OH

ketone level   alcohol level

Figure 5.8 Oxidation levels of carbon groups. In fatty acid (top), glucose (middle), and fructose (bottom), the amount of further oxidation still possible for each group depends on its particular oxidation level in the intact molecule. A ketone group (—C=O) has the same oxidation level as an aldehyde group (—HC=O).

version of ADP to ATP requires on the average 7 kilocalories (kcal; see Chap. 3) per gram-molecular weight (mole) of ADP. A mole is an amount of a substance equal in grams to the molecular weight. Thus, a mole of ADP weighs 406 g. In effect, 7 kcal represents a kind of minimum energy packet in metabolism, and any oxidation that yields less than that will not result in ATP formation. In fuel oxidations the energy yields range from about 5 to about 11 kcal, with an average of about 8 kcal, per mole of fuel oxidized. Most of such oxidations do result in ATP formation (and any energy excess dissipates as heat). But in certain oxidations the energy yield is less than 7 kcal, and in such cases ATP cannot be formed (see also below).

Hydrogen Transfer

Hydrogen removal from fuel molecules is an enzymatic process; a specific *dehydrogenase* is required at each oxidation step. As already noted, moreover, dehydrogenation also requires the presence of a hydrogen acceptor, and the final acceptor in cells usually is oxygen. However, hydrogen from cellular fuels does not combine with oxygen directly. Instead, such hydrogen is first passed along a whole succession of intermediate *hydro-*

gen carriers, and only the last of these finally releases hydrogen to oxygen:

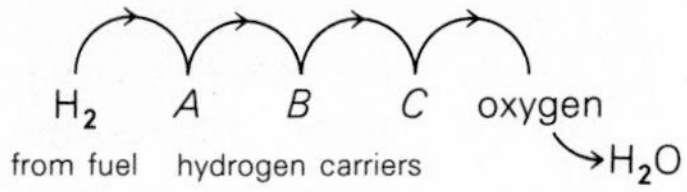

H_2 A   B   C   oxygen

from fuel   hydrogen carriers H_2O

The combination of hydrogen and oxygen is an energy-yielding process. This is well demonstrated by the observation that, when mixed in the right proportions in a test tube, the two gases combine explosively; the energy is released as a single large packet and dissipates almost entirely as heat. Most probably, therefore, H transfer to oxygen in cells occurs stepwise through a succession of carriers for exactly the same reasons that fuel decomposition to CO_2 occurs in small steps; smaller, more numerous energy packets can become available for metabolic use.

Indeed, like fuel decomposition to CO_2, each step in a serial hydrogen transfer is itself a complete redox process. A given hydrogen carrier becomes oxidized when it passes on H_2 to the next carrier in the series, and it becomes reduced when it accepts H_2 from the preceding carrier:

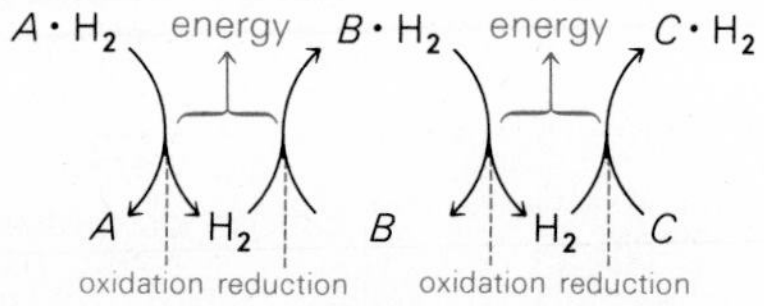

$A \cdot H_2$ energy $B \cdot H_2$ energy $C \cdot H_2$

A H_2 B H_2 C

oxidation reduction   oxidation reduction

Hydrogen carriers operate cyclically and are alternately reduced and oxidized; and $C \cdot H_2$ above is more stable than $B \cdot H_2$, which in turn is more stable than $A \cdot H_2$. The most stable compound appears at the last transfer, when hydrogen combines with oxygen and forms water.

As indicated in the scheme above, H transfer by carriers yields energy at each transfer step. It has been shown that at three of these steps enough energy is released to suffice for ATP formation. Thus, for every H_2 transferred by carriers from fuel to oxygen, a net total of 3 ATP is formed. High-energy phosphates again play a role here, but it is not yet fully known just how the energy actually is trapped in such phosphates. In any case it is clear that there are *two* kinds of ATP sources in respiration. One is fuel decomposition by dehydrogenation; the other is creation of ATP during hydrogen transfer from fuel to oxygen. Of these two sources, the second is often the more important—hydrogen transfer in many cases yields more ATP than that gained through fuel decomposition (Fig. 5.9).

The actual intermediate H carriers are some of the coenzymes already discussed in Chap. 4. Figure 5.9 indicates their names and the sequence in which they function. In this sequence, NAD and FAD are vitamin B–derived coenzymes (see Fig. 4.12). Q is a coenzyme that has not been completely identified as yet, and the cytochromes consist of a family of at least four slightly different coenzymes that operate in succession (see Fig. 4.17).

Aerobic and Anaerobic Transfer

Because H transport requires atmospheric oxygen as the final hydrogen acceptor, this pattern of transfer is said to define an *aerobic* (air-dependent) form of respiration. It is the standard, universal form among animals.

If any one of these aerobic transport reactions were stopped, the whole transfer sequence would become inoperative and the energy it normally supplies would remain unavailable. Reaction blocks actually can occur in a number of ways. For example, *inhibitor* substances of various kinds can interfere specifically with particular transport reactions. Thus, cyanides specifically inhibit the cytochromes, and this is why cyanides are poi-

sons. Another form of reaction block is produced if one of the hydrogen carriers is in deficient supply. For example, inasmuch as riboflavin (vitamin B_2) is a structural part of the FAD molecule (see Fig. 4.12), a consistently riboflavin-deficient diet soon would impair the reactions in which FAD participates.

Although cyanide poisoning and vitamin B deficiencies are not particularly common hazards, animals quite frequently do have to contend with inadequate supplies of atmospheric oxygen. Lack of oxygen is a reaction block of the same sort as cyanide poisoning or vitamin deficiencies, with the consequence that respiration as a whole stops. Whenever hydrogen transport to oxygen is blocked for some reason, animals can respire in a way that does not require oxygen. This *anaerobic* (air-independent) respiration, or *fermentation,* then becomes a substitute or auxiliary source of energy.

The principle of anaerobic hydrogen transport is relatively simple. With the path from NAD to oxygen blocked, another path, from NAD to another hydrogen acceptor, must be used. Such an alternative path is provided by *pyruvic acid,* one of the compounds normally formed in the course of carbohydrate respiration (see below). If enough oxygen is available, pyruvic acid is merely one of the intermediate steps in the decomposition of carbohydrates to CO_2. But pyruvic acid has the property of reacting readily with hydrogen, and if NAD cannot use its normal hydrogen outlet to FAD and oxygen, pyruvic acid is used instead. The acid then ceases to be a fuel and becomes a hydrogen carrier.

When pyruvic acid reacts with hydrogen, the result in animals is the formation of *lactic acid:*

$$NAD \cdot [H_2] \longrightarrow NAD$$
$$\searrow H_2$$
$$CH_3COCOOH \longrightarrow CH_3CHOHCOOH$$

pyruvic acid   lactic acid

This reaction completes anaerobic respiration; lactic acid accumulates in a cell and eventually diffuses into the cellular surroundings.

The energy gained through fermentation is insufficient in most cases to sustain the life of an animal cell. Because oxygen is unavailable, the energy normally harvested by H transfer to oxygen cannot be obtained. Moreover, fuel oxidation has stopped at the pyruvic acid stage, and the potential energy still contained in this acid therefore

Figure 5.9 Coenzymes, hydrogen transfer, and energy sources. Among the coenzymes in respiratory hydrogen transfer from foods, Q is a still poorly known hydrogen carrier and the cytochromes are a series of five successive carriers. Of the two sources of ATP, one is fuel oxidation itself and another is H_2 transfer to oxygen. The latter source yields three high-energy phosphates as shown, hence 3 ATP for every H_2 transferred.

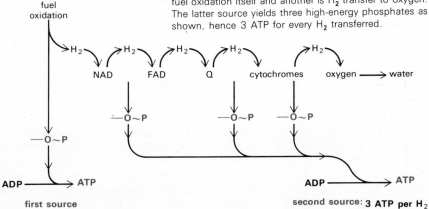

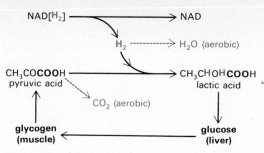

Figure 5.10   Anaerobic respiration. Pyruvic acid accepts hydrogen from NAD and becomes lactic acid. This acid passes via blood to the liver, becomes glucose there, and eventually returns via blood to muscle, where it becomes glycogen. Under later aerobic conditions, the glycogen can be respired completely to CO_2—pyruvic acid in that case need not be a hydrogen acceptor, for H_2 can be transferred to oxygen and form water.

remains untapped and locked in lactic acid. Fermentation actually yields only about 5 percent of the energy obtainable by aerobic respiration (see below). This is too little to maintain the life of animals; as is well known, death occurs within minutes if oxygen is completely absent.

Fermentation nevertheless can *supplement* the aerobic energy gains. Whenever energy demands are high, as during intensive muscular activity, the

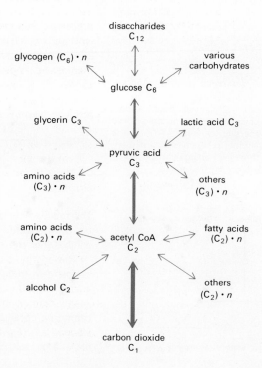

Figure 5.11   Pathways in respiration. Pyruvic acid, acetyl CoA, and carbon dioxide form a main sequence that other pathways join, like branches of a tree.

oxygen supply to the cells tends to become insufficient despite faster breathing. Under such conditions of *oxygen debt,* fermentation can proceed in parallel with aerobic respiration and provide a little extra energy. Lactic acid then accumulates in the muscles. Eventually fatigue becomes so great that the animal must cease its intensive activity. During the later rest period faster breathing at first continues and the oxygen debt is thereby being repaid. Lactic acid is carried by blood from the muscles to the liver, where it is converted to glucose. Returned to muscles, glucose is transformed to glycogen. And with extra oxygen now available, glycogen can be oxidized completely to CO_2. In this indirect way the potential energy still present in lactic acid can be recovered and fatigue then gradually decreases (Fig. 5.10).

Respiration: The Process

In the course of being decomposed progressively, fuel molecules lose their carbons one at a time, until sooner or later the entire molecule is converted to C_1 fragments, or CO_2. If the decomposition sequence is followed backward, the next-to-last stage in fuel breakdown should be a 2-carbon C_2 fragment. This is the case; every fuel molecule sooner or later appears as a 2-carbon fragment, specifically *acetyl* ($CH_3C{=}O$), a derivative of acetic acid. Acetyl does not exist by itself but is attached to a carrier coenzyme called *coenzyme A,* CoA (see Fig. 4.12). Acetyl joined to CoA is *acetyl CoA,* which can be symbolized as $CH_3CO{-}CoA$.

The manner in which the acetyl CoA stage is reached differs for different types of fuel. For example, many carbohydrates are first broken up to 3-carbon compounds. Complex carbohydrates often are built up from 3-carbon units, and their carbon numbers then are whole multiples of 3. This holds, for example, for glucose and all other 6-carbon sugars, for 12-carbon disaccharides, and for polysaccharides such as glycogen. When any of these is used as respiratory fuel, the original 3-carbon units reappear in the course of breakdown. All such C_3 units eventually are converted to *pyruvic acid,* the common 3-carbon stage in respiration. Pyruvic acid then loses one carbon in the form of CO_2 and is converted to acetyl CoA. Fatty acids and related molecules do not break up to 3-carbon units but become 2-carbon units directly, and these eventually appear as acetyl CoA. Amino acids break down partly to pyruvic acid (which subsequently becomes acetyl CoA), partly to acetyl CoA directly. This holds also for many other substances that happen to be used as fuel.

Thus, the overall pattern of aerobic fuel combustion can be likened to a tree with branches or to a river with tributaries (Fig. 5.11). A broad main channel is represented by the sequence pyruvic acid → acetyl CoA → carbon dioxide. Numerous side channels lead to this sequence, some funneling to the C_3 (pyruvic acid) step, others to the C_2 (acetyl CoA) step. In the end the flow from the entire system drains out as C_1 (carbon dioxide). It is best to discuss this common last step first.

Formation of CO₂

The breakdown of the C_2 acetyl fragment in the mitochondria of cells takes the form of a *cycle* of enzymatic reactions. Acetyl CoA is funneled in at one point of the cycle, two carbons emerge at other points as CO_2, and the starting condition is eventually regenerated. The whole sequence is known as the *citric acid cycle*, after one of the participating compounds, or as the *Krebs cycle*, after its discoverer.

A simplified and condensed representation of the cycle is given in Fig. 5.12. The figure shows that acetyl CoA first interacts with water and with *oxaloacetic acid*, a C_4 molecule normally present in the mitochondria of a cell (step 1 in Fig. 5.12). The result is the formation of free CoA and the C_6 compound *citric acid*. In a following series of reactions (step 2), this acid loses 1 H_2 and 1 CO_2 and becomes *ketoglutaric acid*, a C_5 compound. The hydrogen loss here is oxidative: carbon 5 is at a hydrocarbon level in citric acid and at a level equivalent to an aldehyde in ketoglutaric acid. In a further series of reactions (step 3), ketoglutaric acid loses CO_2 in its turn, yielding the C_4 compound *succinic acid*. This step, too, yields H_2 and is oxidative: carbon 5 has become oxidized to the acid level in succinic acid. Through a final sequence (step 4), succinic acid eventually becomes rearranged to oxaloacetic acid, the same C_4 compound that started the cycle. This segment yields 2 H_2 and oxidizes carbon 4, from a hydrocarbon level in succinic acid to a level equivalent to an aldehyde in oxaloacetic acid.

In the whole sequence, the raw materials are one molecule of acetyl CoA and three molecules of water. The endproducts are 2 CO_2 (which represent the two carbons fed into the cycle at the start as acetyl), free CoA, and 4 H_2. These hydrogen pairs are accepted as soon as they appear by NAD and then are transferred via FAD and the cytochromes to oxygen. In these transfers ATP is created, three molecules for every H_2. Thus the citric acid cycle yields a net total of 12 ATP. In summary:

$$CH_3CO \cdot CoA \qquad 2\ CO_2$$
$$\diagdown \ \ CoA \cdot H$$
$$3\ H_2O \qquad \diagup \ \ 4\ H_2$$
$$\vdots$$
$$\text{to oxygen: } 4\ H_2O.\ 12\ ATP$$

As pointed out earlier, acetyl groups generally form by respiratory decomposition of various original foods. How do such foods become acetyl, and how are they thereby funneled to the citric acid cycle?

Figure 5.12   The citric acid cycle. The nine actual steps are condensed and simplified here as four. In step 1, citric acid is formed by combination of acetyl CoA and oxaloacetic acid (small black numbers marking carbon positions in oxaloacetic acid correspond to colored numbers in citric acid). In step 2, carbon 3 is oxidized from an acid level (—COOH) to an anhydride level (CO₂) and H₂ is released. Also, carbon 5 is oxidized from a hydrocarbon (—CH₂—) to a ketone (—C═O), equivalent to an aldehyde level. In step 3, carbon 6 is oxidized from an acid to an anhydride level and H₂ is again released. In step 4, carbon 4 converts from a hydrocarbon to a ketone, 2 H₂ are released, and oxaloacetic acid is regenerated. In this acid the carbons now assume new positions for the start of the next turn of the cycle (colored numbers change to black numbers). Note therefore that the carbons 1 and 2 of acetyl fed into the cycle in one turn do not become oxidized (as carbons 5 and 6) until the following turn.

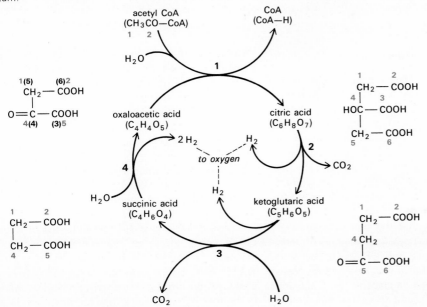

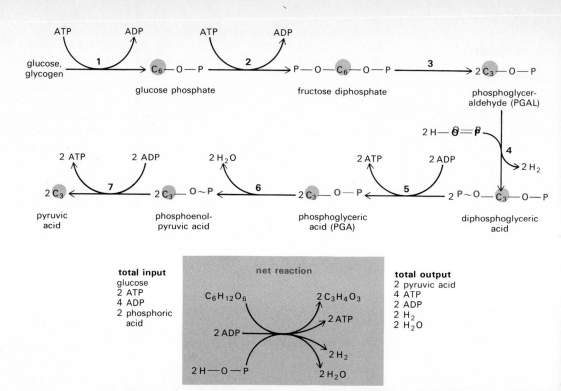

Figure 5.13   Glycolysis. The nine actual steps are shown here simplified and abbreviated as seven, and the structure of the participating compounds is indicated only in the form of carbon numbers. In step 4, the —O—P added by phosphoric acid becomes a high-energy phosphate, which is then transferred to ATP (step 5). Step 4 also yields 2 H_2, which is transported by carriers to oxygen. Through steps 4 and 5 an aldehyde (phosphoglyceraldehyde, PGAL) becomes an acid (PGA). In step 6 oxidative removal of water produces another high-energy phosphate, transferred again to ATP (step 7). In these two reaction steps an alcoholic carbon group (in PGA) is oxidized to a ketone group (in pyruvic acid). The summary below the reaction sequence assumes that free glucose is the starting fuel and that 2 ATP therefore must be expended in the preliminary phosphorylations (steps 1 and 2) for every 4 ATP gained at the end.

Carbohydrate Respiration

Carbohydrate fuels are first converted to pyruvic acid, a common C_3 stage as already noted. In this conversion, often called *glycolysis,* polysaccharides such as glycogen initially are split up and phosphorylated by ATP to glucose-phosphate. If free glucose is used as fuel, it, too, becomes glucose-phosphate through phosphorylation by ATP. Glucose-phosphate therefore is a common early stage of glycolysis. Its further transformation to pyruvic acid is outlined in Fig. 5.13.

This figure shows that the (enzyme-requiring) reaction sequence includes first, a series of additional phosphorylations: —O—P groups become attached at *each* end of the C_6 chain of glucose-phosphate. Then the chain splits to two C_3 chains (step 3 in Fig. 5.13), and more —O—P groups

become attached to the free ends of these C_3 compounds. Next, in successive oxidations, all —O—P groups now present are transformed to —O~P groups. And last, these high-energy phosphates are transferred to ADP. What is then left of the original fuel is a C_3 compound, pyruvic acid. Note that the carbon atoms of the original C_6 fuel are largely at an *alcohol* level of oxidation (see Fig. 5.8), and that the reactions of glycolysis yield first *aldehydes* and then *acids.* Thus the oxidation sequence follows the general stepwise pattern of respiratory decomposition.

If free glucose is considered to be the original raw material, glycolysis includes four phosphorylations, two by H—O—P (phosphoric acid) and two by ATP. Each of the four added phosphates eventually becomes a high-energy phosphate. And of the four ATP molecules then formed, two ''pay

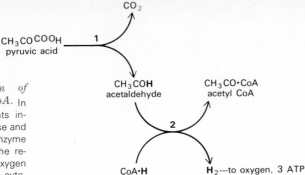

Figure 5.14 Conversion of pyruvic acid to acetyl CoA. In step 1, required participants include the enzyme carboxylase and the vitamin B_1–derived coenzyme cocarboxylase. In step 2 the released H_2 is transferred to oxygen via NAD, FAD, and the cytochromes. The acetyl fragment left after H_2 removal is accepted by coenzyme A, and acetyl CoA is so formed.

back'' for the two expended at the start of glycolysis, while two represent the net gain. The fate of the atoms in glucose is described by the equation

$$C_6H_{12}O_6 \longrightarrow 2\ C_3H_4O_3 + 2\ H_2$$

The net loss of atoms from glucose therefore amounts to 2 H_2, and these are accepted by NAD.

If respiration occurs under anaerobic conditions, the pyruvic acid now must serve directly as the final hydrogen acceptor. Carbohydrate oxidation in this case stops with the formation of lactic acid, and the ATP molecules gained represent the net energy yield of the entire process. But if conditions are aerobic, two advantageous consequences follow. First, the 2 H_2 held by NAD can be passed on to oxygen, a transfer that yields three additional ATP molecules per H_2, or a total of six ATP molecules. Second, since pyruvic acid need not serve as a hydrogen carrier,, it can be oxidized further and more of its chemical energy can become usable.

Pyruvic acid transforms to acetyl CoA through a series of enzymatic reactions in which both hydrogen and CO_2 are removed from pyruvic acid (Fig. 5.14). Removal of CO_2 takes place first, and this step requires (apart from enzymes) a coenzyme called *cocarboxylase*. It is a derivative of *thiamine*, or vitamin B_1, and it serves as a kind of carrier of CO_2. Through CO_2 removal pyruvic acid is converted to *acetaldehyde*, a compound that next becomes dehydrogenated. In this step NAD must be present as a hydrogen acceptor (and transfer of H_2 to oxygen then yields 3 ATP). Another required ingredient is CoA, which serves as the carrier of what is left of acetaldehyde after H_2 has been removed. This remnant is acetyl, and acetyl CoA arises in this manner. In the sequence as a whole, pyruvic acid is transformed first to an *aldehyde* and then to the *anhydride* CO_2, again in line with the general pattern of oxidative decomposition.

Acetyl CoA later can be respired via the citric acid cycle, and we already know that 12 ATP are gained from one acetyl group. If it is therefore assumed that the original starting fuel is free glucose and that it is respired completely to CO_2, then the aerobic energy yield can be shown to total 38 ATP, as outlined in Fig. 5.15. Such a net gain contrasts sharply with an anaerobic yield of only 2 ATP, or about 5 percent of the aerobic yield.

As pointed out earlier, 1 mole of ATP represents energy equivalent to about 7 kcal, on the average. Glucose respiration produces 38 ATP, hence 38 moles of ATP amount to an energy gain of 266 kcal per mole of glucose respired. If 1 mole of glucose is burned in a furnace, it releases energy equivalent to about 686 kcal. Thus, carbohydrate respiration has an average efficiency of $266/686$, or about 40 percent. In other words, some 40 percent of the potential fuel energy actually is harvested as ATP, and the remaining 60 percent escapes as heat. This escaping energy represents the main internal heat source of animals. It does not become trapped chemically, but it is nevertheless exceedingly useful simply as heat. For example, it can serve in counteracting low environmental temperatures. A ''burning'' efficiency of 40 percent is comparable to that of the best fuel-using engines man can construct.

Where glycogen is the chief carbohydrate fuel, as in muscle, carbohydrate respiration occurs almost entirely via glycolysis and the citric acid cycle, as described. In many animal tissues, how-

Figure 5.15 Glucose respiration: summary.

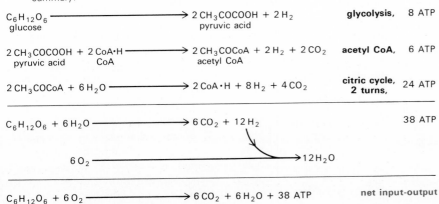

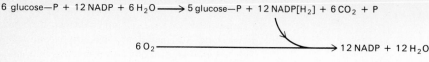

$$6 \text{ glucose—P} + 12 \text{ NADP} + 6 H_2O \longrightarrow 5 \text{ glucose—P} + 12 \text{ NADP}[H_2] + 6 CO_2 + P$$

$$6 O_2 \longrightarrow 12 \text{ NADP} + 12 H_2O$$

$$1 \text{ glucose—P} + 6 O_2 \longrightarrow 6 CO_2 + 6 H_2O + P \qquad \text{net input-output}$$

Figure 5.16 The hexose shunt: summary. In a cyclic reaction sequence one out of six glucose-phosphate molecules is oxidized to CO_2 and H_2O.

ever, the main carbohydrate fuel may be—and in brain must be—glucose, not glycogen. In such cases an important alternative way of carbohydrate respiration exists that involves neither glycolysis nor the citric acid cycle, nor indeed even the mitochondria. Known as the *hexose shunt,* this oxidation process occurs in the free cellular cytoplasm, and it too has the form of a complex cycle of reactions.

It is summarized in a highly condensed manner in Fig. 5.16. The raw materials are the starting fuel glucose-phosphate, water, and NADP (a coenzyme that differs from NAD only in having one more phosphate group). The endproducts are CO_2, free phosphate, H_2 held by NADP, and one less molecule of glucose-phosphate than the number present at the start. In effect, one glucose-phosphate has become decomposed to CO_2 and H_2, and the 12 hydrogen pairs are passed on to oxygen by NADP. This hydrogen transfer yields 36 ATP, of which one must "pay back" for the ATP used up in the original conversion of free glucose to glucose-phosphate. Thus the net energy yield is 35 ATP per molecule of glucose—only slightly less than the 38 ATP obtainable by glycolysis.

Lipid Respiration

A first step in the respiration of a fat molecule is decomposition to glycerol and fatty acids, with lipase as the specific enzyme. This process occurs inside a cell and is equivalent chemically to digestion in animal alimentary systems. Glycerol then can be phosphorylated by ATP and oxidized to

phosphoglyceraldehyde, PGAL (Fig. 5.17). This reaction requires 1 ATP for the phosphorylation but yields 3 ATP in the transfer of H_2 from NAD to oxygen. PGAL, a normal intermediate in carbohydrate decomposition (see Fig. 5.13), then can be respired by glycolysis and the citric acid cycle, processes that yield 20 ATP per molecule of PGAL. Complete aerobic respiration of one molecule of glycerol thus produces a total net gain of 22 ATP.

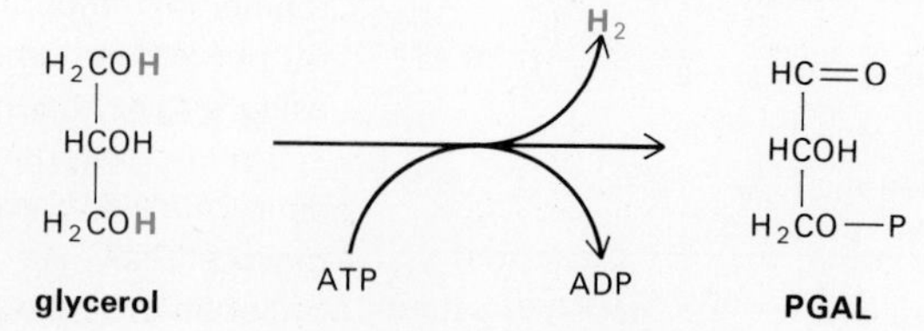

The decomposition of fatty acids is known as *β-oxidation;* the second, or $β$, carbon of the acid (counted after the —COOH group) undergoes oxidative changes. Figure 5.18 summarizes the main steps. In a first reaction CoA becomes linked to the acid end of the fatty acid, with the aid of energy from ATP. A dehydrogenation occurs next, one H atom being removed from each of the $α$ and $β$ carbons. The hydrogen is accepted by FAD (and NAD is bypassed here). In a third step another dehydrogenation takes place, in this case at the $β$ carbon. This hydrogen is carried by NAD. Finally, another molecule of CoA is added, at the $β$ carbon in this instance. Two fragments are thereby produced. One is acetyl CoA, and the other is a CoA-carrying fatty acid that is two carbons shorter than the original acid. This shorter acid now can undergo $β$-oxidation in turn, and consecutive acetyl CoA units can be cut off in this way.

Hydrogen transfer from FAD to oxygen yields 2 ATP (not 3, since NAD is bypassed), and transfer from NAD to oxygen yields 3 ATP. Thus, 5 ATP are gained for every acetyl CoA unit formed. If, for example, an actual starting fuel is assumed to be stearic acid, a C_{18} fatty acid very common in animal fats, then $β$-oxidation of this acid can occur successively eight times, yielding acetyl CoA each time and leaving a ninth acetyl CoA as a remainder. At 5 ATP per $β$-oxidation, the yield is $5 \times 8 = 40$ ATP, minus 1 ATP expended in the first reaction of the stearic acid starting molecule. One C_{18} fatty acid therefore yields a net of 39 ATP and 9 acetyl CoA. The latter generate 9×12, or 108 ATP in the citric acid cycle, so that the total energy gained from the complete respiration of stearic acid is 147 ATP.

By way of comparison, we already know that one glucose molecule yields 38 ATP. Since stearic acid is a C_{18} compound whereas glucose is a C_6 compound, the fatty acid might be expected to yield three times as much ATP. However, three glucose molecules yield only 114 ATP, which means that an 18-carbon fatty acid actually pro-

Figure 5.17 Conversion of glycerol to phosphoglyceraldehyde.

glycerol		PGAL
H_2COH		$HC=O$
$HCOH$		$HCOH$
H_2COH		$H_2CO—P$
	ATP ADP	
	H_2	

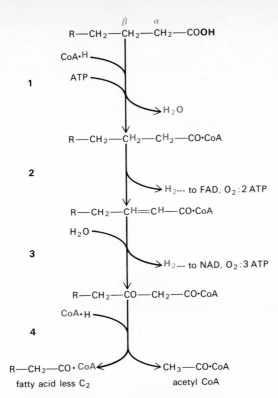

Figure 5.18 β-oxidation of fatty acids. In step 1 a fatty acid is activated by combination with CoA and concurrent removal of water. Step 2 is dehydrogenation, FAD being the H acceptor. Step 3 is an addition of water and represents the actual *β*-oxidation; the carbon group, originally at a hydrocarbon oxidation level, becomes a ketone group (—CO—). In step 4 the 2-carbon fragment acetyl is split off, resulting in acetyl CoA and a fatty acid shorter by two carbons than the original one. The summary shows that 5 ATP are gained through H transport to oxygen. C_x here stands for a fatty acid with x numbers of C atoms.

Figure 5.19 Deamination and transamination. A, oxidation of an amino acid yields a free ammonia molecule, which is excreted, and a keto acid, so named for its ketone (—CO—) in place of an amino group. *B,* interaction of an amino acid *A* and a keto acid *B* can result in a substitution of the amino group by a ketone group in *A* and by a reverse substitution in *B.*

duces *more* than three times as much—almost four times as much—ATP than a 6-carbon carbohydrate. Fatty acids evidently are a richer source of usable energy than equivalent quantities of carbohydrates. The reason is that the carbons in a fatty acid still are largely at the hydrocarbon level, whereas those in a carbohydrate are already at the alcohol or aldehyde level of oxidation from the outset (see Fig. 5.8).

The figures above suggest a reason why fats are the chief storage foods in animals and why animal metabolism is highly fat-oriented in general: energy stored in the form of fat *weighs* less than if it were stored in the form of carbohydrate or protein. Fat storage therefore makes for less bulk, important in a moving animal; every ounce of excess food stored in the form of fat would be equivalent to more than two ounces of excess carbohydrates or proteins. It is interesting that clams, for example, which move very little, do store their foods largely as carbohydrates, like rooted plants, whereas plant seeds, which are adapted for dispersal through air, store foods largely as lipids, like moving animals.

Apart from the greater energy content of fats, the efficiency of fat respiration nevertheless is roughly equivalent to that of carbohydrates. For example, if the common fat tristearin is burned in a furnace, its energy potential can be shown to be about 8000 kcal. Tristearin consists of three C_{18} fatty acids (stearic acid) and one C_3 carbohydrate (glycerol). Its complete respiration in cells therefore will yield 3 × 147 plus 1 × 22, or 463 ATP. At about 7 kcal per mole of ATP, the respiratory energy gained then will be about 3241 kcal, and the efficiency of this respiration will be 3241/8000, or about 40 percent. Fats evidently give rise to the same proportions of ATP and heat as carbohydrates.

Amino Acid Respiration

Amino acids can enter respiratory pathways in two general ways:

One way is made possible by *deamination,* a reaction in which the amino group (—NH₂) is removed from an amino acid (Fig. 5.19). In vertebrates this process takes place particularly in liver cells, where any amino acid excess supplied by eaten food is deaminated. In other animals deaminations can occur in a large variety of cell types. The resulting free ammonia is highly alkaline and

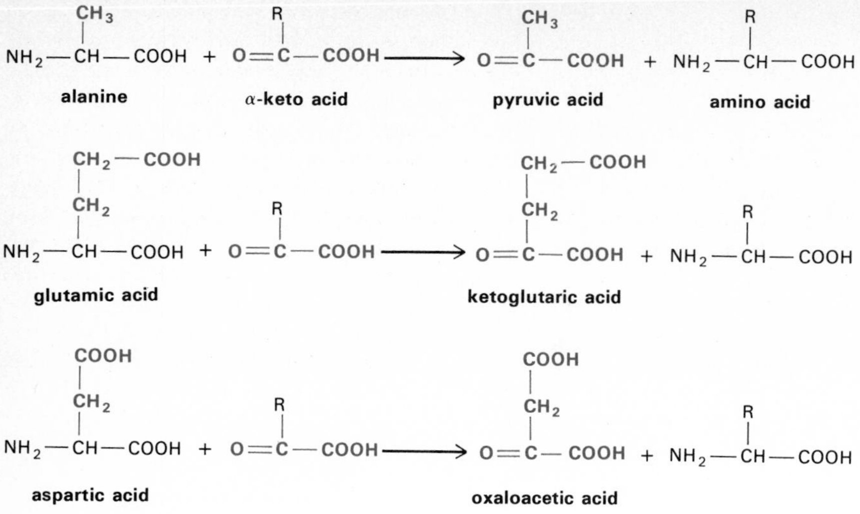

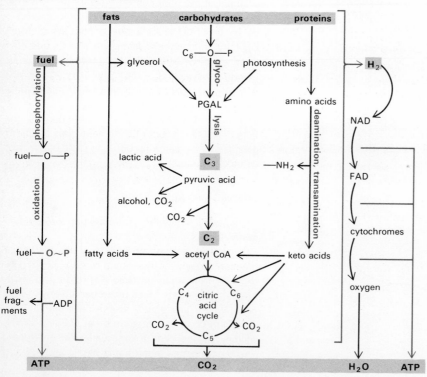

$$CH_2-\!\!\underset{\underset{alanine}{}}{\overset{\overset{CH_3}{|}}{NH_2-CH-COOH}} \;+\; \underset{\underset{\alpha\text{-keto acid}}{}}{\overset{\overset{R}{|}}{O\!=\!C-COOH}} \;\longrightarrow\; \underset{\underset{pyruvic\ acid}{}}{\overset{\overset{CH_3}{|}}{O\!=\!C-COOH}} \;+\; \underset{\underset{amino\ acid}{}}{\overset{\overset{R}{|}}{NH_2-CH-COOH}}$$

$$\underset{\underset{glutamic\ acid}{}}{\overset{\overset{CH_2-COOH}{|}\;\overset{CH_2}{|}}{NH_2-CH-COOH}} \;+\; \overset{\overset{R}{|}}{O\!=\!C-COOH} \;\longrightarrow\; \underset{\underset{ketoglutaric\ acid}{}}{\overset{\overset{CH_2-COOH}{|}\;\overset{CH_2}{|}}{O\!=\!C-COOH}} \;+\; \overset{\overset{R}{|}}{NH_2-CH-COOH}$$

$$\underset{\underset{aspartic\ acid}{}}{\overset{\overset{COOH}{|}\;\overset{CH_2}{|}}{NH_2-CH-COOH}} \;+\; \overset{\overset{R}{|}}{O\!=\!C-COOH} \;\longrightarrow\; \underset{\underset{oxaloacetic\ acid}{}}{\overset{\overset{COOH}{|}\;\overset{CH_2}{|}}{O\!=\!C-COOH}} \;+\; \overset{\overset{R}{|}}{NH_2-CH-COOH}$$

Figure 5.20 *Three transamination reactions* important in amino acid respiration. The keto acids resulting from transamination happen to be participants in glycolysis and the citric acid cycle. Therefore, through the reactions shown, amino acids can enter the same final respiratory pathways as carbohydrates and fatty acids.

Figure 5.21 *Respiration:* overall summary. The center panel outlines the main pathways of the respiratory breakdown of carbohydrates, fats, and proteins. The left-hand panel summarizes the general nature of any of the respiratory reactions that occur in the center panel, and the right-hand panel similarly summarizes the process of hydrogen transport. The two sources of ATP gain are also indicated.

therefore toxic, and it is ultimately excreted. Excretion often occurs in the form of unchanged ammonia, as in most aquatic types. Or ammonia can undergo various chemical transformations and be excreted in the form of, for example, *uric acid,* as in insects and birds, or *urea,* as in mammals (see Chap. 9). The other product of deamination is a *keto acid.* In most keto acids the —R groups are chemically similar to carbohydrates or to fatty acids, and a keto acid then is actually respired either like a carbohydrate or like a lipid.

A second pattern of amino acid respiration results from the circumstance that each amino acid has a structurally corresponding keto acid. Conversely, if to any given keto acid an amino group is added, a corresponding amino acid will be formed (see Fig. 5.19). Through such *transamination* reactions, any amino acid can be transformed to any other provided the appropriate keto acid is available. Transaminations can occur in all cells, and they permit any amino acid to be respired via the citric acid cycle; for it happens that certain transaminations produce keto acids that are normal components of the citric acid cycle. For example, if the amino acid glutamic acid participates in a transamination, this amino acid becomes ketoglutaric acid, an intermediate in the citric acid cycle (Fig. 5.20). Similarly, when the amino acid aspartic acid is transaminated, the corresponding keto acid formed is oxaloacetic acid, another citric acid cycle compound. Moreover, if alanine transaminates with some other keto acid, alanine becomes pyruvic acid. Therefore, any amino acid can, by transamination, become glutamic acid or aspartic acid or alanine; and by a further transamination these three amino acids can become ketoglutaric or oxaloacetic or pyruvic acid, respectively, which in turn can then participate in the citric acid cycle.

The ATP yields of protein and amino acid respiration vary according to the specific reaction pathways followed, but the efficiency again is roughly 40 percent, equivalent to that of carbohydrate or fat respiration. However, oxidation of proteins, far more than that of carbohydrates or fats, is accompanied by many other energy-requiring and therefore heat-producing reactions. For example, transaminations, deaminations, and urea or uric acid synthesis occur in parallel with amino acid respiration, and many of these processes consume ATP and release heat to the environment. It has been shown that for every 100 kcal supplied by protein, heat to the extent of some 30 kcal is generated by accompanying reactions (as compared with only about 5 kcal for every 100 kcal supplied by fat or carbohydrate). Thus the net usefulness of protein as an energy source is only about 70 percent of its potential value, and this probably is one reason why carbohydrates and fats are primary energy sources. The large heat generation accompanying protein degradation is known as the *specific dynamic action* of protein.

An overall summary of respiratory reactions is outlined in Fig. 5.21. These reactions take place exceedingly rapidly in cells. In vertebrates, moreover, respiratory rates are influenced greatly by the thyroid hormone thyroxin, which accelerates respiration in proportion to its concentration. Most animals are not vertebrates, however, and their respiration is not under thyroxin control. Nevertheless, respiratory decompositions still occur extremely rapidly; a glucose molecule can be respired in less than 1 sec. Very efficient enzyme action probably is one condition that makes such speed possible. Another undoubtedly is the close, ordered arrangement of all required ingredients in the submicroscopic spaces of the mitochondria. Just as a well-arranged industrial assembly line turns out products at a great rate, so do the even better arranged mitochondria.

The fate and function of the chief product of respiration, ATP, is the first subject of the next chapter.

Review Questions

1. How does cellular nutrition take place? What is active transport? Describe the uptake of glucose and oxygen by cells.

2. Compare and contrast a fire with respiration. What do they have in common? What is different? Which materials are fuels in respiration? What general types of event occur in respiration?

3. What are redox reactions? Define oxidation and reduction. Show how both electron-transfer and electron-sharing reactions can be oxidative. What energy relations exist during redox reactions?

4. Show how progressive changes of oxidation level occur during the transformation of hydrocarbons to anhydrides.

5. What is dehydrogenation? Where does it occur, and what role does it play in respiration? Under what conditions does it take place? How is hydrogen transferred to oxygen? Describe the sequence of carriers and the specific role of each during aerobic hydrogen transport.

6. Distinguish between aerobic and anaerobic respiration. Under what conditions does either take place? How and where can aerobic respiration become blocked? How is lactic acid formed? How and under what conditions is lactic acid respired? What is an oxygen debt, and how is it paid?

7. Describe the role of adenosine phosphates in respiration. What is a high-energy bond? How and where are such bonds created in fuels? During H transfer?

8. How much energy is required for ATP formation? Show how ATP is both an endproduct and a specific raw material for respiration. Compare the ATP gain in aerobic and anaerobic respiration.

9. Describe the general sequence of events in the citric acid cycle. Which steps are oxidative and what changes in oxidation level take place? What is the total input and output of the cycle? How much ATP is gained and through what steps?

10. Review the sequence of events in glycolysis. Which steps are oxidative and what changes in oxidation level take place? How much energy is obtained? Which classes of nutrients pass through a pyruvic acid stage in respiration?

11. Review the conversion of pyruvic acid to acetyl CoA. What are the functions of cocarboxylase and coenzyme A? How much energy is gained and where? What classes of nutrients pass through an acetyl CoA stage in respiration?

12. Summarize the events of the hexose shunt. What respiratory role does this process play and where does it occur? How does the energy gain compare with that of glycolysis?

13. Describe the process of β-oxidation. Which steps are oxidative? How much energy is gained, and where? Distinguish between deamination and transamination. By what metabolic pathways are amino acids respired?

14. How much potential energy does each of the main classes of nutrients contain? What fraction of this energy is actually recovered as ATP during respiration? Calculate the efficiency of respiration for each of the main classes of nutrients.

15. Review and summarize the overall fate of one molecule of glucose during complete respiratory combustion. What is the total net input and output? What happens to the individual atoms of glucose? What is the total ATP gain, and how much is gained during each of the main steps of breakdown?

Collateral Readings

Baker, J., and G. Allen: "Matter, Energy, and Life," Addison-Wesley, Reading, Mass., 1965. A paperback containing good introductions to the chemical aspects of nutrition, respiration, and metabolism in general.

Giese, A. C.: Energy Release and Utilization, in "This Is Life," Holt, New York, 1962. A good, concise review of respiratory processes.

Green, D. E.: The Metabolism of Fats, *Sci. American,* Jan., 1954. An article on the respiration of fats.

————: Biological Oxidation, *Sci. American,* July, 1958. An account on the process as a whole.

Holter, H.: How Things Get into Cells, *Sci. American,* Sept., 1961. A good general discussion of processes and forces of cellular nutrition.

Lehninger, A. L.: Energy Transformation in the Cell, *Sci. American,* May, 1960. The article includes a discussion of respiratory enzymes from both structural and functional standpoints.

————: How Cells Transform Energy, *Sci. American,* Sept., 1961. The photosynthetic role of chloroplasts and the respiratory role of mitochondria are compared.

————: "Bioenergetics," Benjamin, New York, 1965. This paperback contains discussions of the molecular aspects of biological energy transformations, including those of respiration and photosynthesis. Slightly more advanced than the presentation in this chapter, but not difficult. Recommended for those interested in greater detail.

Loewy, A. G., and P. Siekevitz: "Cell Structure and Function," Holt, New York, 1963. A paperback containing good accounts on all general aspects of respiratory reactions.

McElroy, W. D.: "Cellular Physiology and Biochemistry," Prentice-Hall. Englewood Cliffs, N.J., 1961. Like the preceding reference, a paperback that includes a general review of respiration.

Racker, E.: The Membrane of the Mitochondrion, *Sci. American,* Feb., 1968. The inner of the two mitochondrial membranes is taken apart (and then put together again) in studies of respiratory enzymes.

Siekevitz, P.: Powerhouse of the Cell, *Sci. American,* July, 1957. The structure and function of mitochondria are discussed.

Solomon, A. K.: Pores in the Cell Membrane, *Sci. American,* Dec., 1960. The mechanical aspects of cellular absorption are reivewed.

————: Pumps in the Living Cell, *Sci. American,* Aug., 1962. A discussion of active transport, specifically for sodium ions.

Stumpf, P. K.: ATP, *Sci. American,* Apr., 1953. An article on the cellular roles of this energy carrier.

6

CELL OPERATIONS:
SYNTHESIS,
SELF-PERPETUATION

Energy must be expended in all physical and chemical processes that maintain the metabolism and self-perpetuation of a cell. The most important physical roles of energy are to produce heat, to some extent also light and electricity, and, above all, movement of cells and cell parts. The chief chemical roles of energy are maintenance of respiration itself and, most particularly, of cellular synthesis. In synthesis energy is one requirement, structural building blocks in the form of nutrients are another. And the function of synthesis is to offset the respiratory decomposition and the wear and tear of existing cell components, to make possible cellular repairs after injury, and to maintain growth and development.

The self-perpetuation of a cell is based on these metabolic activities.

Biophysical Metabolism

Movement

Probably the most important and widespread physical functions of animal cells are those that produce movement, resulting in either locomotion of whole animals or motion of parts of animals. All such movements are ATP-dependent.

Certain nonlocomotor movements occur in all cells. Every cell moves nutrients through its boundaries, and we already know that respiratory energy is expended in the absorption of, for exam-

ple, glucose. Compounds also move inside cells, partly by diffusion, partly by cyclosis. The role of ATP is less clear here, but that it plays some role, even if very indirectly, seems almost certain; for if respiration stops, cyclosis stops. Just as incompletely understood are the movements of chromosomes during cell division (see below), the motions of flagella, cilia, and pseudopodia, and the migrations of cells or cell groups in the course of growth and embryonic development. All that can be said at present is that if ATP becomes unavailable the various movements cease.

Somewhat better known today are the events of *muscular* movement. These are among the most important in animals generally; animals carry out few functions that do not include muscular contraction. Moreover, muscles often are the most abundant tissue of an animal, particularly a vertebrate, and a proportionately large amount of energy must be expended to keep muscles operating. Even during "inactive" periods like sleep, for example, the muscular system maintains not only posture and shape but also vital functions such as breathing, heartbeat, and blood pressure. Mainly because of muscular movement the energy requirements of animals are far greater pound for pound than those of any other kind of organism.

The functional units of muscles are long, thin filaments called *myofibrils* (Fig. 6.1). Each actually is a bundle of many ultrathin parallel strands, composed mainly of five kinds of materials: water, inorganic ions, ATP, and two proteins called *actin*

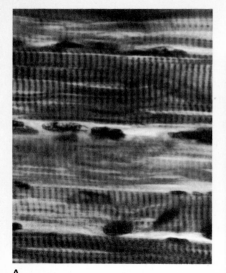

A

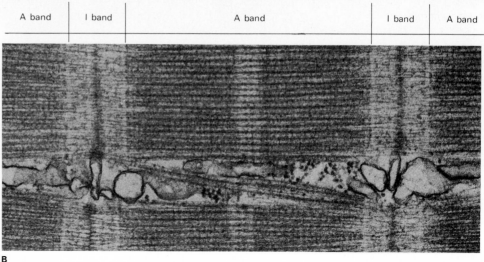

A band | I band | A band | I band | A band

B

Figure 6.1 Skeletal muscle. A, whole muscle fibers. Note the cross striations, the faintly visible internal longitudinal myofibrils, and the nuclei, which appear as dark patches (see also Color Fig. 3). *B*, electron micrograph of portions of two horizontal myofibrils (in the tail muscles of a frog tadpole), separated by a layer of cytoplasmic material (endoplasmic reticulum). The cross striations seen in *A* appear here as A and I bands. The distribution of myosin and actin in these bands is shown in Fig. 6.2.

and *myosin*. Together these form the basic contraction apparatus.

That this is so has been demonstrated dramatically by experiment. With appropriate procedures actin and myosin can be extracted from muscle, and it can be shown that neither actin nor myosin alone is able to contract. But by mixing actin and myosin together artificial fibers of *actomyosin* can be made. If to these are added water, inorganic ions, and ATP, it is found that the fibers contract forcefully. They can lift up to 1,000 times their

I band | A band | I band

I band | A band | I band

Figure 6.2 Muscle activity. Myosin, colored bars; actin, black lines. Top, contracted state; bottom, extended state. The A band and the portions of the adjacent I bands of a muscle fiber shown here are intended to represent a working unit repeated horizontally many times in a whole fiber.

own weight, just as they do in a living muscle. And it is also found that, in a contracted fiber, ATP is no longer present but low-energy phosphates are present instead.

Experiments of this sort provide clues as to how contraction might be brought about in a living muscle. Muscle activity appears to be at least a two-step cycle involving alternate *contraction* and *extension*. These movements are believed to be a result of a sliding of actin fibers back and forth past stationary myosin fibers (Fig. 6.2). Energy is used up at some point or points in such a cycle. One view is that the energy makes possible the contraction of a muscle, like compressing a spring. Extension then is thought to be essentially an automatic recoil. According to an alternative view, energy must be expended to extend a muscle, as in stretching a rubber band. Contraction then would be automatic. A good deal of evidence appears to favor this second hypothesis, but the first cannot be ruled out. Indeed there are strong indications that muscle could require energy for both contraction and extension.

The energy donor in muscle activity is ATP, which, together with actin and myosin, forms a so-called actomyosin-ATP complex. During a contraction-extension cycle, the ATP of actomyosin-ATP gives up its energy. To prepare a muscle for a new contraction-extension cycle new energy must be supplied. The characteristic fuel of muscle is glycogen, and respiration of glycogen is the ultimate source of muscle energy. But glycogen

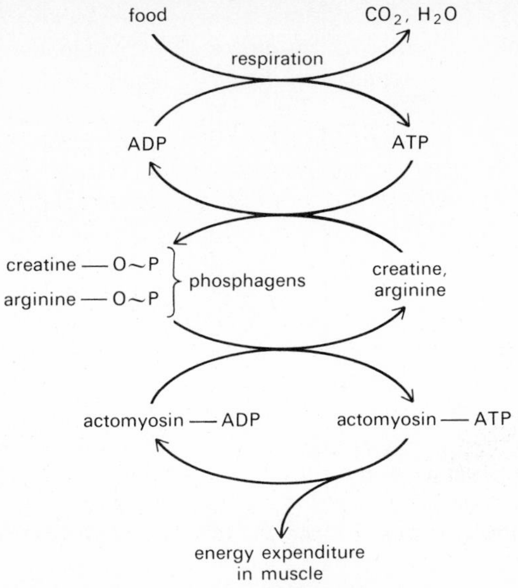

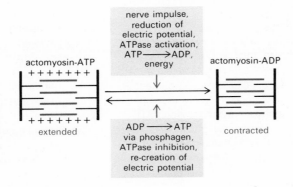

Figure 6.3 Energy and muscle activity. Respiration supplies energy for muscles via the phosphagen stores.

Figure 6.4 Unit action of muscle: summary of probable events. Processes associated with contraction are listed in the box above the horizontal arrows; processes associated with extension, in the box below the arrows.

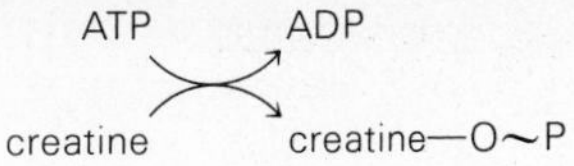

In other words, creatine can accept —O∼P from ATP and become creatine-phosphate. Compounds of this type that carry high-energy phosphates are known as *phosphagens.* The amino acid *arginine* is another of several kinds of compounds that, like creatine, serve as phosphagens in various animals.

Thus, if during rest fuels supply more —O∼P than can be harvested as ATP, then ATP "unloads" —O∼P to phosphagens and becomes ADP. Phosphagens accumulate in this manner. When a muscle later becomes active and must be reenergized rapidly, the immediate energy sources for this "recharging" are the phosphagens. They give up their —O∼P and later are reenergized themselves by respiration. Respiratory ATP therefore replenishes the phosphagen stores slowly, and these re-create actomyosin-ATP rapidly while a muscle is active (Fig. 6.3).

The specific trigger for the contraction of a muscle is a nerve impulse. Also, it is known that inorganic ions, notably Mg^{++} and Ca^{++}, are attached to or in some other way associated with the actomyosin-ATP complex. By virtue of such positive charges the complex has an electric potential over its surface; and one of the known effects of nerve impulses is to bring about a reduction of electric potentials (see Chap. 8). Like several other tissues, furthermore, muscles contain ATPases, enzymes that promote conversion of ATP to ADP. Muscle ATPase either is identical with myosin itself or is so closely linked with myosin that available techniques are unable to separate the two. In effect, the actomyosin-ATP complex contains not only built-in potential energy in the form of ATP but also the necessary built-in enzyme that can make this energy available. The enzyme remains inhibited in some unknown way before arrival of a nerve impulse; directly or indirectly, the impulse appears to be the necessary stimulus for activation of ATPase.

These considerations provide clues to the sequence of events that might occur during muscle action (Fig. 6.4). At first the actin and myosin components of actomyosin-ATP would be stretched apart and the muscle would be extended. Such a condition might be maintained by the electric charges: since like charges repel each other, contraction of the actomyosin-ATP complex

breakdown is not the immediate source; it is far too slow to supply the ATP required by an active muscle. A glycogen molecule might be respired within a second, but in that second the wing muscle of an insect can contract up to 100 times and use up energy far faster than could be supplied directly by fuel oxidation.

Unlike most cells, muscles contain compounds other than ATP that can store relatively large amounts of energy in high-energy phosphate bonds. Among such compounds is, for example, *creatine,* a nitrogen-containing substance found in the muscles of most vertebrates and some invertebrates. The role of creatine can be described as follows:

would be prevented by electric repulsion of actin and myosin. But when a nerve impulse arrives, it would reduce the electric potentials and thereby remove the obstacle to contraction. The nerve impulse also would activate the ATPase. ATP would be split, and the potential energy of acto-myosin-ATP would become actual. As noted, it is still not clear whether this energy brings about contraction or reextension after contraction. In either event new potential energy is supplied thereafter by phosphagen. In some way reextension of muscle also must be associated with re-building of electric potentials, sliding apart of the actin and myosin components, and inhibition of ATPase.

Whatever the actual details of the action cycle eventually prove to be, cycles of this sort clearly take place fast enough to propel a cheetah, for example, at speeds of 50 mph; and they are pow-erful enough to permit many animals to lift objects weighing more than the animals themselves.

Heat, Light, Electricity

One source of cellular heat is the external environ-ment, which supplies heat in varying amounts. Another, most important source is food. As shown in Chap. 5, respiration yields up to 60 percent of the energy content of foods directly as heat. A third heat source is phosphorylation; when ATP is used as a donor of low-energy phosphates, any excess energy of —O~P becomes heat. Still an-other internal heat source is ATP-energized move-ment, for friction of moving parts generates heat. Moreover, conversion of the chemical energy of ATP to the mechanical energy of motion is not 100 percent efficient and is accompanied by a loss of energy in the form of heat. Whatever its source, heat maintains the temperature of an animal and offsets heat lost to the environment by evaporation and radiation. Heat also creates tiny convection currents in cells and thereby assists in diffusion and cyclosis. Above all, heat provides adequate operating temperatures for enzymes and all other functional parts of cells.

In birds and mammals heat production is bal-anced dynamically against heat loss. These ani-mals are "warm-blooded" (*homoiothermic*, or *endothermal*), and their body temperature is kept constant (Fig. 6.5). In all other animals the inter-nal temperature by and large matches that of the external. Animals of this (*poikilothermic*, or *ecto-*

thermal) type are sometimes referred to as being "cold-blooded"—a poor term since the blood could be hot or cold and since many of these animals lack blood to begin with. Such animals cannot survive if the environment is either too cold or too hot. But within these extremes food com-bustion and ATP create internal heat that to some extent counteracts low external temperatures; and the cooling effect of evaporation reduces internal heat and thereby counteracts high external tem-peratures.

Light can be emitted by some members of virtually every major animal group; most phyla include marine or terrestrial representatives that are *bioluminescent*. In all these forms the main components of the light-generating mechanism are two substances called *luciferin* and *luciferase*. They can be extracted from light-producing cells, and they are nonluminous on their own. If ATP is added to luciferin, a luciferin-ATP complex is formed. And if in the presence of oxygen a solution of the enzyme luciferase is now added, the mixture emits light. This light soon disappears, but if then more oxygen and more ATP are added, light is generated again. Light production evidently is an oxygen-requiring, ATP-dependent process (Fig. 6.6).

Bioluminescent animals emit light either con-tinuously or in intermittent flashes. Light emission depends on nervous stimulation of specialized cells in light-producing organs, and the light generated can be red, yellow, green, or blue. The actual color probably is determined by the particular chemical makeup of luciferin. Some animals contain two or more kinds of luciferin and can light up in several colors. Little heat is lost during light production, hence living light often is designated as "cold" light. Also, the intensity of the light is remarkably high; it compares favorably with that of modern fluorescent lamps.

Bioelectricity is a byproduct of all processes in which ions play a part—in other words, virtually all cellular processes. A highly specialized capacity to produce electricity has evolved in certain eels and rays, which have *electric organs* composed mainly of modified muscles. Production of elec-tricity in such organs depends on ATP and a sub-stance called *acetylcholine*. Electricity is generated when acetycholine splits into separate acetyl and choline fractions. The two then recombine as acetycholine with energy from ATP. An electric eel can deliver a shock of up to 400 volts, enough

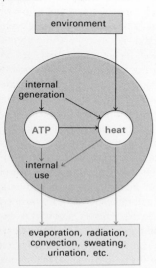

Figure 6.5 Heat gain and loss. Gains (gray) arise by absorp-tion from the environment and by internal generation (mainly respi-ration, movement, friction, and phosphorylations). Losses (color) are partly direct (evaporation or radiation, for example), partly in-direct (excreted materials carry heat with them). In birds and mammals heat gains are actively balanced against losses, and such "warm-blooded" animals can maintain a constant internal tem-perature.

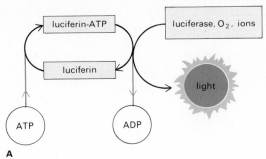

A

B

Figure 6.6 Bioluminescence.
A, the pattern of light production in organisms. *B,* a school of luminescent deep-sea squids.

to kill another fish or to jolt a man severely or to light up a row of electric bulbs wired to the tank that contains the eel. Nervous stimulation of the electric organ triggers the production of electricity.

It is still unknown just how the chemical energy of ATP actually is converted to light or electric energy. But that ATP is the key is clearly established, and this versatile compound thus emerges as the source of all forms of animal energy, chemical as well as physical.

Synthesis Metabolism

Patterns

The general function of synthesis is to produce those chemicals that a cell does not obtain directly as prefabricated environmental nutrients or as secretions from other cells. Such missing ingredients include most of the critically necessary compounds for cellular survival: nucleic acids, structural and enzymatic proteins, polysaccharides, fats, and numerous other groups of complex organic substances. In most cases such synthesis reactions are endergonic and ATP-requiring. A cyclic inter-

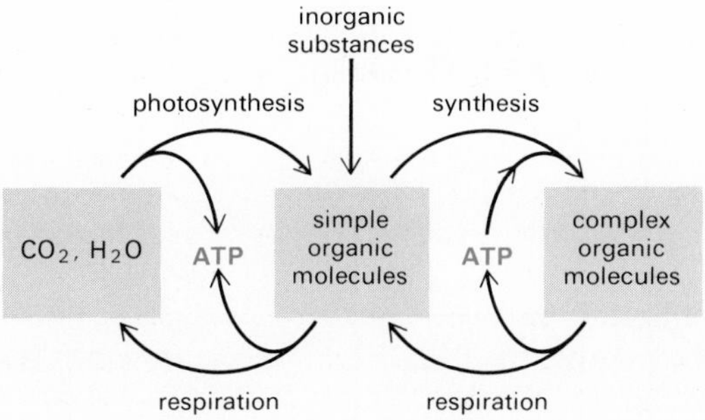

Figure 6.7 Metabolic balance.
All synthetic and other creative processes of metabolism sometimes are collectively referred to as *anabolism* and all respiratory and other destructive processes, as *catabolism.*

relation is therefore in evidence. On the one hand, breakdown of organic compounds leads to a net buildup of ATP through respiration. On the other, breakdown of ATP leads to a net buildup of organic compounds through chemical synthesis (Fig. 6.7).

Synthesis of cellular components and breakdown occur simultaneously, all the time. As pointed out in the preceding chapter, breakdown can affect any cellular component regardless of composition or age. A certain *percentage* of all cellular constituents is decomposed every second, and what particular constituents actually make up this percentage largely is a matter of chance. Such randomness applies also to synthesis. Regardless of the source of materials, a certain percentage of available components is synthesized every second to finished cell substances. If synthesis and breakdown are exactly balanced, the net characteristics of a cell remain unchanged. But continuous turnover of energy and materials occurs nevertheless, and every brick in the building is sooner or later replaced by a new one. Thus the house always remains "fresh."

Yet even when the two processes are balanced exactly, they cannot sustain each other in a self-contained, self-sufficient cycle. Energy dissipates irretrievably through physical activities and through heat losses in chemical reactions; and materials dissipate through elimination, evaporation, and friction. Just to maintain a steady state, therefore, a cell must be supplied continuously with energy and raw materials. Moreover, the rate of supply often must exceed the rate required for mere maintenance, for net synthesis frequently exceeds net breakdown, as in growth or repair after injury.

In a very general sense, synthesis reverses the result of respiratory decomposition; the products formed at successive steps of respiration can be

inorganic substances

photosynthesis   synthesis

CO_2, H_2O ATP   simple organic molecules   ATP   complex organic molecules

respiration   respiration

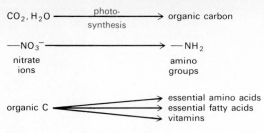

Figure 6.8 Animal dependence on plants: five classes of compounds that green plants can synthesize and that the cells of most animals cannot synthesize. Animals depend on green plants for these products.

the starting materials in synthesis. However, only green, photosynthetic cells as in plants can reverse the final step of respiration. Only they can "fix" the respiratory endproduct CO_2 and create organic carbon (in the form of PGAL) from this inorganic source. But once organic compounds are available, *all* cells, those of animals included, can add CO_2 to these compounds and thereby increase the lengths of carbon chains.

In addition to supplying original organic carbon, green plants also must be the ultimate source of several other categories of raw materials. First, animal cells are unable to manufacture certain required fatty acids, and such "essential fatty acids" must be obtained through plant food. Second, animal cells cannot make 8 or 10 of the 23 required amino acids, and here again plants must

be the original suppliers of these "essential amino acids." Third, plants are able to manufacture usable amino nitrogen ($-NH_2$) from mineral nitrates (NO_3^- ions), but animals cannot utilize such inorganic sources. Animals also cannot make metabolic use of aerial nitrogen (see Chap. 9). Thus they depend on one another and ultimately on plants for their usable nitrogen supply. Finally, plants must provide most vitamins. Animals vary greatly in their vitamin-synthesizing capacities, but almost no animal is able to produce all required vitamins in required amounts (Fig. 6.8).

Such differences between plants and animals and among animals undoubtedly are a result of mutation and evolution. In all probability, the ancestors of animals—as of plants—were able to synthesize all needed cellular compounds on their own. In the course of time, random mutations must have led to a loss of various synthesizing abilities in different organisms. If the affected organism was photosynthetic it must have become extinct, for it could not have obtained the missing ingredients in any other way; and all plants surviving today still must synthesize all needed ingredients on their own. But if the affected organism was an animal it could survive readily, for it could obtain any missing compound from plants or another animal by way of eaten food. That mutations actually can destroy cellular synthesis capacities can be demonstrated experimentally.

Carbohydrates, Lipids, Amino Acids

In animal cells carbohydrates are usually synthesized from glucose. This sugar in turn can be formed from C_3 units such as PGAL, essentially by reverse of part of glycolysis (see Fig. 5.13). Alternatively an animal cell can obtain glucose as a prefabricated nutrient by transport in blood from gut or liver. After phosphorylation of glucose at the cell surface, glucose-phosphate then can interact inside a cell with UTP, uridine triphosphate, a nucleotide derivative functionally equivalent to ATP. The result of this interaction is a *UDP-glucose* complex, the actual starting material in many glucose-requiring syntheses. For example, UDP-glucose participates in the manufacture of monosaccharides such as galactose, disaccharides such as sucrose, and polysaccharides such as glycogen (Fig. 6.9).

Glycogen stored in animal cells normally is the most abundant carbohydrate. Also, it is the usual starting carbohydrate in respiration and in most

Figure 6.9 Main pathways in carbohydrate synthesis. The compounds shown in color represent key intersections of numerous reaction sequences. UTP is uridine triphosphate, a uracil derivative functionally similar to ATP. UDP-glucose is uridine diphosphate–glucose. Interconversions occur readily and frequently according to pathways indicated by the arrows. Note that carbohydrate metabolism is interconnected with fatty acid and amino acid metabolism.

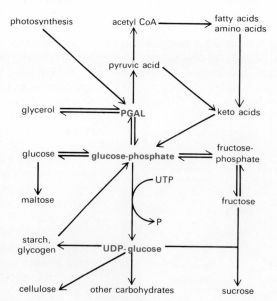

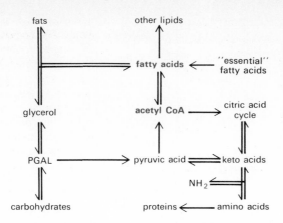

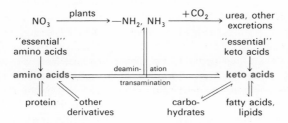

Figure 6.10 Main pathways in lipid synthesis. The key compounds acetyl CoA and fatty acids are interconverted by β-oxidation or its reverse. Note that lipid metabolism is interconnected with carbohydrate and protein metabolism.

Figure 6.11 Main pathways in amino acid synthesis. Arrows do not necessarily imply single reaction steps. See also Fig. 4.18.

syntheses. When used as fuel, glycogen is phosphorylated by ATP and converted to glucose-phosphate; and when used as building material, the polysaccharide is reconverted to UDP-glucose. Glucose-phosphate and UDP-glucose themselves are interconvertible, and glucose-phosphate also can be exported from a cell in the form of free glucose.

In lipid synthesis the simplest starting material is the 2-carbon acetyl group (Fig. 6.10). From acetyl CoA can be synthesized, for example, fatty acids of any length ("essential" fatty acids ex-

cepted, as noted above). This process is almost exactly the reverse of β-oxidation: successive C_2 groups in the form of acetyl CoA are added to acetyl CoA itself or to already existing longer carbon chains. The main steps can be traced by reading Fig. 5.18 in reverse. Acetyl CoA can also enter the citric acid cycle, and various participants of this cycle then can become transformed to pyruvic acid and carbohydrates generally. This is a common pathway in lipid-carbohydrate interconversions.

In the synthesis of fats, fatty acids are one type of required building material, glycerol is the other. Glycerol is usually formed from C_3 compounds such as PGA (phosphoglyceric acid) or PGAL—read Fig. 5.17 in reverse. In fat synthesis, one molecule of glycerol is joined enzymatically to three molecules of fatty acids (see Fig. 4.5). Also, fatty acids and therefore fats can arise from carbohydrate starting materials, and indeed amino acids can contribute to fat formation as well (Fig. 6.11 and see Fig. 6.10).

Amino acid synthesis in cells ordinarily takes place by transamination (see Fig. 5.19). Thus, any amino acid can be formed if an amino group (—NH₂) interacts with an appropriate keto acid. Amino groups can be derived from other available amino acids or, as noted, ultimately from plants. With respect to keto acids, 8 to 10 are the "essential" ones that must be obtained in prefabricated form from plants, and the others can be manufactured directly in animal cells from lipid or carbohydrate raw materials. The principal metabolic pathways that interconnect carbohydrates, lipids, and amino acids are outlined in Fig. 6.12.

Once carbohydrates, lipids, and amino acids are available in a cell, these compounds serve as raw materials for thousands of other manufacturing processes. Among these the most important single type of synthesis is protein manufacture, a process associated intimately with nucleic acid synthesis.

Proteins and Nucleic Acids: The Genetic Code

The Pattern

The synthesis of proteins is bound to differ in at least one basic respect from the manufacture of other cellular compounds: a cell cannot use just any newly made proteins but only *specific* proteins.

Figure 6.12 Synthesis metabolism: overall summary. Many of the reaction pathways shown here are reversible and can lead either to synthesis or to respiratory degradation.

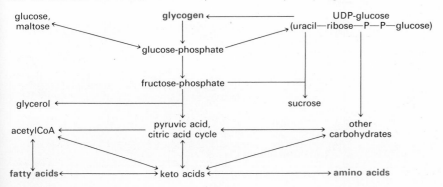

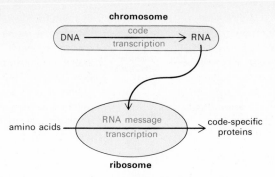

chromosome

DNA ⟶ RNA
code
transcription

amino acids ⟶ RNA message
transcription
⟶ code-specific proteins

ribosome

Figure 6.13 The pattern of protein synthesis. The genetic "message" in the chromosomal DNA of a cell is transcribed to RNA and from RNA to specific amino acid sequences in proteins manufactured in the ribosomes.

For only if new proteins are exactly like earlier ones can a cell maintain its own special characteristics. Such a specificity problem does not arise with most other kinds of compounds. A glycogen molecule, for example, is composed of identical glucose units and is structured more or less like any other glycogen molecule that contains the same number of glucose units. But a protein molecule is composed of *different* kinds of amino acid units, and a merely random joining of such units would make one polypeptide chain quite different from every other. Protein synthesis therefore requires *specificity control;* a "blueprint" must provide instructions about the precise sequence in which given numbers and types of amino acids are to be joined as a protein.

Ultimately such specificity control is exercised by the genes of a cell, the DNA of the chromosomes. *The primary function of genes is to control specificities in protein synthesis.* It has been found that the particular sequence of the nitrogen bases in DNA represents a coded chemical message that specifies a particular sequence of amino acids. The different genes in a cell carry different messages, and a cell can manufacture proteins only as these genetic instructions dictate.

Genes are housed in the chromosomes in the nucleus, but the "factories" where proteins are actually put together are the ribosomes in the cytoplasm. Genetic instructions are transmitted to the ribosomes by RNA. Chromosomes produce RNA, and in the process the chemical message of DNA becomes incorporated, or *transcribed,* in the structure of RNA. The new RNA molecules then leave the chromosomes and diffuse to the cytoplasm, where they eventually reach the ribosomes. Here amino acids are joined together as proteins in accordance with the genetic instruction supplied by the RNA molecules (Fig. 6.13).

Evidently, genes essentially are passive information carriers. All they do, or allow to be done to them, is to have their code information copied by RNA molecules, which then serve as information carriers in turn. Genes thus can be likened to important original "texts" carefully stored and preserved in the "library" of the nucleus. There they are available as permanent, authoritative

Figure 6.14 The pattern of protein synthesis. Specific messenger RNA (*m*RNA) manufactured in chromosomes becomes attached at ribosomes (and the *r*RNA there perhaps aids in the attachment). Amino acids entering a cell as food become joined to specific transfer RNA (*t*RNA) molecules, and the *t*RNA–amino acid complexes attach at specific, code-determined sites along the *m*RNA. The amino acids there link together through formation of successive peptide bonds (as in Fig. 4.7). The completed, code-specific amino acid chains then become free.

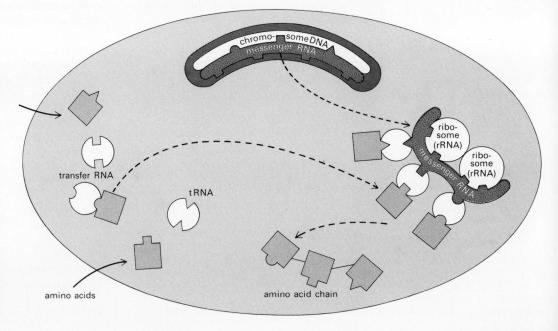

"master documents" from which expendable duplicate copies can be prepared. RNA passing to the cytoplasm actually is expendable and comparatively short-lived; soon after it has exercised its function as code carrier it is degraded by respiration. New RNA from the nucleus then is required if repeated protein synthesis is to occur. The genes on the contrary persist and are protected from respiratory destruction by the nuclear boundary. A nuclear membrane therefore appears to be an important evolutionary adaptation that promotes preservation of gene stability. It is advantageous also that the permanent message center and the manufacturing center are at different locations in a cell. If both were in the nucleus together, the manufacturing center might be too distant and isolated from the energy sources and raw material supplies; and if both were in the cytoplasm together, the message center would be subject to rapid respiratory destruction.

Inasmuch as the RNA formed by the chromosomes carries chemical messages to the ribosomes, it is called *messenger* RNA, or *m*RNA. Two additional types exist in a cell. One is *ribosomal* RNA, or *r*RNA, a normal structural component of the ribosomes. This type of RNA might be derived from RNA stored in the nucleoli inside a cell nucleus, and nucleoli thus might be assembly sites for ribosomal RNA. The other type is *transfer* RNA, or *t*RNA which functions as *amino acid carrier*. A cell contains about 60 different kinds of *t*RNA, roughly three times as many as there are different kinds of amino acid. When a particular kind of amino acid is used in protein synthesis, a corresponding kind of *t*RNA becomes attached to the amino acid and carries it to the ribosomes. Here the *t*RNA "delivers" its amino acid at a particular place along the *m*RNA chain already present. Other *t*RNA carriers similarly deliver their amino acids at other specific locations along the *m*RNA chain. In this manner large numbers of amino acids become lined up along *m*RNA in a particular sequence. As will become apparent presently, the nature of this sequence has been determined by the chemical message in *m*RNA. The "correctly" stationed amino acids then become joined to one another, and a polypeptide chain with a gene-determined specificity results (Fig. 6.14).

Figure 6.15 Code transcription from DNA to *m*RNA. (*A, T, G, C, U*, purine and pyrimidine bases; *R*, ribose; *P*, phosphate.) In a first step part of the DNA double chain unzips. In the second step the nucleotide triphosphate raw materials ATP, GTP, CTP, and UTP become bonded to appropriate nitrogen bases along a single DNA chain (color), and the two terminal phosphates of each of these raw materials split off in the process. In the third step a linked ribose-phosphate chain is formed under the influence of RNA polymerase. The finished *m*RNA then has a nitrogen-base sequence specifically determined by that of DNA. After *m*RNA becomes detached from DNA, the spiraled double chain of DNA is re-formed.

Code Transcription: DNA → Protein

We already know that DNA is a spiraled double chain of nucleotides (see Fig. 4.13). In it, the genetic information is coded by a four-letter alphabet of nitrogen bases, and in the double chain these bases occur in four pair combinations, $A \cdot T$, $T \cdot A$, $G \cdot C$, and $C \cdot G$. The genetic code then consists of a specific succession of such pairs. If the code of a given segment of a DNA chain is to be transcribed, the first requirement is an unspiraling and "unzipping" of this segment of the double chain into two separate single chains. Assume that, as in Fig. 6.15, part 1, the nitrogen-base sequence *CAATGA* of one of the single chains is to be transcribed to RNA.

The first nitrogen base in the DNA sequence to

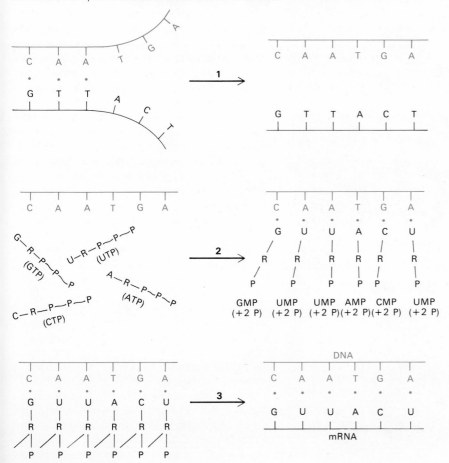

be transcribed is *C*, cytosine. We know that such a base can link up specifically with the nitrogen base *G*, guanine, and form a *C · G* pair. If therefore the available raw materials in a chromosome include a guanine-containing nucleotide, then such a molecule can bond to the *C* of DNA. The raw material actually required is GTP, guanosine triphosphate (G—R—P~P~P in Fig. 6.15, part 2). In the process of bonding enzymatically to the DNA chain, the two terminal phosphates split off as byproduct, the split high-energy phosphate provides bonding energy, and the molecule that then remains attached to DNA is GMP, guanosine monophosphate (G—R—P in Fig. 6.15, part 2). Similarly, an *A* of DNA can bond to UMP derived from a UTP (uridine triphosphate) raw material, a *T* can bond to AMP derived from ATP; and a *G* can bond to CMP derived from CTP (cytidine triphosphate). In other words, the DNA sequence serves as a *template*, or mold, along which raw material molecules can become attached in specific sequence. To form RNA it is then necessary only that the attached nucleotides become linked together as a chain (Fig. 6.15, part 3). This last process occurs not all at once but progressively, and it is catalyzed by the enzyme *RNA polymerase*:

$$(N\text{-base}—R—P\sim P\sim P)_I \xrightarrow{\text{RNA polymerase}} (P\sim P)_I + (N\text{-base}—R—P)_I$$

Here *x* symbolizes the number of ribonucleotide units linked together.

The finished RNA chain is *m*RNA, which separates from the DNA template and eventually reaches a ribosome. Note that the specific DNA code is imprinted in *m*RNA in somewhat the same way that a photographic negative shows light objects as dark areas or that a plaster cast shows elevated objects as depressions. Such ''inverted,'' negative codes in *m*RNA represent the actual working blueprints for protein synthesis.

The particular chromosome regions where *m*RNA is synthesized at any given time sometimes can be seen under the microscope. For example, the salivary gland cells of various insects contain thick giant chromosomes that consist of ordinary chromosome filaments joined together in bundles. It was first demonstrated in such giant chromosomes that *m*RNA synthesis occurs specifically in regions where the chromosomes exhibit conspicuous enlargements, or *puffs* (Fig. 6.16). The puffed regions give the impression that the DNA chains there became separated and loosened like the strands of a frayed string, as if the chains exposed themselves as fully as possible to surrounding raw materials. Puffs wax and wane at any one region, and they appear at different chromosome regions at different times and at different developmental stages of the animal. These observations suggest that not all genes of a cell are active simultaneously in *m*RNA synthesis—as indeed would be expected from what is known about gene function (see below). That chromosome puffs actually do represent regions of active *m*RNA synthesis is indicated fairly conclusively by various experiments. For example, the antibiotic actinomycin is known to be a specific inhibitor of RNA synthesis—and it has also been found to inhibit the formation of puffs. Similarly, uracil is a nitrogen base that occurs specifically in RNA and is required in RNA synthesis; and if radio-labeled uracil is introduced into a cell, the compound accumulates specifically

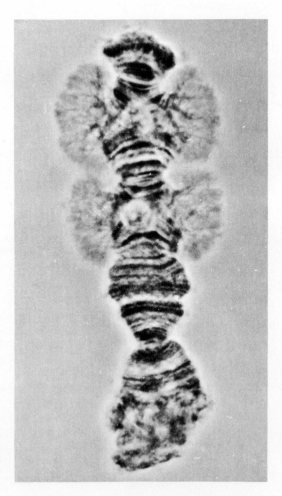

Figure 6.16 Chromosome puffs. High-power view of a stained salivary gland chromosome of a midge, with conspicuous puffs in upper part of photograph.

in the chromosome puffs, indicating that RNA synthesis occurs only there.

Just what does the genetic code in *m*RNA actually say? We know that it must somehow "spell out" in chemical terms an identification of the 23 different amino acids, and that it must do so with a four-letter alphabet. Assuming that nature is as concise as it can be, how are 23 different identifying "words" constructed out of four letters, such that each word contains as few letters as possible? If the code consisted of one-letter words, a four-letter alphabet *A-B-C-D* would allow only four different identifications: *A, B, C, D.* If the code were made up of two-letter words, there could be 4^2, or 16 different letter combinations: *AA, AB, AC, AD, BB, BA, BC,* etc. Yet 16 combinations are still too few to specify 23 words. However, if the code contained three-letter words, there could be 4^3, or 64 different letter combinations, more than enough to spell out 23 words (and roughly as many as there are different kinds of *t*RNA).

On the basis of such reasoning, it was hypothesized and later actually confirmed that the genetic code "names" each amino acid by a sequence of three "letters," or three adjacent nitrogen bases. In such a *triplet code,* 23 triplets would be "meaningful" and would spell out amino acid identities.

Which nitrogen-base triplets identify which amino acids? The answer has been obtained through ingenious experiments. It is possible to extract RNA polymerase from bacteria and to add to this enzyme in the test tube known nitrogen-base raw materials. In the first experiments of this kind, for example, only UTP was added. From such a reaction system could be obtained artificial RNA that consisted entirely of a sequence of uracil-containing nucleotides. To such RNA then could be added a mixture of different amino acids, and a chain of linked amino acids could be synthesized

in the test tube. Analysis of these chains showed that they contained only a single kind of amino acid, *phenylalanine.* In other words, the artificial RNA "selected" only phenylalanine from the many different kinds of amino acids available and controlled the formation of a polypeptide chain that consisted of phenylalanine only. From this it could be concluded that the triplet code for phenylalanine must be *UUU.*

In later work many different kinds of artificial RNA could be prepared from various combinations of the raw materials ATP, GTP, CTP, and UTP. Such artificial RNA of known composition could then be employed as above in a search for the triplet codes of various amino acids. These searches have been successful, and the code triplets for all amino acids have been identified (Table 4). The exact sequence of the three bases in each triplet is still not clear in all cases. For example, the code triplet for aspartic acid contains *G, A,* and *U,* but it is not established yet whether the triplet reads *GAU* or *AGU* or *GUA* or some other sequence of these three bases. Also, it has been found that certain amino acids are coded by more than a single triplet. For example leucine is coded as *AUU, GUU,* and *CUU.* Because of such multiple codings the genetic code is said to be *degenerate.* A limited degree of degeneracy is of considerable adaptive advantage, for despite a change of one code letter—by mutation, for example—the meaning of the whole triplet still can be preserved.

In the sample *m*RNA in Fig. 6.15, therefore, the six-base sequence *GUUACU* consists of two consecutive triplets that could represent the code for the amino acid combination valine-histidine. How does such an *m*RNA code in the ribosomes control the formation of a valine-histidine portion of a polypeptide? To answer this, attention must turn to *t*RNA.

The linking of an amino acid with its *t*RNA carrier in the cytoplasm represents *amino acid activation.* This process requires ATP as energy donor and a specific enzyme that attaches to the amino acid and then joins the acid to *t*RNA (Fig. 6.17). Each *t*RNA is a comparatively short nucleotide chain in which both ends of the molecule play a critical role. One end can be regarded as the "carrier" region; the amino acid becomes attached there. This carrier end consists of the nitrogen-base triplet *ACC* in all types of *t*RNA, and thus is nonspecific; all amino acids are joined to and carried by *t*RNA in the same way. However, the other end of *t*RNA functions in amino acid

Figure 6.17 Amino acid activation. A, joining of an amino acid and a *t*RNA specific for that acid. In a first step the amino acid becomes activated by linking to a specific enzyme with the aid of energy from ATP. In a second step *t*RNA specific for both the enzyme and the amino acid links to the acid, and the enzyme becomes free. *B,* each *t*RNA has a nonspecific carrier triplet of nitrogen bases at one end (*ACC*), where attachment to an amino acid is achieved. At the other end is a specific recognition triplet. *CAA* in the case of valine-specific *t*RNA, *UGA* in the case of histidine-specific *t*RNA. This triplet will attach to an appropriate nitrogen base along the *m*RNA on the ribosomes (see Fig. 6.18).

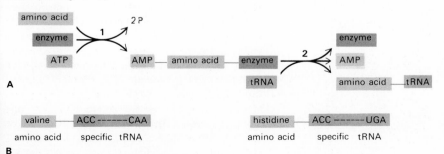

Table 4. *Triplet codes for amino acids in mRNA (A, adenine; U, uracil; G, guanine; C, cytosine)*

AMINO ACID	CODE TRIPLET	AMINO ACID	CODE TRIPLET
alanine	CCG, CGU	leucine	AUU, GUU, CUU
arginine	CCG	lysine	AAU, AAA
asparagine	CAU, AAU, AAC	methionine	GAU
aspartic acid	GAU, CAA	phenylalanine	UUU
cysteine	GUU	proline	CCU, CCC
glutamic acid	GAU, GAA	serine	CUU, CCU
glutamine	AAC, AGG	threonine	AAC, ACC
glycine	GGU	tryptophan	GGU
histidine	CAU, CAC	tyrosine	AUU
isoleucine	AUU	valine	GUU

"recognition"; present there is a nitrogen-base triplet that spells out a positive, DNA-like code for a particular amino acid. In the case of valine, for example, the specific *t*RNA would carry the terminal recognition triplet *CAA*, the inverse of *GUU*; and in the case of histidine, the recognition triplet would be *UGA*, the inverse of *ACU*. It is the specific enzyme in the activation reaction that "recognizes" and promotes interaction of a given amino acid and its corresponding *t*RNA. The enzyme ensures that a particular acid is combined with the "correct" type of *t*RNA. More than one type of *t*RNA can be "correct" in this sense for those amino acids that are identified by more than one code triplet.

When a *t*RNA carrier then arrives at a ribosome, the positive recognition triplet of *t*RNA will be able to bond only to a corresponding inverse code triplet along *m*RNA. Thus, as in our example, a valine-carrying *t*RNA with the recognition triplet *CAA* will be able to bond to a *GUU* triplet along *m*RNA; and a *t*RNA with the triplet *UGA* can bond to an *ACU* triplet of *m*RNA. In this way amino acids become stationed along *m*RNA in a code-determined sequence (Fig. 6.18).

The final joining of amino acids as polypeptide chains occurs through formation of *peptide bonds* between adjacent amino acids. This bonding requires the participation of energy donors such as ATP and is accomplished enzymatically by removal of one molecule of water from two amino acids (see Fig. 4.7). A finished chain of amino acids disengages step by step from its RNA connections, and then constitutes a protein or a part of a protein built according to gene-determined instructions.

Figure 6.14 now can be consulted again for a summary representation of the whole pattern of protein synthesis. In accordance with this scheme, and with extracts of RNA polymerase, *m*RNA, *t*RNA, and ribosomes from bacterial cells, it actually has been possible to synthesize in the test tube several (enzymatic) proteins that are otherwise manufactured only in living cells. But although formation of a specific primary (polypeptide) structure of a protein can now be explained, little precise knowledge is available as yet about how specific secondary, tertiary, or quaternary structures are determined. If two proteins differed, for example, in tertiary structure but not in primary structure, then the same *m*RNA would provide the code for the primary structure of both. What then governs the higher structural specificities?

Newly formed proteins become part of the structural and functional makeup of a cell. Some proteins might be incorporated in various fibrils, membranes, mitochondria, chromosomes, ribosomes, or indeed any other cellular organelle. Other proteins might, by virtue of their particular specificities, come to function as specific enzymes and thus determine at what rates the reactions in a cell can take place. Each of the many genes present in a cell controls the manufacture of a different kind of protein, and the totality of the proteins formed then determines and maintains the nature of that cell.

Metabolic Maintenance

The total metabolic effort of a cell can be considered to serve two general functions. One is *interior maintenance*, or activities that promote chiefly the survival of the cell in which the activities take place. Among physical activities here are, for example, heat generation and internal motions. Chemical activities consist of procurement of substances required for survival and putting such substances to use. Collectively all such processes of interior maintenance counteract normal decomposition and wear and tear, make possible the re-

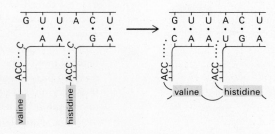

Figure 6.18 tRNA and mRNA. The link-up between *m*RNA and the recognition ends of amino acid-carrying *t*RNA is shown. Adjacent amino acids then join through peptide bonding (as illustrated in Fig. 4.7).

placement of cellular parts after injury, and permit a cell to grow and reproduce.

The other general function is *exterior mainte-nance,* processes that contribute chiefly to the survival of other cells or of the animal as a whole. On a physical level, for example, the contraction of a muscle cell often is far less significant for the survival of that cell itself than for the survival of the entire animal. On a chemical level, every cell exports compounds that it produces but that often are waste in its own metabolism—CO_2 or H_2O, for example. Yet what is waste for one cell well might be an essential metabolite in another.

Moreover, many cells manufacture more or less special substances that are exported and that in many cases play special roles in other cells or regions of an animal. For example, such secretions can be *nutritive,* like glucose exported from liver cells; or *digestive,* like enzymes secreted in ali-mentary systems; or *supportive,* like bone sub-stance secreted by skeleton-forming cells; or *re-productive,* like scents secreted by many animals; or *protective,* like secreted irritants and poisons. Actually there is hardly any function in any animal in which secretions exported from some cell or cells do not play a role.

Any cell or group of cells *specialized* for the manufacture of secretions can be called a *gland.* Thus, cells are not usually regarded as glands if they merely export water or salts as part of their regular housekeeping metabolism; *all* cells actually carry out such exports. Animal glands largely are of two general types. So-called *exocrine* glands empty their secretions into ducts or free spaces. Included in this category are all the digestive

glands (liver, pancreas, salivary glands), skin glands (sweat-, oil-, and wax-secreting glands), and numerous others associated with repro-ductive, excretory, and other systems. The second category comprises the *endocrine* glands, which are ductless and secrete into the blood. The char-acteristic products of these glands are *hormones* (see Chaps. 8 and 9).

In summary, therefore, a cell obtains the mate-rials it requires in three general ways. Some com-ponents, like water, mineral ions, and a number of organic materials must come in prefabricated final form from the external environment. Some others, like hormones, must come as prefabricated secre-tions from other cells in the animal. And all other materials must be synthesized from foods inside the cell itself. Collectively these various chemicals then maintain the body of a cell (Fig. 6.19).

But it must not be imagined that this multitude of ingredients just happens to arrange itself as new living substance. A mere random mixture of such ingredients would only form a complex but lifeless soup. As has long been appreciated, *omnis cellula e cellula*—all cells arise from preexisting cells; new life must arise from preexisting life. New compo-nents become living matter only if already existing living matter provides a framework. The house can be added to and its parts can be replaced or modified, but an altogether new house cannot be built. That apparently occurred only once during the history of the earth, when living cells arose originally.

How then can cell metabolism continue as an ordered, properly functioning series of events rather than as a mere jumble of random activities? The answer is *self-perpetuation:* processes that control, integrate, and coordinate metabolic reac-tions and thereby convert the merely active cellular system to a living one.

Self-perpetuation

Like metabolism, self-perpetuation at the cellular level includes processes that differ from those at the level of the whole animal. For example, on the level of the whole animal the self-perpetuative function of adaptation includes sex, heredity, and evolution. But apart from actual reproductive cells, other cells of an animal are concerned with neither sex nor evolution (although indirectly they contrib-ute to both). In general, *cellular* steady-state control means maintenance of optimal operating condi-

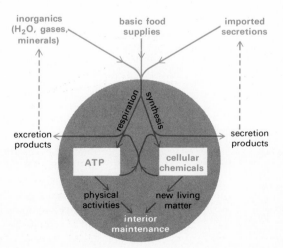

Figure 6.19 Metabolism: over-all summary of the main process.

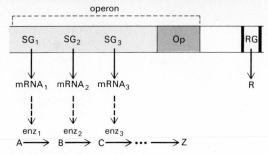

Figure 6.20 Operon structure. SG, structural gene; *Op,* operator gene; *RG,* regulator gene; all three of these gene types are located in a chromosome. *R,* protein product formed by *RG* and influencing *Op* in ways to be shown in Figs. 6.21 and 6.22. *A* to *Z* represents a reaction sequence in which a succession of specifically required enzymes (*enz*) is manufactured under the control of SG and *m*RNA.

tions inside a cell; *cellular* reproduction means cell division and growth in cell numbers; and *cellular* adaptation means long-range cell maintenance, or faithful transmission of cell traits through successive cell generations.

Steady States: DNA and Operons

Regardless of how a cell is stimulated, the stimulus usually affects one or more *metabolic* reactions—particular processes of respiration or synthesis or physical activities such as movement are likely to be speeded up or slowed down. Similarly, regardless of how a cell responds, the response ultimately is produced by metabolic reactions—acceleration back to normal of those that had been

Figure 6.21 Operon functioning: repression. R by itself does not affect *Op,* but if *R* combines with reaction endproduct *Z,* the complex *RZ* inhibits the operator gene *Op* (transverse double bar denotes inhibition). As a result, *Op* also prevents the *SG's* from functioning, enzymes will not be produced, and the reaction sequence *A* to *Z* will cease. Thus the endproduct *Z* eventually represses the continuation of its own manufacture.

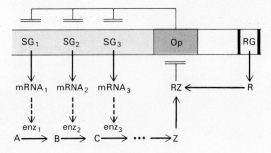

slowed down or deceleration of those that had been speeded up.

In short, steady states can be maintained if, in response to particular stimuli, a cell can adjust and readjust the pattern of its chemical reactions. Reaction patterns are determined by enzyme contents, and these in turn depend on protein synthesis. Ultimately, therefore, cells must maintain steady states by *adjusting the rates of protein synthesis.* Since protein synthesis is under gene control, rate changes will occur if the activity of each gene in a cell somehow can be "turned on and off."

The means by which genes actually might be turned on and off has been suggested by the chemical phenomena of *repression* and *induction.* It has long been known that in certain reaction sequences in cells, the endproduct of the sequence often tends to inhibit some earlier point of the sequence. For example, in a serial transformation of compound *A* to compound *Z,* the endproduct *Z,* once formed in given amounts, might inhibit some earlier step such as $B \rightarrow C$:

$$A \xrightarrow{\text{enz}_1} B \xrightarrow{\text{enz}_2} C \xrightarrow{\text{enz}_3} \cdots \xrightarrow{\text{enz}_z} Z$$

Such "endproduct inhibitions," or repressions (symbolized by the transverse double bar above) prevent more endproduct from being formed.

Conversely, for some reaction sequences in cells it has been found that addition of excess amounts of reactants leads to a rapid formation of excess amounts of specific enzymes. For example, in a serial transformation of *A* to *Z,* an excess of *A* can induce, or lead to an increase in, the amount of enzyme that converts *A* to *B,* hence to a more rapid formation of *Z:*

$$A \xrightarrow{\text{enz}_1} B \xrightarrow{\text{enz}_2} C \xrightarrow{\text{enz}_3} \cdots \xrightarrow{\text{enz}_z} Z$$

Based on research on microorganisms, a widely accepted hypothesis shows how both repressions and inductions actually might operate through a switch mechanism that affects gene activity and thus protein synthesis. This so-called *operon* mechanism (Fig. 6.20) is based on the existence of two types of gene, *regulator genes* (*RG*) and *structural genes* (*SG*). Regulator genes are thought to control the manufacture of protein products *R,* which specifically affect the activity of structural genes. These are genes that control protein syn-

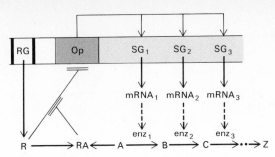

Figure 6.22 Operon functioning: induction. R inhibits *Op*, hence the *SG*'s do not operate and Z is not formed. But if starting material A is introduced into the system, R combines with A and the complex RA inhibits the inhibition exerted by R on *Op*; in effect, RA permits *Op* to function again. The *SG*'s then will function, too, enzymes will be formed, and A can be converted to Z. Thus A removes a repression (it derepresses *Op*), hence it promotes its own conversion to Z.

thesis in general; they transcribe their codes to *m*RNA, and thus they control the synthesis of, for example, enzymes that promote transformation of A to Z. All the structural genes controlling the successive steps of a particular reaction sequence are thought to be located close together on a chromosome. Also, this chromosome region is believed to contain an *operator gene* (*Op*), which must be active if the nearby structural genes are to be active. The whole region, including operator and associated structural genes, is said to form an *operon*.

In a repression (Fig. 6.21) the product R of the regulator gene is assumed not to affect the operator gene *Op*, which then is active. This permits the structural genes to be active as well, resulting ultimately in the formation of endproduct Z. However, Z now is believed to combine with R, and the complex RZ becomes attached to the operator *Op* and thereby inhibits it. The structural genes consequently become inactive, too, transcription ceases, and the reaction sequence A to Z is soon halted. Endproduct Z then is no longer formed. The repression lasts as long as the concentration of Z stays above a certain critical level. Thereafter *Op* becomes free again, and formation of Z can recur—temporarily.

In an induction (Fig. 6.22), the product R of the regulator gene is an inhibitor of *Op*, and endproduct Z therefore cannot be manufactured. But if raw material A is introduced into the cell, A is believed to combine with R, and the complex RA is postulated to abolish the inhibition of *Op*. The structural genes then would become active and

A could be transformed to Z. After all of A is used up, *Op* becomes inhibited again and the enzymes necessary for formation Z are no longer synthesized. Induction therefore amounts to removal of a repression (*derepression*).

Clearly, control through operons saves a cell energy and materials in considerable measure. Enzymes are synthesized only when they are actually needed—when raw materials are actually available and when endproducts are not already present to excess. Quite as important, operon control permits a cell to be responsive to its environment. Compounds that enter a cell as food can be the specific stimuli for their own utilization, by enzyme induction; and compounds that accumulate as finished products can be the specific stimuli halting their own manufacture, by enzyme repression. Adjusting its metabolic activities in this manner, a cell can exercise effective steady-state control.

Reproduction: DNA and Mitosis

Genes govern not only the production of proteins but also their own synthesis. Each gene controls the manufacture of new genes exactly like it; DNA is ''self''-replicating.

New formation of DNA has many features in common with message transcription from DNA to *m*RNA (Fig. 6.23). A DNA double chain first separates progressively into two single chains, and each such single portion links to itself appropriate nucleotide raw materials. In this case, these are deoxyribotides (ATP, GTP, CTP, and TTP in Fig. 6.23). After attaching to appropriate places along a single DNA chain (and becoming monophosphates in the process), such nucleotides become joined together as a new chain. The enzyme *DNA polymerase* catalyzes this linking.

Evidently, DNA synthesis is equivalent to DNA *reproduction*; one double chain gives rise to two identical double chains. One of the two single chains in each ''daughter'' DNA has existed originally in the parent DNA, and the other single chain has been newly manufactured. In this manner the genetic information is inherited by successive DNA generations.

Replication of DNA usually takes place after more or less extended periods of cell growth. Thereafter, some—so far unknown—stimulus brings about a change in chromosome activity: DNA manufactures not only *m*RNA but also new DNA. One set of genes and chromosomes so

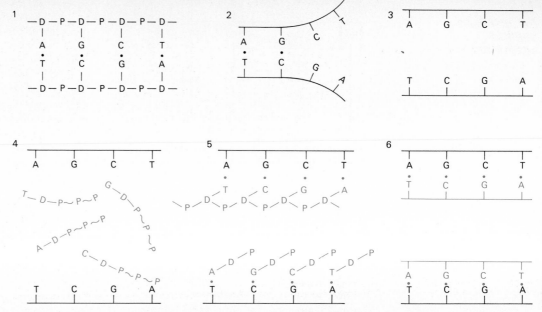

Figure 6.23   DNA duplication. (A, G, C, T, nitrogen bases; D, deoxyribose; P, phosphate.) If a double DNA chain (1) is to be duplicated, it separates progressively (2) into two single chains (3). Nucleotide triphosphate raw materials (4) then become attached to each free nitrogen base as monophosphates (5, top), and the deoxyribose and phosphate parts of adjacent nucleotides become joined with the aid of DNA polymerase (5, bottom). Two DNA double chains thus result (6), identical to each other as well as to the original parent chain. Note that in each newly formed double chain one single chain served as the code-specific template in the manufacture of the (colored) new single chain. In certain respects this message transcription from DNA to DNA resembles that of transcription from DNA to *m*RNA (compare with Fig. 6.15).

becomes two sets. Chromosome reproduction then appears to be the trigger—again in an unknown manner—for cell division.

This event consists of two separate processes: cleavage of the cytoplasm into usually two parts, or *cytokinesis,* and reproduction of the nucleus, *karyokinesis* or *mitosis.* In animal cell division cytokinesis and mitosis normally occur more or less simultaneously, and both together therefore can be called "mitotic division." Four successive, not sharply separated stages can be distinguished, *prophase, metaphase, anaphase,* and *telophase* (Fig. 6.24).

One of the first events of prophase is the division of the centriole, just outside the nucleus. Daughter centrioles then behave as if they repelled each other and migrate toward opposite sides of the cell nucleus. Concurrently portions of the cytoplasm transform to fine gel fibrils (or *microtubules*). Some of these radiate away from each centriole like the spokes of a wheel and form so-called *asters.* Other gel fibrils develop between the two centrioles and produce a *spindle,* with a *spindle pole* marked at each end by the centriole (Fig. 6.25).

The nuclear membrane usually dissolves, the

Figure 6.24   Mitosis: summary. The assumption here is that cytoplasmic cleavage accompanies mitosis. Note that a "resting" cell is resting only from the standpoint of reproductive activity. In all other respects it is exceedingly active.

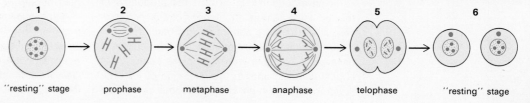

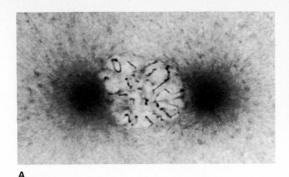

A

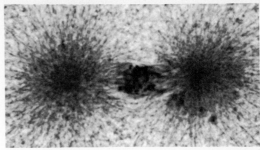

B

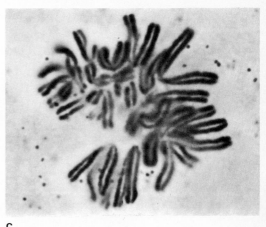

C

Figure 6.25 Mitosis: prophase. A, early prophase. The nuclear membrane is just dissolving and chromosomes are already visible. To either side of the nuclear region is a darkly stained centriole area, from which fine aster fibrils are beginning to radiate out. *B,* late prophase. Asters already conspicuous, spindle fibers present between asters and chromosomes. Chromosome migration to metaphase plate under way. *C,* close-up view of prophase chromosomes, which have already duplicated. Each member of such a pair is a chromatid, and each pair is still held together at one point, the centromere.

nucleoli disintegrate, and distinct chromosomes become visible. Each chromosome has produced a mathematically exact double shortly before prophase, and this doubled condition now becomes clearly apparent. In each such pair the two chromosomes (here called *chromatids*) lie closely parallel. They are joined to each other at their *centromeres*, single specialized regions at corresponding locations along the two chromatids. Spindle fibrils form between the centromeres and the spindle poles.

The metaphase starts when the paired chromatids begin to migrate. If a line from one spindle pole to the other is considered to mark out a spindle axis, then the chromatid pairs migrate to a plane at right angles to and midway along this axis. Here the chromatid pairs line up in a *metaphase plate*. At this stage (or sometimes earlier) the centromeres separate and the two chromatids of each pair thereby become unjoined chromosomes. Such twin chromosomes now begin to move apart; one set migrates toward one spindle pole, and the identical twin set migrates toward the other. This period of chromosome movement

represents the *anaphase* of mitosis (Fig. 6.26).

As a set of chromosomes now collects near each spindle pole, spindle fibrils and asters disappear. A new nuclear membrane soon forms around each chromosome set, and the chromosomes manufacture new nucleoli in numbers characteristic of the particular cell type. Two new nuclei form in this manner, the *telophase* of mitosis. If cytoplasmic cleavage accompanies mitosis, this process occurs in conjunction with nuclear anaphase and telophase. Cleavage begins with the appearance of a *cleavage furrow* in the plane of the earlier metaphase plate. The cleavage furrow is a gradually deepening surface groove that cuts through the spindle fibrils and eventually constricts the cell into two daughters. In each daughter nucleus the genes now resume control of RNA manufacture and a new growth cycle follows.

The mechanical forces responsible for the chromosome movements in division cannot yet be identified precisely. It is clear, however, that DNA is the key to cellular multiplication. It controls cell growth and, through its own periodic reproduction, triggers the reproduction of the whole

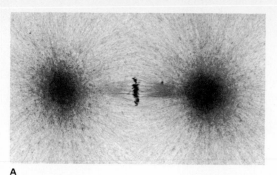

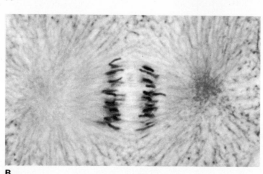

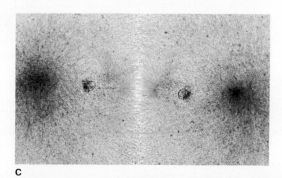

Figure 6.26 Mitosis: metaphase, anaphase, telophase. A, metaphase. Note asters, spindle, and the metaphase plate, halfway along and at right angles to the spindle axis. Fibrils join the chromosomes in the metaphase plate with the spindle poles. *B,* anaphase. The two chromosome sets are migrating toward the spindle poles. *C,* telophase. Asters are subsiding, nuclei are re-forming, chromosome threads have become indistinct, and cytoplasmic cleavage is under way in the same plane as the earlier metaphase plate.

cell. The net result is the formation of two cells that match each other (and the parent cell) precisely in their DNA contents and that contain approximately equal amounts and types of all other components. This identity of the DNA content is the key to cellular and ultimately to all *heredity;*
to the extent that genes are stable, each new cell generation inherits the same genetic codes, hence the same structural and functional potentials, that were present in the previous cell generation.

Adaptation: DNA and Evolution

On occasion, parts of the DNA codes do become altered. Such gene *mutations* can arise, for example, as accidental errors during DNA duplication; for like every other process in nature, gene reproduction undoubtedly is not completely error-free. If a single nitrogen base of old DNA happens to be transcribed incorrectly in new DNA, one code triplet will be altered and one of the protein types synthesized will contain a ''wrong'' amino acid at some point. Mutations also are produced by certain physical and chemical agents, including X rays and other forms of high-energy radiation, both natural and man-made. External agents of this sort change either the position or the chemical nature of at least one nucleotide in DNA, resulting in an alteration of at least one code triplet. Every time a particular protein is then manufactured, it will contain a ''wrong'' amino acid. This can change or abolish the enzymatic effect of the protein, and a particular cell trait can become changed in this manner (see also Chap. 12).

To be sure, some amino acids normally are coded by more than one triplet, and an error in one nitrogen base in such cases need not lead to a trait mutation. Moreover, DNA itself is among the most stable of all organic compounds. Indeed, unless it were relatively stable it would be useless as a dependable code carrier. In addition to this inherent stability several other safeguards ensure that the genetic messages of DNA are not lost or altered. One such safeguard is the nucleus itself. The evolution of distinct membrane-bounded nuclei in ancestral cell types actually can be regarded as an adaptation that shields genes from destructive metabolism of the cytoplasm. Another safeguard is *redundancy;* to ensure that a message is not lost or altered it can be made redundant, or repeated several times. Indeed, the genetic messages are stored in more than one place. First, animal cells ordinarily contain two complete sets of genes, one set inherited from each of two parents; and even if one member of a gene pair then mutates, the other member still carries the original code. Second, each cell type usually is represented by many like cells. Even if some cells die, therefore, the DNA of the remaining cells still contains the specific information characteristic of that cell type.

Yet despite inherent stability, protected existence, and redundancy, DNA nevertheless is subject to mutations, with an estimated average frequency of about one per one million replications of a given gene. Since many animals consist of trillions of cells, each such animal will carry several million mutated genes.

Mutations appear to be completely random events. Any gene can mutate, at any time and in unpredictable ways. It can mutate several times in rapid succession, then not at all for considerable periods. It can mutate in one direction, then back to its original state or in new directions. Every gene today undoubtedly is a *mutant* that has undergone many mutations during its past history (Fig. 6.27).

The effect of a mutation on a trait is equally unpredictable. Some are "large" mutations that affect a major trait in a radical, drastic manner. Others are "small," with but little effect on a trait. Some mutations have *dominant* effects that override the activity of their normal gene partners and produce immediate alterations of traits. Other mutations have *recessive* effects, the normal gene partners here masking the activities of mutated genes and thereby protecting the cell from actual trait changes.

In view of the structural and functional complexity of a cell it might be expected that nearly any permanent change in cell properties would be disruptive and harmful. Indeed, most mutations are disadvantageous, and if they have dominant effects they tend to impair cellular functions. Most mutations with dominant effects actually tend to be eliminated as soon as they arise, through death of the affected cell. In some cases, however, a dominant effect of a mutation (particularly a "small" dominant effect) can become integrated successfully with cellular functions. Such a cell then survives with an altered trait. Yet most of the mutations in surviving cells have recessive effects that remain masked by the normal gene partners.

A small percentage of mutants produces advantageous traits or new traits that are neither advantageous nor disadvantageous. In man, for example, many trillions of cells compose the body and, in view of the average rate of mutations, several million mutations are likely to occur in each individual. Many of these are lethal to the cells in which they occur, and many others remain masked by normal dominants. But some produce nonlethal dominant traits. Such new traits that arise in individual cells then are transmitted to all cells formed by division from the original ones. For example, "beauty spots" probably develop in this manner.

Gene changes that occur in body cells generally are *somatic mutations*. They affect the heredity of the cell progeny—a patch of tissue at most—and usually have little direct bearing on the heredity of the whole animal. A whole individual is likely to be affected only by *germ mutations*, stable genetic changes in immature and mature reproductive cells. Such mutations will be transmitted to all cells that ultimately compose the offspring. To the extent that germ mutations are not masked by normal genes, the offspring then can come to differ from its parents in certain major or minor ways. If the difference happens to be advantageous in the environment in which the animal lives, the animal will be well *adapted* and produce many offspring in turn, passing on its well-adapted characteristics in the process. In this manner, an alteration of a single code triplet in a single cell

Figure 6.27 Mutant types in mice. A, the effects of the mutation "eyelessness." B, the effects of the mutation "hairlessness." Each such alteration in structure is associated with a single mutant gene, and the alterations are stable and inheritable.

A

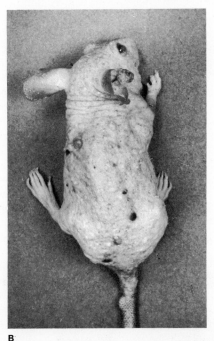

B

eventually can affect a whole animal population. After numerous generations the final result can be a change we recognize as *evolution*.

In effect, therefore, the crucial significance of genes is not simply that they are inherited. Rather, by governing the synthesis of new genes and of specific proteins, DNA plays so strategic a role that ultimately it controls the whole nature and the very life of every cell in every animal. First, since proteins make up the basic organic framework of cells, DNA determines the fundamental *structure* of every cell (including structural differences between cells and in individual cells at different times). Second, genes determine what kinds of *metabolic processes* are possible in a cell, for virtually all such processes are enzyme-dependent, hence protein-dependent, hence DNA-dependent. Third, DNA is the ultimate maintainer of *steady states*, for through its control over metabolism it also governs all control agents in cells, including DNA itself. Fourth, by governing synthesis generally and production of new DNA specifically, DNA directs growth, development, and the *reproduction* of cells. By being exchanged among cells and pooled in cells, DNA becomes the basis of sex (see Chap. 10). By duplicating itself and being inherited by offspring cells, DNA becomes the basis of *heredity*. And through its property of undergoing occasional mutations, DNA becomes the key to *evolution*.

In summary, then, DNA serves in just one primary role: it allows its specificities to be copied. Three indirect secondary roles emerge from this:

DNA controls protein specificities; DNA controls the specificities of new DNA; and, to the extent that DNA stability is imperfect, DNA can change its specificities. Through these three secondary activities DNA has indirect derivative effects that govern every aspect of living. For by controlling all metabolism and all self-perpetuation, DNA governs cell structure, cell function, and cell development. And by controlling the life of cells, DNA governs the life of all animals. Genes started life, genes still continue it, and, by their failure or absence, genes ultimately end it (Fig. 6.28).

In this context it is worth noting that recent experiments with DNA and RNA already have contributed much toward an eventual test-tube synthesis of living matter. It is now possible, for example, to create wholly artificial, man-made genes in the laboratory. For example, the protein hormone insulin consists of 51 known amino acids joined in known sequence (see Fig. 4.10), and the DNA code triplets for each of these amino acids are also known. Hence it is possible to prepare, first, basic organic raw materials such as sugars and amino acids from simple inorganic compounds (water, ammonia, and methane; see Chap. 14). Then, artificial nucleotide triplets can be produced that correspond to those of the DNA for insulin, and these triplets can be linked together. The result is an ''insulin gene''—a DNA chain with the exact genetic information for the control of insulin manufacture.

Moreover, some artificially prepared genes have been used to control conversion of amino acid raw materials to specific test-tube proteins, and such man-made proteins then function enzymatically and catalyze particular chemical reactions. Further, some artificial genes have been made to duplicate themselves in the test tube. Such systems are still far from being alive, to be sure But inasmuch as they operate in precisely the same way as in a living system, the test-tube systems represent major elements of living units.

Clearly, experimental capabilities of this sort pave the way for artificial creation of potentially any gene, for incorporation of such genes into cells, and thus for ''genetic engineering,'' or biochemical manipulation of traits directly at the level of the gene. This prospect is desirable for medicine, but it can also be alarming considering the many possibilities of social misuse. In any event, we seem to be only a few steps away from being able to create test-tube systems that not only

Figure 6.28 The pattern of gene action. Through the fundamental action of transferring their coded specificities, genes control cellular metabolism and all phases of cellular self-perpetuation.

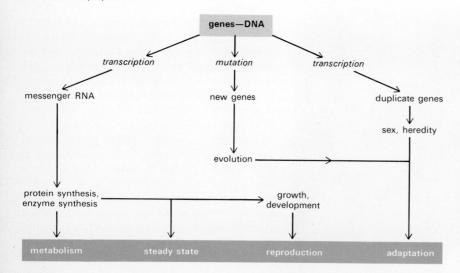

metabolize but also perpetuate themselves, even if only in limited fashion. And although experimental creation of complete living cells still is very far in the future, much of the research now being conducted does provide important data, often as an incidental byproduct, for a possible future laboratory synthesis of life.

Our survey of cellular operations now is substantially completed, for as a cell metabolizes and perpetuates itself, it lives. We are now ready to examine how large groups of living cells cooperate structurally and functionally, and indeed how they form whole animals and all higher levels of animal organization.

Review Questions

1. Describe the internal fine structure of a muscle cell. What and where is actomyosin? What are the roles of ATP in muscle? In what specific ways is the ATP supply maintained? Describe the energetics of a unit cycle of muscle activity.

2. In what ways does an animal obtain and produce heat? What are the functions of heat in metabolism? How do animals produce bioluminescence? How do the properties of living light compare with those of nonliving light? How and by what animals is bioelectricity produced?

3. Review the patterns of carbohydrate and fat synthesis. How is a C_{18} fatty acid synthesized from acetyl CoA? How are glycogen and fats synthesized?

4. How does a cell synthesize amino acids? Show how the following interconversions could occur: (*a*) carbohydrate to fat; (*b*) fat to carbohydrate; (*c*) carbohydrate to amino acid or vice versa; (*d*) fat to amino acid or vice versa.

5. In what metabolic respects are animals dependent on plants, and what are the reasons for such dependence? What are the essential differences between plant and animal patterns of nutrition?

6. Review the chemical structure of DNA. Review also the structure of RNA and distinguish between *m*RNA, *r*RNA, and *t*RNA. Where in a cell do each of these occur, and what are their functions?

7. What is the genetic code, what are code triplets, and on the basis of what reasoning has a triplet code been postulated for amino acid specification?

8. In what sense is the genetic code "degenerate," and what are the functional consequences of this degeneracy?

9. Describe the mechanism of amino acid activation. How do amino acids along *m*RNA become polypeptides?

10. What are chemical repression, induction, operator genes, regulator genes, structural genes? Review the operon mechanism and show what role it plays in accounting for cellular steady-state control.

11. How does DNA synthesis take place? What is the role of DNA polymerase? How does a cell as a whole reproduce? Distinguish between cell division, cytokinesis, karyokinesis, mitosis. Describe the succession of events in cell division.

12. What are mutations, and what are their general effects in a cell? What kind of change qualifies as a mutation? Distinguish between small and large mutations and between somatic and germ mutations.

13. How is the stability of genes safeguarded? What is the importance of gene stability? In what way is genetic information redundant?

14. Review the pattern of gene function as a whole. Which function may be regarded as primary? Which indirect secondary functions derive from this, and which tertiary functions result from the secondary ones in turn?

15. Name cellular activities that serve in internal and external maintenance. What is a gland, and what roles can glandular secretions play? Review the various ways by which a cell obtains all the ingredients it requires for its survival.

Collateral Readings

Allfrey, V. G., and A. E. Mirsky: How Cells Make Molecules, *Sci. American*, Sept., 1961. The role of DNA in protein synthesis is discussed.

Barry, J. M.: "Genes and the Chemical Control of Living Cells," Prentice-Hall, Englewood Cliffs, N.J., 1964. A recommended paperback, covering all aspects of DNA structure and function and the subject of protein synthesis.

Beerman, W., and U. Clever: Chromosome Puffs, *Sci. American*, Apr., 1964. Regional enlargements in giant insect chromosomes probably represent the places where code transcription from DNA to *m*RNA actually occurs.

Clark, B. F. C., and K. A. Marcker: How Proteins Start, *Sci. American*, Jan., 1968. On the initiation of polypeptide synthesis.

Crick, F. H. C.: The Genetic Code, *Sci. American*, Oct., 1962. A review of the triplet code by one of the discoverers of DNA structure; see also the later article by Nirenberg cited below.

————: The Genetic Code: III, *Sci. American*, Oct., 1966. A sequel to the earlier article.

Dawkins, M. J. R., and D. Hull: The Production of Heat by Fat, *Sci. American*, Aug., 1965. A discussion of heat generation by respiration of various fat deposits in young and hibernating animals.

Gamov, G.: Information Transfer in the Living Cell, *Sci. American*, Oct., 1955. The arithmetic from which the nature of the genetic code has been predicted is discussed in this article.

Hayashi, T.: How Cells Move, *Sci. American*, Sept., 1961. The molecular basis of flagellar, amoeboid, and muscular motion is discussed.

Holley, R. W.: The Nucleotide Sequence of a Nucleic Acid, *Sci. American*, Feb., 1966. The first determination of the genetic code of a whole protein is described.

Hurwitz, J., and J. J. Furth: Messenger RNA, *Sci. American*, Feb., 1962. A good discussion of the role of this type of RNA. See also the article by Rich, below.

Huxley, H. E.: The Mechanism of Muscular Contraction, *Sci. American*, Dec., 1965. A good review of the fine structure of muscle cells and the function of their operating units.

Kornberg, A.: The Synthesis of DNA, *Sci. American*, Oct., 1968. The test-tube synthesis of artificial but active DNA is described by the Nobel-prize-winning investigator who first achieved such a synthesis.

Mazia, D.: Cell Division, *Sci. American*, Aug., 1953. Experiments are described in which the entire mitotic apparatus is isolated from dividing cells.

————: How Cells Divide, *Sci. American*, Sept., 1961. A review of the nature of the mitotic process.

McElroy, W. D., and H. H. Seliger: Biological Luminescence, *Sci. American*, Dec., 1962. A comprehensive review of bioluminescence in various organisms and of the chemistry of the process.

Mirsky, A. E.: The Discovery of DNA, *Sci. American*, June, 1968. A review of the 100-year-old history of research on nucleic acids.

Nirenberg, M. W.: The Genetic Code: II, *Sci. American*, Mar., 1963. Experiments that have uncovered the nature of the triplet code are described by one of the Nobel-prize-winning investigators in the field; a sequel to Crick's first article cited above.

Porter, K. R., and C. Franzini-Armstrong: The Sarcoplasmic Reticulum, *Sci. American*, Mar., 1965. The ultrastructure of muscle is examined by means of high-resolution electron microscopy.

Ptashne, M., and W. Gilbert: Genetic Repressors, *Sci. American,* June, 1970. The isolation of the first actual inhibitor substances produced by regulator genes is described.

Rich, A.: Polyribosomes, *Sci. American,* Dec., 1963. An analysis of the role of ribosomes in protein synthesis.

Sinsheimer, R. L.: Single-stranded DNA, *Sci. American,* July, 1962. The discovery of DNA consisting of only one nucleotide chain helps to clarify the duplication mechanism of this nucleic acid.

Spiegelman, S. L.: Hybrid Nucleic Acids, *Sci. American,* May, 1964. Experiments with DNA composed of two different but complementary nucleotide chains throw light on information transfer in cells.

Taylor, J. H.: The Duplication of Chromosomes, *Sci. American,* June, 1958. A description of experiments with radio-labeled chromosomes showing that one of the two strands of just-formed DNA preexisted and that the other strand was newly manufactured.

Watson, J. D.: ''The Molecular Biology of the Gene,'' Benjamin, New York, 1965. A paperback on the nature of the gene and its role in protein synthesis. Strongly recommended.

Yanofsky, C.: Gene Structure and Protein Structure, *Sci. American,* May, 1967. An interesting review of how the correspondence between the genetic code and the amino acid sequence in polypeptides was demonstrated.

Animals obviously are put together in different ways; the anatomies of sponges, earthworms, clams, starfishes, insects, sharks, and men appear to have little in common. Yet despite their highly varied makeup, animals nevertheless are alike in fundamental respects; all are constructed out of cells, in most cases the cells are joined as tissues and organs, and these in turn often form a maximum of 10 organ systems: integumentary, skeletal, muscular, nervous, endocrine, circulatory, breathing, excretory, alimentary, and reproductive systems.

The three chapters of this part concentrate on the ways in which different animal types are put together from these same kinds of building blocks.

PART 3
ANIMAL ORGANIZATION

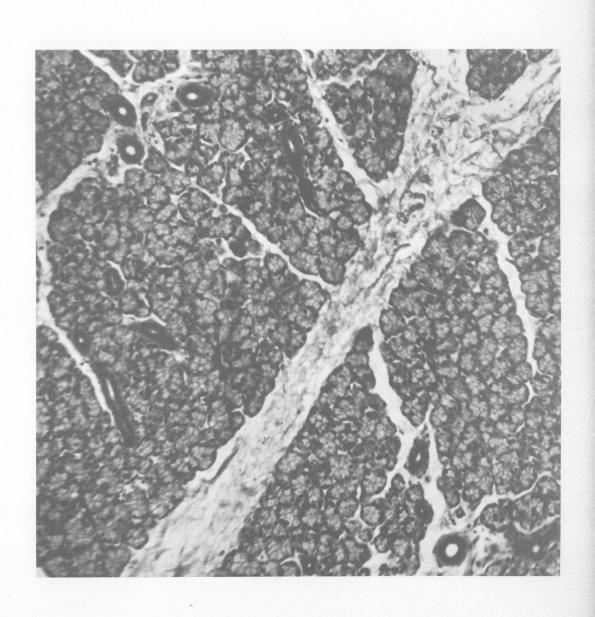

7

BODY TYPES AND BODY PARTS

The structural components of animals differ in anatomic arrangement, and it is partly on the basis of such variations of body form that animals are classified as belonging to different categories or types; *classification* and *animal form* are the first two topics of this chapter.

Particular anatomic arrangements are produced by cells, tissues, organs, and systems. The kinds of arrangements produced by *tissues* and *organs* are examined in this chapter, and those produced by systems, in the following two.

Animal Types

Taxonomic Classification

Animal classification is the special concern of the zoological subscience of *taxonomy*, or *systematics*. The present method of classifying types of animals was originated by Carolus Linnaeus, a Swedish naturalist of the early eighteenth century. This Linnaean system of taxonomy has since become greatly elaborated and now is in universal use. It is based on the proposition that if certain animals can be shown to have similar body construction they can be regarded as members of the same classification group. Moreover, an evolutionary inference is also made: the more closely two animals resemble each other, the more closely are they likely to be related. Thus, taxonomy deals with the structural makeup of animals directly and with their evolutionary histories indirectly.

In a given classification group it is often possible to distinguish several subgroups, each containing animals that have even greater similarity of body structure and, by inference, evolutionary history. Such a subgroup often can be subdivided still further, and a whole hierarchy of classification groups can be established in this fashion. The progressively lower levels in this hierarchy represent *taxonomic ranks*, or *categories*, and each is named; in succession from highest, or most inclusive, to lowest, or least inclusive, the main categories are *kingdom*, *branch*, *grade*, *phylum*, *class*, *order*, *family*, *genus*, and *species*.

Intermediate ranks sometimes interpolated between two main levels are identified by the prefixes *sub-* or *super-*; for example, *subgrade, superclass, subgenus*. The actual animals included in a given rank-category are referred to technically as *taxa*. For example, the sponges form a taxon of phylum rank; mammals are a taxon at the class rank.

In the hierarchy as a whole, progressively lower ranks consist of progressively more but smaller groups. The animal world makes up one kingdom, some two dozen phyla, and at least $1\frac{1}{2}$ to 2 million species. Also, the groups at successively lower ranks have increasingly similar body forms and evolutionary histories. Thus the members of a class resemble each other to a great extent, but the members in one of the orders of that class

A

B

C

Figure 7.1 Resemblance of related taxonomic groups. The three members of the vertebrate class of mammals shown here are alike in, for example, having fur and nursing their young with milk. However, the snowshoe hare (*Lepus*) belongs to the order Lagomorpha, whereas both the jaguar (*Felis*) and the timber wolf (*Canis*) belong to the order Carnivora. The two orders resemble each other greatly, and they also have similar evolutionary histories. But they differ in, for example, tooth structure, eating habits, and locomotion. Of the two carnivores shown, the jaguar belongs to the cat family Felidae and the timber wolf to the dog family Canidae. The two resemble each other more in anatomy and evolutionary history than either resembles the hare.

resemble each other to an even greater extent. A corresponding relation holds for evolutionary histories (Fig. 7.1).

That taxonomy provides direct and inferential information about two kinds of data, body structure and evolutionary history, should be clearly kept in mind. Animals have other characteristics as well, notably functions and ways of life. However, these play only a limited role in defining taxonomic types. Because metabolism and self-perpetuation are broadly the same in all animals, such functions are not very useful as distinguishing traits. Moreover, both the ways of life and the detailed ways of performing functions can become modified greatly. Such characteristics therefore tend to be less permanent than body architecture and evolutionary history.

Structural resemblances of presently living animals can be studied readily, and indeed they are well known for the most part. But studies of evolutionary histories—through fossils, for example—represent an independent line of investigation, and the amount of information available here varies greatly. In general, evolutionary knowledge is less precise the higher the taxonomic rank. For many high rank-groups evolutionary information actually is quite incomplete or lacking altogether, and in such cases classification must be based almost wholly on studies of body structure. Conclusions regarding evolutionary histories then are correspondingly uncertain. In instances of this sort, classification is said to be artificial, or "unnatural," to greater or lesser degree.

The most unnatural grouping in the taxonomic hierarchy in effect is the kingdom. By tradition that goes back to Linnaeus and even to earlier times, the living world as a whole has been classified into two kingdoms, the plant kingdom and the animal kingdom. However, in view of vastly increased knowledge of both the structure and the evolution of organisms, it is highly questionable whether this 250-year-old tradition still is justifiable today. There now are excellent reasons for categorizing the highest taxonomic groups in a different way, a point examined critically in Chap. 14.

In passing from the highest level to lower taxonomic ranks, classification tends to become progressively more "natural": evolutionary knowledge is more complete, and structural and evolutionary data come to dovetail more and more. At the lower levels taxonomy thus tends to indicate an actual, real interrelation of animals. For example, some notable exceptions notwithstanding (see Unit 2), a phylum generally can be defined as the largest group of animals for which a common ancestry has been demonstrated reasonably well and which is characterized by a common, basically unique body construction.

Table 1 in Chap. 2 indicates that most phyla represent more or less familiar categories of animals, often named for one of their most distinctive anatomic features. For example, the phylum *Chordata* includes all animals that have an internal skeleton, the *notochord,* at least as embryos. This phylum contains a subphylum of *vertebrates,* identified by the presence of a vertebral column. In this subphylum one of the classes comprises the *mammals.* These animals share the possession of a vertebral column with all other vertebrate classes. But mammals are identified uniquely by hair and by their nursing young with milk. Every other class similarly has its own distinguishing traits.

The most natural of all taxonomic categories is the species. It is defined as an *interbreeding* group: the members of a species normally interbreed only with one another, not with members of a different species. A species thus encompasses all animals of the same particular kind (see also Chap. 17). In this instance, therefore, classification is based on a characteristic that undoubtedly results from a very close structural and evolutionary similarity of the animals in question. For example, all men now in existence are members of the same single species.

According to Linnaean tradition and internationally accepted rules, a species always is identified by *two* technical names, in Latin or latinized.

For example, the species of grass frogs is known throughout the world as *Rana pipiens;* the species to which we belong is *Homo sapiens.* Such species names are always underlined or printed in italics, and the first name is capitalized. This first name always identifies the genus to which the species belongs. Thus the human species belongs to the genus *Homo,* and the genus *Rana* contains *Rana pipiens* and many other frog species as well.

A complete classification of an animal tells a great deal about the nature of that animal. For example, if we knew nothing else about men except their taxonomic classifications, then we would know that the zoological traits of mankind are as outlined in Table 5. Such data already represent a substantial detailing of the body structure. We would also know by implication that the evolutionary history of men traces back to a common chordate ancestry.

In a number of cases the taxonomic ranks assigned today to certain groups of animals are only tentative, and sometimes a particular group is placed at different rank-levels by different authorities. In general, taxonomic agreement among zoologists increases with the lower, more natural rank-categories and decreases with the higher, less natural ones. Indeed, the higher categories are being reshuffled more or less continually. But this is as it should be, for as our knowledge of evolutionary histories improves, the rankings of the animals must be adjusted accordingly.

On one point agreement is universal, however. Taxonomic studies clearly suggest, and evolutionary evidence fully confirms, that the interrelations among animals have the pattern of a greatly branching *bush.* All presently living animals are *contemporaries,* at the uppermost branch tips of the bush. Ancestral types, most of them long extinct, appear lower down, where branches join. Thus a particular common ancestor can give rise to *several* different types of descendant, each inheriting the characteristics of the common ancestor and evolving innovations of its own. And a particular descendant living today can become a common ancestor of new and different types living tomorrow (Fig. 7.2).

Taxonomic Comparisons

In a discussion of animal types it is often desirable to make comparative statements about two or more body forms and/or evolutionary histories.

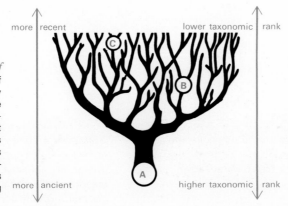

Figure 7.2 The bush pattern of evolution. The uppermost tips of the branches represent currently living forms, all at the same time level, the present. Branches terminating below the top represent extinct forms. Fork points such as *B* and *C* are ancestral types. *B* is more ancient and of higher taxonomic rank than *C. A* represents the archancestor of all living types.

more | recent

lower taxonomic | rank

more | ancient

higher taxonomic | rank

Table 5. *A partial taxonomic classification of man*

RANK	NAME	CHARACTERISTICS	RANK	NAME	CHARACTERISTICS
phylum	Chordata	with notochord, dorsal hollow nerve cord, and gills in pharynx at some stage of life cycle	infraclass	Eutheria	embryos attached to, and nourished by, maternal placenta; young born in fully developed condition; males with urethra and sperm ducts joining into single duct through penis; females with urethra and vagina opening at separate orifices
subphylum	Vertebrata	notochord supplemented or substituted by bony or cartilaginous vertebrae in adult; body with head, trunk, tail, basically segmented; skull enclosing brain			
superclass	Tetrapoda	terrestrial; trunk appendages are walking legs; gills embryonic only, breathing by lungs	order	Primates	basically tree-dwelling; usually with fingers, flat nails; sense of smell reduced
			suborder	Anthropoidea	flat face, eyes forward, with stereoscopic and color vision
(group)	Amniota	land-adapted eggs and embryos surrounded by water sac (amnion); heart four-chambered; excretion by kidneys (metanephros); 12 pairs of cranial nerves	superfamily	Hominoidea	arm freely movable in socket; hands and feet similarly specialized; tail internalized; menstrual cycles
class	Mammalia	young nourished by milk glands; skin with hair; body cavity divided by diaphragm; aortic arch only on left; red corpuscles without nuclei; constant body temperature; 3 middle-ear bones; brain with well-developed cerebrum	family	Hominidae	upright, bipedal locomotion; living on ground; hands and feet differently specialized; family and tribal social organization
			genus	Homo	large brain, speech; life span extended, with long youth
subclass	Theria	eggs not laid, young born; teeth specialized (incisors, canines, premolars, molars)	species	Homo sapiens	prominent chin, high forehead, thin skull bones; spine double-curved; body hair sparse

The taxonomic system and its evolutionary implications then prescribe what kinds of comparisons are legitimately possible. Several sets of contrasting terms are employed, and it is important to use them correctly.

Simple—Complex These terms indicate comparative degrees of structural and/or functional elaboration of animals and their parts. Degrees of elaboration can be judged by the *number* of components present and by the number of *interrela-*

tions among the components (Fig. 7.3). The terms thus describe organizational and operational attributes of animals, not evolutionary attributes. If we say that one animal (a mammal, for example) is more complex than another (an amoeba, for example), we mean only that it has comparatively more cells, tissues, or other units, and comparatively more interrelations among these units. But we do *not* make evolutionary comparisons. A higher level of organization, in the sense defined in Chap. 2, does signify "more complex." Note also that it is quite meaningless to speak of "simple" cells or animals; all units of life are exceedingly complex, and the terms "simple" and "complex" have zoological meaning only if they are used in a *comparative* manner.

Primitive—Advanced   These adjectives are often—and wrongly—employed as equivalents of "simple" and "complex." Unlike the latter, "primitive" and "advanced" do have primarily historical, evolutionary connotations. If animals or their parts are structured according to ancient, ancestral patterns, such patterns can be said to be primitive. Newer, more modern patterns superimposed on the ancient ones then are advanced, again by comparison only (see Fig. 7.3). Evolutionary "earliness" and "lateness" and similar time-contrasting terms are roughly equivalent to "primitive" and "advanced." An animal or a part of it can be more primitive yet at the same time more complex than another (for example, protozoan cells are more primitive and probably more complex than most mammalian cells); more advanced yet simpler than another (for example, a human skull is more advanced but simpler—with far fewer bones—than a fish skull); and also more primitive as well as simpler, or more advanced as

Figure 7.3 Structural and evolutionary comparisons. A, photomicrographic cross section through a hydra, a cnidarian, and *B,* through an earthworm, an annelid. On the basis of the number of tissue layers present in each, the hydra can be considered to be simpler in structure than the earthworm. Hydra also happens to be more primitive—the evolutionary history of the cnidarian phylum goes back farther than that of the annelid phylum. *C,* the bones of a fish skull, and *D,* of a human skull. On the basis of the number of bones present in each, the fish skull is structurally more complex, yet it is more primitive from an evolutionary standpoint than the human skull. Thus, a given degree of structural complexity does not necessarily or automatically signify a comparable degree of evolutionary advancedness. (*A,* Carolina Biological Supply Company.)

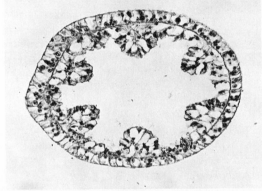

A

B

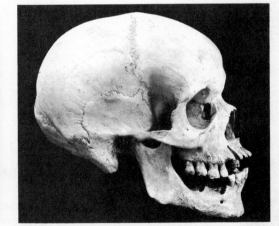

C

D

A

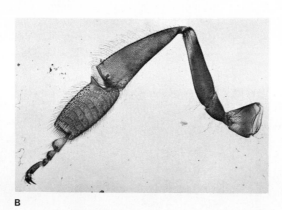

B

Figure 7.4 Generalized and specialized structures. A, the leg of a grasshopper, a generalized leg type within the broad scope of insect structure (even though such a leg is already specialized considerably for, for example, jumping). B, the leg of a honeybee, a comparatively more specialized type: a subterminal segment is enlarged as a bristly pollen basket, adapted specifically for pollen transport, and the leg as a whole is equipped abundantly with bristles, a feature that facilitates adherence of pollen.

Figure 7.5 Homology and analogy. The bird, bat, and insect wings diagrammed here are all analogous—they all serve the same function of flying. Bird and bat wings also are homologous, since they develop in similar fashion and have similar structure. But insect wings are not homologous to either of the other two.

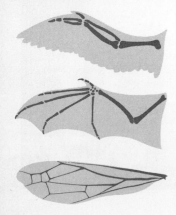

well as more complex (for example, jellyfishes are more primitive and simpler than vertebrate fishes).

Generalized—Specialized These terms have both structural *and* evolutionary connotations. A pattern is generalized (=unspecialized) if it can be and actually has been further modified, through simplification or complication, for example; and it is specialized if it is already modified. Thus the segmented body of an annelid (as exemplified by an earthworm) is generalized by comparison with the segmented body of an arthropod (as exemplified by an insect); the arthropod body is a modified annelid body (Fig. 7.4). Generalized structures originate earlier than the specialized structures they later give rise to. Hence generalized patterns usually are more primitive than specialized ones (but they are not always simpler).

Homologous—Analogous As noted, function and way of life play only relatively small roles as taxonomic criteria. It happens often, therefore, that the various subgroups within a taxonomic rank differ greatly in their ways of life and their methods of

executing functions. Consider, for example, the different functions of fish fins and human arms and the different ways of life of a fish and a man. Nevertheless, the lateral fins of a fish and the arms and legs of a man are basically the same kinds of structures; they have evolved from one common ancestral type of body appendage.

Whenever body parts in an animal or in different animals have evolved from a common ancestral starting point, as in the example just cited, and whenever they also have the same structure at least at early developmental stages, then such components are said to be *homologous*. Homology thus indicates similarity of history and structure, without reference to function. Indeed, homologous structures may or may not function the same way. For example, fish fins and human arms are homologous but do not function in the same manner. By contrast, whenever two structures do function in like fashion, regardless of history or structure, they are *analogous*. Bird wings and bat wings are analogous. They also happen to be homologous, but analogous structures are not always homologous as well. For example, bird wings and insect wings are analogous inasmuch as both are used for flying, but they are not homologous (Fig. 7.5). Essentially, therefore, "homology" is a single term for the two main criteria on which taxonomy rests, structure and history; and a study of homologies is the basis of animal classification.

Higher—Lower These adjectives have a proper role only in reference to taxonomic and organizational levels and in a literal sense. A phylum is higher in the taxonomic hierarchy than a class, a tissue is higher in the organizational hierarchy than a cell, and a bird can be higher off the ground than a worm. In other respects the terms are either erroneous or meaningless if they are used in a structural or evolutionary sense—as they often are. Several decades ago it was believed that animal evolution occurred in a scalelike or ladderlike pattern, one animal type giving rise to another in serial fashion. The terms "lower" and "higher" were used to indicate "rung" positions on the evolutionary "ladder." Also, since "highest" generally meant "man," human vanity could be pleasantly satisfied by regarding all other animals as being "lower."

However, it is now clearly established that evolution is not ladderlike but, as noted above, has the general pattern of a greatly branching *bush* (see Fig. 7.2 and also Chap. 14). In this evolutionary bush all types now in existence occur at the same time level, the present, and the various animals are only *different,* with different specializations. Some of these animals have longer evolutionary histories than others, and some are structurally more complex than others; yet each, man included, today stands only just as "high" (or just as "low") on the evolutionary bush as *every* other animal now living.

The terms thus have lost zoological significance, but old terminologies, like old habits, unfortunately disappear slowly. Even professional zoologists still frequently speak of "higher" or "lower" animals, when the actually intended reference is to advanced and/or more complex ones or to primitive and/or simpler ones. Indeed the terms are not only erroneous but also superfluous. For example, in a statement such as "man is a higher animal than a dog inasmuch as man has a more complex brain," the zoologically significant information is fully expressed by the statement "man has a more complex brain than a dog," and the addition of "higher" is only a self-serving, unscientific value judgment without zoological information content. In the same vein, the sense of smell is developed far better in a dog than in a man, yet this does not make the dog "higher" either. The point is that the terms should be avoided if scientifically meaningful comparisons of body types are to be made.

Animal Form

That generalized traits appear earlier than specialized traits holds true not only in animal evolution but also in embryonic development. For example, a human embryo develops vertebrate characteristics (such as rudiments of a vertebral column) before it develops mammalian characteristics (such as rudiments of hair). The latter arise before human traits (for example, rudiments of a chin). And only at the last moment does the embryo acquire the special personal traits that will distinguish the later adult uniquely from other human adults. Similarly, all other animals by and large develop general traits first and specific or unique traits progressively later (Fig. 7.6).

The more general a trait, the higher, more inclusive a classification group it defines. For example, the general trait of having a vertebral column defines a higher taxonomic level (a subphylum) than the more specialized trait of having hair (which identifies a class). Therefore, since animal development produces a sequence of progressively less generalized traits, this sequence can be used to define a taxonomic hierarchy of body forms characteristic of the major animal groups.

Basic Anatomy

Developmentally the first and most generalized trait of animals is their *level of organization* (here also referred to as *grade of construction*). All animals typically begin life as single cells. Some never develop beyond this stage, but others do develop

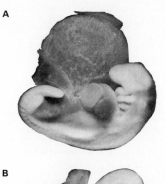
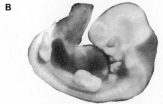
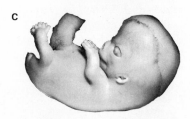
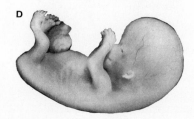

Figure 7.6 The human embryo. A to D, four successive stages: 25 days, 33 days, 6 weeks, and 8 weeks after fertilization, respectively. The series indicates that chordate features arise first (e.g., dorsal skeletal supports, gill pouches, as in *A*); that vertebrate features develop next (e.g., anteroposterior segments, paired limb buds, tail, as in *B*); that tetrapod and mammalian traits appear later (e.g., four legs, umbilical cord, as in *C*); and that distinctly human traits appear last (e.g., arm-leg differences, flat face, individualized facial expression, as in *D*).

to higher levels. On this basis, the Linnaean kingdom of animals traditionally has been subdivided into two *subkingdoms: Protozoa,* identified by a cellular level of construction, and *Metazoa,* identified by various higher levels (Fig. 7.7).

Within the Metazoa, some animals are constructed almost entirely on the tissue level, whereas others pass beyond this level during their embryonic development and become predominantly more complex. Accordingly, the metazoan subkingdom can be considered to include two *branches.* In the branch *Parazoa,* the highest level of organization is the tissue. This branch contains just one phylum, the *sponges.* The second metazoan branch comprises the *Eumetazoa,* defined by the presence of permanent organs and particularly

also of organ systems. The organ level is highest in two phyla, the *cnidarians* and *ctenophores.* All other Eumetazoa pass beyond the organ level during embryonic development and come to exhibit a conspicuous system level of construction. This structural difference between organ and organ-system levels offers a further taxonomic distinction, and indeed it is reinforced by an additional one, the different *symmetries* of the body.

The trait of symmetry has a very high degree of generality; it appears in animal development very soon after levels of organization are established. Four basic types of symmetry exist: *spherical, radial, bilateral,* and *asymmetric.* As they occur among animals none of these symmetries is geometrically precise, but the designations never-

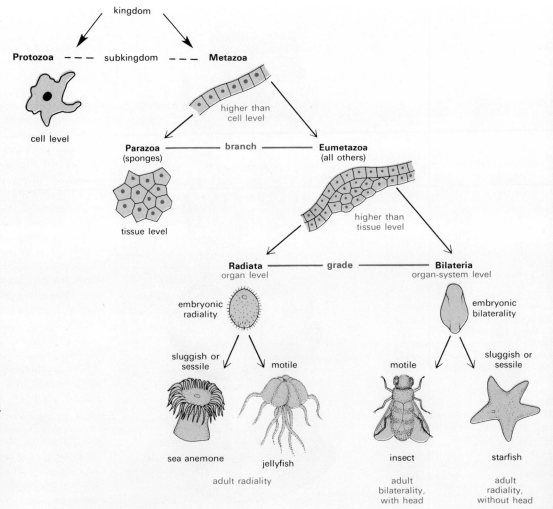

Figure 7.7 Taxonomy, level of organization, and symmetry. Radiates have embryonic as well as adult radial symmetry. Bilateria have a primary embryonic bilateral symmetry. Adults are bilateral if they are motile but often secondarily radial if they are sessile.

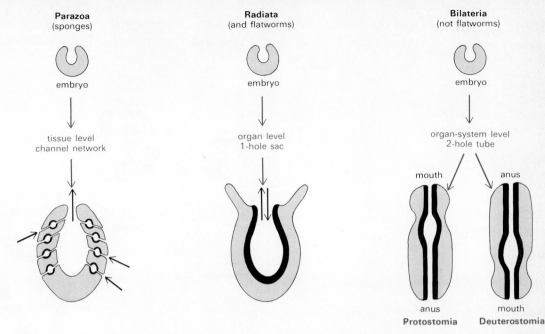

Figure 7.8 Alimentary patterns. The position of the alimentary cells is shown as dark layers (endoderm in Radiata and Bilateria). In radiates the single alimentary opening of the embryo becomes a joint mouth-anus in the adult. In bilaterial types the embryonic alimentary opening becomes the mouth in Protostomia, the anus in Deuterostomia. In each case a second alimentary opening develops later at the opposite end.

theless are useful for descriptive purposes. Many animals maintain the same symmetry throughout development and in the adult stage as well. In other cases adult symmetry often is greatly modified, usually as a specialization for a particular mode of life. It is therefore important to distinguish between *primary* symmetry, exhibited by embryos and larvae, and *secondary* symmetry, exhibited by adults. Where the two are different, only the primary symmetry represents a generalized trait.

Protozoa are variously radial, bilateral, and asymmetric, and a few also approach a more or less spherical form. Among Metazoa, all embryos are radially symmetric at very early stages; they are solid or hollow balls composed of a few cells. Sponge embryos then form adults that exhibit a variety of symmetries, asymmetries most particularly. In the Eumetazoa some groups retain the original radiality throughout embryonic life, and this symmetry later carries over more or less unchanged to the adult stage as well. In all other groups the original radiality changes very rapidly to bilaterality, and the remaining embryonic and adult stages then typically are bilateral.

On this basis, the branch Eumetazoa can be subclassified into two taxonomic *grades*. The grade *Radiata* includes the cnidarians and the ctenophores. These animals are identified by both an organ level of construction and a pronounced radial symmetry at all stages. All other Eumetazoa exhibit a system level of construction, and they are primarily bilateral. These form the grade *Bilateria*. Most adults in this group have an elongated, bilateral body, and a head typically is present. In numerous Bilateria, however, the adult is specialized for a sluggish or sessile way of life, and in such cases the bilateral embryos tend to give rise to adults that exhibit secondary radiality or asymmetry. A head then is usually absent as well (for example, starfishes, barnacles: see Fig. 7.7).

After symmetry, the next most generalized trait is the form of the *alimentary structures;* once symmetry is established in the embryo, the architecture of the alimentary system is among the first to become elaborated. Three major alimentary patterns are encountered among Metazoa (Fig. 7.8). One, unique to the sponges, can be described as a *channel network* pattern. Alimentary channels branch out extensively throughout the body and lead to the outside through openings in

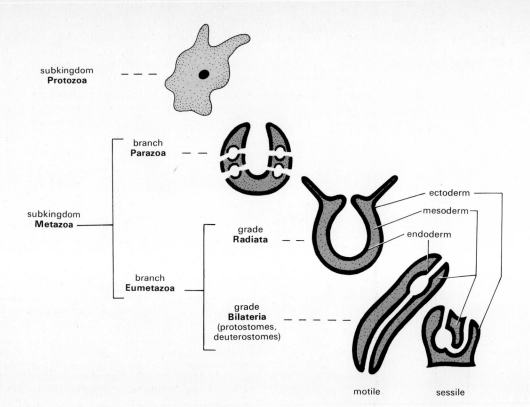

subkingdom
Protozoa

branch
Parazoa

subkingdom
Metazoa

branch
Eumetazoa

grade
Radiata

grade
Bilateria
(protostomes,
deuterostomes)

ectoderm

mesoderm

endoderm

motile   sessile

Figure 7.9 The fundamental body plans of animals. Note the three primary germ layers in eumetazoan animals.

the body wall. The channels are lined by flagellate digestive cells that function in cooperative groups. As would be expected in a sponge, therefore, the alimentary apparatus exhibits a primitive tissue level of construction.

The second type of alimentary architecture has a *one-hole sac* pattern. A single opening in the sac functions as both mouth and anus, and the sac itself is formed by a single tissue layer that contains digestive cells. Such an alimentary organization is encountered in all Radiata and one phylum of Bilateria, the flatworms.

The third alimentary pattern charcterizes all other Bilateria and can be described as a *two-hole tube* system. Developmentally such a tube arises from a sac. The embryos of the Bilateria first develop a one-hole sac, like the embryos of the radiates and flatworms. But whereas the latter retain the sac construction permanently, the bilaterial embryos soon acquire a second opening opposite the first. One opening then specializes as a mouth, the other as an anus, and the tube interconnecting them becomes the alimentary tract through which food passes in only one direction, from mouth to anus.

Of the two openings formed in the embryo,

which becomes the mouth and which the anus? In one group of Bilateria the original opening of the one-hole sac forms the mouth, and the later second opening becomes the anus. In a second group the pattern is reversed, the first opening forming the anus. This distinction can be used to define two general, descriptive categories among the Bilateria. In the *Protostomia,* the first opening is the mouth, and in the *Deuterostomia,* the second opening is the mouth (see Fig. 7.8). Protostomia include most of the wormlike animals as well as mollusks, arthropods, and numerous other phyla. There are fewer phyla among the Deuterostomia, their best-known members being the echinoderms and the chordates.

In bilaterally symmetrical adults of either category, the alimentary tract by and large lies along the longitudinal axis of the body. But in secondarily radial or asymmetric adults, the course of the alimentary tract usually is modified as well. In many sessile, attached animals, for example, the tract has a U shape, the bottom of the U marking the region where the animal is attached to the ground. Mouth and anus then lie close together and as far away from the ground as possible.

Taken together, the level of organization, the

symmetry, and the alimentary pattern provide a broad outline of the fundamental body form of any animal. Alimentary pattern and symmetry specify the basic interior and exterior architecture, respectively, and the organizational level specifies the complexity of the architectural building blocks. This information permits a description of all animal organizations in their most generalized form (Fig. 7.9).

More detailed characterizations can be obtained by studying the interior structure of animals, the solid and hollow regions between the alimentary layer and the skin.

Body Cavities

Figure 7.9 indicates that Metazoa can be considered to exhibit a basic three-ply construction; an outer layer is integumentary, an inner layer is alimentary, and a middle layer includes all other body parts. Such layering is comparatively indistinct in the sponges but is elaborated conspicuously in all Eumetazoa. In these animals the adult body is formed from well-defined embryonic germ layers, already referred to in Chap. 2. The integumentary layer of the adult develops from an embryonic ectoderm, and the alimentary layer matures from an embryonic endoderm. Later a middle mesoderm arises by ingrowth of cells from either of the first two layers.

In the Radiata this middle layer remains a relatively uncomplicated tissue, and it often becomes quite bulky as a result of jelly secreted by the mesoderm cells. In the Bilateria the mesoderm cells accumulate far more extensively than in the Radiata, and the cells also become organized as

several tissues, organs, and systems. Indeed, the bulk of the body comes to have a mesodermal origin. In different cases the mesoderm arises either from the ectoderm or from the endoderm or from both of these primary layers. A distinction can therefore be made between *ectomesoderm* and *endomesoderm* (Fig. 7.10). Ectomesoderm characteristically begins to form by migration of loose cells from the ectoderm. Such cells later multiply and become organized as cohesive layers. Endomesoderm usually is a cohesive tissue from the start and it too gives rise to organs and systems later. In a few cases mesoderm is purely ectomesodermal or purely endomesodermal. Far more commonly, the middle layer develops from both of the other layers. Frequently the mesoderm of the larva is partly or largely ectomesodermal, and the mesoderm of the adult then is predominantly endomesodermal.

In the most primitive Bilateria (for example, the flatworms), the mesoderm of the adult completely fills the region between the ectoderm (skin) and the endoderm (alimentary system). Animals so constructed form a *subgrade Acoelomata* within the grade Bilateria (Fig. 7.11).

In all other Bilateria, the middle body layer does not form a solid accumulation of body parts. Instead, as the mesoderm develops in the embryo, a more or less extensive free (fluid-filled) space is left between the ectoderm and the endoderm. This space later becomes the main *body cavity* of the adult. The original evolution of such cavities has been a major event in animal history and has been mainly responsible for the present success of animals as a whole.

The first advantage of a body cavity probably

Figure 7.10 Mesoderm derivations. The colored layers represent embryonic ectoderm and endoderm; the black cells are mesoderm. Mesenchymal ectomesoderm, characteristic of Radiata and many Bilateria, typically originates as a series of loose cells. Epithelial endomesoderm, as in most Bilateria, usually forms as cohesive layers of cells from the start. Both kinds of mesoderm often develop in bilaterial animals.

ectomesoderm:
largely
mesenchymal

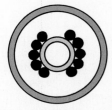

endomesoderm:
largely
epithelial

Figure 7.11 Mesoderm and coelom in Bilateria. The main body cavity is a pseudocoel in the Pseudocoelomata and a peritoneum-lined coelom in the Coelomata. In coelomates a dorsal and ventral mesentery (formed by two mesodermal layers) supports the alimentary tract.

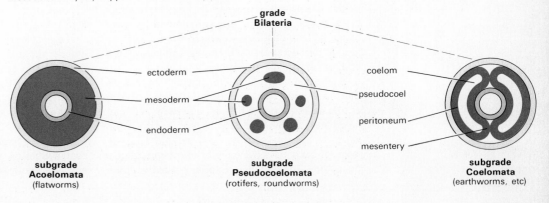

grade
Bilateria

ectoderm

mesoderm

endoderm

coelom

pseudocoel

peritoneum

mesentery

**subgrade
Acoelomata**
(flatworms)

**subgrade
Pseudocoelomata**
(rotifers, roundworms)

**subgrade
Coelomata**
(earthworms, etc)

was that it led to a then new way of animal life, burrowing into sand or mud. Elongated wormlike animals typically make burrows by powerful, wave-like (*peristaltic*) contractions of the circular muscles in the body wall. Acoelomate types can produce peristaltic contractions, too; but because the body here is solidly filled with tissues, the contractions must be shallow and thus cannot generate either the force or the leverage against sand or mud that burrowing requires. By contrast, in animals with fluid-filled body cavities the body wall is mechanically independent from the interior, and it *can* produce the deep peristaltic waves necessary for burrowing. Moreover, the fluid in the body cavity acts like an incompressible ''hydraulic skeleton'' that spreads out and diffuses—and thereby renders harmless—the inward pressure generated by body wall contractions.

The possibility of making a burrow in turn opens numerous new modes of survival. Microscopic organisms and detritus in sand or mud bottoms can provide new food sources, and in a burrow an animal can be protected against becoming food itself. Besides, if the food-collecting front end of the animal protrudes out of the burrow, even the open-water foods become available while the animal still is safe in its burrow. Such ways of life actually appear to have been adopted by the first animals with body cavities, and a vast array of descendants still lives this way today.

Furthermore, a body cavity introduces many important secondary advantages. The cavity permits an animal to attain considerable size, for the hydraulic fluid skeleton provides internal support against gravity. The fluid also can aid in transporting nutrient molecules, wastes, and respiratory gases to and from deep-lying body parts, which in a large animal would otherwise not be accessible merely by direct diffusion from integument and alimentary tract. Moreover, with an internal transport fluid available, breathing and excretory structures can become localized in restricted body regions. As later chapters will show, such structures tend to take up space (and energy) throughout the body of an animal that lacks a body cavity, but they are compact and more efficient in animals that have body cavities. Also, the very presence of free internal spaces permits a gradual accumulation and temporary storage of reproductive cells, which in turn makes possible a simultaneous release of such cells by many or all members of a population. The reproductive potential of the group

then tends to be increased by such synchronous breeding (and it is quite possible that specialized endocrine mechanisms for control of breeding periods might have evolved in parallel with body cavities).

Different groups of Bilateria develop body cavities in different ways. In some (rotifers and roundworms, for example), the embryonic mesoderm accumulates only in particular limited regions between the ectoderm and endoderm. The body cavity left free in this manner is bounded on the outside directly by the body wall and on the inside directly by the alimentary system. Animals so constructed form a subgrade of Bilateria, the *Pseudocoelomata* (see Fig. 7.11).

In all remaining Bilateria, the body cavity arises when, during development of mesoderm, part of it (the *somatic* mesoderm) comes to lie along the inner surface of the ectoderm, while another part (the *splanchnic* mesoderm) comes to surround the alimentary tract. In such animals the body wall therefore contains both ectodermal and mesodermal layers, and the alimentary wall contains both mesodermal and endodermal layers. Both parts of the mesoderm later give rise to various tissues and organs, and particularly also to a cellular membrane that comes to enclose the free space between the outer and inner mesodermal parts. This mesodermal membrane is a *peritoneum*. Vertical portions of it called *mesenteries* suspend the alimentary tract from the body wall. The free space enclosed by the peritoneum now represents the main body cavity.

A body cavity bounded completely by mesodermal components, and especially by a peritoneal membrane, is known as a *coelom*. Accordingly, animals that have a coelom form a bilaterial subgrade *Coelomata* (see Fig. 7.11). The meaning of the terms *acoelomate* and *pseudocoelomate* now becomes clear. Acoelomates are animals without coelom and indeed without body cavity of any kind. Pseudocoelomates have a *pseudocoel*, ''false'' coelom—a body cavity lined mainly by ectoderm and endoderm, perhaps partly by mesoderm, but in any event not by a peritoneum. The cavity thus resembles a true coelom superficially.

Among the coelomate Bilateria, which form by far the largest groups of animals, further subgroups can be distinguished on the basis of *how* the coelomic cavities develop (Fig. 7.12). In one such subgroup, exemplified by mollusks, annelids, and arthropods, the adult mesoderm arises in the

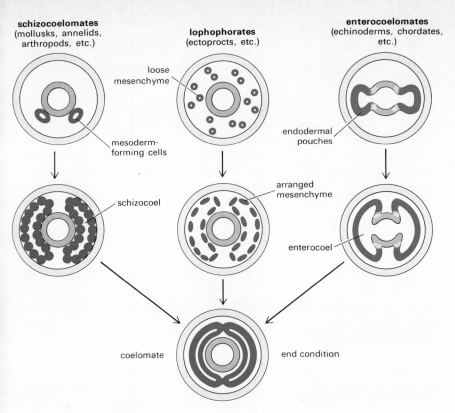

schizocoelomates
(mollusks, annelids, arthropods, etc.)

loose mesenchyme

mesoderm-forming cells

schizocoel

lophophorates
(ectoprocts, etc.)

arranged mesenchyme

enterocoelomates
(echinoderms, chordates, etc.)

endodermal pouches

enterocoel

coelomate

end condition

Figure 7.12 Coelom formation in the subgrade Coelomata. The pattern shown for lophophorates is but one of several known to occur in that group.

mainly by echinoderms and chordates and again constituting a superphylum, the mesoderm arises in the embryo as paired lateral pouches that grow out from the endoderm. The pouches later become disconnected from the endoderm, but their inner portions remain as a layer that surrounds the developing alimentary system. The outer portions become applied against the developing body wall. The final condition is essentially quite similar to that in schizocoelomates. However, since the mesoderm and the coelom here are derivatives of the future gut, or *enteron,* the body cavity is called an *enterocoel.* Animals having such cavities therefore are known as *enterocoelomates.*

In a third subgroup or superphylum, various other patterns of coelom formation are encountered. In one, for example, loose mesoderm cells of the embryo migrate in amoeboid fashion and simply arrange themselves as a continuous peritoneal layer. Coeloms developed in this and similar ways have not been given any special technical names. Animals in this subgroup can be referred to collectively as the *lophophorates.* Including the phyla of phoronids, ectoprocts (moss animals), and brachiopods (lamp shells), these animals are not particularly abundant today. But their ancestors may have been among the most ancient coelomate animals from which both the schizocoelomate and enterocoelomate superphyla later evolved.

Among Bilateria as a whole, only the enterocoelomate superphylum happens to be deuterostomial; the lophophorates and schizocoelomates are protostomial, as are the Acoelomata and Pseudocoelomata (Fig. 7.13).

Body Divisions

Many Bilateria have tough and often stiff epidermal cuticles that provide excellent protection but that also restrict motion—particularly in Pseudocoelomata and Coelomata, where the fluid pressure in the body cavity also makes the animals turgid, somewhat like sausage skins blown up and filled with water. In many of these animals, bending, shortening, or lengthening of the body is made easier by series of permanent ringlike creases in the cuticle and the body wall. Called *superficial annulation,* or superficial segmentation, such ectodermal creasing has little additional significance; it occurs more or less haphazardly among bilateral groups (Fig. 7.14).

embryo from two endoderm-derived cells, one on each side of the future gut. These so-called *teloblast* cells then proliferate as a pair of *teloblastic bands* of tissue. At first the bands are solid cellular masses, but later each mass splits into outer and inner mesodermal sublayers. Therefore, because the coelom here forms by a splitting of mesoderm, it is called a *schizocoel.* Animals characterized by body cavities of this type can be designated as *schizocoelomates,* a grouping taxonomically roughly equivalent to a superphylum.

In another coelomate subgroup, represented

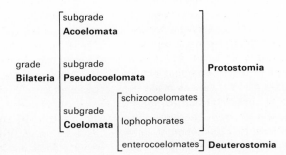

grade **Bilateria**

subgrade **Acoelomata**

subgrade **Pseudocoelomata**

subgrade **Coelomata**: schizocoelomates, lophophorates — **Protostomia**

enterocoelomates — **Deuterostomia**

Figure 7.13 Bilateria. The major groups and their interrelations.

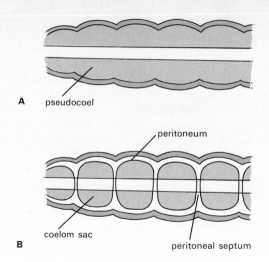

A   pseudocoel

B   coelom sac   peritoneum   peritoneal septum

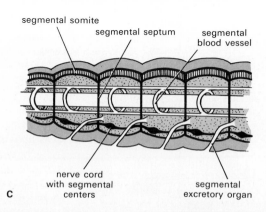

C   segmental somite   segmental septum   segmental blood vessel   nerve cord with segmental centers   segmental excretory organ

Figure 7.14 Segmentation. A, superficial segmentation; only the integument exhibits ringlike creases. B, metameric segmentation; both ectodermal and mesodermal derivatives are arranged segmentally, and the coelom is subdivided into metameric anteroposterior compartments. C, diagram of the segmental arrangement of organ systems in metameric animals. The excretory organs shown are so-called *metanephridia,* essentially tubes that lead from the coelomic body cavity to the outside.

Of much greater importance are various kinds of internal subdivisions in the body of many coelomate animals. Such subdivisions appear first in the embryonic coelom, and later they often also affect mesodermal and ectodermal organ systems. Formed in various ways, the early subdivisions arise as *peritoneal septa,* transverse partitions in the coelom that are part of the peritoneum. On each side of the body the coelom thus comes to be represented by an anteroposterior series of *coelom sacs.* These sacs may or may not interconnect, and if they do they usually can be closed off from one another by muscles in the septa. The evolution of partitioned coeloms was as significant in its own way as the earlier evolution of body cavities themselves. Coelomic cavities made possible new ways of life based on burrow digging; and partitioned coeloms made possible vastly more new ways of life based on motion of limited parts of the body only. Among the ultimate results

were all the advanced, "modern" modes of existence exhibited by the most successful animals today.

Whereas animals with undivided body cavities can make burrows, they cannot readily use the peristaltic contractions of their body walls for either sustained burrowing or outright locomotion. Movement or displacement in any part of the hydraulic skeleton necessarily means movement or displacement in all of it. Also, because the longitudinal muscles of the body wall extend uninterruptedly from front end to hind end, lengthwise contraction in one part of the animal necessarily involves contraction of the whole animal. The body therefore cannot obtain fixed anchorage against the ground, as would be required in continued tunneling or in locomotion based on alternate extension and telescoping of the body. With transverse partitions, however, the necessary purchase on the ground can be obtained: the lengthwise body muscles are interrupted at some or many septa, and a given section of the body thus can contract, expand, or remain stationary independently of other sections; and local pressure changes in one coelom sac are prevented from spreading to other sacs by the septa. In effect, one body part can be used as a stationary anchor on the ground while other parts extend or contract (Fig. 7.15).

In the lophophorate animals the coelom basically has two anteroposterior divisions, and in primitive enterocoelomates there are three. These groups typically still comprise burrowers, but considerably advanced ones. Lophophorates and the earliest enterocoelomates use their anterior body divisions for *filter feeding:* they protrude a ciliated tentacle apparatus from their burrows or tubular housings and strain microscopic food out of the water environment. However, this tentacled forepart can be withdrawn and reextended as often as necessary, while the hind part of the body remains independently stationary and anchored inside the burrow or housing. Some groups of primitive enterocoelomates actually have ceased to be sedentary filter feeders; they bore extensive tunnels through sand or mud by using their anterior body divisions as digging organs and their posterior ones as anchoring levers.

In these animals burrowing thus has become sustained, continuous locomotion. A much greater advance in locomotor capacity is in evidence among coelomates that develop not just two or three but numerous anteroposterior subdivisions of the coelom. Such animals are said to exhibit true segmentation, or *metamerism*. Here typically all organ systems except the alimentary tract form as segmental, metameric series of parts. The muscular system develops in segmental blocks called *somites*, and each segment, or *metamere*, generally also has its own circulatory, excretory, reproductive, and nervous organs, with interconnections between adjacent segments (see Fig. 7.14).

The most conspicuously metameric animals are the segmented worms, or annelids. They too include filter-feeding sedentary types (feather-duster worms, for example; see Color Fig. 1), but also among them are highly efficient tunneling types such as earthworms, and, in addition, types that not only burrow but also swim for extended periods. The important advances here lie not so much in new forms of motion and locomotion as in new speeds and durations: an earthworm or a clamworm is a far faster, stronger, and more active animal than any coelomate evolved before the annelid type, and it can maintain its activity for far longer periods. Annelids in effect might be called the first "lively" coelomates and also the first long-distance migrants, capable of self-powered propulsion well beyond their most immediate surroundings. This capacity of relinquishing a sedentary existence is a major adaptive advantage; it results from an increased operating efficiency of the body, itself a consequence of the segmental construction.

Far-reaching further advantages emerge when a segmented body is combined with a hard skeleton. Such a body organization is encountered in two major, unrelated groups, arthropods and vertebrates. In arthropods, descended from annelid-like ancestors, the originally similar segments of the body have become elaborated and specialized in an endless number of different ways. Moreover, a jointed segmental skeleton has evolved that consists of hard external plates and covers over appendages. The skeleton provides rigid surfaces for muscle attachment and therefore far more efficient locomotor leverage than can be obtained through the compartmented hydraulic system of the annelidlike ancestors (see Fig. 7.15). Arthropods actually have become the largest, most diversified single group of all living organisms, and they exhibit virtually all known forms of locomotion, including flying most particularly.

Vertebrate segmentation is derived indirectly from a repeated subdivision of the most posterior of the three coelom sacs of early enterocoelomates. As later chapters will show, evolution of a segmented body here was associated originally with the development of swimming locomotion by means of tails, as in fishes. Also, a jointed, segmental, internal skeleton arose that, as in arthropods, provided hard attachment surfaces for a segmental musculature. The ultimate result was the emergence of the fastest, strongest, liveliest, and most active animals of all time, with just as many diverse forms of locomotion as among arthropods. Note that, in both arthropods and vertebrates, the presence of a leverage-providing skeleton makes the earlier partitioned coelom and hydraulic system essentially superfluous. Indeed, both groups have coelom sacs only during embryonic stages; as the adult stage is reached the sacs largely disappear. This circumstance strongly supports the generalization that coeloms, and body

Figure 7.15 Advantages of segmentation. A, in an unsegmented animal, contraction of the longitudinal body-wall muscles leads to shortening of the whole animal. Locomotor leverage on the ground therefore is difficult to obtain. B, in a segmented animal, the segmental longitudinal musculature permits independent contraction of given body segments. Such a construction permits an animal to obtain great locomotor leverage on the ground. C, hard endoskeletons (left) or exoskeletons (right) provide firm attachment surfaces and leverage for muscles, hence in both cases the ancestral "hydraulic" fluid skeleton in the body cavity becomes more or less superfluous for locomotion.

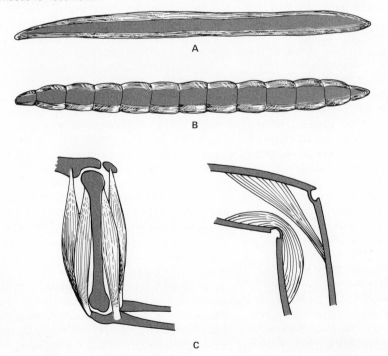

cavities of all kinds, are significant primarily and originally as locomotor adaptations (see Fig. 7.15).

Clearly then, metamerism specifically and coelomic body divisions generally are important features of animal form; they characterize body organization at the level of groups of phyla and individual phyla. The next lower level of organizational detail describes phylum units and subunits within phyla, and it is one of the functions of Unit 2 of this book to provide such descriptions. Attention now can turn to two basic levels of organization inside individual animals, the tissue and the organ.

Tissues and Organs

As an animal matures, most of its cells specialize in various ways; they become more or less diversified in external appearance and internal structure, and they develop the capacity to perform some function or functions especially well. Such characteristics tend to become fixed and irreversible. Once a cell has become specialized—particularly highly specialized—in one way, it normally cannot change and respecialize in another way.

A *tissue* is an aggregation of cells in which each cooperates with all others in the performance of a particular group function. In a *simple tissue* all cells are of the same type. Two or more different cell types are present in a *composite* tissue. The cells of a tissue need not necessarily be in direct physical contact, but this is actually the case in many instances. Tissues can be highly or less highly specialized, according to the degree of specialization of the component cells.

Most tissues of most animals can be classified generally as *connective tissues* and as *epithelia*. Not normally included here are three specific tissues, nerve, muscle, and blood; these are discussed in the following chapters in the context of the organ systems of which they are a part.

Connective Tissues

These are identified by comparatively widely separated cells, the spaces between the cells being variously filled with fluid and solid materials. Another identifying characteristic is the relatively unspecialized nature of the cells. With appropriate stimulation they can transform from one connec-

tive tissue cell type to another. Variants of connective tissues are distinguished on the basis of the kinds of intercellular deposits present and the relative abundance and arrangement of the cells.

The most fundamental of the connective tissues is *fibroelastic tissue* (Fig. 7.16). Conspicuous in it are large numbers of threadlike fibers, some tough and strong, others elastic. These are suspended in fluid and form an irregular, loosely arranged meshwork. The cells of the tissue are dispersed throughout the mesh, and they secrete materials that give rise to the fibers outside them.

The cells are of various types. Many are *fibrocytes,* generally spindle-shaped and believed to be the chief fiber-forming cells. Others, the *histiocytes,* are capable of amoeboid motion and of engulfing foreign bodies (such as bacteria in infected regions). Also present are *pigment cells* (*chromatophores*), *fat cells,* and above all, *mesenchyme cells.* These are undeveloped and relatively quite unspecialized. They form all the connective tissues of an embryo and later give rise to the various cell types of adult connective tissues. Thus it is possible to define connective tissues as all those that have a mesenchymal origin in the embryo.

In many adult animals, mesenchyme cells ''left over'' from embryonic stages play an important role in healing and regeneration. The cells can migrate to injured body regions and contribute to the redevelopment of lost body parts and to formation of scar tissue. Most of the adult cell types of fibroelastic tissue likewise can transform into one another. For example, a fibrocyte might become a fat cell, then perhaps a histiocyte, then a fibrocyte again, and then a pigment cell. The specializations of any of these cells evidently are not fixed.

By virtue of its cellular components, fibroelastic tissue thus functions in food storage and in body defense against infection and injury; and by virtue of its fibers, the tissue is a major binding agent that holds one body part to another. For example, fibroelastic tissue connects skin to underlying muscle. The tough fibers provide connecting strength, yet the elastic fibers still permit the skin to slide over the muscle to some extent.

The relative quantities of cellular and fibrous components vary greatly, and on the basis of such variations other types of connective tissue can be distinguished (Fig. 7.17). For example, *tendons* are dense tissues that contain only fibrocytes and

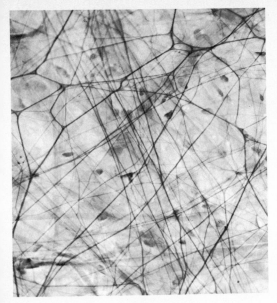

A

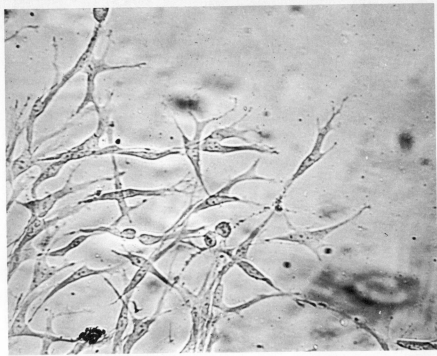

B

Figure 7.16 Connective tissues. *A,* fibroelastic tissue. Fibers form a network, and the cells (small dark dots) secrete the fibers and are embedded in the network (×500). *B,* mesenchyme cells in tissue culture (×5,000). *C,* pigment cells (from skin connective tissue of flounder) in various states of expansion (×5,000).

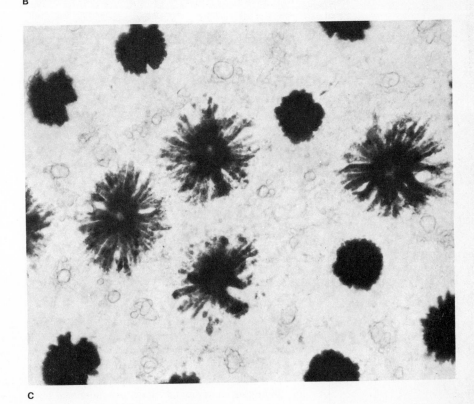

C

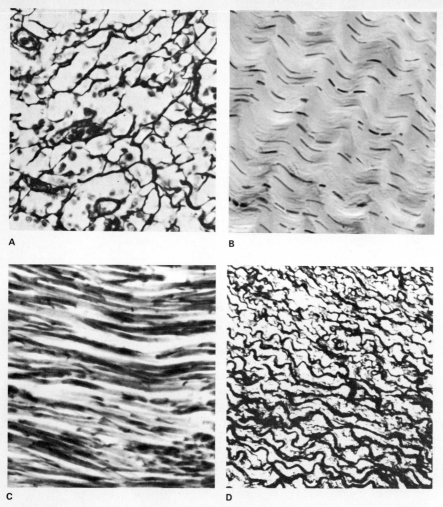

A

B

C

D

Figure 7.17 Variants of fibroelastic connective tissue. A, reticular tissue, composed of fibrous and cellular components. This type of tissue is found in lymph nodes and lungs and in preadult stages is also a forerunner of mature fibroelastic tissue. *B,* tendon. *C,* ligament. *D,* elastic tissue, as in the walls of the large arteries. Elastic fibers predominate here.(*D,* courtesy Carolina Biological Supply Company.)

Many animals contain *jelly-secreting* connective tissues. The cells in such cases are mesenchymal, and as they secrete gelatinous substances the cells become separated from one another more and more. Mesenchymal jelly tissue can become quite bulky, as in the middle body layer of the Radiata, for example. Such tissues also occur abundantly in numerous other groups, either as parts of whole body layers (flatworms, for example) or as parts of organs (around and in the eyes of vertebrates, for example).

In some connective tissues the cells secrete organic and especially inorganic materials that form a solid precipitate around the cells. Such cells therefore appear as islands embedded in hard intercellular deposits. The chief variants of this tissue type are *cartilage,* encountered in several invertebrate groups and in all vertebrates, and *bone,* characteristic of vertebrates. Both cartilage and bone arise from mesenchymal cells, and both function in support and protection (Fig. 7.18).

As a group, the connective tissues serve largely in forming the structural scaffolding of the animal body. By contrast, the primarily functional parts are formed chiefly by the epithelial tissues.

Epithelia

An epithelium is a tissue in which the cells adhere directly to one another. Such cell groups can occur as single-layered sheets, as many-layered sheets, or as compact, irregularly shaped masses. All three embryonic germ layers give rise to such tissues.

Epithelia that form sheets generally rest on *basement membranes,* flat networks of collagen fibers that are secreted by the cells as supporting fabrics. Sheets consisting of single layers of cells are called *simple* epithelia (Fig. 7.19). Distinctions among them are made mainly on the basis of cell shape. If the cells are flat and joined along their edges, the tissue is a "pavement," or *squamous* epithelium. Many tissue membranes and the surface layer of the skins of many animals are of this type. If the cells have the shape of cubes, the tissue is a *cuboidal* epithelium. The walls of ducts and glands frequently consist of such tissues. If the cells are prismatic and are joined along their long sides, a *columnar* epithelium is formed. In many animals this type occurs in, for example, the innermost (digestive-juice-secreting) layer of the intestine or the outermost (exoskeleton-secreting) layer of the skin.

tough collagen fibers, the latter arranged as closely packed parallel bundles. Tendons typically connect muscles to parts of the skeleton. A *ligament* is similar to a tendon, except that both collagen and elastic fibers are present and that these are arranged more or less irregularly. Another variant of fibroelastic tissue is *adipose tissue,* in which fat cells are the most abundant components. Each fat cell contains one or more large fat vacuoles that fill almost the entire cellular space. A collection of such cells has the external appearance of a continuous mass of fat. Still other variants of fibroelastic tissue are known.

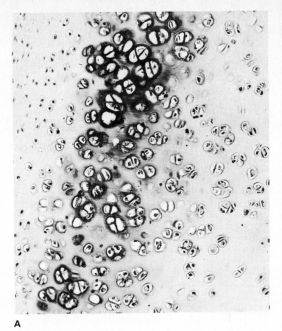
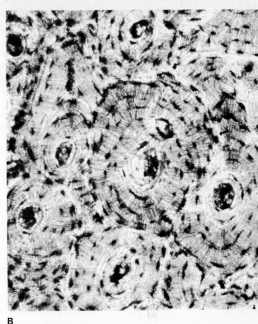

Figure 7.18 Cartilage and bone. A, the many cartilage-forming cells are surrounded by their own secretions. B, bone-forming cells are in the dark patches, arranged in concentric patterns. Hard bone substance appears light. A unit of concentric bone layers represents a so-called Haversian system; a cell-filled Haversian canal is in the center of such a system. See also Fig. 8.8.

If several epithelial layers are stacked as a multilayered sheet, they are said to form a *stratified epithelium*. Such complex epithelia can be stratified squamous, stratified cuboidal, or stratified columnar. The mammalian epidermis, or outermost tissue of the skin, is a good example of a mixed stratified epithelium; the cells are squamous along the outer surface and become increasingly cuboidal with increasing distance from the surface (Fig. 7.20). In contrast with the connective tissues, the epithelia all are fairly highly specialized.

Once their cells are mature, they do not thereafter change in their basic structural and functional characteristics.

Some animal tissues cannot be classified strictly as either connective tissues or epithelia, and they often share certain of the characteristics of both. For example, blood generally is like a connective tissue in that it contains cellular components and intercellular deposits, fluid in this case; but although some of the blood cells have a mesenchymal origin, some do not. Muscle tissue

Figure 7.19 Simple epithelia. A, squamous epithelium, diagram of cross section and photo of surface view of frog epidermis. Note close packing of cells, flat shapes, and angular outlines produced by mutual cell pressure (×1,000) B, cuboidal epithelium, diagram and photo of cross section of lining of tubule in kidney (×1,400). C, columnar epithelium, diagram and photo of inner lining of frog gut. Note surface cilia (×1,500). (B, C, courtesy Carolina Biological Supply Company.)

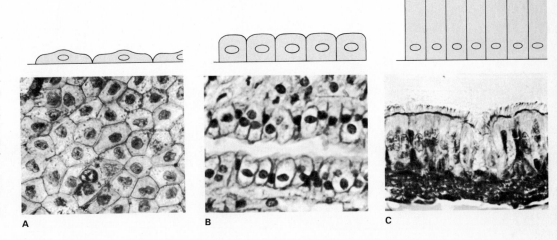

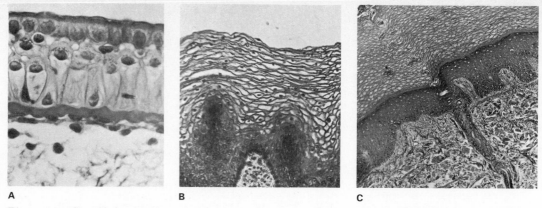

A   B   C

Figure 7.20 Stratified epithelia and skin. *A,* section of skin of frog tadpole, showing stratified epidermis (dark tissue) and connective tissue dermis under it. Note progressive flattening of epidermal cells toward skin surface (see also Fig. 7.19) (×1,500). *B,* section through lining of human uterus. This lining tissue resembles the epidermis of skin. Again note flattening of cells toward surface (×50). *C,* section through mammalian skin. Note stratified epidermis near top of photo and dermis near bottom. Parts of the duct of a sweat gland can be seen meandering from the dermis through the epidermis to the skin surface (×50).

Figure 7.21 Organs. *A,* section through the intestine (at the level of the duodenum), to show the alternating layers of epithelia and connective tissue. The innermost layer is the highly folded mucosa, an epithelium. See Fig. 9.13 for structural details. *B,* section through the liver (of a pig), a compact organ. The photo shows a few of the epithelial lobules, separated from one another by layers of the connective tissue stroma. Branches of the hepatic portal vein in the stroma carry blood to a lobule. Blood then passes freely through the canal-like spaces between the strands of lobule cells, and it is eventually carried off by a branch of the hepatic vein, seen as a large clear space in the center of a lobule. See also Fig. 9.15.

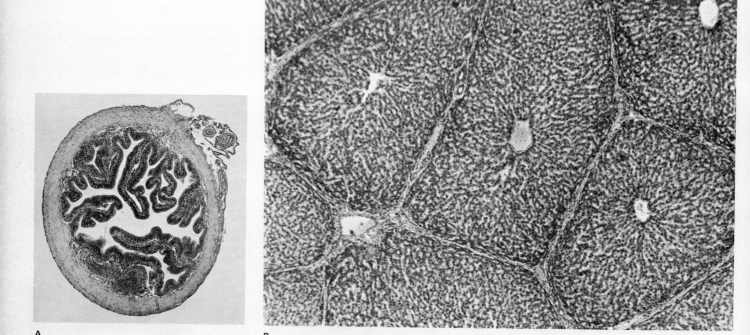

A   B

does have a mesenchymal origin, yet adult muscle resembles an epithelium more than a connective tissue. Nerve tissue has an epithelial origin, and it resembles an epithelium in some respects; but in others—its frequent netlike arrangement, for example—it does not. Similarly mixed characteristics are found in a variety of tissues that occur uniquely in particular animal groups. Such tissues are considered later, in the context of the specific animals that contain them.

Organs

An *organ* is an aggregation of two or more tissues that cooperate in the performance of a common group function. Organs typically consist of one or more epithelia and one or more connective tissues. The epithelia carry out the characteristic specialized functions of the organ, and the connective tissues serve in the necessary auxiliary roles. More specifically, the connective tissues maintain the shape and the position of the organ as a whole,

and they lead nerves, blood vessels, and other ducts to and from the epithelia.

In sheetlike organs such as skin and in tubular organs such as intestine, epithelia and connective tissues form adjacent layers (Fig. 7.21 and see Fig. 7.20). In compact three-dimensional organs such as the liver, the connective tissues form an external enveloping layer and they also extend into the interior as partitions. The connective tissues thus make up a *stroma*, a supporting framework that subdivides the liver into islands of epithelial cells (see Fig. 7.21). Such islands usually represent complete functional units of the organ, and the traffic of materials to and from the units is carried by the surrounding stroma. Smaller or larger groups of such islands often form anatomically recognizable *lobules* or *lobes*.

In their turn, organs often are linked together as *organ systems*, organizational units definable as cooperating aggregations of organs. The construction and operation of the animal organ systems are the subjects of the following chapters.

Review Questions

1. Review the structure of the Linnaean taxonomic system. What are the principal ranks? Name and define them and cite animal groups of each. How is this system related to the size and the numbers of groups at each rank?

2. On what characteristics of animals is the taxonomic system based? Why are other possible characteristics not used? Why are animals not simply classified alphabetically or by some system equivalent to book cataloguing in libraries?

3. What is meant by natural and unnatural taxonomic classifications? Give examples. What rules govern the naming of species? Review the taxonomic classification of man, with attention to the definition of each taxon.

4. Describe the proper zoological use of the following contrasting sets of terms: simple—complex; primitive—advanced; generalized—specialized.

5. Criticize and correct the following statements: "Evolution consists of a progression from simple to more complex animal types." "In the evolutionary scale, higher animals such as vertebrates have descended from lower forms such as protozoa."

6. Define homology, analogy. Give specific examples of each. Why are homologies more important in taxonomy than analogies?

7. Name the subkingdoms, branches, grades, and subgrades of animals. What criteria define each of these ranks? What animals are included in each?

8. List five or six of the most generalized features of animal structure, and show on what basis they are considered to be more generalized than others.

9. Define Protostomia, Bilateria, Parazoa, coelom, schizocoel, enterocoel, endomesoderm, germ layer, peritoneum, mesentery, metamerism, somite, lophophorate, teloblast, mesenchyme.

10. Show how Bilateria are subdivided taxonomically on the basis of mesoderm and coelom formation and show how coeloms arise in different ways in different groups. What is the adaptive advantage of a coelom? Does a pseudocoel offer similar advantages?

11. Distinguish between superficial and metameric segmentation and give their characteristics. In which animal groups is each encountered? What are the adaptive advantages of metamerism?

12. Draw up a comprehensive chart listing the classification of animals down to subgrade, and define each category so listed as fully as possible. Using this chart, state the characteristics of the following animal groups: cnidarians, arthropods, echinoderms, flatworms, sponges, mollusks, ectoprocts, chordates.

13. What is a tissue? A connective tissue? Describe the makeup of several types of connective tissue and state their general functions in animal organization.

14. What is an epithelium? Describe the structural and functional characteristic of an epithelium and list several variants of this type of tissue. Give specific examples of each.

15. What is an organ? How are tissues joined as organs? What is a stroma and what is its role? Define organ system.

Collateral Readings

Bloom, W., and D. W. Fawcett: "Textbook of Histology," Saunders, Philadelphia, 1962. This or any similar standard text can be consulted for further information on *histology,* the zoological subscience that deals with tissue and organ structure.

Clark, R. B.: "Dynamics of Metazoan Evolution," Clarendon Press, Oxford, 1964. An important book that, among other topics, documents the significance of body cavities as locomotor adaptations. The argument presented will be appreciated more fully after Chap. 14 (this text) has been studied.

Hyman, L.: "The Invertebrates," vol. 1, chaps. 1 and 2; vol. 2, chap. 9, pp. 18 ff., McGraw-Hill, New York, 1940 and 1951. The indicated selections from this most comprehensive six-volume English-language treatise on invertebrate animals include discussions of general animal characteristics.

Mayr, E.: "Principles of Systematic Zoology," McGraw-Hill, New York, 1969. One of the important standard texts on taxonomy, well worth consulting for further information on this subscience of zoology.

Sokal, R. R.: Numerical Taxonomy, *Sci. American,* Dec., 1966. Classification by means of computers is now becoming a feasible and useful technique.

8

ANIMAL SYSTEMS: SUPPORT, MOTION, COORDINATION

Every organ system of an animal contributes to the proper operation of every other, and all systems ultimately are equally essential in maintaining all metabolic and self-perpetuative processes. However, some systems contribute to certain functions more directly than others. Thus, five systems play a particularly direct role in relating an animal to its external environment; they provide protection and support against gravity, produce movement, and regulate body functions in line with given environmental conditions. These five are the *integumentary, skeletal, muscular, endocrine,* and *nervous* systems. They are examined in this chapter.

Integumentary Systems

As boundary between the interior of an animal and the physical environment, the integument probably displays more variation than any other system. Its functions variously include protection, locomotion, support, breathing, sensory reception, steady-state control, nutrient uptake, and excretion. Moreover, the integument often plays an important role in mate selection, in expressing behavior, and in permitting recognition of other animals. The system in effect performs a direct or indirect function in virtually every process of metabolism and self-perpetuation.

The basic organ of the system is the *skin,* composed of one or two main layers. An outer layer, the *epidermis,* arises from embryonic ecto-

derm. In most invertebrates the epidermis is the only skin layer present and often represents the entire body wall. In other cases the wall also includes one or more underlying layers of (mesoderm-derived) muscle. In some invertebrates and all vertebrates the skin is composed of two distinct layers, an epidermis and an inner, mesoderm-derived *dermis.* The dermis in turn is underlain by muscle layers (Fig. 8.1).

The invertebrate epidermis is an epithelium generally only one cell layer thick and syncytial in some cases. It often is ciliated and then represents the main or only organ of locomotion. Exchange of respiratory gases, water, and mineral ions usually can occur directly through such a skin, which thus functions as the main breathing and excretory organ. At various places single epidermal cells or small groups of them can be glandular and secrete mucous coats, cementing substances, or protective irritants and poisons. Other epidermal cells often are specialized as sensory structures that connect with nerve fibers.

In numerous cases the epidermal cells manufacture noncellular protective *cuticles* that can become quite thick and acquire the functions of *exoskeletons.* In most invertebrates such coverings consist largely of horny *scleroprotein,* and in arthropods and a few other groups the main component is *chitin.* Some groups manufacture calcareous *shells* and in others the epidermis produces skeletal secretions that do not remain in contact with the skin but form a housing within which the

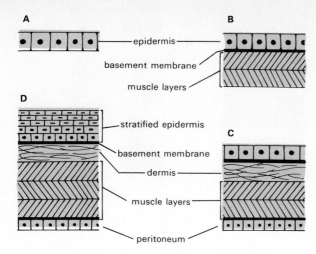

Figure 8.1 Skin and body-wall types. A, as in radiates, some noncoelomates. B, as in numerous invertebrates, both noncoelomate and coelomate (in the latter case a peritoneum underlies the muscle layers). C, as in some coelomate invertebrates and also some chordates. D, as in most vertebrates. In all groups, the muscle layers are longitudinal or circular if only one is present, and both longitudinal and circular if two are present; any additional layers are variously diagonal. The outside-inside sequence of the muscle layers varies considerably for different groups.

animal can move freely. In various wormlike animals, for example, such housings are leathery or calcareous tubes.

Among vertebrates, the epidermis of fishes and many amphibia is thin and without cuticles or cilia. In most other vertebrate groups the epidermis is a *stratified* epithelium, several to many cell layers thick (see Fig. 7.20). The basal layers here continually produce new cells that displace older cells outward. These older cells manufacture *keratin,* an insoluble protein that *cornifies* the cells: it makes them horny and impervious to water. The oldest cells, near and at the body surface, are completely cornified and are actually dead. As this exterior layer keeps wearing off, new layers produced in the basal regions of the epidermis move outward, cornify, and replace those lost.

Various ingrowths and outgrowths from the vertebrate epidermis give rise to numerous special integumentary structures. For example, ingrowths form a variety of glands, among them *mucus glands* and the excretory *sweat glands* and oil-secreting *sebaceous glands* of mammals. Outgrowths can have the form of localized thickenings of the cornified layers, as in the *thumb pads* of

male frogs and the *foot pads* of many mammals. More elaborately formed thickenings are represented by *beaks, claws, hoofs,* and *nails,* and by the outer horny layers of turtle *shells* and of cattle and sheep *horns.* Epidermal outgrowths also produce the horny *teeth* of lampreys, the *scales* of snakes, lizards, and armadillos, and the *rattles* of rattlesnakes (Fig. 8.2).

The most conspicuous epidermal outgrowths are *feathers* and *hairs,* which represent modified epidermal scales. Both types of structures grow out from epidermal pits that penetrate deep into the dermis. At the bottom of such a pit is a *follicle,* composed of a blood- and nerve-supplying *dermal papilla* and a surrounding jacket of epidermal cells. Feather or hair shafts that grow upward from the follicle are dead cornified cells. Most birds and mammals shed (*molt*) their feathers and hairs from time to time. Reptiles too periodically molt the cornified layers of their epidermis.

The epidermis extends from the body surface into various openings and cavities in the animal. For example, the outer skin layer lines the mouth cavity and the terminal portions of the excretory and reproductive tracts. In such places the epidermis remains largely uncornified and therefore differs in texture from external skin.

The dermis is a fibroelastic connective tissue that contains an abundance of tough and elastic fibers (see Fig. 7.20). Its primary function is to carry blood and nerves to and from the epidermis and to endow the whole skin with strength, toughness, and elasticity. The cellular components of the tissue include pigment cells (*chromatophores*), which give the skin its coloration pattern. The dermis also contains the receptor organs for the cutaneous senses of touch, heat, cold, pressure, and pain. Nervous adjustment of the amount of blood flowing through the dermis plays a major role in the regulation of body temperature. A localized permanent increase in the blood supply produces dermal modifications such as the combs and wattles of certain birds.

Most important, the dermis has a general tendency to produce calcareous deposits. For example, the skull plates of all bone-possessing vertebrates are *dermal bones.* Many ancient, now extinct fishes also had armor plates of dermal bone over much of the body, and the modern bony fishes still have overlapping bony *scales.* Such plates and

Figure 8.2 Epidermal derivatives. A, scales and claws on a bird foot. *B, C,* hair and its insertion in skin. The root structure of a feather is quite similar (see Fig. 30.11). *D,* section through a fingertip to show nail; the tissue arrangement is similar in a claw. *E,* section through a hoof. (*D, E,* after Boas.)

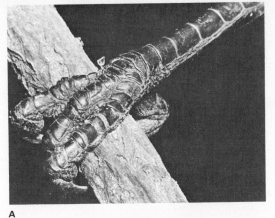

A

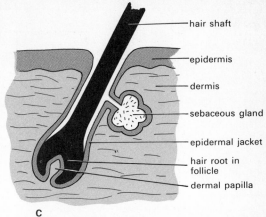

C

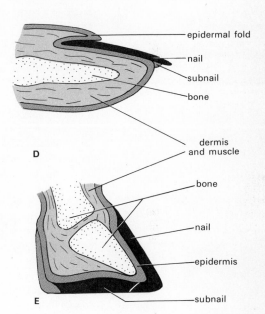

D

E

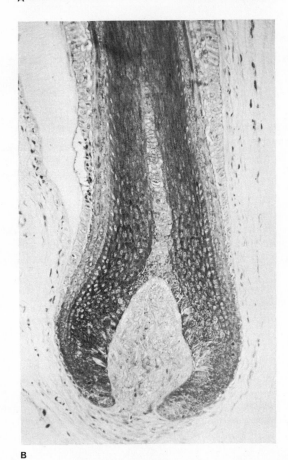

B

scales are overlain by the epidermis. Sharks and related fishes have, instead of scales, dermal *den-*

ticles, composed of an outer layer of *enamel,* an inner layer of *dentine,* and a core of fibroelastic *pulp.* Denticles are structured like the *teeth* of most vertebrates, and teeth are essentially larger dermal variants of denticles. Evidently, teeth as well as scales occur in two structural varieties, one variety epidermal, the other dermal (Fig. 8.3). In all animals, vertebrate or invertebrate, the epidermis regenerates readily. The dermis does not, however. A deep wound that penetrates the dermis usually fills with abnormally aligned dermal fibers, and epidermal cells grow over the region.

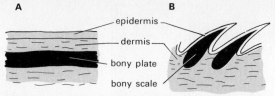

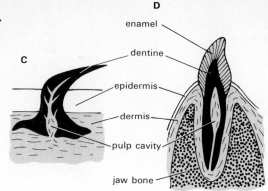

Figure 8.3 Dermal derivatives. A, dermal bony plate, as in echinoderms and vertebrate skull bones. *B*, dermal scale, as in bony fishes. *C*, placoid scale, or denticle, as in shark skin. *D*, mammalian tooth. Note that *C* and *D* are fundamentally alike in structure. The pulp cavities arise from dermal papillae (as in hair) and are filled with connective tissue (in which blood vessels and nerve fibers are present).

But because of the abnormal dermis the healed area differs from normal skin and remains recognizable permanently as a scar.

Skeletal Systems

Many animals, including comparatively large ones such as giant earthworms, do not have hard skeletons. Others are equipped with *exoskeletons* formed from epidermal secretions (see Fig. 2.11). In sessile animals like corals and in slow-moving forms like snails and clams, the exoskeletons are entirely or almost entirely rigid. Made of thick calcareous deposits, the heavy shells afford excellent support and protection in essentially stationary ways of life. As such an animal grows, a new, larger section of shell usually is secreted in continuity with older portions.

Rapid locomotion becomes possible only if an animal has a mobile, flexible skeleton and one that can be carried about with a minimum of effort. Both conditions are realized in the chitinous exoskeleton of arthropods. Chitin is much lighter than lime, and at specific regions the chitin shell is thin and pliable and forms *joints* that permit free body movement. In insects chitin also forms the wings. However, the all-enveloping inert shell would make growth in size impossible if arthropods did not *molt* periodically and enlarge in spurts at such times. At each molt the exoskeleton breaks open along the back and the soft, defenseless animal extricates itself from the old shell and enlarges rapidly during the next few days (see Color Fig. 28). A

new and larger exoskeleton is secreted by the epidermis, and when this new shell has fully hardened growth in size ceases again (although the body weight continues to increase).

An important limitation of all kinds of exoskeletons is that, with increasing size of the encased animal, deep-lying body parts obtain progressively less support. Animals with exoskeletons actually are comparatively small. Moreover, if the exoskeleton is calcareous its proportionate weight (if not also its external position) restricts motion; and if the shell is chitinous it is comparatively less sturdy, since calcium salts are the stronger material. Furthermore, either an external skeleton must have open regions to allow for body growth, in which case support and protection are limited; or, if the shell is all-enveloping, the animal must molt at least during its early development, in which case support and protection are periodically lacking altogether.

Such disadvantages are absent from *endoskeletons*, which support the body both near the surface and in deep parts. The animal thus can be large and still be buttressed adequately. Moreover, endoskeletons make possible continuous growth. Among several invertebrate groups with endoskeletal components, sponges have *spicules*, needle-shaped deposits formed in the interior of certain specialized cells (see Fig. 20.6). Squids and octopuses have mesoderm-derived *cartilage* supports around the brain and in several other body regions. Squids also have a horny or calcareous leaflike supporting structure, the *pen*, underneath the upper body wall (see Fig. 23.23). In

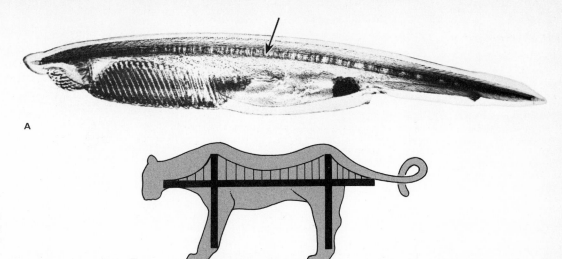

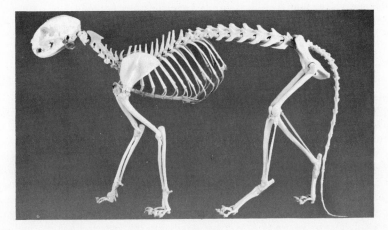

Figure 8.4 The chordate endo-skeleton. A. the notochord (arrow) of amphioxus, an early chordate. *B,* in advanced chordates such as mammals, the skeleton supports the body on the principle of a suspension bridge. See also Fig. 8.5.

Figure 8.5 Skeleton of a cat, illustrating the general skeletal organization of vertebrates. Note skull, axial skeleton (vertebral column and tail), and appendicular skeleton (limbs and limb girdles). The suspension-bridge skeleton of four-footed vertebrates is turned upright in man (see Fig. 2.11).

echinoderms a well-developed endoskeleton is present. It consists of many small calcareous plates formed in the dermis and set together like tiles. The skeleton thus seems to form an external shell, yet it is covered by the epidermis and therefore is a true endoskeleton (see Fig. 27.15).

Elaborate and unrestricting endoskeletons occur only in the chordates, vertebrates in particular. During the embryonic period all chordates develop a dorsal *notochord,* a stiff, elastic supporting rod along the midline of the body (Fig. 8.4). As pointed out earlier, the chordate phylum is named after this structure. In some chordates the notochord persists throughout life as the main or only skeletal support, but in most cases the rod is supplanted in late embryos by a *vertebral column.* Additional supports develop as well, and the adult then comes to have a complex mesodermal skeletal system.

This vertebrate skeleton can be considered to consist of two parts: an *axial* skeleton, comprising skull, vertebral column, and rib cage; and an *appendicular* skeleton, comprising the pectoral and pelvic girdles and the limb supports (Fig. 8.5). In lampreys, sharks, and related fishes, the whole skeleton is cartilaginous throughout life. In all other vertebrates the skeleton is very largely bony, bone developing in two ways. Most of the skull and part of the pectoral girdle arise as *dermal bones,* directly from the connective tissue of the dermis. All other skeletal parts develop as *replacement bones;* they are first laid down in cartilage, which later is replaced by bone substance.

In the case of dermal bones (Fig. 8.6), mesenchyme cells in the dermis first produce a tissue sheet, the *periosteum.* If a bony plate is being formed two parallel sheets usually develop, some distance apart. These bud off new cells into the

dermal connective tissue

ossifying fibers

developed periosteum

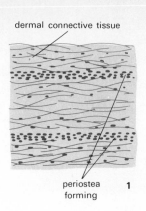
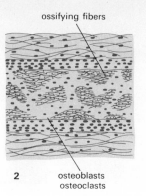
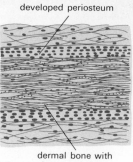

periostea forming   **1**

2   osteoblasts osteoclasts

3   dermal bone with large and small cell islands

Figure 8.6 Dermal bone for-mation. A periosteum (1) is a layer of connective tissue that produces bone-forming and bone-destroying cells [*osteoblasts* and *osteoclasts,* respectively, (2)], and that also covers a bone after it is formed (3).

space between them. Called *osteoblasts,* such cells secrete a meshwork of fibers around them-selves that becomes impregnated with hard min-eral deposits and *ossifies* in this manner. In the spaces left by these deposits are blood cells, fat cells, more mesenchyme cells—in short, the usual constituents of connective tissue. Some of the mesenchyme cells later give rise to *osteoclasts,* bone-destroying cells capable of resorbing the bone substance near them. Operating together, osteoblasts and osteoclasts can reshape and remold bone in response to changes of mechanical stress and the growth of the animal as a whole.

Replacement bone similarly arises from mesen-chyme cells in connective tissue (Fig. 8.7). Such cells first form a *perichondrium,* a layer that buds off cartilage-forming *chondroblasts* and cartilage-destroying *chondroclasts.* Through their activity, cartilage in the shape of a future long bone is produced. Later the perichondrium specializes as a periosteum, and from then on bone formation occurs essentially as in dermal bones; as chondro-clasts resorb cartilage, newly specialized osteo-blasts secrete bone substance in substitution. Car-tilage eventually remains only in two places: at the surfaces of both ends of a long bone, where carti-lage provides smooth, friction-reducing joint pads, and some distance behind each end, where a layer of cartilage permits continuing bone growth in length. In the mature animal these growth centers ultimately ossify as well and elongation of bone

then ceases.

In the outer region of a bone shaft, bone sub-stance is laid down as *compact* bone, in dense concentric patterns that form so-called *Haversian systems* (see Fig. 7.18). Each contains a central canal and concentrically placed islands filled with blood cells and other cellular components of bone. In the core and at the ends of a shaft, bone is *spongy,* with irregular channels filled with *marrow.* As a long bone matures, osteoclasts gradually resorb the spongy bone in the core of the shaft and a distinct *marrow cavity* appears.

Skull bones grow until they meet and then they fuse along the edges. Other bones form movable articulations, or *joints.* A joint is encapsulated by a strong ligament, and the interior of such a cap-sule is filled with a watery fluid that lubricates and facilitates movement.

In its overall construction the vertebrate skele-ton resembles a suspension bridge; the appen-dicular skeleton supplies the vertical supports and the axial skeleton forms the horizontal parts (see Fig. 8.4). Limb bones exhibit a common structural plan (though not always the same shape). The upper part of a limb is supported by one bone, the lower part by two. Wrist and ankle bones and the bones of the digits likewise have basically the same arrangements among vertebrates. Limbs as a whole connect to sockets formed by the appen-dicular girdles. In most vertebrates the pelvic girdle is fused directly to the vertebral column, whereas

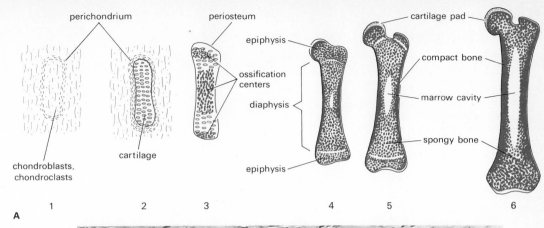

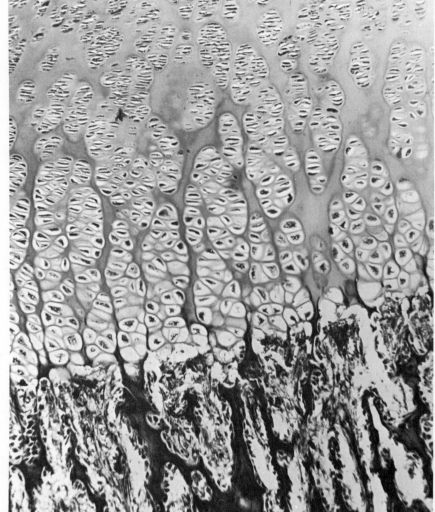

Figure 8.7 Replacement bone.
A, formation. A cartilage rod is first shaped by cartilage-forming and cartilage-destroying cells (*chondroblasts, chondroclasts*) that have been produced from a perichondrium (1, 2). The latter then becomes a periosteum, and osteoblasts produced by it begin secretion of bone substance at three ossification centers (3). Spongy bone replaces all cartilage except in regions near joints (white bands); compact bone begins to form at surface of shaft, and marrow cavity, in center of shaft (4, 5). In a mature bone (6), cartilage layers near joints have been replaced by bone, the marrow cavity is large, and spongy bone largely is replaced by compact bone (solid black). *B*, section through ossification center; cartilage (top) in process of resorption, bone (bottom) in process of formation. Bone deposits are dark. See also Fig. 7.18.

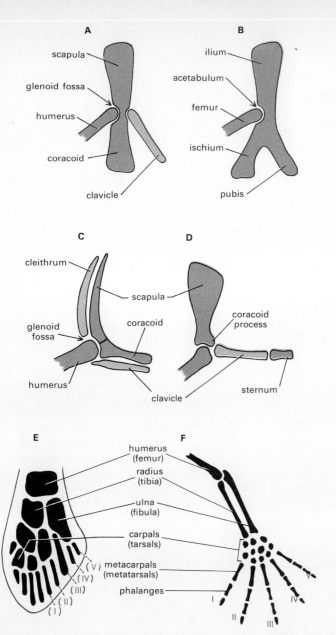

Figure 8.8 Limb girdles and limbs. *A* and *B,* the basic plan of the vertebrate pectoral and pelvic girdles, respectively; note the fundamental similarity of construction. The glenoid fossa is the socket for the humerus, corresponding to the acetabulum for the femur. *C,* primitive vertebrate pectoral girdle as in fishes, showing dermal bones (light color) and replacement bones (dark color). *D,* pectoral girdle as in mammals; note reduction of coracoid to a bony process and disappearance of cleithrum. *E,* fin of lobe-finned fish, indicating the basic ancestral bony elements of all vertebrate walking limbs. *F,* basic plan of bones in vertebrate limb; labels in brackets refer to names of bones in hind limb corresponding to those of forelimb. In *E,* roman numerals in parentheses refer to positions where the metacarpals and phalanges of *F* are added. In evolution from *E* to *F,* wrist and ankle bones (carpals and tarsals) have become reduced in number. (Partly after Woodruff, Holmgren.)

the pectoral girdle is held in place by ligaments and muscles (Fig. 8.8).

From head to tail, the vertebral column consists of five regions: *cervical* (neck), *thoracic* (chest), *lumbar* (lower back), *sacral* (hip), and *caudal* (tail). In each region the individual *vertebrae* are shaped in characteristic ways, but all vertebrae have a common basic structure (Fig. 8.9). The main portion is the *centrum,* and projecting from it are various bony outgrowths called *processes.* Vertebrae articulate with one another by means of *articular* processes, and this jointing leaves room for *intervertebral disks,* pads of cartilage between adjacent vertebrae. Dorsal processes from the centrum form a *neural arch,* a bony enclosure for the spinal cord. The trunk region of the vertebral column bears *ventral ribs* in bony fishes and *lateral ribs* in other bony vertebrates. Lateral ribs extend around to the ventral side of the body, where they articulate with a *sternum,* or breastbone, and form a rib cage.

In invertebrates the skeleton-forming materials are inert deposits that do not contain embedded cells. By contrast, vertebrate endoskeletons have living tissue directly inside the hard supporting substance. The skeleton therefore exhibits living qualities that entail major advantages: the skeleton can grow continuously along with the rest of the animal, and it can adjust to changing mechanical stresses and loads far better than a completely solidified, wholly passive support. Moreover, whereas a broken piece of clam shell, for example, must stay broken, a cracked piece of vertebrate bone can heal by renewed activity of the cells in the interior.

Muscular Systems

With the exception of protozoa, sponges, and some cnidarians, all animals have distinct mesoderm-derived muscle tissues. Two basic types are *smooth* muscle and *striated* muscle (Color Fig. 3). In the smooth variety the cells are elongated and spindle-shaped, and each contains a single nucleus. Contractile myofibrils are aligned longitudinally, and contraction shortens and thickens the cell. Striated muscle is made up of syncytial units. Each such unit, a *muscle fiber,* develops through repeated division of a single cell, and in this process the boundaries between daughter cells disap-

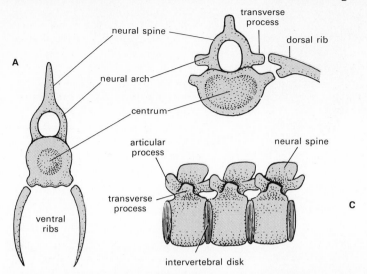

Figure 8.9 Vertebrae. A, trunk vertebra of bony fish, showing ventral ribs. *B,* generalized trunk vertebra of a land vertebrate, showing dorsal ribs. These curve to ventral side of body and articulate with sternum. *C,* side view of mammalian lumbar vertebrae; the spaces between the centra and the articular processes provide paths for the exit of spinal nerves. (Partly after Kingsley.)

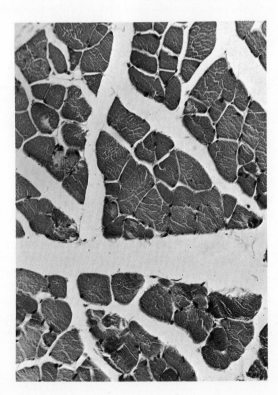

Figure 8.10 Striated muscle, transverse section. The broad white spaces mark the position of layers of connective tissue that envelop and separate individual muscle bundles. Within each bundle, narrower white spaces mark the connective tissue surrounding the individual muscle fibers. Black dots in the fibers are nuclei. See also Fig. 6.1 and Color Fig. 3.

pear. In this type of muscle the longitudinal myofibrils exhibit microscopic cross striations, hence the name (see Fig. 6.1).

Smooth muscle occurs in all animals that have a musculature, and in most groups smooth muscle is the only type present. Characteristically, the muscle tissue forms layers, with cells aligned longitudinally or circularly. Such layers are found, for example, in the body wall and in the walls of the alimentary tract and the larger blood vessels. Striated muscle occurs primarily where great locomotor speed is of advantage; it is encountered (in addition to smooth muscle) mainly in mollusks (particularly squids), in arthropods, and in vertebrates. The striated muscles here typically form a "skeletal" musculature connected to parts of the skeleton; the muscles are fastened to the inner surfaces of arthropod exoskeletons and to the outer surfaces of vertebrate endoskeletons.

A group of striated muscle fibers usually makes up a *muscle bundle* and is enveloped by a layer of connective tissue. Several such bundles form a *muscle,* an organ enclosed in a connective tissue sheath of its own. At either end this sheath gradually becomes a *tendon* and connects to some part of the skeleton (Fig. 8.10). In vertebrates striated muscles are under voluntary nervous control, smooth muscles are not. The hearts of vertebrates are composed of a specialized variant of striated muscle called *cardiac muscle.* The syncytial fibers are fused to each other in intricate patterns, and the whole vertebrate heart is a continuous multinucleate mass of contractile living matter. This muscle is not under voluntary control (see Color Fig. 3).

All striated muscle fibers innervated by a single nerve fiber form a *motor unit* (Fig. 8.11). Hundreds of such units can be present in a whole muscle. Each motor unit operates in *all-or-none* manner; either it contracts fully or not at all. Nervous stimulation of a motor unit is followed in succession by a brief *latent period,* a *contraction period,* and a *relaxation period.* Such a unit cycle of activity is a *simple twitch.* If a second stimulus arrives during the latent period of a twitch, this stimulus will be without effect. But if a second stimulus arrives just after the latent period, the motor unit will relax incompletely or not at all and begin a second twitch. If now a third stimulus again arrives during the contraction phase, a third contraction can follow without intervening relaxa-

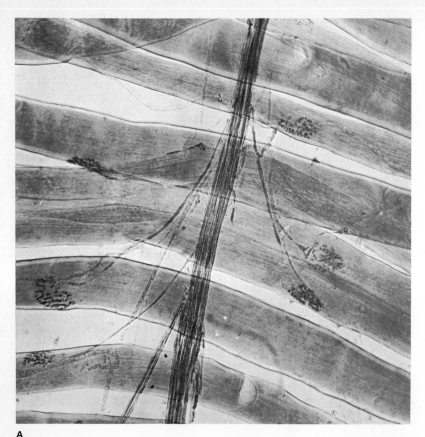

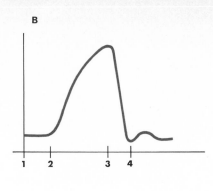

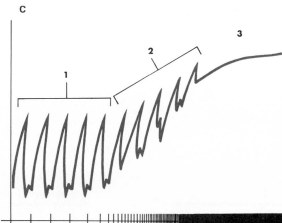

Figure 8.11 Muscle stimulation. A, motor end plates, the knobbed terminals of nerve branches on individual muscle fibers. The group of muscle fibers so innervated by a single nerve fiber and its terminal branches is a motor unit, a set of muscle fibers that functions together ($\times$ 2,000). *B,* muscle contraction in a simple twitch. 1, Stimulation applied. 2, Beginning of contraction (hence interval between 1 and 2 represents latent period). Interval 2–3, contraction period; interval 3–4, relaxation period. *C,* tetanic muscle contraction. 1, Well-spaced stimuli are applied (as marked along bottom line), resulting in separate simple twitches. 2, Summation occurs when frequency of stimuli is increased. 3, Sustained tetanic contraction, resulting from exceedingly high stimulus frequency.

tion. Through such *summation* of a rapid (but not too rapid) succession of stimuli, a sustained contraction can ensue. Known as *tetanus,* the sustained contraction lasts until the motor unit fatigues.

Most muscular activity is tetanic in nature. The many motor units in a whole striated muscle work in relays, different ones contracting tetanically or being at rest at any given moment. Also, a whole muscle can produce a graded response; its contraction will be the stronger the more motor units are active. Muscles actually are never relaxed completely but remain in a partially contracted state even during periods of rest. Little energy is then expended. Such mild contractions maintain

tonus, or muscle "tone." Through it the muscular system preserves posture and the shape of body parts, and provides mechanical support in general. Only stronger contractions, above and beyond tonic ones, result in outright movement of parts (and in pronounced energy expenditure).

Striated muscles operate far more rapidly than smooth muscles, and they can produce faster, more abruptly alterable, and more finely adjustable motions. But the smooth musculature requires comparatively less energy, and its slower, more sustained motions serve particularly in prolonged activities such as gut-wall contractions during digestion.

Whereas many muscles function solely in producing internal movements, the bulk of the muscular system also contributes to the *locomotion* of the whole animal. All animal locomotion is based on the lever principle; a part of the body acts as a more or less rigid lever, and as muscles exert pull against it at one point, another point of the lever exerts push against the environmental medium. The locomotor muscles are usually arranged in opposing sets, one set producing a *flexion* or *adduction* of a body part, the other set producing an *extension* or *abduction*. Through such activities animals propel themselves on solid surfaces by creeping, sliding, and walking; in water by paddling, lashing, and jetting; and in air by various forms of flying. In view of the importance of locomotion in animal life generally, it is not surprising that, in forms such as vertebrates, the musculature has become the largest organ system of the body.

Endocrine Systems

All animals probably contain chemical agents that, after being produced in one body part, have specific regulatory or coordinating effects in other parts. Such substances generally are referred to as *humoral* agents. For example, CO_2 qualifies as a simple humoral agent in mammals; among other effects it exerts a controlling function over breathing (see Chap. 9).

In certain instances humoral agents are distinct hormones, produced by specialized *endocrine* cells and discharged not through ducts but directly to the body fluids. For example, nerve cells are endocrine since, as shown in the next section, they secrete hormones at their ends. Indeed in some animals all endocrines present are *neurosecretory* cells: modified or unmodified nerve cells that secrete a variety of hormones having a variety of functions outside the nervous system. In other instances endocrine cells are not part of the nervous system and often are components of elaborate endocrine organs. Several such organs and any neurosecretory cells present constitute an endocrine system.

Hormones vary greatly in chemical composition. Some are proteins, a few are amino acids, and the rest are various other simple or complex compounds. A few can be synthesized in the laboratory, a few have known chemical structure, and

the remainder are known only through the abnormal effects produced by hormone deficiency or hormone excess.

To date, the presence of endocrine cells has been demonstrated more or less definitely only in animals such as nemertine worms, certain segmented worms, mollusks, most arthropods, and chordates, including tunicates and all vertebrates. Cells of the neurosecretory type occur in all these groups, but nonnervous endocrine cells are conspicuous only in arthropods and vertebrates. The known functions of the endocrine secretions differ greatly among the groups named. In some worms the hormones play a role in growth and regeneration. In squids and octopuses neurosecretory hormones appear to control mainly the expansion and contraction of chromatophores in the skin. By such means the animal can change its coloration to blend with environmental backgrounds or in response to external stimuli (for example, color changes in an "excited" octopus).

Regulation of the activities of pigment cells also is one function of arthropod hormones. Among others are control of breeding and mating behavior and, as discussed in Chap. 26, regulation of molting during development. Apart from such specific direct effects on given tissues and processes, the endocrines of arthropods also affect one another; hormones synthesized by particular glands stimulate or inhibit the hormone production of other glands. It is through interactions of this kind that the glands become an integrated endocrine system, even though they are unconnected structurally.

Vertebrate endocrine systems are by far the most complex. In vertebrates the number of glands is much greater than in other animals, and indeed many of these glands are anatomic composites of two or more distinct endocrine organs, each producing its own hormones (Fig. 8.12). In a vertebrate hardly any function occurs that is not influenced at least in part by hormones. Endocrine control usually operates in conjunction with nervous control, and in many instances the nervous system supplies information about the external environment while the endocrine system regulates the internal response to this information.

Most vertebrate hormones are required in most or all cells of the body. Accordingly, labels such as "sex hormones," for example, are somewhat misleading. To be sure, sex hormones are manu-

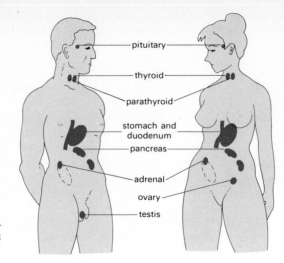

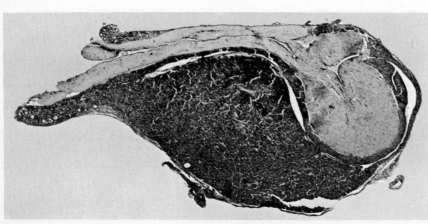

A

Figure 8.12 Vertebrate endocrine organs. See also Fig. 8.13 and Color Fig. 4.

Figure 8.13 The pituitary. A, section through a pituitary gland. The left side of the photo points in the direction of the face. Note the anterior lobe in the left part of the gland and the intermediate and posterior lobes in the right part. The posterior lobe continues dorsally as a stalk that joins the whole gland to the brain. *B,* the stimulative effect of thyrotropic hormone on the thyroid gland, and the inhibitive effect of thyroid hormone on the hypothalamus and the pituitary. Through such control cycles, the output of tropic pituitary hormones is automatically self-adjusting.

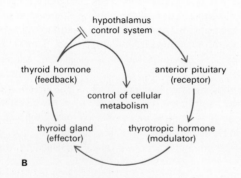

B

factured in sex organs and the hormones contribute to the proper functioning of these organs. However, sex hormones also contribute to the functioning of virtually every other organ of a vertebrate. It happens that the effect of deficiency or excess of a given hormone often reveals itself first or most obviously in a particular body part. For convenience the hormone then can be named according to this body part, but it cannot be concluded that the hormone functions only there.

Apart from their other controlling roles in cells, some vertebrate hormones perform an additional special function: like some arthropod hormones, they control each other. Of particular importance in this respect are the *tropic* hormones of the pituitary gland. These stimulate the endocrine activity of the thyroid glands, the adrenal glands, and the sex organs. In turn, the hormones of the latter inhibit the activity of the pituitary (Fig. 8.13). Through such automatically self-adjusting "feedback" controls, the hormone concentrations are maintained at relatively steady levels. For example, if the amount of thyroid hormone rises unduly, the hormone inhibits the pituitary and the output of thyrotropic hormone then decreases. As a result the thyroid gland now is stimulated less, and reduced secretion of thyroid hormone will follow (Color Fig. 4).

The functions of the vertebrate endocrine glands are summarized in Table 6. Specific roles of most of these glands are discussed in later chapters, in the context of particular body activities.

Nervous Systems

Like all other living control operations, nervous activity is based on five kinds of structures: *receptors,* or stimulus receivers; *sensory pathways,* or transmitters of incoming signals; *modulators,* or signal interpreters and selectors; *motor pathways,* or transmitters of outgoing signals; and *effectors,* or response executors (Fig. 8.14).

The neural receptors are specialized *sensory cells* that in many cases are parts of sense organs. Sensitive to environmental or internal stimuli, receptors initiate nerve impulses that convey information related to the stimuli. These impulses are transmitted over *sensory nerve fibers* to modulators such as ganglia, nerve cords, and brains. Modulators "interpret" sensory impulses, often store

Table 6. **The main vertebrate endocrine glands and their hormones**

GLAND	HORMONES	CHIEF FUNCTIONS	EFFECTS OF DEFICIENCY OR EXCESS
pituitary, anterior lobe	TSH (thyrotropic) ACTH (adrenocorticotropic) FSH (follicle-stimulating) LH (luteinizing) prolactin (lactogenic) growth	stimulates thyroid stimulates adrenal cortex stimulates ovary (follicle) stimulates testes in male, corpus luteum in female stimulates milk secretion, parental behavior promotes cell metabolism	dwarfism; gigantism
pituitary, midlobe	intermedin	controls adjustable skin-pigment cells (for example, in frogs)	
pituitary, posterior	at least five distinct fractions	controls water metabolism, blood pressure, kidney function, smooth-muscle action	increased or reduced water excretion
thyroid	thyroxin calcitonin	stimulates respiration; inhibits TSH secretion lowers blood Ca	goiter; cretinism; myxedema
parathyroid	parathormone	controls Ca metabolism raises blood Ca	nerve, muscle abnormalities; bone thickening or weakening
adrenal cortex	cortisone, other steroid hormones	controls metabolism of water, minerals, carbohydrates; controls kidney function; inhibits ACTH secretion; duplicates sex-hormone functions	Addison's disease
adrenal medulla	adrenaline, noradrenaline	alarm reaction, for example, raises blood pressure, heart rate, blood-sugar level	inability to cope with stress
pancreas	insulin glucagon	glucose $\longrightarrow$ glycogen conversion glycogen $\longrightarrow$ glucose conversion	diabetes
stomach	gastrin	stimulates gastric juice secretion	
duodenum	secretin	stimulates bile and pancreatic juice secretion	
testis	testosterone, other androgens	promote cell respiration, blood circulation; maintain primary and secondary sex characteristics, sex urge; inhibit FSH secretions	atrophy of reproductive system; decline of secondary sex characteristics
ovary: follicle	estradiol, other estrogens		
ovary: corpus luteum	progesterone	promotes secretions of oviduct, uterus growth in pregnancy; inhibits LH secretions	abortion during pregnancy
thymus gland		aids antibody production by lymphocytes in young mammals	
pineal body		aids control of pigment dispersion in amphibian chromatophores contributes to light responses in circadian rhythms of rats	

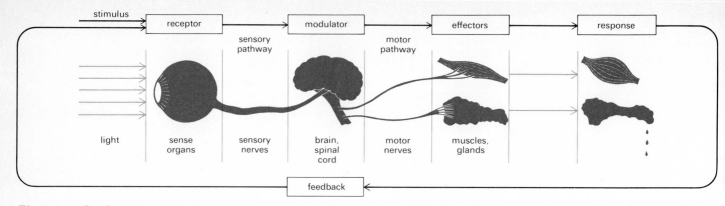

Figure 8.14 Steady-state control: reflex action. The components of any kind of steady-state-maintaining device are indicated along the top row; those of a nervous reflex, immediately underneath. A stimulus (such as light) will produce a response such as muscle contraction and / or glandular secretion. The response itself then becomes a feedback, or new stimulus, through which the modulator is "informed" whether or not the response has produced an adequate reaction to the original stimulus. The feedback stimulus then may or may not initiate a new reflex.

Figure 8.15 Nerve tissue. *A*, general structure of a neuron with a myelinated axon. *B*, micrograph of a stained (multipolar) motor neuron. *C*, nonmyelinated and myelinated fibers. Left, cross-sectional view of a fiber without myelin, showing a Schwann cell surrounding an axon almost completely. Right, cross-sectional view of a developing myelinated fiber. A myelin sheath grows as an extension of the Schwann cells, each of which wraps around the axon several times (more often than shown) and forms an envelope of many layers. The inner layers collectively represent the myelin sheath. *D*, high-power view of a single neuroglia cell. Note the elongated extensions from the main part of the cell. Such cells provide structural and metabolic support for neurons and represent the main nonneural components of nervous systems (×1,500).

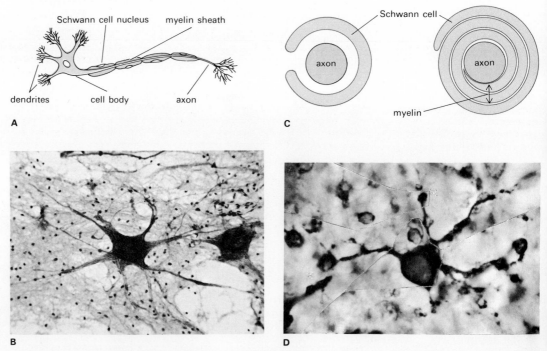

incoming information as memory, and send out motor impulses that usually are associated in some way with the incoming information. Such impulses pass through *motor nerve fibers* to effectors, which are either *muscles* or *glands*. These then carry out responses that normally bear some more or less direct relation to the original stimuli and that in the long run tend to preserve steady states. Many of these responses contribute to recognizable forms of *behavior*.

Neural Pathways

The main cell type in a nervous system is the nerve cell, or *neuron*. Each typically consists of a nucleus-containing *cell body* and one or more filamentous fibers, outgrowths that extend away from the cell body (Fig. 8.15). Impulses usually originate at the terminals of fibers called *dendrites*, which then carry the impulses toward the cell body. Impulses travel away from a cell body through fibers called *axons*. Many dendrites and axons are short, but others have lengths of over a yard (for example, axons from the base of the spinal cord to the toes in man).

In vertebrates the long axons and dendrites are enveloped by a *Schwann sheath*, a single-layered tube of thin flat cells that supplies nutrients to a fiber and that also provides a pathway when a cut fiber regenerates. In certain cases the cells of a Schwann sheath wind around the nerve fiber several times, and the fatty contents of these wrap-

ping layers then form a *myelin sheath*. A myelin wrapping is believed to serve as a kind of insulation, comparable perhaps to an insulating rubber envelope around an electric wire (see Fig. 8.15).

The fibers of two adjacent neurons are not in direct contact; their terminals come close together at a *synapse*, a microscopic space through which impulses are carried by chemical means (Fig. 8.16). In crude terms a whole nervous system can be envisaged as an intricate network of neurons, with fibers interconnecting functionally at numerous synapses. Many pathways in such a network form *reflex arcs*, composed of a sequence of neurons with specific functions. Thus, *sensory* (or *afferent*) neurons in such an arc transmit impulses from a receptor to a modulator, and *motor* (or *efferent*) neurons transmit from a modulator to an effector. Neurons inside a modulator are *interneurons*. Groups of nerve fibers frequently traverse a body region as a single collective fiber bundle, or *nerve*. Nerves are designated as sensory, motor, or mixed, depending on whether they contain sensory fibers, motor fibers, or both.

The most primitive type of neuron arrangement known is a *nerve net* (Fig. 8.17). Representing the only neural structures in, for example, cnidarians, such nets also form at least part of the nervous system of most other animals. In vertebrates, for example, nerve nets occur in the walls of the alimentary tracts. Most animals in addition contain *nerve cords*, condensed regions of nets, and *ganglia*, dense accumulations of neurons and fiber terminals. Ganglia can be sensory, motor, or mixed according to the kinds of fibers they connect with. Large ganglia usually contain functional subdivisions called *nerve centers*, specialized groups of interneurons that regulate specific activities. Very large ganglia usually also store information as memory and control intricate forms of behavior. Large ganglia or groups of ganglia that integrate the main sensory inputs and motor outputs of animals constitute *brains*.

In some animals the main part of the nervous system often is only a single ganglion that forms the hub of an array of sensory and motor pathways radiating to and from it. If several ganglia are present they usually are interconnected by distinct nerves or nerve cords. Among active, motile invertebrates, the main part of the nervous system frequently has the form of a ladder. Such a *ladder-type system* typically consists of two parallel, cen-

Figure 8.16 Neural pathways. Afferent fibers (color) conduct impulses from receptors to modulators, where interneurons transmit the impulses to efferent fibers. The latter send impulses to effectors. Neurons interconnect functionally across synapses. Collected bundles of neuron fibers form nerves. The photo at right depicts a cross section through such a nerve. Note the many individual nerve fibers.

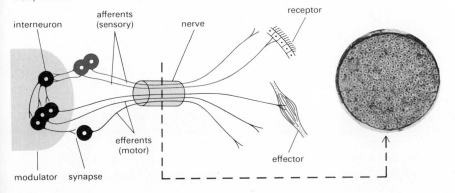

interneuron

afferents
(sensory)

nerve

receptor

efferents
(motor)

effector

modulator   synapse

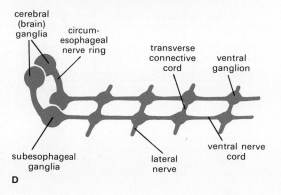

A   B

C

Figure 8.17 Nerve nets, cords, and ganglia. A, portion of a nerve cord, represented as a dense concentration of parts of a nerve net. *B*, schematic representation of a ganglion; a mixed ganglion is indicated here, with sensory and motor neurons as well as interneurons. *C*, section through a spinal (sensory) ganglion of a mammal; note the many cell bodies and also the nerve fibers, some seen in cross section, some in longitudinal section. *D*, schematic representation of a ladder-type nervous system; most such systems have the components shown, though not necessarily in exactly the indicated arrangement.

cerebral
(brain)
ganglia
    circum-
    esophageal
    nerve ring

transverse
connective
cord

ventral
ganglion

subesophageal
ganglia

lateral
nerve

ventral nerve
cord

D

trally located nerve cords, with ganglia along their course and transverse connecting cords between them. At the head end are several large ganglia, usually arranged as a ring around the alimentary tract. The dorsal parts of this ring are the *brain ganglia* (see Fig. 8.17).

By far the most complex nervous systems are those of vertebrates. The main parts develop from a hollow dorsal *neural tube,* formed in the embryo as an ingrowth from the ectoderm (see Chap. 11). The anterior portion of the tube enlarges as a

brain, the posterior portion becomes the *spinal cord,* and nerves grow out from both. The fluid-filled space in the tube forms *brain ventricles* anteriorly and a *spinal canal* posteriorly. In the mature brain the major divisions are the *forebrain,* the *midbrain,* and the *hindbrain* (Fig. 8.18).

Of two main subdivisions of the forebrain, the more anterior one is the *telencephalon.* It contains paired *olfactory lobes,* the centers for the sense of smell. In birds and mammals the telencephalon is greatly enlarged as a pair of *cerebral hemispheres* that cover virtually all the rest of the brain. These hemispheres contain the centers for the most complex sensory integration and for voluntary motor activities, and they also play the key roles in control of memory and intelligence. A conspicuous set of nerve tracts, the *corpus callosum,* interconnects the two hemispheres.

Behind the telencephalon, the (unpaired) *diencephalon* contains the *thalamus* and *hypothalamus.* These lateral regions control numerous involuntary activities, and they also affect consciousness, sleep, food intake, and emotional states. Ventrally the *pituitary* gland and dorsally the *pineal* body project from the diencephalon. The pineal body forms a third eye on the top of the head in lampreys and the reptile *Sphenodon* but is hidden under the cerebral hemispheres in birds and mammals.

The midbrain, or *mesencephalon,* contains dorsally located *optic lobes.* In all vertebrates except the mammals these lobes contain the centers of vision, and the nerve tracts from the eyes terminate there. In mammals the optic nerve tracts continue to visual centers located posteriorly in the cerebral hemispheres. The original optic lobes here are little more than relay stations for visual nerve impulses.

The hindbrain consists of two subdivisions, an anterior *metencephalon* and a posterior *myelencephalon,* or *medulla oblongata.* Dorsally the metencephalon includes the *cerebellum,* a comparatively large lobe that coordinates muscle contractions as smoothly integrated movements. For example, locomotion and balancing activities are regulated from this lobe. Ventrally the metencephalon contains a conspicuous bulge, the *pons,* in which the nerve tracts between brain and spinal cord cross from the left side to the right side. Because of this crossover the left side of the brain controls activities on the right side of the body and vice versa.

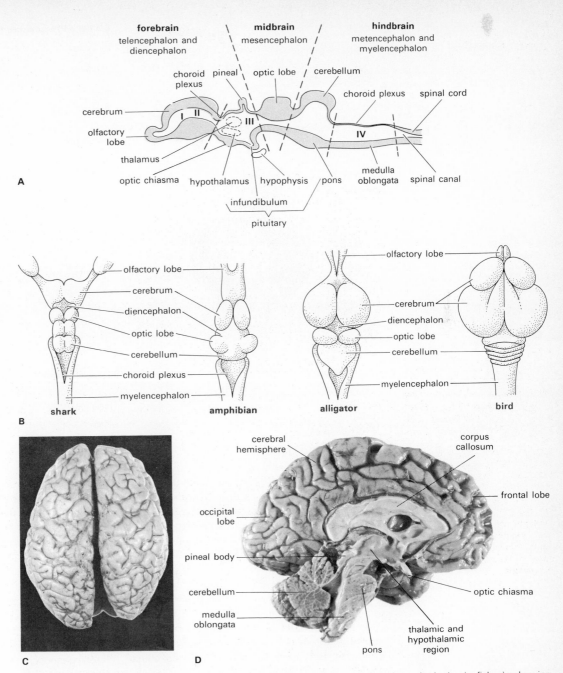

Figure 8.18 The vertebrate brain. A, median section through primitive vertebrate brain (as in fishes), showing basic structural plan. Roman numerals refer to ventricles. Thalamus and hypothalamus regions are indicated in broken lines, since these brain parts lie on each side of the median plane (as do ventricles I and II). The hypophysis forms as an outpouching from the roof of the mouth cavity; it becomes the anterior lobe of the pituitary, the infundibulum becoming the posterior lobe. The two choroid plexuses are membranous regions in which blood vessels are carried. B, dorsal views of various vertebrate brains. Note the progressive enlargement of the cerebrum in a posterior direction (and the parallel reduction of the olfactory lobes). C, dorsal view of human cerebrum. The left cerebral hemisphere is slightly larger than the right, a usual condition in right-handed persons. D, median section through human brain. The cerebrum has become so large that, in dorsal view, it covers the cerebellum posteriorly.

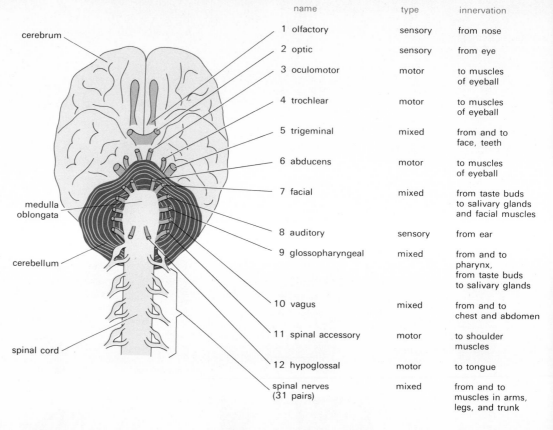

name	type	innervation
1 olfactory	sensory	from nose
2 optic	sensory	from eye
3 oculomotor	motor	to muscles of eyeball
4 trochlear	motor	to muscles of eyeball
5 trigeminal	mixed	from and to face, teeth
6 abducens	motor	to muscles of eyeball
7 facial	mixed	from taste buds to salivary glands and facial muscles
8 auditory	sensory	from ear
9 glossopharyngeal	mixed	from and to pharynx, from taste buds to salivary glands
10 vagus	mixed	from and to chest and abdomen
11 spinal accessory	motor	to shoulder muscles
12 hypoglossal	motor	to tongue
spinal nerves (31 pairs)	mixed	from and to muscles in arms, legs, and trunk

Figure 8.19 The main nerves.
Left, underside of brain and part of spinal cord of man, showing the origin of cranial and spinal nerves. *Right,* the names and functions of these nerves. In the spinal nerves, the dorsal roots (with ganglia) are sensory, the ventral roots, motor.

labels on figure: cerebrum, medulla oblongata, cerebellum, spinal cord

The medulla oblongata, which contains the nerve centers for control of heartbeat and breathing, continues posteriorly as the spinal cord. Twelve pairs of *cranial nerves* (only ten pairs in fishes and amphibia) emerge from the brain, and most of them lead away from the medulla oblongata (Fig. 8.19). The spinal cord gives rise to segmental *spinal nerves* (31 pairs in mammals), which pass to the trunk and the appendages. In each spinal nerve, sensory fibers from the body enter the dorsal part of the spinal cord and motor fibers leave from the ventral part. The cell bodies of the sensory fibers lie just outside the spinal cord, in *spinal ganglia* (Fig. 8.20).

Vertebrates have a well-developed *autonomic* subdivision of the nervous system (ANS, distinct from the remainder, called the *central* subdivision, or CNS). Controlling all involuntary activities and containing only nonmyelinated nerve fibers, this autonomic system has its nerve centers in the spinal cord and the brain and in a series of small peripheral ganglia. To the centers lead sensory nerve fibers from all body parts that are not under voluntary control. And from the centers lead away *two* functionally different sets of *autonomic motor fibers* (Fig. 8.21). Fibers from the brain and the most posterior part of the spinal cord represent a *parasympathetic outflow* of the autonomic system; and fibers from the midportion of the spinal cord form a *sympathetic outflow.*

Each organ of the body that is not under voluntary control receives fibers from both outflows, and these generally have opposing effects. If parasympathetic fibers inhibit a particular organ, sympathetic fibers stimulate it; or vice versa. For example, the inhibitory fibers to the heart (which travel in the vagus nerves, the 10th cranials) are part of the parasympathetic outflow, and the accelerating fibers to the heart belong to the sympathetic outflow (see Fig. 9.5). All other organs that

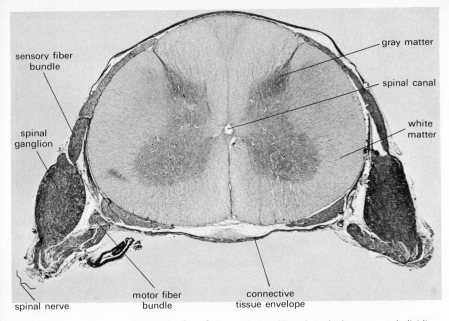

Figure 8.20 *The mammalian spinal cord:* cross section. Note the spinal nerves, each dividing into two fiber bundles. The motor bundle connects with the cord ventrally, and the sensory bundle passes through a spinal ganglion and connects with the cord dorsally. The spinal cord itself is a dense meshwork of neurons, the cell bodies of which are aggregated around the center and form so-called *gray matter.* The axons and dendrites of these neurons form white matter around the gray matter. The central spinal canal contains lymphlike spinal fluid.

function involuntarily likewise are equipped with both braking and accelerating controls.

Each autonomic motor path to a given organ consists of at least two consecutive neurons that synapse in an *autonomic ganglion* located somewhere along the path to the organ. The nerve fiber leading to this ganglion is said to be *preganglionic,* and the fiber from the ganglion to the organ is *postganglionic.* In the sympathetic outflow the autonomic ganglia lie just outside the spinal cord, and on each side of the cord they are interconnected as an *autonomic chain* (Fig. 8.22). These chain ganglia also have interconnections with the spinal ganglia. Parasympathetic ganglia are more dispersed and are not arranged as chains.

The involuntary operations of the ANS are interrelated closely with the voluntary ones of the CNS. For example, within certain limits a man can alter his breathing rate voluntarily through CNS control, even though this rate is basically under involuntary ANS control. Conversely, numerous autonomic changes affect voluntary behavior—for example, ANS-controlled hiccuping limits CNS-regulated speaking.

Figure 8.21 *Autonomic nervous system:* some of the motor pathways. In spinal cord (center), parasympathetic centers are shown in light color, sympathetic centers are darker. The column to the left of spinal cord represents the sympathetic ganglion chain on that side. Each neural path shown occurs pairwise, one on each side of the body. For simplicity only one side is indicated in each case, but every organ is innervated by both sympathetic and parasympathetic nerves.

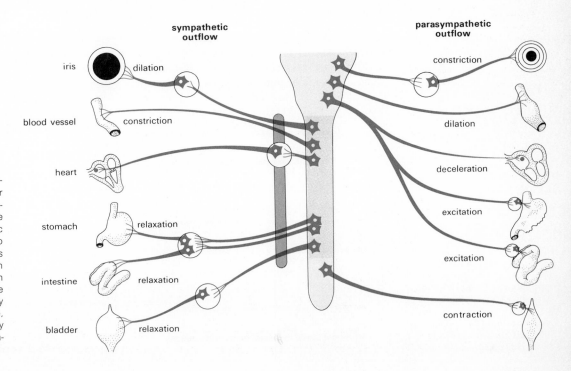

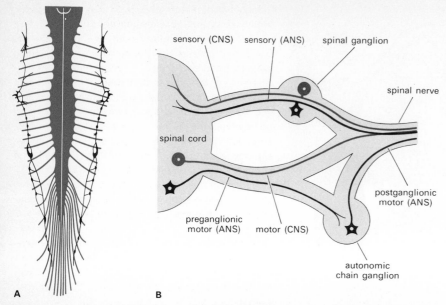

sensory (CNS) sensory (ANS) spinal ganglion

spinal nerve

spinal cord

postganglionic
motor (ANS)

preganglionic
motor (ANS) motor (CNS)

autonomic
chain ganglion

A B

Figure 8.22 Autonomic chains. A, spinal cord and nerves in color, autonomic chains black. *B*, interconnections of autonomic chain ganglia and spinal ganglia. CNS neurons in color, ANS neurons black. The cell bodies of all sensory ANS neurons lie in the 31 pairs of spinal ganglia that also contain the cell bodies of sensory CNS neurons. Sensory fibers of all kinds enter the spinal cord dorsally; motor fibers of all kinds leave ventrally.

Neural Impulses

A nerve impulse consists of a sequence of electrochemical reactions; during the passage of an impulse a wave of electric *depolarization* sweeps along a nerve fiber. After an impulse has passed,

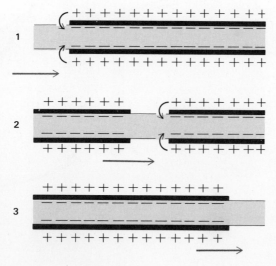

1

2

3

Figure 8.23 The nerve impulse. Passage of an impulse through a nerve fiber involves local depolarization of the fiber membrane, an effect that is propagated like a wave through successive portions of the fiber. After an impulse has passed a given region, the original polarization is reestablished.

the reaction balance returns to the original state, readying the fiber for a new impulse.

A resting, nonstimulated neuron is electrically positive along the outside of its surface membrane and electrically negative along the inside. These electric charges are carried by ions that are part of, or are attached to, the two sides of the cell membrane of the neuron. This membrane is so constructed that in the rest state it prevents the positive and negative ions from coming together. As a result an *electric potential* is maintained through the cell membrane; the membrane is said to be *polarized* electrically. When an impulse sweeps along a nerve fiber, the permeability of the membrane changes at successive points along the fiber. As this happens at any one point, an avenue is created through which the positive and negative ions of an adjacent point can pass, thus depolarizing that region. In other words, one depolarized region drains electric charges from adjacent parts of the membrane, causing depolarization there, too (Fig. 8.23). In this manner the impulse itself produces the necessary conditions that allow it to advance farther, and it travels wavelike along a fiber. Some short time after an impulse has passed a given point, the membrane at that point reacquires both its original permeability state and its polarization.

If fine wires and electric measuring equipment are connected to a nerve, the passage of an impulse is recorded in such equipment by a flow of current of certain characteristics. By studying these action currents, or *action potentials*, of different nerves, it has been found that impulses differ in speed, strength, and frequency. In some nerve fibers, like those from the heart-rate center to the heart, impulses are fired continuously in rapid succession. Adjustment of heart rate occurs through frequency modulation: heart rate changes with alterations in impulse frequency. In motor fibers to many glands, by contrast, the fibers normally are at rest and carry impulses only when secretions are to be produced. Each type of fiber has its own characteristic pattern of impulse transmission, and it has been found also that impulse speeds tend to be directly proportional to the thickness of a nerve fiber. Moreover, speeds are influenced by the presence or absence of myelin sheaths. Myelinated CNS fibers conduct impulses at speeds up to 100 yd per sec, whereas nonmyelinated ANS fibers of comparable thickness conduct at about 25 yd per sec at most.

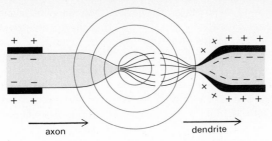

axon →   dendrite →

Figure 8.24 Chemical impulse transmission. In a neural synapse, hormonal transmitter substances are released and spread locally (colored rings) from the axon terminal of one fiber to the dendrite terminal of another. Impulses are transmitted across synapses by such chemical means.

How does an impulse get across a synapse? In certain cases it can be shown that when an impulse reaches an axon terminal, the terminal acts as an endocrine structure and secretes minute amounts of a hormonal transmitter substance. This hormone diffuses through the synapse and some reaches dendrite terminals of adjacent neurons. There the hormone can depolarize the dendrites in such a way that new impulses are initiated in them (Fig. 8.24).

Four hormonal substances that function in this manner have been identified in vertebrates: *serotonin, acetylcholine, adrenaline* (*epinephrine*), and *noradrenaline* (*norepinephrine*). One or the other of the last two is secreted by sympathetic postganglionic fibers, and probably also by at least some of the interneurons in brain and spinal cord (for example, those in the ANS centers of the hypothalamus). All such neurons are said to be *adrenergic.* Serotonin or acetylcholine is produced by sympathetic preganglionic fibers, all fibers of the parasympathetic system, and probably also by CNS fibers and CNS centers in brain and spinal cord. Such neurons are *cholinergic.*

The secretion pattern of these hormonal substances in the brain is not yet known very precisely, largely because of the so-called *blood-brain barrier,* a selective metabolic block between blood capillaries and nerve tissue. This barrier lets only oxygen and basic nutrients such as glucose and amino acids pass into the brain, and it lets only waste products leave. The hormonal transmitter substances normally cannot diffuse in or out, hence investigation of their activity in the brain has proved difficult. Even so, through research done in conjunction with tranquilizers and drugs such as LSD, it has been found that the hormonal trans-

mitters play a key role in regulating moods and emotional states, both normally and in mental illness. Thus, elevated moods and feelings of well-being are associated directly with high levels of noradrenaline and serotonin in the brain, whereas depressed states are associated with low concentrations of these substances. Some findings further suggest that normal and certain abnormal changes in mental state might result from internally generated alterations in the chemical reaction patterns through which the transmitter substances are manufactured in the brain.

Synaptic impulse transmission by chemicals also has important consequences outside the brain. For example, nerve fibers as such rarely fatigue, but their transmitter-secreting terminals "tire" fairly easily. Moreover, since only axon terminals secrete hormones and only dendrite terminals are sensitive to these substances, impulse conduction becomes unidirectional.

Neural Centers

In general, the activity of all types of neural modulator is based on two kinds of information: genetically inherited information and newly acquired information obtained through the sensory system.

Certain genetically determined neural pathways and patterns of neural activity are already established once an animal has completed its embryonic development. Even in a just-formed nervous system, therefore, relatively simple sensory inputs to the modulators often can evoke fairly complex outputs to the effectors, and behavior can be correspondingly complex. In many cases, indeed, sensory inputs are not required at all: neural centers often are spontaneously active. This is particularly true for those that control the most vital rhythmic processes of an animal. For example, the breathing center in the medulla oblongata sends out rhythmic motor impulses spontaneously (see Chap. 9). This rhythm can be shown to persist even if the center is isolated surgically. Similarly spontaneous activity takes place in the heart-rate center and in many—possibly most—other neural control centers as well.

In effect, the genetic endowment ensures that, as soon as an animal is completely developed, it has fully functional neural controls for at least those motor activities that are basically necessary for survival. To the extent that these activities are spontaneous, neural operations evidently do not involve complete reflexes; receptors and sensory paths do not participate directly.

To a greater or lesser degree, however, inherited neural activity usually is modified by sensory experience. Information about the current status of the external and internal environment is acquired by receptors and is transmitted to the neural centers in the form of more or less complex sensory impulses. In such instances nervous activity is based on complete reflexes. But here the modulating centers do not merely relay incoming sensory impulses to outgoing motor parts; even the simplest modulators usually are capable of *reflex modification*. The nature of this modification depends in some cases on inherited factors, in others on information acquired through past experience and stored as memory.

Among the simplest and most basic forms of modification are *suppression* and *augmentation* of reflexes. Suppression occurs when certain neurons inhibit others. Some synapses are so organized that, when impulses arrive in incoming nerve fibers, the production of impulses in outgoing fibers becomes harder rather than easier. The opposite effect, *augmentation,* can be brought about by a *summation* of impulses. In a synapse receiving many incoming impulses, each impulse individually might be too weak to produce an impulse in an outgoing fiber. However, the small quantities of chemical transmitters produced by the many incoming fibers can add together and become sufficiently powerful to initiate impulses in an outgoing fiber (Fig. 8.25).

A more complex form of modulator activity, based largely on summation and inhibition, is *channel selection*. Even a simply organized modulator selects among many possible outgoing fibers and sends out impulses only over certain specifically chosen paths. Normally only *appropriate*

effectors then will receive motor commands. As a result, the effector response of an animal can be adaptively useful and actually contribute to steady-state maintenance. Little is known as yet about the mechanism by which certain neural channels are selected in preference to others. In most cases preferred circuits become established during the embryonic development of the nervous system. Thereafter, given sets of sensory impulses to a modulator result in more or less fixed, predictable sets of motor impulses to effectors. Such neural activities are among those that have a largely genetic basis, and they govern most of the internal operations of most animals.

Preferred channels include, for example, sets of interneurons that are arranged as *oscillator circuits,* in which an impulse travels continuously over a circular route. Each time such an impulse passes a given synapse, a motor impulse to an effector might be initiated. The rhythmic heartbeat of an insect is controlled by an oscillator circuit of this type, in which nine circularly arranged neurons are embedded directly in the heart. Many other rhythmic, automatic activities are known to be governed by oscillator circuits.

A modulator often reacts to an incoming sensory impulse by sending out not just one but several selected motor commands. Here a simple external stimulus can lead to the completion of several or many simultaneous responses, all occurring as a single, integrated pattern of activity, or *programmed behavior*. A good example is the startle response in man. An unexpected blow directed at the head leads to a closing of the eyes, a lowering of the head and assumption of a crouching stance, and a raising of hands to the face. These several dozen separate reflexes occur simultaneously, as a unified "program." Most programs of this type are largely inherited, and in most animals, invertebrates in particular, behavior is very largely programmatic in this sense. Moreover, the execution of such programs involves most of the neurons present in the nerve centers. In comparatively large-brained animals, the majority of neurons again probably form fixed inborn circuits, yet substantial numbers seem to remain available for the later development of new or modified circuits and new or modified programs. Such animals, notably (but not only) cephalopod mollusks, arthropods, and vertebrates, are capable of *learning* and of storing learned experiences as memory.

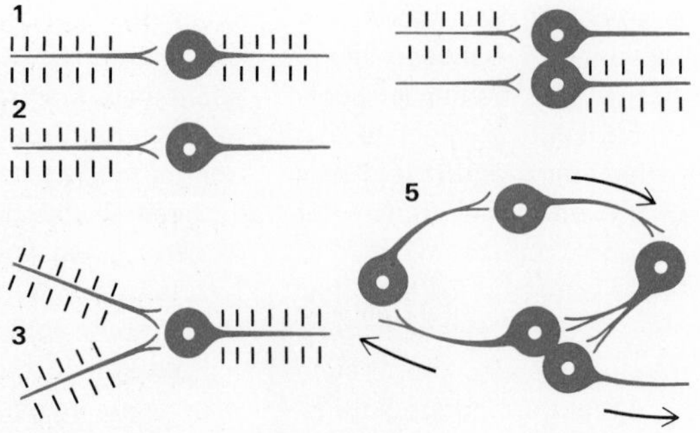

Figure 8.25 Modulator activities. Colored lines symbolize passage of impulses. 1, Simple reflex relay. 2, Suppression of impulse conduction. 3, Augmentation of outgoing impulse by summation of weak incoming impulses. 4, Channel selection. 5, In certain cases information is believed to be stored by continued impulse conduction in an oscillator (cyclic) circuit.

The simplest form of learning probably is *habituation,* or progressive loss of responsiveness to repeated stimulation. For example, most animals carry out avoidance or escape activities in response to mild stress stimuli. However, if such stimuli are repeated many times in succession the response can gradually subside and ultimately often disappear. It is not well known as yet how such habituation comes about. (Indeed, organisms without nervous systems can habituate to environmental stimuli, too; hence learning of this type probably operates on an intracellular, chemical level.)

A more complex form of learning depends on *conditioned reflexes.* In so-called *classical conditioning,* two or more stimuli are presented to an animal simultaneously and repeatedly, until the animal learns to execute the same response to either stimulus. For example, if bright light is directed into the eyes the pupils will contract reflexly. If the light stimulus is given repeatedly and is accompanied each time by food, sound, or some other stimulus, then pupillary contraction eventually can be made to occur by such a stimulus alone, without the bright light. Evidently, the animal now associates the second stimulus with light, and it comes to have not only the inherited neural circuit to the eyes but also an additional one acquired by learning through experience. Either one alone or both together then can produce the pupillary response.

Indeed, pupillary retraining in man is probably possible even if the second stimulus is no more than a single spoken word, and even if this word is spoken by the test subject himself. The stimulus word might not even need to be spoken but could be merely an unvoiced thought. Such a self-conditioned individual then would be able to contract his pupils at will. The occasionally recorded feats of human self-control over pain and other normally nonvolitional responses undoubtedly are based on self-conditioning of this sort.

Learning also can occur through *operant conditioning,* which differs from the classical type in that the animal participates actively and deliberately in the learning process. Thus if a certain activity at first happens to be carried out by chance and if this activity happens to have desirable consequences, then the animal can re-create these consequences by deliberately repeating the activity. Learning by this means becomes particularly effective if it is *reinforced,* that is, if repetition of an activity entails material or psychological rewards. Much of the learning of vertebrates—mammals and man in particular—is based on operant conditioning of this sort (see also Chap. 16).

Undoubtedly the most complex modulator activities are those that are involved in intelligence, personality, ability to think abstractly, and capacity to manipulate and control the environment. Depending extensively on memory and learning, such functions are developed to any substantial degree only in the most advanced mammals. But note again that, regardless of their relative complexities, all modulators depend on adequate *information;* apart from whatever built-in, inherited information they have available, modulators can act only on information that the neural receptors supply in the form of sensory impulses.

Neural Receptors

Receptor cells are either *epitheliosensory* or *neurosensory* (Fig. 8.26). The first type is a specialized, nonnervous epithelial cell that receives stimuli at one end and is innervated by a sensory nerve fiber at the other. The second type is a modified sensory neuron that carries a dendritelike stimulus-receiving extension at one end. At the other is an axon that synapses with other neurons. Both types of receptor cell can occur in clusters and together with accessory cells form *sense organs.* All receptor cells of invertebrates are of the neurosensory type. Vertebrates contain both neurosensory and epitheliosensory receptors.

Nerve impulses generated by receptors become perceptions only in the neural centers. In effect, eyes do not see and ears do not hear; eye-brain complexes are required for seeing, ear-brain complexes for hearing. In some instances the percep-

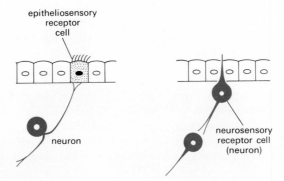

epitheliosensory
receptor
cell

neuron

neurosensory
receptor cell
(neuron)

Figure 8.26 Receptor cells. The two kinds are shown.

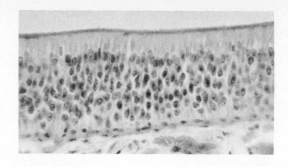

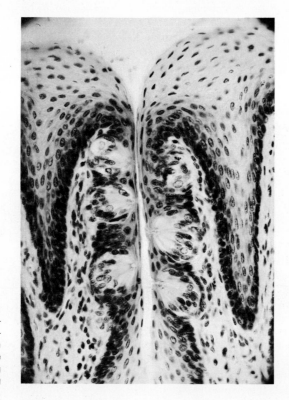

Figure 8.27 Chemoreceptors. A, section through epitheliosensory nasal epithelium of vertebrates. *B,* portion of section through tongue showing taste buds; the buds are located along the deep, narrow channels leading into the tongue from the surface.

tions become conscious, but more often they do not. When nerve impulses reach the brain from a blood vessel, for example, sensing takes place, but in this case the sensation does not become conscious.

Chemoreceptors These receptors give information about environmental *chemicals*. They include receptors that mediate the senses of *smell* and of *taste*. In most animals the receptor structures are located in the skin, and they usually are free sensory nerve endings that can be stimulated directly

by particular chemicals. The ability to sense common environmental chemicals has particular significance for many invertebrates, aquatic ones especially. This sense permits such animals to detect the presence of irritants or poisons and the chemical exudates of enemies, prey, food, and mates.

Smell is mainly a distance sense. Aquatic animals smell traces of chemicals in solution, terrestrial types smell vaporized chemicals or traces of chemicals that adhere to the ground. The sense of taste conveys information mainly about the general chemical nature of potential food substances. In most cases the receptors are localized in and around the mouth (Fig. 8.27), but animals such as flies and moths taste with their legs and head appendages, the receptors being in bristles on these body parts.

Chemical perceptions of all kinds are subjective sensations, not objective properties of chemicals. For example, sugar is not inherently sweet; it merely has the property of stimulating particular taste buds. Sweetness and all similar sensory qualities then are purely subjective interpretations. At least one substance is known (*phenylthiocarbamide*) that one person might not taste at all but that tastes sweet to another, bitter to a third, salty to a fourth, and sour to a fifth. Individual differences of this sort trace back to differences in heredity.

Mechanoreceptors These structures register stimuli of touch and of mechanical pressure generally. The receptors often are free nerve endings that in many cases are highly branched or elaborated as meshworks. Such receptors occur abundantly in the skin and in muscles, tendons, and most connective tissues. The skin of vertebrates also contains distinct nonnervous touch-registering organs innervated by sensory fibers (Fig. 8.28).

Mechanoreceptors are stimulated by mechanical displacement or by changes in the mechanical stresses that affect surrounding parts. For example, a bending or stretching of part of the skin or of an internal organ usually results in receptor stimulation. An animal thereby receives information both about contacts with external objects and about movements of any body part. In mammals the base of each hair is surrounded by a meshlike terminal of a mechanoreceptor fiber, and a nerve impulse is initiated if a hair is touched even lightly. Internal mechanoreceptors are called *stretch re-*

A

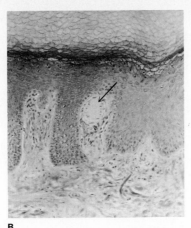

B

Figure 8.28 Mechanoreceptors. *A,* section through a Pacinian corpuscle, a pressure receptor in mammalian skin. *B,* location of a Pacinian corpuscle in the skin.

ceptors, or *proprioceptors*. They contribute to the control of the relative positions and movements of body parts and thus play an important role in maintaining posture and balance.

Statoreceptors   Present in many invertebrate groups and in all vertebrates, these organs are special types of mechanoreceptor for the sense of body orientation in relation to gravity (Fig. 8.29). Most commonly a statoreceptor is a small, fluid-filled *statocyst,* a sac that contains a cluster of ciliated *hair cells.* Attached to or resting against

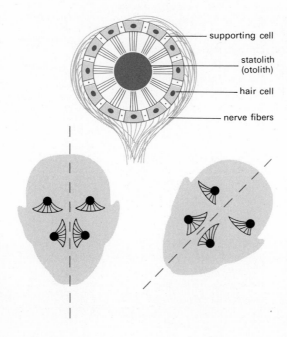

supporting cell

statolith (otolith)

hair cell

nerve fibers

Figure 8.29 Statoreceptors. The statocyst at top shows how hair cells support an ear stone (statolith) in the center of the cyst. Lower figures show position of the receptor organs in relation to the head, and the effect of tilting the head.

the hairs is a *statolith,* a grain of hard, often calcareous material (a sand grain in crayfish). When an animal moves the statolith shifts position under the influence of gravity and its own inertia, and it then presses against a somewhat different set of hair cells. Such a change in the pressure pattern produces a corresponding change in the pattern of nerve impulses that travels away from the sensory cells. In this manner the brain receives information about altered orientations or accelerations of the body. Body position also is sensed independently by information from mechanoreceptors, as noted, but the statoreceptors must function if an animal is to maintain normal orientation in relation to gravity.

In most animals statoreceptors are located in the head or in head appendages such as antennae. In vertebrates they are part of the inner ear. Indeed, the vertebrate ear has evolved primarily as a statoreceptor and only secondarily as an organ of hearing. Hair cells with statoliths (*ear stones*) are located at several places along the walls of two inner-ear chambers, the *saccule* and the *utricle* (Fig. 8.30). These receptors are organs for *static* body balance. They give information about head positions and permit perception of up, down, side, front, and back, even when visual stimuli and sensory data from muscles fail to provide such information.

Vertebrates also have *semicircular canals,* which provide information about head movements and through this about the *dynamic* balance of the body. Three semicircular canals in each ear loop from the utricle back to the utricle. The canals are placed at right angles to one another in three planes of space (Fig. 8.31). At one end of each canal is an enlarged portion, or *ampulla,* which contains a cluster of hair cells. When the semicircular canals move as the head is moved, the fluid in the canals "stays behind" temporarily as a result of its inertia and "catches up" only after the head has stopped moving. This delayed fluid motion bends the hairs of the receptor cells. Different impulse patterns then reach the brain according to the direction and intensity of fluid motion in the three pairs of canals.

Sudden accelerations, particularly in a vertical direction, or rapid spinning, or also uneven warming or cooling of the fluid in the semicircular canals, initiate reflexes through the receptors of dynamic balance that in man often lead to well-known symptoms of dizziness, nausea, and gastric

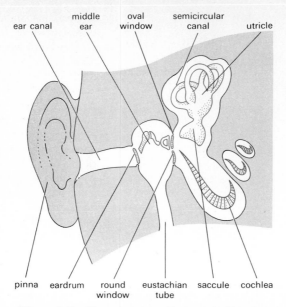

Figure 8.30 *Phonoreceptors: the mammalian ear.* Note ear bones in the middle-ear cavity and attachment of semicircular canals to the utricle. Statoreceptors are in the utricle and the saccule.

Figure 8.31 Semicircular canals of the left ear. Top of diagram is anterior; right side is toward median plane of head. The three canals are set at right angles to one another, hence only the horizontal canal reveals its curvature in this view. Both ends of each canal open in the utricle. The hair cells in the ampullae function as receptors for the sense of dynamic balance. When the head is moved, fluid in the canals bends the hair cells, thereby initiating sensory nerve impulses.

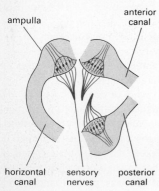

upsets. Seasickness has the same cause, as does the discomfort produced by jerky motions of elevators and airplanes.

Phonoreceptors These organs are sensitive to pressure vibrations in water or air. Organs of hearing are particular kinds of phonoreceptors.

Receptors for nonauditory vibrations, largely free nerve endings and hair cells, are located in or near the body surface. Most aquatic vertebrates have highly specialized phonoreceptors in the head and trunk skin. These regions contain *lateral-line systems*, essentially series of water-filled canals that communicate through pores with the external medium (Fig. 8.32). Movement of the water in the canals produces nerve impulses in hair cells. The lateral-line system is capable of detecting water turbulence created by moving objects or by a fish itself, and in a sense the system is comparable to the ''listening'' component of submarine sonar.

A distinct sense of hearing is restricted to some arthropods (certain crustacea, spiders, and insects) and to vertebrates; by and large only animals that make sounds also can hear them. In arthropods the receptor organs are located in various body regions—antennae (mosquitoes), forelegs (crickets), thorax (cicadas), or abdomen

(grasshoppers). Such arthropod ''ears'' usually are sensitive to certain sound frequencies only. For example, a male mosquito can best hear the sounds that have the same frequency as those produced by the wings of flying females (Fig. 8.33).

Among vertebrates, many fishes have sound receptors in the inner ear (Fig. 8.34). In addition to the utricle and saccule, the inner ear of such fishes contains a *lagena*. This third chamber connects with the swim bladder either directly or via a chain of small bones (*Weberian ossicles*). External sound waves generate vibrations in the air in the swim bladder, and such vibrations are transmitted to the lagena, where a statolith and hair cells produce impulses in auditory nerve fibers.

In terrestrial vertebrates, an *outer ear* carries sound to a *middle ear*, a cavity closed off from the external environment by a membranous *eardrum*. Tiny middle-ear bones (one in amphibia, three in reptiles, birds, and mammals) form an adjustable sound-transmitting bridge from the eardrum across the middle-ear cavity to the *inner ear*. Two membranes, the *round window* and the *oval window*, close off the middle-ear cavity from the

Figure 8.32 *Phonoreceptors: the lateral-line system.* A, general plan of the system in relation to the body surface of fishes. B, schematic cross section through a lateral-line canal, showing supporting cells and sense organ, the latter with phonoreceptive hair cells embedded among supporting cells.

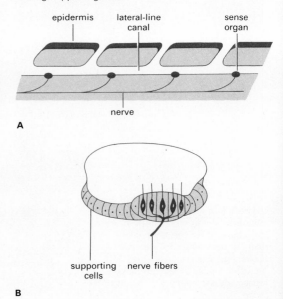

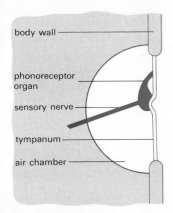

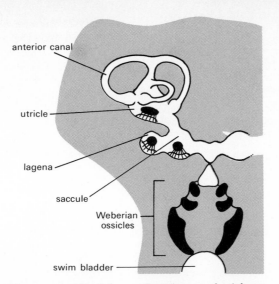

inner ear. The stirrup bone is anchored on the oval window (see Fig. 8.30).

A lagena again is the sound receptor in amphibia, but in reptiles, birds, and mammals the lagena is developed as a coiled, highly elaborate *cochlea* (Fig. 8.35). It contains a *basilar* membrane with fibers of graded lengths stretched across the cochlear tube (an arrangement similar to that of the strings of a piano). Attached to these fibers are innervated hair cells, and each fiber–hair–cell complex is selectively sensitive to a particular vibration frequency. Sound waves transmitted to a cochlea produce fluid vibrations of given frequencies, and the fibers that are sensitive to such frequencies move up and down as a result. As the hair cells attached to the fibers now move up and down, too, they come into contact with an overhanging *tectorial membrane*. It is presumably these contacts that initiate nerve impulses. Impulses from different sets of hair cells are interpreted in the hearing centers of the brain as sounds of different pitch. Basilar membrane, hair cells, and tectorial membrane together represent an *organ of Corti*.

Photoreceptors The light-sensitive receptors of animals are intracellular organelles that contain

Figure 8.33   Phonoreceptors: the arthropod ear. An organ of this type is found, for example, on grasshopper legs. The tympanum corresponds functionally to the eardrum of mammals.

Figure 8.34   The fish ear. This diagram of a left ear shows the lagena and its relation to the Weberian ossicles and the swim bladder. Utricle, saccule, and lagena contain ear-stone receptors; the ampullae of the semicircular canals similarly contain hair-cell mechanisms (not shown). The mammalian cochlea has evolved as an extension of the lagena.

Figure 8.35   The cochlea. A, the coils of the cochlea and a cochlear cross section with the parts of the organ of Corti. B, section through the organ of Corti.

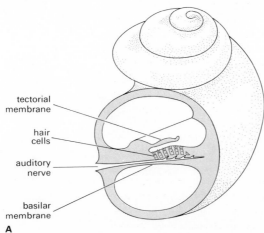

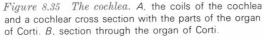

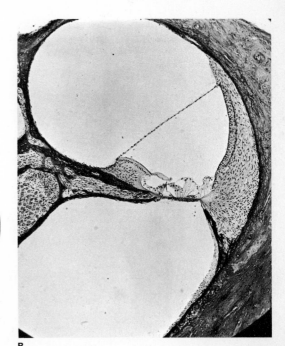

A

B

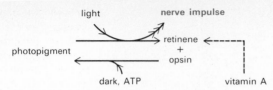

Figure 8.36 Photoreception. The photopigment is iodopsin in cone cells, rhodopsin in rod cells. Note that vitamin A can replenish the supply of retinene.

special photopigments. These pigments are chemically quite similar in all animals, and each consists of two joined molecular parts. One is a variant of *retinene,* a derivative of vitamin A; the other is a variant of *opsin,* a protein. Light splits retinene from opsin, and in the process light energy is in some way converted to chemical energy. Further chemical and electrical reactions then produce nerve impulses in the nerve fibers. At the same time retinene is rejoined to opsin through a series of ATP- and enzyme-requiring reactions, and the

Figure 8.37 Eyes. A, B, C, cnidarian eyes. *A,* spot ocellus; *B,* cup ocellus with lens; *C,* cup ocellus of inverted type: light must pass nerve fibers (part of epidermis, not shown) before reaching photosensitive cells. *D,* cup ocellus of planarian flatworm, inverted type; light passes the neurons first before reaching the flared photosensitive ends of the neurons. *E,* cup ocellus of the chambered *Nautilus,* converted type: light reaches the retina directly, the neurons being behind the photosensitive layer. *F,* vesicular ocellus (eye) as in many snails and annelids, converted type. *G,* eye of the scallop *Pecten,* with double retina; the outer retinal layer is inverted, the inner layer is converted and receives light bounced back from the reflecting layer. *H,* eye of squid, converted type, structurally as complex as vertebrate eye. (After Hesse, Borradaile, and other sources.)

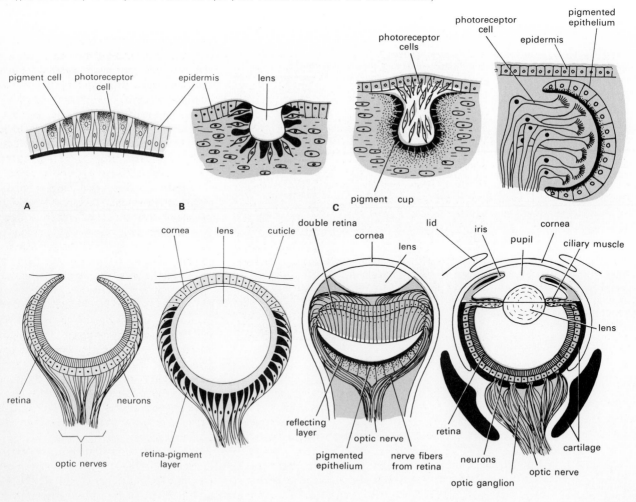

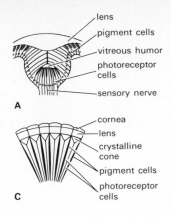

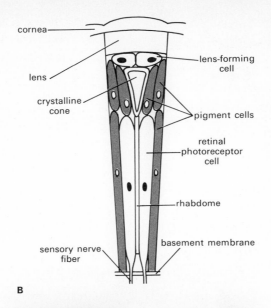

sects, some reptiles, most birds, and monkeys, apes, and man, also have receptor cells that detect color. In vertebrates with color vision, these cells, called *cones*, are structurally different from rods and their photopigments are various kinds of *iodopsin*. These pigments are combinations of retinene and forms of opsin that are different from those in rhodopsin. Functionally, cone cells appear to be of three kinds, sensitive respectively to red, blue, and yellow wavelengths of light. Since all colors are produced from combinations of these three primary colors, the three kinds of cones together can detect and analyze light of any color.

Clusters or layers of photoreceptor cells form *retinas*, the main components of seeing organs. Some of these organs are relatively simple constructed *eyespots* (*spot ocelli*), flush with the body surface. Others are *eyecups* (*cup ocelli*), and the most complex are more or less spherical *eyes* (*vesicular ocelli*). Most types of eyes are equipped with various accessory structures, including light-screening pigment layers and transparent, light-concentrating lenses (Fig. 8.37).

Arthropods have *simple* and *compound* eyes. Each compound eye consists of up to hundreds or thousands of complete visual units, or *ommatidia*. The lenses of adjacent ommatidia form hexagonal *facets* on the eye surface (Fig. 8.38). The eyes of vertebrates and the remarkably similar eyes of certain squids are the most complex photoreceptors. Apart from lenses, such eyes contain transparent *corneas,* which aid in focusing; *pupillary muscles,* which adjust the amount of light admitted into an eye; and a muscular *ciliary body* around the lens of each eye, which varies the curvature of the lens surface and thereby contributes greatly to focusing (Fig. 8.39).

An external object is "pictured" on the vertebrate retina as a series of points, like the points of a newspaper photograph. Each point corresponds to a rod or a cone. Impulses from these points are transmitted to the optic lobe in each cerebral hemisphere according to the pattern illustrated in Fig. 8.40. Each optic lobe thus receives impulses from *both* eyes, and normally the visual interpretations of both lobes give a single, smoothly superimposed picture of the external world. Sometimes a smooth superimposition fails to occur (under the influence of alcohol, for example), in which case one "sees double."

The image of an external object is projected on the retina in an inverted position, just as an

Figure 8.38 Arthropod eyes. A, simple eye (ocellus), as in insects. B, longitudinal section through a single visual unit (*ommatidium*) of a compound eye. C, section through several adjacent ommatidia. D, photo of compound eye in insect. (A, B, after Berlese, Snodgrass.)

photopigment is thereby regenerated (Fig. 8.36).

Numerous invertebrate groups and all vertebrates have photopigments in elongated receptor cells that enable the animals to perceive different black-white intensities of light. Such cells serve mainly as illumination and motion detectors. In vertebrates these cells are called *rods*, and their photopigment is "visual purple," or *rhodopsin*. In addition, a few animal groups, notably some in-

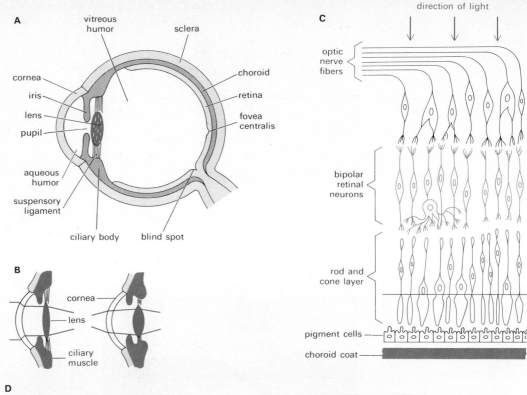

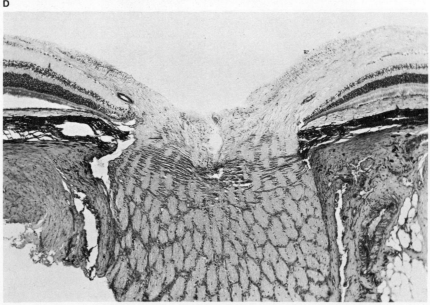

Figure 8.39 Mammalian eyes. A, structure. *B,* focusing. Left, far object, flat lens. Right, near object, curved lens. Lens curves out when the muscles of the ciliary body contract. *C,* the retina, simplified sectional view. This is an inverted retina, with light passing through the neuron layers before reaching the photosensitive rods and cones. *D,* retina, blind spot, and optic nerve. Note the neuron layers at the surface of the retina (top of photo) and the merging of the neuron fibers at the depression of the blind spot, forming the optic nerve (leading down in the photo).

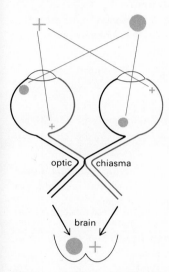

optic chiasma

brain

Figure 8.40 Eye and brain. An object in the left field of vision registers on the right halves of both retinas, and impulses are transmitted to the right half of the brain.

image is inverted on the film of a camera. But such images are not perceived inverted because the optic centers in the brain give visual experiences correct orientations. Space orientation actually depends not only on vision but also on perception of gravity through the statoreceptors; and the brain apparently interprets orientations in the visual world in accordance with its interpretation of gravity stimuli.

It is known also that visual interpretations in the optic lobes are assisted by the retinal neurons, which carry out a considerable amount of "data processing" before impulses are sent to the brain. It has been found, for example, that the retina of a frog generates four distinct sets of impulses to the brain. One set provides the animal with a black-and-white "outline drawing" of the stationary content of the visible world. A second set similarly provides an outline drawing, but only of illuminated objects that move across the visual field. A third set informs of objects that blot out illumination in part of the visual field and that become rapidly larger—undoubtedly indicating the approach of potentially dangerous animals. And a fourth set informs of moving objects that blot out illumination but that remain small—probably an insect-detecting device of special importance to a frog. These four sets of impulses are neatly superimposed in the brain and appear to be interpreted as a single "picture." A frog evidently sees the world in a unique, subjective manner, and a similar conclusion undoubtedly holds for light-sensitive animals of all kinds.

Other Receptors Numerous animal groups contain *temperature receptors* in the integument. In vertebrate skin are found separate heat and cold receptors, for example. Man—and by inference presumably other animals as well—can sense pain, sexual excitation, hunger, thirst, and sleepiness; and man can also discriminate between sensations of burning, tickling, stinging, and limbs "falling asleep." Separate receptors need not necessarily be associated with each of these senses. For example, some of the sensations just mentioned result when pressure, temperature, and pain receptors are stimulated simultaneously and in different combinations.

Other animals well might have comparable and also quite different senses, yet the presence of a sensory structure by itself often is insufficient for a determination of the nature of a particular sense. Does an earthworm experience pain, for example? Most animals do react sharply to stimuli that, by analogy with man, might be assumed to be painful. Yet we cannot really be sure, for sense perceptions must be communicated in some way that we can interpret. And in many cases we cannot make such interpretations. It is fairly certain nevertheless that in many instances receptors for so far unidentified senses do exist, particularly among invertebrates. The environments of such animals often produce stimuli quite different from those we would expect in terrestrial or in human surroundings. Invertebrate senses actually have so far been studied far less extensively than those of vertebrates.

Review Questions

1. Distinguish between epidermis and dermis and describe the basic structure of each. What derivatives does each layer produce in different animal groups? What are cornified cells? Name examples of tooth and scale types produced by the epidermis and dermis. What is the structure of a hair?

2. Which animal groups have exoskeletons? Endoskeletons? What are the comparative advantages and disadvantages of each? List structural variants of exoskeletons. What is a notochord? Distinguish between dermal and replacement bone, and describe the formation of each.

3. Review the organization of the vertebrate skeleton. Name and describe the basic parts of the vertebral column, the limb girdles, and the appendages. Which vertebrates have cartilaginous skeletons? What is the structure of cartilage?

4. What are the functions of a muscular system? Distinguish between smooth, striated, and cardiac muscle. In which animals and where in the body does each type occur?

5. What is a motor unit? Describe the characteristics of a simple twitch. What are summation, tetanus, and tonus?

6. What are humoral agents, hormones? Which animal groups have (*a*) humoral agents, (*b*) endocrine cells, (*c*) endocrine systems? What are some of the known functions of hormones in different invertebrates?

7. What are tropic hormones and what is their functional significance? Review the structure of the vertebrate endocrine system and describe the functions of various hormones.

8. Review the structure of a neuron. What are nerves, nerve nets, nerve cords, and ganglia? What is a ladder-type nervous system? What is the difference between a ganglion and a brain?

9. Describe the structure of the vertebrate brain. List the main subdivisions and the main components of each, and indicate their general functions.

10. Distinguish between spinal and cranial nerves. List the mammalian cranial nerves and review the functions of each. What are myelinated and nonmyelinated fibers? Where in the nervous system does each kind occur?

11. Review the structure and functioning of the autonomic nervous system. Distinguish structurally and functionally between the sympathetic and parasympathetic subdivisions of that system. What are sympathetic chain ganglia?

12. What are preganglionic and postganglionic fibers? Name the general components of a reflex arc and describe the detailed course of a reflex in (*a*) the central nervous system, (*b*) the autonomic nervous system.

13. What is a nerve impulse? How is an impulse propagated in a nerve fiber? What electrical events take place during impulse transmission? What is an action potential of a nerve fiber? How is an impulse transmitted across a synapse? Distinguish between cholinergic and adrenergic fibers.

14. Describe the activities of neural centers. What are reflex modification, summation, oscillator circuits, and inherited neural programs? Distinguish between learning by habituation, by classical conditioning, and by operant conditioning.

15. Distinguish between neurosensory and epitheliosensory cells. Describe various kinds of chemoreceptors and their functions. What is the difference between smell and taste? Are smells and tastes inherent in given substances? Discuss.

16. What are mechanoreceptors and what are their functions? Where in an animal can such receptors be found? Review the structure and function of statoreceptors in general.

17. Describe the organization of the mammalian ear and show where statoreceptors are located. Distinguish between static and dynamic body balance.

18. What is a phonoreceptor? Describe the structure and function of a lateral-line system. Which animals can hear, and by means of what kinds of receptors? Describe the structure of such receptors. What is a lagena? A cochlea? An organ of Corti?

19. Review the chemistry of vision. What are rods and cones? What is a retina? Which animals have color vision? Distinguish between spot, cup, and vesicular ocelli and between simple and compound eyes.

20. What structures does light traverse before it reaches the rods and cones in different kinds of eyes? Describe the eye structure of (*a*) various mollusks, (*b*) arthropods, (*c*) vertebrates.

Collateral Readings

General Background:

Prosser, C. L., and F. A. Brown, Jr.: "Comparative Animal Physiology," 2d ed., Saunders, Philadelphia, 1961.

Ramsay, J. A.: "Physiological Approach to the Lower Animals," 2d ed., Cambridge University Press, Cambridge, 1968.

Schmidt-Nielsen, K. S.: "Animal Physiology," 2d ed., Prentice-Hall, Englewood Cliffs, N.J., 1964.

On Integumentary, Skeletal, and Muscular Systems:

Hayashi, T.: How Cells Move, *Sci. American,* Sept., 1961. The molecular basis of flagellary, amoeboid, and muscular motion is discussed.

Hoyle, G.: How is Muscle Turned On and Off?, *Sci. American,* Apr., 1970. The flow of calcium ions is shown to be associated intimately with muscular contraction and relaxation.

Huxley, H. E.: The Contraction of Muscle, *Sci. American,* Sept., 1958. A good review of present knowledge of the fine structure and function of muscle cells.

McLean, F. C.: Bone, *Sci. American,* Feb., 1955.

Montagna, W.: The Skin, *Sci. American,* Feb., 1965. A good review of the structure and function of this organ system in man.

On Endocrine Systems:

Davidson, E. H.: Hormones and Genes, *Sci. American,* June, 1965. Preliminary evidence indicates that hormones probably exert reaction control by influencing the transcription of genetic codes.

Etkin, W.: How a Tadpole Becomes a Frog, *Sci. American,* May, 1966. A description of the feedback system between the pituitary and thyroid glands that controls amphibian metamorphosis.

Funkenstein, D. H.: The Physiology of Fear and Anger, *Sci. American,* May, 1955. A study of the role of adrenaline in human emotion.

Levine, R., and M. S. Goldstein: The Action of Insulin, *Sci. American,* May, 1958. The cellular and general biological role of this hormone is discussed.

Li, C. H.: The Pituitary, *Sci. American,* Oct., 1950. A general examination of this gland and its hormones.

————: The ACTH Molecule, *Sci. American,* July, 1963. The function of this pituitary hormone is outlined in relation to its structure.

Rasmussen, H.: The Parathyroid Hormone, *Sci. American,* Apr., 1961. The function of this hormone in man is described.

On Neural Operations:

Agranoff, B. W.: Memory and Protein Synthesis, *Sci. American,* June, 1967. The relation between memory and proteins is examined by experiments on goldfish.

Baker, P. F.: The Nerve Axon, *Sci. American,* Mar., 1966. The functioning of these fibers is examined experimentally. Recommended.

Brazier, M. A. B.: The Analysis of Brain Waves, *Sci. American,* June, 1962. Computers can now be used in the study of electrical brain phenomena.

Eccles, J.: The Synapse, *Sci. American,* Jan., 1965. An analysis of synaptic functions by electron-microscopic studies of structure.

French, J. D.: The Reticular Formation, *Sci. American,* May, 1957. A discussion of the important brain region that functions as "volume" control and governs sleep and wakeful states.

Gazzaniga, M. S.: The Split Brain in Man, *Sci. American,* May, 1967. When the two cerebral hemispheres are separated surgically, each becomes an independent center of consciousness.

Hydén, H.: Satellite Cells in the Nervous System, *Sci. American,* Dec., 1961. The probable chemical basis of memory is explored, and special attention is given to neuroglia cells, which occur together with neurons.

Jouvet, M.: The States of Sleep, *Sci. American,* Feb., 1967. Different biochemical mechanisms appear to be involved in light and deep sleep.

Katz, B.: How Cells Communicate, *Sci. American,* Sept., 1961. An examination of the nature of nerve impulses.

Kennedy, D.: Small Systems of Nerve Cells, *Sci. American,* May, 1967. Small ganglia of invertebrates are used in studies of programs of behavior.

Keynes, R. D.: The Nerve Impulse and the Squid, *Sci. American,* 1958. The physiology of nerve impulses is studied with the aid of the giant axons of squids.

Luria, A. R.: The Functional Organization of the Brain, *Sci. American,* Mar., 1970. The functional organization of the speech and writing centers is examined.

Peterson, L. R.: Short-term Memory, *Sci. American,* July, 1966. An examination of the separate mechanisms of short- and long-term retention of information.

Snider, R. S.: The Cerebellum, *Sci. American,* Aug., 1958. Modern experiments on the functions of this brain part are described.

Sperry, R. W.: The Growth of Nerve Circuits, *Sci. American,* Nov., 1959. An excellent account on how complex neural pathway networks become established in the embryo.

————: The Great Cerebral Commissure, *Sci. American,* Jan., 1964. Fascinating experiments on the corpus callosum are described.

Walter, W. G.: The Electrical Activity of the Brain, *Sci. American,* June, 1954. The patterns and possible meaning of brain waves are examined.

Wooldridge, D. E.: ''The Machinery of the Brain,'' McGraw-Hill, New York, 1963. One of the best and most stimulating recent paperbacks on brain function. Strongly recommended.

On Senses and Sense Perception:

Bekesy, G. von: The Ear, *Sci. American,* Aug., 1957. A highly recommended account.

Dowling, J. R.: Night Blindness, *Sci. American,* Oct., 1966. The role of vitamin A in the mechanism of vision is examined.

Fender, D. H.: Control Mechanisms of the Eye, *Sci. American,* July, 1964. Target tracking by the eye is analyzed.

Haagen-Smit, A. J.: Smell and Taste, *Sci. American,* Mar., 1952. A study of the sensitivity and discriminating ability of these chemoreceptors.

Hodgson, E. S.: Taste Receptors, *Sci. American,* May, 1961. The general mechanism of tasting is examined through experiments on receptors in the blowfly.

Hubel, D. H.: The Visual Cortex of the Brain, *Sci. American,* Nov., 1963. Responses of individual interpreter neurons in the optic lobes of cats are examined.

Lissman, H. W.: Electric Location by Fishes, *Sci. American,* Mar., 1963. Some fishes explore their environment by sensing changes in electric fields they have produced.

Loewenstein, W. R.: Biological Transducers, *Sci. American,* Aug., 1960. The conversion of external stimuli to nerve impulses is examined in the Pacinian corpuscle.

MacNichol, E. F.: Three-pigment Color Vision, *Sci. American,* Dec., 1964. The subject is adequately described by the title.

Melzach, R.: The Perception of Pain, *Sci. American,* Feb., 1961. Factors such as past experiences and cultural backgrounds are shown to affect the sensation of pain.

Miller, W. H., F. Ratcliff, and H. K. Hartline: How Cells Receive Stimuli, *Sci. American,* Sept., 1961. The properties of specialized sensory cells are examined, with special reference to the photoreceptors of horseshoe crabs.

Muntz, W. R. A.: Vision in Frogs, *Sci. American,* Mar., 1964. Data processing in the frog retina is examined.

Rock, J., and C. S. Harris: Vision and Touch, *Sci. American*, May, 1967. The dominance of vision over touch is demonstrated by experimental means.

Rosenzweig, M. R.: Auditory Localization, *Sci. American,* Oct., 1961. How the brain interprets sensory information received from the two ears of an animal.

Thomas, E. L.: Movements of the Eye, *Sci. American,* Aug., 1968. The small-scale movements of the eye during object fixation are described.

9

ANIMAL SYSTEMS: TRANSPORT, SUPPLY, REMOVAL

Four organ systems are examined in this chapter, those of *circulation, alimentation, breathing,* and *excretion.* These four represent the internal ''service'' systems for cellular metabolism. As the circulation passes by all cells of the body, it supplies them with nutrients and oxygen and carries off their wastes; and as the circulation passes through the alimentary, breathing, and excretory systems, it delivers the wastes to all three and picks up fresh nutrients from the first and oxygen from the second.

Circulatory Systems

Pathway Patterns

Animals characteristically contain *lymph,* a body fluid that fills free spaces between cells and tissues. In ancestral animals the internal fluid probably was little different from sea water, and modern animals have inherited a form of ''sea water,'' or lymph, as the universal internal medium of their bodies.

Lymph functions chiefly in maintaining *water, salt, pH,* and *osmotic* equilibria between the interior and exterior of cells. Secondarily lymph also provides a medium for the diffusion and transport of foodstuffs, respiratory gases, waste materials, in some cases hormones, and any other substances that pass from one body region to another. In comparatively simply constructed animals (in-

cluding particularly the sponges, radiates, and acoelomates), the free spaces between cells and tissues are minimal and the quantities of body fluid are small. Also, these fluids do not circulate in any specialized manner but are merely redistributed to some extent by body movements.

In animals with pseudocoelomic or coelomic body cavities lymph not only permeates all tissues but also fills the cavities. The movements of the fluid here are readily discernible; as the animal bends and twists, lymph ebbs and flows haphazardly but nevertheless quite conspicuously. This motion has some resemblance to the movement of a blood, and lymph accordingly is then called *hemolymph.* The hemolymph-filled space itself is a *hemocoel* (Fig. 9.1).

In many animals hemolymph is kept in motion in a regulated manner by means of specialized, mesoderm-derived flow channels and pumping organs. These enclose part or all of the hemocoel and the hemolymph contained in it. Any hemolymph confined partly or wholly in mesodermal channels is a *blood,* and the channels themselves are *blood vessels.* Vessel systems occur mainly in the coelomate animals. Groups that lack vessel systems are without blood, though many of such animals do contain hemolymph.

Blood vessels form circulatory systems that are either *open* or *closed.* In an open system the vessels have open ends, and blood therefore flows partly through vessels, partly through free hemocoelic spaces (here usually called *blood sinuses*).

If these spaces are large, the coelom is crowded out and greatly reduced in size. Open systems of this type are encountered in, for example, mollusks (squids and octopuses excepted) and arthropods.

In a closed system, the blood vessels form a complete, self-contained circuit. The only access to and exit from such a system is *through* the walls of the vessels. Free hemocoelomic spaces outside the system usually are absent or greatly reduced in such cases, and the coelom can become large (see Fig. 9.1). The blood in the vessels then is the main or only representative of the hemolymph. However, the whole body still is permeated with lymph as such, and this fluid readily exchanges materials with blood through the vessel walls. Closed systems are found in, for example, annelids, squids, and vertebrates.

The organs of a vessel system usually are *arteries, veins, hearts,* and, in most closed systems, also *capillaries.* Arteries carry blood away from a heart; veins, toward a heart. Capillaries are vessels of microscopic diameter that interconnect the narrowest arteries and veins. In closed systems the capillaries represent the regions where materials are exchanged between blood and lymph. The only tissue component of a capillary is a one-cell-thick *endothelium,* a simple squamous epithelium (Fig. 9.2). Vessels with larger diameters have additional tissues outside the endothelium, mainly layers of fibroelastic connective tissue and muscle. Arteries, which carry blood under the greatest pressure, have thicker walls than veins. At intervals along the larger veins often are found internal *valves* that open toward the heart and prevent backflow of blood.

Vertebrates also have *lymph vessel systems* composed of lymph capillaries and lymph veins (Fig. 9.3). The blind-ended capillaries receive body lymph, which is conducted to lymph veins. These join each other and the widest then empty into certain veins of the blood circulation. As lymph thus passes into the blood, body tissues obtain new lymph by outflow of fluid through the blood capillaries. Present along the course of the lymph vessels are glandular *lymph nodes.* In frogs and some other vertebrates certain regions of large lymph vessels are enlarged as pulsating *lymph hearts.* These aid in driving lymph to the blood circulation.

In both open and closed system, hearts usually lie in a lymph-filled *pericardial cavity.* This space is a part of the coelom and it is enclosed by the *pericardium,* a portion of the peritoneal membrane (Fig. 9.4). A heart itself is either tubular, as in arthropods, or more compact, as in most other animals. In many cases it contains two types of chamber, one or two relatively thin-walled *atria* (or *auricles*), which receive blood, and one or two thicker-walled *ventricles,* which pump blood out. Heart *valves* between atria and ventricles (AV valves) prevent backflow of blood. Some animals have secondary hearts in addition to a main, or *systemic,* heart. For example, squids and octopuses have *gill hearts* that aid in driving blood through the gill capillaries (see Fig. 23.24). Similarly, the bases of the legs of certain insects contain secondary hearts that aid in forcing blood through the narrow hemocoels of these appendages.

In invertebrates, hearts typically lie along the *dorsal* midline of the body. If the system is open, blood usually is pumped out toward both ends of the body. Blood then flows back ventrally from both ends and returns to the heart roughly at

Figure 9.1 Lymph, hemolymph, blood, and vessels. A, the relation of the body fluids to the internal body spaces of animals. *B,* open and closed circulations. 1, the open system as in mollusks. Blood flows partly in vessels, partly in free hemocoelic sinuses. The coelom is reduced to a paricardial space (white). 2, the closed system as in annelids. Blood flows forward dorsally, the pump being the dorsal or the anterolateral vessels. The coelom typically is large. 3, the closed system as in vertebrates. The heart is ventral and blood flows backward dorsally. A portion of the coelom is partitioned off as a pericardial cavity.

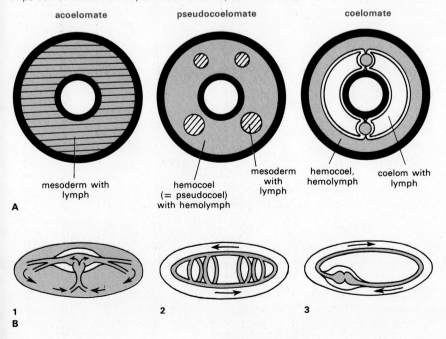

acoelomate   pseudocoelomate   coelomate

mesoderm with lymph

hemocoel (= pseudocoel) with hemolymph

mesoderm with lymph

hemocoel, hemolymph

coelom with lymph

A

1   2   3

B

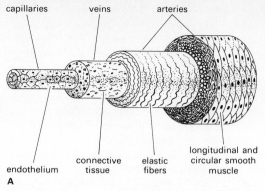

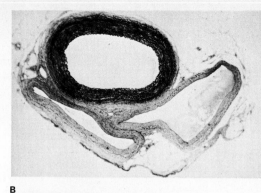

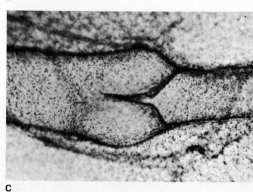

A

B

C

Figure 9.2 *Circulatory vessels.* *A,* the progressively greater thickness and tissue complexity of capillaries, veins, and arteries. The single cell layer of squamous endothelium is continuous throughout a vessel system. Additional tissues do not necessarily occur in such neat layers as sketched here. *B,* section through an artery and two veins. Note the thicker wall of the artery and the many elastic fibers (dark wavy lines) in this wall. *C,* longitudinal section through a lymph vessel showing an internal valve. Such valves prevent backflow. Valves very much like this are present also in the larger veins.

Figure 9.3 *The lymph system.* *A,* the general plan of the blood and lymph circulations. Oxygenated blood, dark color; venous blood, light color. Oxygenation occurs in gills or lungs. An artery carries blood away from the heart; a vein, toward the heart (regardless of the state of oxygenation of the blood carried). Fluid escaped from blood capillaries to surrounding tissues enters the lymph system through the walls of the lymph capillaries. *B,* section through a lymph node.

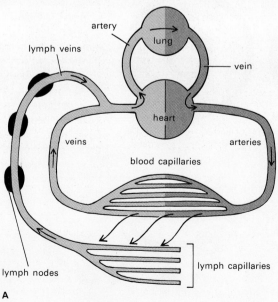

A

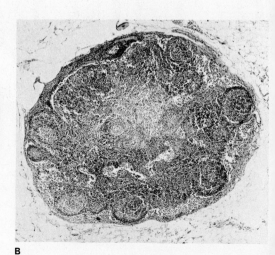

B

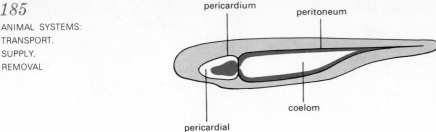

A

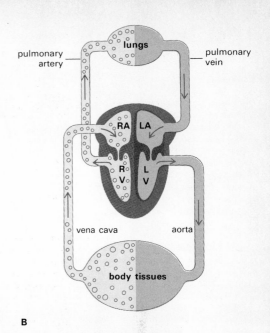

B

Figure 9.4 Heart and circulation in mammals. A, the relation of heart, pericardium, and coelom in vertebrates. The pericardial cavity is a subdivision of the coelomic body cavity. *B,* the mammalian blood circulation. Arterial blood is in the left side of the circulatory system (right side of diagram), venous blood in the right side (left side of diagram). *C,* the human heart. The large blood vessel stump is the aorta. The atria are partly hidden by the aorta. The size of your fist is very nearly the actual size of your heart. *D,* the human heart cut open to show the interior of the left ventricle. Note the strands of tissue attached to the two flaps of the biscuspid valve. These strands prevent the valve from opening into the atrium (white area above the ventricle).

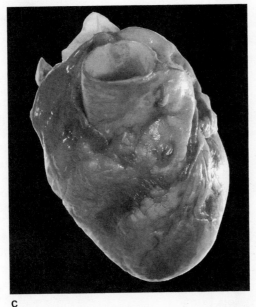

C

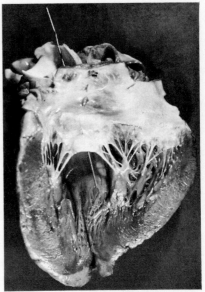

D

mid-body. If the system is closed, blood by and large flows toward the head end dorsally and toward the hind end ventrally. In vertebrates, by contrast, the heart is situated on the *ventral* side and blood is pumped toward the head end ventrally, toward the hind end dorsally (see Fig. 9.1).

The beat of a heart can originate *neurogenically* or *myogenically*. In a neurogenic beat, exemplified by the hearts of most arthropods, rhythmic contraction is initiated by a neural ganglion situated in or close to the heart. In a myogenic beat, contraction is initiated by a *pacemaker,* a special node of modified or unmodified heart muscle typically located in an atrium (the right atrium in mammals, Fig. 9.5). The pacemaker sends rhythmic impulses through the whole heart muscle. In mammals a

Figure 9.5 Innervation of the mammalian heart. The heart-rate center receives messages through many sensory nerves. Some of them originate in the vena cava and the aorta. Such messages can initiate a stretch reflex: the center can send command signals to the pacemaker via motor nerves that either inhibit or accelerate the heart. Note that sensory impulses from the vena cava lead to accelerating motor signals, and impulses from the aorta lead to inhibiting motor signals. Impulses from the pacemaker then stimulate the atria (which contract as a result) and then the atrioventricular (AV) node, which in turn sends contraction signals to the ventricles through the bundle of His.

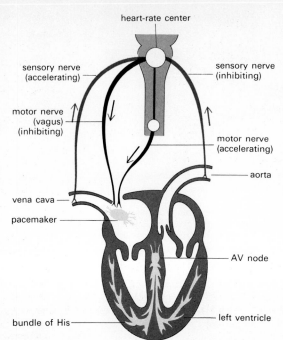

Figure 9.6 Heart action. When atria contract (left), they force blood into the relaxed ventricles. The AV valves are open, but the pressure of blood closes all other exits. When ventricles contract (right), they force blood into the pulmonary artery and the aorta. The atria are relaxed at the same time and fill with blood in preparation for the next beat.

strand of specialized muscle fibers, the *bundle of His,* conducts impulses through the heart. A heart thus has an *intrinsic beat* that can be maintained even when the organ is separated from the body. But inside the body the beat is under nervous control. Heart ganglia or pacemakers are innervated by motor nerves from the neural centers of an animal; and impulses through these nerves override the intrinsic beat frequency and can

thereby adjust and vary the heart rate (Fig. 9.6). An interesting situation exists in the myogenic hearts of tunicates, where the location of the pacemaker alternates every minute or so from one end of the tubular heart to the other. The heart in effect produces an "alternating current," with blood forced in one direction by one series of beats, in the opposite direction by the next series.

The force of the heartbeat is a major determinant of *blood pressure.* In open systems this pressure diminishes greatly once blood has left the vessels and flows free in the blood sinuses. Blood here returns to the heart in part by a "massaging" action resulting from contractions of the body musculature, in part by back-suction of blood to the heart after this organ has emptied with each beat.

In closed systems, blood pressure is determined also by the total volume of blood and by the space available inside the vessel system. In many vertebrates (man not included) blood volume can be adjusted to some extent by the spleen, which serves as a blood-storing organ; when it contracts it squeezes stored blood to the circulation. The blood space can be varied nervously. So-called *vasomotor* nerves innervate the muscles in the walls of arteries. Contraction or relaxation of these muscles leads to *vasoconstriction* or *vasodilation,* a narrowing or widening of a vessel. Blood pressure thus can increase or decrease, either locally or throughout the circulatory system (Fig. 9.7). Nevertheless, overall blood pressure falls gradually with increasing distance from the heart, for the total cross-sectional area of all capillaries is far greater than that of the arteries leaving the heart. Return of blood to the heart here is achieved by the (low) pressure of blood in veins, by prevention of backflow through the valves in the veins, and again by muscular massaging of veins and back-suction of blood to the heart.

The most essential parts of a circulation are the capillary beds in a closed system and the blood sinuses in an open system. It is in such regions that body tissues draw from and add to blood, and that blood in turn draws from and adds to the alimentary, breathing, and excretory systems.

Blood

Blood is a tissue that consists of two components: fluid *plasma* and loose blood *cells* suspended in the plasma.

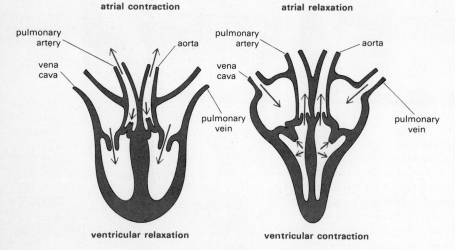

atrial contraction

pulmonary artery

vena cava

aorta

pulmonary vein

ventricular relaxation

atrial relaxation

pulmonary artery

vena cava

aorta

pulmonary vein

ventricular contraction

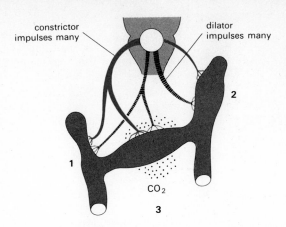

constrictor impulses many

dilator impulses many

2

1

CO_2

3

Figure 9.7 Vasomotion. Vaso-constriction as at 1 occurs when a blood vessel receives many constrictor impulses and few dila-tor impulses from the vasomotor center. Vasodilation as at 2 occurs when constrictor impulses are few and dilator impulses are many. Locally produced CO_2 as at 3 can override the vasoconstriction or-dered by the vasomotor center and can bring about a local vaso-dilation.

Plasma The *water* component of plasma main-tains an aqueous environment around all tissues, and, as pointed out, its presence in a given volume also influences blood pressure. Plasma water also is the transport vehicle of foods and other metabo-lites, including waste materials, hormones, and respiratory gases. Carbon dioxide in particular is carried from tissue cells to the breathing system, largely in the form of bicarbonate ions:

$$CO_2 + H_2O \underset{\text{in breathing system}}{\overset{\text{in body tissues}}{\rightleftharpoons}} H^+ + HCO_3^-$$

The dissolved *mineral components* of plasma contribute to *salt, pH,* and *osmotic* balance be-tween plasma and lymph and between lymph and the interior of tissue cells. Water and mineral ions tend to be kept at constant concentrations, quan-tity control being exercised by the alimentary, breathing, and excretory systems. Some of the foods, gases, and wastes transported in plasma

are likewise held at constant levels. The concen-trations of most metabolites usually vary, however, in line with changing rates of supply and utiliza-tion. Exchange of such components between plasma and lymph and between lymph and tissue cells is accomplished by diffusion and active trans-port.

Blood plasma differs importantly from lymph in that it contains *plasma proteins*. These too tend to remain at constant concentrations and, in closed systems, normally do not leave the circula-tory vessels. Most studied in vertebrates, the plasma proteins of these animals are manufactured mainly in the liver and they serve a variety of nonnutritional roles. First, like the mineral ions, all these proteins contribute to maintaining the os-motic pressure and pH of blood. Particularly im-portant in osmotic regulation are the *albumins,* blood proteins named after their chemical resem-blance to egg white. Second, some of the proteins are ingredients in the *clotting* reaction (see below). Third, many of the proteins are active enzymati-cally. One of them, *prothrombin,* is a clotting factor. Others are enzymes such as are found in tissue cells generally and even in the intestine. Their catalytic functions in plasma are still largely obscure. Fourth, proteins called *globulins* are the basis of differences in the *blood types* of animals, a consequence of the general phenomenon of protein specificity (see also Chap. 15). And last, some of the globulins serve as defensive *anti-bodies,* which destroy or render harmless infective agents such as bacteria. Animals thereby can be-come *immune,* an important aspect of steady-state control that ultimately is a consequence of protein specificities, too (Fig. 9.8).

Apart from its protein content, plasma in many animals differs from lymph also in another respect, the presence of *respiratory pigments.* However, in a good many animal groups such pigments occur not in plasma but in blood cells.

Blood Cells The bloods of most animals contain *nonpigmented* cells of various kinds, and many also contain *pigmented* cells. Both kinds are mesodermal derivatives and they usually are so specialized that they do not divide (Color Fig. 5).

Most of the nonpigmented cells are capable of amoeboid locomotion. Called *amoebocytes* gener-ally and *white* cells in vertebrates, such cells can squeeze in between the cells that form a capillary vessel and leave the circulatory system in this

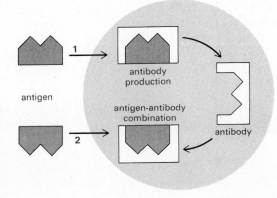

antigen

1

antibody production

antigen-antibody combination

2

antibody

Figure 9.8 Antibodies. A for-eign protein introduced into an animal is an antigen. It elicits the formation of antibodies (1), which "fit" precisely the surface con-figuration of the antigen. If later these same antigens invade the animal (2), the specific antibodies already present can combine with the antigens and render them harmless (immune reaction). Anti-bodies typically are globulins, a group of blood proteins.

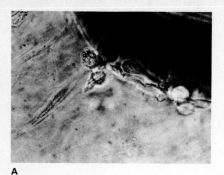

A

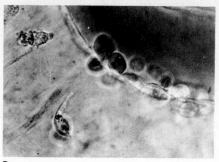

B

Figure 9.9 Blood cell migration. In each photo a blood-filled capillary is in upper right portion. In *A* two white blood cells have just penetrated through the capillary wall into surrounding tissues. In *B* the white blood cells have migrated farther into the tissues.

manner. White cells are of two main types, *leukocytes* and *lymphocytes*. In leukocytes the amoeboid habit is developed particularly well. Once leukocytes are out in the body tissues, they can migrate toward sites of infection and engulf the bacteria present there (Fig. 9.9). Such *phagocytic* activity generally is characteristic also of the nonpigmented blood cells of invertebrates. An accumulation of phagocytes, bacteria, and cellular debris in wounded areas represents *pus*.

Lymphocytes, too, serve in body defense. By transforming to mesenchyme cells and fibrocytes, lymphocytes contribute importantly to wound healing and scar-tissue formation after injury. Lymphocytes also serve as lymph-purifying agents. Particles of dust, smoke, and other materials frequently get into lungs, and lymph then usually carries such particles to the lymph nodes. The lymphocytes there engulf the particles and retain them permanently. Lymphocytes are manufactured mainly in the lymph nodes; leukocytes originate in the liver and spleen during embryonic stages but in the marrow of long bones in the adult.

Bone marrow also is believed to be the main generating tissue of the *platelets* of vertebrate blood (see Color Fig. 5). These bodies are cell fragments, often without nuclei, and when they rupture on the rough edges of torn blood vessels,

they initiate blood *clotting*. Broken platelets release an enzymatic substance, *thrombokinase*. This substance interacts in plasma with calcium ions and the blood protein *prothrombin*. The protein is an inactive precursor of the catalyst *thrombin*. In the presence of calcium ions and thrombokinase, prothrombin becomes thrombin. The latter then reacts with *fibrinogen*, another of the blood proteins, yielding fibrin (Fig. 9.10). This insoluble coagulated protein is the blood clot. It is a yellowish-white meshwork of fibers in which pigmented blood cells are trapped, hence the color of the clot. Invertebrate bloods too can clot. The detailed reactions are less well known, but they likewise lead to formation of coagulated protein meshworks.

Pigmented blood cells of vertebrates are manufactured in the liver and spleen of embryos and in the bone marrow of adults. In mammals the nuclei of the cells disintegrate as they mature. The resulting *red corpuscles* remain in the circulation for limited periods only, in numbers that are maintained constant: the spleen and to some extent also the liver destroy corpuscles, while bone marrow manufactures them. The rate-controlling factor here is the amount of oxygen in the environment. A high O_2 content in the air slows the production rate (and increases the destruction rate), whereas a low O_2 content (as at high altitudes) has the opposite effect. Red cells are the most abundant cellular components of vertebrate blood. For example, a cubic millimeter of human blood contains some 5 million red corpuscles but only some 8,000 white cells and about 250,000 platelets.

Pigmented blood cells derive their color from *respiratory pigments* dissolved in their cytoplasms. The pigments serve primarily in oxygen transport, but CO_2 can be carried to some extent too. All respiratory pigments are linked to proteins, which make the reaction with oxygen reversible:

$$\text{protein-pigment} + O_2 \underset{\text{in body tissues}}{\overset{\substack{\text{in breathing} \\ \text{system}}}{\rightleftarrows}} \text{protein-pigment}-O_2$$

Respiratory pigments are largely of four different types: *hemoglobin, hemerythrin, chlorocruorin,* and *hemocyanin.*

Hemoglobin (Hb) is by far the most widespread. Its presence has been demonstrated in all animal phyla except the sponges and the radiates. How-

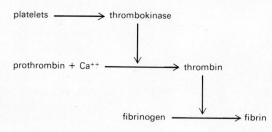

Figure 9.10 Blood clotting. A simplified reaction sequence is shown.

ever, the pigment is not always a blood constituent. For example, Hb occurs in protozoan cells, in the eggs of many animals, and in vertebrate skeletal muscle, where it is responsible for the redness of flesh and the color of "dark meat" in cooked condition. In such noncirculatory locations Hb nevertheless serves as a supplier of oxygen, just as it does in blood.

Hemoglobin consists of the iron-containing pigment *heme* (chemically similar to the cytochromes) and the protein *globin* (see Fig. 4.17). The nature of this protein varies for different animal groups, and each such form of Hb endows blood with a unique oxygen-carrying capacity. Table 7 shows that Hb actually is the most efficient of the respiratory pigments. The table also indicates that oxygen-carrying efficiency is related closely with the level of activity an animal is capable of.

Hb combines to some extent with CO_2, and it joins carbon monoxide (CO) in preference to O_2. Consequently, CO is an oxygen-displacing poison in animals that contain Hb. When red blood cells of vertebrates are destroyed in the liver, the iron of heme is salvaged for renewed use and the rest of the heme molecule is converted to green *biliverdin* and red *bilirubin*. These pigments are eliminated both in the feces (via bile) and in urine (via blood and kidneys), hence the characteristic colors of these elimination products. Robins deposit the heme remnants in their egg shells, which become blue as a result.

Iron is present also in hemerythrin, a colorless pigment that turns red when combined with oxygen. It occurs only in the blood cells of one group of annelids, some echiuroid worms, and the brachiopod *Lingula.* Hemocyanin too is colorless, but it turns blue in combination with oxygen. It differs also in that it is a copper compound and in that it does not combine with CO. The pigment occurs in plasma and it is encountered only in mollusks (squids, octopuses, and certain snails) and in arthropods (crustacea most particularly). Snails that have hemocyanin in blood often have Hb in the muscles.

Chlorocruorin is a green pigment with a reddish color in a concentrated state. Like Hb it is an iron-containing heme protein. It is found only in certain sedentary marine annelids, where it occurs in plasma. One genus of these animals (*Serpula*) has Hb as well as chlorocruorin in plasma, the only known instance where two different blood pigments are found together. In another genus, one species has chlorocruorin and green blood, a second has Hb and red blood, and a third no pigment

Table 7. *Respiratory pigments and oxygen-carrying capacity of blood*

PIGMENT	ANIMAL GROUP	cm³ O_2 CARRIED PER 100 cm³ BLOOD	OCCURRENCE IN BLOOD
hemocyanin:	some snails	2	plasma
colorless to blue	squids	8	plasma
Cu	crustacea	3	plasma
hemerythrin:	some annelids and	2	cells
colorless to blue	related worms		
Fe			
chlorocruorin:	some annelids	9	plasma
green to reddish			
Fe			
hemoglobin:	some mollusks	2	plasma
purple-red	some annelids	7	plasma or cells
to orange-			
red	fishes	9	cells
Fe	amphibia	11	cells
	reptiles	10	cells
	birds	18	cells
	mammals	25	cells

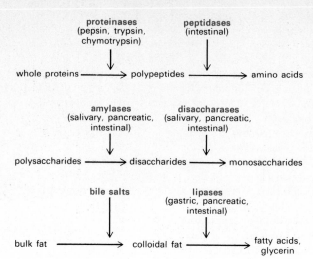

Figure 9.11 Digestion: summary. Polypeptides are products of partial protein breakdown. Bile salts act physically, not chemically; they emulsify fats and thereby reduce large fatty masses to tiny droplets of colloidal dimensions.

and colorless blood. The distribution of the pigments evidently does not follow any clearcut evolutionary pattern.

An interesting group of blood pigments with a possible respiratory function occurs in tunicates. The blood of these marine chordates contains oxides of vanadium, combined with protein. The pigments are orange, green, and blue, and they are in the blood cells. Hb is not present (and indeed none has as yet been demonstrated in any chordates other than the vertebrates). The vanadium pigments combine readily with O_2 and they release the gas in acid media. Since tunicate blood cells actually contain free sulfuric acid (!), a respiratory role of the pigments is probable though not fully proved.

Alimentary Systems

If an animal could obtain all the nutrients it requires in the form of immediately usable ions and molecules, it could simply absorb such nutrients from the environment directly through its cell surfaces. This actually is the nutritional pattern in many parasitic animals, in which alimentary structures often are highly reduced or even absent. But, apart from water and minerals dissolved in water, directly usable nutrients are largely unavailable to

free-living animals; and it is the function of an alimentary system to permit *ingestion* of nutrients that are available, *digestion* of these to directly absorbable ions and molecules, and *egestion* of all remaining materials.

The key process, digestion, is accomplished by mechanical and chemical means. Mechanical digestion subdivides ingested food to fine, often colloidal particles suspended in water. Chemical digestion then reduces these particles to molecular dimensions. This process requires specialized *digestive enzymes*, secreted by various parts of an alimentary apparatus. The enzymes speed up hydrolysis of food molecules, or dissolution by water:

$$food + H_2O \xrightarrow{enzyme} food\ components$$

Complete hydrolysis of a food compound usually occurs in a series of steps, each of them a hydrolytic reaction catalyzed by a specific digestive enzyme. The actual digestion pattern of the main classes of foods is summarized in Fig. 9.11.

The combined result of mechanical and chemical digestion is an aqueous food solution. In simply constructed animals, absorption and distribution of the usable nutrients in this solution is accomplished entirely by *short-distance transfer*—diffusion and active transport directly from cell to cell. In all more complexly constructed animals, *long-distance transfer* by lymph and the circulatory system occurs as well.

In different animals alimentary processes take place by *intracellular* alimentation, by *extracellular* alimentation, or by a combination of both (Fig. 9.12). The first occurs mainly in protozoa and sponges. An individual cell here functions as a complete alimentary apparatus. It ingests food particles by some form of phagocytosis (for example, amoeboid engulfment, with or without the aid of cilia or flagella), and thereby acquires an internal *food vacuole*. The cytoplasm then secretes digestive enzymes into the vacuole and usable nutrients later diffuse out of it. Indigestible remains are egested from the vacuole.

A combination of intracellular and extracellular alimentation occurs mainly in radiates and flatworms, where the alimentary apparatus is a one-hole sac. The digestive cells that line the sac secrete enzymes, and these promote a preliminary extracellular decomposition of foods that have been brought into the sac by ingestion. Small food particles then are engulfed by the cells of the sac

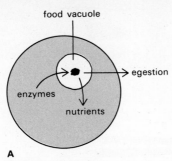

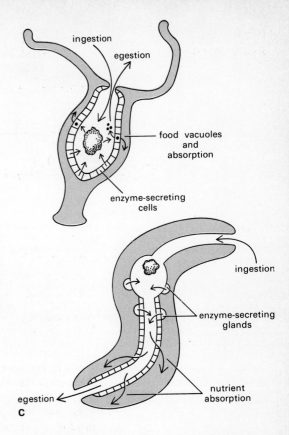

Figure 9.12 Forms of digestion. A, the intracellular pattern, in which enzymatic digestion occurs in a food vacuole. After usable nutrients are absorbed into the cell, the unusable remains of the original food are egested from the vacuole and the cell surface. *B,* the joint intracellular and extracellular pattern. Enzymes secreted into the alimentary cavity accomplish a preliminary extracellular phase of digestion, and small food particles engulfed in vacuoles of digestive cells then are digested intracellularly. *C,* the extracellular pattern. Digestive glands secrete enzymes to the alimentary cavity and nutrient molecules then are absorbed through the gut wall.

Figure 9.13 Mammalian intestine: cross section through a portion of the intestinal wall. The mucosa, adjacent to the lumen (gut cavity), consists of columnar epithelium, connective tissue, and thin layers of muscle; the submucosa, of connective tissue, nerves, and blood vessels; the muscularis, of thick layers of inner circular and outer longitudinal muscles; and the serosa, of connective tissue and squamous epithelium, the latter continuous with a membrane (mesentery) that holds the whole intestine in place. See also Fig. 7.21.

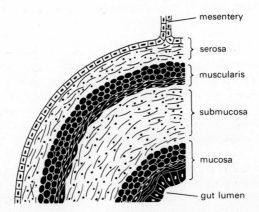

lining and are digested intracellularly, as in protozoa or sponges.

Where the alimentary apparatus is a two-hole tube, purely extracellular alimentation is the rule in most cases; enzymes produced by specialized digestive glands are secreted into the cavity of the alimentary tube, where complete digestion takes place. The main tissue of the alimentary tube is the *mucosa,* an endodermal layer one cell thick,

often ciliated over part or all of its surface. In many cases the mucosa represents the whole alimentary wall, but in most animals this layer is surrounded by various additional (mesodermal) tissues. These include layers of fibroelastic connective tissue, which carry networks of blood and lymph vessels; muscle layers, which by contracting aid in moving food along the alimentary tract; and typically also (ectodermal) nerve nets, which provide the stimuli for the muscle contractions. In coelomate animals the outermost layer is part of the peritoneal lining of the coelom (Fig. 9.13).

Animals with tubular alimentary tracts ingest foods of three general forms: fluids; solids of microscopic or near-microscopic dimensions; and solids of greater bulk. Many animals live entirely on plant or animal juices. The ingestive organs in such cases are *sucking* devices of some kind. Suction generally is produced by muscular distension of the mouth cavity, the pharynx, or the stomach. Numerous aquatic animals live on microscopic solid food. Ingestion here often takes the form of *filter feeding.* In this process ciliated structures such as tentacles strain food particles from

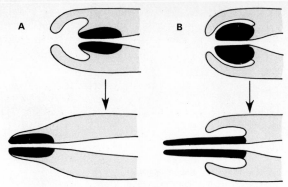

Figure 9.14 Pharyngeal ingestion. A, eversible pharynx, rest condition (top) and extended condition (bottom). *B,* protrusible pharynx, rest condition (top) and extended condition (bottom). The walls of the pharynx are in black. In eversion, the whole pharynx slides out through the mouth; in protrusion, the pharynx narrows and elongates out of the mouth.

water. This type of ingestion is particularly widespread among burrowing and sessile animals (see Color Fig. 1). Microscopic food also can be obtained with the aid of filelike *raspers* of various kinds, which scrape off minute fragments from larger food masses. The intake of bulkier food usually is accomplished by *tearing, biting,* and then *grinding* or *chewing.* Such processes also aid in mechanical digestion and they are carried out by *jaws* and *teeth* of various types. Some animals obtain bulk food with the aid of the *pharynx,* the portion of alimentary tract just behind the mouth cavity. In such cases the thick muscular walls of the pharynx can be either *protruded* or *everted* through the mouth and food is then gripped by the projecting muscular tube and pulled back (Fig. 9.14).

Inside the alimentary tract, food is propelled along both by cilia of the mucosa and by *peristalsis,* wavelike contractions of the alimentary wall. In some animals the alimentary tract is without conspicuous modifications along its course, but in others the tract is specialized regionally as a series of distinct organs. Apart from mouth cavity and pharynx, such organs usually include *esophagus, stomach,* and *intestine.* In vertebrates the intestine consists of a long, comparatively narrow *small* intestine and a shorter, wider *large* intestine (or *colon*). All such subdivisions are specialized for some phase of the alimentary process (Fig. 9.15 and Table 8).

Mechanical digestion often begins in the mouth

and is continued by the grinding action of a stomach or specialized grinding organs in the pharynx of some animals. In vertebrates, moreover, mechanical subdivision of food is accelerated greatly by strong HCl secreted from the stomach wall. This acid dissolves the cementing substance between cells and makes a food mass literally fall apart. Chemical digestion is the specialized function of various digestive glands present along the alimentary tract. Some glands are embedded in the alimentary wall; others lie outside it but connect to it by ducts. In vertebrates such external glands are the *salivary glands,* the *pancreas,* and the *liver.* Many invertebrates likewise have salivary and pancreatic organs, but none has an organ quite like the liver (even though certain ones are sometimes called "livers"). The vertebrate liver performs most of its functions *after* digestion and absorption have already taken place (see below). One liver function does contribute to digestion directly, manufacture and secretion of bile. This fluid, which can be stored temporarily in a *gallbladder,* contains salts that emulsify fats in the gut. The fats thereby become subdivided to colloidal droplets, and the digestive enzyme lipase in the surrounding water then can act on the extensive surfaces of the fat droplets.

After digestion, absorption of usable nutrients occurs in the mucosa of the intestine and, in invertebrates, also in various glands ("livers") and pouches that connect with the intestine. In many animals the absorptive surfaces are enlarged greatly by the presence of intestinal loops or branches, or by folds in the mucosal lining. The vertebrate mucosa in addition contains millions of microscopic fingerlike projections called *villi,* which give the intestinal lining a fine carpetlike texture and enormous surface area (Fig. 9.16).

From the mucosa absorbed nutrients are distributed to all parts of the body by short-distance and long-distance transport. In invertebrates this distribution is always direct, from intestine to utilizing cells. In vertebrates, by contrast, distribution is largely indirect; most foods first pass from the intestine to the liver. Some colloidal fats usually escape digestion in the intestine, and these, along with some water, are absorbed by intestinal lymph channels that bypass the liver. All other usable nutrients are absorbed by blood vessels that lead to the liver.

Figure 9.17 shows that liver cells act on all main classes of food compounds. Incoming carbohydrates from the intestine are stored as glycogen, any excess is converted to fat, and certain amounts are released to the general circulation as glucose. Within narrow limits the liver thus maintains a constant glucose concentration in blood (Fig. 9.18). Lipid nutrients from the intestine are partly stored in the liver, partly converted to carbohydrates, and partly released to blood for utilization in cells or for storage in adipose tissues elsewhere.

Figure 9.15   The mammalian alimentary tract. Photo inserts: *A*, section through a salivary gland. Note the connective tissue stroma (light areas in photo) traversing the gland and binding groups of gland cells together. Note also the several small salivary ducts (dark rings). *B*, section through a portion of stomach wall. Note folded mucosa near top of photo. *C*, section through liver, showing parts of a few lobules injected to reveal the blood channels (dark). Blood brought by the hepatic portal vein to a lobule passes to the hepatic vein in the center of the lobule (see also Fig. 7.21). *D*, section through pancreas. The large round space is a branch of the pancreatic duct. *E*, section through the wall of the duodenum. The cavity of the gut is toward the top. Underneath the folded inner surface tissues note the glandular layer. Its secretion is discharged to the gut cavity and contributes to the composition of intestinal juice. *F*, section through a mucosal fold of the large intestine. Note the many mucus-secreting goblet cells in the mucosal lining.

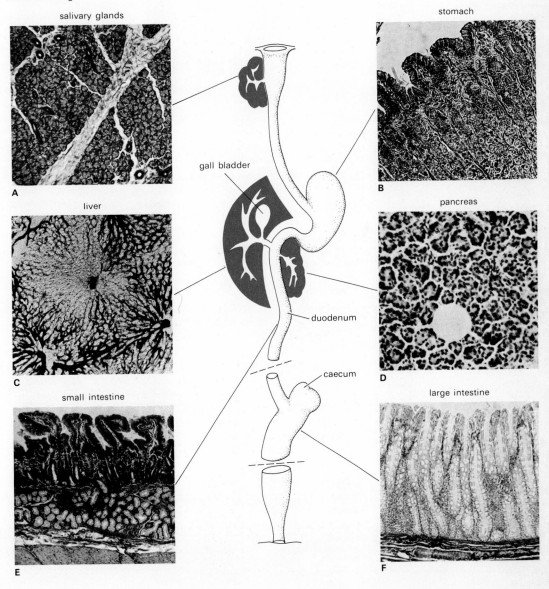

Table 8.   *The composition and action of digestive juices in mammals*

	SALIVA	GASTRIC JUICE	INTESTINAL JUICE	PANCREATIC JUICE	BILE
source	salivary glands	stomach wall	duodenal wall	pancreas	liver, gall-bladder store
pH	neutral	highly acid	alkaline	alkaline	alkaline
secretion started by	food in mouth reflex; thought of food	food in mouth; emotions; food in stomach ⟶ gastrin hormone, stimulates stomach wall	food contact in duodenum	secretin hormone from intestinal juice	secretin hormone from intestinal juice
carbohydrases	amylase, for polysaccharides maltase, for maltose		amylase, for polysaccharides disaccharases (maltase, sucrose, lactase), for disaccharides	amylase, for polysaccharides	
lipases		lipase (brought in from duodenum)	lipase	lipase	
proteinases		prorennin $\xrightarrow{\text{HCl}}$ rennin ⟶ curdles milk protein (caseinogen ⟶ casein) pepsinogen ⟶ pepsin, for proteins	amino-peptidases, for products of partial protein breakdown (polypeptides)	trypsinogen ⟶ trypsin,* for proteins chymotrypsinogen ⟶ chymotrypsin,* for proteins carboxy-peptidases, for products of partial protein breakdown (polypeptides)	
other components (all contain water, mucus, mineral ions)		HCl, macerates food, activates gastric proteinases	enterokinase, activates trypsinogen, chymotrypsinogen* secretin hormone, stimulates pancreas and liver secretions		bile salts, emulsify fats into colloidal drops bile pigments, excretion products from hemoglobin breakdown in liver

* The pancreatic proteinases are secreted in inactive form, which protects pancreatic tissues from being digested. Enterokinase from intestinal juice converts trypsinogen to active trypsin. The latter then transforms more trypsinogen to trypsin, and it also converts chymotrypsinogen to active chymotrypsin.

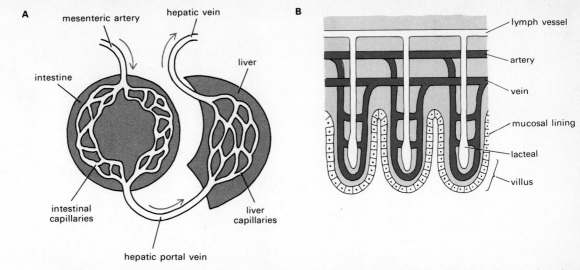

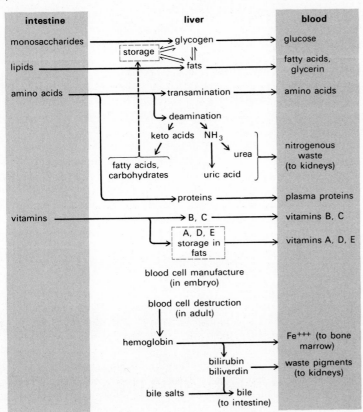

Figure 9.16 Circulatory paths in the intestine. A, the hepatic portal circulation. Blood reaches the intestine through a branch of the aorta, the mesenteric artery. Intestinal capillaries join and form a hepatic portal vein, which enters the liver. Blood leaves the liver through a hepatic vein and carries nutrients to all parts of the body. *B,* diagram of the blood and lymph circulation through the villi of the vertebrate intestine. Colloidal whole fat enters the blind-ended lymphatic lacteals, passes from there to the lymph circulation of the body, and so bypasses the liver. All other organic nutrients from the intestine are absorbed directly to the blood circulation of the villi and are transported from there to the liver.

Figure 9.17 Liver functions: summary.

Amino acids are partly released to blood for distribution, partly transaminated and deaminated. Any resulting keto acids are converted to carbohydrates or fats, and any resulting free amino groups appear as ammonia. This compound is eventually excreted, in many cases after prior conversion to urea or uric acid (Fig. 9.19). Liver cells also store fat-soluble vitamins and, as Fig. 9.17 indicates, they perform various functions not primarily nutritional in nature.

In effect, whereas the vertebrate alimentary tract makes foods available, it is the liver that determines what is to happen to them: liver cells regulate the kinds and quantities of foods sent out to body tissues. Representing the largest "gland" of the vertebrate body, the liver has been estimated to perform well over one hundred different functions. The adaptive advantage of these functions is important: whereas the metabolism of other animals reaches peaks just after food has been eaten, the metabolism of vertebrates can remain at a continuously steady level regardless of when or how much food is consumed.

Materials that cannot be or simply have not been digested eventually move to the posterior part of the alimentary tract. In this region permanent populations of bacteria live on the indigestible

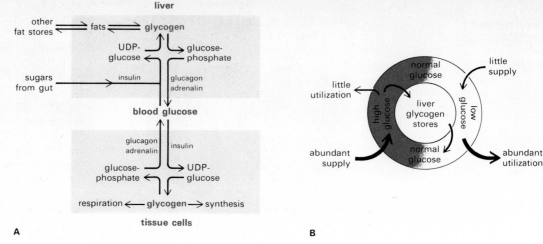

Figure 9.18 Carbohydrate balance in vertebrates. A, the pathways of interconversion between liver glycogen, blood glucose, and tissue glycogen. The hormones promoting given reactions are indicated in smaller type. Any excess liver glycogen can be converted to fat. Tissue glycogen is a main respiratory fuel as well as a starting compound in synthesis. *B,* the maintenance of blood-glucose balance by the liver. If much sugar is supplied to blood from food and little is used, then the blood-glucose concentration will tend to be high (left). Under such conditions the liver withdraws glucose and stores it as glycogen, thus establishing a normal glucose level (top). But if much glucose is used up and little is supplied, then the blood-glucose concentration will tend to be low (right). The liver then adds glucose to blood from its glycogen stores and so reestablishes the normal glucose level (bottom). By these means the liver maintains a constant blood-glucose concentration.

remains and often also excrete various substances useful to the animal host. For example, an animal frequently obtains some of the vitamins it requires from the byproducts of bacterial metabolism. Another result of bacterial activity is the *decay* of the intestinal remains, which are thereby transformed to *feces* ready to be egested.

In many animals the intestine continues without further structural modifications directly to the anus. In other cases a short elimination tube (*rectum*) connects intestine and anus. The excretory ducts, reproductive ducts, or both often open into the most posterior portion of the intestine, in which case this region is called a *cloaca*.

Breathing Systems

The basic component of any breathing system is a *breathing surface,* where O_2 can diffuse in and CO_2 can diffuse out (Fig. 9.20). Only in coelomate animals are the breathing surfaces part of specialized breathing systems. In all other animals (and indeed also in coelomates such as annelids), gas exchange takes the form of *skin breathing;* O_2 and CO_2 diffuse directly through all body surfaces ex-

posed to the environment—the skin mainly, but also the alimentary surfaces to some extent. Internal distribution in such cases is accomplished by gas diffusion from cell to cell and by transport through lymph, hemolymph, or blood. The skin also can serve as an accessory breathing organ in animals with specialized breathing systems (frogs, for example).

Such specialized systems function either in an aquatic or an aerial environment. In both cases the actual breathing surfaces are one-cell-thick epithelia exposed to the environmental medium on one side and the internal circulation on the other. Where the circulation is open, blood sinuses usually extend right up to the breathing epithelium and blood is in direct contact with it. In closed circulations, networks of capillaries cover the inside surfaces of the breathing epithelium.

Breathing systems in most invertebrates are *ectodermal* derivatives. Aquatic invertebrates have feathery, leaflike, or filamentous *gills* in various body regions, formed as outgrowths from the integument. Terrestrial invertebrates breathe through ingrowths from the integument. For example, land snails have saclike *lung chambers* under the forward edge of their shells (see Fig. 23.12); scorpions and spiders have chitin-lined chambers on the underside in which leaflike breathing plates form *book lungs;* and insects breathe by means of branching chitin-lined *tracheal tubes* that originate on the body surface and pipe air directly to all interior regions (see Fig. 9.20).

In chordates, by contrast, both gills and lungs are *endodermal* derivatives of the pharynx (Fig. 9.21). In the embryo a series of pouches grows out from each side of the pharynx, and a corresponding series of pouches grows in from the skin on each side of the head. The two sets of pouches on each side then meet and fuse and form *pharyngeal gill slits*. These channels are lined with a ciliated breathing epithelium. The cilia draw water through the mouth, strain food out of the water,

and pass it back to the esophagus. The water itself is diverted to the outside through the gill slits. The tissues under the breathing epithelium contain a rich supply of capillary blood vessels, and as water passes through the slits gas exchange takes place.

Ancestral vertebrates also evolved lungs. These animals are believed to have lived in freshwater environments that dried out periodically. The adaptive response to such hazards was the evolution of a pouch that grew out ventrally from the pharynx. This pouch was supplied with blood by a branch of the vessel (*aortic arch*) that also supplied the most posterior gill slit. Air could then pass through the mouth into the pouch and gas exchange could take place there. The pouch in effect was a lung, and the duct connecting it to the pharynx was a windpipe, or *trachea* (Fig. 9.22).

These ancestral vertebrates gave rise to two main groups of descendants. In one, the modern bony fishes, the air pouch is used as a *swim bladder*. In some of these fishes the connecting duct to the pharynx still exists, but in other cases it has degenerated. The second group of descendants continued to use the air pouch as a lung, and from this group of air-breathing fishes the land vertebrates evolved. The original single lung later became a paired sac, and an air passage from nose to pharynx developed as well. Air then could pass from nose to lung *across* the food channel, and food could pass from mouth to esophagus across the air channel; the pharynx came to be a common segment in the food and air channels.

Amphibia still develop pharyngeal gills during their aquatic larval stages. In most cases these gills later degenerate and the adults breathe through lungs instead. In reptiles, birds, and mammals gill pouches begin to form in early embryos but they never reach a functional slit stage. Nevertheless, the pattern of the blood circulation between heart and lung in the land vertebrates still indicates its derivation from the original aortic gill arches (Fig. 9.23 and see Fig. 9.22).

The tracheal duct from pharynx to lungs is lined with a ciliated, mucus-secreting epithelium. The tube is prevented from collapsing by C-shaped cartilage rings in its wall. At its lower end the trachea divides into a left and right *bronchus*, and after some distance each subdivides into *bronchioles* that branch repeatedly in turn. The diameters of these channels become progressively smaller and their walls become thinner. Only the ciliated

Figure 9.19 Metabolism of amino nitrogen. Urea production through the ornithine cycle in the vertebrate liver is outlined at left. The net input and output are in the color box under the cycle. NH_3 is also used in the synthesis of nitrogen bases, and later stepwise degradation of such bases yields the series of excretion products shown in the vertical sequence at right. Uric acid is a major excretory endproduct in, for example, insects, birds, and some reptiles; allantoin, in turtles and some mammals; allantoic acid, in some bony fishes; urea (produced in two ways as this diagram indicates) is excreted in most vertebrate groups not named here; and NH_3 is excreted in most invertebrate groups. Animals generally excrete mixtures of various endproducts but one or the other substance tends to predominate.

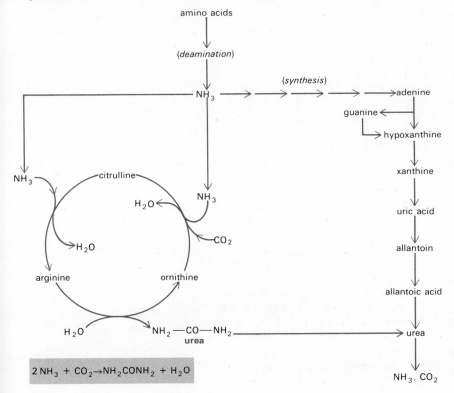

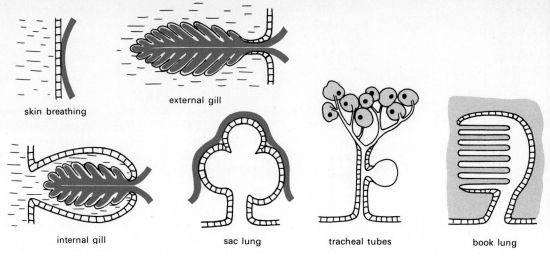

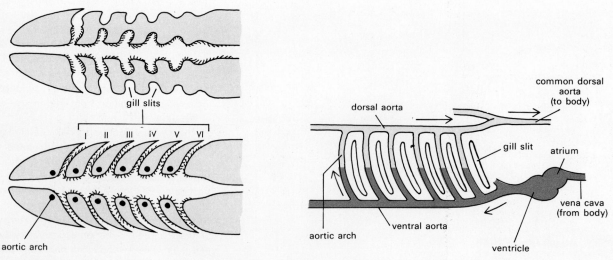

Figure 9.20 Breathing in invertebrates. Epidermis shown as layer of cells, blood and vessels in dark color. As indicated, all such breathing organs are outfolded or infolded integumentary derivatives. External and internal gills differ in principle only in whether or not they are tucked into chambers. Tracheae occur in terrestrial arthropods, fine air tubes here reaching individual cells; muscle-enveloped sacs folded out from a main tube (as shown) often serve as air reservoirs and air pumps. Book lungs occur in most spiders, the gas-carrying medium here being blood in blood sinuses (color). Free blood rather than blood vessels also transports gases in skin, gill, and lung systems of animals in which vessel circulations are absent. Some aquatic arthropods have book gills, constructed like book lungs but operating in water.

Figure 9.21 Vertebrate gills and their circulation. Top left, gills develop as endodermal pouches from the pharynx that meet corresponding pouches from the integument. Successive stages are indicated in anteroposterior sequence. *Bottom left,* the six basic pairs of ciliated vertebrate gill slits. Food is strained into the esophagus (toward right), while water flows out through the slits and in the process oxygenates blood in the aortic arches. *Right,* side view of the basic gill circulation on the left side of the body. Venous blood, dark shading; oxygenated blood, light shading.

lining layer and some connective tissue continue to the microscopic terminations of the branch system. Each such terminus is an *alveolus*, a raspberry-shaped sac, and the sum of all alveoli represents a lung. A thin layer of connective tissue around the alveoli carries nerves and a dense network of blood capillaries. These vessels lead venous, oxygen-poor blood from the pulmonary arteries to the lungs and arterial, oxygen-rich blood to the pulmonary veins away from the lungs (see Figs. 9.4 and 9.23).

In birds and mammals the breathing system also contains a voice-producing organ. Song birds have a *syrinx*, a cartilaginous chamber located where the trachea divides into bronchi. The comparable organ in mammals is the *larynx*, just behind the pharynx. The larynx contains *vocal cords*, a pair of membranes with a slitted opening between them. Cartilages in the wall of the larynx can be moved by muscles and in this manner the tension of the vocal cords can be altered voluntarily. Sounds of different frequencies are produced when exhaled air passes through the slit between the cords (see Fig. 9.23).

Figure 9.22   Gills and hearts of vertebrates. A, the basic six pairs of aortic arches and gill slits, the nasal chamber with external nares, and the primitive air pouch and duct to the pharynx. *B,* the first gill slit is reduced to a spiracle, and the first aortic arch has degenerated (broken line). *C,* the first two aortic arches are absent and the air pouch is a swim bladder. *D,* the air pouch functions as a lung, with a pulmonary artery branching to it from the sixth aortic arch. *E,* essentially the lungfish condition. *F,* with the disappearance of gills, some of the aortic arches degenerate. The heart is three-chambered and the ventricle extends posteriorly, which brings the two atria to an anterior position. *G, H, I,* a four-chambered heart is present, with venous blood confined to the right atrium and ventricle. Only the right aorta remains in birds, only the left in mammals.

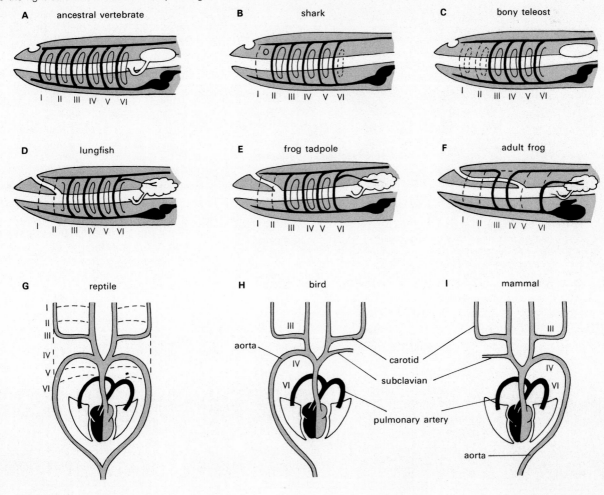

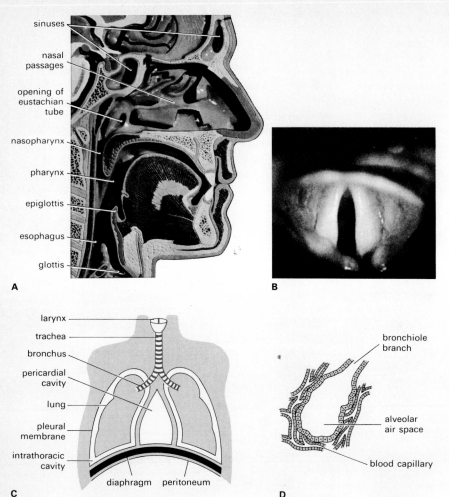

A

B

larynx
trachea
bronchus
pericardial cavity
lung
pleural membrane
intrathoracic cavity
diaphragm   peritoneum

C

bronchiole branch
alveolar air space
blood capillary

D

Figure 9.23 The mammalian breathing system. A, the upper parts of the human breathing system. In the pharynx the air path (from nasopharynx to glottis) crosses the food path (from mouth cavity to esophagus). *B,* the vocal cords of man. The view is from above, looking into larynx and trachea. *C,* the lower parts of the breathing system. Note that the intrathoracic cavity is sealed off. The dark bars in the trachea symbolize the cartilage rings that prevent this channel from collapsing. *D,* an alveolus of the lung, surrounded by blood capillaries. Connective tissue (not shown) envelops the entire alveolar system.

sinuses
nasal passages
opening of eustachian tube
nasopharynx
pharynx
epiglottis
esophagus
glottis

The mammalian breathing system includes a *diaphragm,* a muscular partition that seals the chest cavity off from the abdominal cavity. When the dome-shaped diaphragm contracts it flattens out and enlarges the chest cavity. The lungs are thereby "sucked" open and air rushes into them. Relaxation of the diaphragm permits elastic recoil of the lungs and expulsion of air. Breathing is mainly under nervous control. A breathing center in the medulla oblongata spontaneously generates

rhythmic impulses, with a basic frequency that ultimately is determined genetically. These impulses travel over motor nerves to the diaphragm, which contracts as a result and produces an inhalation (Fig. 9.24). As a result the lungs become inflated and stretched, and this stretching in turn stimulates sets of stretch receptors in the walls of the lungs. Impulses from there then temporarily *inhibit* the breathing center from sending more signals to the diaphragm. This muscle consequently relaxes and exhalation occurs. As the lungs now recoil, the stretch receptors there cease to be stimulated and the breathing center therefore ceases to be inhibited. At this point the center sends out motor impulses again, and a new cycle starts.

The breathing center is sensitive to the CO_2 concentration in blood, and when this concentration is high the rate of breathing increases. Thus, breathing becomes faster during intense physical or emotional activity, when rapid respiration raises CO_2 levels in blood. The gas here hastens its own removal through the lungs and faster breathing at the same time increases the oxygen supply, just when the tissues require more oxygen. If an animal holds its breath deliberately the accumulating CO_2 soon stimulates the center so strongly that breathing *must* be resumed, even against the most intense will; the automatic controls ensure that breathing does not occur too slowly.

Conversely, breathing slows down during rest or sleep, when respiration and CO_2 production are minimal. The extreme is the *hyperventilated* condition, produced, for example, when breathing is intentionally made as deep and as rapid as possible. Carbon dioxide then is exhaled so fast that its concentration in blood becomes abnormally low. In that case the breathing center usually ceases to operate temporarily, a "blackout" ensues, and breathing will remain stopped until the CO_2 concentration again has built up to a normal level.

Fresh inhaled air contains some 20 percent oxygen and 0.03 percent carbon dioxide. Air exhaled by man includes only 16 percent oxygen but as much as 4 percent carbon dioxide. A fifth of the available oxygen thus is retained in the body, and more than 100 times the amount of CO_2 is expelled. This exchange is governed by differences in gas pressures between blood and lung (Fig. 9.25). Venous blood flowing to the lungs from the body is comparatively oxygen-poor, but air in the

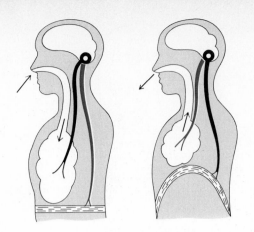

Figure 9.24 Breathing control. Left, the breathing center sends motor impulses to the diaphragm, leading to inhalation. *Right,* sensory impulses from the inflated lung inhibit the breathing center, resulting in exhalation.

lungs is oxygen-rich. A pressure gradient therefore leads from the lungs *to* blood, and more oxygen diffuses into the blood capillaries than in the reverse direction. At the same time, venous blood coming to the lungs is almost saturated with CO_2, whereas the air in the lungs contains a far lower concentration of this gas. A pressure gradient here leads *out* of the blood capillaries, and more CO_2 diffuses to the lungs than in the opposite direction. As a result blood ceases to be venous, and the incoming oxygen makes it arterial.

As in the lungs, gas exchange between blood and the body tissues likewise is governed by pressure gradients. Tissue cells use up oxygen and the higher pressure of this gas in blood therefore drives oxygen into the cells. Blood then ceases to be arterial. At the same time respiration in cells builds up a higher CO_2 pressure than in blood, and CO_2 therefore diffuses out of the cells. Blood here becomes venous.

The breathing systems of animals not only supply oxygen, but they also *excrete* CO_2 and water and, in many cases, other substances as well.

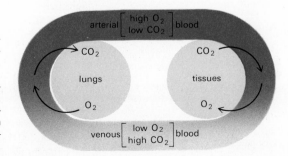

Figure 9.25 Gas exchange between lungs and blood and between body tissues and blood. Oxygen enters blood in the lungs, is carried as HbO_2 in red corpuscles, and leaves in the tissues. Carbon dioxide enters in the tissues, is carried as HCO_3^- in plasma (see section on circulation), and leaves in the lungs.

Excretory Systems

These systems are not named very adequately, for elimination of wastes is only one of their functions; they also retain and adjust. Indeed what is or is not "waste" at any given moment is precisely what an excretory system must determine.

Animal life originated in the sea, and the primitive members of most groups still live there. Correspondingly, the interior of most marine invertebrates is isotonic to sea water. When such an animal ingests food, it invariably takes in some sea water as well. If then an internal steady state is to be maintained, this excess salt and water must be eliminated. Moreover, the main byproducts of nutrient metabolism are CO_2 and H_2O from respiration and NH_3 from interconversions among nitrogenous compounds. These byproducts must be eliminated, too. From marine animals evolved descendants that live in fresh water, a far more dilute medium. A permanent osmotic differential exists here that draws water from the hypotonic freshwater environment into the hypertonic interior of an animal. To avoid internal flooding, therefore, such animals must eliminate far more water, and more continuously, than their marine relatives.

Thus the basic problem of excretion is to eliminate NH_3, CO_2, excess salts, and excess water. This means that any excretory structure must be able to discriminate between waste and nonwaste, for water and salts are both waste and nonwaste simultaneously; the difference is one of amount, not kind. Ammonia is a toxic waste in almost any amount, and CO_2 is largely waste; it is excreted primarily through the breathing surfaces.

If all or most of the cells of an animal are exposed directly to the environment, each cell can be its own excretory apparatus; waste substances can be eliminated to the external medium by diffusion and active transport across the cell boundary. Such unicellular excretion occurs in all protozoa, sponges, and radiates. Most marine members of these groups are without excretory structures of any kind (other than the cell membranes themselves), and the freshwater members have *contractile vacuoles* specialized for excretion of large quantities of water. Such cells often must excrete an amount of water equal to the volume of the whole cell every 4 min (Fig. 9.26).

If most cells of an animal are not exposed directly to the external environment, the interior

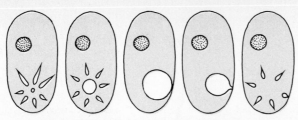

Figure 9.26 Contractile vacuole activity. From left to right, cytoplasmic water channels empty into a growing vacuole, which then expels water through a point on the cell surface. In the meantime water accumulates again in channels and a new vacuole begins to form.

cells can excrete only to lymph, hemolymph, or blood. Specialized excretory systems then are required, and these must operate in close association with the body fluids; the excretory structures must act as screening devices that retain valuable substances in the body fluids and collect only wastes for removal to the outside. In all systems this excretory function appears to be performed through one or more of three basic processes: *filtration, reabsorption,* and *secretion* (Fig. 9.27).

Filtration takes place between the body fluids and the interior space of an excretory structure. The pressure of the fluids—blood pressure in vertebrates, for example—supplies the force necessary for filtration. Cells and proteins in blood or hemolymph normally cannot pass through the filter, but most other components can. Thus the filtrate that collects in the excretory structure is essentially lymph, here called *initial urine.* A separation of waste from nonwaste then occurs when initial urine flows through another region of the excretory system on its way to the outside. Reabsorption takes place there: excretory cells in contact with initial urine remove substances "judged" to be valuable and return them to the body fluids. When salts or other dissolved mate-

rials are reabsorbed, the remaining urine becomes more dilute, or hypotonic to the body fluids; when water is reabsorbed, the remaining urine becomes hypertonic and more concentrated than the body fluids.

The third process, secretion, can take place either in the same general region of the excretory system or in another body part altogether (gills, for example). Secretion transfers materials from the body fluids to urine or the external environment directly. Secretion and reabsorption therefore operate in opposite directions, but the substances reabsorbed often differ greatly from those being secreted. The fluid then present, *final urine,* is discharged to the outside either continuously or, after accumulating in a *bladder,* intermittently. In many cases the excretory system opens directly to the outside. In others it empties into a cloaca or the outgoing ducts of the reproductive system.

Among the simplest excretory systems are *protonephridia.* A protonephridium consists of branching tubules that lead to the outside of the body through one or more openings (*nephridiopores;* Fig. 9.28). The inside end of each branch tubule is closed off either by a *flame bulb* or by *solenocytes.* A flame bulb is a hollow, flask-shaped structure, often formed from a single cell. The outer surface of the flask is in contact with the body fluids, and on the inner surface is a tuft of beating cilia suggestive of a flickering flame. A solenocyte too is a hollow tubular cell, but instead of cilia a long, beating flagellum is present. In protonephridia of both types the cilia or flagella maintain a flow of wastes from the body fluids to the outside. Reabsorption and secretion can take place along the ducts, but it is not known too well to what extent these processes actually occur. Protonephridial systems develop as ingrowths from embryonic ectoderm, and they occur quite widely in acoelomate and pseudocoelomate animals and also in a few coelomate groups.

Another type of excretory structure, found particularly in invertebrates with large coeloms (annelids, for example), is a *metanephridium.* It too is a duct that opens at a nephridiopore, but the inner end of the duct is open to the coelom, at a ciliated, funnel-shaped *nephrostome.* The cilia propel coelomic fluid through the duct, where reabsorption and secretion take place. Filtration has occurred earlier, in the peritoneum that lines the coelom; as lymph passes from body tissues to the coelom, the fluid is filtered by the peritoneal

Figure 9.27 Excretory processes. In an excretory system a filtrate of blood (or lymph) forms initial urine, from which some materials are reabsorbed and to which other materials are secreted. These functions are performed by the cells lining the system. Final urine is the ultimate product.

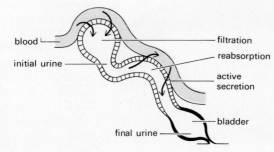

blood

initial urine

filtration

reabsorption

active secretion

bladder

final urine

cells. Near the nephridiopore a tubule can be enlarged as a bladder, and tubules can open to the outside individually or in groups with a single exit. Metanephridia probably form in part as mesodermal outgrowths from the coelomic lining, in part as ectodermal ingrowths from the integument.

Invertebrates with small coeloms have *renal glands* of various types. These often are evolutionary derivatives of metanephridia, and they again are essentially tubes that filter the body fluids and that reabsorb and secrete through the walls. Insects must cope with a problem common to all terrestrial animals, reduced availability of external water. Thus, whereas a marine animal and even more so a freshwater animal is burdened with an internal water excess, a terrestrial animal has the very urgent opposite problem of conserving as much internal water as it can. The problem is solved in part by water-retaining, evaporation-resistant integuments, in part by the excretory system: land animals have excretory systems in which water is reabsorbed, not excreted. As a result, the urine becomes highly concentrated and hypertonic to the body fluids.

A concentrated urine creates a further problem, however, for NH_3 is toxic and must be diluted by

Figure 9.28   Excretory systems of invertebrates. A, protonephridia of the flame-bulb type; single-celled flame bulb at left (note flamelike tuft of cilia), branch pattern of entire system at right. *B,* protonephridia of the solenocytic type; single-celled solenocyte at right (note flagellum), branch pattern of entire system at left. *C,* the structure of a metanephridium; body fluids between the body wall and the peritoneum are filtered into the coelomic cavity. *D,* the excretory organ of a clam. The nephridiopore opens to a water channel that leads out of the animal. *E,* the excretory organ of an aquatic arthropod (lobster). Antennal, or "green," glands such as this often contain excretory giant cells (as shown), with two or more nuclei and cytoplasmic excretory channels. *F,* the excretory Malpighian tubules of terrestrial arthropods (insects).

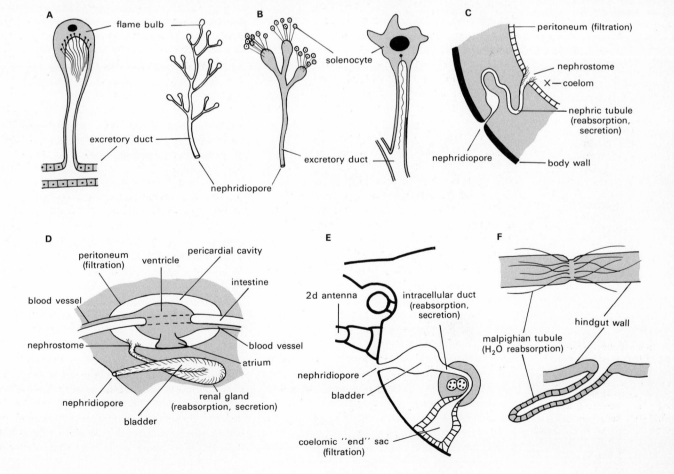

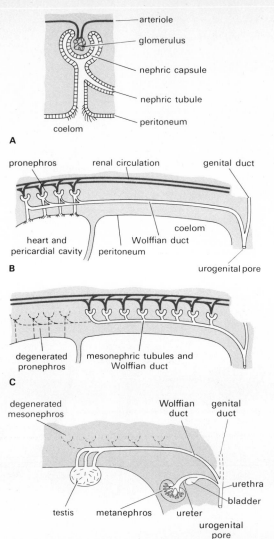

Figure 9.29 The vertebrate kidney. A, a primitive nephron unit. Note ciliated funnel (*nephrostome*) opening to coelom. B, the pronephric kidney. Note ciliated openings into the pericardial (coelomic) cavity. More nephron units actually occur than are sketched here. C, the mesonephric kidney. The pronephros has degenerated, and new nephrons have developed along the Wolffian duct. Note absence of ciliated openings to coelom. D, the metanephric kidney. Mesonephric nephrons have degenerated, and a new large collection of nephrons has developed from the hind part of the Wolffian duct. In males this duct also makes new connections with the testes and becomes the sperm duct (as shown). In females the Wolffian duct degenerates, too, but the old genital duct becomes the egg channel from the ovaries to the outside. See also Fig. 10.16.

relatively large amounts of water. The land animal cannot spare such water. Instead, it relies on converting NH_3 to another, less toxic compound. One such compound is *urea*; another is *uric acid* (see Fig. 9.19). Either can be excreted in much smaller amounts of water. Indeed, uric acid can be excreted in virtually solid form. Insects, birds, and most reptiles excrete uric acid, and their urine is almost solid; the hindgut reabsorbs virtually all the water still left in urine. Mammals excrete a concentrated liquid urine containing mainly urea.

In the vertebrate *kidney*, the functional unit is a *nephron*. In its primitive form a nephron consists of three main parts (Fig. 9.29): a *glomerulus*, a tiny ball of blood capillaries; a *nephric capsule*, a cup-shaped outgrowth from the dorsal coelomic lining that partially envelops the glomerulus; and a *nephric tubule*, a relatively short duct that leads from the nephric capsule to a *nephrostome* opening located along the roof of the pericardial coelom over the heart. On each side of the body the tubules from all nephrons also connect with a common channel, the *Wolffian duct*, which passes posteriorly and empties to the outside at a urogenital pore. Such an excretory system is a *pronephros*, or "head kidney." In it the nephric capsules filter both the blood and coelomic fluid.

A pronephros occurs in the larvae of all fishes and in the adults of a few, but in most fishes and in amphibia such an excretory system is a transient, embryonic kidney. Before these animals become adults all parts except the Wolffian duct degenerate and a new kidney, the *mesonephros*, develops. This structure again forms along the roof of the coelom, but it extends from just behind the position of the earlier pronephros right to the level of the anus. The mesonephros too consists of sets of nephrons, yet nephrostomes are no longer present; the nephric capsules filter only the blood. Tubules again carry this filtrate to the original Wolffian duct.

Being kidneys of vertebrates that originally evolved in fresh water, both the pronephros and mesonephros are adapted mainly for elimination of water; they largely filter, but they are not constructed to effect much reabsorption of either water or salt. Consequently, the urine of freshwater fishes is more dilute than the body fluids. Salt loss through these kidneys is compensated for by absorption of salts from the environment, typically by means of special *salt-absorbing glands* located in the gills (Fig. 9.30).

Ancestral freshwater fishes have given rise to all the marine fishes now in existence. These live in a hypertonic environment, hence marine fishes tend to *lose* water osmotically; and they are in danger of dehydrating internally even though (because, really) a vast ocean is around them. The *bony* marine fishes solve this dilemma in two ways. First, they reabsorb as much water as their short nephric tubules will permit. In some marine bony fishes, moreover, the glomeruli of the mesonephros degenerate altogether. This makes the kidney *aglomerular* and unable to filter much water from blood to begin with. A copious urine then cannot form, and the little that does form remains nearly isotonic to blood. Second, to compensate for the water lost by osmosis, marine bony fishes swallow sea water deliberately; but they excrete the salts of the swallowed water through specialized *salt-excreting glands* in the gills (see Fig. 9.30).

A different solution to the osmotic problem has evolved in the *cartilage* fishes such as sharks. These marine animals convert NH_3 to urea and they retain this compound in all tissues and in blood. Urea concentrations thereby become high enough to make the body interior slightly hypertonic to the external sea water. Such fishes therefore actually acquire some water osmotically, and their mesonephric kidneys can excrete hypotonic urine, like those of freshwater fishes (see Fig. 9.30).

Fishes are able to "make do" with a mesonephros mainly because ample water—even if it is sea water—is always available. But land verte-brates lack this convenience. As noted, a basic excretory requirement for all terrestrial forms is efficient retention of internal water and production of a highly hypertonic urine. Yet inasmuch as a mesonephros cannot produce such a urine in fishes, it can be expected to be unable to do so also in land vertebrates. Indeed, reptiles, birds, and mammals no longer depend on a mesonephros but on a newly evolved *metanephros*. This type of kidney is capable of forming a hypertonic urine, and it is because a metanephros actually has evolved that some vertebrates have been able to become efficient land animals.

In the embryos of the land vertebrates, a pronephros and later a mesonephros develop as transient structures only. They degenerate usually well before hatching or birth, and in their place the metanephros arises as the adult kidney. This organ forms behind the mesonephros (see Fig. 9.29). One portion of it develops as a branch duct that grows out from the hind region of the Wolffian

Figure 9.30 Water and salt balance in vertebrates. A, freshwater fish, medium hypotonic. Osmotic water uptake is balanced by salt uptake through gills and excretion of copious hypotonic urine. B, marine bony fish, medium hypertonic. Osmotic water loss is balanced by salt and water ingestion, salt secretion from gills, degeneration of mesonephric glomeruli to reduce water loss in urine, and excretion of nearly isotonic (only slightly hypotonic) urine. C, marine cartilage fish. Urea retained in body makes body hypertonic to sea water. Osmotic water uptake then is balanced by excretion of hypotonic urine. D, land vertebrate. Water is conserved by metanephric reabsorption and excretion of hypertonic urine.

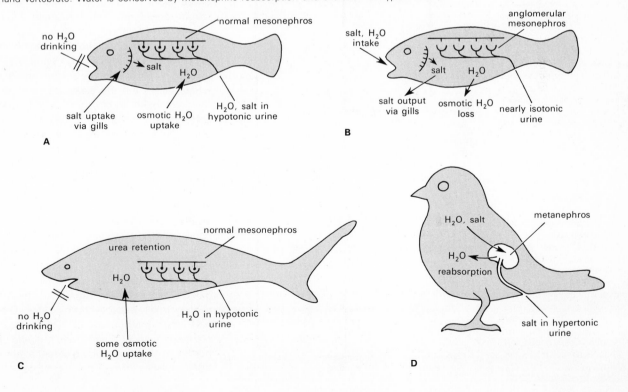

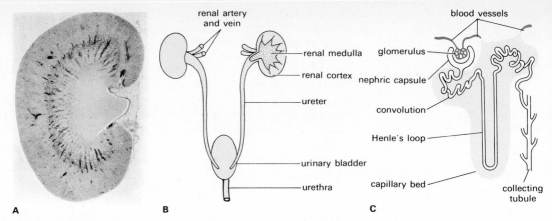

Figure 9.31 Kidney structure. *A*, section through rat kidney. Note outer renal cortex, inner renal medulla. *B*, the kidneys, their ducts, and the bladder (kidney on right in section). In males the urethra also carries sperms from the sperm ducts (not shown, but see Fig. 10.16). *C*, a nephron unit, showing the convoluted portions of the tubule and Henle's loop. The capillary bed that envelops the coiled parts of a nephron is indicated as the tinted area.

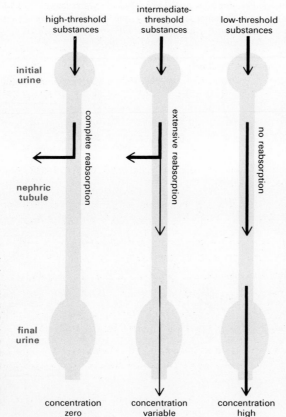

Figure 9.32 Selective excretory action of the cells of a nephric tubule. The color areas symbolize nephric capsules, tubules, and bladders of excretory systems. High-threshold substances like glucose are reabsorbed more or less completely (left). Low-threshold substances like urea are not reabsorbed at all (right). And intermediate-threshold substances like water and many mineral ions are reabsorbed in part, depending on the amounts already present in blood (center).

duct. This branch becomes the *ureter* of the adult system; it divides repeatedly at its free end into many tiny *collecting tubules*. The second portion consists of numerous nephrons that develop around the terminals of these collecting tubules. These nephrons differ from those in a mesonephros in one fundamental respect: their nephric tubules are far longer (Fig. 9.31).

Each metanephric tubule contains two highly coiled (*convoluted*) regions, connected by an extensive loop. This *Henle's loop* and the convoluted regions are the essentially new features of the metanephric kidney; they function as a *water-reabsorbing* segment, and their presence permits the metanephros to produce a highly concentrated, hypertonic urine. The two kidneys of man, for example, filter the equivalent of about 160 qt of blood a day (5 qt once every 45 min), but only about $1\frac{1}{2}$ qt of fluid per day is actually released in the form of urine; some 99 percent of the water in the blood filtrate is reabsorbed through the convoluted portions and the loops of Henle.

Among substances normally reabsorbed completely is glucose. Initial urine contains glucose in the same concentration as in blood. But all this glucose tends to be returned to the blood, and none normally escapes in final urine. Other materials undergoing more or less complete reabsorption include amino acids, lipids, vitamins, hormones—in short, all the essential nutrients and other usable supplies in transit to cells. Such materials are *high-threshold* substances; they are reabsorbed entirely unless their concentrations in blood are excessively high (Fig. 9.32).

So-called *low-threshold* substances on the contrary are not reabsorbed by the cells of the nephric tubules. Among these substances are nitrogenous byproducts such as urea, pigmented blood-breakdown products, and other materials the tubule cells "determine" to be outright wastes. They become highly concentrated as water is withdrawn from initial urine. In man, for example, final urine contains some seventy times more urea than an equal volume of initial urine. As a result of such increases in concentration the osmotic pull of urine becomes greater. But despite this force, which tends to draw water *from* blood *to* urine, tubule cells nevertheless continue to transport more water from urine to blood. In counteracting this prevailing osmotic gradient tubule cells expend large amounts of energy.

A third group of materials comprises *intermediate-threshold* substances. They are or are not absorbed depending on whether or not their blood concentrations are at optimal levels. Mineral ions belong to this category. For example, if blood already contains normal concentrations of sodium chloride, then additional salt intake with food will be followed by salt excretion. But if the internal supply is low, the tubule cells will reabsorb salt and will not reduce the concentration in blood any

further. Water itself qualifies as an intermediate-threshold substance.

In reptiles and birds, the ureters from the kidneys empty into a cloaca along with the reproductive ducts. Most mammals lack a cloaca, and in males the ureters and reproductive ducts join and lead through a common *urethra* to an independent *urogenital orifice*. Female mammals have separate genital openings and urine here passes through the urethra to an independent *urinary orifice* (see Fig. 9.29). Many reptiles and most mammals have a *bladder* near the posterior end of the ureter, but this organ is absent in birds (as would be expected in animals whose urine is semisolid; it mixes with the feces in the cloaca).

Although a kidney or its equivalent is functionally the most essential component of an excretory system, it is not the only component. As already noted, major excretory functions are performed also by the breathing organs, the large intestine, and the skin of many animals. For example, the sweat glands of mammals virtually represent tiny kidneys, and their water- and salt-excreting activities ease the work load of the main kidneys considerably. Additional excretory functions are carried out by structures such as salivary glands, nasal epithelium, tear glands, and liver. Indeed, any body part that opens to the outside either directly or indirectly contributes to excretion at least to some extent.

Review Questions

1. Define lymph, hemolymph, pseudocoel, hemocoel, blood, pericardium. What are the basic functions of lymph and blood?

2. Distinguish between open and closed circulations. Relate each to coelom size. Which animal groups have open and which have closed circulations? Which are without blood circulations?

3. What organs are the main components of a circulatory system? A lymph system? What is an endothelium? How do arteries differ from veins and capillaries (a) structurally, (b) functionally?

4. What is the structure of different types of heart? What are lymph hearts? Gill hearts? Lymph nodes? Review the course of blood and lymph circulation in a vertebrate.

5. How is the beat of a heart controlled? What is a pacemaker? How is blood pressure controlled? What is meant by vasomotion? Through what forces does blood return to the heart?

6. Describe the composition of blood plasma. Name some of the plasma proteins and state their functions. What is immunity and how is it produced? Name the different kinds of blood cells and state their function.

7. Review the clotting process. Name respiratory pigments, state in which animal groups each kind is encountered, and describe their functions. How is CO_2 transported in the body fluids?

8. Distinguish between short-distance and long-distance transport of nutrients and show how each is accomplished. Describe the patterns of intracellular and extracellular alimentation. In which kinds of animals is each encountered?

9. What is the typical tissue structure of an alimentary tract? Describe some of the different ways by which ingestion and mechanical digestion take place. What is peristalsis? An eversible pharynx?

10. How and where does chemical digestion occur and what are the detailed chemical results? What digestive functions are performed by each of the different portions of an alimentary tract? How and where does nutrient absorption take place? What is the role of lymph in nutrient absorption?

11. Describe the structure and function of the vertebrate liver. Show how this organ processes each of the main classes of foods. How are urea and uric acid produced?

12. What different breathing processes and systems are encountered among invertebrates? How does each of these systems operate? What type of breathing system is characteristic of chordates and vertebrates? Review the pattern of the basic vertebrate gill circulation.

13. Show how the gill and lung circulations have changed in the course of vertebrate evolution. How did the vertebrate lung evolve originally? Review the pattern of blood circulation between lung and heart.

14. Describe the makeup of the mammalian breathing system and show how breathing movements are produced and controlled.

15. What is the basic structure and function of an excretory system? Show how a contractile vacuole operates. What does it excrete? Distinguish between excretory filtration, secretion, and reabsorption.

16. Describe the structure and operation of protonephridia and metanephridia. What other types of excretory structures occur in invertebrates?

17. Show how the pattern of water and salt excretion varies according to whether an animal lives in a marine, a freshwater, or a terrestrial environment. Which animals produce hypotonic, isotonic, or hypertonic urine?

18. Describe the structure of the vertebrate pronephros, mesonephros, and metanephros. How do they differ functionally? What is a nephron and how does it operate?

19. Describe the excretory patterns of different kinds of fishes and land vertebrates. Which groups of fishes do and which do not swallow water and why? What is an aglomerular kidney?

20. What is Henle's loop? A ureter? A urethra? Which kinds of vertebrates have salt-absorbing and which have salt-excreting glands? What body parts other than the excretory system play a role in excretion?

Collateral Readings

Adolph, E. F.: The Heart's Pacemaker, *Sci. American*, Mar., 1967. Recommended reading for further background.

Botelho, S. Y.: Tears and the Lacrimal Gland, *Sci. American*, Oct., 1964. A study of the secretion of the tear glands and of the functions of tears.

Burnet, M.: How Antibodies Are Made, *Sci. American*, Nov., 1954. A hypothesis regarding this still unsolved problem is described.

————: The Mechanism of Immunity, *Sci. American*, Jan., 1961. The article examines the antibody-antigen interplay in immune reactions.

Chapman, C. B., and T. H. Mitchell: The Physiology of Exercise, *Sci. American*, May, 1965. The adaptive changes that occur in the breathing and circulatory systems during exercise.

Comroe, J. H., Jr.: The Lung, *Sci. American*, Feb., 1966. A recommended review of the biology of this organ.

Dougherty, E. C.: "The Lower Metazoa," University of California Press, Berkeley, 1963. The emphasis is on comparative biology and physiology.

Fox, H. M.: Blood Pigments, *Sci. American*, Mar., 1950. The title describes the contents adequately.

Hong, S. K., and H. Rahn: The Diving Women of Korea and Japan, *Sci. American*, May, 1967. The breath-holding capacity of these divers is examined.

Irving, L.: Adaptations to Cold, *Sci. American*, Jan., 1966. The circulatory redistributions of blood in the body play an important role in counteracting external temperature changes.

Kylstra, J. A.: Experiments in Water-breathing, *Sci. American*, Aug., 1968. How air breathers can obtain adequate amounts of oxygen under water.

Mayer, J.: Appetite and Obesity, *Sci. American*, Nov., 1956. A noted nutritionist discusses the control of food intake in man.

Mayerson, H. S.: The Lymphatic System, *Sci. American*, June, 1963. The functions of this system are described.

McKusick, V. A.: Heart Sounds, *Sci. American*, May, 1956. The production and meaning of these sounds are examined.

Nossal, G. J. V.: How Cells Make Antibodies, *Sci. American*, Dec., 1964. Experiments on single cells in culture show how genes appear to control antibody production.

Perutz, M. F.: The Hemoglobin Molecule, *Sci. American*, Nov., 1964. An account on the detailed configuration of the four polypeptide chains that compose hemoglobin.

Ponder, E.: The Red Blood Cell, *Sci. American*, Jan., 1957. A discussion of the significance of red corpuscles in circulation and breathing.

Porter, R. R.: The Structure of Antibodies, *Sci. American*, Oct., 1967. The chemistry of this class of proteins is examined.

Prosser, C. L., and F. A. Brown, Jr.: "Comparative Animal Physiology," 2d ed., Saunders, Philadelphia, 1961. Chapters 4 and 5 of this text contain detailed discussions of alimentary functions in various animal types.

Ramsay, J. A.: "Physiological Approach to Lower Animals," 2d ed., Cambridge University Press, Cambridge, 1968. A small book, well worthwhile consulting for additional data.

Schmidt-Nielsen, K.: Salt Glands *Sci. American*, Jan., 1959. An examination of specialized glands in marine birds and reptiles that play an important role in maintaining the salt balance of these seawater-drinking animals.

Scholander, P. F.: The Master Switch of Life, *Sci. American*, Dec., 1963. The article describes the vasomotor mechanism of blood distribution in relation to the oxygen-carrying role of blood.

Smith, H.: The Kidney, *Sci. American*, Jan., 1953. A recommended article on the function of the mammalian kidney.

Speirs, R. S.: How Cells Attack Antigens, *Sci. American*, Feb., 1964. The role of defensive cells in protecting against antigens is described.

Surgenor, D. M.: Blood, *Sci. American*, Feb., 1954. A good general discussion.

Wiggers, C. J.: The Heart, *Sci. American*, May, 1957. The structure of the heart muscle and the quantitative work of the human heart are described.

Wood, J. E.: The Venous System, *Sci. American*, Jan., 1968. An excellent article on the structure and functions of these vessels.

Wood, W. B., Jr.: White Blood Cells vs. Bacteria, *Sci. American*, Feb., 1951. A good account on the infection-combating activity of white cells.

Zweifach, B. J.: The Microcirculation of the Blood, *Sci. American*, Jan., 1959. An examination of the steady-state functions of the capillary circulation.

The chapters of this part deal with the three kinds of processes that enable animals to survive on a long-term basis: *reproduction*, a mechanism of self-perpetuation that increases the number of animals and permits formation of successive animal generations; *development*, the means by which new generations are actually molded from parts of the old; and *heredity*, a mechanism of adaptation that controls the forms and functions produced by development and plays a role in fitting a new generation to its environment. Reproduction implies and includes both development and heredity, yet all three processes contribute separately to the continuity of animals in time and their expansion in space.

PART 4
ANIMAL CONTINUITY

10

REPRODUCTION

Reproduction can be defined broadly as an extension of living matter in space and time. The self-perpetuative importance of this process is clear, for the formation of new living units makes possible replacement and addition at every level of organization. Among molecules or cells, among whole organisms or species, replacement offsets death from normal wear and tear and from accident and disease. *Healing* and *regeneration* are two aspects of replacement. Apart from such maintenance functions, net addition of extra units results in four-dimensional *growth,* or increase in the amount of living matter.

The creation of new units requires raw materials; reproduction at any level depends on ample nutrition. Duplication of a particular living unit also implies prior or simultaneous duplication of all smaller units in it. Reproduction therefore must occur on the molecular level before it can occur on any other.

Forms of Reproduction

Reproductive Methods

Animal propagation includes at least two steps. First, a *reproductive unit* forms from a parent animal. Second, an offspring arises from the reproductive unit through *development.* Reproductive units are of two kinds, *vegetative bodies* and *gametes.*

A vegetative body can be almost any portion of the parent animal; such a reproductive unit is not specialized exclusively for reproduction and can range in size from the whole parent body to a minute fragment of that body. Moreover, the unit develops into an offspring directly and independently. The main forms of vegetative reproduction are *binary fission, multiple fission, fragmentation,* and *budding* (Fig. 10.1).

Binary fission is essentially mitotic cell division. In most protozoa cell division is equivalent to reproduction of the whole individual, and this process actually represents the *only* form of reproduction in such types. In Metazoa binary fission contributes to *cell replacement,* as in regeneration or wound healing, or adds to *cell number,* as in growth. Multiple fission is a variant form of cell division, in which a parent cell divides into several smaller cells simultaneously. Thus, numerous offspring cells, each with its own nucleus, form from one parent cell at the same time. Multiple fission occurs in various groups of protozoa.

In fragmentation, a parent animal spontaneously splits up into two or more fragments, each then regenerating the missing body parts. For example, certain sea anemones and various types of worms occasionally fragment into two or more units (Fig. 10.2). Fragmentation also occurs when an animal becomes split or cut through injury. In many cases each resulting piece can then regenerate into a whole new animal. Earthworms, starfishes, and sponges are good examples of animals

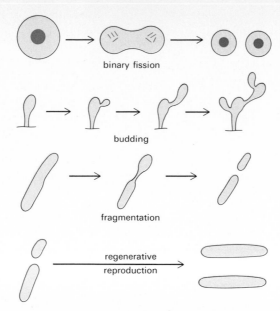

binary fission

budding

fragmentation

regenerative reproduction

Figure 10.1 Vegetative reproduction: summary of the main variants.

Figure 10.2 Vegetative reproduction. A, fragmentation. A sea anemone fragments lengthwise into two offspring organisms. *B,* regenerative reproduction. An arm of a starfish regenerates all missing parts and becomes a whole animal. *C,* budding in *Hydra.* When such buds are mature they separate from the parent and pursue an independent existence.

separate from the parent or remain attached and later bud in turn. Large *colonies* of joined individuals can form in this fashion. Budding is particularly widespread in sessile animals; it occurs, for example, among sponges, cnidarians, flatworms, ectoprocts, and tunicates (see Fig. 10.2).

The chief adaptive advantage of all forms of vegetative reproduction is that the process can occur without dependence on other animals. The only requirements are favorable environmental conditions and ample food supplies. Thus, vegetative reproduction becomes particularly useful if an animal is sluggish or sessile and therefore relatively isolated from contact with other members of the species. Moreover, the method requires little tissue or cell specialization, and it also offers obvious advantages in the form of regenerative reproduction. However, vegetative reproduction entails a major limitation that can be circumvented only by the second basic method of multiplication, reproduction through gametes.

The smallest parental unit that contains the genetic information and the operating equipment representative of a whole animal is a single cell. Accordingly, the minimum unit for the construction of an animal should be one cell. This is actually the universal case. Regardless of whether or not it can also reproduce vegetatively, every metazoan is capable of reproducing through single reproductive cells, or *gametes.* Such cells are *specialized* for reproduction, they are usually formed in specialized reproductive tissues or organs of the parent, and, in contrast to vegetative units, they

with great regenerative capacity. This capacity varies with the species, and in many animals it is highly limited. Salamanders can regenerate a whole limb, but a limb cannot regenerate a whole salamander. In most vertebrates the regeneration potential is not even as great as in salamanders but, as in man, is limited to the healing of relatively small wounds.

In reproduction by budding, small groups of cells form *buds* in different body regions of parent animals. A bud develops into an adult that can

A

B

C

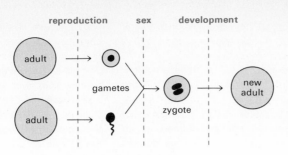

reproduction   sex   development

gametes

adult

adult

zygote

new adult

Figure 10.3 Gametic propagation. The three basic phases: reproduction (gamete formation), sex (gamete union), and development (zygote growth).

usually cannot develop directly. Instead, they must first undergo a *sexual process*, in which two gametes fuse. The cells therefore are also called *sex cells*, or *germ cells*. Male gametes are *sperms;* female gametes are *eggs*. A mating process makes possible the pairwise fusion, or *fertilization*, and the fusion product is a *zygote*.

Development of gametes to adults normally cannot occur until fertilization has taken place. Where sex occurs, therefore, it is interpolated between the two basic phases of the reproductive sequence, formation of reproductive cells and development of such cells into adults. In a strict sense, evidently, reproduction should not be labeled as being "asexual" or "sexual," as is often done. Sex is *not* a "reproductive" process but one of *gamete fusion*. The fundamental "reproductive" event in all forms of multiplication is formation of reproductive units. The rest is development, and it is this developmental phase that may or may not require sexual triggering. Consequently, whereas *development* can start sexually or asexually, reproduction as such, or formation of reproductive units, is always "asexual," even in the case of gametes. After gamete formation sex must *follow*, and the term "gametic reproduction" can be used

to distinguish this type of multiplication from the vegetative type (Fig. 10.3).

Gametic reproduction entails serious disadvantages. The process depends on chance, for gametes must meet and very often they simply do not. Meeting requires locomotion, moreover, but neither eggs nor many animals can move. Above all, gametic reproduction requires a water medium. In air, gametes would dry out quickly unless they had evaporation-resistant shells. But if two cells were so encapsulated they could then not fuse together. As will become apparent below, terrestrial animals actually can circumvent this dilemma only by means of special adaptations.

However, all these various disadvantages are relatively minor compared to the one vital advantage offered by gametes. This advantage is a result of sex.

Sex and Reproduction

The role of sex is revealed most clearly in protozoa, in which sexual processes and reproduction do not occur together. These organisms reproduce vegetatively when food is plentiful, when a population is not too crowded, and when environmental conditions are optimal in general. On the contrary, sexual processes take place under unfavorable conditions, such as overcrowding or lack of food (Fig. 10.4).

One kind of sexual response is *syngamy;* two whole protozoa come to function as gametes and fuse as a zygote. Another involves *conjugation;* two mating cells fuse partially by forming a cytoplasmic connecting bridge between them. The nucleus of each cell then gives rise to two *gamete nuclei,* and one of these migrates through the bridge to the other cell. After such a nuclear exchange, the two gamete nuclei now present in each conjugating individual fuse as a zygote nucleus and the two mating cells then separate.

Thus the sexual process is fundamentally quite distinct from reproduction. Protozoa do not increase in numbers through sex—if anything quite the contrary; two cells become one in syngamy or remain two as in conjugation. In all Metazoa sex and reproduction are equally distinct, even though in most cases the two processes do occur together. Moreover, note that sexual activity tends to take place particularly during periods of environmental stress. Indeed, among most animals in the temperate zone, sex typically occurs during spring

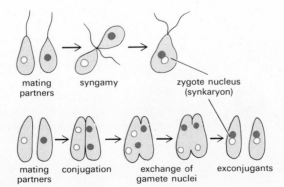

mating partners   syngamy   zygote nucleus (synkaryon)

mating partners   conjugation   exchange of gamete nuclei   exconjugants

Figure 10.4 Sexual patterns, as in protozoa. *Top*, syngamy, the fusion of two gamete cells. *Bottom*, conjugation, or exchange and fusion of gamete nuclei.

or autumn. The process here is a response to the stress conditions of the preceding season and it anticipates those of the following season.

Just how are sexual processes effective against conditions of stress? Events in protozoa supply the general answer: every cell resulting from the sexual process contains the genes of *both* parental cells. Basically, therefore, sex can be defined as an accumulation, in a single cell, of genes derived from two relatively unrelated cells. One method by which such a gene accumulation is achieved is syngamy, or fusion of two whole cells, as in some protozoa and in all Metazoa after various mating processes; another method is exchange of duplicate nuclei, as in conjugating protozoa.

Sex evidently counteracts stress conditions on the principle of "two are better than one." If the self-perpetuating powers of two relatively unrelated parent animals are joined through union of their genes, then the offspring produced later can acquire a greater survival potential than that of either parent alone (Fig. 10.5). Moreover, a sexual pooling of genes leads to a still poorly understood "rejuvenation" in the offspring, on a biochemical, metabolic level. For example, if in certain protozoa sex is experimentally prevented during many successive vegetative generations, the vigor of the line eventually declines. The organisms ultimately die, even under optimal environmental conditions. Genetic malfunctions and imbalances have been found to accumulate in aged vegetative generations, and only a genetic restoration through sex then can save the line from dying out.

Reproduction thus is a "conservative" process, which faithfully passes on parental characteristics unchanged. As long as the external and internal environments remain favorable, succeeding gener-

ations survive as well as preceding ones. Sex, by contrast, is a "liberalizing" process that can facilitate survival under new or changed conditions. By combining the genes of two parents sex introduces *genetic change* in the later offspring. And to the extent that such change is advantageous for survival in new environments or under new conditions, sex has *adaptive* value. That is the key point. Sex is one of the chief processes of adaptation; it is *not* a process of reproduction.

In many protozoa all gametes, though functionally of two different types, are structurally alike (*isogamy*). In another group the gametes of a species are visibly of two different types; one kind is distinctly smaller than the other, but in other structural respects they again are alike (*anisogamy*). Still other protozoa and all Metozoa exhibit *oögamy,* a special form of anisogamy: one gamete type is flagellate and small, the other is nonmotile (or amoeboid) and large. The small types are *sperms,* the large types, *eggs* (Fig. 10.6).

If the two gamete types are produced in different individuals, as in many protozoa and most Metazoa, the sexes are said to be *separate.* If such separately sexed organisms also exhibit isogamy or anisogamy, the terms "male" and "female" are not strictly applicable. For example, isogamous protozoa are neither male nor female. Instead the two sex types, or *mating types,* are customarily identified by distinguishing symbols such as + and −. True male and female sexes are recognized only in cases of oögamy—in Metazoa, where distinct sperms and eggs are produced. These animals often exhibit a variety of other sex differences, too. Thus, sperms and eggs usually are produced in differently constructed sex organs; sperms arise in *testes,* eggs in *ovaries.* Collectively called *gonads,* these organs largely belong to differently structured male and female reproductive systems, and numerous *secondary sex characteristics* often provide external distinctions as well. Where the sexes are separate, therefore, the degrees of sex distinction can vary considerably. At one extreme are the isogamous protozoa, where visible differences between sex types are zero. At the other extreme are the advanced animals, in which nearly every part of the body, even every cell, usually exhibits characteristics of maleness or femaleness.

In numerous animals both gamete types are produced in the same individual. Known as *hermaphroditism,* this condition is believed to be more primitive than that of separate sexes; hermaphroditism could have evolved to separate sex-

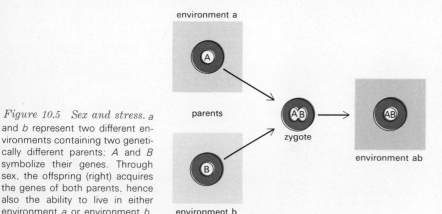

environment a

parents

environment b

zygote

environment ab

Figure 10.5 Sex and stress. a and *b* represent two different environments containing two genetically different parents; *A* and *B* symbolize their genes. Through sex, the offspring (right) acquires the genes of both parents, hence also the ability to live in either environment *a* or environment *b*.

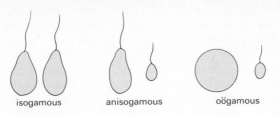

Figure 10.6 Gamete types. In the oögamous pattern, the large gamete is a nonmotile egg, the small gamete a motile sperm.

isogamous   anisogamous   oögamous

uality by suppression of either the male or the female potential in different individuals. For example, all vertebrates have both potentials in the embryonic state, but only one potential later becomes actual in a given individual (see also below). Most conjugating protozoa are hermaphroditic, each individual producing a gamete nucleus of each sex type. Among Metazoa hermaphroditism occurs in some groups of almost every phylum, and it is sometimes encountered as an abnormality in vertebrates, man included.

In most cases hermaphroditism appears to be a direct adaptation to ways of life that offer only limited opportunities for geographic dispersal, as

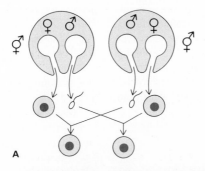

A

B

Figure 10.7 Hermaphrodites. Fertilization pattern in cross-fertilizing hermaphrodites A, the symbols ♀ and ♂ identify female and male reproductive systems, respectively. The symbols ♀ identify whole hermaphroditic organisms. If sperms fertilize eggs from the same organism, the pattern is self-fertilizing. B, copulation of the land snails. Helix, which are cross-fertilizing hermaphrodites. See also Fig. 24.11.

in sluggish, sessile, and parasitic forms. Since every hermaphrodite can function both as a "male" and a "female," a mating of two individuals often is not even required and self-fertilization can take place. But most hermaphrodites must carry out mutual cross-fertilization, such that two mating partners fertilize each other (Fig. 10.7). In these species some kind of block normally prevents self-fertilization. Among certain protozoa, for example, the block operates through so-called compatibility genes. More often the block results from the anatomy of the body, as when the two reproductive systems of a hermaphrodite open to the outside in different surface regions. Moreover, as in mollusks, for example, sperms and eggs frequently are manufactured at different times during a breeding season. Some species produce ovaries first and testes next (protogyny; the same sex organ often switches function and produces both eggs and sperm in succession). Other species form testes first and ovaries later (protandry).

The general adaptive advantage of hermaphroditism is that potentially fewer reproductive cells are wasted: sperms from one individual can meet eggs in any other individual, for every hermaphrodite produces eggs. In separately sexed types, by contrast, many sperms are wasted through chance misdistribution to the wrong sex. Similarly, if cross-fertilizing hermaphrodites are capable of some locomotion, like land snails or earthworms, then fertilization becomes possible whenever any two individuals meet. Sluggish individuals do not meet very frequently, yet every such meeting can result in fertilization.

In any animal, separately sexed or hermaphroditic, it often happens that gametes fail to find compatible partners. Most of such unsuccessful gametes disintegrate very soon, but in exceptional cases single gametes can begin to develop and form normal adults. This phenomenon is parthenogenesis, "virginal development" of an egg without fertilization. A natural form of parthenogenesis occurs in, for example, rotifers, bees and other social insects, certain crustacea, and sporadically also in birds such as turkeys and chickens. In some species artificial parthenogenesis can be induced by experimental means. For example, an egg of a sea urchin, a frog, or even a rabbit can be made to develop before it has become fertilized by pricking its surface with a needle (in conjunction with other treatments). The puncture simulates the entrance of a sperm, and development then begins. But a sexual process has not taken place.

Meiosis and Life Cycles

One consequence of every sexual process is that a zygote formed from two gametes contains twice as many chromosomes as a single gamete. An animal developed from such a zygote would consist of cells having a doubled chromosome number. If the next generation again is produced sexually the chromosome number would quadruple, and this process of progressive doubling would continue indefinitely through successive sexual generations.

Such events do not happen, and chromosome numbers do stay constant from one life cycle to the next. This constancy is maintained by a series of special nuclear divisions known as *meiosis*. It is the function of meiosis to counteract the chromosome-doubling effect of fertilization by reducing a double chromosome number to half. The unreduced doubled chromosome number, before meiosis, is called the *diploid* number, symbolized as $2n$; the reduced number, after meiosis, is the *haploid* number, symbolized as n (Fig. 10.8).

Excepting only a few protozoan groups, animal meiosis takes place just before fertilization, during the production of gametes. These cells are haploid, and when two of them fuse the resulting zygote is diploid. This diploid condition persists into the adult stage. When the adult then manufacturers gametes of its own, the gamete-forming

cells in the gonads undergo meiosis. Mature gametes therefore are haploid. Thus the only haploid phase in the life cycle is the stage of the mature sperm and egg. A life cycle of this type, characterized by diploid adults and gametic meiosis, is a *diplontic* cycle. It is distinguished from other kinds encountered in many microorganisms, in plants, and in certain protozoa, in which meiosis takes place at different points of the life cycle (see Chap. 19).

Meiosis occurs through two *meiotic cell divisions*. The gonads contain diploid, gamete-forming *generative cells* (*oöcytes* in ovaries, *spermatocytes* in testes). When meiosis takes place a generative cell undergoes two successive cytoplasmic cleavages, which transform the one original cell to four cells. During or before these cleavages, the chromosomes of the diploid cell duplicate once. As a result, $2n$ becomes $4n$. And of these $4n$ chromosomes, one n is incorporated in each of the four cells formed. Thus, *one diploid* cell becomes *four haploid* cells (Fig. 10.9).

In man, for example, a diploid generative cell in the gonads contains 46 chromosomes (or 23 pairs, one set of 23 being of maternal and the other set of 23 of paternal origin). During meiosis the chromosome number doubles once, to 92, and in the process the cell divides twice in succession. Four cells result and among them the 92 chromosomes are distributed equally. Each mature gamete therefore contains 23 chromosomes, a complete haploid set.

Meiosis has many features in common with mitosis. As in mitosis, each meiotic division passes through prophase, metaphase, anaphase, and telophase. Moreover, spindle formation and other nonchromosomal events are as in mitotic divisions. The critical difference between mitosis and the *first* meiotic division lies in their metaphases. In mitosis all chromosomes, each of them already duplicated and forming paired chromatids, migrate to the metaphase plate where all the centromeres line up in the same plane. In the first meiotic division the $2n$ chromosomes similarly duplicate during or before prophase. These $2n$ pairs of chromatids then migrate to the metaphase plate, but now only n pairs assemble in one plane. The other n pairs migrate to a plane of their own, closely parallel to the first. Moreover, every pair of chromatids in one plane comes to lie next to the corresponding type of chromatid pair in the other plane. The metaphase plate therefore is made up of *paired chromatid pairs*, or *tetrads* of like chromatids lying

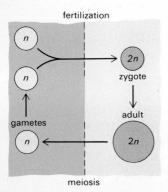

Figure 10.8 Life cycle and chromosome numbers. Animals have a diplontic cycle, with a prolonged $2n$ (adult) phase and meiosis during gamete formation. The haploid n phase is restricted to the gametes only.

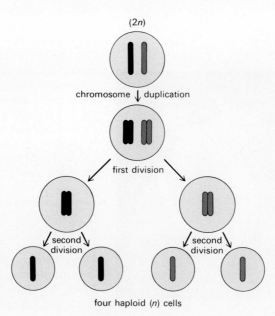

Figure 10.9 Meiosis. The assumption here is that $2n = 2$. During the meiotic divisions each member of a chromosome pair is referred to as a chromatid, as in mitosis.

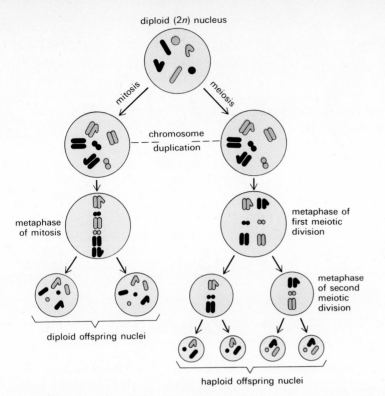

Figure 10.10 Mitosis and meiosis. The assumption is that $2n = 6$. The key difference between the two processes is the way the pairs of chromatids line up in the first metaphase.

Figure 10.11 Meiosis. A, in males all four haploid cells formed become functional sperms. In females one cell formed by the first meiotic division is small and degenerates and becomes the first polar body. Also, one cell formed by the second meiotic division becomes the second polar body. Thus only one cell matures as a functional egg. B, C, polar-body formation. B, section through the edge of an immature whitefish egg, showing the extremely eccentric position of the spindle and the chromosomes during a meiotic division. Chromosomes are in anaphase. Cleavage will occur at right angles to the spindle axis and will therefore produce an extremely large and an extremely small cell. C, cytoplasmic cleavage under way. The small cell formed will degenerate, and its remnant will be a polar body.

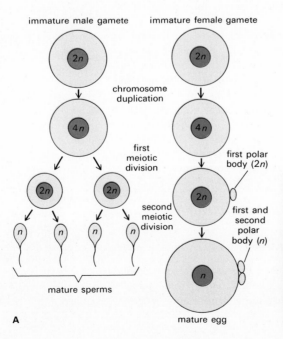

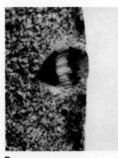

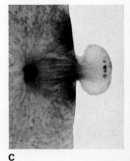

side by side. And there are n of these tetrads in the whole plate (Fig. 10.10).

During the following anaphase, two chromatids of each tetrad migrate to one spindle pole, two to the other. The first meiotic division thus produces two cells, each with n pairs of chromatids. In the metaphase of the second meiotic division these n pairs line up in the same plane, and n single chromosomes eventually migrate to each of the poles during anaphase. At the termination of meiosis as a whole, therefore, four cells are present, each with n chromosomes.

Meiosis represents the *nuclear* phase of gamete maturation. In a male, a diploid generative cell in a testis undergoes both meiotic divisions in fairly rapid succession, and all four resulting haploid cells are functional sperms. In a female, a generative cell in the ovary undergoes a first meiotic division and produces two cells. Of these one is small and soon degenerates; called the *first polar body,* it remains attached to the other cell. When this cell passes through the second meiotic division, one of the two resulting cells becomes the egg, and the other again is small and degenerates. It forms the *second polar body,* which like the first remains attached to the egg. Each original generative cell thus gives rise to only one functional egg (Fig. 10.11).

In the eggs of some animals—cnidarians and echinoderms, for example—both meiotic divisions

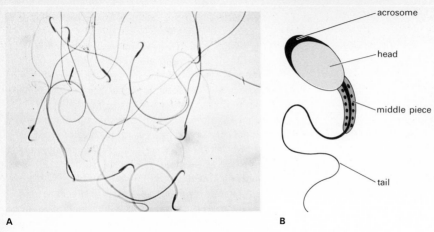

acrosome

head

middle piece

tail

A   **B**

Figure 10.12 Sperms. A, rat sperms. Note sperm head, with hooklike acrosome faintly visible at forward end; sperm tail; and middle piece darkly stained. *B,* sperm structure.

occur in rapid sequence (as happens in the sperms of all animals). Eggs so formed are haploid, mature, and ready to be fertilized. In most other animals, however, vertebrates included, eggs are ready for fertilization as soon as the first meiotic division has occurred. Such eggs remain in a state of meiotic arrest until a sperm enters, which provides the stimulus for the completion of meiosis.

In parallel with the nuclear maturation of gametes cytoplasmic maturation takes place. In sperm-forming cells much of the cytoplasm degenerates. The nucleus enlarges as an oval *sperm head,* and the mature sperm retains only three structures that have a cytoplasmic origin: a long posterior *sperm tail,* which serves as locomotor flagellum; a *middle piece,* which contains energy-supplying mitochondria; and an *acrosome,* an organelle at the forward end of the sperm head that will make contact with an egg (Fig. 10.12). Having lost most of the cytoplasm, mature sperms are among the smallest cells in the body. But mature eggs are among the largest; their cytoplasms accumulate *yolk,* which contains most of the raw materials for the construction of embryos.

Patterns of Reproduction

Reproductive Systems

Reproductive systems typically form from embryonic mesoderm. Gonads usually are globular organs with gamete-forming *generative epithelia.*

Mature gametes formed by them often accumulate in chambers or spaces inside the gonads before discharge. Since the gonads are mesodermal, any such spaces are coelomic in nature (*gonocoels*), even in noncoelomate animals. Eggs in an ovary often become surrounded by nutritive *follicle* cells. These supply raw materials that accumulate in the eggs as yolk. Most animals typically contain one pair of gonads, but in many instances single gonads or several pairs are encountered.

In sponges, radiates, and a few other groups the gonads represent the whole reproductive system; gametes are discharged directly to the outside or into an interior body cavity first. However, most reproductive systems include *gonoducts* that lead to the outside. These are *sperm ducts* in males, *oviducts* in females. Various accessory organs generally are associated with such channels (Fig. 10.13). Sperm ducts often pass through or past *sperm sacs,* chambers that store sperms before discharge. Also, *prostate glands* and *seminal vesicles* often are present that secrete *seminal fluids.* Together with the sperms these fluids make up *semen.* Many male animals have *copulatory organs* at or near the exterior termination of the sperm ducts. In many instances a sperm duct passes right through a copulatory organ, in which case the organ is called a *penis.*

Along the course of oviducts can be present *seminal receptacles,* pouches that store sperms after mating and before fertilization, as well as *yolk glands, shell glands,* and other glands that produce nutritive or protective layers around fertilized eggs. Near its exterior termination an oviduct can be enlarged as a *uterus,* a chamber where egg development takes place. In copulating animals the terminal section of the oviduct that receives the male copulatory organ during mating is called a *vagina.*

Gonoducts generally open to the outside through independent *gonopores.* However, in many cases the ducts discharge into a cloaca and in others they join with the excretory ducts and empty through a common *urogenital pore.* Hermaphroditic animals have separate gonopores for each of their two reproductive systems or a common single gonopore for both.

In noncoelomates the gonads usually connect directly with the gonoducts. By contrast, the gonads of coelomate animals generally discharge gametes into the coelom, and the gonoducts then collect and transport them from the coelom to the exterior. The opening from coelom to gonoduct is a ciliated *gonostome*. In such cases the gonoducts resemble metanephridia. Some coelomate animals actually lack gonoducts and their metanephridia serve as both excretory and reproductive channels; and, conversely, some coelomates virtually are without excretory ducts, their gonoducts serving as common exit channels (Fig. 10.14).

Vertebrate gonads arise as a pair of pouches that project into the coelom near the anterior part of the mesonephros (Fig. 10.15). The space in each pouch fills loosely with strands of meso-

nephric tissue. Gamete-producing cells, called *primordial germ cells,* arise not in the developing gonad itself but in the head mesoderm of the embryo. From there the cells migrate to the embryonic gonad and disperse inside it. At this stage the gonad is sexually undetermined as yet and can develop in either a male or a female direction. The factors that determine the actual direction are examined in Chap. 12.

If the gonad becomes a testis, its outer tissue, the *cortex,* develops very little more but the mesonephric core, or *medulla,* proliferates and becomes

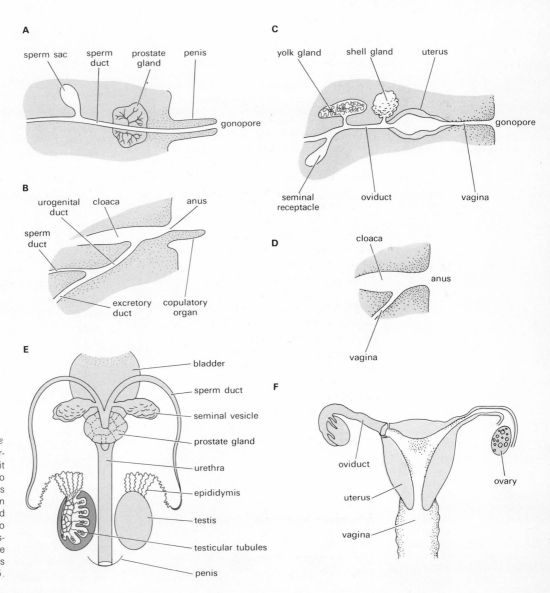

Figure 10.13 Reproductive systems. A and *B,* common organs of male systems, and exit modifications according to whether or not a cloaca or a penis is present. *C* and *D,* common organs of female systems, and exit modifications according to whether a cloaca is or is not present. *E* and *F,* the male and female systems in man. In *E,* left testis in section. See also Fig. 30.15.

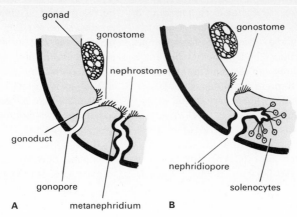

Figure 10.14 Reproductive and excretory systems in coelomate invertebrates. A, basic condition of coelomic gonads and separate reproductive (fine line) and excretory (heavier line) systems. In many cases the gonoducts degenerate and the metanephridia then serve as both reproductive and excretory channels. *B,* nephromixium (as in some annelids), with gonoduct joined to solenocytic protonephridium and both having common exit. In a variation of this pattern, degeneration of solenocytes produces a joint system composed partly of gonoduct and partly of excretory duct and serving both reproductive and excretory (metanephridiumlike) functions.

the bulk of the mature testis. In it arise interconnected sperm chambers, or *testicular tubules.* Their inside lining is a sperm-producing epithelium formed from the primordial germ cells of the embryo. After sexual maturity the cells of this layer divide mitotically, and the new cells so formed accumulate in the testicular tubules and mature as sperms. As noted, meiosis occurs in the process.

If the immature gonad develops as an ovary, it is the medulla that undergoes little further

growth, whereas the cortex enlarges greatly. Also, the primordial germ cells give rise to an egg-producing generative epithelium on the surface of the cortex. This layer produces successive batches of new cells, and in each batch all but one of the cells become nutritive follicle cells. After sexual maturity the remaining cell can undergo meiosis and develop as a functional egg. An embryonic ovary of a human female is estimated to contain some 400,000 primordial germ cells; but only about 400 functional eggs are actually formed during adult life.

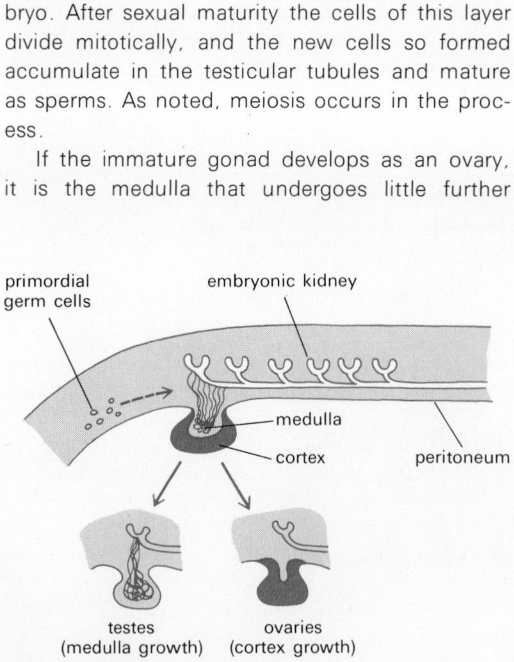

In jawless fishes, the most primitive vertebrates, gametes pass from the gonads of either sex into the coelom and from there to the outside via small pores in the posterior region of the Wolffian ducts (Fig. 10.16). In all other vertebrates, the *females* have oviducts (*Muellerian ducts*) that carry the eggs from the coelom. The Muellerian ducts typically join the excretory ducts posteriorly and form a common *urogenital sinus.* This channel then empties to the outside through the cloaca. In most female mammals, however, separate exits develop; excretory ducts open via a urethra at a urinary pore, reproductive ducts open via a vagina at a gonopore, and both types of duct are independent of the alimentary system. Only the *left* ovary and oviduct persist in most birds; the right side of the reproductive system degenerates in the embryo, an adaptation to the production of comparatively huge eggs in these animals.

In *male* vertebrates (other than jawless fishes), the Muellerian ducts degenerate in the embryo and sperms are carried to the outside by the Wolffian ducts. Fish and amphibian testes retain the embryonic connection with the mesonephros, and in the adults the Wolffian ducts then carry both sperms and urine. In male land vertebrates the Wolffian ducts discharge sperms only, urine flowing from the metanephric kidneys through the ureters. As in females the reproductive and excretory ducts form a common urogenital sinus, and in most male mammals this sinus is a urethra that opens independently of the alimentary tract at a separate urogenital orifice. Simultaneous ejection of sperms and urine here is prevented reflexly; muscles close off one set of ducts when the other is discharging.

In comparatively primitive mammals such as opossums, bats, and whales, testes remain permanently at their original locations inside the body (where ovaries are in females). In a second group, which includes elephants and many rodents, the

Figure 10.15 Gonad development. Top, sectional side view of part of vertebrate embryo, indicating place of origin of primordial germ cells in head region in relation to developing gonad, embryonic kidney, and coelom. *Bottom,* differential growth of the medullary (kidney-derived) tissue results in testes, and differential growth of the cortical (peritoneum-derived) tissue results in ovaries.

testes leave the trunk when the breeding season begins and migrate to a *scrotum,* a skin sac between the hind legs. After sperms cease to be produced at the end of the breeding season, the testes migrate back to their original positions. In a third group, exemplified by rodents such as mice and rats, the testes descend to a scrotum when the animals reach sexual maturity, and from then on the gonads remain there permanently. And in a fourth group, of which man is a member, the testes are internal only during embryonic stages; the organs migrate to a scrotum just before birth and then remain in this sac throughout life.

It is known that the temperature in a scrotum is up to 7°C lower than inside the trunk. It is also known that lower temperatures tend to promote

Figure 10.16 Reproductive ducts in vertebrates. Ducts of the left side of the body are diagrammed. In lampreys, note the pore in the Wolffian (mesonephric) duct for sperm or egg exit from the coelom. In fishes and amphibia, the Wolffian ducts carry urine in females but largely sperms only in males, urine here traveling mostly through accessory excretory ducts. In reptiles and birds, as also in mammals, the Wolffian ducts are sperm ducts exclusively, urine being carried through the ureters in males as well as females. Females have reproductive Muellerian ducts in all vertebrates except the lamprey group. The uterus and vagina of placental mammals are derived from the original Muellerian ducts. Such ducts form also in early embryos of males but degenerate soon after. See also Fig. 30.15.

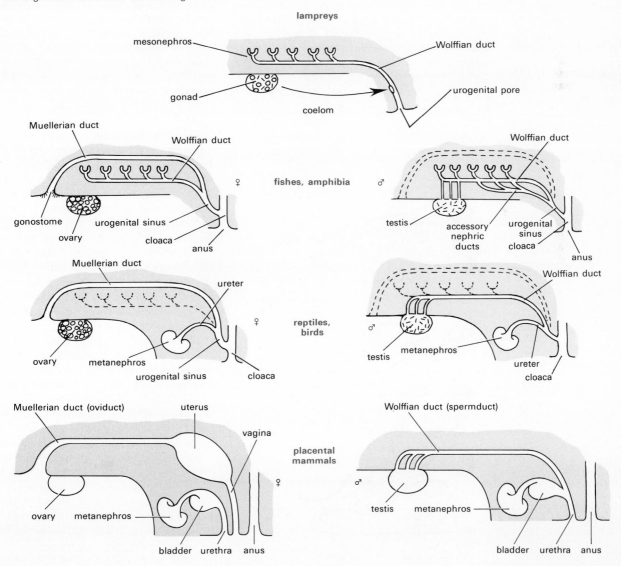

and higher temperatures to inhibit sperm production. Temperature, testis location, and continuity of sperm manufacture therefore appear to be interrelated.

Reproductive Behavior

In all animals, sperms must fuse with eggs in an aqueous environment. This water requirement has led to the elaboration of two basic mating patterns:

In *external fertilization,* mating partners are or come close to each other in natural bodies of water and both simultaneously *spawn;* they release sperms and eggs directly into the water, where many fertilizations then take place. This pattern is characteristic of most aquatic animals and also of terrestrial types such as certain insects and amphibia, which migrate to permanent bodies of water for reproduction.

The second pattern is *internal fertilization.* Mating partners here come into physical contact and *copulate;* by various means males transfer sperms directly to the reproductive systems of females. Fertilization then takes place in the oviduct, as sperms move up through the duct and

meet eggs coming down. The internal tissues of the female here provide moisture for the sperms, and the need for external water is thereby circumvented. Internal fertilization is characteristic of most terrestrial animals, but the process occurs also in numerous aquatic groups (for example, in many fishes).

Among several variant forms of internal fertilization, numerous animal groups produce, not loose sperms in seminal fluid, but compact sperm packets, or *spermatophores.* Squids and octopuses then use their tentacles to transfer spermatophores to females. Certain salamander females use cloacal lips to enfold spermatophores deposited on the ground by the males. Males of some other amphibia transfer spermatophores by mouth. In certain spiders sperms are placed on an anterior body appendage, which is then inserted directly into the female reproductive system. Some animals transfer sperms into females by a process akin to hypodermic injection, through any part of the skin.

In all cases of external and also in some of internal fertilization, development of zygotes to adults takes place externally, in natural bodies of water. Such animals, in which eggs are shed to the outside either in an unfertilized or a fertilized state, are said to be *oviparous* (Fig. 10.17). Among vertebrates, for example, many fishes are oviparous and externally fertilizing, whereas all birds are oviparous and internally fertilizing. The developing eggs eventually *hatch,* as larvae or as miniature, immature adults. If development takes place in water, the zygotes often have coats of jelly around them (for example, frog eggs) but are otherwise protected very little. Zygotes developing on land usually are protected against evaporation by cocoons or shells.

Most animals are oviparous as above. Among the rest, one group is *ovoviviparous.* Fertilization here is internal and the zygotes are retained inside the female reproductive system, specifically a uterus. Development then occurs there. However, beyond providing a substantial measure of protection the female body does not otherwise contribute to zygote development; as in oviparous types, food is supplied by the yolk stored in each egg. Ultimately the young are *born* rather than hatched, and the females release fully formed animals, not eggs. Among vertebrates some of the fishes, amphibia, and reptiles are ovoviviparous.

A third group of animals comprises *viviparous*

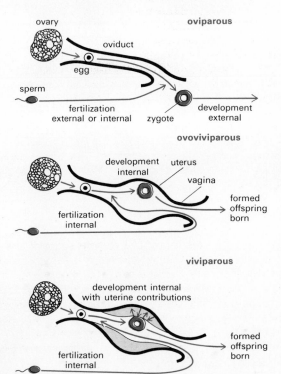

Figure 10.17 Fertilization and development in relation to the environment and the maternal body.

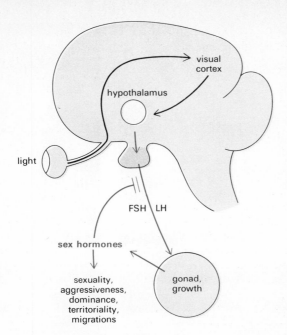

Figure 10.18 Neuroendocrine control of reproductive functions. Light (and many other external) stimuli, as well as internal psychological states, are communicated to the hypothalamus over neural pathways (black arrows). Chemical signals (color arrows) then stimulate the pituitary, which (superimposed on an intrinsic rhythm of activity) secretes gonadotropic hormones (*FSH, LH*). These induce gonad growth and sex hormone production. The hormones in turn bring about the onset of breeding seasons and all associated reproductive functions, as well as various sex-associated behaviors. When sex-hormone concentrations exceed certain levels, they inhibit the pituitary (double bar), and the reproductive and associated activities then decline.

types. Here fertilization again is internal, zygotes are retained in a uterus, and the young are born as developed animals. However, the female body now provides not merely protection but also food, and it contributes to offspring metabolism in other vital ways. The females in such cases are *pregnant*. Viviparous vertebrates include, for example, some fishes, some snakes, and the majority of the mammals (see Fig. 10.17).

Unlike vegetative reproduction, which can occur at any favorable time, gametic reproduction takes place in most animals only during specific *breeding seasons*. Such seasons are largely annual, most of them occurring in spring or fall. But many animals have two or more breeding seasons per year; for example, there are two in dogs. In some cases breeding seasons last only a single day (or night, as in clamworms and many other marine invertebrates), but in others, as in monkeys, apes, and men, reproduction can occur the year round. Breeding seasons tend to be continuous also where environmental conditions remain uniformly favorable throughout the year,. as in domesticated cattle, chickens, and rabbits, and in laboratory mice and rats. Even in such cases, however, fertility is usually greatest during the spring.

Breeding periods tend to be synchronized with seasonal rhythms in day lengths and temperatures.

Thus, breeding seasons in vertebrates usually are initiated after the animals have been exposed to a certain total amount of illumination, counted from the winter solstice in spring breeders and the summer solstice in fall breeders. Such *photoperiodic* stimuli affect the reproductive system through the eyes and the neural-endocrine mechanism outlined in Fig. 10.18 (see also Chap. 16). The anterior lobe of the pituitary thereby becomes stimulated to secrete increasing amounts of gonadotropic hormones, and all parts of the reproductive system increase in size as a result and become functional. The higher concentrations of sex hormones then produced by the gonads elicit enhanced reproductive behavior and sexuality. After a breeding season, the hormone output of the anterior pituitary declines again and the reproductive system becomes quiescent and reduced in size.

Behavioral processes can be shown to be just as significant for reproduction as internal functional ones. For example, if mature pigeons or mice are prevented from having social contact with other members of the species, they usually fail to come to breeding condition, even at the right season. However, pigeons can be induced to lay eggs if they are at least permitted to see themselves in a mirror, and female mice come into heat when exposed at least to the smell given off by males. In many zoo animals reproduction will not occur even when a male and a female are allowed to live together. Such animals evidently require considerably more elaborate social settings if their neural and endocrine controls of reproductive processes are to function normally.

Furthermore, mating in most vertebrates and many invertebrates is functionally dependent on preliminary *courtship*. Consisting of a series of more or less precisely programmed displays and movements, courtship serves to advertise the presence of sexually receptive individuals and to forestall or inhibit aggression by a prospective mate. The last is particularly necessary if the mate is a predator, an individual of high social status, or the owner of a territory. Thus, courtship by either sex usually includes token or actual submissive behavior, prominent display of sex recognition signs, or other signals indicating nonaggressive intent (Fig. 10.19).

Most important, courtship often is required to bring both mating partners to reproductive readiness simultaneously. In some vertebrates, for ex-

A

B

Figure 10.19   Courtship. A, the tail display of male peacocks is one of the familiar courtship behaviors. *B,* male Siamese fighting fish displaying for female.

ample, the internal reproductive mechanism does not become operational unless visual, olfactory, or other stimuli from a mate provide a trigger. Frequently a whole sequence of successive triggers is needed to initiate a corresponding sequence of internal hormonal processes. For example, courtship by male ringdoves elicits production of sex hormones in females, which then are induced by the hormones to build nests. This activity in turn is the specific stimulus for mate acceptance and egg laying. In female canaries, similarly, a progressive sequence of hormonal changes and a succession of steps of nest construction have been found to be reciprocally necessary stimuli. In some instances, as among cats and rabbits, male-female coordination through courtship dovetails so finely that eggs are not released from the ovary until a mating pair is actually copulating.

Courtship (like aggressive and submissive displays, see Chap. 16) probably has evolved as a highly ritualized version of combat. In both sexes courtship usually includes behavioral elements that are very similar to those of attack and flight, and that in some instances involve real attack and flight. In sticklebacks and gulls, for example, intermittent bursts of male attack and female flight

continue right up to mating. In finches it is the female that attacks the male with increasing vigor until copulation actually takes place. In most cases, however, the ritual elements of courtship appear to differentiate reproductive intent from real hostility. Males use aggressive displays for actual threats but ritualized versions of them to attract females; and whereas females are submissive to discourage real attack, their stylized submission in courtship is used to invite attention from males.

Mating normally terminates when sperms and eggs have been shed. In many vertebrates sexual activity ceases not primarily because the male has released sperms, but because he may be inhibited from further activity by the sight of shed eggs (as in certain fishes) or because he may have become habituated to a particular female. In guinea pigs and cattle, for example, a female normally is inseminated several times in succession during a single mating sequence. The sequence ends when male interest wanes. However, a male that has become inactive with one female will readily and repeatedly copulate with a succession of other females. In the case of bulls this prolonged copulatory capability has become commercially important; prize bulls can be induced to deliver maximal

226

amounts of semen (for use in artificial inseminations) by presenting them with a series of dummy cows.

Reproductive behavior often extends well beyond the breeding season. Among vertebrates particularly, many guard and tend eggs at least until hatching, and many also nourish, care for, and train their young after hatching or birth. Such parental behavior is induced partly by the sex hormones and specifically by *prolactin,* one of the pituitary gonadotropic hormones (see Chap. 8, Table 6). In male vertebrates prolactin promotes protective attitudes toward mates and offspring, intensive food gathering for the family, and expressions of paternalism in general. In females, similarly, the hormone induces maternalism, including broodiness, strong emotional attachment to the offspring, and the drive to bring the young to a state of self-sufficiency (Fig. 10.20). In mammals prolactin also initiates milk secretion in the mammary glands (hence the alternative name *lactogenic* hormone).

Parental behavior does not have an exclusively endocrine basis, however, but is subject also to complex psychological controls. In mammals, for

Figure 10.20 Parental behavior. Among less-than-common examples is this black-chinned mouthbrooder female (*Tilapis mossambica*), which incubates its eggs in the mouth. Even after hatching the fry retreats into the mother's mouth if danger threatens. Experiments show that fry will seek refuge in any dark opening and that surrogate hiding places are accepted as readily as the parental mouth.

example, development of the emotional bond between mother and offspring, normally firmly established within an hour or two after birth, depends in large part on maternal licking, cleaning, and handling of offspring. If a just-born offspring is cleaned by an experimenter or is handled by another animal, the mother often will refuse to acknowledge this offspring as her own. Similarly, many female mammals will retrieve young that have rolled out of the nest, and since this activity is undertaken even by nonbreeding animals it, too, appears to be independent of hormonal states. Continued milk secretion likewise requires not merely hormonal but also external stimuli, particularly tactile stimulation of the nipples and withdrawal of already accumulated milk by suckling. Moreover, as shown in Chap. 16, adequate social settings and social behavior learned during young stages are essential for normal parental behavior and establishment of proper parent-young relations.

Reproductive Mechanisms

Males produce sperms continuously during a breeding season and often also to a reduced extent between such seasons. In male vertebrates, pituitary control is exercised chiefly through LH (the luteum-producing hormone; see below), one of the pituitary gonadotropic hormones. (A male function for the follicle-stimulating hormone FSH, a second such hormone, has not been demonstrated, and the male function of prolactin has been described above.) Pituitary LH is the specific hormone that stimulates the testes to produce *androgens,* the male sex hormones (see Fig. 4.16). These are manufactured in the tissue partitions between the sperm-producing tubules of a testis (Fig. 10.21). The concentration of *testosterone,* the most potent of the androgens, rises sharply at the start of a breeding season under the influence of LH. Sperms then are produced actively, sex urge increases, and secondary sex traits such as mating colors in skin and plumage become pronounced. If they are present in blood in excessive concentrations, androgens have an inhibitory effect on the pituitary. This gland then produces less LH and androgen secretions therefore decline as well. Through such feedback control, androgen concentrations remain at fairly steady levels during the breeding season. Sperm production begins at sex-

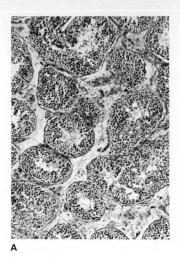

A

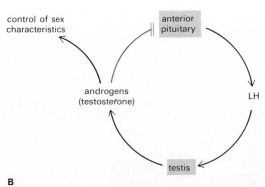

B

Figure 10.21 The testis and its hormonal control.
A, section through a mammalian testis. Sperms are
produced in the tubular chambers, and mature sperms
accumulate in the central spaces of the tubules. The
tissue between the tubules contains interstitial endocrine
cells that manufacture androgens, the male sex hormones, under the stimulus of LH from the pituitary. *B,*
the control of androgen secretion. LH is one of the
gonadotropic hormones of the pituitary. Arrow tipped
with transverse double bar (color) signifies inhibition.

ual maturity, or *puberty,* and in man can continue
for life.

Females produce one or several successive
batches of eggs during a breeding season. Periods
of egg maturation are called *estrus cycles* in mammals, and female dogs, for example, come into
estrus just once per breeding season (*monoestrous*
types). Horses and sheep are among *polyestrous*
types, which produce several batches of eggs

during a breeding season. However, egg manufacture ceases after one batch is fertilized. The hormonal controls of egg production in vertebrates
parallel those of sperm production. At the onset
of a breeding season the pituitary secretes gonadotropic hormones. Of these, FSH, the follicle-
stimulating hormone, initiates increased ovarian
activity.

More specifically, the follicle cells around an
egg are stimulated to produce *estrogens,* the female sex hormones (see Fig. 4.16). Corresponding
functionally to the male hormones, they promote
growth of the follicle, pronounced development of
secondary sex traits, and an increase in sex urge
(Fig. 10.22). Certain amounts of estrogen accumulate in the fluid-filled cavity of a follicle, which
becomes larger as it develops. At maturity a follicle
lies just under the ovary surface, where it produces
a conspicuous bulge.

When estrogen concentrations in blood exceed
a certain level, the hormones exert an inhibitory
effect on FSH production in the pituitary. At the
same time also the estrogens stimulate the pituitary to produce the gonadotropic hormone LH. As
a result LH concentrations begin to rise just when
FSH concentrations begin to fall. These shifting
hormone balances are the specific stimulus for
ovulation; the ovary surface and the follicle wall
both rupture, and the mature egg escapes to the
coelom and the oviduct. In some animals, among
them cats, rabbits, and squirrels, ovulation also
requires nervous triggering through the act of
copulation.

An immediate consequence of ovulation is that
the ruptured and eggless follicle remaining in the
ovary loses its fluid and collapses. Another consequence is that, since FSH production by the pituitary now has ceased, the remnant of the follicle
ceases to manufacture estrogen. Instead, under
the specific influence of LH, the follicular remains
transform to a yellowish body, the *corpus luteum*
(hence the name "LH," which stands for
"luteum-producing hormone"). Under the continuing influence of this hormone the corpus luteum
begins to secrete a new hormone of its own,
progesterone. Chemically this hormone is similar
to estrogen and testosterone (see Fig. 4.16).

In oviparous vertebrates progesterone induces
the oviducts to secrete jelly coats or shells around
the eggs passing through. When the concentration
of progesterone eventually exceeds a certain level,

the hormone inhibits LH production in the pituitary. The corpus luteum then ceases to manufacture progesterone, but by this time the eggs have already been shed. A new FSH-initiated egg-growth cycle then can begin if the breeding season has not come to a close.

The pattern of events is somewhat different in the viviparous mammals. Progesterone here stimulates growth of the uterus in preparation for pregnancy: the wall of the uterus thickens greatly and develops numerous glandular pockets and extra blood vessels. If a fertilized egg reaches the uterus it becomes firmly embedded in the thickened uterine wall (Fig. 10.23). The maternal blood in this wall then supplies the egg with nourishment and oxygen. But if an egg is not fertilized in the oviduct, it soon disintegrates and the uterus will have been made ready for nothing. In that case progesterone again inhibits pituitary LH production eventually, and the corpus luteum then ceases to manufacture progesterone. Without this "pregnancy hormone," however, the ready condition of the uterus cannot be maintained and the wall soon reverts to its normal thickness.

In Old World monkeys, apes, and human females, the preparations for pregnancy in the uterus are so extensive that, if fertilization does not occur and progesterone production then ceases, the inner lining of the uterus actually disintegrates. Tissue fragments separate away and some blood escapes from torn vessels. Over a period of a few days all this debris is expelled to the outside through the vagina, a process called *menstruation* (see Fig. 10.23).

A *menstrual cycle* in such animals lasts about 28 days. Follicle maturation occurs during the first 10 to 14 days under the control of FSH and estrogen, and this *follicular phase* of the cycle terminates with ovulation. During the next 14 to 18 days the uterus grows in preparation for pregnancy under the influence of LH and progesterone. If pregnancy does not begin during this *luteal phase*, menstruation takes place in the course of the first few days of the next menstrual cycle. A new cycle starts when, in the absence of LH and progesterone, the pituitary resumes FSH production and a new follicle matures in the ovary.

Evidently, a menstrual cycle is governed by two successive control cycles with built-in feedbacks. FSH and estrogen are components of one control cycle, LH and progesterone of the other; and the termination of one is the specific stimulus for initiation of the second. Menstrual cycles begin to occur at puberty and continue to the time of *menopause* in middle age, when the sex-hormone control system gradually ceases to operate.

Estrogen is not absent altogether during the luteal phase, nor is progesterone completely absent during the follicular phase of a menstrual

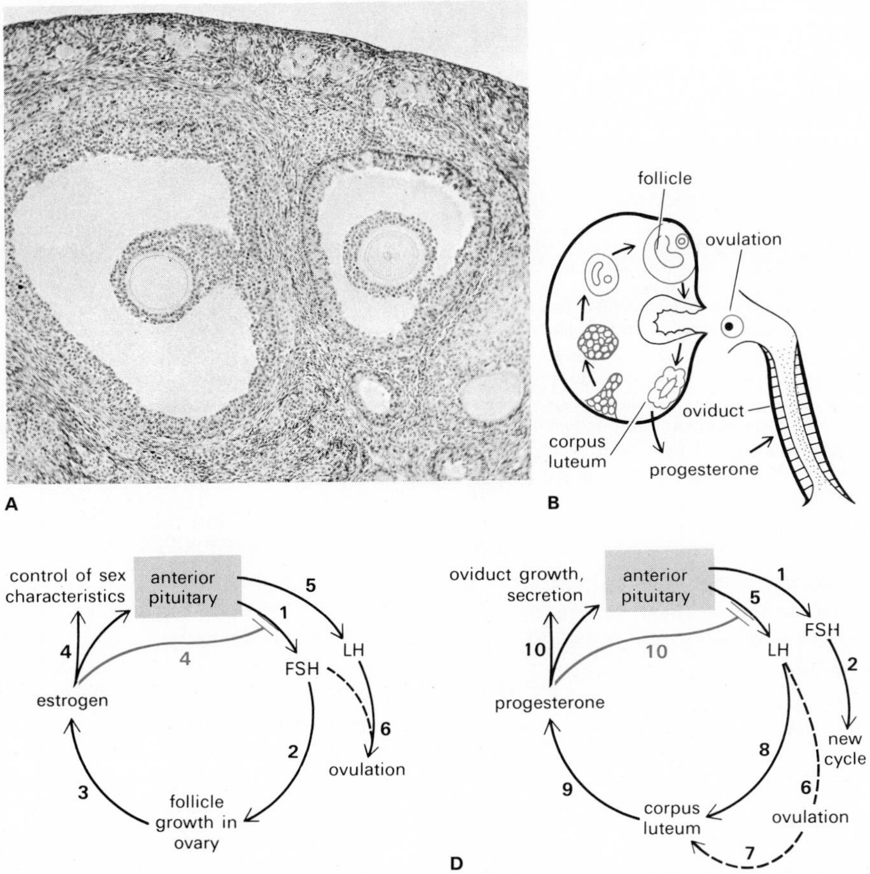

Figure 10.22 *Egg growth in oviparous vertebrates.* A, section through an ovary. Note the two large follicles, the follicular cavities, and the large egg cell in each follicle embedded in a mass of cells along the follicular wall. Endocrine cells secrete estrogens into the follicular cavity. When the eggs are mature they will ovulate and escape by rupture of follicle and ovary walls. Along the edge of the ovary (top) the relatively large cells are immature eggs that will mature later, in follicles yet to be formed. B, progesterone induces the oviduct to secrete jelly coats or shell-forming substances around eggs. C, hormonal changes during the follicular phase, leading to ovulation. Numbers indicate sequence of steps. Black arrows, stimulation; colored double-bar arrows, inhibition; broken lines, decreasing concentrations. D, hormonal changes after ovulation (luteal phase) in oviparous vertebrates. The sequence begins at steps 5 and 6, equivalent to the terminal steps in C, and ends at step 10. Steps 1 and 2 then repeat the first events of C.

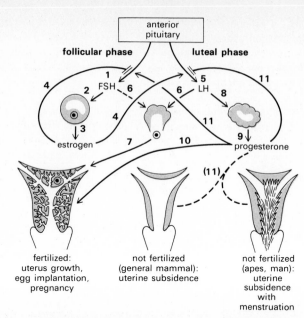

Figure 10.23 Egg growth in viviparous vertebrates. Steps 1 to 9 correspond to the follicular and luteal phases in oviparous vertebrates (see Fig. 10.22). Thereafter events differ according to whether fertilization does or does not take place. If it does, pregnancy ensues (step 10); if it does not, progesterone production declines (via step 11), leading to tissue resorption in the uterus wall in most mammals and to menstruation in monkeys, apes, and man. The cycle then repeats (step 1).

Figure 10.24 Fertilization, as in aquatic invertebrates. A sperm enters an egg at an egg cone that comes to surround the sperm (*C*). A fertilization membrane lifts off the egg surface after a sperm has made contact, preventing additional sperms from entering (*C, D*). The sperm tail is left at the egg surface, and the sperm head (nucleus) alone migrates into the egg cytoplasm, where it fuses with the egg nucleus. An egg is fully fertilized only after sperm and egg nuclei have fused.

temperature increases somewhat during the follicular phase, then falls during the luteal phase. Sex drive is likely to be more pronounced during the follicular phase, since estrogen maintains it. And inasmuch as estrogens, like androgens in males, affect many aspects of behavior, monthly hormonal and behavioral fluctuations in some cases tend to be interrelated.

The production of sperms and eggs sets the stage for fertilization. Regardless of whether this process occurs in free water after spawning or in an oviduct after copulation, the first step always is the entrance of a sperm into an egg. The union is usually achieved through a fusion of the cell membranes of the sperm and the egg. The acrosome at the sperm tip is specialized to adhere to the egg, and sperms hitting an egg head on therefore are most likely to remain attached. Also, the eggs of many animals are amoeboid and assist actively in sperm entrance. Normally only a single sperm can enter, for as soon as a first sperm has made head-on contact the egg undergoes metabolic changes that make it unreceptive to additional sperms. In certain aquatic invertebrates, moreover, a *fertilization membrane* has formed during egg maturation and this membrane lifts away from the egg surface after a sperm has made contact (Fig. 10.24).

During sperm entrance the tail of the sperm usually drops off. At this point the egg is *activated;* its development has been started. A mature egg is ready and able to develop, but this ability remains latent until a specific stimulus starts development. Sperm penetration normally serves as this stimulus. Recall in this connection that, in parthenogenesis, eggs become activated by natural or experimental means without sperms. Under more usual circumstances, a sperm nucleus inside an activated egg migrates toward the egg nucleus, and this meeting of the two haploid nuclei completes fertilization. The nuclear membranes dissolve, the now diploid chromosomes line up in a metaphase plate, and the zygote undergoes a first mitotic division. The development of a new animal is launched in this manner.

cycle. The hormones merely attain definite peak concentrations at particular times. These fluctuations have important consequences. Under the influence of progesterone, for example, which promotes growth of the duct system in the mammary glands, a slight swelling of these glands generally occurs during the luteal phase. Body

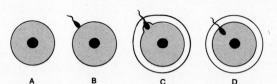

Review Questions

1. How does reproduction contribute to steady-state maintenance? To self-perpetuation in general? Name and define the different forms of vegetative reproduction and give examples of each. Under what circumstances does vegetative reproduction occur?

2.   Distinguish between reproduction, development, and sex. What is gametic reproduction? What are the advantages and disadvantages of vegetative and gametic reproduction?

3.   What are the most basic events of every sexual process? Under what conditions does sex tend to occur? In what way is sex of adaptive value? Illustrate by examples among protozoa.

4.   Distinguish between syngamy and conjugation. Define mating, fertilization, zygote, gamete, isogamy, oögamy, and parthenogenesis.

5.   What is hermaphroditism? In which organisms does it occur, generally and specifically? What is its adaptave value? Distinguish between self-fertilization and cross-fertilization.

6.   What is the basic function of meiosis and what makes such a process necessary? Where and when does meiosis occur? Define haploid, diploid. How many *pairs* of chromosomes does a diploid cell contain? Of these, which and how many are maternal and which and how many are paternal?

7.   How many chromosome duplications and how many cell duplications occur during meiosis? In what respects are mitosis and meiosis alike? What is the essential difference between the metaphase of mitosis and the metaphase of the first meiotic division? Describe the complete sequence of events during both divisions of meiosis.

8.   What are the first and second polar bodies? Are they found in males as well as females? Explain. What is the general structure of a mature sperm and a mature egg?

9.   In which animals is fertilization (*a*) external, (*b*) internal? Define oviparity, ovoviviparity, viviparity. In which vertebrates does each occur?

10.   Describe the general structure of the reproductive system in male and female animals. How does this structure usually vary for externally and internally fertilizing animals? What are spermatophores?

11.   Show how the reproductive and excretory systems of some invertebrates are interrelated structurally and functionally. What is a gonostome? A gonocoel?

12.   What are primordial germ cells? Show how an immature vertebrate gonad develops as either a testis or an ovary.

13.   Describe the evolutionary development of the vertebrate reproductive system. What is the Muellerian duct? Trace the anatomic interrelations of the reproductive and excretory systems in vertebrate females and males. What is a urogenital sinus?

14.   Describe the relation of testis location and breeding season in different mammals. How are vertebrate breeding seasons initiated, maintained, and terminated?

15.   Describe various aspects of vertebrate reproductive behavior and its hormonal controls. What are the behavioral characteristics of animal courtship?

16.   Outline the hormonal controls of sperm production in vertebrates. What are monoestrous and polyestrous mammals? What are the roles of prolactin in male and female vertebrates?

17.   Describe the hormonal controls and the process of follicle growth in vertebrates up to the time of ovulation. What events take place during ovulation? After ovulation, what happens to (*a*) the egg and (*b*) the follicle?

18.   Outline the interrelation of ovulation, corpus luteum formation, progesterone production, and the role of the pituitary. Review the entire pattern of egg-growth cycles in vertebrates generally and viviparous vertebrates specifically.

19.   Which animal groups have menstrual cycles? Describe the events in the uterus up to the time of menstruation. What happens during menstruation in the ovary and the uterus?

20.   How and where does fertilization occur? Describe the process of fertilization. What is an activated egg?

Collateral Readings

Berrill, N. J.: "Sex and the Nature of Things," Dodd, Mead, New York, 1953. A beautifully and interestingly written paperback on the significance and processes of sex in various organisms.

Bullough, W. S.: "Vertebrate Reproductive Cycles," Wiley, New York, 1961. An account on breeding behavior and the hormonal controls of vertebrate reproduction.

Csapo, A.: Progesterone, *Sci. American,* Apr., 1958. A good review of the function of this hormone.

Edwards, R. G.: Mammalian Eggs in the Laboratory, *Sci. American,* Aug., 1966. Eggs raised in culture media and fertilized there produce embryos outside the body that can be studied directly.

Farris, E. J.: Male Fertility, *Sci. American,* May, 1950. The fertilizing capacity of the male depends on normally structured sperms and certain proportions of them in semen.

Lehrman, D. S.: The Reproductive Behavior of Ring Doves, *Sci. American,* Nov., 1964. External stimuli, the behavior of the mating partner, and hormones form a sequence of triggers that initiate successive steps of reproductive activity.

Levine, S.: Sex Differences in the Brain, *Sci. American,* Apr., 1966. Masculine behavior in both males and females appears to be a consequence of the effect of testosterone on the brain of just-born offspring.

Monroy, A.: Fertilization of the Egg, *Sci. American,* July, 1950. The fine details of the process are examined.

Nelsen, O. E.: "Comparative Embryology of the Vertebrates," McGraw-Hill, New York, 1953. Vertebrate reproductive processes are reviewed in detail in this large text.

Pincus, G.: Fertilization in Mammals, *Sci. American,* Mar., 1951. Fertilization and early development can be studied in isolated mammalian eggs.

Ross, R.: Wound Healing, *Sci. American,* June, 1969. The three-step process of this regenerative event is examined.

Tinbergen, N.: The Courtship of Animals, *Sci. American,* Nov., 1954. A noted student of animal behavior discusses the requirements for mating.

Tyler, A.: Fertilization and Antibodies, *Sci. American,* June, 1954. Experiments showing that fertilization in sea urchins is chemically equivalent to an antigen-antibody reaction.

11

DEVELOPMENT

In very general terms, *any* part of living history is a part of development—development in fact creates living history, and together with structure and function it represents the third fundamental dimension of life. However, in specific association with animal reproduction and the progression of life cycles, development customarily signifies just those events of living history that relate to the formation of single whole animals and their parts.

Processes of Development

The most comprehensive event in animal development is the elaboration of a complex multicellular individual from a single cell—vegetative unit or zygote. Such a transformation is brought about by groups of processes among which *morphogenesis* and *differentiation* are two of the most essential.

One component of morphogenesis is increase in size, or *growth*. It can occur through an increase in the number of parts, the size of parts, the spacing between parts, or any combination of these. Molecular growth is the basic prerequisite for growth at any other level; animals grow from their molecules on up. On the cellular level, the most significant growth process usually is increase in cell number by mitotic division. Fully grown animals differ from undeveloped ones mostly in cell number, not in cell size or spacing.

Growth of any kind generally does not proceed randomly in all directions. How does it happen, for example, that net growth stops just when the nose, the brain, the liver, and all other body parts are of the ''right'' proportional size and shape? Or that the different parts of the fully grown adult *retain* correct proportions and shapes? Or that, when the limb of a salamander is cut off, regenerative growth stops just when the new limb has the size and the shape of the original one? Thus, apart from growth as such, establishment of species-specific body *form* is a second component of morphogenesis.

The main aspects of form are *polarity* and *symmetry*. If they are given, a great deal about the general appearance of an object is already specified. A structure is polarized if one of its three dimensions of space is in some way dominant. For example, the head-tail axis in most animals is longer than the other two. This is the chief axis around which the whole animal is organized, and such types are said to be polarized longitudinally. Symmetry indicates the degree of mirror-image regularity. A structure can be symmetric in three, two, one, or no dimensions and thus it can be spherical, radial, bilateral, or asymmetric.

Every animal exhibits a certain polarity and a certain symmetry—the earliest and most permanent expressions of form that appear during development. Many traits of an animal can be changed by experimental means, but its original polarity and symmetry can hardly ever be changed. Millions of years later, long after the animal has become a fossil, polarity and symmetry often are still recog-

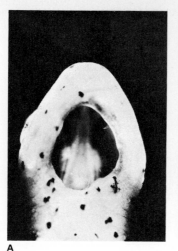
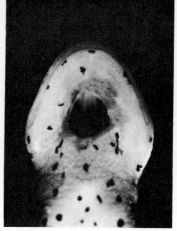
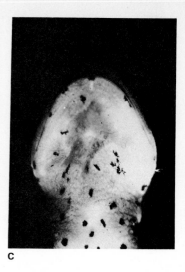

A   B   C

Figure 11.1 Differential growth. The floor of the mouth of this salamander has been cut out (left). The photos show successive stages in the regeneration of a new floor. Evidently, the regenerating tissues grow differentially far faster than the rest of the animal.

nizable even if all other signs of form have disappeared. It is a fairly general principle of development that the earlier a particular feature appears, the later it disappears.

One form-producing process is *differential growth*. Here the amounts and rates of growth are unequal in different body parts, or they differ in different directions (Fig. 11.1). A second such process is *form-regulating movement*, which produces shifts and outright migrations of growing parts relative to one another. Through these processes the form of a body part or a whole animal is first blocked out in the rough, through establishment of polarity and symmetry, and then it becomes progressively more refined in regional detail. Thus the organ system is delineated ahead of its organs; the tissue acquires definitive shape in advance of its cells, and the molecules are last to assume final form. Evidently, form develops as in a sculpture, from the coarse to the fine. Whereas an animal grows from its molecules on up, it forms from gross shape on down.

The net result of growth, differential growth, and form-regulating movements is the emergence of an appropriately shaped and sized anatomy. These processes of morphogenesis thus represent an organizational, architectural component of development.

The second group of processes, differentiation, is an operational, functional component. Development does not produce simply a collection of many identical cells but a series of widely different cells—for example, some become nerve cells, some liver

cells, some skin cells. Yet all arise from the same zygote, and all inherit the same genes from this zygote. Nevertheless, a multitude of different cells is formed (Fig. 11.2).

Moreover, in the course of development body parts often change their operational characteristics; a part frequently carries out functions that are not yet in existence at earlier stages and that no longer exist at later stages. Evidently, a developing part need not necessarily grow and it need not necessarily change form, but by the very meaning of development, it must differentiate. Through differentiation, structural units become functionally *specialized* in various ways and at various times.

To be differentiation, operational changes must have a certain degree of permanence. We can make an animal vitamin-deficient, for example, and many of its cells then will behave differently. But if we now add the missing vitamin to the diet normal cell functions probably will be resumed very promptly. Here cellular capacities have not been changed in any fundamental way. Only their expression has changed temporarily in response to particular conditions. Such easily alterable, reversible changes are *modulations*. On the contrary, differentiation implies a more or less fundamental, relatively lasting alteration of operational potentials. A vitamin-deficient cell that, after addition of the missing vitamin, *maintained* its altered characteristics would have differentiated.

Cell differentiation might come about in three ways. First, the developing differences among cells might be a result of progressive changes in gene action. In a given cell some genes might become active at certain developmental stages, whereas others might become inactive. The operon mechanism (Chap. 6) suggests how such switching of gene activities might occur. Activity patterns might change differently in different cells, and this might contribute to differentiation. Or, second, gene actions might remain the same, but the operations of the cytoplasm could become altered progressively. For example, one round of cytoplasmic reactions might use up a certain set of starting materials, and in the later absence of these, similar reactions then could no longer take place. A next round of reactions would proceed with different starting materials and therefore would produce different endproducts. The net result could be progressive differentiation. Or, third, both nuclear and cytoplasmic changes might occur in reciprocal fashion. This is probably the likeliest possibility, and much current research is devoted to a study of this very complex problem.

Like growth, differentiation occurs from the molecule on up; the operations of any living level are based on those of subordinated levels. It is this that makes the problem of analysis so enormous. For if differentiation is as complex as the totality of molecular interactions in cells, then it cannot be any less complex than the very process of life itself.

Morphogenesis and differentiation are two of the forces that drive development. A third is *metabolism*. There could be no development if energy were not available and if molecular synthesis did not occur. To be sure, there could be no metabolism if morphogenesis and differentiation did not develop it. At no point in the life cycle of any animal is metabolism more intensive and development more rapid than during the earliest stages. Both then decline in rate, until the zero point is reached at death; the metabolic clock is wound only once, at the beginning.

This circumstance introduces a number of major problems. For example, respiration in reproductive cells necessitates gas exchange through the cell surfaces. This requirement limits the size of such a cell, however, for diffusion could not be effective in too large a cell mass. But the requirement of smallness in turn limits the amount of food that can be stored in reproductive cell, which puts a time limit on the amount of development possible. Clearly, the developmental consequences of so "simple" a requirement as oxygen supply are quite far-reaching.

The molecular equipment for energy production is inherited complete by all reproductive units and is functional from the start. This is an absolute necessity for survival. But only relatively few kinds of metabolic syntheses are possible at first, for synthesis must itself develop. Endproducts of a first round of synthesis must become the starting materials for a second, more complex round. In this manner synthetic capacities must be increased and broadened progressively. Synthesis metabo-

Figure 11.2 Differentiation. Mouse embryo tissue pieces from the region of the future salivary gland were put together in a culture (*A*). These pieces grew and interacted (*B*) and eventually differentiated as secretion pockets and ducts characteristic of normal salivary glands (*C*).

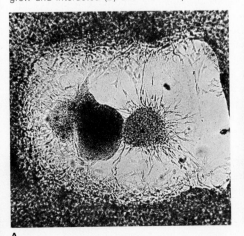
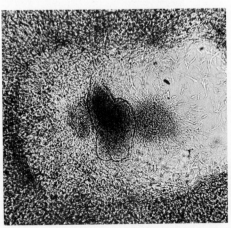
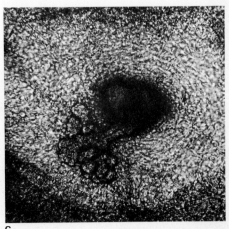

A   B   C

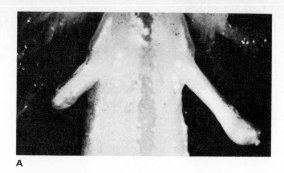

A

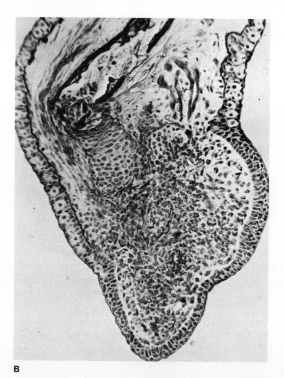

B

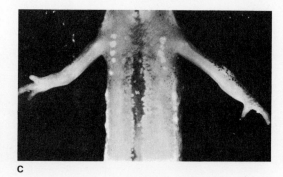

C

Figure 11.3 Wholeness. If the arms of a salamander are amputated (*A*), tissues in the stumps gradually reorganize and produce a regenerating cell mass (*B*). Eventually whole new limbs are re-formed, 31 days later in this case (*C*). Development evidently tends to produce or restore wholeness in an organism.

lism clearly is as much a result of development as it is a prerequisite; it is one aspect of differentiation.

How do morphogenesis, differentiation, and metabolism mesh together to produce a sensibly functioning whole? A zygote or a bud does not yet have any of the features of the adult. How then does it happen to give rise to just one head end and one tail end, not two of either, yet in a man also two arms and two legs, not one of each? The developing system evidently behaves as if it "knows" its objectives precisely, and it proceeds without apparent trial and error. For normally there is no underdevelopment, no overdevelopment, and there are no probing excursions along the way. Development is directed straight toward wholeness (Fig. 11.3).

This tendency toward wholeness is manifested at every level of organization and in any developing unit. It pinpoints the fourth universal component of development, *control.* Ultimately the control of development is undoubtedly genetic, as it is for every living process. But such a generalization is not very informative and actually is little more than a restatement of the problem. *How* do genes control development? More specifically, how does a particular gene, through control over a particular enzyme or other protein, regulate a particular developmental occurrence? Answers to such small problems are just beginning to be obtained. The collective larger issue, the controlled, directed emergence of wholeness in an entire animal, remains a matter of future research.

Another genetic aspect of development is of significance in this context. The course of development varies considerably according to whether the starting point is a zygote or some form of vegetative body. Zygotic development passes through several distinct phases: fertilization, embryonic period, typically also larval period, and then adulthood. In sharp contrast, all other forms of development are exceedingly direct. In the development of vegetative units of any type there is no sex, hence no fertilization; there is no embryo, hence also no hatching; there is no larva, hence also no metamorphosis. Instead, the reproductive unit becomes an adult in a smoothly continuous single developmental sequence (Fig. 11.4).

This marked difference between sexual and asexual patterns of development undoubtedly is a result of the presence or absence of the sexual process itself. Unlike a vegetative body, an egg

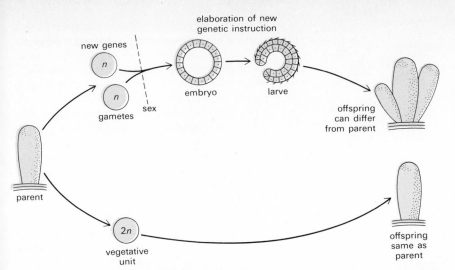

Figure 11.4 Sexual and asexual development. In sexual development (top) new genetic instructions are introduced in the zygote by the gametes, and later these instructions are elaborated in the embryo and the larva. Hence the mature offspring can differ to greater or lesser extent from the parent. In asexual development (bottom) new genetic instructions are not introduced, and the offspring therefore resembles the parent fully.

cuting such instructions therefore are not needed. Correspondingly, embryos and larvae are absent here.

Moreover, to the extent that the genetic instructions received by an egg are different from those received by earlier generations of eggs, embryos and larvae permit introduction of *evolutionary* changes into developmental histories. Indeed, evolutionary change is achieved not by alteration of already developed, fully differentiated adults, but primarily by modifications of eggs, embryos, and larvae. These incompletely developed stages still are plastic and unformed enough to be capable of executing new genetic instructions.

Embryonic Development

An embryo represents a first developmental phase in which, through morphogenesis and differentiation, all basic structures and functions of an animal become elaborated in at least rough detail. The embryonic period begins with *cleavage,* the subdivision of the fertilized egg.

Cleavage and Blastula

Mitotic cleavage divisions cut an egg into progressively smaller cells called *blastomeres* (Fig. 11.5). The divisions occur so rapidly that the blastomeres have little opportunity to grow, hence the egg subdivides without enlarging as a whole. Cleavage typically continues until the blastomeres have attained the size of adult cells. During cleavage different surface regions of an egg differentiate as a series of *organ-forming zones,* areas that later will give rise to the various body parts of the adult. Cleavage *segregates* these zones to different cells and cell groups. On the basis of how soon such zones become established, two categories of eggs can be distinguished: *mosaic* (or *determined*) eggs and *regulative* (or *undetermined*) eggs.

The first type is encountered in most acoelomates, pseudocoelomates, and schizocoelomates. The fate of every egg part here becomes fixed unalterably before or at the time of fertilization. Therefore, the head-tail, dorsal-ventral, and left-right axes already are firmly established in the zygote. After cleavage in such an egg, it is possible to separate the cells from one another experimentally. Each isolated blastomere then develops

is more than simply a reproductive unit; it is also the agent for sex and therefore an *adaptive* device. Through fertilization the egg acquires new genes, which often endow the future offspring with new, better-adapted traits. However, before any new traits actually can be displayed they must be *developed* during the transition from egg to adult. Embryonic and larval periods appear to be the outcome. These stages provide the means and the necessary time for translating genetic instructions acquired sexually by the zygote into adaptively improved traits of the adult. Vegetative units do not acquire new genetic instructions through sex, and equivalent developmental processes for exe-

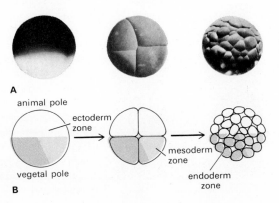

Figure 11.5 Cleavage and organ-forming zones. *A,* egg, four-cell, and later cleavage stage in frogs. *B,* organ-forming zones. The main egg axis is marked by so-called *animal* and *vegetal poles* at opposite areas of the egg. As cleavage progresses, a given zone of the egg becomes segregated in progressively smaller but more numerous blastomeres.

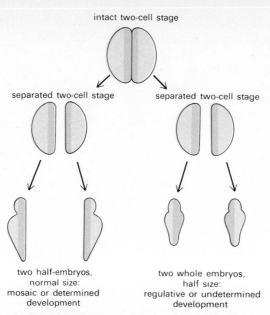

intact two-cell stage

separated two-cell stage   separated two-cell stage

two half-embryos,
normal size:
mosaic or determined
development

two whole embryos,
half size:
regulative or undetermined
development

Figure 11.6 Mosaic and regulative development. If
the cells of early cleavage stages of mosaic eggs are
isolated experimentally (left), then each cell develops as
it would have in any case. The inference is that the fates
of cytoplasmic regions in such eggs (dark central parts)
are determined very early. But if the cells of early cleaving
regulative eggs are isolated (right), then each cell develops as a smaller whole animal. The inference here is
that the fate of the cytoplasm still is undetermined. Thus
if the two-cell-stage blastomeres are separated, the central cytoplasm, which normally would form central body
parts, actually forms left structures in one case, right
structures in the other.

Figure 11.7 Egg size, cleavage rate, and yolk: effects on development.

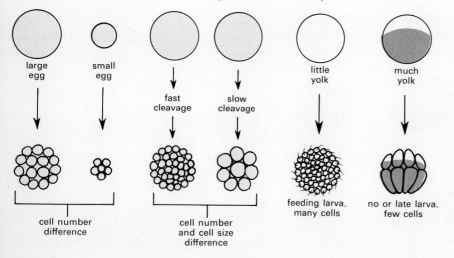

large
egg

small
egg

fast
cleavage

slow
cleavage

little
yolk

much
yolk

cell number
difference

cell number
and cell size
difference

feeding larva,
many cells

no or late larva,
few cells

as a *partial* embryo; it produces the same portion
of the embryo it would have produced if the cleaving egg had been left intact. The egg once determined thus behaves like a quiltwork, a mosaic, in
which the fate of the egg parts cannot be altered
after fertilization (Fig. 11.6).

By contrast, the eggs of most enterocoelomates, vertebrates included, are of the regulative
variety. In these the fate of various egg portions
becomes fixed much later, usually not until three
cleavages have produced eight cells. If blastomeres are isolated before that time, each develops
as a *whole* embryo, not as a partial one. During
the undetermined phase any cell thus can substitute for any other cell and can develop into any
structure, including a whole animal. Developmental determination is a form of differentiation,
and the underlying mechanism still is completely
unknown.

Formation of two or more whole animals from
separated blastomeres is equivalent to the production of identical twins, triplets, quadruplets, or
larger sets. Natural *twinning* undoubtedly occurs
through similar separations. However, the forces
or accidents that actually isolate blastomeres in
nature are not understood. If the blastomeres are
incompletely separated, Siamese twins result. And
if separation occurs after the time of developmental determination, each blastomere forms only
a partial embryo, like a part of a mosaic.

Apart from the mosaic or regulative characteristics of an egg, three additional properties
influence development greatly: egg *size,* cleavage
rate, and amount and distribution of *yolk* (Fig.
11.7). Depending on the original size of an egg,
cleavage will continue for a longer or a shorter
period until the cells are considerably smaller.
Initial egg size thus influences the number and size
of the blastomeres that will be present at the end
of cleavage, and this in turn affects the later architectural development of the embryo greatly (see
below).

The rate of cleavage likewise influences the
number and size of cells available later. For example, if an egg divides successively seven times
during a given period, it will yield 128 cells. But
if another, equally large egg cleaves eight times
during the same period, it will yield 256 far smaller
cells, a number fully 100 percent greater than in
the first case. Clearly, even slight differences in
cleavage rates can produce large differences in cell
numbers and sizes.

The quantity of yolk in an egg determines how long an embryo can develop without external food sources. As a general rule, yolk-poor eggs pass through the embryonic phase quickly and become feeding larvae very soon, whereas animals with highly yolky eggs have long embryonic periods and late larvae or no larvae at all.

With respect to the distribution of yolk, four main egg types can be distinguished (Fig. 11.8). In *isolecithal* eggs, yolk is dispersed rather evenly and comparatively little is usually present. Sponges and mammals are among animals with such eggs. In *centrolecithal* eggs, the amount of yolk is large and it is collected compactly in the center of the egg cell. Eggs of this type occur in some cnidarians and most arthropods. In *telolecithal* eggs, the amount of yolk again is large but is collected excentrically in one region of the egg. Annelids, mollusks, and most amphibia have eggs of this kind. In *discoidal* eggs the amount of yolk is so enormous that the nonyolky part of the cell forms a *blastodisc*, a mere microscopic spot atop the yolk mass. Such eggs occur in squids and octopuses, in fishes, and in reptiles and birds.

A large amount of yolk interferes physically with the subdivision of an egg during cleavage. Thus when centrolecithal and discoidal eggs cleave, their yolk does not even become included in the blastomeres. Instead, a layer of cells forms that surrounds or lies on top of the undivided yolk mass (*superficial* and *dicoidal* cleavage, respectively, or *meroblastic* cleavage collectively; see Fig. 11.8). In isolecithal and telolecithal eggs the cleaving blastomeres do include the yolk, but yolky cells divide more slowly and remain larger than yolk-free ones (*equal* and *unequal* cleavages, or *holoblastic* cleavage collectively).

Where yolk does not interfere with cleavage unduly, the blastomeres are arranged in various symmetry patterns. Undivided eggs are radially symmetric around an axis that passes from an *animal pole* in one (nonyolky) egg hemisphere to a *vegetal pole* in the other (yolky) hemisphere. This animal-vegetal axis establishes the primary longitudinal polarity of an egg, and cleavage patterns are oriented with reference to this axis. Three patterns are particularly common, *radial, bilateral,* and *spiral* ones (Figs. 11.9 and 11.10). In all three the first two division planes pass through the animal-vegetal axis at right angles to each other. The result is a quartet of blastomeres, four cells lying in the same plane.

In both radial and bilateral cleavage patterns, later divisions typically occur in such a way that blastomeres come to lie one directly over another. Also, the embryo develops either several mirror-image planes, as in the radial pattern, or only one, as in the bilateral pattern. In spiral cleavage, the quartet of blastomeres formed by the first two divisions are called *macromeres*. These later bud off a succession of four or five quartets of *micromeres*, all stacked one on top of the other. However, the first micromeres do not lie directly over the macromeres but in the valleys *between* the macromeres. Viewed from the animal pole, the shift of position typically is to the right. When the macromeres then cut off a second quartet of micromeres, these are shifted to the left in relation to the macromeres. The third quartet of micromeres shifts to the right again (and thus comes to lie directly under the first micromeres). These alternating shifts of the successive quartets are the identifying feature of spiral cleavage. However, notwithstanding the early spiral pattern, the embryo develops an overall bilateral symmetry. This type of cleavage occurs in various protostomial groups, notably acoelomates, annelids, and mollusks.

Many animals have other, *irregular* cleavage patterns that cannot be classified neatly. Moreover, the occurrence of particular egg types and cleavage patterns is not always taxonomically tidy. Related phyla or closely related groups in a phylum often have distinctly different forms of cleavage. For example, tunicates have small, isolecithal,

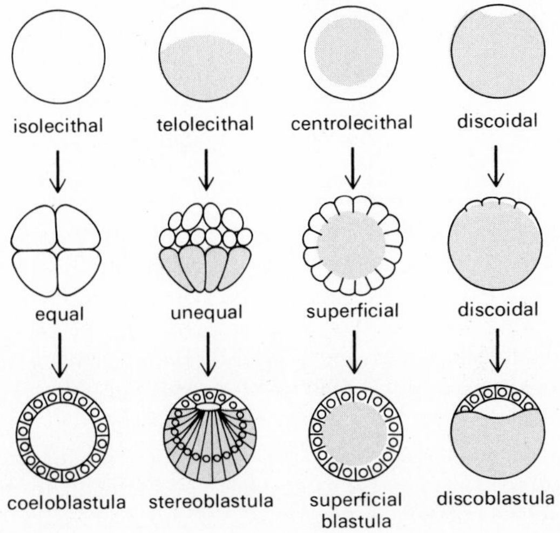

Figure 11.8 Eggs, cleavage, and blastulas. Top row, egg types, named according to yolk content and distribution (indicated by colored areas). *Center row,* cleavage patterns. Note that yolk influences the nature of a given pattern. The two patterns at left are *holoblastic,* the two at right, *meroblastic* (cleavage does not cut off complete cells). *Bottom row,* blastula types. Any free space in a blastula is a blastocoel.

isolecithal   telolecithal   centrolecithal   discoidal

equal   unequal   superficial   discoidal

coeloblastula   stereoblastula   superficial blastula   discoblastula

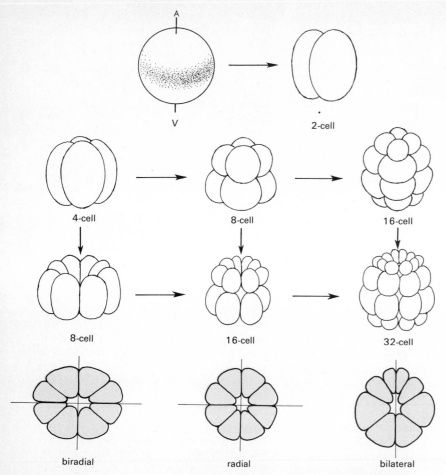

Figure 11.9 Cleavage patterns. First three rows, animal half (A) toward top, vegetal half (V) toward bottom. *Bottom row,* sections as seen from animal pole. *Left column,* biradial (or dissymmetric) cleavage; *middle column,* radial cleavage; *right column,* bilateral cleavage. Note that the 4-cell stage is similar for all three patterns shown, and that the 8- and 16-cell stages can be attained in different ways. The 32-cell stage often is more strongly bilateral than shown here if the pattern is a bilateral one.

the forerunner of the adult pseudocoel. The size of this space is governed by the number and size of the blastomeres present, hence ultimately by the original egg size, the cleavage rate, and the yolk content. The blastocoel is small in a *stereo-blastula* and comparatively large in a *coeloblastula*. In meroblastic embryos, the blastocoel either is filled completely with yolk (*superficial* blastula) or forms by a lifting away of the blastodisc from underlying yolk (*discoblastula*).

Gastrula and Induction

After a blastula has developed, the next major event is the establishment of the three primary germ layers of the embryo—*ectoderm* exteriorly, *endoderm* interiorly, and *mesoderm* between. The processes that transform a blastula to such a developmental stage collectively are called *gastrulation,* and the resulting embryo itself is a *gastrula.*

There are almost as many different methods of gastrulation as there are animal types. In many radiates and some other groups the blastula is a solid morula, and gastrulation often consists of little more than arrangement of the outer cells as a distinct ectoderm layer. The inner cells then form endoderm, and cells that later migrate in from the ectoderm become mesoderm.

In the majority of animals the prospective regions that will produce the three germ layers are already marked out in the organ-forming zones on the original egg surface. Gastrulation therefore involves an *interiorization* of the endoderm- and mesoderm-forming zones of the blastula. The prospective endoderm usually is brought to the interior first, generally by means of two processes: *epiboly,* or overgrowth of the ectoderm-forming regions around the endoderm-forming ones; and *emboly,* or ingrowth of the endoderm-forming regions under the ectoderm-forming ones. Different animals gastrulate by either one of these processes or by a combination of both (Fig. 11.11).

Emboly can occur in at least four different ways: by *ingression,* or inward migration of individual endoderm-forming cells from the blastula surface; by *delamination,* or establishment of a whole sheet of endodermal cells that are budded off toward the inside of the surface layer of the blastula; by *involution,* or inrolling of a sheet of endoderm cells from the blastula surface, followed by a spreading of this sheet underneath the surface layer; and by *invagination,* or indenting of the

mosaic eggs; but vertebrates, with the same ancestry as tunicates, typically have large, discoidal, regulative eggs, in adaptation to life in fresh water and later on land. Mammals have reverted to small isolecithal (yet still regulative) eggs, in conjunction with the viviparous mode of development.

At the end of cleavage, the embryo consists of a ball of cells in holoblastic types and a layer of cells over yolk in meroblastic types. This developmental stage is a *blastula* (see Fig. 11.8). In holoblastic embryos the blastula is either solid or hollow. If it is filled with cells, it is called a *morula.* If it is hollow, the internal space is a *blastocoel,*

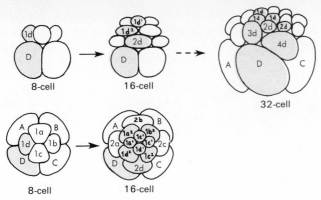

Figure 11.10 Spiral cleavage. Top row, side views; *bottom row,* view from animal pole. Blastomeres in the *D* quadrant in color, to indicate the alternating right and left spiraling of successive tiers of blastomeres. Capital letters identify macromeres, lowercase letters, micromeres. In the 16-cell stage, *D* has produced 2d (a micromere of the second quartet), while 1d (of the first quartet) has divided into 1d^1 and 1d^2; similarly for the *A, B,* and *C* quadrants. Subsequent cleavages do not all occur synchronously. The numbers before the letters indicate the quartet a blastomere belongs to; exponent numbers indicate sequence of blastomere origin. In the 32-cell stage, the 4d cell has just cut off from *D.* This 4d micromere is the source of the later adult mesoderm. All other micromeres will develop most of the ectoderm, and the macromeres, most of the endoderm.

Figure 11.11 Gastrulation. A, gastrulation by ectodermal overgrowth (epiboly). *B,* three forms of gastrulation by endodermal ingrowth (emboly). *C,* gastrulation by invagination, another form of emboly. *D,* early starfish development. 1, Blastula. 2, Embolic invagination, early gastrula. 3, Late gastrula, beginning of mesoderm formation.

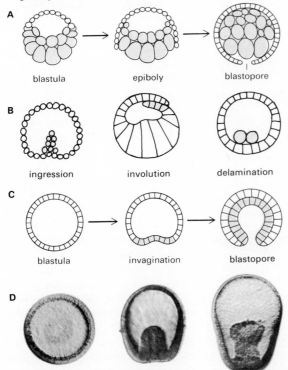

endoderm-forming part of the blastula surface, as when a balloon is pushed in with a fist in one region.

Whether epiboly or one or more variants of emboly occur is determined in large part by the size and number of blastula cells, by the amount of yolk present, and by the size of the blastocoel. For example, embolic ingrowth—particularly by invagination—is most feasible mechanically in a coeloblastula, one with many small cells and a large blastocoel.

By contrast, invagination is mechanically difficult or impossible where the blastocoel is small and where blastula cells are large, yolky, and few in number. Gastrulation then is achieved by epiboly or one or more forms of noninvaginative emboly. Thus, stereoblastulas usually gastrulate by epiboly or involution or both. Morulas often gastrulate by epiboly, delamination, ingression, or by combinations of these. And meroblastic blastulas acquire interior layers by delamination or by involution.

The common result of all methods of gastrulation is a basically two-layered gastrula having ectoderm on the outside, endoderm on the inside, and usually a *blastopore,* an opening in the vegetal half. This perforation later becomes the single alimentary opening of radiate animals and flatworms, the mouth of the protostomial groups, and the anus of deuterostomial groups. A second alimentary opening arises later as a perforation at the opposite side of the embryo. In a gastrula, also, the endoderm either has an internal space from outset or, if the interior is originally solid, soon develops such a space. This endodermal cavity is a *gastrocoel* or *archenteron,* the embryonic alimentary cavity. It communicates with the outside through the blastoporal opening (see Fig. 11.11).

Development of the third germ layer, the mesoderm, usually begins as soon as endoderm has formed. Mesoderm cells are budded off from ectoderm, endoderm, or both, and they migrate to the space between the ectoderm and the endoderm. If, early in gastrulation, most of the mesoderm-forming regions of the original blastula do not become interiorized but remain part of the ectoderm, then the mesoderm arising in the late gastrula will be largely an *ectomesoderm.* By contrast, if the original mesoderm-forming regions do become interiorized during gastrulation, then the later mesoderm will be largely an *endomesoderm* (Fig. 11.12).

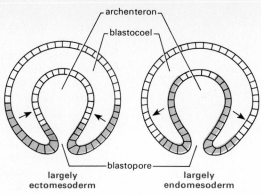

Figure 11.12 Ectomesoderm and endomesoderm. If little of the future mesoderm-forming zone (color) is interiorized during gastrulation, as at left, then most of the mesoderm eventually formed will be ectoderm-derived ectomesoderm (small arrows pointing into blastocoel). But if most of the mesoderm-forming zone is interiorized by gastrulation, as at right, then mesoderm will be largely endoderm-derived endomesoderm. Although the diagram depicts gastrulation by invagination, the principle regarding the source of mesoderm applies to any gastrulation pattern. Primary (embryonic and larval) mesoderm often is ectomesodermal, adult mesoderm, usually endomesodermal.

Figure 11.13 The vertebrate gastrula. In this diagrammatic side view are shown the adult organ systems formed by each of the primary germ layers. These layers often give rise to other organ systems in other animal groups.

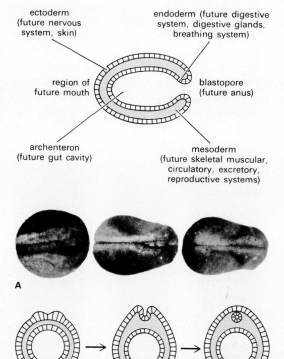

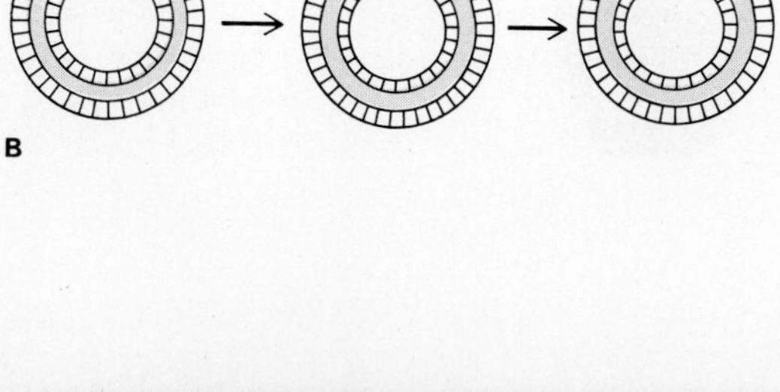

Figure 11.14 Early development of vertebrate nervous system. A, left to right, dorsal views of progressive stages in neural-tube formation in frogs. The anterior ends of the embryos are toward the right. *B,* diagrammatic cross sections corresponding to the stages shown in *A.*

In most animals both the ectoderm and the endoderm contribute to mesoderm formation, though often at different stages. In most coelomates, for example, the larval mesoderm is largely ectomesodermal whereas the adult mesoderm is predominantly endomesodermal. Thus, as already shown in Chap. 7 (see Fig. 7.12), adult enterocoelomates acquire mesoderm as pouches that grow out from the endoderm. Similarly, the adult mesoderm of schizocoelomates has an endodermal origin. More specifically, a pair of *teloblast* cells buds off from the endoderm and these cells proliferate as teloblastic mesoderm bands on each side of the embryo. In spirally cleaving types the teloblast cells arise from the 4d blastomeres, in the fourth quartet of micromeres (see Fig. 11.10).

Thus, according to how mesoderm arises and becomes arranged later, the embryo will come to exhibit an acoelomate, a pseudocoelomate, or a variously coelomate body structure. The late gastrula evidently represents a key stage in embryonic development. At that time the germ layers are established in proper positions, the future front and hind ends of the animal are marked, in many cases the top-bottom and left-right axes are already determined, and the body cavities have developed in rudimentary form. In effect, the fundamental architecture of the future animal has become elaborated in rough outline (Fig. 11.13).

Note, moreover, that these early developmental events produce the progressively less generalized features of animal construction discussed in Chap. 7. Eggs and cleavage establish the level of organization and the symmetry of an animal; the early gastrula sets the alimentary pattern; and the late gastrula delineates the basic architecture of the mesoderm and the nature of the body cavities. All later development consists essentially of sculpturing of detail—formation of organ systems and organs with specialized tissues and cells.

The actual ways in which the germ layers develop as well-defined body parts differ vastly for different animals. In general, however, such transformations take place by outfolding or infolding, outpouching or inpouching, of portions of the three layers of the gastrula. For example, the nervous system of vertebrates forms by an infolding of a tube of ectoderm along the dorsal midline of the embryo (Fig. 11.14). Limbs arise by combined outpouchings from ectoderm and mesoderm. Lungs and digestive glands develop as outpouchings at various regions of the endoderm. All other

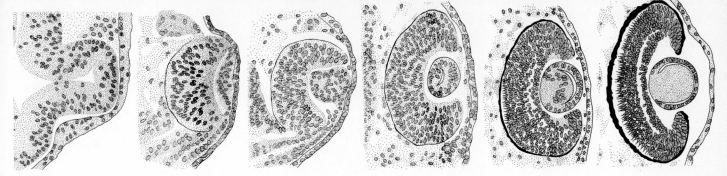

Figure 11.15 Vertebrate (amphibian) eye development. This series of diagrams shows the successive outgrowth of a pocket from the brain, contact of this pocket with the outer body ectoderm, formation of an eyecup, gradual formation of a lens from the outer ectoderm, and development of the pigmented and other tissue layers of the eyeball.

body parts develop in similar fashion. The ultimate result of these processes of morphogenesis and differentiation is a fully formed embryo, clearly recognizable as a young stage of a particular species.

Experiments have revealed some of the controls that ensure orderly sequences of development. Eye development in vertebrates provides a striking example. An eye begins to form when a pocket grows out from the side of the prospective brain (Fig. 11.15). This pocket has a bulbous tip that soon invaginates as a double-layered cup, the future eyeball. The rim of this cup eventually comes into contact with the outer, skin-forming ectoderm of the body. Just where the eyecup rests against it, skin ectoderm begins to thicken as a ball of cells, and the ball then is nipped off toward the inside. It fits neatly into the eyecup and represents the future lens. The lens cells and the overlying skin ectoderm later become transparent, and in this manner the basic structure of the eye is established.

In experiments on amphibian embryos, the stalked eyecup can be cut off before it has grown very far and can be transplanted to, for example,

a region just under the belly ectoderm of another embryo. Under such conditions the patch of belly ectoderm that overlies the transplanted eyecup soon thickens, a ball of cells is nipped off toward the inside, and a lens differentiates. Moreover, lens and overlying skin become transparent. In effect, the transplanted structures have caused the formation of a structurally normal eye in a highly abnormal location (Fig. 11.16).

A common conclusion emerges from this and many similar types of experiments. One embryonic tissue interacts with an adjacent one, and the latter is thereby induced to differentiate, to grow, to develop in a particular way. This developed tissue then interacts with another one and induces it to develop in turn. In such fashion one tissue provides the stimulus for development of the next. This phenomenon of *embryonic induction* therefore provides an explanation for the orderly sequence in which body parts are normally elaborated. However, the nature of such inductive processes in terms of reactions in and among cells is still obscure in many respects.

Postembryonic Development

Larvae

In oviparous animals a larval period begins when the embryo hatches from its enclosing membranes and protective coats and assumes a free-living existence.

Larvae serve mainly in one or both of two capacities. In many animals, sessile ones in particular, the larvae are the chief agents for geographic distribution of the species. In some cases such species dispersers still are without fully formed

Figure 11.16 Eye transplantation. If an embryonic eyecup is excised from a donor embryo and transplanted to an abnormal location in a host embryo, then a structurally perfect eye can develop at that abnormal location. The photo shows a larva of a salamander (*Amblystoma*) with two supernumerary eyes grafted to abnormal locations. The photo was taken 43 days after the transplant operation.

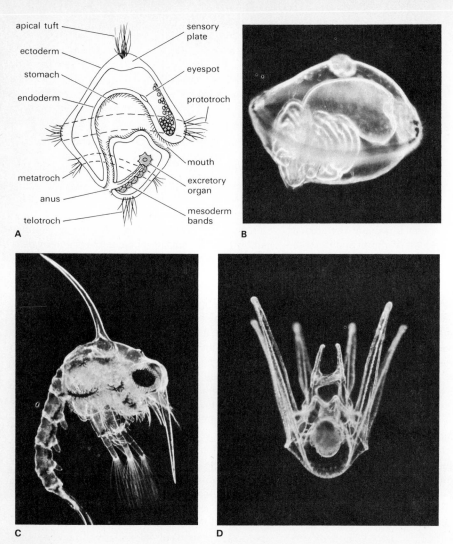

Labels in figure A: apical tuft, ectoderm, stomach, endoderm, metatroch, anus, telotroch, sensory plate, eyespot, prototroch, mouth, excretory organ, mesoderm bands

A

B

C

D

Figure 11.17 Larval types. A, sagittal section through a trochophore larva, characteristic of both mollusks and annelids. Mesodermal structures in color. The prototroch and metatroch circlets of long cilia are indicated by broken lines. A metatroch or telotroch is not necessarily present in all types of trochophores. B, late trochophore of the archiannelid *Polygordius.* The mouth is toward the right. C, late zoaea larva of the shore crab *Carcinus.* D, pluteus larva of a sea urchin (*Echinus*). The ciliated arms endow such a larva with large surfaces that increase buoyancy.

mary function of larvae here appears to be nutritional; the organisms are transitional feeding stages that accumulate food reserves required for the construction of a more complex adult. In numerous animals a larva is both a dispersal and a feeding device.

Larvae occur in virtually all phyla, but some of the subgroups in a phylum often are without them. In most cases animals with larvae tend to be the comparatively primitive members of a phylum; those without larvae, the more advanced or specialized members. When larvae do occur, some resemble the adults greatly whereas others are structurally quite distinct from the adults. For example, fish larvae leave little doubt that these animals will become fishes, but it is not nearly so obvious that tadpoles become frogs or that caterpillars become butterflies. Furthermore, where the resemblance is great, metamorphosis to the adult condition usually is a very gradual, barely perceptible process. But where a resemblance is lacking, metamorphosis generally involves a drastic and abrupt reorganization of the whole body. What accounts for such differences in the durations, forms, and terminal events of larval life?

The answer has three parts. One is illustrated well by the larvae of crustacea (Fig. 11.18). Like arthropods generally, crustacea develop after germ-layer formation by elongating at the posterior end. In the process the body segments are laid down in anteroposterior succession and paired appendages grow from each segment. A sequence of stages therefore can be recognized, each with several more segments than the preceding stage. These stages are named, in sequence, *nauplius, metanauplius, protozoaea, zoaea, mysis,* and *adult.* A nauplius has only the most anterior head segments in reasonably developed condition, and a preadult mysis has all but the most posterior abdominal segments.

The important consideration is that different crustacea become larvae at different points in the developmental series. For example, the brine shrimp *Artemia* develops as an embryo up to the nauplius stage, and then it hatches as a nauplius larva. After some time the nauplius molts and becomes a metanauplius larva, and later molts give rise to a sequence of larvae corresponding to the series above.

By contrast, the fairy shrimp *Branchipus* passes the nauplius stage as an embryo and hatches only later, as a metanauplius larva. The mantis shrimp

alimentary systems (as in ascidian tadpoles), but locomotor structures always are well developed. Thus, tadpoles have strongly muscled tails, and the swimming larvae of many invertebrates have greatly folded and ciliated surfaces, with bands and tufts of extralong cilia in given regions (Fig. 11.17). In many other animals the larvae can disperse geographically far less well than the adults (insect caterpillars, for example). The pri-

Squilla passes both the nauplius and metanauplius stages during the embryonic phase and hatches only as a protozoea larva. Still later hatching occurs in many crabs, which emerge as zoaea larvae; in lobsters, which do not become larvae until the mysis stage; and in certain shrimps and prawns, which hatch only as fully formed adults.

Clearly, then, the earlier the time of hatching, the more larval development must take place before the adult condition is attained; and the less therefore does the young larva resemble the adult.

Conversely, the later hatching occurs, the less larval development is necessary to produce the adult and the more will larva and adult resemble each other. Moreover, if hatching takes place very late, a larval phase can be absent entirely and the emerging animal will be an adult.

These relations hold not only for crustacea but for other animals as well. Among insects, for example, some are early hatchers and their larvae are more embryonic than adult. All caterpillar-producing types belong to this category. Other insects, grasshoppers among them, pass through the wormlike caterpillar stage as embryos, and when they then hatch comparatively late they already resemble the adults greatly. Still other insects, exemplified by silverfish, hatch comparatively even later and their "larvae" in effect are young adults.

Thus the durations and forms of larval periods vary according to the developmental stage the animals have attained at the time they *hatch*. This generalization answers in part why some larvae resemble the adults more than others and why larvae do not necessarily occur in all members of a group.

Another part of the answer is illustrated in the development of brittle stars. In these echinoderms the eggs have considerably different sizes and yolk contents. At one extreme eggs are small and isolecithal; at the other, they are large and highly telolecithal. The small eggs develop as outlined in Fig. 11.19. A long-lived pluteus larva forms, with a large, ciliated swimming surface folded into conspicuous arms. In the course of larval life the number of these arms increases from two to four to six and finally to eight. After several months the pluteus metamorphoses into a small adult.

By contrast, development is quite different in brittle stars with larger eggs. The yolkier the egg, the fewer arms develop in the larva. Some plutei form only six arms, others only two, still others none. Moreover, the fewer the arms, the shorter the larval period and the earlier the metamorphosis. In the extreme condition, the large original amounts of yolk first change embryonic development to a pattern that is very unusual for echinoderms, and then a larva does not develop at all; soon after gastrulation the embryo becomes an adult directly. Brittle stars of this type are viviparous; their eggs develop in the maternal parent, which then releases not eggs but young adults.

Evidently, the nature of the egg influences the

Figure 11.18 Crustacean larvae. A through *E*, series of larvae hatched at successively later developmental stages (segmental stage attained is indicated below each figure). Thus, a just-hatched brine shrimp nauplius still must develop a great deal before the adult condition is attained, but a just-hatched mysis of a lobster already resembles the adult considerably. If hatching occurs very late, as in some shrimps and prawns, a larval phase is absent. (After Claus, Faxon, Herrick, Ortman, Weisz.)

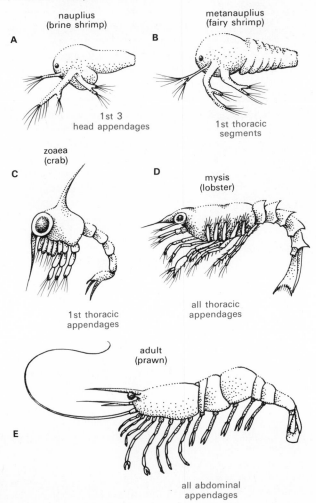

A nauplius (brine shrimp)
1st 3 head appendages

B metanauplius (fairy shrimp)
1st thoracic segments

C zoaea (crab)
1st thoracic appendages

D mysis (lobster)
all thoracic appendages

E adult (prawn)
all abdominal appendages

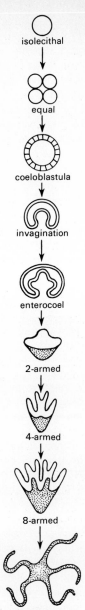

isolecithal

equal

coeloblastula

invagination

enterocoel

2-armed

4-armed

8-armed

adult

telolecithal

unequal

stereoblastula

epiboly

delamination

schizocoel

adult

Figure 11.19 Extremes in brittle star development. Left column, pattern starting with small yolkless egg leads through many-armed larval stages. *Right column,* pattern starting with large telolecithal egg leads to absence of larval stages and viviparous adults.

pattern of all further development—a point already made earlier. Second, the later metamorphosis occurs, the more body parts develop that have purely larval significance: the arms of a pluteus are larval only, not adult, and the longer the adult condition is delayed, the more pluteus arms can form. Conversely, early metamorphosis prevents development of larval structures. Indeed if meta-

morphosis occurs very early a larva is suppressed entirely and the embryo becomes an adult directly.

These relations again apply quite generally. For example, numerous mollusks and annelids develop from small, slightly telolecithal eggs, their larvae are long-lived, free-swimming *trochophores,* and metamorphosis occurs comparatively late (see Fig. 11.17). By contrast, mollusks such as squids and octopuses and annelids such as earthworms develop from large, very yolky eggs. In these groups metamorphosis occurs so early that in effect it does not occur at all: the stages corresponding to the trochophore are suppressed and the embryos become adults directly.

Thus the durations and forms of larval periods vary according to the developmental stage given animals have attained at the time they *metamorphose.* This generalization shows why larval periods can have different durations and again why larvae do not necessarily occur in all members of a group. Note that if a larval period is long, for example, it is so either because hatching occurs early or because metamorphosis occurs late or for both these reasons. Also, absence of a larva can be due either to a forward extension of the embryonic phase up to the adult or to a backward extension of the adult phase down to the embryo. The result is the same but the developmental mechanism is not (Fig. 11.20).

The timing of metamorphosis also influences the general nature of that transformation. As just noted above, larvae are *composite* organisms; some of their parts serve during larval life only, others serve during adult life only, and still others serve in both the larva and the adult. Thus the surface structures of plutei and trochophores are purely larval, the immature reproductive structures are purely adult, and the alimentary apparatus functions in both the larva and the adult. Similarly, a frog tadpole has gills and tail for larval life, the beginnings of lungs and limbs for adult life, and nervous, alimentary, and other parts for both larval and adult life.

Accordingly, metamorphosis usually includes three kinds of changes: resorption or disintegration of purely larval components, rapid development of purely adult components, and continuation of growth of all other components (Fig. 11.21). The *rate* of development of these various body parts determines how metamorphosis will affect them.

For example, assume an extreme condition in which the developmental rates of larval and adult

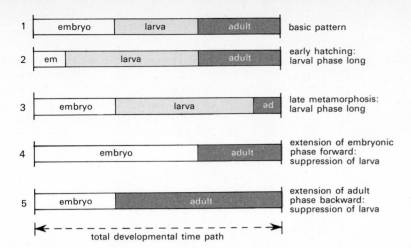

1 | embryo | larva | adult | basic pattern

2 | em | larva | adult | early hatching: larval phase long

3 | embryo | larva | ad | late metamorphosis: larval phase long

4 | embryo | adult | extension of embryonic phase forward: suppression of larva

5 | embryo | adult | extension of adult phase backward: suppression of larva

← total developmental time path →

Figure 11.20 Larvae and developmental time. Larval differences due to variations in hatching time (as in pattern 2) are illustrated by crustacea; those due to variations in time of metamorphosis (pattern 3), by brittle stars. Early hatching and late metamorphosis yield the same result, prolonged larval periods. Pattern 4 results from extremely late hatching (as in some crustacea), pattern 5 from extremely early metamorphosis (as in some brittle stars); both again produce the same result, complete suppression of larval periods.

components are nearly equal. All parts of a larva then mature simultaneously, and a distinction between "larval" and "adult" components in effect cannot be made; the whole larva becomes the adult. Metamorphosis here will not include drastic tissue destruction but will occur smoothly and gradually. Such so-called *incomplete* metamorphoses are encountered in, for example, crustacea, silverfish insects, and the planula larvae of cnidarians. In these cases all tissues of the larva are incorporated in the adult (Fig. 11.22).

Assume next an intermediate condition in which purely larval components develop substantially faster than future adult components. The larval tissues then develop so rapidly that they are already senile when the adult tissues just begin to mature. And because they are senile the larval tissues die off during metamorphosis, whereas the young adult tissues grow rapidly. For example, when frog tadpoles become adults the senile gill and tail tissues disintegrate but the immature lung and limb tissues proliferate greatly. Most metamorphoses are of this type, partly smooth transitions, partly drastic reorganizations; they are another variety of incomplete metamorphosis.

A

Figure 11.21 The nature of metamorphosis. In these photos of amphibia before and after metamorphosis, some parts of the animal degenerate (e.g., tail), some parts proliferate (e.g., legs), and some parts undergo relatively little change (e.g., skin). *A,* tadpole, hind legs developing. *B,* hind legs at advanced stage, forelegs already visible underneath skin. *C,* forelegs have broken through, tail is degenerating. *D,* froglet, limbs well developed, tail resorbed almost completely.

B

C

D

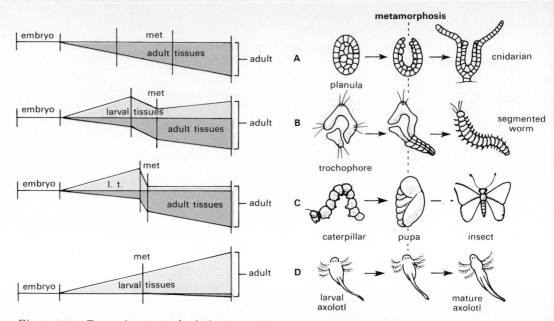

Figure 11.22 Types of metamorphosis. Graphs on left indicate comparative durations and degrees of body reorganization during metamorphosis (met), and also the relative amounts of transient larval and prospective adult tissues (color areas) composing a larva and later adult. Specific examples are sketched at right. *A,* if the larva consists entirely of tissues that are essentially part of the later adult, metamorphosis will be so gradual and prolonged as to be almost unnoticeable. *B,* if the larva consists partly of purely larval and partly of future adult tissues, metamorphosis will be somewhat less gradual than in *A* and will involve a certain amount of body reorganization (dissolution of larval tissues, proliferation of adult tissues). *C,* if the larva consists very largely of purely larval tissues, metamorphosis will be a very abrupt and drastic transformation. *D,* if the larva is entirely without prospective adult tissues (opposite of *A*), the larva in effect will become permanent and the "adult" state will require only a later development of reproductive organs (neoteny).

A far more drastic transformation occurs if the developmental rate of the adult components is exceedingly slow in relation to that of the larval components. In such cases almost the whole larva consists of purely larval tissues; and at a time when all larval parts are already senile, adult parts often have not even begun to develop. Metamorphosis then entails an almost total dissolution of the larva, as in insect caterpillars, for example. These are composed almost entirely of purely larval tissues. Future adult tissues are represented only by small, undeveloped islands of cells, so-called *imaginal disks,* located in various regions of the larval body. At metamorphosis a caterpillar forms a cocoon around itself and in this *pupa* virtually the whole caterpillar disintegrates. With raw materials from the disintegration products the imaginal disks then develop rapidly into the adult flying insect. Abrupt and drastic reorganizations of this sort are called *complete* metamorphoses.

In the most extreme situation, adult tissues develop so slowly that their rate in effect is zero. Such tissues then do not mature at all, metamorphosis does not occur, and the larval condition becomes permanent: the adult stage is suppressed (see Fig. 11.22). This direct opposite of the total suppression of a larva is encountered in, for example, axolotls. In these salamanders the thyroid mechanism that controls amphibian metamorphosis does not become operational, and the animals remain aquatic larvae permanently, with gills but without lungs. Later, one adult component, the reproductive system, does escape the growth suppression, which permits the animals to breed despite their otherwise larval state. That axolotls are in fact *larval* can be proved in one species by injection of thyroid hormone. Such individuals metamorphose immediately.

Permanent retention of larval or youthful traits is called *neoteny.* The axolotl example suggests that neoteny can provide a means of creating essentially new types of animals. Indeed, as will be shown in Chap. 14, neoteny is believed to have been an exceptionally important mechanism in the evolution of major new categories of animals.

Land Eggs and Pregnancy

Vertebrates are good examples of a large group in which the early members—fishes and amphibia—have well-defined larval periods, whereas in the later members—reptiles, birds, and mammals—free larval life has been replaced by a long-lasting embryonic phase. This change in advanced vertebrates is an adaptation to living on land; the embryo develops inside a shelled egg capable of surviving on land or inside the uterus of the maternal parent. In either case the embryonic phase continues until the offspring has developed far enough to survive out of water, in a terrestrial environment.

Ancestral reptiles were the first vertebrates without free-swimming aquatic larvae. These animals evolved not only shelled eggs that could be laid on land, but also four so-called *extraembryonic membranes* inside the eggs that made egg survival on land possible. All modern reptiles, birds, and mammals have inherited these membranes.

Early development and formation of the membranes in reptiles and birds is sketched in Fig. 11.23. Gastrulation takes place primarily by involution. After endoderm is interiorized as a layer underneath the ectoderm, mesoderm involutes at a highly elongated, slitlike blastopore, here called the *primitive streak*. The mesoderm soon forms a coelom-enclosing layer, and the embryo now is a *blastodisc*, a flattened double-layered vesicle on top of the large yolk mass. The margins of the blastodisc later proliferate and give rise to the extraembryonic membranes. One of them grows over and around the yolk and becomes a *yolk sac*. The second, or *amnion*, arises from upward folds along the margin of the blastodisc and a later fusion of the folds above the embryo. The third, or *allantois*, forms as a downward fold from the hind region of the blastodisc. And the fourth, or *chorion*, becomes an outer enclosure around the embryo and the three other membranes.

In the mature reptile or bird egg, the chorion lies just inside the shell and prevents excessive evaporation of water through the shell. The yolk sac holds the yolk and gets smaller gradually as yolk is used up during embryo growth. Eventually the empty sac becomes the floor of the alimentary tract. The allantois, lying against the eggshell just inside the chorion, is the breathing structure of the embryo; gas exchange occurs between the numerous blood vessels in the allantois and the air outside the shell. Also, the allantois serves as an embryonic urinary bladder in which metabolic wastes are stored up to the time of hatching. The amnion surrounds the developing embryo everywhere except on its ventral side. This membrane holds lymphlike *amniotic fluid,* which bathes the embryo in a "private pond." The fluid represents the substitute of the actual ponds in which the larvae of ancestral aquatic vertebrates developed. Because reptiles, birds, and mammals all have an amnion, these vertebrates collectively are also called *amniotes* (and the other vertebrates are *anamniotes*).

In mammals, certain primitive forms still lay shelled eggs like reptiles and birds (see Chap. 30), but most mammals become pregnant and the eggs develop in a uterus. Such eggs are very small and almost yolkless. Each develops into a spherical blastula with an *inner cell mass* (Fig. 11.24). The cell layer around the fluid-filled blastocoel, called the *trophoblast,* is equivalent to the chorion. An amnion soon develops in the inner cell mass as a cavity, and some cells of this mass also proliferate and spread as a sheet around the inside of the trophoblast. This sheet represents the yolk sac, which does not contain any yolk, however. By this time the inner cell mass has formed a double-layered disk composed of an ectodermal layer continuous with the amnion and an endodermal layer continuous with the yolk sac. This disk represents the embryo proper, and in it mesoderm later involutes at a primitive streak as in reptiles and birds. An allantois similarly arises from the posterior endoderm.

But although mammals acquire extraembryonic membranes in ways that are basically similar to those in reptiles and birds, the mammalian membranes largely function in new ways. After a mammalian egg is fertilized in the oviduct, it reaches the uterus as an embryo that already has a well-formed trophoblast. The embryo then *implants* in the uterus wall, which, as noted in Chap. 10, has already become prepared for pregnancy under the influence of progesterone from the corpus luteum (see Fig. 10.23). The chorionic trophoblast next develops numerous branching, fingerlike outgrowths that erode paths through the uterus wall. In this manner the tissues of chorion and uterus

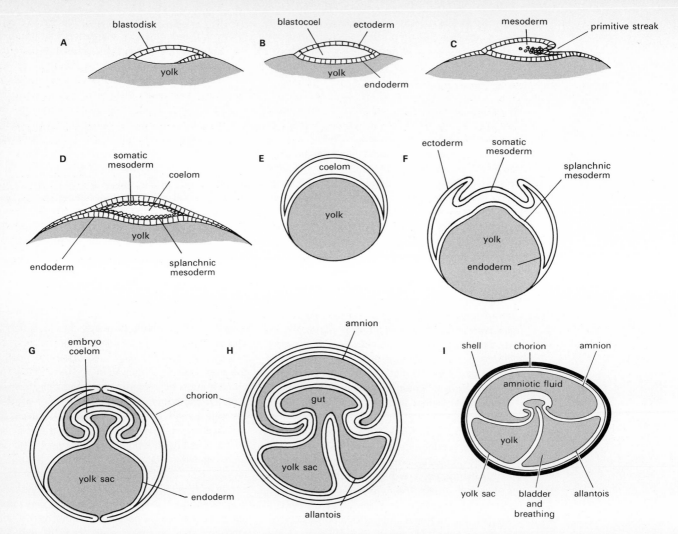

Figure 11.23   Early development in reptiles and birds. All figures represent longitudinal sections, with the head end toward left. *A,* beginning of endoderm formation between outer blastodisc layer and yolk. *B,* blastocoel established. *C,* ingression and involution of mesoderm from primitive streak. *D,* proliferation of mesoderm as somatic and splanchnic layers, and establishment of coelom. *E,* through *I,* progressive growth of ectoderm, somatic and splanchnic mesoderm, and endoderm. Each of these layers is shown as a single line. The embryo proper lies in the area directly above the yolk and gradually becomes covered by amniotic-cavity-forming folds. Note the establishment of the four extraembryonic membranes. Chorion and amnion are ectodermal, yolk sac and allantois are endodermal. Each is accompanied by a mesodermal layer (but these double-layered membranes are shown as single lines in *I*).

become attached to each other firmly. These interfingering and interlacing tissues represent the *placenta.* When fully formed, the placenta functions both as a mechanical and a metabolic connection between the embryo and the maternal body: the viviparous mammals are placental mammals, and it is because a placenta has evolved that most mammals have been able to become viviparous.

The developing allantois still serves as in reptiles and birds as an embryonic lung, except that now gas exchange occurs in the placenta, between the blood vessels of the allantois and the maternal blood of the uterus. The embryonic and maternal bloods do not mix in the placenta; the chorion always separates the two circulations (Fig. 11.25). Maternal blood also carries off embryonic wastes and provides nutrients. If a raw material is in low supply in the maternal circulation, it is usually in still lower supply in the embryonic circulation. Needed metabolites therefore tend to diffuse to embryonic blood even if this produces a pronounced deficiency in the prospective mother. In this sense the embryo is parasitic on maternal metabolism. Conversely, waste substances pass preferentially to maternal blood, which then is usually rebalanced fairly quickly by increased maternal kidney activity. The maternal circulation also supplies defensive antibodies to the embryo. An offspring thereby acquires much of his mother's immunity for the first few months of life. Such a period usually just suffices to allow the young to manufacture his own antibodies in response to exposure to infectious agents.

Because the maternal circulation removes wastes from the embryo and supplies food and

Figure 11.24 Early development in mammals. A, the ectodermal trophoblast (later chorion) and the inner cell mass. *B,* establishment of the amniotic space by cavitation in the inner cell mass and growth of endoderm layer from inner cell mass around blastocoel. *C,* growth of mesoderm and coelom from inner cell mass and growth of amnion and empty yolk sac. *D,* formation of allantois; this stage is comparable to diagram *H* in Fig. 11.23. *E,* growth of allantois and chorion to form placenta, enlargement of amnion, and collapsed yolk sac (heavier lines between extraembryonic membranes symbolize double layer of mesoderm). *F,* placenta and embryo. The yolk sac is rudimentary and collapsed, and the allantois functions as a breathing organ, via blood vessels it carries from placenta to embryo. The amnion and chorion are ectodermal, the yolk sac and allantois are endodermal, and the white areas correspond to mesodermal regions.

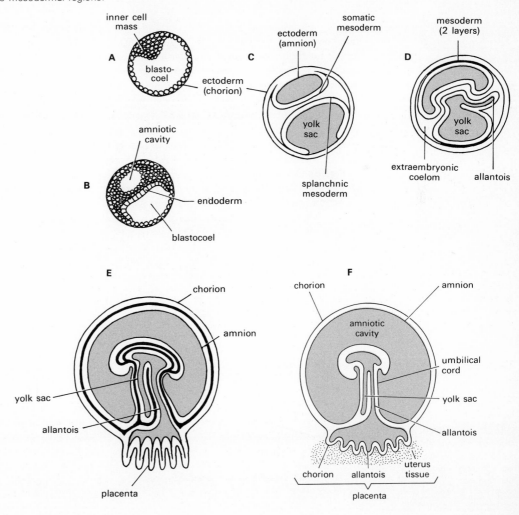

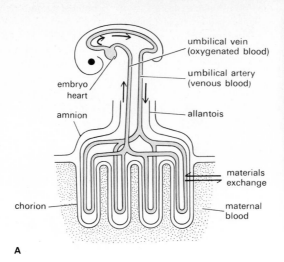

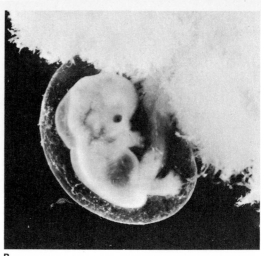

Figure 11.25 Placenta and embryo. A, the embryonic blood circulation in the placenta. Embryonic and maternal bloods do not mix, being separated by the chorionic and allantoic membranes. *B*, embryo about 8 weeks after fertilization, obtained after surgical removal of portions of the reproductive system of a female patient. The chorion is pushed to the side, revealing the amniotic sac. Note umbilical cord.

other necessary materials to it, the allantois no longer functions as a urinary bladder and the yolk sac no longer contains yolk. Both these membranes are collapsed, empty sacs. They later become enveloped by connective tissue and skin, the whole representing the *umbilical cord.* The blood vessels in the allantoic membrane make the umbilical cord a lifeline between placenta and embryo; it leaves a permanent mark in the offspring in the form of the navel. The amnion still forms as in reptiles and birds, and its fluid again serves as a ''private pond.'' It is this enlarging amnion, more than the embryo itself, that produces the bulging abdomen of the pregnant female (see Fig. 11.25).

The presence of placenta and embryo inhibits further egg production (and menstruation) during

the period of pregnancy. The pituitary continues to produce LH, and the corpus luteum at first continues to secrete progesterone. Thus the thickened wall of the uterus can be maintained and the developing embryo can remain implanted in it. Moreover, the placenta itself soon specializes as an endocrine organ. It manufactures slowly increasing amounts of estrogen and progesterone, and the progesterone output eventually becomes far greater than that of the corpus luteum. This body degenerates at some stage during pregnancy (roughly the twelfth week in the human female). From that time on the placenta provides the main hormonal control of pregnancy, and through its progesterone output it maintains its own existence (Fig. 11.26).

This period when the *luteal phase* of pregnancy changes over to the *placental phase* generally is rather critical. For if the corpus luteum should degenerate a little too soon and the placenta should reach full development a little late, then the amount of progesterone available during this gap is likely to be inadequate. In the absence of enough hormone, however, placental tissues could not be maintained. The uterine lining then would disintegrate just as during menstruation, and the embryo, no longer anchored securely, would be aborted. In man miscarriages occur frequently near the end of the third month of pregnancy. Such mishaps can be prevented by hormone injections.

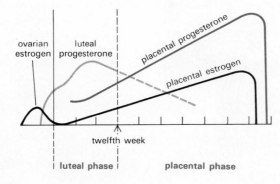

Figure 11.26 Hormones in pregnancy. Curves indicate the amounts and sources of sex hormones during pregnancy in human females.

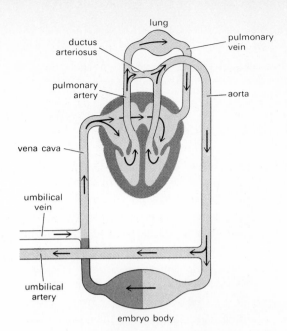

Figure 11.27 Embryonic circulation in mammals.
Oxygenated blood, white (only in umbilical vein, from placenta); venous blood, dark color (from embryo body to vena cava); mixed blood, light color (in vena cava, heart, lung, and aorta). The embryo lung is nonfunctional, and blood can pass directly from the pulmonary artery via the ductus arteriosus to the aorta. Note also the open passage between the two atria. The ductus arteriosus is a portion of the sixth aortic arch of the vertebrate gill circulation (see Fig. 9.22). Determine the effect if the umbilical vessels, the ductus, and the atrial passage all disappear and the lungs become functional, as at birth. Compare with adult circulation, Fig. 9.4.

At the time of the hormonal phase change, the species of a developing mammal usually can be identified. From then on one speaks of a *fetus* rather than an embryo. In parallel with the development of the fetus the amnion gradually enlarges and the uterus stretches. Also, the mammary glands enlarge markedly under the stimulus of the sex hormones from the placenta and lactogenic hormone from the pituitary. Numerous ducts form in the interior of the glands, and in later stages the milk-secreting cells mature. For the first day or so after birth their product is not milk but *colostrum,* a lymphlike fluid that has some laxative action in the offspring.

Birth begins when the chorion and amnion rupture and the amniotic fluid escapes to the out-

side. Labor contractions of the uterine muscles occur with increasing frequency and strength, pressing against the offspring and pushing it out through the vagina. Concurrently the placenta loosens from the wall of the uterus and the connection between mother and offspring is thereby severed. An important result is that CO_2 produced by the offspring must accumulate in its own circulation. However, within seconds or minutes the concentration of the gas becomes high enough to stimulate the breathing center of the offspring.

In conjunction with this switchover from placental to lung breathing a structural change occurs in the heart (Fig. 11.27). Before birth the dividing wall between the right and left atria is incomplete, and a movable flap of tissue provides a passage between these chambers. Once lung breathing is initiated the blood-pressure pattern in the heart changes and the tissue flap is pressed over the opening between the atria. The flap eventually grows in place, and the left and right sides of the heart then are separated permanently. Another structural change involves an embryonic blood vessel (*ductus arteriosus*), which before birth leads from the pulmonary artery to the aorta and thus shunts blood around the nonfunctional lung. At birth a specially developed muscle in this vessel constricts and never relaxes thereafter. Blood is thereby forced to pass through the lungs. The muscle degenerates to scar tissue, and the blood vessel as a whole degenerates soon after birth.

The loosened placenta, still connected to the umbilical cord, is expelled to the outside as the *afterbirth* within an hour or so after the offspring is expelled. Mammalian mothers (modern human ones excepted) bite off the umbilical cord and eat the cord and the placenta. Even herbivorous mothers do so, though they are vegetarians at all other times. Normal egg-producing cycles generally are not resumed as long as milk production and nursing continue. (In this context it is perhaps appropriate to point out that, contrary to a surprisingly wide belief among urban people, cows do not give milk at just any time; they must have been pregnant first.) After the offspring is weaned, FSH is formed again by the pituitary and a new follicle then begins to mature in the ovary.

It should be kept in mind that the reproductive continuity of animals, as outlined in the preceding, depends critically on genetic continuity. This subject is examined in the next chapter.

Review Questions

1. Define morphogenesis, differential growth, form-regulating movements, polarity. Through what kinds of growth processes does an animal enlarge in size? Explain the meaning of the phrase ''animals grow from their molecules on up.''

2. What different symmetries are exhibited by living units? In what ways do polarity and symmetry circumscribe the form of an animal? What is the role of differential growth in the development of form?

3. Define and distinguish between differentiation and modulation. What is the relation between differentiation and specialization? Give examples of differentiation on the level of molecules, cells, whole animals, and societies.

4. What role does metabolism play in development? How do metabolic rates vary during the developmental history of an animal? Illustrate the tendency of development to produce ''wholeness.'' Which metabolic capacities are in existence in a zygote, and which are not?

5. Describe and define the developmental phases in the life history of an animal when this history includes a sexual process, and when it does not. What is the significance of the greater number of phases in the first case?

6. Define blastomere, mosaic egg, organ-forming zone. How can it be shown by experiment whether an egg is mosaic or regulative? In which animals do each of these egg types occur?

7. How are twins formed? Distinguish between identical and fraternal twinning. Can identical twinning take place in mosaic eggs? What is a blastula and what major types can be distinguished?

8. Show how development is influenced by egg size, cleavage rate, yolk content, yolk distribution. What events occur during the cleavage of an egg? What types of holoblastic and meroblastic cleavage patterns can be distinguished?

9. Describe the patterns of radial, bilateral, and spiral cleavage. Which animal groups exemplify each pattern? What is a 4d cell? A macromere? A micromere?

10. In what general ways does gastrulation occur? What are blastocoels and gastrocoels? How do the characteristics of an egg influence the later gastrulation in a mechanical sense?

11. Show how the primary germ layers develop in vertebrates, and name the adult body parts that arise from each layer. Review the processes of mesoderm and coelom formation in different animal groups.

12. By what general processes of morphogenesis do the primary germ layers develop into adult structures? Illustrate this in the development of the vertebrate eye. What role does induction play in such transformations?

13. What are the general functions of larvae? Name and describe some of the larvae of different animals. What is direct and indirect development?

14. Describe the larvae of crustacea. Show how the relative time of hatching influences the nature of a larva. Show similarly how the relative time of metamorphosis influence the nature of a larva.

15. What general events occur during metamorphosis? What developmental factors determine whether metamorphosis will be complete or incomplete? What is neoteny?

16. Review the events of early development (including gastrulation) in amniote vertebrates. Describe the formation, location, and function of the extraembryonic membranes in reptiles and mammals.

17. What happens to a mammalian fertilized egg after it arrives in the uterus? When and how is a placenta formed, and what are its functions?

18. Describe the hormonal controls during the luteal and the placental phase of pregnancy.

19. Review the structure of the human placenta, with attention to embryonic and maternal blood circulation through it. Describe the whole pathway of the embryonic circulation. What is a fetus?

20. What events take place in the reproductive system of a pregnant mammal during birth of offspring? What changes take place in the blood circulation of the offspring at birth? How is milk production initiated and maintained?

Collateral Readings

Barth, L. J.: "Development," Addison-Wesley, Reading, Mass., 1964. This paperback deals with selected developmental topics, with special emphasis on cellular and early embryonic development.

Berrill, N. J.: "Growth, Development, and Patterns," Freeman, San Francisco, 1961. Numerous special aspects of embryonic and later development are discussed in this highly recommended book.

Dahlberg, G.: An Explanation of Twins, *Sci. American*, Jan., 1951. Obviously pertinent in the context of this chapter.

Ebert, J. D.: The First Heartbeats, *Sci. American*, Mar., 1959. An analysis of early chemical differentiation in heart development.

————: "Interacting Systems in Development," Holt, New York, 1965. A strongly recommended analytical paperback on some of the mechanisms at work in developing units.

Fischberg, M., and A. W. Blackler: How Cells Specialize, *Sci. American*, Sept., 1961. Progressive differentiation from the egg stage onward is discussed.

Gurdon, J. B.: Transplanted Nuclei and Cell Differentiation, *Sci. American*, Dec., 1968. A description of experiments showing how genes might influence differentiation during embryonic development.

Hadorn, E.: Transdetermination in Cells, *Sci. American*, Nov., 1968. Changes in cellular differentiation are examined.

Konigsberg, I. R.: The Embryological Origin of Muscle, *Sci. American*, Aug., 1964. The differentiation of muscle is described.

Moscona, A. A.: How Cells Associate, *Sci. American*, Sept., 1961. A case study of tissue culture, illustrating how separated cells reaggregate.

Patton, S.: Milk, *Sci. American*, July, 1969. The work of the mammary glands and the composition of milk are examined.

Puck, T. T.: Single Human Cells in Vitro, *Sci. American*, Aug., 1957. The article shows how from single starting cells human cells can be cultured like bacteria in artificial media.

Reynolds, S. R. M.: The Umbilical Cord, *Sci. American*, July, 1952. A recommended article in the context of this chapter.

Singer, M.: The Regeneration of Body Parts, *Sci. American*, Oct., 1958. Limb regeneration in salamanders and frogs is contrasted experimentally.

Sussman, M.: "Animal Growth and Development," Prentice-Hall, Englewood Cliffs, N.J., 1960. A partly comparative and partly analytical discussion of developmental processes. Recommended.

Taylor, T. G.: How an Eggshell Is Made, *Sci. American*, Mar., 1970. On the calcium metabolism involved in the synthesis of hen's eggshells.

Waddington, C. H.: How Do Cells Differentiate? *Sci. American*, Sept., 1953. A review of the embryonic and genetic aspects of differentiation during egg development.

————: "Principles of Development and Differentiation," Macmillan, New York, 1966. A recommended paperback on animal morphogenesis, differentiation, and the relation of gene action to development.

Wigglesworth, V. B.: Metamorphosis and Differentiation, *Sci. American*, Feb., 1959. The developmental processes and control mechanisms of insect metamorphosis are examined.

12

HEREDITY

Like sex, heredity has *adaptive* significance, for what an animal inherits in large measure determines its survival potential. But animals do *not* inherit strong muscles, sensitive hearing, red blood, or any other trait. Animals inherit the genes and all the other contents of reproductive cells. Visible traits then *develop* in an offspring, under the control of the inherited genes. The result of such heredity is an adult that exhibits *likeness* to its parents in certain major respects and *variation* from parents in many minor respects. If the variations are not lethal or do not cause infertility, the animal will survive and pass on its genes to following generations.

Modern studies of heredity were begun in the last half of the nineteenth century by the Austrian monk Gregor Mendel. He discovered two basic rules that laid the foundation for all later advances in genetics, the biological subscience dealing with inheritance. Accordingly, this discussion will include an examination of the rules of *Mendelian inheritance* and of some of the main aspects of *non-Mendelian inheritance* brought to light since the time of Mendel.

Mendelian Inheritance

The Chromosome Theory

Most traits of animals occur in two or more variant forms: some traits are exhibited in more or less sharp alternatives, like the different eye colors in man, while others form graded series between extremes, like body height in man. By studying the offspring from matings between such variant animals, the patterns of trait inheritance often can be determined. For example, in the fruit fly *Drosophila*, one of the most widely used animals in genetic research, the trait of body pigmentation is expressed in at least two alternative forms. In one the general coloration of the animal is gray and the abdomen bears thin transverse bands of black melanin pigment. A gray body represents the *wild type*, or predominant form of coloration in nature. By contrast, some flies are pigmented black uniformly all over the skin, a coloration pattern referred to as the *ebony* trait (Fig. 12.1).

If two gray-bodied wild-type flies are mated, all offspring produced are also gray-bodied. Indeed, all later generations again develop only wild-type colorations. Similarly, a mating of two ebony flies yields ebony offspring in all later generations. Gray and ebony body colors here are said to be *true-breeding* traits (Fig. 12.2).

In Mendel's time it was generally supposed that if alternative forms of a trait are crossbred, a *blending* of the traits would result. Thus if gray and black were mixed together, like paints, a dark gray should be produced. And if blending really occurred, dark gray should be true-breeding as well; for mixed traits, like mixed paints, should be incapable of "unblending." In reality, however, the results of crossbreeding are strikingly different.

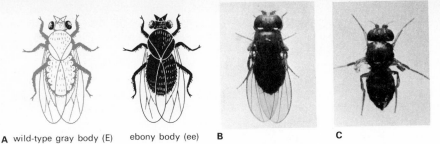

A wild-type gray body (E)   ebony body (ee)   **B**   **C**

Figure 12.1 Traits of fruit flies. A, female gray-bodied wild type on left, ebony-bodied fly on right. *B,* photo of wild-type fly. *C,* vestigial-winged fly, produced by a mutation of a single gene affecting wing structure. The animal is unable to fly.

When a wild-type and an ebony fly are mated (parental generation, *P*), all offspring (first filial generation, *F₁*) are gray-bodied, exactly like the wild-type parent (Fig. 12.3). And when two such gray-bodied *F₁* flies then are mated in turn, some of the offspring obtained are gray-bodied, others are ebony; color mixtures do not occur. Numeri-

cally, some 75 percent of the second generation (F_2) are gray-bodied, like their parents and one of their grandparents; and the remaining 25 percent are ebony, unlike their parents but like the other grandparent.

Evidently, the color traits of the offspring do not breed true; from gray-bodied flies in the F_1 can arise ebony flies in the F_2. Large numbers of tests of this kind have clearly established that, quite generally for any trait, blending inheritance does not occur and traits remain distinct and intact. If they become joined together in one generation, they can again become separated, or *segregated,* in a following generation. Mendel was the first to reach such a conclusion from studies on plants. Moreover, he not only negated the old idea of blending but postulated a new interpretation.

He realized that traits trace back to the gametes that produce an organism, and he suspected that some "factors" in the gametes controlled the later development of traits. For any given trait, he argued, an organism must inherit at least one factor from the sperm and one from the egg. The offspring then must contain at least two factors for each trait. When that offspring in turn becomes adult and produces gametes, each gamete similarly must contribute one factor to the next generation. Hence before gametes are mature, two factors must be reduced to one. Mendel therefore postulated the existence of a factor-reducing process.

With this he in effect predicted meiosis. Near the end of the nineteenth century meiosis was actually discovered, and it was later recognized that chromosome reduction during meiosis corresponded precisely to Mendel's postulated factor reduction. Chromosomes therefore came to be regarded as the carriers of the factors, and the *chromosome theory of heredity* emerged. This theory has since received complete confirmation, and Mendel's factors eventually became the genes of today.

Segregation

On the basis of the chromosome theory the fruit fly data above can be interpreted as follows: A true-breeding wild-type fly contains a pair of gray-color-producing genes on some pair of chromosomes in each cell. These genes can be symbolized by the letters *EE.* Thus the gene content, or *genotype,* is *EE,* and the visible appearance, or

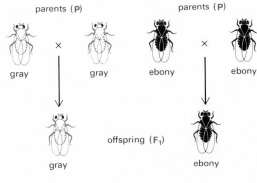

parents (**P**)   parents (**P**)

gray   gray   ebony   ebony

offspring (**F₁**)

gray   ebony

Figure 12.2 True-breeding in Drosophila. If two gray-bodied (wild-type) flies are mated, all offspring will be wild-type (left); and if two ebony flies are mated, all offspring will be ebony (right).

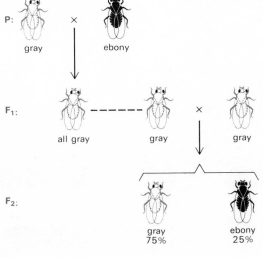

P:

gray   ebony

F₁:

all gray   gray   ×   gray

F₂:

gray   ebony
75%   25%

Figure 12.3 Crossbreeding in Drosophila. If a gray-bodied fly is mated with an ebony fly (*P* generation), all offspring will be gray-bodied (*F₁* generation). And if two of the *F₁* flies then are mated in turn, the offspring will be gray-bodied and ebony in the ratio shown.

phenotype, is gray. When such an animal produces gametes, meiosis occurs. Mature gametes therefore contain only one of the two chromosomes, hence only one of the two genes (Fig. 12.4).

It is entirely a matter of chance which of the two adult chromosomes will become incorporated in a particular gamete. Since both adult chromosomes here carry the same color gene, all gametes will be genetically alike in this respect. That is why *EE* animals are true-breeding, and why a mating of *EE × EE* will produce only gray-bodied offspring. In similar fashion, the genotype of a true-breeding ebony fly can be symbolized as *ee*. A mating of two such flies will yield only black-bodied offspring (see Fig. 12.4).

If now a wild-type and an ebony fly are mated, all offspring will be gray-bodied (Fig. 12.5). In such offspring the *E* and *e* genes are present together, yet the effect of the *e* gene evidently is overridden or masked completely. The single gene *E* by itself exerts the same effect as two *E* genes. By contrast, the single gene *e* by itself is without visible effect; a double dose, *ee*, is required if a visible result is to be produced. Genes that exert a maximum effect in a single dose, like *E*, are said

to produce *dominant* traits. Such genes mask more or less completely the effect of corresponding genes such as *e*, which are said to produce *recessive* traits.

Genes that affect the same trait in different ways and that occur at equivalent (*homologous*) locations in a chromosome pair are called *allelic* genes, or *alleles*. Genes such as *E* and *e* are alleles, and pairs such as *EE, ee,* and *Ee* are different allelic pairs. If both alleles of a pair are the same, as in *EE* or *ee*, the combination is said to be *homozygous*. The combination *EE* is *homozygous dominant*; the combination *ee, homozygous recessive*. A *heterozygous* combination is one such as *Ee*, in which one allele produces a dominant and the other a recessive trait.

Thus the F_1 resulting from a mating of a wild-type and an ebony fly as above is heterozygous, and this F_1 reveals that the wild-type trait is dominant over the ebony trait. That the heterozygous F_1 condition is not a true-breeding blend now is shown if two F_1 flies are mated (Fig. 12.6). After meiosis, each fly will produce two types of gamete: of the genes *Ee*, either the *E* gene or the *e* gene could by chance become incorporated in any one gamete. Approximately 50 percent of the gametes therefore will carry the *E* gene, and the other 50 percent, the *e* gene.

In almost all animals it is wholly a matter of chance which of the two genetically different sperm types fertilizes which of the two genetically different egg types. If many fertilizations occur simultaneously, as is usually the case, then all possibilities will be realized with appropriate frequency. The result is that three-quarters of the offspring are gray-bodied and resemble their parents in this respect. One-quarter is ebony and these offspring resemble one of their grandparents. Evidently, the result can be explained fully on the basis of nonblending, freely segregating genes and the operation of chance. Offspring in ratios of three-fourths : one-fourth (or 3 : 1) usually are characteristic for matings of heterozygous animals as above.

However, not all genes produce sharply dominant and sharply recessive forms of a trait. Many allelic genes give rise to traits that are neither dominant nor recessive. In such cases *each* allele in a heterozygous combination such as *Aa* can exert a definite effect, and the result usually is a visible trait intermediate between those produced by *AA* and *aa* combinations. But even here the intermediate result again is not produced by

Figure 12.4 Genetic effects of meiosis. A and B left, because of meiotic chromosome reduction, allelic gene pairs in adult cells become reduced to single genes in the gametes. *A and B right,* gray-bodied wild-type flies and ebony flies, respectively, breed true because in each case all offspring receive the same kinds of gray-body-controlling or ebony-controlling genes from each of the two parents.

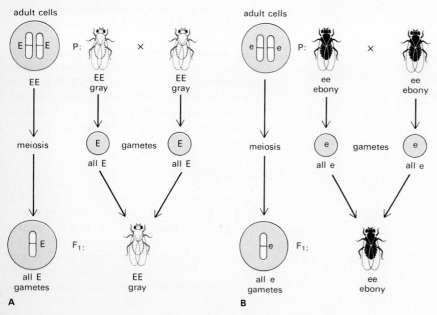

A

B

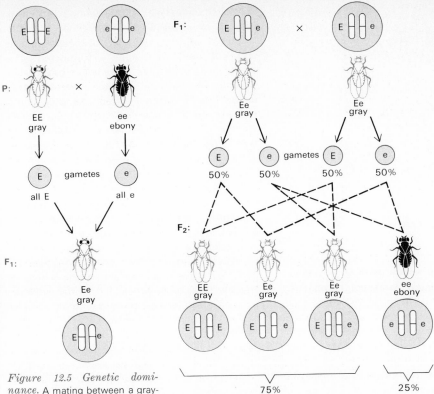

P:

EE
gray

×

ee
ebony

gametes

E
all E

e
all e

F₁:

Ee
gray

Figure 12.5 Genetic dominance. A mating between a gray-bodied and an ebony fly yields all gray-bodied offspring, indicating that the gray-body trait is dominant and masks the recessive ebony trait.

Figure 12.7 Partial dominance. If genes *A* and *a* each have their own definite effect on phenotype, and if neither thus is completely dominant or recessive, a mating of two heterozygous adults will produce an *F₁* with phenotypes and genotypes in the ratios shown. In this *F₁*, only 50 percent of the offspring are like the parents (the other 50 percent resembling two of their grandparents; verify this).

F₁:

Ee
gray

×

Ee
gray

gametes

E
50%

e
50%

E
50%

e
50%

F₂:

EE
gray

Ee
gray

Ee
gray

ee
ebony

75%   25%

Figure 12.6 Dominance and segregation. If two heterozygous gray-bodied (*F₁*) flies are mated, the offspring (*F₂*) will be 25 percent gray-bodied and homozygous like one of their gray parents (*P* in Fig. 12.5), 50 percent gray-bodied and heterozygous like their parents, and 25 percent ebony and homozygous like the other grandparent (*P* in Fig. 12.5).

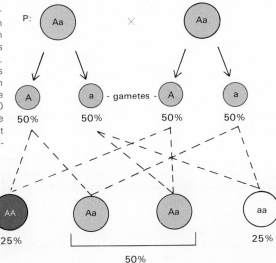

P:

Aa   ×   Aa

A
50%

a
50%

A
50%

a
50%

- gametes -

F₁:

AA
25%

Aa

Aa

50%

aa
25%

blending, for the *Aa* condition is not true-breeding. A mating of *Aa* × *Aa* segregates *AA*, *Aa*, and *aa* in a characteristic phenotype ratio of one-fourth : one-half : one-fourth (or 1 : 2 : 1). The inheritance pattern of the genotypes here is precisely the same as where genes have sharply dominant and recessive effects, and only the phenotype ratios are different (Fig. 12.7). Evidently, when genes are inherited in a particular pattern, the expressions of visible traits can differ according to the particular effects that the genes have on one another and on cell metabolism generally.

In modern terminology, Mendel's first law, the *law of segregation,* now can be stated as follows:

Genes do not blend but behave as independent units. They pass intact from one generation to the next, where they may or may not produce visible traits depending on their dominance characteristics. And genes segregate at random, thereby producing predictable ratios of visible traits among the offspring.

Implied in this law are chromosome reduction by meiosis and the operation of chance in the transmission of genes.

Independent Assortment

Animals do not inherit genes one at a time, but all of them are inherited together. What then will offspring be like with respect to two or more simultaneous traits inherited from particular parents? Mendel discovered a fundamental rule here. Phrased in modern terms, this *law of independent assortment* states:

The inheritance of a gene pair located on a given chromosome pair is unaffected by the simultaneous inheritance of other gene pairs located on other chromosome pairs.

In other words, two or more traits produced by genes located on two or more different chromosome pairs "assort independently"; each trait is expressed independently, as if no other traits were present.

The meaning of the law emerges from an examination of the simultaneous inheritance of, for example, two traits of fruit flies, *body color* and *wing shape.* As already noted, body color can be either wild-type gray or recessive ebony. Wing shape can be either normal or *vestigial.* In the latter condition the wings are reduced in size to such an extent that the animal cannot fly (see Fig. 12.1). Such stunted wings can be shown to de-

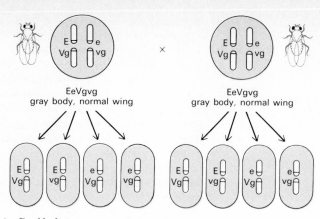

Figure 12.8 Double heterozygotes. If two flies heterozygous for both body color and wing shape are mated (top), each fly will produce gametes as shown at bottom.

velop whenever a recessive gene *vg* is homozygous, *vgvg*. Normal wings represent the dominant wild type, produced by either *VgVg* or *Vgvg* gene combinations. The body-color and wing-shape genes are located on different chromosome pairs of *Drosophila*, and the wing genes, like the color genes, obey the law of segregation.

What now will be the results of a mating between two *EeVgvg* flies, individuals that are heterozygous for both traits simultaneously? After meiosis, each gamete will contain only *one* color gene and only *one* wing gene. But which of each pair—the dominant or the recessive gene? This is a matter of chance. Thus a gamete might contain the genes *E* and *Vg*, or *E* and *vg*, or *e* and *Vg*, or *e* and *vg*. If many gametes are produced, all four combinations will occur with roughly equal frequency (Fig. 12.8).

Fertilization, too, is governed by chance. Consequently *any* one of the four sperm types might fertilize *any* one of the four egg types. Hence there are 16 different combinations that can occur in fertilization. If large numbers of fertilizations take place simultaneously, all 16 combinations will occur with roughly equal frequency. These 16 combinations can be determined from a grid in which the gametes of one parent are put along a horizontal edge and the gametes of the other parent along a vertical edge (Fig. 12.9).

Among the 16 offspring types so formed, some individuals contain *both* dominant genes at least once, some contain one *or* the other of the dominant genes at least once, and some contain none of the dominant genes. A count reveals gray-

normal, gray-vestigial, ebony-normal, and ebony-vestigial to be present in a ratio of 9 : 3 : 3 : 1.

This result proves the law of independent assortment. For if body color is considered *alone*, there are 9 plus 3, or 12 animals out of every 16 that are gray, and 3 plus 1, or 4 that are ebony. But 12 : 4 is a 3 : 1 ratio. Similarly, if wing shape is considered *alone*, again 12 out of every 16 animals have normal wings and 4 have vestigial wings; here, too, the ratio is 3 : 1. Evidently, although the color and wing traits are inherited simultaneously and yield a 9 : 3 : 3 : 1 offspring ratio overall, each trait considered *separately* nevertheless gives a 3 : 1 ratio of offspring. Each trait therefore is inherited as if the other trait were not present; or, in Mendel's phrase, the traits assort independently.

Mendel's second law applies specifically to gene pairs located on different chromosome pairs. The law therefore will hold for as many different gene pairs as there are chromosome pairs in each cell of an animal. Suppose we considered the inheritance of *three* different gene pairs, each located on a different chromosome pair, in a mating of two triple heterozygotes $AaBbCc \times AaBbCc$.

As shown above, a double heterozygote *AaBb* produces *four* different gamete types. Applying the same principles, it can be verified readily that a triple heterozygote produces *eight* different gamete types: *ABC, ABc, AbC, aBC, Abc, aBc, abC,* and *abc*. To determine all possible genotypes of the offspring, we can use a grid 8 squares × 8 squares and place the 8 gamete types of each parent along the edges, as above. The result will be 64 offspring types, of which 27 will express all the three traits in dominant form. The complete phenotype ratio easily can be verified as 27 : 9 : 9 : 9 : 3 : 3 : 3 : 1.

Two quadruple heterozygotes, *AaBbCcDd*, would manufacture 16 gamete types each, and we would need a grid 16 × 16 to represent the 256 different genotype combinations. Evidently, the possibilities rapidly become astronomical once more than a few traits are considered simultaneously.

Animals heterozygous for a large number of gene pairs are called *hybrids*, and one, two, or three heterozygous gene pairs can be referred to as monohybrids, dihybrids, or trihybrids, respectively. In man there are 23 pairs of chromosomes per cell. Mendel's second law therefore will apply to any 23 traits controlled by genes located on different chromosome pairs. What then would be the genotypes resulting from a mating of, for example, two 23-fold hybrids: $AaBb \cdots Ww \times AaBb \cdots Ww$? We know that:

A monohybrid yields $2^1 = 2$ gamete types
A dihybrid yields $2^2 = 4$ gamete types
A trihybrid yields $2^3 = 8$ gamete types
A quadruple hybrid yields $2^4 = 16$ gamete types

Carrying this progression further, a 23-fold hybrid can be shown to produce 2^{23}, or over 8 million, genetically different gamete types. Hence in considering just 23 gene pairs on different chromosome pairs, a grid of 8 million × 8 million would be required to represent the over 64 trillion possible genotypes. This number is far larger than the totality of human beings ever produced, and a good many millions or billions of these genotypes therefore have probably never yet arisen during the entire history of man.

Accordingly, chances are great that every newborn human being differs from every other one, past or present, in at least some genes controlling just 23 traits. And the genetic differences for *all* traits must be enormous indeed. Here is one major reason for the universal generalization that no two animals produced by separate fertilizations are precisely identical.

A chromosome contains not just one gene but anywhere from a few hundred to a few thousand. What is the inheritance pattern of two or more gene pairs located on the *same* chromosome pair? This question leads beyond Mendel's two laws.

Linkage

Genes located on the same chromosome are said to be *linked;* as the chromosome is inherited, so are all its genes inherited. Such genes clearly do not assort independently but are transmitted together in a block. The traits controlled by linked genes similarly are expressed in a block. In fruit flies, for example, the same chromosome pair that carries the wing-shape genes also carries one of many known pairs of eye-color genes: a dominant allele *Pr* produces red, wild-type eyes, and a recessive allele *pr* in homozygous condition produces distinctly purple eyes. If now a normal-winged, red-eyed heterozygous fly *VgvgPrpr* produces gametes, only two types should be expected, *VgPr* and *vgpr*, 50 percent of each (Fig. 12.10). In actuality, however, four gamete types are produced, in the proportions shown in Fig. 12.10.

If these four types occurred in approximately equal numbers, each about 25 percent of the total, then the result could be regarded simply as a case without linkage, governed by Mendel's second law. But the actual results include significantly *more* that 25 percent of each of the expected gamete types and significantly *less* than 25 percent of each of the unexpected types.

To explain odd results of this sort, T. H. Morgan, a renowned American zoologist of the early twentieth century, proposed a new hypothesis. He postulated that, during meiosis, paired chromosomes in some cases might twist around each other and might break where they were twisted. The broken pieces then might fuse again in the "wrong" order (Fig. 12.11). Such occur-

Figure 12.9 The Punnett square. Gametes produced by flies as in Fig. 12.8 and at top of this figure can combine in the 16 combinations shown inside the square. The offspring then have a 9:3:3:1 phenotype ratio.

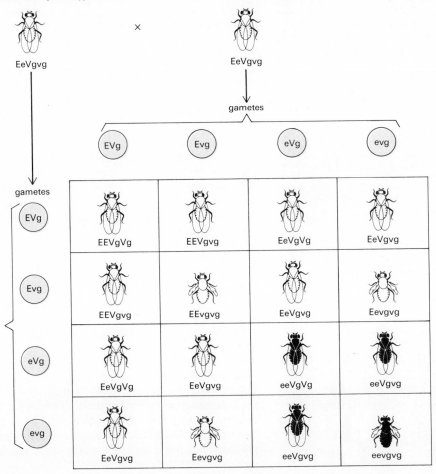

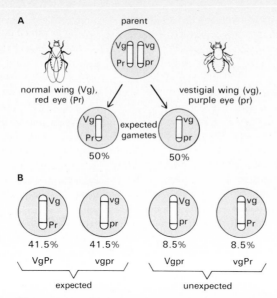

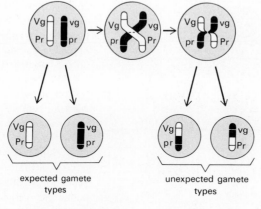

Figure 12.10 Linkage. A, left and right, the phenotypes of wing shape and eye color in *Drosophilia,* two traits controlled by linked genes. Center, the expected gametes of a heterozygous fly *VgvgPrpr. B,* the proportions of the actual gamete types obtained. Expected types appear in smaller numbers than predicted, and considerable numbers of wholly unexpected types appear as well.

Figure 12.11 Crossing over. If during meiosis chromosomes twist, break, and re-fuse at the break points in the wrong order, then unexpected gamete types will appear.

Figure 12.12 Crossing over and gene distances. If two genes are far apart, crossing over between them is likely to occur fairly frequently (top). But if genes are close together, crossing over between them is less likely (bottom). In general, the farther apart given genes are on a chromosome, the more frequent crossing over will be.

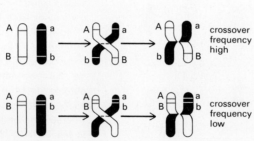

rences could account for the large percentage of expected and the small percentage of unexpected gamete types. The hypothesis was tested by microscopic examination of cells undergoing meiosis, and it could indeed be verified that chromosomal *crossing over* actually takes place. Crossing over now is believed to involve a breaking of DNA chains and subsequent synthesis of new connections between the broken ends of these chains.

The implications of crossing over have proved to be far-reaching. It has been reasoned that the frequency of crossovers should be an index of the *distance* between two genes. If two genes on a chromosome are located near each other, the chances should be relatively small that a twist would occur between these close points. But if two genes are relatively far apart, twists between them should be rather frequent. In general, the frequency of crossovers should be directly proportional to the distance between two genes (Fig. 12.12). Inasmuch as the crossover percentage of two genes can be determined by breeding experiments, it should be possible to construct *gene maps* that show the actual location of particular genes on a chromosome. Indeed, since Morgan's time the exact positions of a few hundred genes have been mapped in the fruit fly. Smaller numbers of genes similarly have been located in mice and various other organisms.

A second implication of crossing over is that genes on a chromosome must be lined up single file. Only if this is the case can linkage and crossing over occur as they actually do occur. This generalization has become known as the *law of the linear order of genes.* It represents the third major rule that governs Mendelian inheritance.

Third, crossing over makes meiosis a *source of genetic variations.* For example, when a diploid cell in a testis undergoes meiosis and produces four haploid sperms, these four do not contain merely the same whole chromosomes as the original cell. Rather, because of crossing over, the chromosomes in each sperm will be quiltworks composed of various joined pieces of the original chromosomes; and the four sperms are almost certain to be genetically different not only from one another but also from the original diploid cell. Moreover, any two genetically identical diploid cells are almost certain to give rise to two genetically different quartets of sperms. Genetic variations thus are produced by both phases of sex: through fertilization, which doubles chromosome numbers, and through meiosis, which reduces numbers and also shuffles the genes of the chromosomes.

The three rules of heredity here outlined describe and predict the consequences of *sexual recombination*—the various results possible when different sets of genes become joined through fertilization and are pooled in the zygote. In other words, sexual recombination of genes leads to Mendelian inheritance. However, a great many hereditary events have been found that do not obey the three basic rules.

Non-Mendelian Inheritance

Mutation

The most common type of non-Mendelian variation is a *mutation*. As pointed out in Chap. 6, a mutation is any stable, inheritable change in the genetic material of a cell. The most frequent form is a *point mutation*, a stable change of one gene. In such cases it is not necessary that all the DNA of a whole gene become altered; a change in not more than

a single pair of nucleotides in a double DNA chain can amount to a mutation. Such a change can alter the genetic code for a single amino acid in a protein, and one different amino acid often suffices to affect the function of the protein. If the protein is an enzyme, for example, a particular metabolic reaction could become altered or even blocked, and the consequences in a cell could be very significant.

The smallest portion of a gene that can produce a mutational effect is called a *muton*; the smallest muton would be a single pair of nucleotides in DNA. A "gene," consisting of numerous mutons in linear series, can therefore be defined as a *unit of mutation:* a section of a chromosome that, after becoming altered in at least one of its mutons, changes just one trait of a cell.

Traits are affected not only by point mutations but also by various *chromosome mutations*. These include, for example, *inversions*—a piece breaks off a chromosome and reattaches itself in inverted position; *translocations*—a piece breaks off a chromosome and attaches itself to another chromosome; *duplications*—a section of a chromosome doubles; and *deletions*—parts of chromosomes break off and become lost. All such chromosome mutations alter the nucleotide sequences of DNA and thus the genetic instructions that govern the nature of a cell (Fig. 12.13).

Mutational changes can be induced by high-energy radiation such as X rays, and the frequency of mutation has been found to be directly proportional to the amount of radiation a cell receives. Some of the naturally occurring mutations probably are produced by cosmic rays and other space radiation and by radioactive elements in the earth. But this unavoidable natural radiation is not sufficiently intense to account for the mutation frequency characteristic of genes generally, about one per million replications of a given gene. Most of these mutations probably represent errors in gene reproduction. Others undoubtedly are caused by man-made radiation, which adds to and increases the natural "background" radiation. Mutations can also be produced experimentally by physical agents other than radiations, and by various chemical agents.

Apart from limitations as to numbers, types, and range of possible effects, mutations occur entirely at random, and most of them are recessive and disadvantageous (see discussion in Chap. 6). Most mutated genes therefore accumulate in a cell without immediate visible effect, their action being

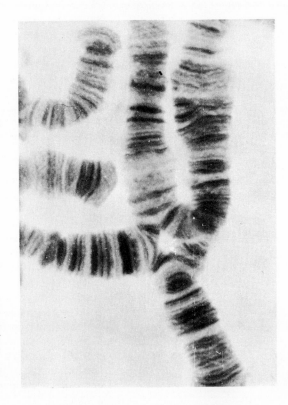

Figure 12.13 Chromosome mutations. The U-shaped portion in the upper part of this photo is a normal (stained) salivary gland chromosome of a midge. The branches leading away from the bottom of the U are parts of other chromosomes that have become translocated and attached here. These translocations have been induced by irradiation with X rays.

suppressed by the dominant wild-type alleles. Traits become altered visibly only if a mutation has a dominant effect from the outset or if a mutation with recessive effect becomes homozygous (by segregation of two mutant genes of the same kind in one offspring). Furthermore, even such mutations affect later generations only if they are *germ mutations*, genetic changes in reproductive cells. Provided new traits so produced are not lethal or do not cause sterility, they will persist as non-Mendelian variations. Mutations therefore can affect the adaptation of an individual as much as sexual recombination of genes.

Sex Determination

Groups of genes in a cell often cooperate in controlling a single composite trait. One of the best illustrations is the trait of sexuality, which in numerous animals is controlled not by individual genes acting separately but by whole chromosomes acting as functionally integrated units.

Each animal is believed to have genes for the production of both male and female traits. Such genes need not necessarily be specialized "sex genes" but could be of a type that, among other

Figure 12.14 Nongenetic sex determination. Left, M and F in zygote and adults indicate that male- and female-determining genetic factors are equally balanced. The actual maleness, femaleness, or hermaphroditic condition of the adult depends on external nongenetic influences acting at some point during development. Right, such influences act at different developmental stages in different types of organism (progressively lower down along the vertical development line of the zygote). Thus, separate sexuality can become manifest in the embryo or, by contrast, not until the adult produces gametes. Up to such an externally influenced determination stage the organism retains a bisexual (hermaphroditic) potential.

effects, also happen to influence sexual development. Animals thus are considered to have a genetic potential for *both* maleness and femaleness, and two categories can be distinguished according to how this genetic potential is translated to actual sexual traits.

In one category, comprising probably the majority of animals, the masculinizing genes are exactly equal in effect, or "strength," to the feminizing genes. In the absence of other influences an animal then will develop as a hermaphrodite. If other sex-determining factors do exert an influence, they are *nongenetic* and *environmental*; different conditions in the external or internal environment affect an animal in such a way that it develops either as a male or as a female. In most cases the precise identity of these environmental conditions has not yet been discovered. The genetic nature of a species determines whether nongenetic influences will or will not play a role (Fig. 12.14).

Also, the genetic nature of the species determines *when* during the life cycle the sex of such organisms becomes fixed by nongenetic means. In hydras, for example, sex determination does not occur until an adult is ready to produce gametes. Up to that time the animal is potentially bisexual. In other animals sex is determined during the larval phase. After metamorphosis, therefore, such animals are already males or females.

In at least one case, the echiuroid worm *Bonellia*, the sex-determining factor is known to be environmental CO_2. The relatively low concentrations of CO_2 in sea water cause free-swimming larvae of this worm to develop as females. But if a sexually still undetermined larva happens to make contact with an adult female or with a larva already determined as a female, then the added respiratory CO_2 produced by that animal causes the undetermined larva to develop as a male; it becomes a small, sperm-producing, structurally simplified parasitic animal, permanently attached inside the excretory organ of the female (see Fig. 24.13).

In still other animal groups, nongenetic sex determination occurs even earlier during the life cycle—for example, after cleavage or in the fertilized egg itself. In general, the portion of the life cycle before the stage of determination always is potentially bisexual, and if the determination occurs no later than the time of gamete production in the adult, the animal will produce either sperms

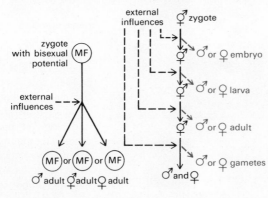

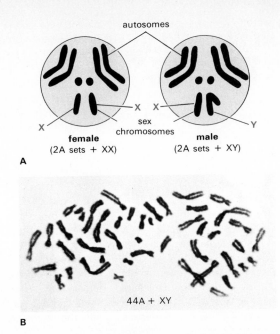

A

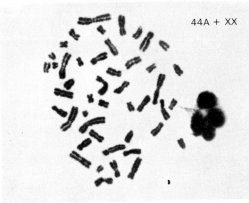

B

C

Figure 12.15 Sex chromosomes and autosomes. A, chromosome types and numbers in the fruit fly, where $2n = 8$. Note male and female differences. *B, C,* isolated chromosomes from cells of a human male and female, respectively. In both cases 46 pairs of chromosomes are present; each of the original 46 chromosomes has duplicated as a preliminary to cell division. (*B, C,* Courtesy Carolina Biological Supply Company.)

or eggs. But if by then a determination has not taken place, the animal will be a hermaphrodite.

An altogether different form of sex determination occurs in a second category of animals, notably insects and vertebrates. In these the masculinizing genes are *not* equal in their effects to the feminizing genes, and the primary determination of sex has a purely genetic basis. Every individual becomes either a male or a female; hermaphroditism does not occur except as an abnormality. Also, sex always becomes fixed at the time of fertilization, and every cell later formed is genetically male or female.

Animals of this type contain special *sex chro-* *mosomes,* different in size and shape from all other chromosomes, the *autosomes.* Sex chromosomes are of two kinds, *X* and *Y.* Each diploid cell contains a pair of sex chromosomes, either *XX* or *XY,* and all such cells are genetically either male or female. For example, *XY* cells are genetically female and *XX* cells genetically male in butterflies, most moths, some fishes, and birds. In these animals, maleness appears to be controlled by the genes of the *X* chromosomes, and femaleness, by the genes of the autosomes (and in part perhaps also by those of the *Y* chromosomes). By contrast, *XY* cells are male and *XX* cells female in flies and mammals. Femaleness here is controlled by the genes of the *X* chromosomes; maleness is known to be determined by the autosomes in fruit flies, but to a large extent by the *Y* chromosomes in mammals (Fig. 12.15).

In man, for example, each adult cell contains 22 pairs of autosomes plus either an *XY* or an *XX* pair. Female cells, $44A + XX$, thus have two female-determining chromosomes, whereas male cells, $44A + XY$, contain one female-determining and one male-determining chromosome. This difference of one whole chromosome lies at the root of the sexual differences between males and females. More specifically, in a female cell the femininizing effect of the two *X* chromosomes outweighs any masculinizing influence the autosomes might have; and in a male cell, the masculinizing effect of the *Y* chromosome (and probably also the autosomes) outweighs the feminizing influence of the single *X* chromosome.

These relations suggest that the sexual nature of an individual might depend on a particular numerical ratio or balance between different chromosomes. That this is actually the case has been shown by experiments in fruit flies. In these animals, in which maleness is controlled by the autosomes, it is possible to vary the normal number of *X* chromosomes and autosomes that occur in sperms and eggs. One can then obtain offspring with, for example, normal paired sets of autosomes but three *X* chromosomes instead of two. Such individuals grow into so-called *superfemales;* all sexual traits are accentuated in the direction of femaleness. Other chromosome balances give rise to *supermales* and *intersexes,* the latter with sexual traits intermediate between those of normal males and females (Fig. 12.16). Paradoxically, supersexes and also intersexes generally are sterile; for as a result of the abnormal chromosome

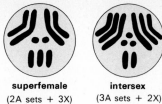

superfemale
(2A sets + 3X)

intersex
(3A sets + 2X)

supermale
(3A sets + 1X)

Figure 12.16 Chromosome balances. The sexual character of a fruit fly is determined by the specific balances of autosomes and *X* chromosomes as shown.

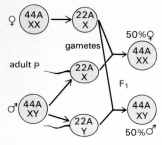

Figure 12.17 Sex determination in man. Males produce two genetically different sperm types, roughly 50 percent of each. Offspring then will be male and female in a 1 : 1 ratio.

numbers meiosis occurs abnormally, and the sperms and eggs then produced are defective.

In the light of such balances, the normal mechanism of sex determination at the time of fertilization becomes clear. For example, human females produce eggs of which each contains $22A + X$ after meiosis (Fig. 12.17). Males produce two kinds of sperm, $22A + X$ and $22A + Y$, in roughly equal numbers. Fertilization now occurs at random, and a sperm of either type can unite with an egg. Therefore, in about 50 percent of the cases the result will be $44A + XX$, or female-producing zygotes; and in the remaining 50 percent the zygotes will be $44A + XY$, or prospectively male. In man therefore it is the paternal

parent who at the moment of fertilization determines the sex of the offspring. When only a single offspring is produced, a 50:50 chance exists of its being a son or a daughter. When many offspring are produced the number of males generally will equal the number of females.

Note that, in females, the recessive effect of a gene located on one *X* chromosome can be masked by an allele with a dominant effect located on the other *X* chromosome. In males, by contrast, genes on the *X* chromosome can exert even recessive effects, since another *X* chromosome with masking dominants is not present. Genes located on *X* chromosomes are said to be *sex-linked*, and they are inherited in a characteristic pattern. For example, red-green color blindness in man is a trait produced by a sex-linked recessive gene *c*. Suppose that a color-blind male, X_cY, marries a normal female, *XX*. In this symbolization an *X* chromosome without subscript is tacitly assumed to contain the dominant gene *C*, which prevents the expression of color blindness. All offspring of such a mating have normal vision, but the daughters carry the recessive gene *c* (Fig. 12.18).

If now one of these daughters marries a normal male, all female offspring will be normal but half the sons will be color-blind. Thus the trait has been transmitted from color-blind grandfather to normal mother to color-blind son. Such a zigzag pattern of inheritance is typical for all recessive sex-linked traits; males exhibit the traits, females merely transmit them. The second *X* chromosome in females prevents expression of recessive sex-linked traits. Among other characteristically male, sex-linked abnormalities is *hemophilia*, a bleeder's disease resulting from absence of blood platelets.

Genetic Systems

The example of genetic sex determination shows clearly that genes of one or more chromosomes can act in concert and control one highly composite trait. Many other illustrations of such gene interactions are known. For example, if genes were simply independently functioning units, then it should not matter if the position of genes relative to one another were rearranged. Yet experiment shows that gene rearrangements actually do produce altered traits in a cell, as in chromosome mutations in which genes are neither added nor lost but only repositioned. Such findings indicate clearly that genes normally interact with their

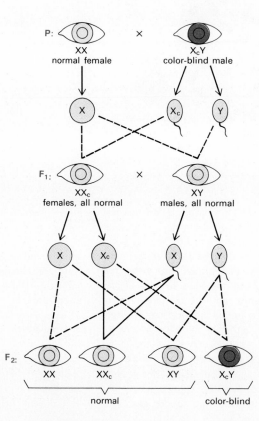

Figure 12.18 Sex linkage. The inheritance of the sex-linked recessive gene for color blindness. Note that a color-blind male (*P*) transmits the gene X_c to all his daughters, who have normal vision, however. Such females then transmit the gene to half their sons, who are color-blind. The abnormality thus is expressed in males and only transmitted by females.

P: XX
normal female × X_cY color-blind male

X X_c Y

F_1: XX_c females, all normal × XY males, all normal

X X_c X   Y

F_2: XX XX_c XY X_cY

normal   color-blind

neighbors. The same conclusion is suggested by paired allelic genes with dominant and recessive effects; for if any trait is to be dominant over another, the genes and gene products that control these traits must certainly interact. Interactions also have been described among the operator, regulator, and structural genes discussed in the context of the operon mechanism (Chap. 6). It appears, indeed, that *all* genes in a cell, collectively called the *genome,* are closely interdependent and interacting.

This generalization is reinforced further by recent studies indicating that genes actually are not the smallest hereditary units. As already pointed out, for example, a mutational change can be produced by alteration of a single muton, a unit far smaller than a whole gene. It has also been found that adjoining mutons act cooperatively and form larger functional blocks called *cistrons.* A gene might contain a single cistron only or it can consist of two or more. Whatever the number, if the cooperating mutons in a cistron happen to become disjoined by a break in the chromosome, then the hereditary function of that cistron is abolished. But whole cistrons in a gene can become disjoined without loss of function. Such data show clearly that, in addition to interaction among whole genes, considerable interaction also occurs among the smaller functional units within genes.

Moreover, genes operate in an environment that includes not only other genes and their subunits but also the cell cytoplasm. It is there that the basic function of genes is exercised, through control of protein synthesis. A second, molecular definition of "gene" can be formulated on this basis: a gene is a chromosomal DNA unit that controls the synthesis of a single type of polypeptide. Such amino acid chains form proteins and thus enzymes or structural components, and these can be regarded as the *primary* traits of a cell.

Various *secondary,* sometimes visible traits then are produced as a result. For example, an enzyme that catalyzes a particular metabolic reaction contributes to the formation of a reaction product, which represents a secondary trait. In certain cases such a reaction product is a visible, final trait—a pigment, for example. In other instances a final trait is a composite of numerous separate reaction products, formed not only in one cell but often in many. For example, disease resistance is a functional property of a whole animal, a composite trait produced by millions of cells. Each of these contributes some particular function to the total trait, and specific genes in each of these cells control each of these functions. Similarly, body size, general vigor, intelligence, fertility, and many others—all, like sexuality, are interaction products of several dozens or hundreds or thousands of different genes.

In all probability, most of such highly composite traits are controlled by the collective action of possibly all genes of an animal, each contributing a tiny effect to the total trait. Such traits then can be expressed in a correspondingly great variety of ways. As is well known, for example, traits like body size or intelligence range from one extreme to another, through enormously varied series of intergradations. But if all genes contribute to the control of numerous composite traits then any one gene clearly must contribute to the control of more than one trait. We are led to the generalization that one trait can be controlled by many genes, and one gene can contribute to the control of many traits (Fig. 12.19).

That critical interactions occur between genes and their cellular surroundings is perhaps best shown by the observation that the same kinds of genes often produce different traits in different cells of an animal. For example, *all* cells of a man contain eye-color genes, but only iris cells actually develop the color. Evidently, the cytoplasms of different cells react differently to the genes they contain, and the various cell differentiations and specializations are the result (Fig. 12.20).

So-called "inherited diseases" likewise must be interpreted in the light of gene-cytoplasm interactions. Certain mental diseases, diabetes, alcoholism, cancer, and many other abnormalities are known to "run in families." What is inherited here is not the disease itself; a child of diabetic ancestry is not automatically diabetic. However, *susceptibility* to disease might be inherited. The genes are present, but before the disease can develop partic-

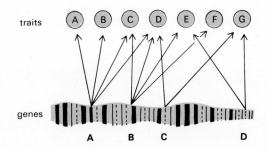

Figure 12.19 Traits and genes. One trait (for example, *D*) often is controlled by many genes, and one gene (for example, *B*) often controls many traits.

same genes
in all cells,
but pigment
only in some

Figure 12.20   Gene expression. Although all cells of an organism have the same kinds of genes, gene action is influenced differently by different cells. The result is a differential expression of traits. Thus all cells of a butterfly can have pigment-producing genes, but an actual pigmentation might develop only in wing-forming cells.

cases, however, the genes controlling muscular development might be the same.

Clearly, then, the actual visible traits of an animal always are an interaction product of genetic heredity *and* environment. Genes supply a reasonable promise, as it were, and the total environment of the genes then permits or does not permit a translation of promise to reality. Furthermore, whereas the pre-Mendelians thought that traits were inherited, and whereas the Mendelian era advanced to the idea that individual factors, or genes, were inherited, the present post-Mendelian era recognizes that actually neither traits nor genes nor even subunits of genes are inherited. Instead, the hereditary units turn out to be whole chromosome sets, coordinated complexes of genes, subtly interacting *genetic systems*. Moreover, as long as a cell is alive such genetic systems never lose their functional integration with the nongenetic parts of cells; and it is biologically almost meaningless to consider genes separately from their immediate or even more distant environment. Ultimately, therefore the smallest real unit of inheritance is one whole cell.

In the individual animal the interplay between sex, heredity, and environment results in various degrees of individual adaptation. In the long reproductive succession of animals this same interplay results in *evolution*.

ular cellular environments must make gene expression possible. Similarly, a person who performs physical exercise regularly will develop strong muscles and so will acquire traits differing from those of a person who does not exercise. In both

Review Questions

1.   What was meant by "blending inheritance"? Describe breeding tests showing that blending inheritance does not occur. What hypothesis did Mendel substitute for the blending concept?

2.   State the chromosome theory of heredity. What is the evidence that genes actually are contained in chromosomes? Define genome, true-breeding, phenotype, genotype, allele, dominant and recessive traits, homozygous, heterozygous.

3.   Review the breeding tests on inheritance of gray and ebony body colors in fruit flies, and explain them on the basis of genes and chromosomes.

4.   What are the quantitative results of the mating *Aa* × *Aa* if (*a*) *A* is dominant over *a*, (*b*) neither gene is dominant over the other?

5.   In your own words, state the law of segregation. If *A* is dominant over *a*, what phenotype ratios of offspring are obtained from the following matings: (*a*) *Aa* × *aa*, (*b*) *AA* × *aa*, (*c*) *Aa* × *Aa*, (*d*) *Aa* × *AA?*

6.   In your own words, state the law of independent assortment. By what kinds of breeding experiments and by what reasoning did Mendel come to discover this law? Interpret the law on the basis of genes, meiosis, and gametes.

7. How many genetically different gamete types will be produced by an animal heterozygous for 10 gene pairs? If two such animals were mated, how many genetically different offspring types could result?

8. Define linkage. Why does inheritance of linked genes not obey Mendel's second law? What were Morgan's observations that led him to the hypothesis of crossing over? Describe this hypothesis.

9. How do crossover data permit the construction of gene maps? State the law of the linear order of genes. Show how crossing over can be a source of genetic variations.

10. Distinguish between chromosome mutations and point mutations and between somatic mutations and germ mutations. What is the relation between mutation frequency and radiation intensity?

11. What are the characteristics of mutations from the standpoint of (*a*) predictability, (*b*) functional relation to normal alleles, (*c*) effects on traits, and (*d*) relative advantage to an animal?

12. Review the patterns of sex determination based on (*a*) nongenetic, (*b*) genetic mechanisms. How is sex determined in (*a*) birds, (*b*) mammals?

13. What is the significance of a given numerical balance between autosomes and sex chromosomes? What are supersexes and intersexes? What are mutons and cistrons?

14. What are sex-linked genes? Describe the inheritance pattern of the sex-linked recessive hemophilia gene *h*, assuming that a hemophilic male mates with a normal female.

15. Cite examples of interactions among genes and between genes and their environment. What contributions are made to the expression of traits by (*a*) genes, (*b*) the environment? What does the phrase "inherited disease" mean?

Collateral Readings

Bearn, A. G., and J. L. German: Chromosomes and Disease, *Sci. American,* Nov., 1961. Visually demonstrable abnormalities in human chromosomes are associated with genetic diseases.

Benzer, S.: The Fine Structure of the Gene, *Sci. American,* Jan., 1962. A review of experiments that have led to the concept of mutons and cistrons.

Bonner, D. M.: "Heredity," Prentice-Hall, Englewood Cliffs, N.J., 1961. A paperback covering general aspects of both Mendelian and non-Mendelian genetics.

Deering, R. A.: Ultraviolet Radiation and Nucleic Acid, *Sci. American,* Dec., 1962. A discussion of mutations induced by this form of energy.

Hollaender, A., and G. E. Stapleton: Ionizing Radiation and the Cell, *Sci. American,* Sept., 1959. The primary and secondary genetic effects of radiation are discussed.

Kormondy, E. J.: "Introduction to Genetics," McGraw-Hill, New York, 1964. A programmed paperback designed for self-instruction in general genetic principles; recommended as a study aid.

Ledley, R. S., and F. H. Ruddle: Chromosome Analysis by Computer, *Sci. American,* Apr., 1966. A computer technique analyzes human cells for genetic abnormalities and their associated chromosome defects.

Mendel, G.: "Experiments in Plant Hybridization," Harvard, Cambridge, Mass., 1941. A translation of the 1865 original, on which modern genetics is based.

Mittwoch, U.: Sex Differences in Cells, *Sci. American,* July, 1963. Differences in sex chromosomes and other cellular traits of males and females are examined.

Müller, H.: Radiation and Human Mutation, *Sci. American,* Nov., 1955. The discoverer of the genetic effects of radiation discusses the influence of mutations on human evolution.

Sager, R.: Genes Outside the Chromosomes, *Sci. American,* Jan., 1965. A sequel to the article by Sonneborn cited below. Recommended.

Sonneborn, T. M.: Partner of the Gene, *Sci. American,* Nov., 1950. A recommended article, showing that genetic control is exercised not only by genes but also by cytoplasmic factors.

Stahl, F. W.: ''The Mechanics of Inheritance,'' Prentice-Hall, Englewood Cliffs, N.J., 1964. An excellent nonintroductory paperback on ''molecular'' genetics; virus and bacterial research is reviewed in detail, and references to original papers are given.

Every animal is *adapted* to its environment. Among thousands of shapes that a fish, for example, could have, its actual shape is well suited for rapid locomotion in water. A bird is cast in a form eminently suited for aerial life, yet its ancestry traces to fish. Over long periods of time, clearly, animals can change their particular adaptations in response to new environments.

Based on steady-state control and reproduction, adaptation is achieved through sex, heredity, and evolution. In this series of chapters we concentrate on the last. The first chapter is an account on the mechanism of evolution—the basic forces that govern and direct evolutionary changes. The following chapters then examine the actual results that this mechanism has produced since life originated on earth.

13

EVOLUTION:
MECHANISMS

No zoologist today seriously questions the principle that species arise from preexisting species. Evolution on a small scale can be brought about by experimental means, and the forces that drive and guide evolutionary processes are understood quite thoroughly.

That evolution really occurs did not become definitely established till the nineteenth century. The first section below outlines this historical *background* of evolutionary thought. Following sections then deal with the *forces* of evolution as understood today, and with the evolutionary *effects* produced by these forces.

Background

The earliest written discussion of organic creation is contained in the Old Testament: God made the world and its living inhabitants in six days, man coming last. Later ideas included those of *spontaneous generation* and of *immutability of species,* which largely held sway until the eighteenth and nineteenth centuries. Each species was considered to have been created spontaneously, completely developed, from dust, dirt, and other nonliving sources. And once created, a species was held to be fixed and immutable, unable to change its characteristics.

In the sixth to fourth centuries B.C., Anaximander, Empedocles, and Aristotle independently considered the possibility that living forms might represent a *succession* rather than unrelated, randomly created types. However, the succession was thought of in an essentially philosophical way, as a progression from ''less nearly perfect'' to ''more nearly perfect'' forms. The historical nature of succession and the continuity of life were not yet recognized.

Francesco Redi, an Italian physician of the seventeenth century, was the first to obtain evidence against the idea of spontaneous generation by showing experimentally that organisms could not arise from nonliving sources. Contrary to notions held at the time and earlier, Redi demonstrated that maggots would not form spontaneously in meat if flies were prevented from laying their eggs on the meat. But old beliefs die slowly, and it was not until the nineteenth century, chiefly through the work of Louis Pasteur on bacteria, that the notion of spontaneous generation finally ceased to be influential.

By this time the idea of continuity and historical succession, or *evolution,* had occurred to a number of thinkers. An important evolutionary hypothesis was that of the French biologist Lamarck, published in 1809. To explain how evolution occurred, Lamarck proposed the two ideas of *use and disuse of parts* and of *inheritance of acquired characteristics.* He had observed that if a body part of an organism was used extensively such a part would enlarge and become more efficient, but that if a structure was not fully employed it would degenerate and atrophy. Therefore, by differential use and disuse of various body parts during its lifetime, an organism would change to some extent and would acquire certain traits. Lamarck then thought that such acquired traits were inheritable and could be transmitted to off-spring.

According to this Lamarckian scheme evolution would come about somewhat as follows. Suppose a given short-necked ancestral animal feeds on tree leaves. As it clears off the lower levels of a tree it stretches its neck to reach farther up. During a lifetime of stretching the neck becomes a little longer, and a slightly longer neck then is inherited by the offspring. These in turn feed on tree leaves and keep on stretching their necks; and so on, for many generations. In time a very long-necked animal is formed, something like a modern giraffe.

This theory was exceedingly successful and did much to spread the idea of evolution. But Lamarck's views ultimately proved to be untenable. That use and disuse do lead to acquired traits is quite correct. For example, it is common knowledge that much exercise builds powerful muscles. However, Lamarck was mistaken in assuming that such acquired (nongenetic) variations were inheritable. Acquired traits are *not* inheritable, since they are effects produced by environment and development, not by genes. Only *genetic* traits are inheritable, and then only if such traits are controlled by genes that are present in the reproductive cells. What happens to cells other than gametes through use and disuse, or in any other way for that matter, does not affect the genes of the gametes. Accordingly, although Lamarck observed some of the effects of use and disuse correctly in some cases, such effects cannot play a role in evolution.

One famous attempt at experimental refutation of Lamarckism was carried out by Weismann, an eminent zoologist of the nineteenth century. The tails of mice were cut off for many successive generations. According to Lamarck, such enforced disuse of tails should eventually have led to tailless mice. Yet mice in the last generation of the experiment still grew tails as long as their ancestors.

The year in which Lamarck published his theory—1809—also was the year in which Charles Darwin was born. During his early life Darwin undertook a 5-year-long circumglobal voyage as naturalist on the naval expeditionary ship *H. M. S. Beagle.* He made innumerable observations and collected a large number of different plants and animals in many parts of the world. He then spent

nearly 20 years sifting and studying the collected data. In the course of his work he found evidence for certain generalizations. Another naturalist, Alfred Russel Wallace, had been led independently to substantially the same generalizations, which he communicated to Darwin. In 1858 Darwin and Wallace together announced a new theory of evolution, which was to supplant that of Lamarck. Darwin also elaborated the new theory in book form. This famous work, entitled *On the Origin of Species by Means of Natural Selection, or the Preservation of Favored Races in the Struggle for Life,* was published in 1859.

In essence, the Darwin-Wallace *theory of natural selection* is based on three observations and on two conclusions drawn from these observations:

Observation Without environmental pressures, every species tends to multiply in geometric progression.

In other words, a population that doubles its number in a first year has a sufficient reproductive potential to quadruple its number in a second year, to increase eightfold in a third year, etc.

Observation But under field conditions, although fluctuations occur frequently, the size of a population remains remarkably constant over long periods of time.

The validity of this point will become apparent in Chap. 17.

Conclusion Evidently, not all gametes will become zygotes; not all zygotes will become adults; and not all adults will survive and reproduce. Consequently there must be a "struggle for existence."

Observation Not all members of a species are alike; considerable individual variation is exhibited.

Conclusion In the struggle for existence, therefore, individuals that exhibit favorable variations will enjoy a competitive advantage over others. They will survive in proportionately greater numbers and will produce offspring in proportionately greater numbers.

Darwin and Wallace thus identified the *environment* as the chief cause of natural selection; the environment would gradually weed out organisms with unfavorable variations but preserve those with favorable variations. Over a long succession of generations, and under the continued selective influence of the environment, a group of animals would eventually have accumulated so many new, favorable variations that a new species would in effect have arisen from the ancestral stock.

Nonprofessionals often are under the impression that this Darwin-Wallace hypothesis is *the* modern explanation of evolution. This is not the case. Indeed, Darwinism was challenged even during Darwin's lifetime. What, it was asked, is the source of the all-important individual variations? How do individual variations arise? Here Darwin actually could do no better than fall back on the Lamarckian idea of inheritance of acquired characteristics. Ironically, the correct answer regarding variations began to be formulated just 6 years after Darwin published his *Origin*, when Mendel announced his rules of inheritance. But Mendel's work remained unappreciated for more than 30 years, and progress in understanding evolutionary mechanisms was retarded correspondingly.

Another objection to Darwinism concerned natural selection itself. If this process simply preserves or weeds out what already exists, it was asked, how can it ever create anything new? As will soon become apparent, natural selection actually does create novelty. The earlier criticism arose in part because the meaning of Darwin's theory was—and still is—widely misinterpreted. Social philosophers of the time and other "press agents" and disseminators of scientific information, not biologists, thought that the essence of natural selection was described by the phrase "struggle for existence." They then coined alternative slogans like "survival of the fittest" and "elimination of the unfit." Natural selection thus came to be viewed almost exclusively as a negative, destructive force. This had two unfortunate results. First, a major implication of Darwin's theory—the creative role of natural selection—was generally overlooked; and second, the wrong emphasis often was accepted in popular thinking as the last word on evolution.

Such thinking proceeded in high gear even in Darwin's day. Many people thought that evolution implied "man descended from apes," and man's sense of superiority was duly outraged. Also, because evolutionary views denied the special crea-

tion of man, they were widely held to be antireligious. Actually, the idea of evolution is not any more or less antireligious than the idea of spontaneous generation. Neither really strengthens, weakens, or otherwise affects belief in God; to the religious person only the way God operates, not God as such, is in question.

Moreover, under the banner of phrases like "survival of the fittest," evolution was interpreted to prove an essential cruelty of nature; and human behavior, personal and national, often came to be guided by the ethic of "jungle law," "might is right," "every man for himself." Only in that way, it was thought, could the "fittest" prevail. Even today, unfortunately, evolution is still commonly—and erroneously—thought to be a matter of "survival of the fittest."

By now, a full century after Darwin and Wallace, it has become clear that natural selection is preeminently a peaceful process that has little to do with "struggle," "weeding out," or "the fittest." Also, natural selection is recognized to represent only a part of the evolutionary mechanism since, like Lamarck, Darwin was unsuccessful in identifying the genetic causes of evolutionary change. In short, Darwin (and Wallace) supplied an incomplete explanation; but as far as it went theirs was the first to point in the right direction.

The current modern theory of evolution is not the work of any one man, though it is the spiritual offspring of Mendel and of Darwin. It evolved slowly during the first half of the current century, and many biologists of various specializations contributed to its formulation.

Forces of Evolution

The Evolutionary Process

As now understood, the mechanism of evolution can be described as *natural selection acting on the genetic variations that appear among the members of a population.*

A population is a geographically localized group of individuals of the same species, in which the members interbreed preferentially with one another and also occasionally with members of neighboring sister populations (see also Chap. 17). The result of this close sexual communication in a population is a *free flow of genes*. Hereditary material present in some portion of a population

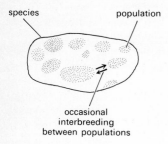

Figure 13.1 The gene pool. In a species genes flow in and between populations. The total gene content of the species thus represents a gene pool to which all members of the species have access. Genes normally cannot flow between the gene pools of two different species.

can in time spread to the whole population, through the gene-pooling and gene-combining effects of sex. Moreover, interbreeding also can interconnect the *gene pools* of sister populations, and the total genetic content of a whole species thus can become shuffled and reshuffled to some extent among the member organisms (Fig. 13.1).

Chapter 12 has shown that genetic variations in a gene pool can arise by *sexual recombination,* by *mutation,* or both. Where reproduction is uniparental, as in vegetative forms of propagation, mutation is the only source of genetic variation; and in biparental reproduction, where animals are produced by gametes from two parents, both mutation and sexual pooling of genes in zygotes are sources of genetic variations.

Thus, in each generation some individuals appear that have new, variant traits as a result of either recombinational or mutational processes. If such variant animals survive and have offspring of their own, then their new genetic characteristics will persist in the gene pool of the population. And in the course of successive generations the genetic novelty can spread to many or all members of the population.

Whether or not such spreading actually takes place depends on natural selection. The real meaning of this term is *differential reproduction:* some members of a population leave more offspring than others. Those that have more offspring will contribute a proportionately greater percentage of genes to the gene pool of the next generation than those that have fewer offspring. Therefore, if differential reproduction continues in the

same manner over many generations, the abundant reproducers will contribute a progressively larger number of individuals to the whole population. And as a result *their* genes will become preponderant in the gene pool of the population (Fig. 13.2).

Which individuals leave more offspring than others? Usually (but not necessarily) those that are best adapted to the environment. Being well adapted, such individuals on the whole are healthier and better fed, can find mates more readily, and can care for their offspring appropriately. To be sure, on occasion comparatively poorly adapted individuals have the most offspring. Yet what counts here is not how well or poorly an animal copes with its environment, but only how many offspring it manages to leave. The more there are, the greater the role that the parental genes will play in the total gene pool of the population. By and large the well-adapted animal contributes most to the gene pool.

It can happen, therefore, that a new trait that originates in one animal spreads by differential reproduction and in time becomes a standard feature of the whole population. This is the unit of evolutionary change. Many such unit changes must accumulate in a population before the animals are altered enough in structure or function to be recognizable as a new species. In any event, evolution operates through this basic two-step process:

1. Appearance of genetic variations by sexual recombination and mutation

2. Spreading of these variations through a population by differential reproduction in successive generations

To the extent that genetic variations originate at random, evolutionary innovations similarly appear at random. But inasmuch as the best reproducers generally are the best adapted, evolution as a whole is directed by adaptation and is oriented toward continued or improved adaptation. It is therefore not a wholly random process.

In this modern view of evolution, evidently, natural selection is fundamentally a *creative* force, one that spreads genetic novelty. And while it does eliminate the *reproductively* "unfit," it does not necessarily eliminate the behaviorally or socially "unfit." The mightiest and grandest animal in the population could be exceedingly "fit" in a behav-

Figure 13.2 Differential reproduction, or natural selection. Assume that a variation arises in one individual of a parental generation (black dot) and that the variant organism is able to leave three offspring. Each nonvariant organism (white dots) on the other hand only manages to leave one offspring. The complexion of the population then will change as shown during subsequent generations; the variant type will represent a progressively larger fraction of the numerical total. Such spreading of variations, brought about by differential reproduction, constitutes natural selection.

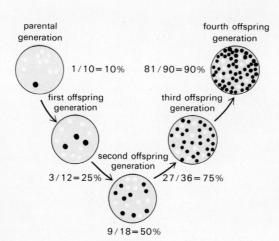

parental generation

1/10 = 10%

first offspring generation

3/12 = 25%

second offspring generation

9/18 = 50%

27/36 = 75%

third offspring generation

81/90 = 90%

fourth offspring generation

ioral sense, but if it happened to be sterile it would be inconsequential in a reproductive and therefore an evolutionary sense. Conversely, a sickly weakling might be behaviorally or socially "unfit" yet have numerous offspring. Natural selection thus operates basically through reproduction, not through struggle for survival. Such struggles certainly occur, often in a very physical sense, and indirectly they can affect reproductive success. To that extent such factors can have evolutionary consequences. But the important issue is neither "struggle" nor "elimination" nor even individual "survival"; all that finally matters is comparative reproductive success.

The Genetic Basis

From the preceding discussion, evolution can be described as *a progressive change of gene frequencies;* in the course of successive generations the proportion of some genes in a population increases and the proportion of others decreases. Clearly, the *rates* with which gene frequencies change will be a measure of the *speed* of evolution. What determines such rates?

Suppose we consider a large population made up of individuals of three genetic categories, *AA, Aa,* and *aa,* in the following numerical proportions:

AA	Aa	aa
36%	48%	16%

Assuming further that the choice of sexual mates is entirely random, that all individuals produce roughly equal numbers of gametes, and that the genes *A* and *a* do not mutate, we can then ask how the frequency of the genes *A* and *a* will change from one generation to the next.

Since *AA* individuals make up 36 percent of the population, they will contribute approximately 36 percent of all the gametes formed in the population. These gametes will all contain one *A* gene. Similarly, *aa* individuals will produce 16 percent of all gametes in the population, and each will contain one *a* gene. The gametes of *Aa* individuals will be of two types, *A* and *a,* in equal numbers. Since their total amounts to 48 percent, 24 percent will be *A* and 24 percent will be *a.* The overall gamete output of the population therefore will be:

PARENTS	GAMETES	PARENTS	GAMETES
36% AA $\longrightarrow$ 36% A		16% aa $\longrightarrow$ 16% a	
48% Aa $\longrightarrow$ 24% A		48% Aa $\longrightarrow$ 24% a	
	60% A		40% a

Fertilization now occurs in four possible ways: two *A* gametes join; two *a* gametes join; an *A* sperm joins an *a* egg; and an *a* sperm joins an *A* egg. Each of these possibilities will occur with a frequency dictated by the relative abundance of the *A* and *a* gametes. Since there are 60 percent *A* gametes, *A* will join *A* in 60 percent of 60 percent of the cases—60×60, or 36 percent of the time. Similarly, *A* sperms will join *a* eggs in 60×40, or 24 percent of the cases. The total result:

SPERMS	EGGS	OFFSPRING
A + A $\longrightarrow$	60×60 $\longrightarrow$	36% AA
A + a $\longrightarrow$	60×40 $\longrightarrow$	24% Aa
a + A $\longrightarrow$	40×60 $\longrightarrow$	24% Aa
a + a $\longrightarrow$	40×40 $\longrightarrow$	16% aa

The new generation in our example population thus will consist of 36 percent *AA,* 48 percent *Aa,* and 16 percent *aa* individuals—precisely the same proportions that were present originally. Evidently, gene frequencies have not changed.

By experiment and calculation it can be shown that such a result is obtained regardless of the numbers and the types of gene pairs considered simultaneously. The important conclusion is that, *if mating is random, if mutations do not occur, and if the population is large, then gene frequencies in a population remain constant from generation to generation.* This generalization is known as the *Hardy-Weinberg law.* It has somewhat the same central significance to the theory of evolution as Mendel's laws have to the theory of heredity.

The Hardy-Weinberg law indicates that, when a population is in genetic equilibrium and gene frequencies do not change, the rate of evolution is zero. Genes then continue to be reshuffled by sexual recombination and, as a result, genetic variations continue to originate from this source. But the overall gene frequencies do not change. Of themselves, therefore, the variations are not being propagated differentially and evolution consequently does not occur. What does make evolution occur are deviations from the "ifs" specified in the Hardy-Weinberg law.

First, mating is decidedly *not* random in most natural situations. For example, it happens often that all genetic types in a population do not reach reproductive age in proportionate measure. Suppose that, in the sample population of *AA, Aa,* and *aa* individuals above, the *AA* genes cause death in one-third of the embryos of the population. Under such conditions 36 percent of all zygotes will be *AA,* but only two-thirds of their number will reach reproductive age. The *Aa* and *aa* individuals then will form a proportionately larger fraction of the reproducing population and will contribute proportionately more to the total gamete output. The ultimate result over successive generations will be a progressive decrease in the frequency of the *A* gene and a progressive increase of the *a* gene. Hence the *effective* mating population correspondingly will become more and more nonrandom.

In an effective mating population, moreover, mating pairs usually are formed on the basis of nonrandom criteria such as health, strength, mentality, external appearance, sexual attraction, or simply availability and geographic proximity of particular males and females. Mating thus becomes

even more nonrandom, and the overall result is an uneven, nonrandom shuffling of genes—in effect, a form of natural selection. As some genes then spread more than others, gene frequencies become altered, and a Hardy-Weinberg equilibrium is not maintained. This represents evolutionary change. Through nonrandom mating a certain *intensity* of natural selection, or *selection pressure,* operates for or against most genes, and in time even a very slight selection pressure substantially affects the genetic makeup of a population.

Second, mutations do occur in populations, and Hardy-Weinberg equilibria change for this reason also. Depending on whether a mutation has a beneficial or harmful effect on a trait, selection will operate for or against the mutated gene. In either case gene frequencies will change, for the mutated gene will either increase or decrease in abundance. Note, however, that the effect of selection on mutations will vary according to whether the trait changes produced are dominant or recessive. A newly originated mutation with dominant effect will influence traits immediately, and selection for or against the mutation will take place at once. But if in a diploid animal a mutation has a recessive effect, it does not influence traits immediately. Natural selection then will not influence the mutation immediately either. This is the case with most mutations, since, as noted in Chap. 6, most mutations produce recessive effects.

A recessive mutant gene nevertheless can spread through a population if it happens to be closely linked to another gene that produces a dominant, adaptively desirable trait. Both genes then are inherited and propagated together. The recessive mutant here simply remains in the gene pool without effect, until two individuals that carry the same mutation happen to mate. Then one-fourth of their offspring will be homozygous recessive; if the mutant gene is *a',* then a mating of *Aa'* × *Aa'* will yield 25 percent *a'a'* offspring. These will exhibit altered visible traits, and natural selection now will affect the frequency of the mutation directly (Fig. 13.3).

The evolutionary role of mutations varies according to how greatly a given mutation influences a given trait. A "large" mutation that affects a vital trait in major ways is likely to be exceedingly harmful and usually will be lethal. But a "small" mutation that has only a minor effect on a trait can persist far more readily. Evolutionary alterations actually occur almost exclusively through an accu-

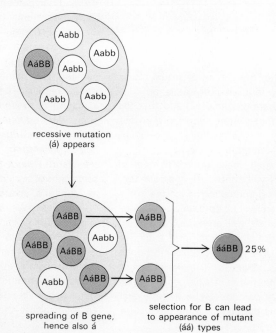

Figure 13.3 Gene spreading. If a recessive mutation *a'* appears in an organism and if that organism also carries a gene *B* that is strongly "selected for," then both *B* and *a'* can spread through a population together. The appearance of mutant phenotypes *a'a'* then becomes rather likely.

recessive mutation (á) appears

spreading of B gene, hence also á

selection for B can lead to appearance of mutant (áá) types

mulation of *many, small* changes in traits, not through single, large changes.

The third condition affecting Hardy-Weinberg equilibria is population size. If a population is large, regional imbalances of gene frequencies that might arise by chance are quickly smoothed out by the numerous matings among the many individuals. The underlying principle here applies to statistical systems generally. In a coin-flipping experiment, for example, heads and tails will each come up 50 percent of the time, but only if the number of throws is large. If only three or four throws are made, it is quite possible that *all* will come up heads by chance alone. Similarly, gene combinations attain Hardy-Weinberg equilibria only if a population is large. In small groups (less than about a hundred individuals), chance alone can lead to *genetic drift,* a random establishment of genetic types that are numerically not in accordance with Hardy-Weinberg equilibria. For example, if *AA, Aa,* and *aa* individuals are expected in certain proportions, chance alone could result in the formation of many more *AA* and many fewer *aa* individuals if the population is small.

Such a genetic drift effect resembles that of natural selection; if several genotypes are possible,

a particular one would likewise come to predominate if there were a selection pressure for it. But because genetic drift is governed by chance, natural selection plays little role. Genes are being propagated not for their adaptive value as in natural selection, but because they happen to be spread by chance. The result is that, in small populations, nonadaptive and often bizarre traits can become established. These actually can be harmful to the population and can promote its becoming even smaller. Or, genetic drift might by chance happen to adapt a small population rather well to a given environment, and such a population later might evolve through natural selection and eventually give rise to a new species. Genetic drift is believed to be a significant factor in the evolution of animals on islands and in small, reproductively isolated groups in general.

Evolution as it actually occurs must be—and indeed can be—interpreted on the basis of the mechanism here described.

Effects of Evolution

Speciation

The key process to be explained is how evolutionary changes in a population eventually culminate in the establishment of new species. As pointed out in Chap. 7, a species can be defined as a reproductive unit in which the members interbreed with one another but not usually with members of other species. In other words, a reproductive barrier isolates one species from another. A more or less free gene flow therefore is maintained within a species but not normally between species. Thus the problem of speciation is to show how reproductive barriers arise.

Primary barriers usually appear through *geographic* changes, as when water comes to intrude on land or vice versa, or when a forest belt grows through a prairie or vice versa, or when mountains or new climatic conditions become interposed across a region. Sister populations of the same species on each side of such barriers then become isolated geographically, and reproductive contact is lost. Most commonly, populations become isolated geographically simply by distance. As a species in time occupies a progressively larger territory, two populations at opposite ends of this territory eventually will be too far apart to permit direct interbreeding (Fig. 13.4).

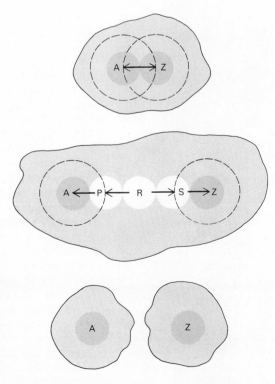

Figure 13.4 Speciation. Top, two populations (*A, Z*) in a parent species have overlapping reproductive ranges (dark circles), hence gene flow can be direct. *Center,* after population growth and territorial expansion the reproductive ranges of *A* and *Z* no longer overlap, and gene flow between them now must be indirect, via intervening populations (*P, R, S*). *Bottom,* if gene flow between *A* and *Z* ceases altogether, the two populations will have become reproductively isolated and will be independent offspring species.

Figure 13.5 Interbreeding.
Platyfish female at top, swordtail male at bottom. These animals belong to different species, and in nature they do not interbreed; but they can and do interbreed in the laboratory.

Figure 13.6 Artificial selection. Red jungle fowl, an example of a wild organism from which man has bred domesticated varieties.

Regardless of how isolation is actually brought about, an interruption of gene flow between sister populations will be followed by independent natural selection in each. In the course of many generations numerous genetic differences will appear, and these are likely to include *biological* barriers to interbreeding. For example anatomic changes in reproductive organs might make mating mechanically impossible; or gametes might become incompatible through altered protein specificities; or the times of the annual breeding seasons might shift; or psychological changes might make mates from neighboring populations no longer acceptable. In effect, two sister populations of the same species in time will become two separate, new species.

At first, newly formed sister species tend to be rather similar structurally and functionally, and in some cases interbreeding still might take place if the species were not isolated geographically. Thus it sometimes happens that, when members of two similar species are brought together under artificial conditions, they can interbreed readily. Evidently, the circumstance that two different species *do* not interbreed in nature does not always mean that they cannot interbreed (Fig. 13.5). But after two sister species have been separated for long periods, interbreeding eventually will become impossible even if contact is provided artificially; the differences sooner or later become pronounced enough to preclude interbreeding.

Speciation by this means is the chief way in which new species evolve. Such a process takes on an average about 1 million years. Consciously or unconsciously making use of this principle of reproductive isolation, man has been and is now contributing to the evolution of many other organisms. The most ancient evolution-directing effort of man is his successful *domestication* of various plants and animals. Darwin was the first to recognize the theoretical significance of domestication, and indeed it was this that led him to his concept of natural selection. He reasoned that if man, by *artificial* selection and isolation, can transform wild varieties of plants and animals to domesticated forms, then perhaps *natural* selection and isolation, acting for far longer periods, can produce even greater evolutionary transformations in nature. The domesticating process actually does involve all the elements of natural evolution: deliberate isolation of a wild population by man, followed by carefully controlled, differential reproduction of those individuals that have traits considered desirable by man. The result is the creation of new strains, subspecies, and even species (Fig. 13.6).

During the last few decades, furthermore, rather rapid, man-caused evolution has taken place among certain viruses, bacteria, insects, various parasites, and other pests. These now live in an environment in which antibiotics and numerous pest-killing drugs have become distinct hazards.

And the organisms have evolved and are still evolving increasing resistance to such drugs.

Clearly then, these examples offer direct proof that evolution actually occurs and is observable, and that it can be made to occur under conditions based on the postulated modern mechanism of evolution. Incidentally, speciation also accounts for the establishment of genera, families, and other higher taxonomic groups: the differences between newly created species or groups of species can be sufficiently extensive that we classify such animals as members of different higher taxonomic categories.

Some zoologists have argued that the distinctions between high-ranking taxonomic categories are far too great to be explainable by gradual accumulation of many small, minor variations in different species. A hypothesis of "large" mutations has been postulated instead. According to it, a major mutation that affects many vital traits simultaneously is assumed to transform an animal suddenly, in one jump, to a completely new type. Such a type would represent not only a new species but also a new high-ranking taxonomic group. In most cases an animal of this sort would not survive, for it would probably be entirely unsuited to the local environment. But it is assumed that in extremely rare cases such "hopeful monsters" might have arisen by freak chance in environments in which they could survive. Only relatively few successes of this sort would be needed to account for the existing classes, phyla, or superphyla of animals.

Few zoologists accept this hypothesis of jump evolution. In studies of natural and experimental mutations over many years it has always been found that sudden genetic changes with major effects are immediately lethal. This is the case not only because the external environment is unsuitable but also, and perhaps mainly, because the internal metabolic upheaval caused by a major mutation is far too drastic to permit continued survival. But even supposing that a hopeful monster could survive, it would by definition be so different from other individuals that it would be structurally, functionally, and behaviorally incompatible as a mate.

Furthermore, although the differences between phyla and other major categories are great, they are not so great that such categories could not have arisen through gradual evolution of different representative species. Indeed, the evidence from

fossils and embryos shows reasonably well in what pattern this evolution might have occurred (see Chap. 14). Also, as the next section will show, important aspects of the evolutionary process cannot be explained by jump evolution but can be explained rather well on the basis of gradual evolution.

Diversification

Even at the species level evolution is an exceedingly slow process. As noted, a very large number of small variations must accumulate before a significant alteration of animals can occur. Moreover, since genetic variations arise at random, animals must *await* the appearance of adaptively useful changes. But there is no guarantee that useful variations will appear in successive regeneration or that they will appear at all. Thus even though evolution might occur, it could occur too slowly to permit successful adaptation to changed environments.

As a rule, the actual *rates* of past evolution have been proportional to the instability of the environment. Terrestrial animals by and large have evolved faster than marine types, land being a less stable environment than the sea. Also, evolution generally has been fairly rapid in times of major geologic change, as during ice ages or periods of mountain formation (see Chap. 15). By contrast, the rate of evolution has been virtually zero for several hundred million years in a few marine groups— horseshoe crabs, brachiopod lamp shells, and some of the radiolarian protozoa, for example. The environments in which such "living fossils" exist evidently have remained stable enough to make the ancient way of life still possible (Fig. 13.7).

Apart from variations in rates, evolution has tended to occur through successive *adaptive radiations*. Just as a parent species can give rise simultaneously to two or more descendant species; so a similar pattern of bushlike *branching* descent characterizes evolution on all levels; a newly evolved type becomes a potential ancestor for many different simultaneous lines of descendants. For example, the ancestral mammalian type has given rise simultaneously to several lines of modern grazing plains animals (horses, cattle, goats), to burrowing animals (moles), to flying animals (bats), to several lines of aquatic animals (whales, seals, sea cows), to animals living in trees (monkeys), to carnivorous predators (dogs, cats),

A

B

Figure 13.7 Speeds of evolution. A, the ascidian *Molgula,* a tunicate chordate that, like tunicates in general, has not changed in any fundamental way from the most ancient ancestral chordates of about 400 million years ago. But during this same time interval another line descended from chordates has given rise to the vertebrates, of which the carp in *B* is a modern representative. The rate of evolution evidently was minimal in the tunicate line but high in the vertebrate line. (*A,* Carolina Biological Supply Company.)

and to many others. Each such line of descent has become specialized for a different way of life, and the sum of all these branch lines leading away from the common ancestral type represents an adaptive radiation.

Further, in the course of its history a branch line can give rise to one or more adaptive radiations of smaller scope. In the line of tree-living mammals, for example, simultaneous sublines and subsublines have evolved that have resulted in contemporary animals as varied as monkeys, lemurs, tarsiers, apes, and men. The important implication here, as also pointed out in Chap. 7, is that evolution is not a "ladder" or a "scale." Man for example did not descend from apes, but both have had a common ancestor and are contemporary members of the same adaptive radiation (see Fig. 7.2). All such radiations generally exhibit *divergence,* or development of more or less dissimilar traits in groups descended from a common ancestor.

Not all the branches on a bush lead right to the top, but some terminate abruptly at various intermediate points. In evolution, similarly, *extinction* has been a general feature. In many cases of extinction the specific causes may never be known. However, in line with the nature of the evolutionary

mechanism, the general cause of all extinctions is inability of animals to adapt rapidly enough to environmental changes. In the past, extinctions have been the more common the lower the taxonomic category. Extinction of species and even of genera has been a nearly universal occurrence, but relatively few orders and still fewer classes have become extinct. And virtually all phyla that ever originated continue to be in existence today. The phylum evidently includes so broad and far-flung an assemblage of different adaptive types that at least some of them have always survived, regardless of how environments have changed. By contrast, species usually are adapted rather narrowly to limited, circumscribed environments, and the chances for extinction therefore are greater (Fig. 13.8).

In conjunction with extinction, *replacement* has been another common occurrence in evolution. Replacement occurs when, after one group of organisms has become extinct, another group evolves that adopts the vacated environment and way of life. For example, pouched marsupial mammals were very abundant in the Americas a few million years ago, but with the exception of forms like the opossum they were replaced in the Western Hemisphere by the competing placental mammals. In this case replacement was more or less

A

B

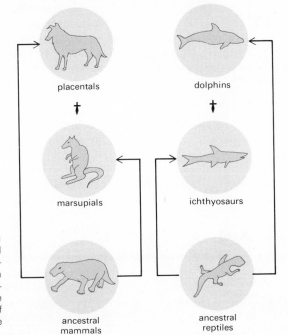

C

D

Figure 13.8 Recent extinctions. A, dodo. B, Irish elk. C, sabertooth cat. D, woolly mammoths. The dodo survived till just a few hundred years ago. Mammoths and sabertooths became extinct some twenty thousand years ago, and Irish elks, some thousands of years before that.

placentals

dolphins

†

†

marsupials

ichthyosaurs

ancestral mammals

ancestral reptiles

Figure 13.9 Replacement. Left, immediate replacement: very soon after placental mammals evolved they replaced the earlier marsupials virtually everywhere except in Australia. Right, delayed replacement: dolphins replaced the ichthyosaurs many millions of years after the latter had become extinct.

immediate, but on occasion millions of years can elapse before a new group evolves into a previously occupied environmental niche. Such *delayed* replacement took place, for example, in the case of the ichthyosaurs. These fishlike marine reptiles became extinct some 100 million years ago. Their particular mode of living then remained unused for about 40 million years, when a newly evolved mammalian group, the whales and dolphins, replaced the ichthyosaurs. Similarly delayed replacement occurred between the flying reptilian pterosaurs and the later mammalian bats (Fig. 13.9).

Replacing animals usually exhibit some degree of evolutionary *convergence*, or *parallelism;* they resemble each other in one or more ways even though they need not be related particularly closely. For example, the development of finlike appendages in both ichthyosaurs and dolphins or of wings in both pterosaurs and bats illustrates evolutionary convergence in replacing types. Since both an original and a replacing group are adapted to the same type of environment and way of life, their evolution has been oriented in the same

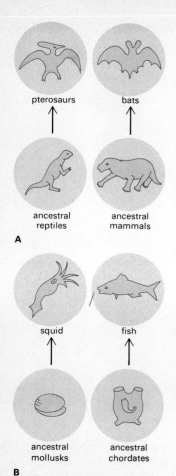

A

B

Figure 13.10 Convergence. Relatively unrelated ancestors give rise to descendants whose ways of life and even superficial appearance are similar in many respects. In *A* the convergence is replacing: bats now occupy the same kinds of niches that pterosaurs did many millions of years ago. In *B* the convergence is nonreplacing: squids and fishes coexist today, each group occupying its own niche.

direction; and the appearance of similar, or convergent, traits is therefore not surprising. But convergence occurs also in nonreplacing types. For example, the eyes of squids and of fish are remarkably alike. Squids and fish are not related directly, and neither replaces the other. However, both types of animal are large, fast swimmers, and good eyes of a particular construction are a distinct advantage in the ways of life of both. Selection evidently has promoted variations that have led to eyes of similar structure, and the observed convergence is the result (Fig. 13.10).

Although the eyes of squids and fish are strikingly alike, they are by no means identical. Convergence leads to *similarity*, not identity. Moreover, neither squids nor fish have a theoretically "best" eye structure for fast swimmers. Similarly, none of the various animal groups that fly has a theoretically "best" wing design. An organ or organism actually need not be theoretically "best" or "most efficient." The structure only needs to be practically workable and just efficient enough for a necessary function. In a way of life based on flying, wings of *some* sort clearly are essential. But most requirements for living can have *multiple* solutions, and so long as a certain solution works at all, it does not matter how the solution is arrived at. The various animal wings do represent multiple solutions of the same problem, each evolved from a different starting point and each functioning in a different way.

Figure 13.11 Evolutionary opportunism. Parts of the lower jaw and the upper bone of the next bony gill arch of ancestral fishes have been the evolutionary sources of the middle-ear bones of man (black). Similarly, parts of the other gill supports of fishes have evolved into the cartilages of the mammalian larynx (dark color), and the gill musculature of fishes has contributed to the muscles of the lower part of the face of mammals (hatched areas).

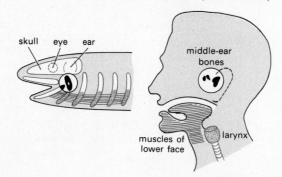

This consideration focuses attention on one of the most important characteristics of evolution, that of *random opportunism.* Evolution has produced not what is theoretically desirable or best, but what is practically possible. There has been no predetermined plan, no striving for set "goals," but only the exploitation of actually available opportunities offered by selection among random hereditary changes. For example, although it might have been adaptively exceedingly useful for terrestrial animals to grow wheels, such a development did not occur because it could not occur; the ancestors simply did not have the necessary structural and functional potential. However, they did have the potential to evolve adequate and workable alternative solutions. Among vertebrates, for example, already existing fins could be reshaped as walking legs.

Evidently, evolution only can remodel and build on what already exists, in small, successive steps. Since, given a long enough time span, every feature of every animal undergoes random variations in many different directions, opportunities for diverse evolutionary changes have been and still are very numerous. Every animal in effect represents a patchwork of good opportunities seized by natural selection at the right time. In man, for example, the bones of the middle ear have arisen opportunistically from pieces of earlier vertebrate jawbones. The musculature of the lower face has evolved from the gill muscles of ancestral fish. The larynx has developed from the gill skeleton of ancient fish (Fig. 13.11).

Such instances of evolutionary opportunism are legion. Animals clearly are not the result of any planned, goal-directed, or predetermined course of creation. Instead they are the result of a cumulative, opportunistic process of piece-by-piece building, based on existing animals and governed entirely by natural selection acting on random variations.

Through past evolution the living mass on earth has increased fairly steadily in individual numbers and types and has seeped into virtually all possible environments. Indeed it has created new environments in the process. For example, the evolution of trees has created new possibilities of life in the treetops, exploited later by very many new animals, including our own ancestors. The evolution of warm-blooded birds and mammals has created a new environment in the blood of these animals, exploited later by many new parasites. The evolution of man has created numerous new environments in human installations, and these have been exploited by a large variety of new animals.

We recognize here yet another general characteristic of evolution: a progressive, creative *expan-siveness* with respect to both living mass and ways of life. The expansion is still under way, faster in some cases than in others, and the end cannot be predicted as yet.

Review Questions

1. Describe the essential points of the evolutionary theories of (*a*) Lamarck, (*b*) Darwin and Wallace. How could the evolution of giraffes from short-necked ancestors be explained in terms of these two theories? What are the weaknesses of each theory?

2. What different kinds of inheritable variation can arise in animals? Do such variations appear randomly or are they oriented toward usefulness? How do noninheritable variations arise, and what role do they play in evolution?

3. Define the modern meaning of natural selection. Show how natural selection basically has little to do with ''survival of the fittest'' or ''struggle'' or ''weeding out,'' and how it is a creative force. How does it happen that natural selection is oriented toward improved adaptation?

4. State the Hardy-Weinberg law. If a population contains 49 percent *AA*, 42 percent *Aa*, and 9 percent *aa* individuals, show by calculation how the law applies. If a Hardy-Weinberg equilibrium exists in a population, what are the rate and amount of evolution?

5. What three conditions disturb Hardy-Weinberg equilibria? For each condition show in what way such equilibria are disturbed and how evolution is therefore affected. How do recessive genes spread through a population? What is genetic drift and where is it encountered?

6. Define species in genetic terms. Describe the process of speciation. What are some common geographic isolating conditions, and what is their effect on gene pools? How do reproductive barriers arise between populations?

7. Review some actual evidence for past and present evolution. Describe the hypothesis of jump evolution. What are its weaknesses and what is the commonly accepted alternative hypothesis?

8. How have rates of evolution varied in the past? What is an adaptive radiation? Illustrate in the case of mammals. What are the general causes of extinction? What has been the pattern of extinction on different taxonomic levels?

9. What is evolutionary replacement? Distinguish between immediate and delayed replacement and give examples. Distinguish between evolutionary divergence and convergence and give examples. In what important way is evolution randomly opportunistic?

10. List five structural and functional features of man and show for each (*a*) how it has evolved opportunistically, and (*b*) that it cannot be labeled as being ''theoretically best.'' Show how evolution has created new environments and therefore new opportunities for evolution.

Collateral Readings

Cavalli-Sforza, L. L.: ''Genetic Drift'' in an Italian Population, *Sci. American,* Aug., 1969. Studies of blood-group frequencies reveal evolutionary changes in an isolated human population.

Crow, J. F.: Ionizing Radiation and Evolution, *Sci. American,* Sept., 1959. A discussion of the effects of radiation-induced mutations on evolution.

Darwin, C.: "The Origin of Species and The Descent of Man," Modern Library, New York, 1948. A reprint of two of Darwin's classic books.

———— and A. R. Wallace: On the Tendency of Species to Form Varieties; and of the Perpetuation of Varieties and Species by Natural Means of Selection, in "Great Experiments in Biology," Prentice-Hall, Englewood Cliffs, N.J., 1955. A reprint of the original 1858 statement of the theory of natural selection.

Dobzhansky, T.: The Genetic Basis of Evolution, *Sci. American,* Jan., 1950. Genetic experiments on evolution are described, including the production of bacterial strains resistant to bacteriocides.

————: "Genetics and the Origin of Species," 3d ed., Columbia, New York, 1951. A well-known and recommended analysis of evolutionary theory.

Dodson, E. O.: "Evolution: Process and Product," Reinhold, New York, 1960. A good text on general principles and specific evolutionary processes and phenomena.

Ehrlich, P. R., and R. W. Holm: "Process of Evolution," McGraw-Hill, New York, 1963. A nonintroductory, recommended book on genetics, populations, and evolutionary principles.

Eiseley, L. C.: Charles Darwin, *Sci. American,* Feb., 1956. A biographical article.

Kettlewell, H. B. D.: Darwin's Missing Evidence, *Sci. American,* Mar., 1959. A study of evolution in progress; the case of body colors of certain species of moths is analyzed.

Lack, D.: Darwin's Finches, *Sci. American,* Apr., 1953. An examination of the famous Galapagos birds on which Darwin based much of his theory.

Lamarck, J. B. P. A. de: Evolution through Environmentally Produced Modifications, in "A Source Book in Animal Biology," McGraw-Hill, New York, 1951. A translation of the original 1809 statement of Lamarck's hypothesis.

Mayr, E.: "Animal Species and Evolution," Harvard, Cambridge, Mass., 1963. An important (nonintroductory) book on speciation and evolutionary theory.

Pasteur, L.: Examination of the Doctrine of Spontaneous Generation, in "Great Experiments in Biology," Prentice-Hall, Englewood Cliffs, N.J., 1955. A translation of the famous 1862 original.

Redi, F.: Experiments on the Generation of Insects, in "Great Experiments in Biology," Prentice-Hall, Englewood Cliffs, N.J., 1955. A translation of the 1688 refutation of the doctrine of spontaneous generation.

Ross, H. H.: "A Synthesis of Evolutionary Theory," Prentice-Hall, Englewood Cliffs, N.J., 1962. A stimulating book on all kinds of evolving systems, from galaxies and stars to living populations on earth.

Simpson, G. G.: "The Meaning of Evolution," Yale, New Haven Conn., 1949, or Mentor Books M66, New York, 1951. A most highly recommended analysis of modern evolutionary theory by one of the principal students of the subject; hard cover and paperback.

Volpe, E. P.: "Understanding Evolution," Brown, Dubuque, Iowa, 1967. A recommended paperback reviewing general evolutionary theory and principles.

Wallace, B.: "Chromosomes, Giant Molecules, and Evolution," Norton, New York, 1966. This paperback deals with the genetic aspects of the evolutionary mechanism.

14

ORIGINS: PHYLOGENY

The actual past events of animal history can be studied through *phylogeny* and *paleontology*. The first seeks to determine the exact pattern of the evolutionary bush, the "genealogy" of life from the moment life first began. Since such determinations must be based largely on built-in historical clues detectable in presently living creatures, phylogeny is rich in hypothesis and speculation. Paleontology is a study of fossils. These provide a very direct, even if spotty, record of the remains of formerly living types. Unfortunately this record does not go back more than about half a billion years, whereas life probably began as long as three billion years ago. Most of living history thus must be deduced from phylogenetic considerations.

Origins of Life

It is thought today that life began through a progressive series of chemical synthesis reactions that raised the organization of inanimate matter to levels of successively greater complexity. Atoms first formed simple compounds, these later formed more complex ones, and the most complex of them eventually became organized as living cells.

The details of these processes are at present known only partly. Some information can be obtained by deducing from viruses, bacteria, and other primitive existing forms what the earliest living systems might have been like. Other clues come from astronomy, physics, and geology, sciences that supply data about the probable physical characteristics of the ancient earth. Important insights are also gained from chemical experiments designed to duplicate in the laboratory some of the steps that could have led to the beginning of life.

What has been learned in this way indicates that living creatures on earth are a direct product of the earth. There also is every reason to believe that living things owe their origin entirely to certain physical and chemical properties of the ancient earth. Nothing supernatural appears to have been involved—only time and natural physical and chemical laws that operate within the peculiarly suitable earthly environment. Given such an environment life probably *had* to happen; once the earth had formed, with particular chemical and physical properties, it then was virtually inevitable that life would later originate on it also. Similarly, if other solar systems have planets where chemical and physical conditions resemble those of the ancient earth, then life is likely to originate on such planets as well. It is now believed strongly that life occurs not only on earth but probably widely throughout the universe.

Chemical Evolution: Cells

According to the most widely accepted hypothesis, the solar system started out some 10 billion years ago as a hot, rotating ball of atomic gas. In it, hydrogen atoms probably were the most abundant, and other, heavier kinds of atoms were present in lesser quantities. The sun was formed when most of this gas gravitated toward the center of the ball. Even today the sun is composed largely of hydrogen atoms. A swirling belt of gas remained outside the new sun. In time this belt broke up into a few smaller gas clouds; these spinning masses of fiery gases were the early planets.

The earth thus probably began, about 4.5 to 5 billion years ago, as a glowing mass of free hydrogen and other elements. These eventually became sorted out according to weight. Heavy ones such as iron and nickel sank toward the center of the earth, where they still are present today. Lighter atoms such as silicon and aluminum formed a middle shell. The very lightest, such as hydrogen, nitrogen, oxygen, and carbon, collected in the outermost layers. In time the temperature of these surface gases became low enough to permit the formation of compounds, and free atoms then largely disappeared.

On the basis of the known chemical properties and the presumed relative abundance of hydrogen, carbon, oxygen, and nitrogen, the surface gases should have given rise to some half dozen different combinations: water (H_2O); methane (CH_4); ammonia (NH_3); carbon dioxide (CO_2); hydrogen cyanide (HCN); and hydrogen molecules (H_2). We have evidence that at least the first three of these compounds actually came into being not only on the early earth but on other planets as well. On Jupiter, for example, water, methane, and ammonia appear to be present today in the form of thick surface layers of permanently frozen solids. These compounds apparently formed there as on earth, but at that great distance from the sun the surface of the planet probably froze before much additional chemical change could occur. On the hot earth, by contrast, the early compounds remained gaseous and could give rise to new compounds later.

Temperatures in the outer layers of the earth eventually became low enough to allow some of the gases to liquefy and some of the liquids in turn to solidify. Thus although to this day the earth contains a hot, thickly flowing center, the middle shell of lighter substances became a solid, gradually thickening crust. And as this crust thickened and cooled, it wrinkled and folded and gave rise to the first mountain ranges. Overlying the crust was the outer atmospheric mantle, which remained gaseous.

When the crust had cooled below the boiling point of water, most of the water in the atmosphere must have fallen as rain and formed oceans. Dissolved in these must have been some of the atmospheric methane and ammonia, as well as salts and minerals that leached out slowly from the solid crust of the earth and that spewed forth from numerous volcanoes. The oceans apparently acquired their saltiness relatively early, and to a small extent they became saltier still during succeeding ages (Fig. 14.1).

The stage was then set for synthesis reactions. Such reactions among water, methane, and ammonia resulted in *organic* compounds that contained *linked* carbon atoms. That simple organic materials can indeed be created by reactions among these gases was demonstrated in the early 1950s through dramatic and now classic laboratory experiments. In them the presumed environment of the early earth was duplicated in miniature. Mixtures containing water, methane, and ammonia were put into a flask, and electricity was discharged through these mixtures for several days to simulate the lightning discharges of the early earth. When the contents of the flask were examined, many amino acids, fatty acids, sugars, and other simple organic compounds were found to be present.

Thus there is excellent reason to think that, with energy from lightning and also from solar radiation, simple inorganic materials could give rise to a variety of organic compounds that accumulated in the ancient seas. Such compounds would have represented the chemical "staples" out of which more complex organic materials could be synthesized later (Fig. 14.2). At some point in the course of this *chemical* evolution, *biological* evolution must have begun: formation of the very complex chemicals so characteristic of life, and development of cells, the first actual living units.

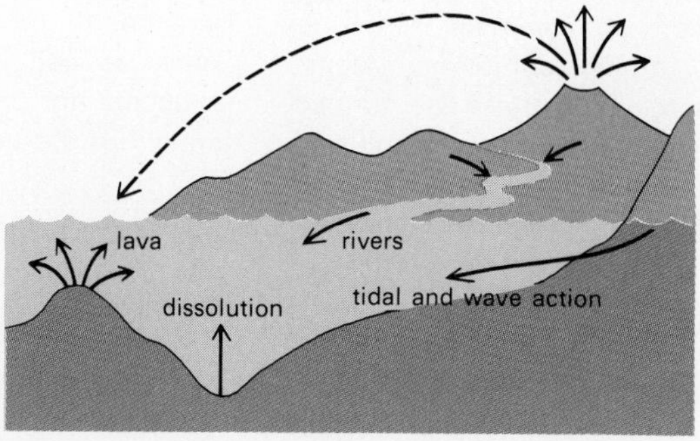

Figure 14.1 Sources of ocean salt. Some came from volcanoes, both submarine and terrestrial; some was dissolved out at the sea bottom; some resulted from tidal action, which crumbled and dissolved the shorelines; and some originated on the land surface, leached out by rain and rivers.

lava

rivers

dissolution

tidal and wave action

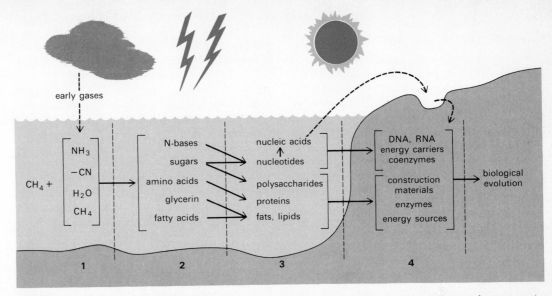

Figure 14.2 Chemical evolution: summary of probable early synthesis reactions. At least three or four successive phases appear to have been involved. The original gaseous raw materials came from the early atmosphere (1), and, with the aid of energy from lightning and the sun, key biological compounds were·synthesized progressively in the ocean (2,3). The later, more complex synthesis reactions perhaps took place in sand pockets along the shore, where required ingredients could become highly concentrated by evaporation of water (4). The outcome was the origin of the first living units, and further biological evolution then followed.

It is physically and chemically most plausible to assume that the critical events in the creation of cells took place not in the open sea, where the necessary ingredients probably would not have remained together, but along the shores of the early ocean. The sand and clay particles there would have provided surfaces to which oceanic molecules could have adhered, and evaporation of water would have increased the concentrations of such molecules. Indeed, microscopic sand pockets containing highly concentrated, nearly dry accumulations of simple organic materials well could have been the birthplaces of cells. This surmise again is reinforced by experimental data. For example, if concentrated mixtures of amino acids are heated under nearly dry conditions, proteinlike complexes are formed. Moreover, these develop surface films resembling cell membranes quite closely. Furthermore, if mixtures of simple starting compounds such as ammonia and certain amino acids are heated under almost dry conditions, products appear that have the characteristics of nitrogen bases.

It is therefore possible that, inside minute membrane-bounded droplets along the shore, some of the proteinlike materials formed at the time could have been active enzymatically. These could then have promoted a combination of nitrogen bases, simple sugars, and phosphates as nucleotides. And nucleotides are only a simple chemical step away from energy carriers such as ATP. Further, once enzymes and ATP are available, a few additional chemical steps can lead to nucleotide derivatives such as coenzymes and to nucleic acids such as genetically active DNA and RNA. In their turn, the DNA molecules in the droplets would have controlled the formation of particular, *specific* proteins, just as they are known to do today.

In the laboratory it is now possible to create DNA, RNA, and specific proteins artificially from simpler organic precursors, and by steps that virtually duplicate those taking place in a living cell (see also Chap. 6). Thus there exists important experimental support for the hypothesis that progressive synthesis reactions, roughly as outlined above, must have been the critical steps in the original development of living matter. In effect, the droplets on the ocean shore would have been distinctly individual units marked off from the surrounding ocean water, and they would have remained individualized even if they absorbed more water and were later washed back to the open ocean. Such units would have been primitive cells (Fig. 14.3).

Undoubtedly numerous trials and errors must have occurred before the first actual cells left their places of birth. Many and perhaps most of the accumulations on the shore probably were ''unsuccessful.'' In given cases, for example, the right kinds of starting ingredients might not have come together; or the right amounts of ingredients might not have accumulated; or the mixture might have dried up completely; or it might have been washed out to sea prematurely and dispersed. Clearly, numerous hazards must have led to many false starts and to many incomplete endings. Yet when an appropriate constellation of materials did accumulate, the formation of a living cell would have been a likely result.

A basic problem in tracing the possible origin of cells (and of later groups of organisms in general) is to decide whether they evolved *monophyletically,* from a single common ancestral stock, or *polyphyletically,* from several separate ancestral stocks. Did early cells arise from a single first cell, an ''archancestor'' of all life, or did numerous ''first'' cells originate independently? The answers to such questions simply are not known.

Monophyletists invoke a *probability* argument: cells are so complex that multiple origins of such complexities are statistically quite unlikely. The argument leads logically to the view that all life started in a single place at a single time, from a single cell formed by ''lucky accident.'' Proponents of polyphyletism counter with a probability argument of their own. They point out that, compared to their later complexities, the original cells really could not have been so complex. At first they probably were little more than accumulations of mixed chemicals surrounded by membranes, and structural complexity must have evolved later, in

gradual steps. Moreover, if the conditions for cell formation could develop in one place at one time, it is statistically most likely that such conditions would develop also in other places and at other times.

In one sense, therefore, chance undoubtedly did play a role in cell formation: many droplets never became cells, and those that did owed their existence to the chance accumulation of the right ingredients. But in another sense cell formation was not simply an enormously ''lucky accident,'' a one-time occurrence of very remote probability. On the contrary, given an early earth so constituted that certain compounds could form, and given these compounds and their special properties, then cell formation *had* to take place sooner or later. The only element of chance here was time; the uncertainty was not a matter of ''if'' but of ''when'' and ''how often.'' The origin of cells therefore was as little an accident as is the eventual appearance of sevens and elevens in a series of dice throws.

Monophyletists also invoke a *similarity* argument: all cells are fundamentally so remarkably alike in chemical composition, microscopic structure, and function that a common origin from a single ancestry seems clearly indicated. Polyphyletists here counter with the observation that all living cells must *necessarily* share certain basic structural and functional characteristics, for those early droplets that did not exhibit such characteristics simply would never have been ''alive,'' by definition of this term. The similarities among cells therefore are not necessarily an indication of common ancestry, but might merely reflect the universal attributes anything living must have regardless of origin.

Best estimates at present suggest that the first cell or cells probably arose perhaps 3 or 4 billion years ago, or about 1 or 2 billion years after the formation of the earth. Yet no one point in the earth's early history really qualifies as a ''beginning'' of life. The cell is the major product of the first few billion years, and we regard this product as being alive. But the earlier organic compounds dissolved in the ocean already had the properties that eventually made life possible. Such compounds in turn did not originate their characteristic properties, but acquired them when they were formed from various simpler compounds. The potential of life clearly traces back to the original atoms, and the creation of life out of atoms was

Figure 14.3 The possible origin of the first cells. Appropriate chemical ingredients might have accumulated by adsorption in microscopic pockets along the seashore (1), and these ingredients could have become concentrated progressively (2). Under relatively dry conditions and perhaps with the aid of ATP, which might have been present, nucleic acids and proteins could have formed (3). Some of the proteins then could have made possible enzymatically accelerated reactions and formation of structural membranes and internal fibrils (4). Finally, primitive cellular compartments might have been washed out to sea (5).

but a step-by-step exploitation of their properties.

In short, life did not burst forth from the ocean finished and ready. Instead it *developed,* and here is perhaps the most dramatic illustration that small beginnings can have surprisingly large endings. Development has been the hallmark of life ever since, and life today still is unceasingly forming and molding. Indeed it will never be finally "finished" until its last spark is extinguished.

Biological Evolution: Premonera

For present purposes the earliest cells collectively can be called *Premonera,* or organisms that preceded the evolution of the Monera (see below).

In later ages, premoneran cells must have become diversified as numerous new cell types in response to a powerful environmental stimulus: the gradual disappearance of free organic molecules from the ocean. For as more and more such molecules were used as food by an ever-increasing multitude of reproducing cells, the global rate of food consumption eventually must have become greater than the rate of food formation from methane, ammonia, and water. In time, therefore, the ocean became a largely inorganic medium, as it still is today. Unless the early cells then could have evolved new ways of obtaining food they would soon have nourished themselves to extinction. Evidently they did not.

One of the first evolutionary responses to dwindling food supplies probably was the development of *parasitism.* If foods could not be obtained from the open ocean, they still could be obtained inside the bodies of living cells. Methods of infecting cellular hosts well could have evolved almost as soon as cells themselves had originated, and para-

sitism must have been an effective new way of life for many of the early cells (Fig. 14.4).

Another new way that required relatively little evolutionary adjustment was *saprotrophism.* Here a cell drew food molecules from the bodies of dead cells or disintegrated cellular material. Organic *decay* of such material was a result and has occurred ever since. Saprotrophic types are so abundant today in all environments that virtually every organic substance begins to decay almost immediately after exposure. A third new process that permitted survival despite dwindling food supplies was *holotrophism,* "eating" other living cells whole. This method of feeding presumably originated through phagocytosis, the general capacity of a cell to engulf microscopic particles (see Chap. 5). Amoeboid pseudopodia and permanent cellular mouths (*gullets*) must have been later elaborations.

But all three of these new food-gathering procedures were ultimately self-limiting. Collectively they represent the *heterotrophic* forms of nutrition, exhibited even today by organisms that require preexisting foods and that therefore depend nutritionally on other organisms. Early heterotrophic nutrition among Premonera thus merely redistributed organic materials that already existed. If totally new food sources had not been obtained, life would have had to cease sooner or later.

The raw materials for new foods still were present in abundance. Water was in inexhaustible supply and, in addition to methane, an even better source of carbon now existed directly inside cells in the form of respiratory carbon dioxide. With CO_2 and water available, organic molecules could be manufactured by cells, provided that new external sources of energy could be found. Internal energy was obtainable from ATP, to be sure, but ATP was itself among the very organic molecules that were in danger of disappearing.

We know by hindsight that some Premonera actually did evolve means of using new external sources of energy. One group included cell types that could absorb sulfur, iron, nitrogen, or one of a number of other mineral materials obtainable in the environment. Such inorganic substances could be made to undergo various exergonic, energy-yielding reactions in the cells, and chemical energy so obtained then could serve in transforming internal CO_2 and water to organic food molecules. Nutrition by such a process of *chemosynthesis* still is encountered today in some of the bacteria (Fig. 14.5).

Figure 14.4 Heterotrophic nutrition. In parasitism one organism obtains food from another living one (a small parasitic cell is shown inside a larger host cell). In saprotrophism food is obtained from dead organisms. And in holotrophism one organism eats another in whole or in part. These three methods are noncreative; they merely redistribute already existing foods and do not generate new supplies.

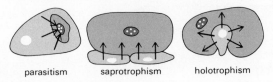

parasitism   saprotrophism   holotrophism

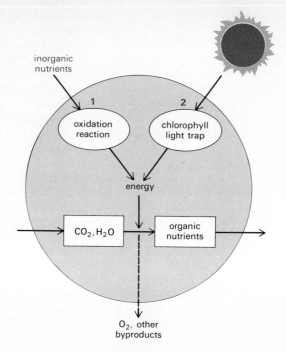

Figure 14.5 Autotrophic nutrition. In chemosynthesis (1) energy is obtained from inorganic raw materials, and with the aid of this energy the organism creates organic materials out of carbon dioxide and water. Various inorganic byproducts result as well. In photosynthesis (2) energy is obtained from the sun by means of energy-trapping molecules such as chlorophyll. The organism then again creates organic nutrients out of carbon dioxide and water. Oxygen is a byproduct here (derived from water, which is split into separate hydrogen and oxygen fractions).

cells could make foods for themselves; holotrophic forms could swallow such cells as well as one another; parasites could invade photosynthesizers or holotrophs; and saprotrophs could find foods in the dead bodies of any of these. Today photosynthesis still supports all living creatures except the chemosynthesizers.

As photosynthesis occurred on an ever-increasing scale, it brought about far-reaching changes in the physical environment. A byproduct of photosynthesis is free molecular oxygen (O_2), a gas that combines readily with other substances. The gas escaped from photosynthetic cells into the ocean and from there to the atmosphere, and it must have reacted promptly with everything it could. A slow, profound "oxygen revolution" then occurred. In the course of it the ancient atmosphere was transformed to the modern one, which no longer contains methane, ammonia, and cyanide. Instead it consists mainly of water vapor, carbon dioxide, and molecular nitrogen, plus large quantities of free molecular oxygen itself (Fig. 14.6).

Moreover, under the impact of X rays and other high-energy radiation from space, oxygen molecules several miles up in the new atmosphere combined with one another and formed a layer of *ozone* (O_3). This layer became an excellent screen against high-energy rays, and ever since it has protected living organisms from excessive amounts of such space radiation. Free oxygen also reacted with the solid crust of the earth and converted most metals and minerals to *oxides,* the familiar ores and rocks now making up much of the land surface. Finally, free oxygen made possible a new form of respiration. The earliest cells respired *anaerobically,* without oxygen, and to a small extent all living organisms still do so (see Chap. 5). But when environmental oxygen became available in quantity, organisms newly evolving at the time developed means of using this gas. Since then the far more efficient oxygen-requiring, or *aerobic,* form of respiration has been perpetuated.

Evidently, the activities of the early organisms greatly altered the physical character of the earth and also the biological character of the organisms themselves. So has it been ever since: the physical earth creates and influences the development of the biological earth, and the biological earth then reciprocates by influencing the development of the physical earth.

However, chemosynthesis probably was far less widespread right from the start than another new nutritional process, *photosynthesis,* evolved by other groups of Premonera. In photosynthesis the external energy source is light, and with the aid of the green photosensitive compound chlorophyll the energy of light is used to transform CO_2 and water to foods. Collectively, chemosynthesis and photosynthesis represent the *autotrophic* forms of nutrition, exhibited by organisms that can survive in an exclusively inorganic environment and that therefore do not depend nutritionally on other organisms.

Thus, once the autotrophic production of new organic compounds was assured, it did not matter that the supply of free molecular foods in the ocean finally became inadequate. Photosynthetic

Figure 14.6 The oxygen revolution. Oxygen from photosynthesis reacted with other materials. A major result was the establishment of a new, modern atmosphere that contained N_2, CO_2, and H_2O, in addition to O_2 itself.

$$CH_4 + 2 O_2 \longrightarrow CO_2 + 2 H_2O$$

$$4 NH_3 + 3 O_2 \longrightarrow 2 N_2 + 6 H_2O$$

$$O_2 + 2 O_2 \longrightarrow 2 O_3, \text{ ozone}$$

$$\text{metals, minerals} + O_2 \longrightarrow \text{ores, rocks}$$

$$\text{organisms} + O_2 \longrightarrow \text{aerobic respiration}$$

Monera and Protista

In parallel with the nutritional evolution of the Premonera an internal structural evolution must have taken place as well. At first the gene-forming DNA molecules probably were suspended free in the cell substance. Occasionally such nucleic acid molecules could have escaped from a cell to the open ocean and by accident could have encountered other cells and entered them. Such transferable nucleic acids conceivably might have been ancestral to the viruses. Modern viruses have similarly transferable nucleic acids, and it is possible that ancestral viruses perhaps were little more than naked nucleic acid molecules. If so, introduction of such molecules from one early cell to another must have had important consequences. For the transferred acids were cellular genes, and as these became shuffled among cells, so did the activities that such genes controlled. Exchanges of nucleic acids thus would have altered the genetic consti-

tution of cells, and this process could have contributed to the evolution of a great variety of new cell types. Certain kinds of modern viruses still transfer genes from one cell to another, a phenomenon known as *transduction*.

Judging from the results today, we know that, of the new structural cell types, two came to have particular significance in later evolution (Fig. 14.7). In one line of descent, the originally freely suspended nucleic acids later must have become joined together as threadlike filaments that formed loose clumps. Such clumps remained embedded in and in direct contact with the rest of the cell substance. The organisms exhibiting this type of internal cellular arrangement collectively form the *Monera* (or *Prokaryota*, organisms preceding the evolution of cell nuclei). They are represented today by the *bacteria* and the *blue-green algae*. As a group they exhibit four of the five methods of nutrition: bacteria are variously photosynthetic, chemosynthetic, parasitic, or saprotrophic; blue-

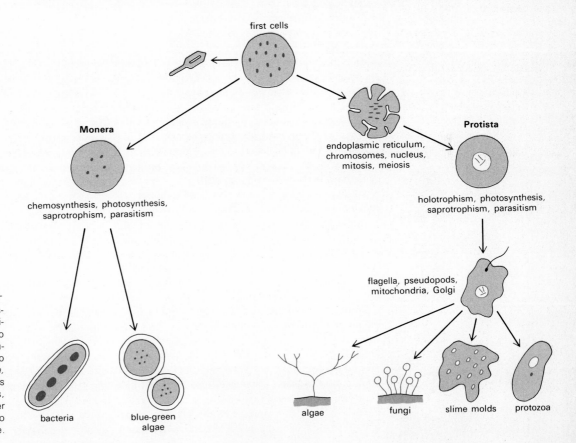

Figure 14.7 Early cell evolution. The earliest cells (with dispersed genetic material, as indicated) probably gave rise to two structural types, one without nuclear membranes and referred to as moneran cells, or *Prokaryota,* the other with nuclear membranes and referred to as protistan cells, or *Eukaryota.* The presumed later evolutionary history of these two cellular stocks is sketched above.

Figure 14.8 Vegetative states in protists. A flagellate cell can become amoeboid, and later that same cell or any of its offspring can revert to a flagellate state. Loss of flagella and development of multinuclearity leads to the nonmotile coccine state, in which, after cell-boundary formation during reproduction, the offspring cells can either remain coccine or resume flagellate or amoeboid existence. Loss of flagella and successive vegetative divisions produce the sporine state. The resulting cells can separate or stay together as colonies, and any of the cells can also assume the flagellate or the amoeboid condition.

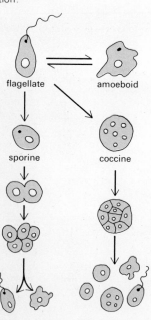

green algae are largely photosynthetic. Evidently, all methods except holotrophic eating occur.

In a second major line of descent, the gene-forming nucleic acids in each cell again condensed as threadlike filaments. However, proteins became an integral structural part of such filaments, and these nucleoprotein organelles represented *chromosomes*. Furthermore, a membrane formed around the chromosomes in a cell, setting off a distinct *nuclear* region from a surrounding *cytoplasmic* region. In the course of probably millions of years then must have evolved all the cytoplasmic organelles we now find in "modern" cells—endoplasmic reticulum, mitochondria, Golgi bodies, centrioles and kinetosomes, various vacuoles, granules, and fibrils, as well as flagella, cilia, and the capacity to form pseudopods. Ribosomes must have been inherited from earlier premoneran ancestors. Moreover, cell division must have become associated with new processes of chromosome duplication, mitosis and meiosis, not encountered as such in the Monera.

Primitive organisms that exhibit such a "modern" cellular organization collectively constitute the *Protista* (or *Eukaryota*, organisms with distinct nuclei). Four groups descended from early Protista represent a major part of the living world today: the *algae* (other than the blue-greens), the *fungi*, the *slime molds*, and the *protozoa*. Collectively they exhibit all nutritional methods except chemosynthesis—photosynthesis, holotrophism, saprotrophism, and parasitism. Indeed, judging from primitive algal types living today, early protists must have had the capacity to obtain food by two or more methods simultaneously. For example, it must have been quite common for a protist to engulf or absorb preexisting food from the external environment *as well as* photosynthesize new food internally. In this respect the ancient protists probably were both plantlike and animal-like at once.

In conjunction with these alternative forms of nutrition and again judging from living types, ancient protists probably also were capable of existing in four alternative vegetative states, two of them locomotor and two nonmotile (Fig. 14.8). The locomotor states are *flagellate* and *amoeboid*, and they are readily interconvertible. An amoeboid state can arise when a flagellate cell casts off its flagellum (but retains the kinetosome) and then moves by means of pseudopodia. If later the amoeboid activity ceases, a new flagellum can grow out.

The two nonmotile states similarly develop by temporary or permanent loss of flagella. One, which we can call the *coccine* condition, is a special type of multinucleate state. The nucleus of a nonmotile cell here continues to divide, but the cytoplasm does not. The result is a progressively more multinucleate but still unicellular organism. Cytoplasmic division occurs only at the time of reproduction, when *multiple fission* takes place: the cell becomes partitioned into numerous offspring cells simultaneously, each containing a nucleus. Such new cells (*spores*) then either become temporarily flagellate or amoeboid or grow into new nonmotile multinucleates directly. In the second nonmotile condition, conveniently called the *sporine* state, a cell does divide regularly during its vegetative life. The resulting nonmotile daughter cells then often remain attached to one another and form a multicellular aggregate. At any time, however, any of the cells can develop flagella or become amoeboid and disengage from the aggregate.

Of these four different states of vegetative existence, the flagellate condition occurs in the most primitive protists known and therefore is believed to be basic; all other states might be derived from it. The adaptive advantage of such multiple alternative states must have been great, particularly in conjunction with the multiple means of nutrition. Thus, by permitting cellular locomotion, the two motile states must have made possible heterotrophic nutrition generally and holotrophic feeding specifically. At night or on the dimly lit bottoms of natural waters, an ancestral protist could actively hunt for food. But in the presence of ample light in the daytime, photosynthesis could occur and locomotor energy could be reduced to zero by assumption of one of the nonmotile states.

Among some modern Protista given individuals still can exist in two or more of these four alternative states, and certain species of unicellular algae can both photosynthesize and feed heterotrophically. However, most living protists exhibit just one particular vegetative state and one method of nutrition more or less permanently. It can be inferred, therefore, that ancestral groups gave rise to modern ones by partial losses of function. Certain ancestral protists could have lost the heterotrophic capacity and thus evolved into more nearly plant-like forms, or they could have lost the photosynthetic capacity and evolved into more nearly animal-like forms. Similarly, different ancestral groups could have become specialized to exist in but one of the four vegetative states and lost the capacity to exist in the others. That such differential losses of function actually can occur is demonstrable in the laboratory. For example, primitive living algae with multiple means of nutrition can be converted by various experimental procedures to either purely photosynthetic or purely heterotrophic organisms. Similarly, experimental removal of, for example, kinetosomes can convert flagellate cells to permanently nonmotile ones.

If we therefore regard specialization in photosynthetic and heterotrophic nutrition as two possible directions of evolution, and if each of these two could at the same time evolve four different vegetative states of existence, then there must have been at least eight possible evolutionary pathways that ancient protists could have followed. Moreover, modern protists include not only unicellular but also multicellular members. Accordingly, we can assume that the available evolutionary paths could be further multiplied by a factor of two, depending on whether the organisms remained unicellular or became multicellular. Thus there must have existed at least 16 different ways in which early protists could have evolved further.

These 16 lines of descent are represented today by the modern Protista (Fig. 14.9). Some unicellular algae now living still display the ancestral traits of being both plantlike and animal-like simultaneously. All other algae are exclusively photosynthetic. They are variously flagellate, amoeboid, coccine, and sporine, and each of these lines include both unicellular and often quite complex multicellular members. The remaining Protista represent the exclusively heterotrophic lines of descent. Fungi exemplify a culmination of the coccine state; all fungi are or become multinucleate organisms that form distinct uninucleate cells only at the time of reproduction. Slime molds exemplify a culmination of the amoeboid state. These largely saprotrophic organisms are creeping multicellular or multinucleate amoeboid masses without permanent shape. Protozoa are holotrophic, parasitic, and in some cases saprotrophic, and the cells are mostly flagellate or amoeboid, though a few groups are coccine. The multicellular condition has developed to only a limited degree, but protozoa appear to have exploited the unicellular way of life perhaps more fully than any other group of protists.

Most of the modern protists, unicellular or multicellular, that remain in one vegetative state

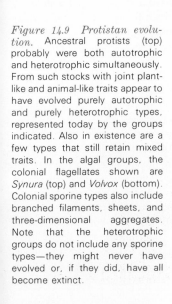

Figure 14.9 Protistan evolution. Ancestral protists (top) probably were both autotrophic and heterotrophic simultaneously. From such stocks with joint plantlike and animal-like traits appear to have evolved purely autotrophic and purely heterotrophic types, represented today by the groups indicated. Also in existence are a few types that still retain mixed traits. In the algal groups, the colonial flagellates shown are *Synura* (top) and *Volvox* (bottom). Colonial sporine types also include branched filaments, sheets, and three-dimensional aggregates. Note that the heterotrophic groups do not include any sporine types—they might never have evolved or, if they did, have all become extinct.

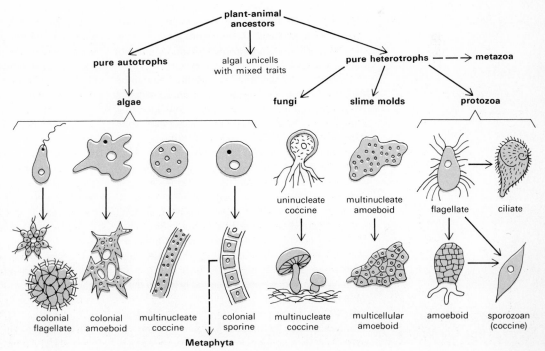

permanently still reveal at the time of reproduction the ancestral potential of developing other states. Thus even where the organisms are nonmotile, the reproductive cells generally are flagellate or amoeboid. Indeed, cells of both kinds often are formed at the same time; flagellate cells become sperms, amoeboid cells become eggs.

Monera and Protista probably were the only organisms in existence for long ages. Eventually, at certain points during their history, the Protista also gave rise to the two largest groups of organisms now in existence, the *Metaphyta* and the *Metazoa*. Metaphyta are green plants, moss plants (*bryophytes*) and vascular plants (*tracheophytes*). Metazoa are the multicellular animals. In both groups the cell structure is typically protistan, and flagellate, amoeboid, and sporine states of cellular existence are common. Thus the protistan origin of the two groups is not in doubt.

But the two groups differ from Protista in important respects: the body is always multicellular and has an organ or organ-system level of construction; the reproductive structures are always multicellular, either tissues or organs; oögamy is universal, the gametes always being distinct sperms and eggs; and development always passes through distinct embryonic stages. The two groups differ from each other in that Metaphyta do not have larval phases whereas Metazoa typically do. Moreover, in Metaphyta all nutritional methods except photosynthesis have been lost, and in Metazoa all but holotrophic (and parasitic) methods have been lost.

The probable ancestry of the Metaphyta can be pinpointed comparatively well. The group appears to have evolved from early sporine green algae that also gave rise to the complex sporine green algae of today (See Fig. 14.9). The tissue-level construction of these algae is repeated in the Metaphyta and is augmented by the organ and organ-system levels. Moreover, such algae and all metaphytes have identical varieties of chlorophyll and other pigments, and their biochemical traits as a whole are virtually identical. Metaphyta did not arise until about 350 million years ago, some 150 million years after the beginning of ample fossil records. As it happens, the early history of the metaphytes is documented reasonably well in these records.

By contrast, metazoan beginnings must be ascertained without fossil evidence and without the aid of built-in biochemical clues such as chloro-

phyll. Thus, although it is reasonably certain that Metazoa originated much earlier than Metaphyta and that they evolved from Protista, it is at present quite impossible to decide which of the protistan groups actually might have been ancestral. Numerous hypotheses have been proposed, and the chief ones are examined in the next section.

Origins of Animals

The First Animals

Protozoa are regarded traditionally as "first animals" and ancestors of the Metazoa. However, before any group can be labeled as animal, first or otherwise, it is of some importance to be clear about the intended meaning of such a label.

As noted in Chap. 7, Protozoa and Metazoa form the traditional Linnaean *animal kingdom*. All other living groups—Metaphyta, Protista other than protozoa, and Monera—form the traditional *plant kingdom*.

In such a Linnaean sense, plants and animals are completely undefinable by biological criteria. Virtually every trait usually regarded as characteristic of one kingdom occurs also in the other. Plants are defined traditionally as organisms that photosynthesize and do not exhibit locomotion, and animals as organisms that exhibit heterotrophic nutrition as well as locomotion. On such a basis, however, many algae and all fungi and slime molds would have to be animals, for all these organisms are nongreen heterotrophs and a good many of them exhibit locomotion as well. Yet sponges, corals, barnacles, tunicates, and other groups could not be strictly included among animals since the adult forms are attached and without locomotion. Moreover, certain primitive algal types would be both plant and animal at once, since, as pointed out above, such organisms can be alternately photosynthetic and nonlocomotor or heterotrophic and locomotor.

Traits other than nutrition or locomotion fail similarly as distinguishing criteria. If animals are defined as organisms with nervous and muscular structures, or at least equivalent components, then many algae would have to be animals; and sponges, which lack such structures, would have to be plants. If plants are defined as organisms with cellulose, then tunicates would have to be plants and a number of algal phyla without cellulose would have to be animals. Actually there does not appear to be a single characteristic that would distinguish the traditional plant and animal kingdoms uniquely.

In effect, some organisms fit in neither the plant nor the animal category and several other groups fit in both. Traditional views notwithstanding, for example, bacteria really have very little in common with either plants or animals, and quite a number of unicellular algae and other types can be regarded equally well as plants *or* animals. To be sure, no one has much difficulty in deciding whether advanced organisms like cabbages and cats are plants or animals. But such a difficulty does exist with many organisms now known to be primitive—those closely related to the ancestral types that gave rise to both cabbages and cats. As shown earlier such ancestral types exhibited both plantlike and animal-like traits *simultaneously,* as is still true of some of their primitive descendants today. And if we go even farther back in time, the very first organisms on earth appear to have had neither plantlike nor animal-like traits at all.

The point is that plants and animals, clearly so recognizable, were not in existence right from the beginning. Instead, plant and animal traits became separated out *gradually,* from a protistan base in which both types of trait were present jointly. Therefore, the Linnaean division of the living world merely into plant and animal kingdoms is too arbitrary and too simple. It does not take into account the slow *evolution* of distinct plant and animal groups, and it allows no place for those primitive organisms that even now are neither "plant" nor "animal" or that are both.

The difficulty disappears if the living world is classified into *four* largest categories—Monera, Protista, Metaphyta, and Metazoa—and if the term "plant" is restricted to Metaphyta, the term "animal" to Metazoa. These four groups *are* definable sharply, and they give every living creature a proper place. On such a basis, protozoa then are not "animals" but protists; and the *first* animals would have been the first Metazoa. Throughout this book, any unqualified reference to animals is restricted to mean Metazoa.

But regardless of whether protozoa are called animals or protists, they are still considered traditionally as the ancestors of Metazoa. Most zoologists agree that Metazoa probably are not monophyletic: sponges are believed to represent one separate line of descent, and Eumetazoa, one or more others. Most zoologists also agree that sponges probably did evolve from protozoa, as complex flagellate colonies just barely at a tissue level of construction (see Chap. 20). Thus the question now remains as to whether or not protozoa are the ancestors of the Eumetazoa. Most traditional hypotheses give an affirmative answer.

In one group of such hypotheses, some of them rather recent, the eumetazoan ancestors are identified as early *ciliate* protozoa. Modern ciliates are multinucleate, and some of their ancient representatives are postulated to have become multicellular by development of internal cell boundaries, in the manner of coccine algae and fungi. The first Eumetazoa so formed are argued to have been bilateral *acoel* flatworms, a group now quite generally believed to represent the most primitive Bilateria (Fig. 14.10).

The "ciliate-flatworm hypothesis" thus rests on two assumptions, that ciliate protozoa became Eumetazoa by *septation,* or internal partitioning, and that these first Eumetazoa were acoel flatworms. The septational part of this view is totally

Figure 14.10 Hypotheses of eumetazoan origins. A, the ciliate-flatworm hypothesis. *B,* the flagellate-planuloid hypothesis. In *A,* multicellularity is assumed to have arisen by septation of a ciliate into an acoel; in *B,* by aggregation of flagellates as a blastulalike form. In *A* bilaterality is assumed to be basic via the acoels, and radiates are taken to be secondarily radial. In *B,* bilaterality is assumed to have arisen during the evolution of radial planuloids into acoels, which thus would make the radiates primitively radial. See also Fig. 14.15.

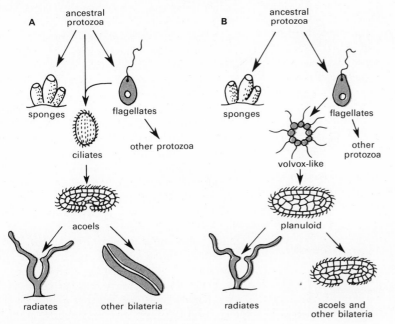

without evidence, even of a suggestive kind. Indeed, only exceedingly few organisms develop a multicellular body by septation. In the vast majority of coccine algae and fungi, the cells formed by internal partitioning are reproductive cells (spores) that do not stay together but disperse geographically; septation here is primarily an adaptive device to make possible just such dispersal, not to produce multicellular organisms. And in the few cases where the cells do stay together and form coccine colonies, the level of structural complexity attained always remains exceedingly low; coccine types of any known kind evidently lack the evolutionary potential to reach even the level of tissues. But an acoel or any other eumetazoan does have tissues. Because of this complete lack of support from living forms, septational views have found comparatively few adherents. The same also holds

for the acoel part of the hypothesis, for reasons to be outlined below.

More widely in vogue are hypotheses that postulate ancestral *flagellate* protozoa as eumetazoan ancestors (see Fig. 14.10). In support of such views it is pointed out that all Metazoa still develop flagellate gametes and numerous other cell types that are flagellate or, by derivation, amoeboid; and that flagellate-amoeboid protists, protozoa included, have a known evolutionary capacity to become multicellular by colony formation, or *aggregation*, just as Metazoa become multicellular when an egg transforms to a "colony" of embryonic cells. Indeed, in these hypotheses it is generally postulated that the first Eumetazoa might have resembled either hollow flagellate colonies such as the algal *Volvox* or solid flagellate colonies such as the algal *Synura*. Such colonies in turn would have resembled the *planula* larvae of cnidarians, and it is generally held today that cnidarians actually were the first Eumetazoa.

This "cnidarian hypothesis" thus consists of two parts, too, the flagellate origin and the planula-like, or *planuloid*, result. The first part is subject to an objection similar to the one raised above. Judging from living Protista, flagellate-amoeboid organisms have exhibited very limited evolutionary resourcefulness only; like the coccine forms, none of these types has attained a tissue level of complexity. Protozoan flagellates very largely have remained unicellular, and among algal flagellates the most complex structures actually evolved have been types such as *Volvox* and *Synura*. It is therefore unlikely that flagellate ancestors could have evolved much beyond a colony level.

Besides, in a eumetazoan body the vast majority of cells is neither flagellate nor amoeboid or even coccine; the bulk is *sporine*—nonmotile cells that divide vegetatively and stick together. And sporine protists also have a proved, clearcut capacity to evolve well beyond the colony level—complex tissue-level algae and all the even more complex plants are among their known derivatives. Accordingly, it is suggested here that Eumetazoa originated from sporine ancestors. It so happens that sporine types do not occur among protozoa or any other heterotrophic protists now known. Therefore, sporine ancestors of Eumetazoa would have had to be "preprotozoan," evolved not *from* presently known protozoa or other colorless protists, but *in parallel* with them as a separate group (Fig. 14.11).

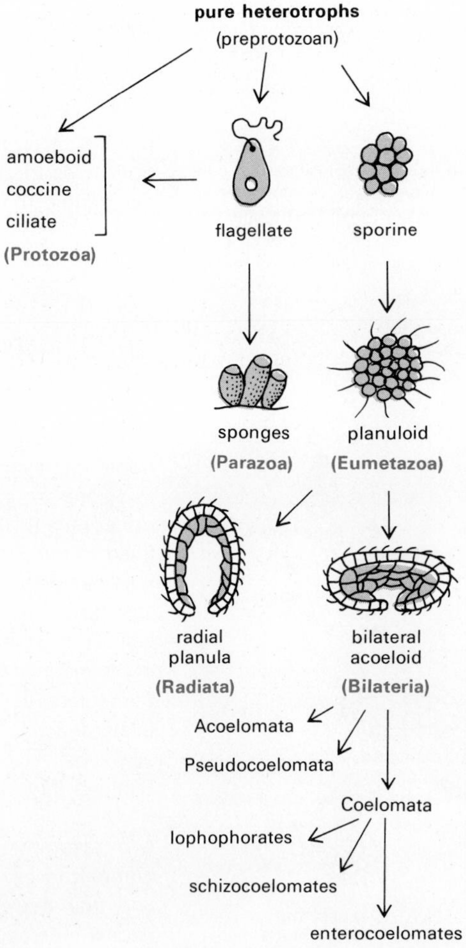

Figure 14.11 A hypothesis of eumetazoan origins. The first eumetazoan here is postulated to have been a morulalike, spherical, sporine form. It is assumed to have evolved not after, but in parallel with, the heterotrophic protists that gave rise to the Protozoa and presumably also the Parazoa. The sporine planuloid type then is envisaged to have produced the Radiata, which were medusalike floaters primitively and were basically radial in adaptation to this mode of life; and the Bilateria, which adopted a creeping mode of life and developed bilaterality in conjunction.

Such ancestors could have given rise to sporine, *nonmotile* colonies that floated passively in shallow water and perhaps settled to the bottom occasionally. Their nonphotosynthetic cells would not have required direct exposure to sunlight and thus need not have been organized as hollow spheres or filaments or sheets, as in sporine algal colonies. Instead the colonies could have been solid balls of cells, resembling the early cleavage stages of many eumetazoan embryos today. Nutrition could have been saprotrophic or holotrophic or both, and digestion could have been carried out intracellularly by all cells. Some or all of the exterior cells sooner or later could have redeveloped flagella, just as cells of sporine algae or of animal embryos can become flagellate. In this way the exterior cells could have assumed locomotor functions, and alimentary and reproductive functions could have become the specialized responsibility of the interior cells.

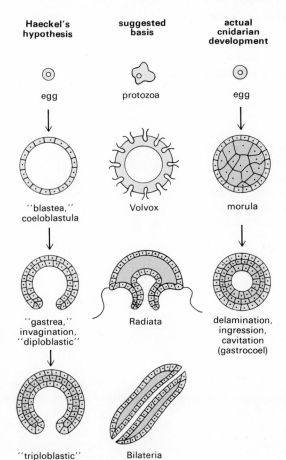

Haeckel's hypothesis	suggested basis	actual cnidarian development
egg	protozoa	egg
"blastea," coeloblastula	Volvox	morula
"gastrea," invagination, "diploblastic"	Radiata	delamination, ingression, cavitation (gastrocoel)
"triploblastic"	Bilateria	

Figure 14.12 Haeckel's "gastrea" hypothesis. Based primarily on cnidarian development, the hypothesis is at variance with the actual developmental patterns typical of most cnidarians (as at right). Thus, most cnidarians have solid, not hollow, blastulas; most do not gastrulate by invagination but by a variety of other processes, as indicated. Moreover, again contrary to Haeckel's hypothesis, the presence of mesoderm in cnidarians makes not only the Bilateria but also the Radiata (hence all Eumetazoa) "triploblastic," or formed from three primary germ layers.

In the end, therefore, the first Eumetazoa again would have been exteriorly flagellate balls of cells, but the original evolutionary starting point would have been a sporine, not flagellate, state. The difference is important, for, as noted, sporine types do and flagellate types do not have a known potential to develop beyond the colony level. Moreover, a sporine origin would make eumetazoan history begin far back in protistan history, at a time when protists still had most of their evolution before them. Eumetazoa therefore could be derived from generalized, primitive ancestors, as is proper, not from already specialized ancestors such as protozoa. On this basis Eumetazoa and Protozoa would have evolved not one after the other but in parallel, as separate lines of an adaptive radiation that started from early colorless protists. And if protozoa later gave rise to sponges, these would thus not have come into existence until *after* eumetazoan history was already well launched. In short, Eumetazoa well might be older than, or at least just as old as, protozoa and sponges.

The exact starting point of the Eumetazoa apart, most current hypotheses do envisage the first of these animals to have been multicellular spheres with exterior locomotor cells. The view that such balls were hollow, like *Volvox,* is now less prevalent than the view that they were solid; and the belief that they represented cnidarian planuloids is accepted widely. These hollow-colony and cnidarian concepts received their strongest original impetus from a famous hypothesis proposed by Ernst Haeckel, a German zoologist of the late nineteenth century. Most of his views now are largely discredited, but they were once so influential that many of them still persist today under various guises. For this reason it may be of some value to review them briefly.

Recapitulation versus Divergence

Haeckel recognized, as did others before him, that animal development typically passes through certain common embryonic stages—zygote, blastula, gastrula, mesoderm formation. Haeckel thought that this succession of embryonic stages mirrored a succession of past evolutionary stages (Fig. 14.12). Thus the zygote would represent the unicellular protistan stage of evolution. The blastula would correspond to an evolutionary stage when animals were, according to Haeckel, hollow one-layered spheres. Haeckel coined the term "blastea'' for such hypothetical adult animals, and he thought that his ancestral blasteas may have been quite similar to currently living algae such as *Volvox*. The gastrula stage would correspond to a hypothetical ancestral adult type that Haeckel called a ''gastrea.'' He believed that gastreas were still represented by the living cnidarians, some of which do resemble early gastrulas in certain respects—in a number of cases the body consists of two layers and the single alimentary opening is reminiscent of the blastopore of a gastrula.

On such grounds Haeckel considered the gastrea to have been the common ancestor of all Eumetazoa. He assumed the two-layered gastrealike condition to represent a *diploblastic* stage in animal evolution, attained by radiates such as the cnidarians. Further evolution then added mesoderm to the gastrealike radiates and so produced a three-layered *triploblastic* condition, as in flatworms and all other Bilateria. By extension the hypothesis also implied that, for example, a caterpillar larva represented an annelid stage in insect evolution, that a frog tadpole represented a fish stage in frog evolution, and that a human embryo, which exhibits rudimentary gill structures at certain periods, represented a fish stage in man's evolution.

Haeckel condensed his views into a ''law of recapitulation,'' the essence of which is described by two phrases: ''ontogeny recapitulates phylogeny'' and ''phylogeny causes ontogeny.'' The first statement means that the embryonic development of an egg (ontogeny) repeats the evolutionary development of the phyla (phylogeny); the second, that *because* animals have evolved one phylum after another, their embryos still pass through this same succession of evolutionary stages.

Therefore, if one wishes to determine the course of animal evolution, he need only study the

course of embryonic development, for, according to Haeckel, evolution occurs by addition of extra embryonic stages to the end of a given sequence of development. If to a protozoon is added cleavage, the protozoon becomes a zygote and the new adult is a blastea. If to a blastea is added gastrulation, the blastea becomes a blastula and the new adult is a gastrea. Similarly, if to a fish are added lungs and four legs, the fish represents a tadpole and the new adult is a frog. And if to such an amphibian are added a four-chambered heart, a diaphragm, a larger brain, an upright posture, and a few other features, then the frog is a human embryo and the new adult is a man.

We must attribute to the lingering influence of Haeckel, not to Darwin, this erroneous idea of an evolutionary ''ladder,'' or ''scale,'' proceeding from ''simple amoeba'' to ''complex man,'' with more and more rungs added on top of the ladder as time proceeds. All such notions are invalid because Haeckel's basic thesis is invalid. Indeed, Haeckel's arguments were shown to be unsound even in his own day, but his generalizations were so neat and they seemed to explain so much so simply that the fundamental difficulties were ignored by many.

For example, it was already well known in Haeckel's time that cnidarians, which should develop most nearly like the postulated gastrea, in actuality develop quite differently (see Fig. 14.12). In most cases the blastulas are not hollow but solid, and gastrulation occurs only very rarely by invagination but most often by delamination and ingression. Furthermore, apart from exceptional forms like the hydras, the radiate animals do not really have two-layered bodies but distinctly three-layered ones, with a mesoderm often highly developed (as in sea anemones, for example). Two-layered animals in effect do not exist, and a distinction between diploblastic and triploblastic types cannot be justified. Thus the conceptual foundation on which the recapitulatory law was based was never valid.

Moreover, new types are not known to evolve by addition of extra stages to ancestral adults. Instead, new evolution occurs for the most part through *developmental divergence;* a new path of embryonic or larval development branches away from some point along a preexisting ancestral path of development (Fig. 14.13). The best example is evolution by larval *neoteny*, a common process by which numerous new groups are believed to

Figure 14.13 Developmental divergence. Top, the pattern of evolution according to Haeckel, by addition of extra stages to a preexisting ancestral path of development. *Bottom*, the actual pattern of evolution, by divergence of new developmental paths (2–5) from various points of a preexisting ancestral path (1). Thus if divergence occurs comparatively late, as in adult 2, its embryonic and most of its larval development can be very similar to that of adult I. But if divergence takes place early, as in adult 5, then almost the whole developmental pattern will be quite dissimilar from that of adult I. Adults 1 and 5 then might represent different phyla, whereas adults 1 and 2 might belong to the same class or order.

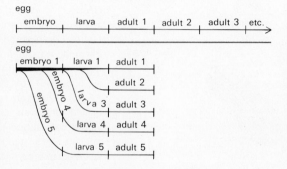

have arisen from the *larvae* of ancestral groups. An ancestral larva here does not metamorphose into the customary adult, but instead develops sex organs precociously and becomes established in this larval form as a new type of adult animal.

As pointed out in Chap. 11, for example, axolotls are neotenous amphibia. Similarly, it is now considered most probable that vertebrates represent neotenous tunicate tadpoles. The tadpoles of certain tunicate ancestors are believed to have retained their tails, with notochord and dorsal hollow nerve cord intact, and to have become reproductively mature in such a larval condition. These permanent larvae then came to represent a new chordate subphylum, the vertebrates (see also Chap. 29).

Numerous other instances of evolution by neoteny are known and as will appear in the next chapter man, too, is a neotenous product. In all these cases the new developmental path branches away sooner or later along the course of the old path. The sooner two such paths do diverge, the more dissimilar will be the two types of resulting adult and the greater will be the taxonomic gulf between the types. For example, the embryos of man and of monkeys resemble each other till relatively late in development, and the developmental paths diverge only then (and lead to superfamily differences). The embryos of man and of tunicates are similar for considerably shorter periods; their developmental paths correspondingly diverge much sooner (and result in subphylum differences Fig. 14.14).

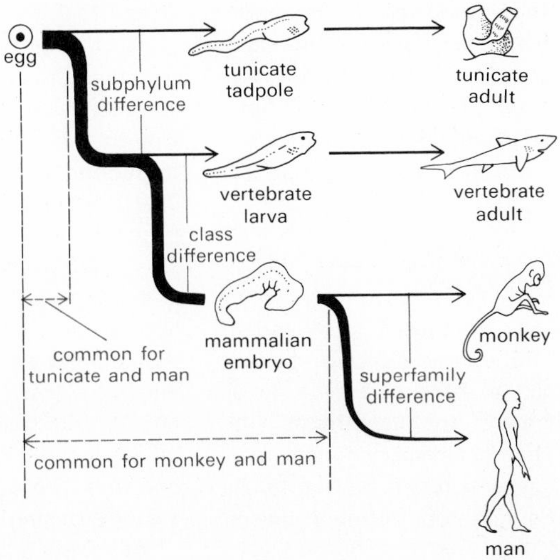

Figure 14.14 Developmental divergence in chordate evolution. Tailed, tadpolelike early stages are characteristic of all chordates, and the earlier the developmental paths diverge from the ancestral one, the more different will be the adults and the taxonomic results. Thus the developmental paths of tunicates and men diverge early, and the adult differences are at the level of subphyla. But monkeys and men develop similarly for far longer periods, and the adult differences here are only at the level of superfamilies. To be sure, chordate eggs are not all the same but are merely similar. Also, "common" developmental stages of two animals are merely similar, not identical.

Such developmental correlations were clearly recognized before Haeckel and even before Darwin. They do have evolutionary meaning, but not in the Haeckelian sense. It is quite natural that related animals descended from a common ancestor should resemble each other in some of their developmental features. Thus, human embryos resemble those of fish and frogs in certain respects not because an egg of man becomes a fish embryo first, a frog embryo next, and a human embryo last. The similarities arise, rather, from the common ancestry of all these three animal types, including their common developmental histories up to certain stages. Beyond these stages each type has modified its developmental processes in its own way.

Moreover, it can be shown that the formation of a particular structure in one type of animal often depends on the existence of an anatomic precursor that evolved in an ancestral type. Some of the evolutionary opportunisms discussed in Chap. 13 illustrate this generalization. As noted, for example, the human larynx has evolved opportunistically as a derivative of skeletal gill supports of ancestral fishes. An important point here is that, from an architectural standpoint, a larynx cannot be put together unless gill-derived skeletal components are available in the human embryo as structural raw materials. In other words, human embryos today still form rudimentary, nonfunctioning gills, not because they "recapitulate" fish evolution as Haeckel supposed, but because gill rudiments must serve as necessary intermediate steps in the building process that leads to larynx construction. Similarly, gill rudiments still are required in human development as intermediate structural steps in the elaboration of the face musculature. In like fashion also, most other supposed instances of "recapitulation" turn out to involve precursor stages still necessary today for the development of other structures.

It should be pointed out, furthermore, that a branching pattern of developmental divergence is fully in line with the known bushlike pattern of evolution, whereas ladderlike end addition is not. It is quite obvious also that mammalian development, for example, does not really represent a successive transformation of an actual protozoan to an actual cnidarian, flatworm, tunicate, fish, etc. Indeed it is hardly conceivable that the billion or more years of animal evolution could be crowded into the few weeks or months of animal development. The common stages in animal development thus give evidence of general similarities only, not of actual identities.

For these various reasons the Haeckelian idea of recapitulation is not tenable; the embryonic stages of animals do not repeat the adult stages of earlier animals. Moreover, phylum evolution also cannot be the "cause" of the progressive stages in animal development. If anything just the reverse probably holds; as pointed out above, developmental stages provide the sources from which new groups can evolve.

What now remains of Haeckel's notions are three ideas. First, the three primary germ layers are probably homologous in all Eumetazoa. Often called the "germ-layer theory," this generalization did not originate with Haeckel but he certainly popularized it. The theory is acceptable today provided an analogy of the layers is not implied. Thus, although the layers of different Eumetazoa are equivalent developmentally and structurally, they do not always serve equivalent functions.

Second, similarities among paths of development do exist, but not for recapitulary reasons. Instead, the development of two individuals or groups on the whole does tend to be similar if the animals are related, and the similarity lasts the longer the closer the relationship. Indeed, as has become apparent in Chap. 7, developmental resemblances are important aids in both classifying animals and elucidating phylogenies—and this insight, too, predates Haeckel.

Third, cnidarians can be hypothesized to be the most ancient Eumetazoa that still have living representatives today. The majority of zoologists accept this view, but they do so largely on a non-Haeckelian or neo-Haeckelian basis. As outlined, Eumetazoa now are usually envisaged to have evolved not from hollow "blasteas" and "diploblastic gastreas," but from solid, planuloid cell colonies that formed three-layered cnidarian animals (see also Chap. 21).

The Later Animals

All major phylogenetic hypotheses today regard the Eumetazoa as a monophyletic group; the first members are assumed to have given rise to all others. However, none of the hypotheses succeeds in producing a fully satisfactory genealogical "tree," and this failure well might be due to the basic assumption of eumetazoan monophyletism.

According to the most common current form of the flagellate-cnidarian hypothesis, the planula-like first Eumetazoa are postulated to have given rise to two groups, the modern radiate phyla (cnidarians and ctenophores) and the flatworms (Fig. 14.15). A planuloid would have become a cnidarian by a hollowing out of the interior into a simple alimentary cavity, and by the development of a single opening to and from this cavity in the blastopore region. Such processes actually occur in the embryos of most modern radiates. Also, a free-swimming, radially symmetric planuloid could have evolved to a flatworm by adopting a creeping mode of existence and at the same time becoming bilateral and flattened down. In such a change of symmetry, the blastopore mouth would have remained at a midventral location. The result would have resembled the *acoels*, the most primitive flatworms now living, and the postulated first Bilateria accordingly are referred to as *acoeloids*. In acoel development a change of symmetry and flattening down actually do take place. Acoeloids then would have given rise to all flatworms and all other Bilateria.

According to the ciliate-flatworm hypothesis, the above events are postulated to have taken place in the reverse order: acoeloids (derived from ciliates) are assumed to have given rise to flatworms and planuloids, the latter then leading to cnidarians. On this basis Eumetazoa would have been bilateral originally and the radial symmetry of cnidarians would be a derived condition. Proponents of this view also point out that cnidarians such as sea anemones actually are bilateral and therefore represent the most primitive radiates (see Chap. 21).

On the contrary, proponents of the cnidarian hypothesis maintain that the fully radial cnidarians are primitive and that those with superficial bilaterality are the more advanced members of the group. In support of this view it can be argued that if bilaterality really were primitive, then the embryos of cnidarians and certainly of acoel flatworms should show signs of this symmetry. Yet all these embryos and all cnidarian larvae are radial (acoels lack larvae). Most zoologists accept the postulate that radiality evolved to bilaterality and that radiate ancestors gave rise to flatworms. A third and equally plausible possibility is that early radial cnidarians and early bilateral flatworms arose not one from the other but each independently from earlier protistan ancestors (see Fig. 14.15).

Although opinion regarding the origin of the

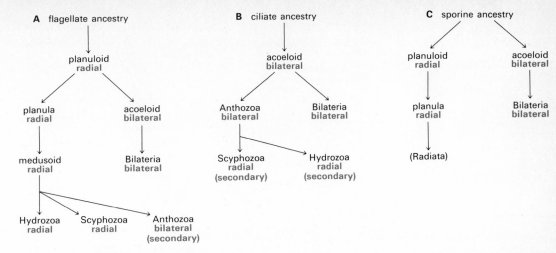

Figure 14.15 Radiate and bilaterial origins. A, according to the "flagellate" hypothesis, radial planuloids are postulated to have given rise to both the Radiata and the Bilateria. Among the Radiata, Hydrozoa then would be the most primitive cnidarians and Anthozoa the most advanced (with secondary bilaterality in some of their features). *B,* according to the "ciliate" hypothesis, bilateral acoeloids are postulated to have given rise to both Bilateria and Radiata. In this case the Anthozoa would be the most primitive cnidarians and the Hydrozoa the most advanced (their radiality then being regarded as secondary). *C,* a third possibility is that radial and bilaterial animals have evolved not one from the other, but independently and in parallel from sporine ancestors. Here, too, Hydrozoa would be the most primitive cnidarians, their radiality being an adaptation to a free-floating way of life. The bilaterial groups, by contrast, would have been bottom creepers originally, with bilateral symmetry most adaptive for that way of life.

first Bilateria thus is divided, it is agreed quite widely that acoelomates produced all other Bilateria. A first basic step would have transformed the near-microscopic acoelomate animals that swam free by means of cilia to types that were larger, bottom-creeping forms. Creeping would have been accomplished initially by the cilia, but later a stronger musculature in the larger animals would have assisted in motion and eventually would have become the main means of motion. The probable next major step was a radical departure from the acoelomate construction: evolution of a body cavity and, as a result, a change from creeping to burrowing ways of life, as outlined in Chap. 7 (Fig. 14.16).

There can be little doubt that body cavities evolved several times—some were pseudocoels, some were true coeloms. Notwithstanding numerous hypotheses on this subject to the contrary, it is impossible at present to decide whether pseudocoelomate or coelomate animals are derived from each other (and in which sequence), or whether or not each group evolved independently from acoelomate ancestors. It is generally agreed that the Pseudocoelomata today are a highly poly-

phyletic assemblage. Coelomata by contrast are widely believed to be monophyletic; yet forcing arguments for this view actually have not been made and, indeed, none of the existing hypotheses on later coelomate evolution is entirely satisfactory. In all likelihood the Coelomata, too, well might be polyphyletic—lophophorates could represent one main independent group, schizocoelomates another, and enterocoelomates a third.

A number of traits are shared in common by pseudocoelomates, lophophorates, and schizocoelomates. All three groups have predominantly mosaic eggs, cleavage is mostly spiral or modified spiral, and the animals are protostomial, the mouth forming at or near the blastopore. Moreover, the adults have broadly common features such as ventral ladder-type nervous systems, protonephridial and metanephridial excretory systems, and main blood vessels or hearts in dorsal locations. However, it is difficult to decide whether any of such correspondences indicate actual evolutionary interrelations or are simply similarities that result from a particular grade of construction and thus lack historical implications. Among coelomates, the lophophorates appear to be the most

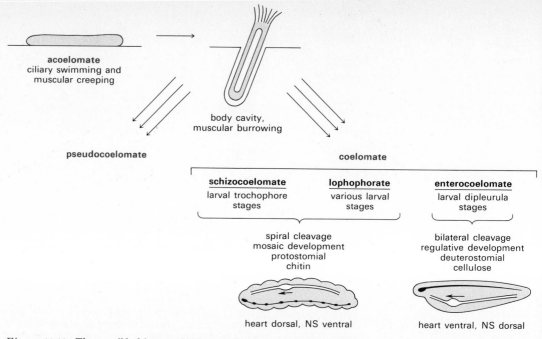

acoelomate
ciliary swimming and
muscular creeping

body cavity,
muscular burrowing

pseudocoelomate

coelomate

schizocoelomate	**lophophorate**	**enterocoelomate**
larval trochophore stages	various larval stages	larval dipleurula stages

spiral cleavage
mosaic development
protostomial
chitin

bilateral cleavage
regulative development
deuterostomial
cellulose

heart dorsal, NS ventral

heart ventral, NS dorsal

Figure 14.16 The possible history of Bilateria. The evolution of body cavities would have led to a transition from creeping acoelomate types to burrowing pseudocoelomate and coelomate types. Coelomates in turn would have evolved as at least three major groups, with different patterns of embryonic development (including methods of coelom formation). In each of these three, some later subgroups also would have acquired new capabilities of locomotion and modes of living, based on, for example, evolution of hard skeletons (see Fig. 7.15). Note that the most characteristic coating of the body surface is chitin in lophophorates and schizocoelomates, but cellulose (tunicin) in most enterocoelomates. Also note that the polarity of the body differs diametrically in schizocoelomates and enterocoelomates.

primitive animals now living. They also suggest possible distant affinities with the schizocoelomates through their protostomial nature, as just noted, and with the primitive enterocoelomates through their anterior and posterior coelomic divisions and their filter-feeding adult organization. Conceivably, therefore, the three living coelomate groups might have evolved from ancestors that themselves had various traits in common.

There is little doubt that the phyla within the schizocoelomates are related quite closely. They share not only the embryonic, protostomial traits already mentioned, but also a life cycle that typically includes a common *trochophore* larva. Early schizocoelomates appear to have been unsegmented, and among their descendants were groups that exhibited a conspicuous metameric segmentation. Similarly, with one probable exception, the enterocoelomate phyla appear to form a fairly closely related assemblage. They typically develop from regulative eggs by bilateral cleavage;

the animals are deuterostomial, the mouth here forming opposite the blastopore; adult mesoderm typically arises as enterocoelic pouches from the endoderm; and (usually) three paired coelomic subdivisions form in the embryos. Most of these groups also pass through a so-called *dipleurula* stage during later development, and in the adults the nervous system tends to be dorsal, the heart ventral. The animals thus exhibit a reversed antero-posterior polarity as compared with the other coelomates (see Fig. 14.16). This circumstance in particular makes it difficult to derive entero-coelomates from other known coelomates, and it reinforces the possibility that enterocoelomates, like lophophorates and schizocoelomates, might have evolved independently from separate coelomate ancestors. The enterocoelomates too produced later members that became segmented.

On this general basis, a comprehensive phylogenetic scheme such as outlined in Fig. 14.17 appears to be most consistent with the very in-

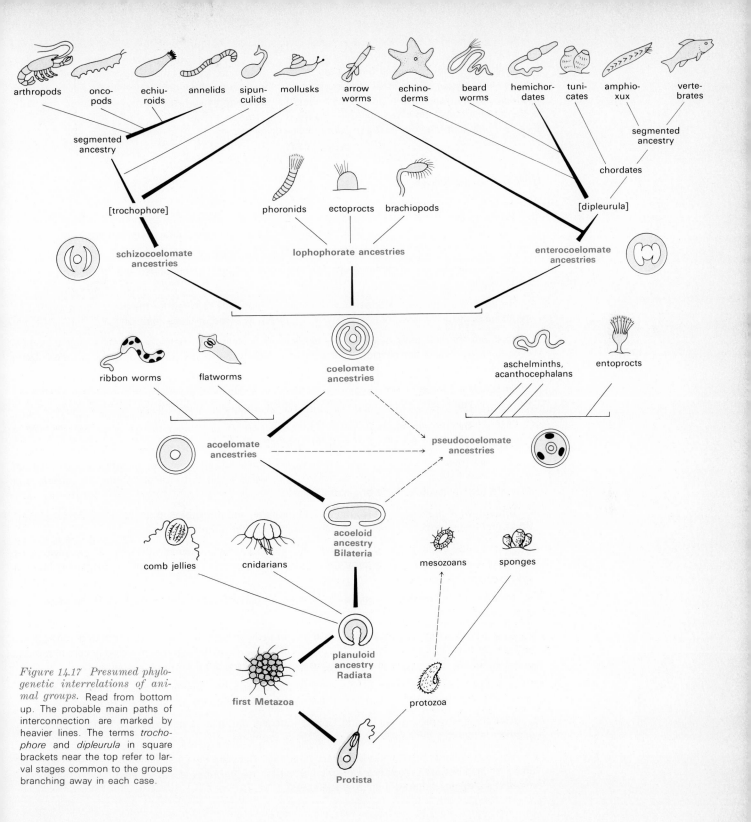

arthropods   onco-pods   echiu-roids   annelids   sipun-culids   mollusks   arrow worms   echino-derms   beard worms   hemichor-dates   tuni-cates   amphio-xux   verte-brates

segmented ancestry

segmented ancestry

chordates

[trochophore]

phoronids   ectoprocts   brachiopods

[dipleurula]

schizocoelomate ancestries

lophophorate ancestries

enterocoelomate ancestries

ribbon worms   flatworms

coelomate ancestries

aschelminths, acanthocephalans   entoprocts

acoelomate ancestries

pseudocoelomate ancestries

acoeloid ancestry Bilateria

comb jellies   cnidarians

mesozoans   sponges

planuloid ancestry Radiata

first Metazoa

protozoa

Protista

Figure 14.17 Presumed phylogenetic interrelations of animal groups. Read from bottom up. The probable main paths of interconnection are marked by heavier lines. The terms *trochophore* and *dipleurula* in square brackets near the top refer to larval stages common to the groups branching away in each case.

complete evolutionary evidence now available. In many instances the interrelations between particular phyla can be deduced much better than those between larger groups, and most of these phylum interrelations are discussed in Unit 2. With regard to the larger groups, however, for the present it is certain only that they did evolve in *some* phylo- genetic pattern, and that these events occurred from roughly 3 or 2 billion to about 500 million years ago. By the latter date all phyla were already well established, and from then on our knowledge of evolutionary histories moves to surer ground by virtue of fossils.

Review Questions

1. Which gases in the early atmosphere of the earth appear to have contributed to the formation of biologically important compounds? What were some of these compounds and what evidence do we have that they could actually have formed?

2. Review the role of (*a*) temperature, (*b*) water, (*c*) organic compounds, and (*d*) enzymes in the origin of life.

3. Review the synthesis reactions through which compounds required for the origin of living systems might have occurred. Describe experiments duplicating some of these reactions.

4. How are the first cells believed to have evolved? Distinguish between chemical and biological evolution. What was the physical character of the earth (*a*) at the time it formed, (*b*) before living systems originated, and (*c*) after living systems originated?

5. Review the events of the oxygen revolution and describe the consequences of this revolution. How many years ago did (*a*) the earth form, (*b*) life originate?

6. Distinguish between monophyletism and polyphyletism as related to the origin of cells. What arguments have been advanced in support of either view?

7. How are moneran and protistan cell types distinguished, and how could both have arisen from the first cells? What living groups belong to the Monera and the Protista? What are viruses, and how are they believed to have originated?

8. Through what processes of evolution could moneran and protistan nutritional patterns have arisen? What processes probably necessitated and promoted such nutritional evolution?

9. Describe the basic events of, and the differences between, chemosynthesis and photosynthesis. What are the basic forms of heterotrophic nutrition and which of these forms occur among Monera and Protista?

10. What states of existence could early protists probably exhibit? Define each of these states and give specific examples.

11. What were the probable methods of nutrition of early protists? In view of the nutritional possibilities and the various states of existence, how many different evolutionary directions could ancestral protists probably follow? Show how the diversity of existing protists supports such assumptions.

12. Which protistan group probably gave rise to the Metaphyta? What is the basis for such an assumption? How are Metaphyta and Metazoa distinguished from each other and from Protista?

13. Why are the terms ''animal'' and ''plant'' inadequate in a strict taxonomic sense? Show how different taxonomic definitions could make these terms precise.

14. Review traditional hypotheses of metazoan origins. Describe the ciliate-flatworm and the flagellate-cnidarian hypotheses of eumetazoan origins. What are the weaknesses of each? What alternative hypotheses might circumvent such weaknesses?

15. Review the recapitulatory hypotheses of Haeckel and show why they are not tenable. What Haeckelian views are still valid today? What is meant by developmental divergence?

16. Review hypotheses designed to account for the origin of the first Bilateria.

17. What are the presumed phylogenetic interrelations between acoelomates, pseudocoelomates, and coelomates? On what evidence are such assumptions based?

18. What are the likely phylogenetic interrelations between schizocoelomates, lophophorates, and enterocoelomates? On what data are such conclusions based?

19. On the basis of current phylogenetic thinking, how many times would (a) the coelom, (b) segmentation have evolved independently? Which groups exhibit metameric segmentation?

20. Without recourse to texts, draw up a comprehensive diagram that indicates the presumed phylogenetic interrelations of all major animal groups. Justify the structure of such a diagram with supporting data.

Collateral Readings

Brown, H.: The Age of the Solar System, *Sci. American*, Apr., 1957. Estimations are based on the radioactivity of meteorites and rocks.

Cairns, J.: The Bacterial Chromosome, *Sci. American*, Jan., 1966. An article discussing the structure and duplication of this DNA-containing organelle.

Clark, R. B.: ''Dynamics in Metazoan Evolution,'' Clarendon Press, Oxford, 1964. Highly recommended, especially the first and last chapters which analyze various phylogenetic hypotheses.

Eglinton, G., and M. Calvin: Chemical Fossils, *Sci. American*, Jan., 1967. On organic compounds found in 3-billion-year-old rocks.

Ehrlich, P. R., and R. W. Holm: ''Process of Evolution,'' McGraw-Hill, New York, 1963. The first chapter of this book is a discussion of the origin of life.

Fox, S. W.: The Evolution of Protein Molecules and Thermal Synthesis of Biochemical Substances, *Am. Scientist*, vol. 44, 1956. The article relates some of the heating experiments referred to in this chapter.

———— and R. J. McCauley: Could Life Originate Now? *Natural History*, vol. 77, 1968, p. 26.

———— K. Harada, et. al.: Chemical Origins of Cells, *Chem. and Engineering News*, June, 1970.

Hyman, L.: ''The Invertebrates,'' vol. 1, chap. 5; vol. 2, chap. 9, McGraw-Hill, New York, 1940, 1951. Various hypothesis on animal phylogeny are reviewed.

Keosian, J.: ''The Origin of Life,'' 2d ed., Reinhold, New York, 1964. A highly recommended review in paperback form.

Kerkut, G. A.: ''The Implications of Evolution,'' Pergamon, New York, 1960. This stimulating little book examines some of the assumptions underlying evolutionary thinking, including the problem of monophyletism versus polyphyletism.

Landsberg, H. E.: The Origin of the Atmosphere, *Sci. American*, Aug., 1953. The history of the atmosphere is traced on the basis of data from various sciences.

Marcus, E.: On the Evolution of Animal Phyla, *Quart. Rev. Biol.*, vol. 33, 1958. A review by a noted zoologist.

Miller, S. L.: The Origin of Life, in ''This Is Life,'' Holt, New York, 1962. This article gives details on the atmosphere experiments referred to in this chapter and also discusses the general problem of the origin of life. References at the end of the article cite all pertinent original sources.

Morowitz, H. J., and M. E. Tourtellotte: The Smallest Living Cells, *Sci. American*, Mar., 1962. Certain microbes smaller than some viruses are described, leading to the question of the smallest unit compatible with life.

Oparin, A. I.: "Life: Its Nature, Origin, and Development," Academic, New York, 1962. This author fathered modern thinking on the subject in a book published in 1936; the present volume is the latest revision of it.

Ross, H. H.: "A Synthesis of Evolutionary Theory," Prentice-Hall, Englewood Cliffs, N.J., 1962. This highly recommended book contains two early chapters on the origin of the universe and of life.

15

DESCENT:
FOSSILS
AND MAN

Any long-preserved remains of animals are fossils. They can be skeletons or shells, footprints later petrified, impressions left by body parts on solidifying rocks, or the remnants of animals trapped in amber, gravel pits, swamps, and other places. Whenever a buried animal or any part of it becomes preserved in some way before it decays it will be a fossil.

The fossil record begins roughly 500 million years ago, when all animal phyla were already in existence. Thus, fossil study, or *paleontology*, deals mainly with evolutionary histories within phyla. Such studies have contributed to knowledge of the animal past as much as, and often far more than, embryologic and anatomic inferences drawn from living animals.

The Fossil Record

Geologic Time

Fossils embedded in successive earth layers provide a time picture of evolution. Deep-lying fossils normally are not accessible, but on occasion a canyon-cutting river or an earthquake fracture exposes a cross section through rock strata. Moreover, erosion gradually wears away top layers and exposes deeper rock (Fig. 15.1).

The age of a rock layer can be determined by "clocks" in the earth's crust: radioactive substances. For example, a given quantity of radium

is known to "decay" to lead in a certain span of time. When radium and lead are found together in one mass in a rock the whole mass presumably was radium originally, when the rock was formed. From the relative quantities of radium and lead present today, one then can calculate the time required for that much lead to form. A similar principle is used in potassium-argon dating and radiocarbon dating. In the first process one measures how much of a naturally occurring unstable form of potassium has decayed to argon. Radiocarbon dating involves measurement of how much of an unstable form of carbon is still present in a rock or fossil sample. The potassium-argon method can be used for dating fossils many millions of years old, but the carbon method is accurate only for fossils formed within the last 50,000 years. Fossils themselves often help in fixing the age of a rock layer. If such a layer contains a fossil known to be of a definite age (*index fossil*), then the whole layer, including all other fossils in it, is likely to be of the same general age.

Based on data obtained from radioactive and fossil clocks, geologists have constructed a *geologic time table* that indicates the age of successive earth layers and provides a calendar of the earth's past history. This calendar consists of five successive main divisions, or *eras*. The last three are subdivided in turn into a number of successive *periods* (Table 9). The beginning and terminal dates of the eras and periods have been made to coincide with major geologic events known to have

Figure 15.1 Rock layers of different ages often are exposed to view. Generally speaking, the deeper a layer in the earth's crust, the older it is.

occurred at those times. For example, the transition from the Paleozoic to the Mesozoic dates the *Appalachian revolution,* during which the mountain range of that name was built up. Similarly, the transition between the Mesozoic and the Cenozoic was marked by the *Laramide revolution,* which produced the Himalayas, the Rockies, the Andes, and the Alps.

The first geologic era, the immensely long

Azoic, spans the period from the origin of the earth to the origin of life. Living history begins with the next era, the Precambrian. The fossil record from this era is exceedingly fragmentary, and it shows mainly that simple cellular life already existed about 3 to 2 billion years ago. From the end of the Precambrian on, a continuous and fairly abundant fossil record is available.

It is a very curious circumstance that rocks older than about 500 million years are so barren of fossils, whereas younger ones are comparatively rich in them. Did the Precambrian environment somehow preclude the formation of fossils? Or were Precambrian organisms still too insubstantial to leave fossilizable remains? We simply cannot be sure. But we *are* reasonably sure that Precambrian evolution must have led to the origin of three of the four present main groups of organisms, the Monera, the Protista, and the Metazoa. Virtually all phyla in these three groups were in existence by the end of the Precambrian, but not a single species of these ancient organisms has survived to the present.

Thus the long Precambrian spanned not only three-quarters of evolutionary time but also three-quarters of evolutionary substance. Nearly all the organisms in existence at the end of the Precambrian were aquatic. With the probable exception of some of the bacteria and some of the protists, the land apparently had not been invaded as yet. After the Precambrian, evolution brought about an extensive diversification in the existing phyla and a repeated replacement of ancient forms by new ones. In each of the three main categories this process also included an evolution of types

*Table 9.   The geologic time table.**

ERA	PERIOD	DURATION		BEGINNING DATE
Cenozoic ("new life")	Quaternary Tertiary	75	1 74	1 75
Mesozoic ("middle life")	Cretaceous Jurassic Triassic	130	60 30 40	135 165 205
Paleozoic ("ancient life")	Permian Carboniferous Devonian Silurian Ordovician Cambrian	300	25 50 45 35 65 80	230 280 325 360 425 505
Precambrian		1,500		3,000
Azoic ("without life")		3,000		5,000

* All numbers refer to millions of years; older ages are toward bottom of table, younger ages toward top.

that could live on land. Among these new forms were the Metaphyta. Very soon after the appearance of this last of the four large groups now in existence, some of the animals began to follow the plants to the land.

The Paleozoic

All animal phyla in existence 500 million years ago still survive today, but most subgroups in these phyla have since been replaced; all the ancient species and genera became extinct long ago. The present brachiopod genus *Lingula* goes back 400 million years and represents the most ancient of the known "living fossils" among animals.

During the very long Cambrian and Ordovician periods, life in the sea was already abundant. A large variety of protists probably existed already long before the Cambrian, and fossil protozoa (*foraminiferans* and *radiolarians*) are known from earliest Cambrian times onward (Fig. 15.2). The fossil history of animals begins with Cambrian

Figure 15.2 The fossil record of animals. The varying width of each graph approximates the changing abundance of a given group in the course of time. Such widths are directly comparable only for different times within a group, not between groups. For example, the absolute Cenozoic abundance of insects would equal or exceed that of all other groups combined. The graphs clearly show the major decline that took place at the end of each era, particularly during the Permo-Triassic transition and on a smaller scale during the Mesozoic-Cenozoic transition.

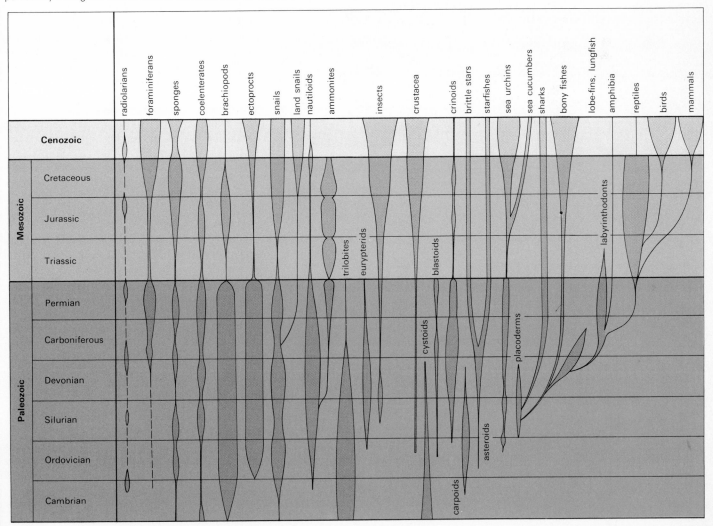

A

B

Figure 15.3 Paleozoic fossils: restorations. *A,* Cambrian seas. Various algae, trilobites (in center foreground), eurypterids (in center background), sponges, jellyfishes, brachiopods, and different types of worms are the most prominent organisms shown. *B,* Ordovician seas. The large animal in foreground is a straight-shelled nautiloid:

Figure 15.4 Graptolites. Left, portion of a colony. *Right,* detail of the exoskeleton of a few individuals (zooids) in a colony, showing their interconnecting black stolon.

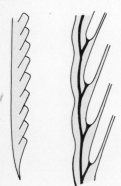

sponges and jellyfishes and Ordovician corals and sea anemones. A variety of tube-forming worms is known from these and even earlier periods, and stalked brachiopods already existed in the Cambrian. The Ordovician marks the beginning of a rich record of entoprocts and ectoprocts (Fig. 15.3).

Also abundant during the first half of the Paleozoic were the *graptolites,* a group of uncertain affinities. At first the animals were thought to be related to cnidarians, but now they are believed to be allied to primitive hemichordate stock. They became extinct by the end of the Paleozoic, and it is unknown if they left any descendants in the form of animals now living (Fig. 15.4).

Echinoderms were amply represented by six archaic groups. Three of these (*cystoids, carpoids, blastoids*) have since become extinct, but the stalked sessile *crinoids* and the ancient *asteroids* and *echinoids* gave rise to the present-day echinoderms (Fig. 15.5 and see Fig. 15.2). Mollusks included archaic clams and snails, as well as *nautiloids,* a group closely related to squids and octopuses and still represented today by the chambered nautilus. Early nautiloids had both coiled and uncoiled shells, and the uncoiled forms probably were the largest animals of the time; their shells were up to 5 or 6 yd long (see Fig. 15.3).

The most ancient arthropod types were the *trilobites,* believed to have been ancestral to all other arthropod groups (Fig. 15.6). Their body was marked into three lobes by two longitudinal furrows, hence the name "trilobites." These animals were already exceedingly abundant when the Cambrian began, and they are among the most plentiful of all fossil forms. A somewhat later group of arthropods comprised the *eurypterids,* large animals that well may have been ancestral to all chelicerate arthropods (horseshoe crabs, sea spiders, and arachnids). The surviving horseshoe crab genus *Limulus* has existed unchanged for the last 200 million years.

Vertebrate history begins in the late Ordovician (Fig. 15.7). The marine tunicate ancestors probably were already present at or near the start of the Paleozoic, and the vertebrates evolved from them as freshwater forms. The first fossil vertebrates are *ostracoderms,* bone-plated members of the class of jawless fishes. Most of them became extinct near the end of the Devonian, and their only surviving descendants are the lampreys and hagfishes. All traces of external bone have been lost in these animals (Fig. 15.8).

The Silurian was the period during which the first tracheophytes evolved, and these land plants were soon followed by land animals: late Silurian land scorpions, probably evolved from earlier sea scorpions, are the earliest known terrestrial animals. Other groups of arthropods invaded the land during the latter part of the Silurian and the beginning of the Devonian. The first spiders, mites, centipedes, and insects appeared during these times. In the sea the nautiloids gave rise to a new molluscan group, the shelled *ammonites,* which were to flourish for long ages (see Fig. 15.8). Among echinoderms the ancient asteroids branched into two descendant groups, the brittle stars and the starfishes. And among vertebrates

Figure 15.5 Paleozoic echinoderms. A, a cystoid. B, a blastoid. C, a crinoid. See Fig. 15.2 for history of first two.

Figure 15.6 Paleozoic arthropods. A, trilobites. B, a eurypterid.

a major adaptive radiation occurred during the Silurian and Devonian (see Fig. 15.7).

During the early Silurian some of the jawless ostracoderms gave rise to a new vertebrate class, the *placoderms.* Bone-plated like their ancestors, these fishes were generally small, but some reached lengths of 12 yd or more. Most placoderms used their jaws in a carnivorous mode of life (see Fig. 2.4). In time the placoderms replaced the ancestral ostracoderms as the dominant animals, and late in the Silurian they produced two descendant lines of fishes that in turn came to replace the placoderms themselves; by the end of the Paleozoic the placoderms had disappeared completely, the only vertebrate class (and one of the few animal classes in general) that has become extinct.

The two new groups of fishes evolved from placoderms were the *cartilage fishes* and the *bony fishes,* each representing a separate class. Both groups arose in fresh water, but the cartilage fishes rapidly adopted the marine habit that sharks and rays still display today. The bony fishes at first remained in fresh water, where they soon radiated into three main subgroups: the *paleoniscoid fishes,* the *lungfishes,* and the *lobe-finned fishes.* During the Devonian, often called the ''age of fishes,'' the paleoniscoids spread to the ocean and became the ancestors of virtually all present bony fishes, both freshwater and marine. The lungfishes declined in later Paleozoic times, and only three genera survive today. The lobe-fins similarly now are almost extinct (Fig. 15.9).

But the Devonian representatives of the lobe-fins included the ancestors of the *amphibia,* the first land vertebrates. Lobe-fins probably lived in fresh waters that dried out periodically, and their air sacs and fleshy fins probably enabled them to crawl overland to other bodies of water or to embed themselves in mud and breathe air through the mouth. It appears likely, therefore, that terrestrial vertebrates arose not because certain fish preferred the land, but because they had to become terrestrial temporarily if they were to survive as fish.

Thus when the Devonian came to a close sharks were dominant in the ocean and bony fishes in fresh water. On land, terrestrial arthropods had become abundant and the first amphibia had made their appearance. Many of these land animals could shelter in the forests of primitive trees then already in existence.

During the Carboniferous and Permian periods

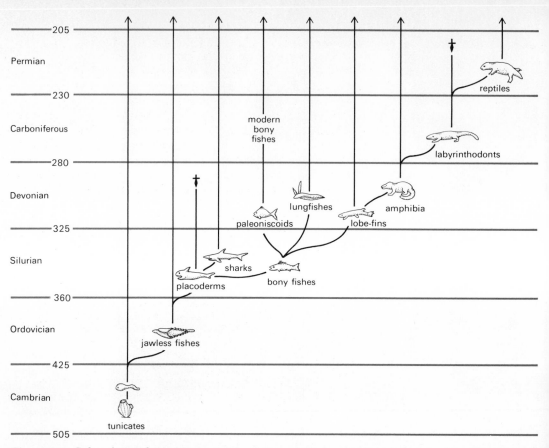

Permian

Carboniferous

Devonian

Silurian

Ordovician

Cambrian

205

230

280

325

360

425

505

reptiles

labyrinthodonts

modern
bony
fishes

amphibia

lungfishes

paleoniscoids

lobe-fins

sharks

placoderms

bony fishes

jawless fishes

tunicates

Figure 15.7 Paleozoic vertebrates. The name of each group is shown roughly at the time level at which this group first appears in the fossil record. Daggers indicate extinction of a group.

Figure 15.8 Paleozoic fossils. *A*, ostracoderm, an ancient jawless fish. Note bony armor. A placoderm is illustrated in Fig. 2.4. *B*, an ammonite, a cephalopod mollusk related to the chambered nautilus (see Fig. 23.26).

A

B

the first crablike and crayfishlike animals and the first land snails appeared, and insects produced extensive adaptive radiations. Some of these ancient insect types had wingspreads of close to a yard. Among vertebrates the early amphibia gave rise to more or less clumsy, often bizarre forms, the *labyrinthodonts* (Fig. 15.10). These became the ancestors of two groups, the modern amphibia and the *reptiles*. This new class was represented at first by the stem reptiles, or *cotylosaurs*, the first vertebrates that laid eggs on land. Cotylosaurs produced a major reptilian radiation during the Permian, which brought about the decline of the labyrinthodonts and also set the stage for an "age of reptiles" during the Mesozoic era (Fig. 15.11).

The Paleozoic era terminated with the Appa-lachian revolution, a geologically unstable time that precipitated a *Permo-Triassic crisis* in the animal world. Archaic forms became extinct and were replaced by rapidly evolving new groups, and the total amount of animal life—particularly in the sea—decreased temporarily (see Fig. 15.2). Brachiopods and ectoprocts became almost extinct. All mollusks passed through a major decline; nautiloids became extinct with one exception, and ammonites were reduced to a small group. Trilobites and eurypterids disappeared altogether. Only a few crinoid types and a single echinoid type survived to the Triassic. Placoderms died out, and extensive replacements occurred among the cartilaginous and bony fishes. Land animals were less affected on the whole, though their numbers did decline temporarily. Labyrinthodonts lingered on to the Triassic, but soon they, too, became extinct. The reptiles on the contrary survived the crisis well, and when the new Mesozoic era opened, they were already dominant.

The Mesozoic

The era as a whole was characterized by a re-expansion of nearly all groups that survived the Permo-Triassic crisis and by extensive replacements within groups (see Fig. 15.2). Protozoa, sponges, and cnidarians underwent major adaptive radiations during the Jurassic, with the result that these forms exist today in greater numbers than every before. Ectoprocts and brachiopods similarly increased in abundance, but the latter became virtually extinct again at the end of the Mesozoic. Clams and snails diversified greatly and became the predominant mollusks from then on. The ammonites also reexpanded during the late Meso-

Figure 15.9 Lobe-fins and lungfish. A, restoration of fossil lobe-finned fishes. B, a coelacanth, a rare lobe-finned fish still surviving today. The pectoral and pelvic fins have internal bony skeletons. Note the diamond-shaped enamel-covered (*ganoid*) scales. C, a lungfish (*Protopterus*), found today in Western Africa.

A

B

C

A

B

Figure 15.10 Late Paleozoic fossils. A, the wing of a Permian insect, two-thirds actual size. Insects larger than this existed in the Permian, but even the owner of the wing shown was far larger than any insect today. *B,* reconstruction of *Diplovertebron,* a labyrinthodont amphibian.

zoic, yet not a single one survived beyond the end of the era. The crinoids managed to linger on as relics, and the more abundant brittle stars and starfishes held their own. But the single echinoid group that survived from the Paleozoic underwent an explosive expansion during the late Mesozoic. In the course of it the modern sea urchins and sea cucumbers evolved. Crustacea gained slowly and steadily in numbers and types. Insects diversified explosively in parallel with the rise of flowering plants, and the present importance of insects traces to this Mesozoic expansion. An extensive radiation occurred also among the bony fishes, which became the dominant animals of the aquatic world from the Cretaceous on.

The most spectacular Mesozoic event was the expansion of the reptiles. These not only evolved into numerous terrestrial types but also invaded the water and the air. As a group they reigned

Figure 15.11 Reptile evolution. A, reconstruction of *Seymouria,* a transitional amphibian type probably related to the stock from which reptiles appear to have evolved *B,* reconstruction of *Labidosaurus,* one of the cotylosaurian stem reptiles.

A

B

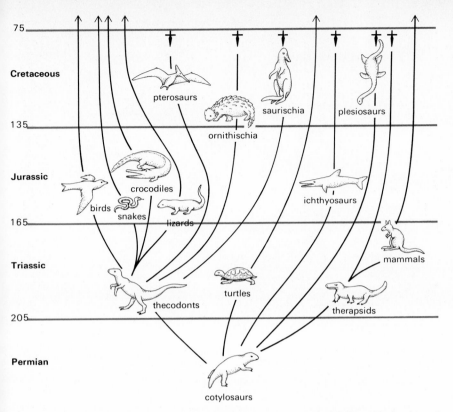

Cretaceous

75

135

Jurassic

165

Triassic

205

Permian

pterosaurs

ornithischia

saurischia

plesiosaurs

crocodiles

birds

snakes

lizards

ichthyosaurs

mammals

thecodonts

turtles

therapsids

cotylosaurs

Figure 15.12 Mesozoic reptiles. Placement of groups corresponds roughly with the time of their greatest abundance. Numbers at left indicate past time in millions of years.

Figure 15.13 Bird evolution: plaster cast of *Archeopteryx.* Note feathered tail, wings. The head is bent back, and the tooth-bearing mouth is not easily visible here.

supreme on earth for 130 million years, longer than any other animal group before or since. When their dominance was eventually broken, they were replaced by two new groups that they themselves had given rise to, the birds and the mammals.

At the beginning of the Mesozoic five major reptilian groups were in existence, all evolved from the Permian stem reptiles (Fig. 15.12). One group, the *thecodonts,* reradiated extensively during the Triassic and gave rise to the following types: the ancestral *birds;* the ancestors of the modern *crocodiles, lizards,* and *snakes;* the flying *pterosaurs;* and two other groups collectively called *dinosaurs.* A second group was ancestral to the modern *turtles.* A third and fourth produced two kinds of marine reptile, the porpoiselike *ichthyosaurs* and the long-necked *plesiosaurs.* The fifth group comprised the *therapsids,* which included the ancestors of the mammals.

These various reptilian types did not all flourish at the same time. The Triassic was dominated largely by early therapsids and thecodonts. The former were four-footed walkers, but thecodonts were rather birdlike, with large hind limbs for walking, an enormous supporting tail, and diminutive forelimbs. During the Jurassic ichthyosaurs became abundant in the ocean, and one of the thecodont lines evolved into birds. This transition is documented beautifully by the famous fossil animal *Archeopteryx* (Fig. 15.13). The reptile had teeth and a lizardlike tail, but it was also equipped with feathers and wings and presumably flew.

Like mammals, birds remained inconspicuous during the whole remaining Mesozoic. They were overshadowed particularly by their thecodont kin, the flying pterosaurs. These animals had their heyday during the Cretaceous, when reptiles as a whole attained their greatest abundance and variety (Figs. 15.14 and 15.15). The unique, long-necked plesiosaurs were then common in the ocean, and the dinosaurs gained undisputed dominance on land. Not all dinosaurs were large, but some were enormous. The group included, for example, *Brontosaurus,* the largest land animal of all time, and *Tyrannosaurus,* probably the fiercest land carnivore of all time.

Mammals arose from therapsids during the late Triassic (Fig. 15.16). Some of the factors that probably promoted this mammalian evolution can be deduced. The ancestors of mammals among therapsid reptiles were generally small, in the size range of mice, and this smallness must have en-

Figure 15.14 Mesozoic reptiles. Reconstruction of plesiosaurs (left) and ichthyosaurs (right).

Figure 15.15 Dinosaurs. Left, Triceratops, a Cretaceous horned herbivore. Right, Tyrannosaurus, a giant Cretaceous carnivore.

tailed perpetual danger in a world dominated by far larger animals. Survival must have depended on ability to escape danger, mainly by running. But the running capacity of early reptiles probably was as limited as that of their modern descendants. Reptilian lungs hold only moderate amounts of air, and breathing movements are fairly shallow. Also, the oxygen-holding capacity of blood is comparable to that of fish and amphibian blood, which suffices for high levels of metabolic activity in water (see Chap. 9, Table 7). On land, however,

such capacities cannot support sustained body activity; most living reptiles actually are active only in brief bursts and otherwise are relatively slow and sluggish.

Living mammals give clear evidence that the evolution of the group from reptiles involved a pronounced elevation of metabolic levels. A newly developed diaphragm increased breathing efficiency; red blood corpuscles became nonnucleated and maximally specialized for oxygen transport; new chemical variants of hemoglobin

Figure 15.16   Mammalian evolution: a Triassic therapsid, a mammal-like reptile.

permitted blood to carry almost three times as much oxygen as before; and along with a change from reptilian scales to mammalian hair, a temperature-control mechanism came into being that permitted maintanance of a constant, uniformly high body temperature. The whole *rate* of metabolism intensified, and mammals thus became perhaps the most active animals of all time. (In part for similar reasons of escaping danger, in part because of the energy-requirements of flight, birds became nearly as active. However, they did not evolve a diaphragm, the oxygen capacity of their blood is only twice that of reptiles, and their red blood cells remain nucleated. Yet body temperature is maintained at an even higher level than in mammals.)

It can be inferred that, as early mammals adapted to a life of running, they must have sought safety in the forests. They became primarily nocturnal and furtive, with a distinct preference for hiding in darkness. Their sense of sight was of limited value in such an existence and, like almost all mammals today, they were color-blind. The sense of smell became dominant instead, and the basic orientation to the environment came to depend on odor cues from the forest floor. Furthermore, in an active life of running and hiding it is unsafe merely to lay eggs and leave them in reptilian fashion. The first mammals probably did just that, and the primitive egg-laying mammals today still do likewise. Later mammals evolved means to carry the fertilized eggs with them, inside the females; they became viviparous. The pouched marsupial mammals and the placental mammals are their modern descendants.

After mammals had originated late in the Triassic, they remained inconspicuous for the rest of the Mesozoic, a period of about 80 million years. But, as the Cretaceous came to a close, most of the reptilian multitude became extinct, probably because of cooler climates following the Laramide revolution. The lower temperatures appear to have caused an extinction of much tropical vegetation, and with their food sources thus reduced the reptiles would have died out as well. Whatever the precise causes may have been, extinction of the Mesozoic reptiles cleared the way for a great Cenozoic expansion of mammals and birds.

The Cenozoic

The progressively cooler climates after the Laramide revolution culminated in the ice ages of the Pleistocene epoch, a subdivision of the Quaternary period of the Cenozoic (Table 10). Four ice ages occurred during the last 600,000 years. In each, ice sheets spread from both poles to the temperate zones and then receded. Warm interglacial periods intervened between the successive glaciations. The last recession began less than 20,000 years ago, at the beginning of the Recent epoch, and it is still in progress; polar regions are still covered with ice. Cenozoic climates played a major role in plant evolution and animal evolution was greatly affected as well. Pleistocene ice in particular influenced human history, modern man in a sense being one of the products of the ice ages.

The adaptive radiations of mammals and birds were the main evolutionary events of the Cenozoic,

Table 10. *The epochs and periods of the Cenozoic era**			
PERIOD	EPOCH	DURATION	BEGINNING DATE
Quarternary	Recent	20,000 years	20,000 B.C.
	Pleistocene	1	1
Tertiary	Pliocene	11	12
	Miocene	16	28
	Oligocene	11	39
	Eocene	19	58
	Paleocene	17	75

* Unless otherwise stated, numbers refer to millions of years.

Figure 15.17 Horse evolution, Cenozoic. This reconstructed sequence begins at left, with the fossil horse *Eohippus,* and proceeds via *Mesohippus, Hyponippus,* and *Neohipparion* to *Equus,* the modern horse at right. The drawings are to scale and show how the average sizes and shapes of horses have changed. Progressive reduction in the number of toes took place, as well as changes in dentition. Note, however, that the animals shown represent a highly selected series, and it should not be inferred that horse evolution followed a straight-line pattern. Here, as elsewhere, a bush pattern is evident. See also Color Figs. 6 and 7.

the "age of mammals." Terrestrial mammals replaced the dinosaurs; aquatic mammals eventually took the place of the former ichthyosaurs and plesiosaurs; and bats, but more especially birds, gained the air left free by the pterosaurs. A total of some two dozen independent mammalian lines came into existence, each ranked as an order (see Fig. 30.18). The fossil record is fairly extensive for nearly all these groups, and it is extremely good for a few, horses and elephants in particular (Fig. 15.17 and Color Figs. 6 and 7).

Each mammalian line exploited either a new way of life available at the time or one left free after the extinction of the Mesozoic reptiles. One mammalian line is of particular interest, for it eventually led to man. The members of this line still made their home in the forests, like much earlier mammalian groups, but they adapted to an *arboreal* life in the treetops. Fossils show that such arboreal mammals of the early Paleocene were the ancestors of two orders, the *Insectivora* and the *Primates.* Some of the shrews now living still have the joint traits of primitive insect eaters and primitive primates, but the later members of the two groups became very different; insectivores now also include moles and hedgehogs, and primates include men.

Nevertheless, there can be little doubt that the distant mammalian ancestors of men were shrew-like, 2 to 3 in. long, with a long snout and a bushy tail (Fig. 15.18). They lived in trees and, like modern shrews, they probably were furtive, quick, color-blind and given to hiding in dark places, as well as ferocious and voracious; they ate anything eatable of appropriate size, insects particularly, and they ate nearly continuously, consuming their own weight every 3 hr or so: a highly active but small body requires comparatively very large amounts of fuel. These early primates then gave rise to several sublines, of which five still survive: *lemuroids, tarsoids, ceboids, cercopithecoids,* and *hominoids* (Fig. 15.19).

Lemuroids include the *lemurs* and *aye-ayes,* now found largely on the island of Madagascar (Fig. 15.20). These animals still have long snouts and tails, but instead of claws and paws they have strong flat nails, a general characteristic of all modern primates. Long nails probably are more

Figure 15.18 Early mammals.
An arboreal squirrel shrew, order Insectivora.

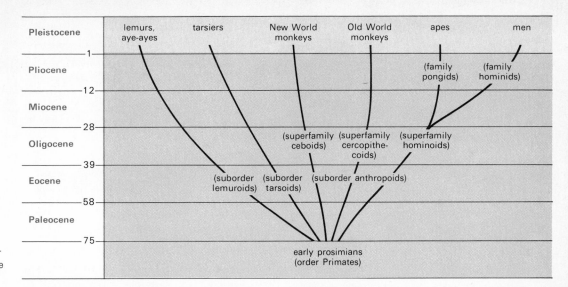

Figure 15.19 The primate radiation. Numbers at left indicate past time in millions of years.

useful than claws in anchoring the body on a tree branch.

In a tree, moreover, smelling is less important than seeing, and in lemuroids an important visual adaptation has evolved: each of the eyes can be directed forward independently, which permits better perception of branch configurations than if the eyes were fixed on the side of the head (as in shrews).

Figure 15.20 Primitive mammals. A, a lemur from Madagascar. *B,* a tarsier from Indonesia. Both these groups are arboreal.

A

B

Tarsoids are represented today by the *tarsiers* of southern Asia and Indonesia. In these animals the ancestral "smell brain" has become a "sight brain"; the olfactory lobes have become small but the optic lobes have increased in size. Reduction of the olfactory lobes also has led to a shortening of the snout, hence to the appearance of a fairly well-defined face. Indeed the eyes have moved into the face, and, although they can move independently, both can be focused on the same point. As a result tarsiers are endowed with stereoscopic vision and efficient depth perception, traits that all later primates also share. Such traits are of considerable adaptive value if balance is to be maintained in a tree. Tarsiers also have independently movable fingers and toes, with a branch-gripping pad at the end of each. Moreover, whereas most mammals produce litters of several to many offspring, tarsiers have a single offspring at a time, a safer reproductive pattern in the branches of a tree.

Ceboids comprise the *New World monkeys,* confined today to South and Central America. These animals have long, strong tails used as fifth limbs. Cercopithecoids are the *Old World monkeys,* found in Africa and Asia. They are identified by tails that are not used as limbs (Fig. 15.21). The two groups represent separate, though remarkably similar, evolutionary developments; monkeys evolved twice independently. In both groups adaptations to arboreal life have developed a good deal farther than in the earlier primates.

Figure 15.21 Monkeys. A, howler monkey, a ceboid (New World) type. Note prehensile tail. *B,* rhesus monkey, a cercopithecoid (Old World) type.

A

B

If tarsoids have a "sight brain," monkeys can be said to have a "space brain." The eyes are synchronized like those of man, and each also contains a fovea centralis, a retinal area of most acute vision. Moreover, monkeys are endowed with color vision, the only mammals other than the hominoids so characterized. Color is actually there to be seen: flowers, foliage, sky, and sun provide an arboreal environment of light and space, far different from the dark forest floor that forced the ancestral mammals literally to keep their noses to the ground. Indeed, the smelling capacity of monkeys is as poor as their sight is excellent. Correspondingly, the cerebrum is greatly enlarged and contains extensive vision-memory areas; monkeys store visual memories of shape and color as we do.

Monkeys also adopt a predominantly *sitting* position at rest, a posture that in a tree is probably safer than lying and that also relieves the forelimbs of locomotor functions. As a direct consequence, monkeys have a new freedom to use hands for touch exploration. Evolution of ability to *feel* out the environment and one's own body has led to a new self-awareness and to curiosity. Ability to touch offspring and fellow inhabitants of the tree has contributed to new patterns of communication and social life, reinforced greatly by good vision, by voice, and by varied facial expressions. Touch-control areas of the brain and centers controlling hand movements are extensive, and the brain as a whole thus has enlarged in parallel with the new patterns of living. The mind has quickened as a result, and the level of intelligence has increased well above the earlier primate average. A ground mammal such as a dog still sniffs its environment, but monkeys and the later primates explore it by sight and by touch. Evolution of intelligence has been associated specifically with improved coordination between eye and hand. Fundamentally, we may note, primate intelligence is an adaptation to an arboreal way of life.

Tree life also provides a basically secure existence. Actually only two kinds of situation represent significant dangers, the hazard of falling off a branch and the hazard of snakes. The first is minimized by opposable thumbs, independently movable fingers on all limbs, and precocious gripping ability in general (displayed also in newborn human babies). The second is countered by the strength of the body, which is larger than in earlier primates. Perhaps partly because of this emancipation from perpetual fear, the life span is lengthened considerably; monkeys live up to four decades, whereas shrews have a life expectancy of a single year. Furthermore, absence of danger and continuously warm climates make feasible a breeding season spanning the whole year, and in Old World monkeys this year-round breeding potential is accompanied by menstrual cycles.

The fifth primate line, the hominoids, evolved important locomotor modifications over and above those attained by the monkeys. Early hominoids developed universal limb sockets and a fully upright posture, and by hand-over-hand locomotion between two levels of branches they became tree *walkers* more than tree sitters; even a monkey cannot match the acrobatic tree swinging of hominoids. Moreover, only hominoids can swivel their hips, and only they have long, strong collar

A   **B**

Figure 15.22 Apes. A, a gibbon from Malaya. *B,* a mountain gorilla from Africa.

main sublines. One of these led to the *pongids,* the family of apes, the other to the *hominids,* the family of man (see Fig. 15.19). Apes are represented today by four genera, gibbons, orangutans, chimpanzees, and gorillas. Gibbons are survivors of an early branch of ape evolution, characterized by comparatively small, light bodies and retention of a fully arboreal way of life. Indeed the gibbon is undoubtedly the most perfectly adapted arboreal primate (Fig. 15.22).

But the other three types of ape, representing later and heavier pongid lines, have abandoned life in the trees to greater or lesser extent. It appears that these later apes ceased to be completely arboreal after their bodies had become so large and heavy that trees could no longer support them aloft. The feet of these apes give ample evidence of the weight they have to support; foot bones are highly foreshortened and stubby, as if crushed by heavy loads. Correspondingly, the agile grace of the arboreal gibbon is not preserved in the ground apes. These do not actually walk, but they *scamper* along in a crouching shuffle gait.

So far as is known the hominid line left the trees right after it split away from the common hominoid stock, just when the adaptations to tree life had finally become perfected. Was this descent, too, prompted by great body weight, as in the case of the large apes? Probably not, or else the human foot would resemble that of the ground apes and man would scamper rather than walk. In actuality the human gait is unique among ground forms, and as a gibbon swings in a tree so a man literally swings on the ground. It is likely, therefore, that the hominid line left the trees when its evolution had progressed to a gibbonlike level and when the body still was comparatively small and light. The walking grace perfected in the trees then could persist on the ground. Early hominids thus appear to have been small, perhaps only 3 to 4 ft tall, and in contrast to the later apes their size probably increased only *after* they had come out of the trees (Fig. 15.23).

But if not body weight, what other conditions could have forced hominids to the ground? The chief cause appears to have been the progressively cooler climate, which led to a thinning out of forests in many regions. Our prehuman ancestors thus may have been forced to travel on the ground if they wished to move from one stand of trees to another. Such forced excursions often must have been fraught with considerable danger, how-

bones and chests that are broader than they are deep. Hominoid locomotion also has become facilitated by a shortening of the tail and its interiorization under the skin between the hindlimbs. In this position the tail skeleton helps in counteracting the internal sag produced by gravity acting on an upright body.

During the early Miocene, some 30 million years ago, the hominoid line branched into two

Figure 15.23 Hominoid size evolution. One group of apes probably developed to a large size in the trees, and modern descendants such as gorillas therefore were already heavy when they adopted life on the ground. Another group, exemplified by the gibbons, remained light and arboreal. Early hominids likewise probably remained light and small. They presumably left the trees as small types and their evolutionary size increase then occurred on the ground.

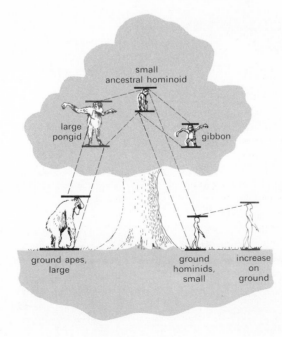

small ancestral hominoid

large pongid

gibbon

ground apes, large

ground hominids, small

increase on ground

ever, for saber-toothed carnivores and other large mammals dominated the ground at those times (see Fig. 13.8). Ability to dash quickly across open spaces would then have had great selective value, and such conditions probably promoted the evolution of *running* feet. Also, strong muscles would be required to move the hindlimbs in new ways. Indeed, a unique trait of the hominid line is the presence of such muscles, in the form of curving calves, thighs, and buttocks. In conjunction with this muscular change the hips have become broad and the waist narrow, and in these respects, too, man differs from apes.

These and other features that now distinguish men and apes came to be superimposed on the traits of the earlier arboreal primates. Clearly then, the modern human type could not have evolved if the ancestral type had not first been specialized for life in trees.

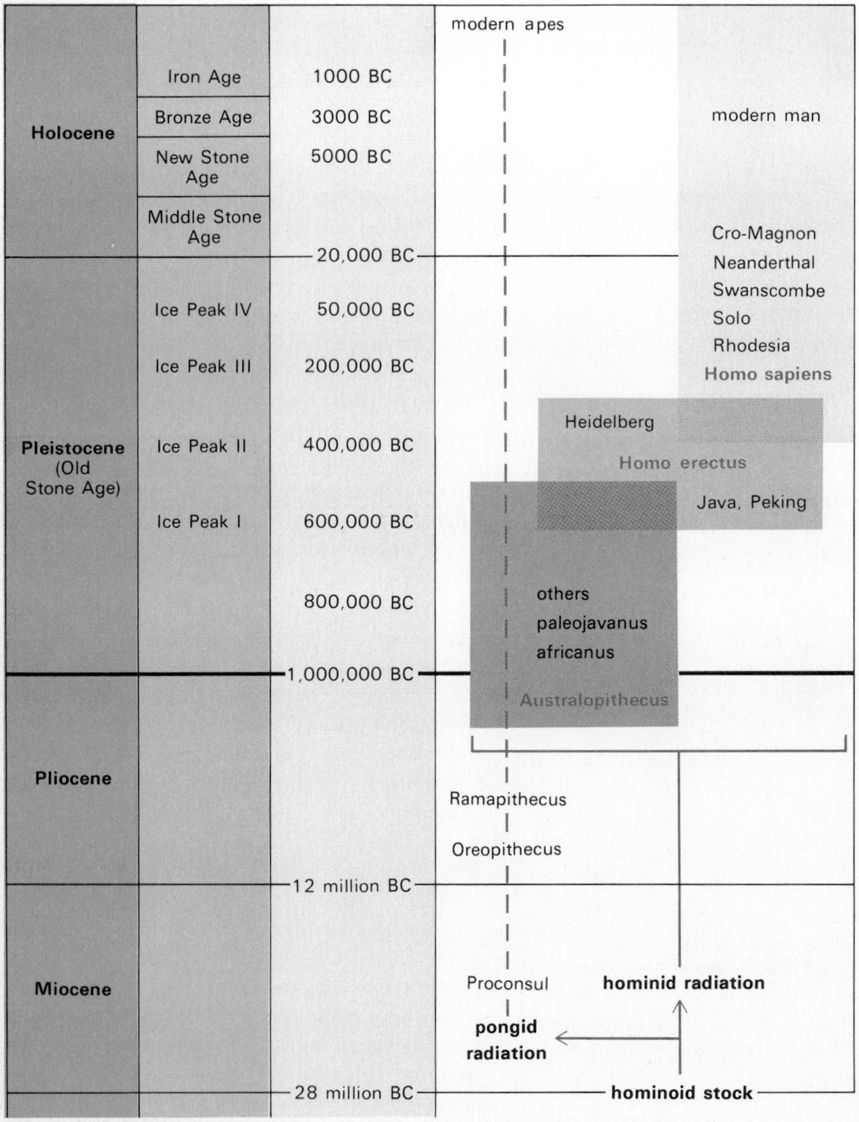

Human History

The fossil evidence of the hominid group is tantalizingly scant; we can trace the recent evolution of almost any mammal far better than our own. Moreover, the exact path of descent of our own species remains undiscovered as yet, and other known hominids are related to us somewhat as uncles or cousins. With the exception of the line leading to ourselves, all other lines of the hominid radiation have become extinct at various periods during the last 30 million years (Fig. 15.24).

Apart from their zoological distinctions, fossil hominids are defined as prehuman or truly human on the basis of cultural achievements. Any hominid that *made* tools in addition to using them can be called a "man." If a hominid only used stones or sticks found ready-made in his environment he is considered prehuman; if he deliberately fashioned natural objects into patterned tools, no matter how crude, he is considered human. By this criterion quite a few hominid types were men.

Prehistoric Man

The earliest fossils acknowledged today to be hominid date back to the late Pliocene, about two or more million years ago. Still earlier ones well might be discovered at any time. The Pliocene hominids generally are assigned to several different species of the genus *Australopithecus* (Fig. 15.25). One such species is *A. africanus* (previously called *Homo africanus* and originally named *Homo habilis*), which dates back roughly 2 million years. About 1¾ million years old is *A. paleojavanus* (originally named *Zinjanthropus*), a type that, like *A. africanus*, was discovered in East Africa. *A. paleojavanus* used wooden clubs and stone hammers as tools and thus was a true man.

Figure 15.24 The hominid radiation. Each main hominid group (in color boxes) is shown roughly at a time level at which it is known to have existed. Within each such group several more subgroups than indicated are known. In most cases the exact period at which the subgroups have lived is not certain.

A

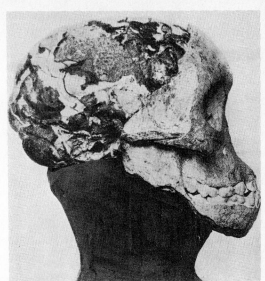

B

Figure 15.25 Australopithecus. A, drawing of *A. paleojavanus* (formerly *Zinanthropus*), the East Africa man. Note the exceptionally low forehead and small braincase. *B,* skull of a later species of *Australopithecus.*

As his large molars indicate, his diet probably was mainly coarse vegetation. The skull reveals that the head was held erect and that jaw muscles were attached as they are in modern man, suggesting that *A. paleojavanus* probably had some form of speech. A forehead was almost absent; the volume of the brain could not have been larger than 600 cm³, comparable to the brain of a modern gorilla. Other species of *Australopithecus* might not have been as far advanced, and at least some of them apparently did not make tools.

Representatives of *Australopithecus* also lived in Pleistocene times, and the genus appears to have persisted till about half a million years ago. Some so far unidentified species of *Australopithecus* is generally believed to have evolved into the genus *Homo,* to which we belong. The time of this transition is not known, nor is the place—many investigators believe it to have been eastern Africa. The earliest known species of *Homo* is *H. erectus* (originally called *Pithecanthropus*), which lived from about 600,000 to about 250,000 years ago. Thus the first representatives of this species probably were contemporaries of late representatives of *Australopithecus* (Fig. 15.26).

H. erectus came to have a wide distribution, with subspecies known from southeastern Asia ("Java man"), China ("Peking man"), Europe ("Heidelberg man"), as well as various regions of Africa. All these groups were true men who made

tools of stone and bone and used fire for cooking. The brain volume averaged 900 to 1,000 cm³, nearly double that of *Australopithecus.* The skull had a flat, sloping forehead and thick eyebrow ridges, and the massive protruding jaw was virtually chinless. *H. erectus* probably practiced cannibalism—some of his fossil remains include skullcaps separated cleanly from the rest of the skeleton; sheer accident does not appear to have caused such neat separations.

Some 500,000 to 250,000 years ago *H. erectus* appears to have given rise to our own species, *Homo sapiens.* The place of this event again is unknown, and if it indeed occurred in just one place *H. sapiens* must have spread rapidly from there to most parts of the world. Early subspecies of *H. sapiens* are known from, for example, Africa ("Rhodesia man"), Java ("Solo man"), England ("Swanscombe man"), and the Eurasian land mass ("Neanderthal man"). The last-named, *Homo sapiens neanderthalis,* is the best known of all prehistoric men (Fig. 15.27).

Neanderthalers probably arose 150,000 years ago or even earlier, they flourished during the period of the last ice age, and they became extinct only about 25,000 years ago, when the ice sheets began to retreat. The brain of these men had a volume of 1,450 cm³, which compares with a volume of only 1,350 cm³ for modern man. The Neanderthal brain also was proportioned differ-

Figure 15.26 Homo erectus. A, skull, and *B, C,* reconstructions of Peking man (formerly *Pithecanthropus*).

A

B

C

ently; the skull jutted out in back, where we are relatively rounded, and the forehead was low and receding. Heavy brow ridges were present, and the jaw was massive and almost without chin. Culturally the Neanderthalers were nomadic cavemen of the *Old Stone Age* (see Fig. 15.24). They fashioned a variety of weapons, tools, hunting axes and clubs, and household equipment. Their territory covered most of Europe and west and central Asia, with fringe populations along the Mediterranean coasts.

Late groups of Neanderthalers were contemporaries of early representatives of our own subspecies, *Homo sapiens sapiens.* One such early representative was *Cro-Magnon man,* who lived in

Europe from about 50,000 to 20,000 years ago. Cro-Magnon man appears to have interbred with the Neanderthalers, and he is also believed to have caused the extinction of the European Neanderthal populations. Cro-Magnon was 6 ft tall on the average, with a brain volume of about 1,700 cm^3 (Fig. 15.28). In addition to stone implements he used bone needles for sewing animal skins into crude garments. The dog became his companion, and he was a cave-dwelling hunter who also painted remarkable murals on cave walls.

Cro-Magnon man was a contemporary of other populations of *H. sapiens sapiens* in different parts of the world. Through evolution, migration, and interbreeding these groups gradually developed

A   B

Figure 15.27   Homo sapiens neanderthalis. A, B, reconstructions of Neanderthal man.

Figure 15.28   Cro-Magnon man: two reconstructions.

A

B

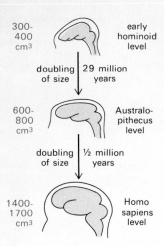

300-400 cm³   early hominoid level

doubling of size | 29 million years

↓

600-800 cm³   Australo-pithecus level

doubling of size | ½ million years

↓

1400-1700 cm³   Homo sapiens level

Figure 15.29 The hominid brain. During the last 29 million years the brain not only has doubled in overall size twice in succession, but also has changed in the relative proportion of parts. The frontal and temporal lobes have increased more than proportionately, which is indicated also by the changes in the contours of the skull.

into the present populations found around the globe. By the time the Pleistocene came to a close, some 20,000 or 25,000 years ago, the ice had started to retreat, and eventually even nontropical man no longer needed to shelter in caves.

For the next 10,000 years or more man was culturally in the *Middle Stone Age,* characterized chiefly by great improvements in stone tools. Man still was a nomadic hunter, however. The *New Stone Age* began about 7,000 to 10,000 years ago, about the time Abraham settled in Canaan. A great cultural revolution took place then. Man learned to fashion pottery, he developed agriculture, and he was able to domesticate animals. From that period on modern civilization developed very rapidly. By 3,000 B.C. man had entered the *Bronze Age* (see Fig. 15.24). Some 2,000 years later the *Iron Age* began. And not very long after

Figure 15.30 Neoteny in man. Two equivalent developmental stages of pig and man are shown, drawn to the same scale. The openings behind the eyes at left mark the ears; note also the rudimentary gill arches, tissues of which contribute to the development of the lower parts of the face. In the time the pig embryo has grown to a stage at which its head is about one-quarter of its total length (right), the human embryo has grown only to a stage at which its head is still roughly one-half its total length. The slopes of the black dashed lines indicate this comparatively slower human growth, and the vertical color arrows mark the time difference in attaining equivalent body proportions. The result is that at birth man is in a neotenous state compared with mammals generally. See also Fig. 15.31.

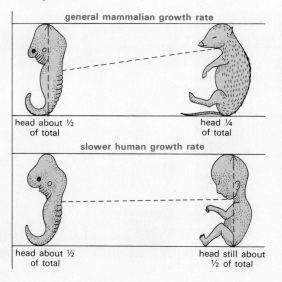

general mammalian growth rate

head about ½ of total   head ¼ of total

slower human growth rate

head about ½ of total   head still about ½ of total

that man discovered steam, electricity, and now the atom and outer space.

Modern Man

Man is an animal with unique attributes. He is fully erect and has a double-curved spine, a prominent chin, and walking feet with arches. He is a fairly generalized type in most respects, being not particularly specialized for either speed, strength, agility, or rigidly fixed environments. However, he has a brain proportionately far larger and functionally far more elaborate than any other animal. And most of the uniquely human traits have their basis in man's brain.

Man acquired his brain by an exceedingly rapid, explosive process of evolution. Judging from Miocene fossils and living apes, the first hominids 30 million years ago might have had a brain volume of about 300 to 400 cm³, comparable to that of a newborn human baby today. From then to the beginning of the Pleistocene brain volume increased to an average of 600 to 800 cm³, as at the *Australopithecus* stage. Thus, in a span of about 29 million years, brain size doubled (and body volume increased by roughly the same factor). But during the first half of the Pleistocene, in only ½ million years, brain size more than doubled again, from the 600 to 800 cm³ range to the 1,400 to 1,700 cm³ range of *Homo sapiens,* while the size of the whole body remained relatively unchanged (Fig. 15.29).

Evidently, the human brain has become considerably larger than might be expected on the basis of general increase in body size alone. The greatest brain growth has occurred in the temporal lobes, which contain speech centers, and especially in the frontal lobes, which control abstract thought. As a result, a basic *qualitative* difference between man and an ape or any other primate is that man has the capacity to think in a new time dimension, the *future.* An ape or other mammal has a mind that can grapple well with the present and the past, but such an animal at best has only a rudimentary conception of future time. It does not have elaborate frontal lobes, which contain the control centers for this dimension of existence.

Man alone therefore is able to plan, to reason out the consequences of actions not yet performed, to choose by deliberation, and to have aims and purposes. Also, only man can think in symbolic terms to any appreciable extent, and only he can generalize, weep, laugh, and envision beauty. Directly or indirectly all such unique attributes of human mentality are based on ability to project abstractly to the future. Man rightly has been called both a "time binder" and the only philosophical animal.

There can be little doubt that the key event that led to this remarkable enlargement of the human brain, and that in fact gave man nearly all his other most characteristically human traits, was a genetic change that slowed down the processes of embryonic development. In all animal embryos the head region begins to develop sooner and faster than more posterior body parts. Thus when a mammalian embryo is about the size of a sand grain, roughly three-quarters of it comprises head structures. But at birth the head usually takes up no more than a sixth or an eighth of the total length. If now the rate of embryonic development were to slow down, the stage of birth would be reached when the head still is relatively large and the rest of the embryo still comparatively undeveloped. In other words, a slowing down of developmental processes leads to neoteny, retention of early developmental features including comparatively larger heads with larger brains (Fig. 15.30). Such a developmental slowdown unquestionably took place during the first part of the Pleistocene,

when *Australopithecus* evolved to *Homo* and when *H. erectus* became *H. sapiens*.

The developmental deceleration also resulted in a host of other neotenous traits in man—traits that in other mammals have a chance to attain a more mature condition before birth. An example is our lack of furry skin—our birth occurs before the stage at which other mammals develop extensive hair. Similarly, man has a smaller jaw with fewer teeth than apes; a smooth round skull without the prominent eyebrow ridges and skull crests characteristic of apes; thinner, more delicate skull plates and skeletal components in general; and several other, less conspicuous anatomic distinctions (Fig. 15.31).

But above all, the deceleration of early development resulted in a substantial lengthening of the whole human life cycle. As a result man became virtually the longest-lived of all animals, and he acquired yet another and most significant characteristic, a proportionately very long *youth*. No other animal passes through such an extended preadult stage; and at an age when a man is just entering adulthood, an ape is already senile and a dog or a horse has been dead for several years. Man therefore has *time* to be young, to learn, to make use of his large brain.

By the same token man's prolonged infancy makes him more helpless for far longer than any other animal. Man thus is critically dependent on fellow individuals when he is young, and he needs to be fed, groomed, protected, trained, and *taught* if he is to survive: man's basic *social* nature is another consequence of his neotenous condition.

All the fundamental and peculiarly human aspects of our "nature" thus arose in one stroke, as it were, as interlocked results of the initial developmental slowdown—"fetal" body, longevity, and large brain; long infancy and dependency; long youth and learning capacity; long maturity to rear and teach infants and to apply learning; and continuous sociality. In these traits lies our basic difference from apes, the essence of our humanity.

To some extent the environmental conditions that promoted the evolution of a long-lived, future-contemplating primate can be identified. The adaptive pressures appear to have been the problems of food and family.

Like apes and most other primates, early hominids at first probably lived in small polygamous bands of perhaps 10 to 30 members. In such a band each male exerted dominance over as many

Figure 15.31 Neoteny in man. In these skulls of a gorilla (left) and a man (right), note that man has a proportionately far larger brain case, a more prominent forehead, much thinner cheek bones, and a less massive lower jaw. See also body proportions of an ape in Fig. 15.22, and compare with those of man.

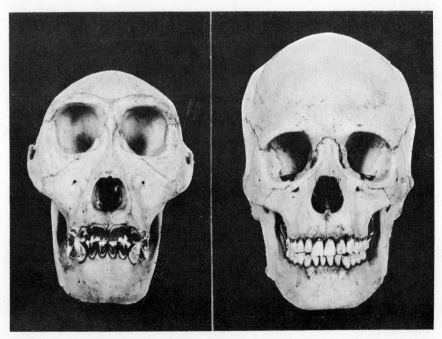

females as he could and aggressively warded off any likely competitors. But unlike other primates, which are herbivorous, hominids eventually became omnivorous and meat-eating. Meat had to be hunted on the hoof and, as today, food procuring probably always was a major function of the male. However, since nearly every desirable food animal was either far stronger or far faster than a hominid, a single individual stood little chance of hunting successfully on his own. Operating in groups therefore must have become a necessity. But group hunting in turn must have depended on a mentality of a very specific kind, one that did not yet exist at the time.

First, the group hunter must be able to produce and comprehend a comparatively complex system of communication; concerted action by a team necessitates frequent and diversified signaling among the participants, over long as well as shorter distances and both in and out of sight of other team members. For maximum effectiveness, such a communications system should be highly symbolic, with signal elements as simple yet as informative as possible. Even at minimum complexity, such a system would have had to be far more complex than any other type of communication then in existence. It is therefore probable that natural selection operated in favor of enlarged temporal lobes, which made possible speech and language, hence efficient communication and teamwork. The genetic potential for language speech could already have been in existence in at least some species of *Australopithecus*.

Further, the group hunter must be able to formulate some sort of hunting plan before going into action, to coordinate the activities of all team members. It is therefore likely that natural selection favored the development of enlarged frontal lobes, which made possible abstract reasoning, foresighted, future-directed conceptualization, and thus again efficient teamwork, food, and survival.

Considerable foresight also would have been required to resolve two major emotional conflicts that group hunting would have entailed. First, it must have become mandatory that every participant subordinate his individuality to the common purpose of the group, if not to actual direction by a leader. Yet subordination would have required a hitherto nonexistent mode of social behavior. The customary mode took the form of dominant self-assertion by each male and aggressiveness toward any other male who ventured too near.

Hence the new motivation for obtaining meat by cooperating with others must have conflicted badly with the ancient motivation to be automatically hostile to others. Simple reason could have reduced this conflict, however, for it would have shown that temporary suppression of hostility might be more desirable than certain future hunger.

Second, a male on a hunt is absent from his home base for often prolonged periods, and his females are left unguarded and subject to the attentions of other males. Thus the male habit of residing in a domain and jealously asserting dominance over females must have conflicted strongly with the desirability of going out in search of meat. Here again reason would have helped to resolve the conflict, for in a choice between "a full house and a full stomach," as one zoologist put it, the foresighted option would be a full stomach and survival—particularly since effective propagation can occur even without a harem.

It could have been pressures of this kind that led to the most significant social development of the hunting age, the modern *monogamous family*. Inasmuch as males and females are born in roughly equal numbers, a society based on polygamy nearly always suffers shortages of females and unrelieved male rivalry. In a monogamous pattern, by contrast, a single female suffices adequately for reproduction sexual competition can decrease substantially, a main cause of male aggressiveness can be minimized, and male teamwork can be enhanced correspondingly. The monogamous family probably was the only feasible social institution compatible with *both* hunting and reproduction.

It appears, then, that the hunting stage in human history was made possible by a major biological change that produced an enlarged brain, and a major social change that introduced the monogamous family. The main behavioral result of the new brain was speech and reasoning, and of the new family, increased male cooperation and reduced aggressiveness. The chance for an exercise of reason was probably enhanced by several structural and functional body changes that made aggressiveness both less necessary and less possible: greater male-female equality in size and strength than among apes; disappearance of externally visible signs of estrus in females and year-round mating readiness; reduction in the size of the canine teeth (by neoteny), adaptive in lessening aggressiveness; and lowered neural and endo-crine responsiveness—man on the whole has far slower reflexes and is emotionally far less "touchy" than almost any other animal.

In the monogamous family the less aggressive male then came to be not merely a biological parent but also a social one; he became a *father* who contributed actively to the care of the young, not only by supplying food but also by being a *teacher*. The offspring could be taught hunting skills and social conduct, and in this way the long period of dependence on adults could be turned from potential liability to vital asset. Indeed this new relation between father-teacher and young came to be the basis of human *culture*. Teaching and learning led to traditions, to accumulation of knowledge, and to transmission of such knowl-edge over successive generations. The vehicle for this cultural transmission was speech, and in effect we deal here with a new kind of evolution. Man is unique in that he evolves not only biologically, through inherited characteristics passed on by genes, but also socially, through an inherited cul-ture passed on by words.

Human Races

H. sapiens sapiens has been the only human group in existence since the late Pleisotcene, and its evolution has continued ever since. That it is still continuing now is shown perhaps most obviously by the changing racial structure of mankind.

The Race Concept

A race is a subunit of a species—subspecies or subsubspecies—that differs from other such sub-units in the *frequency of one or more genes*. Thus the Neanderthalers, Swanscombe men, Solo men and other subspecies of *H. sapiens* mentioned previously were early racial subunits. The subunits of the species today are the various human groups found around the globe. *They are races if, and only if, they differ from one another in the percentage frequencies with which genes occur in them.*

Therefore, racial distinctions are exhibited by *populations*, not individuals. Single individuals cannot differ in the "frequency" of a gene; they can differ only in having or not having the gene, and thus they can exhibit *individual variations*. Frequency differences occur between *groups* of individuals. Also, race is a *statistical* concept; races differ relatively, not absolutely. The differ-ences are of degree, not of kind: distinctions result not from different types of gene sets but from frequency variations in the *same* types. Thus *all* races of a species share the basic species traits, including, for example, full cross-fertility of all members. Animals with different kinds of gene sets are by definition not members of the same species.

Any change that affects gene frequencies also affects racial differences. Such changes can occur through nonrandom mating, through mutation, through genetic drift—in short, all those processes that bring about evolution, as discussed in Chap. 13. Races therefore represent *temporary* collec-tions of genes, transient stages in the evolutionary history of a species. There have been races of man ever since mankind originated, and the ones today are different from earlier ones—and undoubtedly from later ones as well. Races usually evolve when a species covers a large territory and when the populations in different parts of this territory have relatively little reproductive contact. As shown in Chap. 13, evolution then is likely to take different directions in the various localities. Most often the ultimate result of racial evolution is speciation, a splitting of the parent species into two or more new offspring species. Under certain conditions, how-ever, the speciation process can become slowed down or blocked completely. In man, for example, extensive migration and interbreeding so far have prevented formation of separate species, and in many areas have even reduced or obliterated racial differences.

Basically, then, races are temporary geographic variants of a species. Among variants that cover large areas, species traits usually tend to change gradually from one end of the territory to another, in line with gradual variations in flora and fauna, climate, and environmental conditions in general. In such cases large adjacent racial groups are said to form *clines*, or gradual transitions from one to the other: the frequency of a particular gene changes in small steps from one racial median value to another. Most of the large, continental populations of man intergrade clinally in this way. In other instances the transitions are more abrupt; a relatively small population sometimes forms a fairly distinct genetic "island" surrounded by pop-ulations of large racial groups. The isolating factors usually are environmental and ecological and in man also cultural and social. Among examples in man are the Basques and Lapps of Europe, the Ainus of Japan, the Aborigines of Australia, the Bushmen and Pygmies of Africa.

This present statistical and genetic meaning of race is far different from what was believed in the days before modern genetics—and what is commonly believed even today by most laymen. The early hypothesis was a "typological" one. Mankind was assumed to have started out as a structurally homogeneous collection of individuals, which in the course of time became split into a number of pure racial "types"—australoids ("Browns"), negroids ("Blacks"), mongoloids ("Yellows"), caucasoids ("Whites"), amerinds ("Reds"), and several or many more depending on the classifier.

In practice this hypothesis of "racial types" often proved to be clearly untenable, yet in most cases it was psychologically, socially, politically, and even scientifically more expedient not to change the basic typological premise. As a result, in order to make fact fit preconception, it was frequently necessary to invent all sorts of human migrations for which there was—and is—no evidence. For example, to account for the occurrence of black-skinned people in both Africa and Melanesia, it was necessary to postulate that an ancestral group of Blacks migrated from one place to the other or from some intermediate point to both places.

Another consequence of typological thinking was that individuals in one type category could differ more than individuals of different such categories. In the case of skin colors, for example, some "Whites" are far darker than many "Blacks," some are far redder than many "Reds," some are far yellower than many "Yellows." Skins actually run through the *whole* human color spectrum in *each* of the postulated type categories, and a similar conclusion holds for any other racial trait—an excellent illustration that racial traits actually differ only in relative degree, not in absolute kind. Moreover, attempts to classify races were just as inadequate when the distinguishing criteria were several traits simultaneously. Regardless of how simple or how complex the arbitrary type categories were, some individuals always fitted into more than one such category and some fitted into none.

The difficulty here was an inescapable and inherently insoluble boundary problem: no matter how many categories are established, the gaps between them do not disappear. The reason is that, as emphasized earlier, racial differences are statistical population differences in gene frequencies, *not* absolute differences between individuals. But, one might be prompted to ask here, is there not a very evident racial difference between, for example, this particular white-skinned individual and that particular black-skinned one? No—there is an *individual* difference, obviously, but not a "racial" one! The genetic difference between the two well might be less great than between the White and another White or than between the Black and another Black; or the difference might be greater. The important point is that the White and the Black are each members of *populations*—often, indeed, the same population; and unless we compared the genetic characteristics of such populations in statistical terms, we would be back to imposing arbitrary type categories on nature. If it could be shown that the five or so genes believed to control skin color actually differed in frequency in the populations to which our individuals belonged, then we would be describing a real racial distinction—of these *populations, not* of the individuals.

Note also that, although gene-frequency variations do measure racial *differences*, they do not give us a count of the number of races. Because large racial groups usually intergrade clinally, it is generally impossible to draw sharp boundary lines between them. Even if we counted the number of "racial centers," where the gene frequencies for a given trait attained peak values, we would have counted not the number of races but only the number of racial distinctions with respect to *one* trait. And there are thousands of such traits, each with its own distinctive frequency distribution that rarely matches any other. Any "race count" therefore becomes completely arbitrary. In the past, typological classifiers have named as few "races" as 2 or as many as 200. They might as well have identified 2,000 or 200,000; no fixed number corresponds to reality, because preconceived type categories are themselves unreal.

Because the term "race" in human affairs almost invariably implies fixed, outdated typologies, modern investigators generally avoid using the term. They speak of geographic populations instead of artificial "races"; of population differences instead of "racial traits"; and they measure gene frequencies. These concepts do describe human variations as they occur in reality.

The Meaning of Race

What "good" are racial differences? Is a population with a high percentage of whorled fingerprints somehow better or worse off than one with a high percentage of looped prints? Numerous attempts have been made to uncover possible adaptive values of such traits, but in many instances the results have been inconclusive; an adaptive significance in a relatively minor trait is often difficult to demonstrate, if indeed there is any.

For example, it has been suggested that dark skin pigmentation might be a protective adaptation against the ultraviolet radiation of the sun, and perhaps also against the low night temperatures in humid, swampy, tropical regions. By contrast, light pigmentation might facilitate the formation of vitamin D in the skin under the influence of ultraviolet light, particularly in misty, cloudy regions at high latitudes, where the amount of sunlight is reduced.

It is possible, and perhaps even probable, that some of the many hypotheses for various external traits of man actually might be valid or at least come close to being valid. But it is another matter to prove their validity—or prove that they really relate to *racial* distinctions. In nearly all such cases the genetics of the trait in question is virtually unknown, as is the frequency distribution of the genes involved. Moreover, in most instances it is not even known to what extent a trait actually is determined by genes and to what extent its expression depends on environmental or cultural variables. There has been far too little study as yet on the actual nature of numerous human traits. Their possible adaptive significance therefore remains unproved, too—and indeed untested experimentally.

Furthermore, it is quite possible that certain traits might not be adaptive in any case. Mongolian peoples exhibit a high frequency of dry earwax, and black-skinned ones, a high frequency of sticky earwax. If such trait differences were shown to be truly racial, are they necessarily adaptive? No one has yet proposed a plausible hypothesis. Indeed, certain traits well could be adaptively neutral. Yet even such neutral traits could be important. As pointed out in Chap. 12, many genes are known to control more than one trait (so-called *pleiotropic* effects of genes). One such trait might be vital or adaptive in some way whereas another is not, and both could be controlled by the same gene and thus would be expressed simultaneously. Hence the importance of, for example, dry earwax—or slit eyes or even frizzy hair or white skin—might lie not in that trait itself but in another, perhaps internal trait that happened to be controlled through the same genes. Actual pleiotropic correlations of this sort have not yet been made, but they might explain the occurrence of adaptively neutral traits.

The best and so far the only really significant clues about possible adaptive values of racial differences have come from studies of *blood groups*. The genes that control human blood groups have been identified, and the frequency distributions of these genes in nearly all peoples have become known through blood tests. Blood group genes control the manufacture of the blood proteins discussed in Chap. 9. The genes occur in different combinations in different individuals, and people thus can be grouped into several categories according to their blood types.

The most widely studied blood group is the ABO system, which is controlled by three allelic genes. The blood types produced by them are the familiar A, B, AB, and O categories (Fig. 15.32). Well-studied blood groups also include the Rh system and several others—the MN, Kell, Lutheran, Duffy, Diego, and Kidd systems, for example. The blood of every person belongs to a particular type in each of these systems, and human popula-

Figure 15.32 ABO genes and phenotypes. The blood group is indicated by the blood cell antigens, or proteins manufactured under the control of the genes shown. A person of group A contains antigen A in the cells but no anti-A antibody in the serum. Anti-B antibody is present, however, hence such serum will cause clumping of blood cells of groups B or AB (both having the B antigen). Cells of group A will themselves be clumped by serum of O or B donors, both of which contain the anti-A antigen. Group AB blood contains antigens A and B in the cells but neither anti-A nor anti-B antibodies in the serum, hence such serum will not cause clumping of cells of any blood type. Conversely, group O blood contains both serum antibodies but neither cellular antigens, hence such cells will not clump when mixed with serum of any blood type.

genes	blood cell antigens	serum antibodies	serum agglutinates cells of group	cells agglutinated by serum of group
$I^A I^B$	AB	—	none (universal donor)	O, A, B
$I^A I^A$ or $I^A i$	A	anti-B	B, AB	O, B
$I^B I^B$ or $I^B i$	B	anti-A	A, AB	O, A
ii	O	anti-A,B	A, B, AB	none (universal recipient)

tions around the globe differ in the frequencies with which these types are represented. For example, the highest frequency of the gene for type A, up to 80 percent, occurs among the Blackfoot Indians of North America. The B group is most frequent in central and south-central Asia. One of the Rh-positive genes is extremely common in native African populations but rarely exceeds frequencies of 5 or 6 percent in others. Over 50 percent of the Basques are Rh-negative, while this condition is virtually unknown in native Australasian groups. Every other blood group similarly has its own distinctive frequency distribution.

Why are these distributions nonrandom? Is it possible that blood groups are somehow adaptive and are being maintained at various frequencies by natural selection? An affirmative answer is strongly suggested by a landmark study on the so-called *sickle-cell* trait of the blood. The gene for that trait occurs in two allelic forms, *Si* and *si*. Homozygous dominants, *SiSi*, usually die in infancy from sickle-cell anemia; the blood corpuscles are deformed in sickle shapes, and they contain only reduced amounts of hemoglobin that correspondingly reduce the oxygen-carrying capacity. Heterozygotes, *Sisi*, are carriers of the sickle gene; they are essentially normal, with far fewer blood corpuscles affected and only mild or occasional symptoms of anemia. Homozygous recessives, *sisi*, are completely normal. Here evidently is a gene that can be lethal and that therefore should rapidly eliminate itself through the death of the homozygous dominants.

Yet the gene persists. Moreover, when its distribution is examined, an important relation becomes evident. The gene is most frequent in central and western Africa, in many regions along the

Mediterranean coasts, and spottily also in Asian areas bordering the Indian Ocean (Fig. 15.33). In all these places the gene is especially common among populations that inhabit humid, swampy lowlands—where malaria is most abundant and where deaths from malaria occur in the greatest numbers. In view of this matching distribution of sickle-cell anemia and malaria, could there be some interconnection between the two diseases? Indeed, extensive investigations actually have revealed that most malarial deaths take place among normal individuals without the sickle trait (*sisi* homozygotes), whereas malaria is far less lethal in the heterozygotes (*Sisi*)—as if the sickle trait afforded protection. Further studies have shown that the malarial parasite (*Plasmodium*, see Chap. 19) cannot readily parasitize sickled blood corpuscles. And experimental injections of malarial organisms into volunteers of both the *Sisi* and *sisi* types have proved conclusively that the sickle trait actually does protect against malaria. Homozygotes thus die either from abnormal blood or from malaria, but the heterozygotes, who can be slightly afflicted by both diseases, can nevertheless survive them. It is this protective effect of the sickle gene that preserves it at high frequencies in certain populations.

The sickle trait has provided the first clearcut demonstration of *balanced polymorphism*—self-adjusting frequency balances of different genes in a population. Thus, whereas by themselves genes such as *Si* and *si* can have lethal consequences, together they become far less detrimental. Adaptive advantages of this sort have not yet been clearly proved for the ABO and other blood types. But a number of relations have come to light which suggest that balanced polymorphisms might maintain the observed frequencies of the blood group genes too.

For example, some studies have indicated that type A individuals might have the highest incidence of pernicious anemia and of cancer of the stomach and the female reproductive system. Ulcers seem to be associated most frequently with type O, and in such individuals syphilis also appears to be easier to treat than in other groups. Moreover, type O babies might be least affected by bronchopneumonia and by smallpox. In general, the most common selective advantage associated with given blood groups seems to be protection against various diseases, including many of the well-recognized infant killers of the past. If substantiated, such a finding actually would not

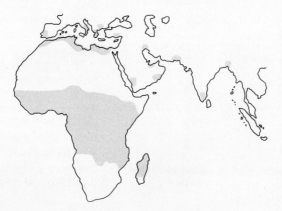

Figure 15.33 The distribution of the sickle-cell gene (color).

be too surprising, for blood groups have their basis in blood proteins, and these are well known as the agents that confer immunity (see Chap. 9).

A tentative but nevertheless highly significant conclusion appears to emerge from these various data: a major selective force—and possibly *the* major selective force—in human racial evolution might have been disease and genetic protection against disease. Compared to the hazard of disease other environmental stresses actually might have been relatively insignificant. After all, even primitive man is known to have lived in virtually all climates and regions; and having had the human potential, he must have adapted quite readily to most local conditions by simple technological expedients like use of fire, more or less clothing, or shelters of various kinds—and without necessarily evolving a new body in the process. Thus it is conceivable that climates and geographic locations of early human populations might have had relatively little direct bearing on whether they evolved black, yellow, or white skins, long or short limbs, lean or chubby physiques, or other external traits.

However, man has been exposed to one environmental hazard that until quite recently he could not cope with by simple technological means: disease. This contingency has entailed not only frequent lethality but quite often also lethality before reproductive age. Genes that protected against the different diseases of various regions thus must have had a strong selective advantage.

And as internal traits such as blood chemistries then underwent evolutionary change, perhaps some of the external traits became altered in parallel—in part through pleiotropic effects, for example. In any case, the best historical assessment now possible appears to be that human races might have evolved as populations that, apart from other distinctions, were characterized by particular collections of disease-resisting traits. Some investigators have pointed out, for example, that resistance to malaria as through sickle genes might have been an important adaptive factor that could have contributed specifically to the evolution of African Blacks.

If such a disease hypothesis should prove to be valid, it would follow that racial evolution should slow or even stop as a result of modern medical techniques; for these technological protections would substitute for the earlier evolutionary ones. An obliteration of racial distinctions is hastened also by the great mobility of modern man, which tends to increase the frequency of interbreeding. However, this trend is often counterbalanced by reproductive barriers that have social and cultural foundations. The racial future of man thus is as unpredictable as the future of mankind as a whole. It is certain only that man today is just as subject to evolutionary forces as he was in the past, with the important difference that modern man creates many of these forces himself—and is usually unaware where they might lead.

Review Questions

1. What is a fossil? How can the age of a fossil be determined? Review the names and dates of the geologic eras and periods. What were the Appalachian and Laramide revolutions?

2. Describe the key events of animal evolution during the (*a*) Cambrian-Ordovician, (*b*) Silurian-Devonian, (*c*) Carboniferous-Permian. Review the course of vertebrate evolution during the entire Paleozoic.

3. Review the evolutionary happenings during the Mesozoic among groups other than reptiles. Make a similar review for reptile evolution. Which reptilian groups exist today, and what ancient groups did they derive from?

4. What group was ancestral to mammals? What factors probably promoted the evolution of mammals from reptiles? What factors appear to have promoted the evolution of birds from reptiles, and in what ways has the direction of bird evolution been different from that of mammals?

5. Describe the main features of the Cenozoic mammalian radiation, with special attention to the origin of primates. Describe the major features and the time pattern of the primate radiation, and name living animals representing each of the main lines. When and from where did the line leading to man branch off?

6. Describe the various adaptations of each of the primate groups to arboreal life. Which structural, functional, and behavioral traits of man trace back specifically to the arboreal way of life of his ancestors? How do hominoids differ from other primates? How does the hominid line differ from the pongid line?

7. Describe some of the characteristics of *Australopithecus* and *Homo erectus*. When and where did these hominids live? Were they men? Roughly when did *Homo sapiens* arise? What traits distinguish Neanderthalers from modern men?

8. Name some of the traits that distinguish man uniquely from other animals. List as many neotenous traits of modern man as you can. What quantitative changes have occurred in the human brain during the last 30 million years?

9. Describe some of the selective pressures that might have promoted the evolution of a large brain in man. What particular significance might future-directed thinking and language ability have had in human evolution? Would such capacities not be equally useful in other animals, and if so why did they not evolve in other groups?

10. Describe some of the selective pressures that might have promoted the development of monogamy in man. What were some of the social and familial consequences of monogamy? What traits tend to make male rivalry less intense in man than in other primates?

11. Define "race" and "racial difference." Why can an individual not be assigned to a race? What is a cline? Why is it not possible to specify a precise number of human "races?" What has so far prevented mankind from evolving into several species?

12. Contrast the outdated typological and the modern genetic definitions of race. What makes typological race concepts no longer tenable? Show how differences in human skin colors were interpreted along typological lines and how they are now interpreted in terms of population genetics.

13. Describe adaptive values that have been ascribed to some external human traits and show how such traits actually might be neither racial nor adaptive. What is pleiotropy, and what is its possible significance in the development of human traits?

14. Describe the biology of the sickle-cell trait and how it has been discovered that it protects against malaria. What is balanced polymorphism?

15. What is the possible significance of the nonrandom frequency-distribution of blood-group genes? What kind of hypothesis about racial evolution is tentatively suggested by studies on blood-group distributions?

Collateral Readings

Abelson, P. H.: Paleobiochemistry, *Sci. American,* July, 1956. A study of amino acids in 300-year-old fossils.

Allison, A. C.: Sickle Cells and Evolution, *Sci. American,* Aug., 1956. Very pertinent in the context of this chapter.

Berrill, N. J.: "Man's Emerging Mind," Dodd, Mead, New York, 1955, or Premier Books, Fawcett, New York, 1957. An extremely fascinating and stimulating account on human evolution. Highly recommended.

Broom, R.: The Ape-men, *Sci. American,* Nov., 1949. A description of *Australopithecus* by the discoverer of the first fossils of this hominid.

Brues, C. T.: Insects in Amber, *Sci. American,* Nov., 1951. Up to 90-million-year-old fossil insects are compared with modern ones.

Colbert, E. H.: The Ancestors of Mammals, *Sci. American,* March, 1949. Therapsids and other Triassic reptiles are examined.

————: "Evolution of the Vertebrates," Wiley, New York, 1955. A good text on fossil vertebrates.

Deevey, E. S.: Living Records of the Ice Age, *Sci. American,* May, 1949. An examination of Pleistocene history in the living world.

————: Radiocarbon Dating, *Sci. American,* Feb., 1952. This technique of age determination in fossils is described.

Dobzhansky, T.: The Present Evolution of Man, *Sci. American,* Sept., 1960. A discussion of how human evolution is influenced by man's control over his environment.

Ericson, D. B., and G. Wolin: Micropaleontology, *Sci. American,* July, 1962. The new field of fossil study by microscope is described.

Glaessner, M. F.: Precambrian Animals, *Sci. American,* Mar., 1961. Recent fossil finds predating the 500-million-year mark are discussed.

Hockett, C. F.: The Origin of Speech, *Sci. American,* Sept., 1960. The article traces the development of human speech from more primitive systems of animal communication.

Howells, W. W.: The Distribution of Man, *Sci. American,* Sept., 1960. An examination of some of the strain differences among geographic variants of man.

————: Homo Erectus, *Sci. American,* Nov., 1966. A good description of this human species and a discussion of its ancestral role for *Homo sapiens.* Recommended.

Hurley, P. M.: Radioactivity and Time. *Sci. American,* Aug., 1949. An examination of the various atomic "clocks" that can be used to date given earth layers.

Millot, J.: The Coelacanth, *Sci. American,* Dec., 1955. An account on a recently discovered living lobe-finned fish, formerly believed to have been extinct for 300 million years.

Moore, R. C., C. G. Lalicker, and A. G. Fischer: "Invertebrate Fossils," McGraw-Hill, New York, 1952. A good source book for further information on extinct invertebrates.

Mulvaney, D. J.: The Prehistory of the Australian Aborigine, *Sci. American,* Mar., 1966. Man appears to have reached Australia very soon after the end of the last ice age.

Napier, J.: The Antiquity of Human Walking, *Sci. American,* Apr., 1967. A million-year-old fossil toe sheds light on the origin of our gait.

Newell, N. D.: Crises in the History of Life, *Sci. American,* Feb., 1963. An examination of the several geologic periods during which organisms died out on a large scale.

Penrose, L. S.: Dermatoglyphics, *Sci. American,* Dec., 1969. On the genetic and anthropological significance of fingerprints and footprints.

Sahlins, M. D.: The Origin of Society, *Sci. American,* Sept., 1960. A subordination of sexual drives appears to have been basis in the development of human society; on this point see also Berrill's book above.

Simmons, E. L.: The Early Relatives of Man, *Sci. American,* July, 1964. An examination of primate fossil history during the last 60 million years.

————: The Earliest Apes, *Sci. American,* Dec., 1967. A 28-million-year-old fossil ape is described and its evolutionary significance is assessed.

Simpson, G. G.: "Life of the Past," Yale, New Haven, Conn., 1953. A very good review of paleontological events; see also the reference by the same author at the end of Chap. 13.

Tanner, J. J.: Earlier Maturation in Man, *Sci. American,* Jan., 1968. A discussion of changes in human growth patterns during the last 100 years, indicating that trait alterations can be substantial without being "racial."

Washburn, S. L.: Tools and Human Evolution, *Sci. American,* Sept., 1960. The article shows how inferences about human evolution can be made from fossil tools.

Weckler, J. E.: Neanderthal Man, *Sci. American,* Dec., 1957. A hypothesis about the relation of modern man and Neanderthalers is discussed in this article.

Wiener, A. S.: Parentage and Blood Groups, *Sci. American,* July, 1954. The nature of blood groups and the genetics of their inheritance are discussed.

Because every animal is specialized to greater or lesser degree, it invariably depends on other organisms and the physical environment for some essential product or process; survival requires *group* association. Interacting, interdependent natural groupings of animals are as ancient as the animals themselves, and as the animals evolved, so did their groupings.

The simplest grouping consists of a single animal and its surroundings, and the observable outcome of interaction here is *behavior*. Several individuals exhibit behavior in relation to not only their environment but also each other, the result in many cases then being a *social* grouping. Social or nonsocial groups make up local *populations*, and many like populations are the components of a *species*.

In a given territory usually live a number of populations of several different species, a grouping that represents a *community*. The communal territory together with the organisms in it forms an *ecosystem*, and all ecosystems on earth collectively make up the *biosphere*. This most inclusive association encompasses the entire inhabited part of the globe and comprises all nonliving and living components.

The chapters of this part examine these progressively more inclusive forms of living interdependence.

PART 6
ANIMAL ASSOCIATIONS

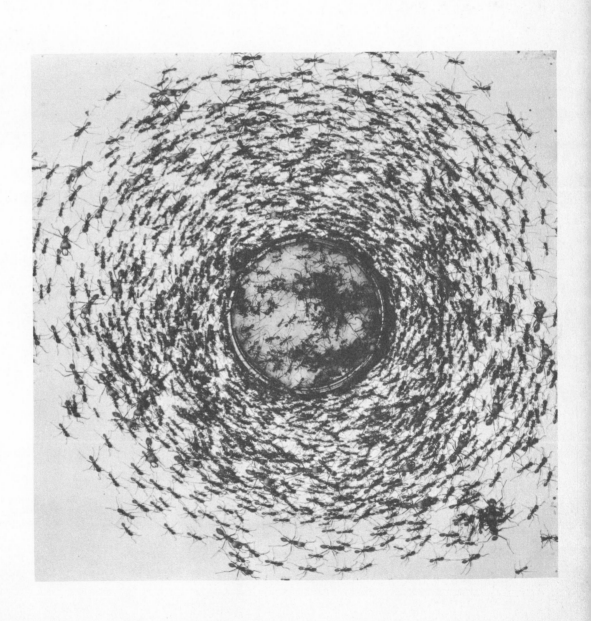

16

Behavior includes any externally observable activity of an animal. In most cases behavior involves some form of movement, either motion of body parts or outright locomotion. But secretions to the outside, color changes, or slow alterations brought about by growth similarly qualify as behavioral events. Indeed, absence or cessation of an overt activity in itself is an expression of behavior.

In all its forms behavior is basically adaptive and has a self-perpetuative function; directly or indirectly it tends to promote survival. Moreover, behavioral acts are just as subject to evolutionary change as their functions; behavior and function evolve together.

Forms of Behavior

Every behavioral process studied closely is now known to include *both* inherited and learned elements. Equivalent to "inherited" in this context are also terms such as "inborn," "innate," or "instinctive;" all of them signify that particular aspects of a behavior are based on the genetic endowment of an animal and are controlled by mechanisms that develop independently of learning. Terms equivalent to "learned" include "acquired," "experiential," or "environmental," and they signify that certain aspects of a behavior are based on external influences and are controlled by nongenetic mechanisms.

However, there can be little question that *all*

properties and activities of animals ultimately are under genetic control. At the same time, *no* property or activity can be exhibited without an environment. It is therefore experimentally impossible to filter out all environmental influences from genetic effects or all genetic influences from environmental effects; genes and environment always form an inseparable continuum. For this reason, behavioral (or other) processes no longer can be categorized—as they usually have been in the past—in sharp alternatives like "instinctive" or "learned."

But although both inherited and learned factors thus play a role in any behavior, the relative contributions of these two sets of determinants vary considerably for different behaviors. Such relative contributions usually can be judged by the degree of modifiability of a behavioral process. A behavior can be largely unlearned and nearly unmodifiable, or partly learned and fairly modifiable, or largely learned and exceedingly modifiable. For present purposes these three general categories can be referred to respectively as *reactive*, *active*, and *cognitive* forms of behavior.

Reactive behavior includes all largely automatic, stereotyped activities. As a group they probably represent the most primitive behavioral level. Reactive responses by and large are fully "programmed" as soon as an animal has completed its embryonic development, and thereafter they occur in relatively fixed all-or-none fashion. Animal *tropisms* are included in this type of behav-

Figure 16.1 Animal tropism. Tropistic behavior, or orientational responses to external stimuli, is elicited in the case of these fish by the diver who takes the picture.

Figure 16.2 Active behavior based on learning. These circus elephants respond to stimuli selectively and display behavior made possible by earlier training and conditioning.

ior. A tropism is an automatic orientational response to an external stimulus. Examples of such responses are the balancing, positioning, and orienting processes that are largely under the control of nervous reflexes (Fig. 16.1). Apart from tropisms, predominantly reactive behaviors of animals also comprise certain routine and recurrent "housekeeping" activities, such as breathing, circulation, pupillary responses, and in general most processes that in vertebrates are controlled primarily by the autonomic nervous system and that in man do not or need not become conscious.

Reactive behavior thus maintains some of the most basic vital functions, and it is a considerable adaptive advantage that such behavior is comparatively unmodifiable. It would be highly detrimental if, for example, a life-preserving balancing or heat-avoiding response were readily modifiable by external factors. Reactive behaviors nevertheless can be modified, even if only to a minor degree in many cases. Most commonly such modifications occur through habituation—progressive loss of responsiveness to repeated stimulation, as outlined in Chap. 8. In some instances more drastic modifications are possible through classical conditioning, as in the retraining of the pupillary response in man (see Chap. 8).

In the category of active behavior, genetic inheritance endows an animal with certain behavioral potentials, but the actual realization of these potentials depends to a substantial degree on learning and the directive influence of the environment. Thus whereas an animal behaving reactively is comparable to a robot, a system that passively delivers preprogrammed responses on command, an animal behaving actively first "assesses" and "judges" stimuli and then "chooses" and adjusts its responses. It does so essentially by "comparing" a present situation with the memorized experiences of similar past situations and then behaving accordingly. The animal therefore is not merely automatically reactive, but selectively active (Fig. 16.2).

This form of behavior requires a fairly elaborate nervous system, and in effect it is limited largely to relatively advanced animals, notably arthropods and vertebrates. The specific behaviors in this category all must become perfected progressively, in part through parental training at immature stages and in part through later experience. Examples are flying and walking, searching for and handling food, grooming and sanitary behavior,

COLOR FIGURE 1 *Motion and feeding.* Animals obtain food either by active locomotion or by trapping small moving food organisms, as among the sessile feather-duster worms shown here. These polychaete annelids, named *Sabella* and related distantly to earthworms, live in attached tubular housings and project their feathery food-trapping crowns from the open ends of the tubes.

COLOR FIGURE 2 *Stages in the life cycle* of amphibians symbolize the main stages in the sexual development of animals generally. *A,* eggs. *B,* embryo. *C,* larva. *D,* adult. The transition from embryo to larva is achieved by hatching; that from larva to adult, by metamorphosis. The photographs are not reproduced to the same scale.

A

B

C

D

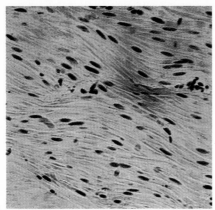

A

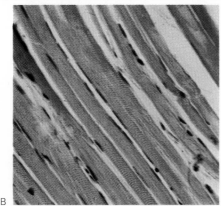

B

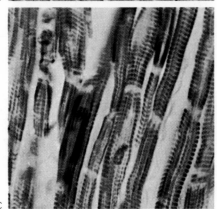

C

COLOR FIGURE 3 *Muscle types.* A, smooth muscle. Note the spindle-shaped cells. B, a few fibers of skeletal muscle. Note the cross striations and the nuclei in each fiber. See also Fig. 6.1 for structural details. C, cardiac muscle. Note the branching fibers, the nuclei, the faint longitudinal fibrils in each fiber, and the cross striations.

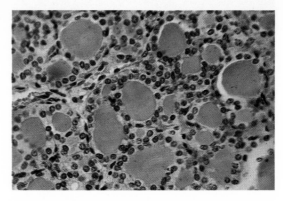

COLOR FIGURE 4 *The thyroid gland.* In this photomicrographic section through a portion of the gland, the spaces filled with red-colored material (''colloid'') are the regions where thyroid hormone accumulates before being transported away by blood.

COLOR FIGURE 5 *Blood.* A, human blood. Most of the round bodies in this smear are red corpuscles, note the absence of nuclei. Three nucleated white cells are in the center of the photo, and a few blood platelets are near the right edge. All these blood components are highly specialized and they do not divide. B, frog blood, red cells. Note the nuclei in the cells of this type.

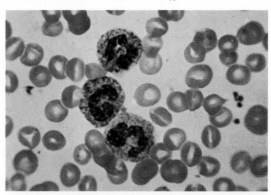

A

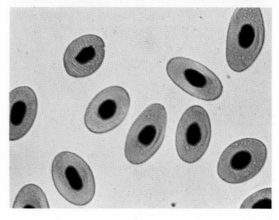

B

COLOR FIGURE 6 *Cenozoic animals:* reconstruction of a 8-ft-high titanothere (*Brontotherium*) from the Oligocene. Tortoises (*Stylemys*) in foreground.

COLOR FIGURE 7 *Pleistocene animals:* reconstruction of a huge ground sloth (*Megatherium*) and two armadillolike glyptodonts (*Glyptodon*).

COLOR FIGURE 8 *Parasitism.* This moth caterpillar carries the pupal cocoons of a parasitic wasp. When the adults emerge from the cocoons they will feed on the tissues of the caterpillar. (Courtesy *Carolina Biological Supply Company.*)

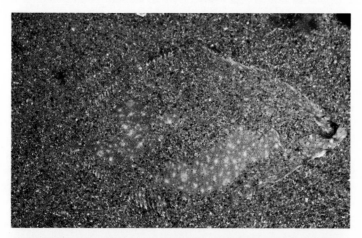

COLOR FIGURE 9 *Camouflage and body color.* Seen against three different backgrounds, this flatfish (a plaice) can adapt its skin-pigmentation to the environment. Information about the environment is communicated to the skin by eyes, nerves, and hormones. Pigment cells in the skin respond to such information by contracting or expanding, thereby altering body coloration.

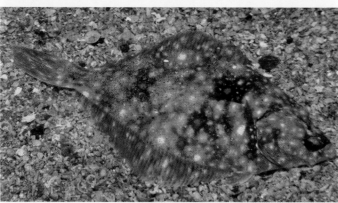

Facing page ▶

COLOR FIGURE 10 *Marine plankton.* Among organisms included here are diatoms, copepods, crustacean larvae, protozoa, animal eggs, and others. The conspicuous animal near center of photo is a copepod (*Calanus*), and toward the right the animal with the large eye is a crustacean (zoaea) larva.

COLOR FIGURE 11 *Plankton. A,* diatoms, common components of both marine and freshwater plankton. *B,* freshwater plankton. The algal components here are largely *Volvox,* globular flagellate colonies, and the animals are crustacean copepods and water fleas. The large-eyed animal near center is the water flea *Daphnia.*

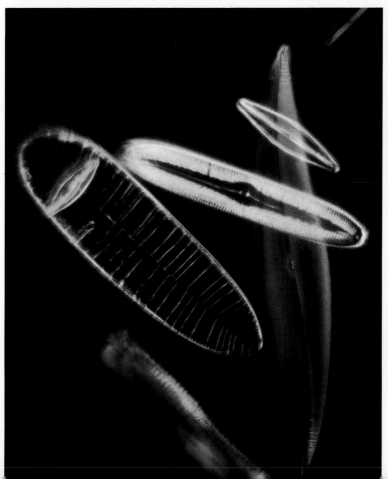

A

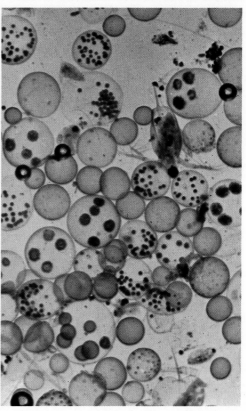

B

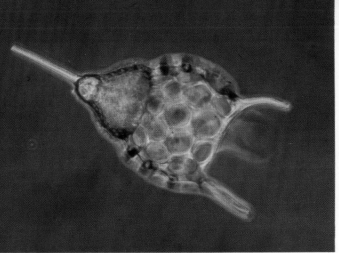

A

COLOR FIGURE 12 *Radiolaria.* A, *Podocystis.*
B, *Saturnulus.*

Facing page ▶

COLOR FIGURE 13 *Amoeboid protozoa.* A–D, sequence showing an amoeba catching and engulfing a paramecium.

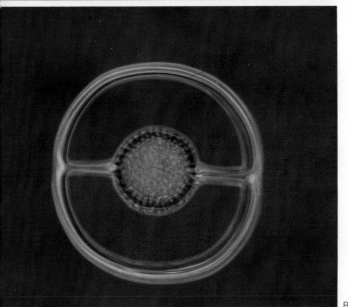

B

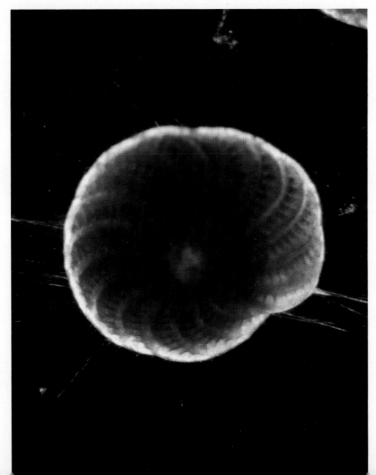

COLOR FIGURE 14 *Foraminifera:* Elphidium
(= *Polystomella*). Note the fine pseudopods protruding
through shell openings.

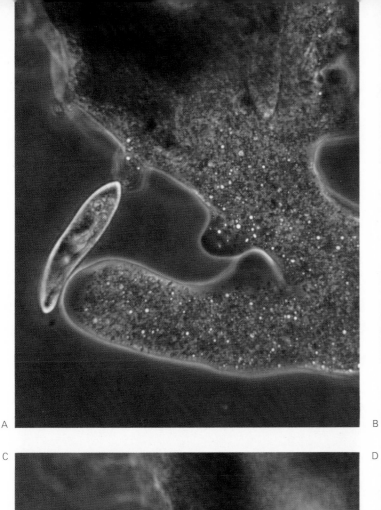

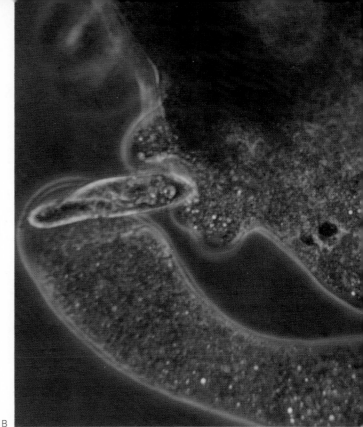

A

B

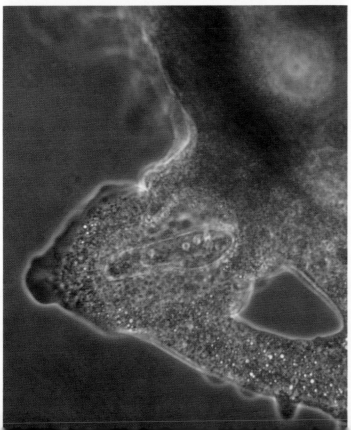

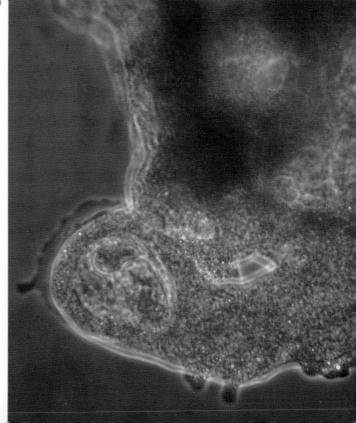

C

D

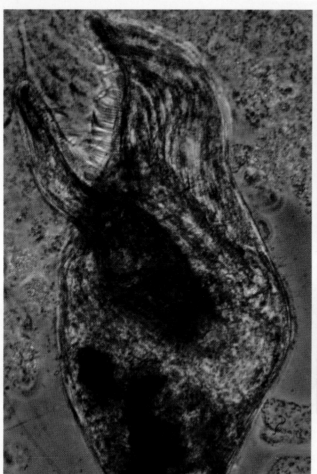

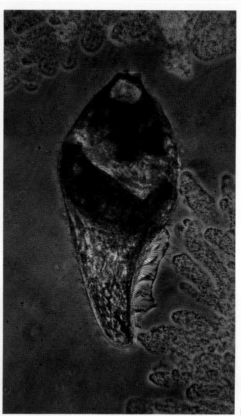

B

A

COLOR FIGURE 15 Ciliate protozoa. A, B, cannibalism in the heterotrich *Blepharisma.* One individual engulfs and digests a fellow member of the species.

COLOR FIGURE 16 Ciliate protozoa. The hypotrich *Stylonychia* is shown. Note the cirri.

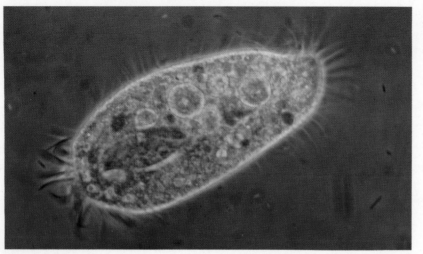

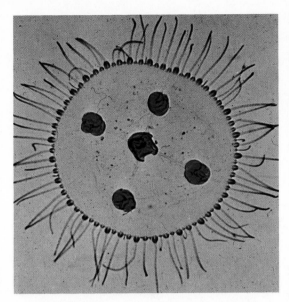

COLOR FIGURE 17 *Obelia.* In this top view of a medusa, note the central manubrium, the sex organs along the four radial canals, the marginal tentacles, and the absence of a velum in this genus. (Courtesy *Carolina Biological Supply Company.*)

COLOR FIGURE 18 *Tubularia:* portion of a polyp colony, from life.

COLOR FIGURE 19 *Physalia,* the Portuguese man-of-war. Each tentacle suspended from the gas-filled float represents a single cnidarian individual. Several different types of tentacle are present, and such colonies thus exhibit a high degree of polymorphism.

COLOR FIGURE 20 *Cyanea,* a scyphozoan jellyfish.

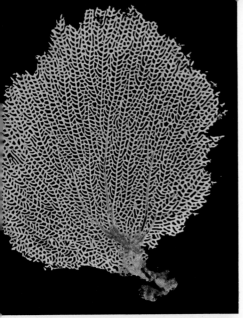

COLOR FIGURE 21 Octo-corallian Anthozoa. A, skeleton of the sea fan *Gorgonia*, a horny gorgonian coral. B, portion of the sea fan *Eunicella*, with expanded polyps. C, skeleton of the organpipe coral *Tubipora*. (A, C, courtesy Carolina Biological Supply Company)

COLOR FIGURE 22 Hexacorallian Anthozoa. A, top view of two sea anemones (*Actinia*). B, *Porites*, a stony madreporian coral.

A

B
C

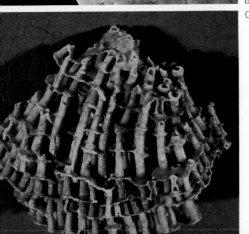

B

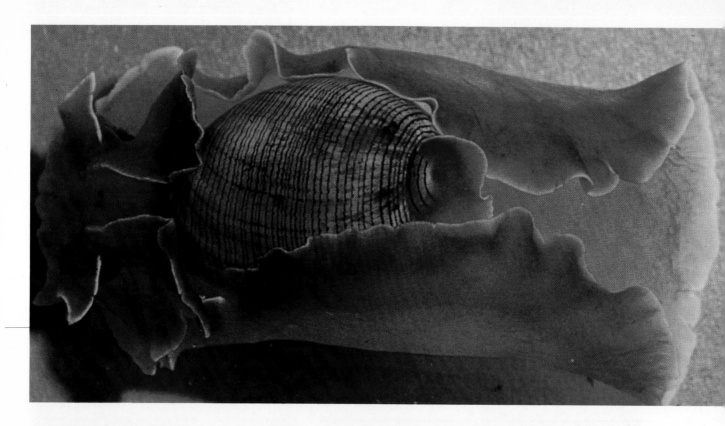

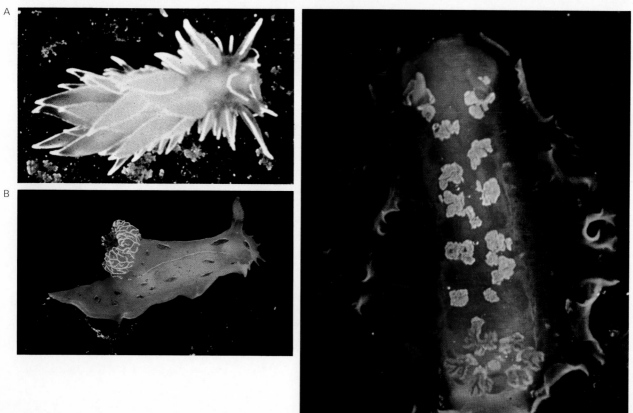

A

B

C

COLOR FIGURE 25 Clams: *Chlamys*, a scalloplike form. Note the small, light-blue ocelli on the fringed mantle along the shell edges. The orange animals at lower right are sea squirts, ascidian tunicates.

◀ Previous page

COLOR FIGURE 23 *Opisthobranch snails:* Hydatina, a sea-hare-like form.

◀ Previous page

COLOR FIGURE 24 *Opisthobranch snails.* A, *Dirona,* a nudibranch. Note cerata and absence of shell. B, *Polycera,* a nudibranch sea slug. Note the breathing rosette dorsoposteriorly. C, *Hexabranchus,* a nudibranch. Note the tentacles anteriorly. (A, courtesy *Carolina Biological Supply Company*.)

COLOR FIGURE 26 Clams: *Pholadidea,* a rock-boring form positioned in rock.

COLOR FIGURE 27 *Polychaetes.* A, a clamworm (*Nereis*) at the breeding stage. The egg-filled hetero-nereis portion is the reddish hind part of the worm. *B*, anterior part of the palolo worm *Leodice*. This animal swarms to the surface of the ocean on a particular night once a year and bursts, releasing its eggs or sperms. The polychaete *Sabella* is shown in Color Fig. 1. (*A*, courtesy *Carolina Biological Supply Company*.)

A

B

C

A

B

C

COLOR FIGURE 28 *Insect molting.* The animal is a cicada, an exopterygote of the order Homoptera. *A,* larva with skin beginning to split along dorsal midline. Note wing buds. *B,* adult emerging from larval skin. *C,* adult and discarded larval skeleton. (Courtesy *Carolina Biological Supply Company.*)

COLOR FIGURE 29 *Insect metamorphosis.* The animal is *Bombyx mori,* the silkworm (order Lepidoptera). *A,* cocoon and pupa inside. *B,* adult emerging, wings still uninflated. *C,* some minutes later, wings attaining mature size.

COLOR FIGURE 30 *Echinoderms.* *A, Antedon,* a crinoid feather star. *B,* a holothuroid sea cucumber. *C, Acanthaster,* an asteroid prickly starfish. *D,* an ophiuroid brittle star. *E, Culeita,* a echinoid cushion star, oral side.

Figure 16.3 Cognitive behavior and manipulative activity. Deliberate manipulation of some part of the environment, as exemplified by the actions of this circus bear, is characteristic of most cognitive forms of behavior.

communicating, courtship and mating, and most activities associated with offense and defense, family life, and sociality. Learning in active behavior often involves conditioned reflexes, particularly those leading to classical conditioning (see Chap. 8). Thus, by conditioning in the course of development internally controlled activities can come to be used in increasingly diverse environmental circumstances.

The most advanced forms of behavior are represented by cognitive activities. In these, genes provide only a very broad and general background for behavioral potentials, and the actual utilization of such potentials depends very largely on external influences. A cognitive activity in effect is an action pattern elaborated in more or less "deliberate" fashion by the animal itself. The animal here does not merely respond to stimuli actively, but it can *invent* its own actions. In the process the animal often also manipulates the environment—an important distinction from merely active behavior, in which such manipulation typically does not occur (Fig. 16.3).

Among the simplest cognitive behaviors, encountered in many arthropods and all vertebrates, are *exploratory* activities through which an animal familiarizes itself with any new condition in its environment. Strange objects are approached and inspected by sight, smell, and touch. Withdrawal usually follows quickly. Inspections then are undertaken repeatedly, but with gradually declining frequency. For any animal studied it has been found that the most complex novelties in the environment always tend to elicit the most exploration; a simple environmental change evidently is not as "interesting" or significant to the animal as an elaborate one. Animals placed in unvaryingly monotonous environments still exhibit exploratory behavior, though in greatly reduced fashion. Also, if mammals are prevented for extended periods from carrying out exploratory activities, their behavior can become highly abnormal. This is often the case with caged animals and with men suffering prolonged imprisonment.

Another type of cognitive behavior is *play*, particularly common in mammals but observed also in most other vertebrates. Play can involve activities more complex and diversified than those of exploration, and in some cases play and exploration can have the same adaptive result: increased familiarization with the environment, including most particularly the social environment. In this respect play among young animals often has an incidental (but nevertheless important) training function for later offensive and defensive behavior, for courtship activity, and for management of social conflicts. In many animals, for example, play involves activities that are but slightly modified forms of aggression or courtship. Also, mammals raised to adulthood in isolation or with insufficient social play later tend to make poor parents and often will ignore or abandon their offspring (Fig. 16.4).

Both exploration and play usually contribute to the development of even more complex cognitive activities, including particularly manipulative, or *instrumental,* behavior. It is characterized by responses that require "insight" and "foresight" and a capacity to analyze and weigh alternative choices. Also, such behavior can include purposeful alterations of environmental conditions and an exercise of skills, creativity, and inventiveness. Behavior of this type is restricted very largely to advanced mammals, particularly to monkeys, apes, and man. Behavioral control here (or in man, at any rate) is fully subject to conscious volition, and the result is conspicuously individual-specific;

Figure 16.4 Exploratory behavior.

each individual must develop his own level of manipulative competence, and in this respect the members of the species can vary greatly.

Most manipulative behavior, as well as cognitive behavior as a whole, depends largely on learning through operant conditioning (see Chap. 8). In the human case the reinforcement generally associated with such learning can be exceedingly subtle and indeed need not be immediate. In many instances, for example, we regard the very execution of a particular activity as its own reward, and we often find sufficient incentive for certain behaviors in a mere expectation of potential rewards, even if they accrue to others after our own lifetimes.

Note in this context that the execution of a behavior itself can sometimes be just as important as the objective of the behavior. For example, animals often stop eating not so much because they have consumed enough food as because they have become satisfied behaviorally; gratification can come primarily from the *act* of eating, and only secondarily from food as such. Thus, men, dogs, and other mammals often will look for food even when well fed, and under certain circumstances men sometimes refuse food despite gnawing hunger. In such cases the actions of an animal appear to be determined by behavioral or psychological states, not by metabolic states. A large body of evidence also indicates that the longer a particular activity has *not* been performed, the more readily it can be evoked; unfulfilled drives (*central excitatory states*) tend to become stronger with time. Such observations suggest that not all of behavior can be explained on the basis of just a few basic motivations, like hunger and sex. Virtually *each*

behavior appears to be motivated by a central excitatory state of its own.

Systems of Behavior

Stimulus and Response

Animals normally respond to *selected* stimuli only; they can filter particular sets of stimuli from the totality of the sensory cues they receive and respond specifically to just these. In many cases the nature of the response depends on the context in which stimuli are perceived, the whole configuration (or "Gestalt") of an environmental situation. As is well known, for example, the same stimuli can elicit diametrically opposite responses in different environmental circumstances. The ability to assess the setting in which stimuli are perceived and to modulate responses accordingly is a function of cognitive behavior and depends on learning and experience.

In many other cases, by contrast, the contexts of stimuli are quite unimportant; selected stimuli often evoke certain responses regardless of the environmental setting. Such stimuli are said to be *symbols*, behavioral *signs*, or *releasers*; they trigger, or "release," certain preprogrammed activities. Behavior in this case is active or even reactive, and the animal needs to learn only how to recognize given sign stimuli. The response then follows in more or less stereotyped fashion.

In herring gulls, for example, young birds peck at the bill of a parent when they wish to receive regurgitated food from the parent. The pecking is directed particularly at a red spot located near the tip of the lower bill of the parent. Experiments with cardboard models show that any pencil-shaped object with a red spot near its tip will elicit pecking. Indeed, if the colored area is designed in the shape of several short red stripes, the young gulls will peck more effectively and more vigorously at this substitute "bill" than at that of the natural parent. Similarly, many birds flee when a flying predator bird comes into their field of vision. Tests with cardboard models show that this standardized escape response can be evoked by any overhead object having the general outline of a long-stemmed cross—suggestive of the short neck, extended wings, and long fuselage typical of most predator birds (Fig. 16.5).

Among animals other than birds, male sticklebacks are prompted to courtship activity by nearly any model with a pronouncedly swollen, silvery underside, suggestive of a female with eggs. A true fishlike shape is otherwise not required, and male attraction will be the greater the more the model bulges downward. Courtship in certain male butterflies has been shown to be elicited not so much by the presence of females as by the speed with which females flutter their wings. Cardboard models that do not resemble females to any great extent but have winglike parts with particularly high flutter rates can arouse males even more than real females. Evidently, greatly exaggerated stimuli often tend to be more effective than normal ones. In man, too, supernormal stimulation is well known to be preferred in a wide variety of activities, and this phenomenon appears to be a general attribute of sign-induced behavior.

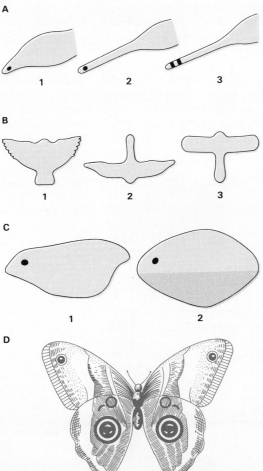

Figure 16.5 Releasers. A, models of bills of herring gulls, effective in eliciting pecking by young birds. Long, thin model bills (as in 2) are more effective as symbols than short, thick ones, and the models are most effective if the red-pigmented areas near the tips have the form of several stripes (as in 3). *B,* the hawk model (1) will elicit an escape response by birds, but the goose model (2) will not. Model 3 will provoke escape if it is moved in the direction of the top of the page but not if it is moved in the opposite direction. In the first case it suggests a hawk, in the second a goose. *C,* a male stickleback will court an inanimate model of a fish having a swollen underside (1) suggestive of a female with eggs. Also, a male will attack a model having a red underside (2), suggestive of another male in breeding condition. *D,* many butterflies have conspicuous pigmented areas on their wings, suggestive of the eyes of an owl. Such eyelike patterns probably serve to protect the insects from predation by owl-fearing birds. (After Tinbergen.)

Signs need not necessarily be visual but can also involve auditory, tactile, or any other sensory modes. In sound-producing animals, for example, auditory symbols are well known to be specific releasers in courtship and mating, in offense and defense, and indeed in most social behavior. Numerous releasers are chemical, mediated by the senses of taste and smell. For example, certain clams exhibit an escape response if placed in water in which starfish have been present. But the same water will attract certain polychaete worms that normally live with starfish. Among mammals, aggressive or submissive behavior evoked by chemical signs is particularly well known and familiar. For example, a cat is alerted to flight by dog smells but to attack by rodent smells.

Specially important are *pheromones,* chemical signals used between members of the same species. For example, bees in a food-rich area secrete pheromones that attract other bees to the location. When wounded, minnows and other fishes release alarm pheromones from the skin that elicit escape responses in other members of a school. Under conditions of stress many types of ant broadcast alarm substances that induce fellow ants to become aggressive. Probably best known are the sex pheromones, by which individuals of either sex attract or evoke reproductive behavior in potential mates. Sex pheromones also can have more subtle effects. For example, caged female mice will become synchronized in their reproductive cycles if the animals are exposed to the scent of a male mouse. Moreover, a pregnant mouse can be induced to abort by exposing her to the odor of a strange male (even if the normal male mate is present). In many animals pheromonal cues also play important roles in the development of parent-young relationships, in recognition of species members, in establishing appropriate spatial distributions among the individuals of a social group, and indeed in influencing most behavioral interactions in the species.

Clearly then, signs as a whole represent various forms of *communication.* Moreover, it need hardly be pointed out that signs are particularly significant in human behavior. Man actually responds to more kinds and more complex kinds of signs than any other animal. His entire language is a sign system, and this language includes not only words but also postural and facial expressions, as well as visual, olfactory, and tactile cues. Not every such communication evokes stereotyped responses, to be sure, but many certainly do. Indeed, the wide occurrence of releasers of this sort is underscored even by common phrases such as "authority symbol," "danger sign," "friendship token," or "sex symbol."

A particularly dramatic demonstration of the importance of releasers is provided by the phenomenon of *imprinting*. Newly hatched chickens, geese, ducks, and other birds develop a strong bond of kinship with their female parent, manifested primarily through a standardized "following response": the young run behind their mother (Fig. 16.6). It has been found that this response develops during a specific period soon after hatching, the so-called critical period, and that it depends on certain releaser stimuli emanating from the mother bird: her overall size, certain aspects of her shape and coloring, the sounds she produces, and her movement away from the young. If such stimuli become "imprinted" in the young during the critical period, the offspring henceforth will accept the adult as their mother and will acknowledge this acceptance by the "following" response. Imprinting normally also provides the recognition signs that in later life prompt selection and acceptance of appropriate mates; a mate must have the same stimulus attributes as the parent (which presumably ensures that the mate will be of the right species).

Figure 16.6 The "following" response, illustrated by the offspring of a Canada goose.

Experiments show clearly that imprinting still can occur if the actual mother is dispensed with. All that is required is another stimulus source that will produce the necessary signs during the critical period. Thus, young birds will readily accept foster mothers of the same species, or adults of another species (man included), or even mechanical models or inanimate objects that have the requisite stimulus properties. The fixation on such "surrogate" mothers becomes just as strong and permanent as on a natural parent. Indeed, birds raised with surrogates may flee from their real parents; and at maturity such birds sometimes reject potential mates of their own species but will try to mate with the surrogates, even if these are inanimate objects. Imprinting probably represents a special kind of learning. It differs from other kinds in that its occurrence is limited to a brief period in life and that it remains permanent thereafter (Fig. 16.7).

Releaser stimuli can also originate inside the animal body. One of the best examples is hunger in mammals, which probably is induced when a low sugar concentration in blood stimulates a hunger center in the brain. The irrelevance of stimulus contexts is shown by the finding that an animal will eat whenever its hunger center is stimulated, regardless of the actual nutritional state of the body. More recently a thirst center has been located that is differentially sensitive to salt concentrations in the body. If this concentration exceeds a certain level an animal will feel thirsty and will drink, regardless of how much water it has already consumed.

Experiments on rats have also revealed the existence of pleasure and displeasure centers in the brain. Test rats were fitted with tiny permanent electrodes that reached to these centers, and such animals were allowed to activate these devices on their own by pressing appropriate control levers. After a rat had stimulated its displeasure center once, it avoided a repeat performance forever. But if such animals stimulated their pleasure centers, they continued this self-stimulation from then on without interruption, not even pausing to eat, until they dropped from exhaustion; and on recovery they promptly resumed the self-stimulation to the exclusion of any other activity.

Under normal conditions, however, responses evoked by sign stimuli are among the most valuable functionally and thus among the most adaptive in the behavioral repertoire of an animal.

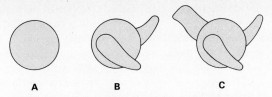

A   **B**   **C**

Figure 16.7 Imprinting. Inanimate models such as shown here are effective in eliciting the "following response" in young chickens. A simple model as in *A* proves to be more effective than a more complex one as in *C*. Also, blue-colored models are more effective than red ones, and the latter are more effective than yellow ones. Effectiveness is measured by the relative distance a chick will follow a model.

Time Orientation: Biological Clocks

Much of animal behavior is regularly recurrent in cyclic fashion, and such behavioral rhythms often are synchronized with some cycle in the physical environment. For example, numerous behaviors have a 24-hr *circadian* rhythm, geared to the solar cycle of day and night and the accompanying cycle of temperature. Most coastal animals display behavioral periodicities synchronized with *tidal* cycles. Many aquatic and terrestrial animals exhibit *lunar* rhythms. And behavior in nearly all animals also occurs in annual *seasonal* rhythms. Reproductive activity in particular tends to be seasonally cyclic.

In an attempt to explain such behavioral rhythms, a first obvious hypothesis might be that the rhythms are causally related to the recurrent environmental conditions. If so, it should be readily possible to change a behavioral rhythm by experimentally changing the associated environmental cycle. For example, what would happen to a circadian (or other kind of) rhythm if an animal were maintained artifically under uninterruptedly *constant* conditions of lighting, temperature, humidity, and other relevant environmental variables. Numerous experiments of this sort have actually been carried out, and the results obtained are exceedingly uniform and clearcut: an original behavioral rhythm always tends to *persist,* despite constant environmental conditions (Fig. 16.8).

It is usually observed in such cases that the successive cycles gradually drift out of phase from an exact 24-hr (or other) periodicity. The rhythm then stabilizes at some shorter or longer cycle interval, yet this new interval nevertheless remains fairly close to the original rhythm. Such results show clearly that rhythmic behavior can occur independently of environmental cycles. Animals therefore are considered to have built-in *biological clocks,* according to which the timing of recurrent activities is programmed.

Much has been learned about the operation of biological clocks by relocating animals geographically. Animals are normally adapted to the geographic latitudes and longitudes at which they live, and their behavior follows the rhythm of local time. But it has been found that if animals are transported to different geographic locations, their activities nevertheless continue in synchrony with the original local time. For example, Atlantic fiddler crabs flown in dark chambers to Pacific waters exhibit cyclic color changes according to the tides at their Atlantic point of origin. Bees trained in Paris to feed at a fixed time of day remain on Paris time after being airfreighted to New York. And fall-breeding sheep transported from the northern to the southern hemisphere continue to reproduce at the time of the northern fall, when the southern hemisphere is in the spring season.

However, relocation experiments also show that after some interval biological clocks invariably do reset to new local times. Thus the crabs, bees, and sheep referred to above eventually synchronize behaviorally with their new longitudes and latitudes. Such gradual adjustment also is the common experience of people who take up residence in new time zones. In general, internally controlled

Figure 16.8 Circadian rhythms. In these two examples, the (color) curves indicate the changing intensity of the activities symbolized in column at left. Note the regular repeats during the 3-day intervals shown here. The daily activity curve applies specifically to rats but holds also for many other organisms. (After F. A. Brown, Jr.)

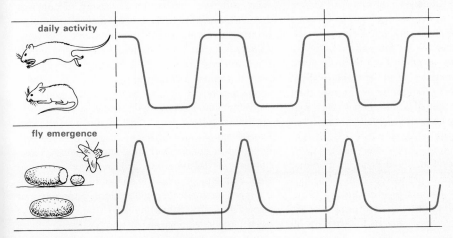

daily activity

fly emergence

rhythms are said to become *entrained* to local external cycles.

Entrainment proves to be possible only if a new periodicity imposed on an animal does not differ too greatly from that of a natural environmental cycle. In mice, for example, circadian rhythms can be entrained to artificial periodicities as short as 21 hr or as long as 27 hr. But entrainment cannot be readily achieved if an artificial cycle deviates by more than 3 hr from the 24-hr norm. Similarly, the sleeping-waking rhythm in man has repeatedly proved to be resistant to artificial night-day cycles that deviate substantially from 24-hr periods. In general, cycles significantly shorter than 24 hr tend to add together and those longer than 24 hr tend to subdivide, with the result that the actual behavioral rhythms follow a natural 24-hr cycle as closely as the experimental conditions will allow.

In both circadian and seasonal rhythms, the most common stimuli by which animals clock the passage of time appear to be light direction, light intensity, and light duration. For example, many animals tell daily time by the progressive change in the position of the sun (see next section). Animals are also sensitive to the daily and seasonal cycles of light intensity, and many of their rhythmic activities can be reset or modified by experimental alterations of such intensities. Moreover, animals tell seasonal time most particularly by assessing changes in day lengths, or *photoperiods*.

For example, the onset of a breeding season is generally clocked by the changes in the day lengths during the preceding months (see Chap. 10). The important photoperiodic stimulus here is the total time during which an animal is exposed to illumination—as though the animal could count the hours of light it receives. In spring breeders, for example, sex organs become active when a certain total amount of illumination has been received in the course of the progressively longer days after the winter solstice. By artificially giving an animal the necessary total amount of light within a shorter or a longer time interval than nature does, the onset of a breeding season can be correspondingly hastened or delayed.

In arthropods and vertebrates the clocks that control seasonal reproductive rhythms are well known to have their functional basis in rhythmic neural and endocrine processes (see also Fig. 10.18). However, nervous or endocrine components are present neither in protozoa nor in sponges, yet such organisms exhibit behavioral

rhythms nevertheless. Also, any single animal exhibits numerous behavioral cycles simultaneously, all with their own specific periodicities. An animal therefore must be considered to contain many different biological clocks, with different rhythms, locations, and functional mechanisms.

Even so, the general conclusion is probably warranted that control of cyclic behavior is achieved by interaction of at least two sets of factors. One is internal and is represented by inherited biological clocks that, if left to themselves, keep their own intrinsic time. The other set is external and is represented by cycles in the physical environment. Within definite and relatively narrow limits, the settings of the internal clocks can become adjusted or readjusted until synchrony with the external cycles is established. In this way an animal interacts with its physical environment and can adapt its behavior to the cyclic time changes in its local habitat.

Space Orientation: Directional Motion

The most basic orientational behaviors are tropisms, which can maintain even sessile animals in optimum positions relative to their immediate surroundings. Motile animals in addition can move to optimum locations in their nearby space through various *kineses,* or movements that intensify in proportion to the strength of particular stimuli. For example, flatworms stay relatively quiescent in darkness but are stimulated to locomotion by light, the more so the stronger the light. As a result, the worms will congregate in the darkest places available to them, as under stones in nature, where locomotor activity will be at a minimum. Animals also exhibit kineses in response to stimuli other than light. For example, zoned environmental variations in temperature, humidity, salinity, oxygen content, and many other conditions are well known to elicit locomotor kineses that distribute animals to the optimum habitat zones available (Fig. 16.9).

Many animals move well beyond their immediate home surroundings, and in the process they usually remain fully "aware" of their position and direction in space. A particularly significant navigational aid is *sun compass orientation,* ability to steer by the position of the sun and to compensate course directions according to the apparent motion of the sun across the sky. This process has been studied extensively in honeybees, in which

A

B

C

Figure 16.9 Kinesis, or increase in activity with increased stimulation. In the example illustrated, the stimulus of the toy becomes stronger as the cat's interest increases. Activity then increases correspondingly.

direction finding plays an especially important role: survival of a colony depends on scouting bees that must inform other members of a hive about the exact location of food-yielding fields of flowers.

It has been found that a scouting bee returning to its hive communicates with fellow bees by means of a *waggle dance*. This is a side-to-side wiggle of the abdomen, carried out while the bee moves along the upright surface of the honeycomb. The intensity of the waggle (and also sounds made by the bee) is an indication of the distance to the food source. Actually bees do not communicate distances as such, but elapsed time on the outward flight from the hive. For if a scouting bee has flown against a headwind its waggling will signal prolonged time in flight, as if it had traveled greater-than-actual distances. Similarly, if a bee is made to *walk* to a food source only a few feet from the hive, its waggling will erroneously imply travel over very great distances.

The angle between the vertical on a honeycomb and the direction of a bee's dancing path indicates the angle between the sun's position and the food source in relation to the hive (Fig. 16.10). The direction of the dancing path therefore differs according to the time of day the dance is performed; it shifts progressively counterclockwise relative to the vertical, in correspondence with the gradual change in the sun's position. If the sky is overcast, polarized and ultraviolet light passing from the sun through the clouds still provides the bee with adequate data on the sun's position. Moreover, bees also have been found to utilize visual clues from landmarks in the environment. During the night waggle dances continue to shift in direction, in line with where the sun would be stationed if it were visible. Thus the orientation mechanism is clearly dependent on the biological clock, and, whether the sun is actually visible or not, a bee "knows what time it is."

Ability to orient by the sun is influenced greatly by learning. If inexperienced bees are trained to search for food only during afternoons but subsequently are allowed to scout only during morning hours, the insects communicate flight directions incorrectly at first. Apparently they have not yet learned to assess the entire daily course of the sun. However, after some days of morning experience the bees do communicate correctly.

Sun compass orientation is known to occur also in horseshoe crabs, certain ants, beetles, spiders, and in several other arthropod groups. Among

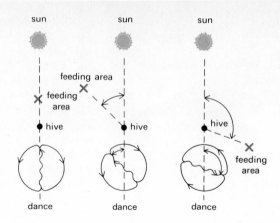

Figure 16.10 Waggle dances in bees. The angle of the dancing path of a honeybee relative to the vertical on the surface of a honeycomb (wavy lines in lower diagrams) indicates the direction of the feeding area in relation to the position of the sun as seen from the hive. (After von Frisch.)

vertebrates the process has been demonstrated to date in some fishes, turtles, and birds. In contrast to arthropod navigation, that of vertebrates appears to require actual sun sightings. At night, under overcasts, or even under clear skies with the sun not directly visible, the orientation of these animals becomes confused or less precise.

In certain cases sun compass mechanisms appear to play some role in navigational guidance

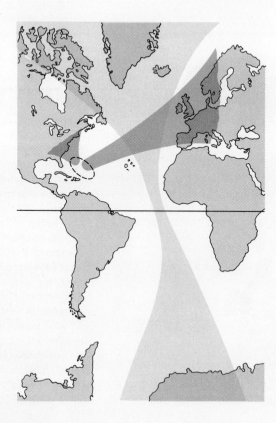

Figure 16.11 Migrations. The migratory paths of arctic terns are shown in light color, those of Atlantic eels in dark color. American and European eels spawn in the Sargasso Sea (indicated by a dashed line).

during seasonal migrations and in homing. For example, certain sea turtles undertake annual migrations of several thousand miles to lay eggs at specific beaches. Correct navigation here appears to depend on the sun at least in part; as in other vertebrates, location finding becomes confused or less precise if the sun is not visible. Homing pigeons, too, orient by the sun. By entraining these birds to artificial light-dark cycles, it has been found that the flight directions taken in homing are adjusted to solar time and are influenced by the biological clock. However, other migrating birds do not appear to navigate specifically by the sun. Indeed such migrants actually travel more by night than by day, and experimental evidence suggests that they probably orient by star sightings.

Seasonal migrations serve to bring animals to fixed breeding or feeding grounds. Apart from birds, migrant animals include certain butterflies and some other arthropods, eels, salmon, and other fishes, some turtles, and a large variety of mammals. The record for long-distance travel is held by the arctic tern, a bird that makes an annual round trip of some 22,000 miles between the Arctic and the Antarctic (Fig. 16.11). The methods by which most of these migrant animals navigate remain largely unclear. Although sun and star orientation might serve as navigational aids in certain cases, such mechanisms are insufficient to account for all aspects of migratory behavior. In aquatic types, for example, optical cues can have only limited significance at best. In some cases, as in salmon, navigation appears to be based on *chemotaxis*, or ability to distinguish different kinds of waters by relative oxygen content, amount of silting, and more subtle chemical characteristics that can be smelled or tasted. But for many other animals, aquatic, terrestrial, or aerial, the sensory basis of navigation still is obscure.

The problem is puzzling enough where animals make the same trip every year, like seals and birds. If nothing else, a remarkable memory for landmarks, prevailing winds, or ocean currents appears to be indicated. The problem becomes even more puzzling, however, when correct navigation takes place without optical cues of any kind and without demonstrable prior experience. For example, migrant birds generally can maintain their course despite zero visibility, as in flight through dense fog, and they can correct course after displacement by winds, even over the open sea where the view is the same in all directions. And although young birds generally fly with older ones, experiments show that young birds still navigate correctly when they migrate alone and for the first time.

Some aspects of how long-distance travel is initiated have been clarified for birds, and a similar mechanism presumably operates in other vertebrates. In these animals migrations are associated with the neural and hormonal changes of the annual breeding cycle. Birds typically migrate during the period preceding their breeding season. It has also been demonstrated that day lengths and total light hours received are the specific external stimuli that elicit breeding migrations, just as they are the stimuli for the breeding season itself. For example, migratory behavior can be changed experimentally by resetting the environmental light clock. It is known, moreover, that photoperiodic stimulation leads not only to sexual but also to adaptive metabolic changes in migrating birds: the animals deposit increasing amounts of storage fat in their bodies, and they can then subsist without regular food intake during migrations (see Fig. 10.18).

For animals other than vertebrates the controls for the onset of migrations are largely unknown as yet, as are those for the control of return migrations, in which the end result is not breeding but most often feeding. Migratory behavior thus still represents one of the most enigmatic phenomena in all zoology.

Social Behavior

Localized groupings of members of the same species are common throughout the living world, but true social populations are not. Individuals of a species group together in locations favorable to them, and men also aggregate temporarily at places of special interest. By contrast, the members of a population form a society only when they interact with each other directly.

Social Groupings

In all social animals the members of a group are interdependent and variously specialized in function, and their survival depends on cooperation in the group. In most cases the member animals are also *polymorphic,* or specialized structurally. For example, colonial cnidarians, ectoprocts, and tunicates often consist of aggregations of different kinds of polymorphic individuals. Because their survival requires mutual cooperation in activities such as feeding, locomotion, protection, or reproduction, colonies of this type qualify as simple societies. Polymorphism also is pronounced in the far more complex societies of insects and vertebrates. Insect societies include structurally different castes, and social vertebrates usually exhibit *sexual dimorphism,* or structural distinctions between males and females (and their young; Fig. 16.12).

The adaptive advantages of social living appear to be many. First, an interacting group often is more effective than separate individuals in obtaining food, in finding mates, and in protective and defensive activities. Indeed, a behavior unique to societies is individual self-sacrifice for the sake of the group, as among worker bees, which die after stinging, and among men, who often are prompted to give their lives for others. Second, metabolism and growth in many cases take place more effectively in a social milieu. In a slightly unfavorable environment, for example, goldfish and other aquatic animals survive better in social groups than as separate individuals, apparently because normal body secretions (mainly salts) neutralize toxic substances in the water and thereby "condition" the medium. Also, some secretions of social animals are known to have stimulating (or inhibiting) effects on growth, and the collective body heat generated by a compact group can reduce the level of metabolism that each individual must maintain. Third, social living substantially facilitates learning and imitative behavior, an important factor in individual survival. Above all, by regulating and controlling the interactions among individuals, social life reduces aggression and competition, keeps conflict and combat on largely nonlethal levels, and in general provides standards of behavior that tend to promote survival of both the individual and the whole species.

Advanced societies occur in animals with advanced evolutionary status—insects and vertebrates. Social life has evolved quite independently in these two groups, and the organization of their societies also is quite different. It is interesting nevertheless that certain external forms of their social behavior are superficially rather parallel. Among all animals, only some social insects and

Figure 16.12 Polymorphism, a common phenomenon in social groups. *A,* the Portugese man-of-war *Physalia.* This cnidarian is a floating colony of numerous individuals of various structural types—feeding, reproductive, protective, and float-forming polymorphs. The unit qualifies as a primitive social formation. *B,* ants (as well as bees and other social insects) exhibit polymorphism in that queens (shown here) are winged whereas other individuals of a species are wingless. *C,* sexual dimorphism, a special form of polymorphism, is illustrated by the fur seals shown here. The male at right is far larger and more colorful than the female at left.

some social vertebrates pursue agriculture, domesticate other organisms, practice slavery, engage in war, and commit suicide.

In each of four insect groups—termites, ants, bees, and wasps—different species form societies of different degrees of complexity. All typically build *nests,* and all contain structurally and functionally different social *castes;* each member is adapted from the outset to carry out specific social functions. For example, a population of honeybees contains three structurally distinct types, a *queen,*

tens or hundreds of male *drones,* and from 20,000 to 80,000 *workers.* The queen and the stingless drones are fertile, and their main functions are reproductive. The smaller-bodied workers are all sterile females. They build the hive, protect the colony from strange bees and enemies, collect food, feed the queen and the drones, and nurse the young (Fig. 16.13).

Polymorphic castes of this sort develop from basically the same kinds of eggs. Drones arise by natural parthenogenesis, from eggs that remain unfertilized. Fertilized eggs develop either into queens or into workers, depending on the type of food the larvae receive from their worker nurses. Larvae to be raised as workers are fed a "regular" diet of plant pollen and honey. Queens form when the larvae receive an especially rich *royal jelly,* containing pollen, honey, and comparatively huge amounts of certain vitamins (particularly pantothenic acid).

Figure 16.13 Social insects: honeybees. Worker at left, queen in middle, drone at right.

Figure 16.14 Social insects: termites. A, worker. B, soldier. C, winged king. D, portion of nest. In central chamber note queen, her abdomen swollen with eggs, being cared for by workers. Winged king in lower right corner, larval queen in upper left corner.

In other social insects polymorphic distinctions can be considerably more diversified. For example, many species of ants and termites include sterile wingless workers as well as sterile wingless *soldiers*. The latter are strong-jawed, heavily armored individuals that accompany work crews outside and keep order inside the nest. Soldiers in many cases cannot feed themselves and are cared for by workers. Besides a winged fertile queen and one or several winged fertile males (kings), ant and termite societies often maintain structurally distinct lesser "royalty." Such individuals probably develop when larvae are fed more than enough to produce workers but not enough to produce queens (Fig. 16.14).

Among insect societies in general, well-ordered group living is a result of the early and permanent subordination of each individual to the group. The stereotyped nature of this subordination is well shown in, for example, tropical army ants. If a column of such ants is made to travel in a circle they will maintain the circle endlessly, each individual continuing to follow the one before it. Behavior evidently is so automatic that an ant is incapable of thinking itself out of an unproductive situation (Fig. 16.15).

In effect, the essence of the behavioral system of social insects is that an individual is never free to seek its place in the society on its own, but that its place is already predetermined as soon as life begins. Intrasocial competition is thereby forestalled, and the responsibilities of the individual to the group and of the group to the individual remain fully defined at all times. This social predetermination is nowhere more complete than in reproduction; social conflict is reduced most particularly by the circumstance that reproduction is virtually removed from the field of competition. Reproductive activity revolves around a single fertile female, and the whole social population consists substantially of her own immediate family—which is sterile. All social institutions therefore rest on the genes of this one female, and her genes alone provide continuity from one generation to the next. As a result, internal social conflicts can hardly begin to become an issue.

Vertebrate societies are organized quite differently in this respect. Apart from the sexual dimorphism of males and females, the members of vertebrate populations are structurally more or less alike, at least during the early stages of life. Most important, *all* members of a vertebrate society are potential reproducers. Later specializations and social stratifications are largely functional and are based on differences in size, strength, intelligence, emotionality, skill, and the comparative effectiveness with which the animals apply these and other personal traits to the problems of living. Unlike a social insect, a social vertebrate then does have to seek a place in the society on its own, and it must usually do so in direct competition with the other members of the group. The competition revolves primarily around securing personal living space, food, and reproductive opportunity.

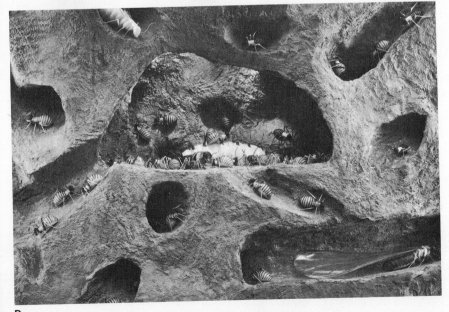

A   B   C

D

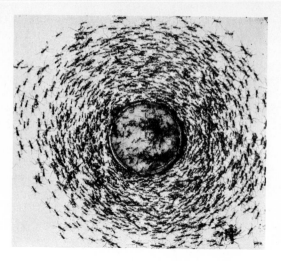

Figure 16.15 Social insects: army ants. If a column of army ants is made to travel in a circle, as in the photo, then the animals will continue to circle endlessly. Evidently this behavior is governed so completely by internal factors that the insects are incapable of modifying their response.

of *aggressive behavior*. On the other hand, the group is equipped to prevent its own disruption, and thus is equipped to survive as an adaptive unit, by controlling and limiting the aggressive behavior of its members through a number of powerful *social constraints*.

Aggression

Through aggressive (or *agonistic*) behavior an individual often asserts his membership in the group and obtains his share of space, food, and mates. The means of aggression include *threat displays* and actual *fighting*.

By variously warning, bluffing, or scaring a potential opponent, aggressive displays serve to forestall real fights. In many cases such displays are effective even against members of other species. For example, a prominent presentation of weapons, a fixed gaze leveled directly at an opponent, an evident muscular tensing, a slow wary stalk accompanied in some cases by fright-inducing sounds, or a size-increasing fluffing up

The vertebrate society therefore features a perpetual interaction of two opposing forces. On the one hand, each individual is equipped to function as a more or less effective competer, and thus is equipped to survive in a group, by being capable

Figure 16.16 Aggressive and submissive displays. A, sticklebacks assume a threatening stance by pointing the head downward (left), a submissive stance by lying in horizontal position or by pointing the head upward (right). *B, top,* a male finch in an aggressive posture, with head extended forward and feathers sleeked down tight against the body; *bottom,* a female finch in a submissive posture, with the body in sitting position, the head retracted, and the feathers fluffed up. *C,* dogs signal threat and submission by the familiar postures indicated at right and left, respectively *D*, in this confrontation, the aggressive protagonist at left bares its teeth, points its ears up, stares at the opponent, raises its tail, and assumes a threat posture. The submissive animal meanwhile crouches down, lays its ears back, closes its mouth, lowers its tail, and carefully looks away from the opponent.

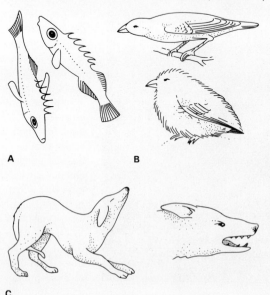

A

B

C

D

or puffing up of the body, all are widely recognized among vertebrates as signs of threat. In most instances a series of exchanges of such displays between opponents suffices to elicit flight or sub-

Figure 16.17 Redirection and displacement. A, male baboon in foreground is being threatened by male in background. However, avoiding both aggressive and submissive responses, the threatened male gives a redirected response instead: it focuses attention on uninvolved bystanders and walks over to the grooming couple nearby. *B,* under threat by male baboon at left, the male at right gives a displacement response: displaying neither aggressive nor submissive behavior, it carries out an activity not pertinent to the confrontation and grooms its mate instead.

A

B

mission by one (Fig. 16.16). Aggressive (and also submissive) displays have probably evolved as symbolic, less dangerous offshoots of real fights; the actions and movements of displays often are quite similar to those of combat. However, they have become *ritualized*, by increases or decreases in the speeds, intensities, or durations of particular movements, and by behavioral changes that exaggerate or draw attention to specific releaser signs of aggression or submission.

When an animal finds itself in a threatening social situation, the necessity of deciding on either fight or flight often creates psychological conflict—in many cases aggression and submission can be equally undesirable. In such circumstances the animal can frequently avoid becoming involved in either aggression or submission by minimizing the attention it creates; it might "freeze" to immobility or feign sleep. Sometimes psychological conflict can also be resolved through *redirection:* directing an aggressive response not at the opponent but at an extraneous third entity—some nearby inanimate object, for example, or an "innocent bystander." Or the animal can carry out a *displacement* response, one not pertinent to the threat situation at all—excessive grooming, for example, or sudden preoccupation with some irrelevant feature of the environment. Responses of this sort are well known in human behavior, and they are also encountered widely in other mammals and in birds (Fig. 16.17).

Outright combat occurs comparatively rarely, and indeed only when it cannot be avoided by other means. Also, notwithstanding some known exceptions (man is one), very few animals seek out fights deliberately. In general, fighting tends to occur mainly when the risks of fighting become less significant than the motivation to compete for space, food, mates, or social status. Thus it has been shown for most vertebrates studied that the frequency of fighting increases with rising population density. Similarly, many animals require a free personal space immediately around them and will fight if this space is violated (Fig. 16.18). Fighting also increases with progressive food scarcity and at the approach of the breeding season. A good many fights occur when an individual is prevented from completing a goal-directed activity, a situation equivalent to frustration in man.

Whatever the provocation, vertebrates rarely fight to the death (man sometimes being a notable exception). If an exchange of aggressive displays

Figure 16.18 Personal space. These nesting cormorants illustrate how each individual of a group often requires, and defends, a certain amount of space around it.

does not establish the dominance of one of the opponents, such a dominance usually becomes apparent fairly quickly once combat has started. The fighting then terminates, and in most cases it terminates in any event as soon as one of the protagonists is wounded. The objective of animal combat is not destruction of an opponent, but merely elimination of the threat that his presence represents—at minimum cost of effort and risk to the threatened animal (Fig. 16.19).

The capacity to behave aggressively or submissively is inherited, but the execution of such behavior is largely learned. Experience teaches not only the techniques of aggressive display and the tactics of combat, but also the acceptable limits of the possible advantages and risks. Similarly significant is learning by imitation and, in predatory types, learning how to obtain food: predation and aggression involve many similar behavior patterns, even though killing is the objective in the first case. Notwithstanding the genetic variations among different individuals, early training to a large extent determines the threshold levels for aggressive responses and the intensities with which an individual will exhibit such responses. As has been shown for mice and as is well known for men, an individual can be *taught* to respond more or less readily and more or less intensely to a given provocation, and thus to be an aggressive or a submissive "type."

In vertebrates, aggressive behavior is under the control of at least two neuroendocrine mechanisms. Both are brought into play by a large variety of stimuli that can be variously visual, auditory,

olfactory, tactile, or combinations of these. In one of these mechanisms, nerve impulses are relayed from sense organs to the brain, which in turn stimulates the adrenal medulla. This gland then secretes increased amounts of adrenaline and noradrenaline, hormones that initiate the "alarm reaction" (see Chap. 8, Table 6). This reaction facilitates the sharply intensified activities that are required under most conditions of emotional or physical stress, including especially those calling for aggressive responses. Adrenaline specifically has been shown to raise emotional tension (manifested by red-faced anger in man), whereas noradrenaline specifically spurs overt action (accompanied by white-faced determination in man).

In the second mechanism, which can operate by itself or together with the alarm response, sensory impulses are transmitted to the hypothalamus in the brain. This region in turn transmits hormonal transmitter substances (particularly noradrenaline) to the pituitary gland, which secretes increased amounts of gonadotropic hormones as a result (see Fig. 10.18). These now induce the reproductive organs to secrete larger quantities of male and female sex hormones. Acting through effects on the nervous system (and especially the hypothalamus), the male hormones are particularly potent in triggering aggressive behavior. As pointed out in Chap. 8, both sexes produce *both* male and female hormones, though in different relative quantities; and in either sex the male hormone promotes aggressive behavior. That this is so is clearly shown by experimental hormone injections. Female hormones either are without significant effect or, in most vertebrates (man included), reduce aggressiveness and correspondingly increase submissiveness.

In effect, vertebrate aggressiveness waxes and wanes in synchrony with sexuality and the annual breeding cycle. However, aggressive behavior is not completely dependent on sex hormones; castrated animals do not become totally submissive, and in some cases castration reduces aggressiveness only very slightly. Chronic aggressiveness also has nonreproductive effects, all ultimately traceable to the endocrine and neural mechanisms just discussed. For example, growth is often inhibited and resistance to infections is lessened. Man in particular is subject to a distinct "stress syndrome," characterized by an increase in allergic sensitivity, in the frequency of headaches and constipation, in the incidence of ulcers, and in other debilities typically associated with prolonged tension.

It is important to recognize, however, that notwithstanding the undesirable connotations of the idea of "aggression" in human affairs, occasional aggressive (or at least self-assertive) individual behavior is probably essential in vertebrate society, human society included. To be sure, a potential to behave aggressively might not even need to exist if every cause of competition and every source of frustration could be removed from all individuals in a social group. But competitive fac-

Figure 16.19 Fighting is one of the behaviors (feeding is another) in which four successive action components can be recognized, as in the encounter of the lesser black-backed gulls shown here. *A*, elicitation phase: two males sight each other at boundary between their territories. *B*, appetitive phase: the animals approach each other warily and "examine the situation." *C*, consummation phase: a fight occurs. *D*, satiation phase: grooming after disengagement.

A

B

C

D

tors play at least some role in all vertebrate groups, for the environmental supply of space and food is inherently limited. Moreover, loss of aggressive potentials in a species probably would lead to its destruction: the vertebrate mechanism for aggressive behavior is so intimately connected with other vital functions that abolition of one would amount to abolition of all. As noted, for example, reproduction, breeding cycles, learning through play and imitation, in some cases search for food and migratory behavior, all are based to greater or lesser degree on the same processes that also generate aggressive behavior. In addition, these same neuroendocrine processes regulate numerous internal metabolic reactions, as well as the social constraints against aggression (see below); and in man at least they contribute substantially to derivative motivations such as love and the drive to create, explore, control the environment, or be active in a very general sense. Without this constellation of capacities, aggressiveness necessarily incuded, a vertebrate would cease to be a functioning animal.

However, if aggressiveness—or submissiveness—is exaggerated unduly it ceases to be adaptive; like any other biological process, assertiveness or its lack becomes detrimental when carried to excess. In general, then, the *potential* of behaving aggressively can be regarded as an important safeguard for survival, evolved for the self-preservation of the individual under the massive—and not necessarily malicious—impact of the group. But if actual aggressiveness is misapplied and its use is mislearned, its survival function is lost and it becomes a destructive force instead.

Social Constraints

Every society operates within a framework of rules that serve to reduce individual aggression and that thereby prevent disruption of the group. Like aggression, such constraining rules have a built-in, inherited basis, but their application is largely learned. Two kinds of constraint are characteristic of vertebrate societies, *dominance* hierarchies and *territoriality*. Both are based on a structured, organized utilization of the aggressive and submissive tendencies of individuals.

In a dominance hierarchy the members of a group are ranked according to a scale of relative status or superiority. Whereas in a caste system an individual inherits a fixed rank, in a dominance

hierarchy he must seek status on his own. Moreover, this status can become higher or lower in the course of time. The most widespread basis of superiority scales is physical size and strength, or at least display features that suggest size and strength. However, more subtle, psychological attributes of individuals often can be quite as significant. As is well known, in human society factors such as intelligence, economic level, or cultural background can be just as important as size and strength, or even more so.

Social dominance is established more or less automatically if the members of a group differ greatly and obviously. Thus, large vertebrates usually dominate without contest over smaller ones, older individuals dominate over younger ones, and males largely dominate over females. Among individuals of more or less equal apparent status, the actual status is determined by sham or real contests. Losers then usually indicate their acceptance of subordinate status by various forms of submissive behavior. Submission does not always guarantee cessation of a contest. Yet in most cases, once the relative status of two individuals has become determined, it tends to have a measure of permanence; in a subsequent encounter the two individuals are not likely to recontest their comparative rank. In this way dominance is effective in reducing social conflict.

Among more than two individuals, hierarchic *peck orders* become established through pair contests carried out in round-robin fashion. In chickens, for example, the females of a flock become ranked by pecking contests. A given hen then can peck without danger of reprisal all lower-ranking birds, and she in turn can be pecked by any higher-ranking bird. If a new hen is introduced, she is subjected to a pecking contest with each fellow hen. Winning here and losing there, she soon acquires a particular rank. The peck order here has a simple linear pattern, but in other animals triangular, monarchic, or oligarchic relations can develop. Coalitions, factions, and alliances can be formed as well, and all such interrelations can change under changed conditions. As in human society, moreover, an individual can be a member of several different peck orders at the same time, in which case he usually maintains a different rank in each (Fig. 16.20).

The effectiveness of dominance hierarchies in reducing conflict depends greatly on easy rank identification: each individual must be able to recognize on sight the social rank of every other individual he encounters (as in the military organizations of man). Dominance systems actually are most common in comparatively small social units, where every member can become a familiar "acquaintance" of all others. In such units recognition is aided by frequent greeting, play, sniffing, and also by the coloration patterns, which tend to be showy and quite individual-specific in their fine details. Moreover, males tend to be distinctly larger than females, and they usually have flashy display devices. Such males typically contribute little or nothing to the parental duties of building nests and caring for offspring. Because easy recognition and familiarity have great importance in social units of this type, the members tend to exhibit ingroup clannishness and considerable antagonism to outsiders.

Individuals of high status (technically often named by the first letters of the Greek alphabet; for example, "alpha hen," top-ranking; "delta fish," fourth-ranking) generally are favored in almost all respects. They can claim the best territory, the largest amounts of the choicest food, and the preferential services of members of the opposite sex. Indeed, in the presence of dominant animals low-ranking individuals often exhibit "psychological castration," or inability to mate. However, animals of lower status usually do not surrender their privileges permanently and passively, and the dominant individuals are frequently chal-

lenged to reassert their high rank. Dominance thus only reduces conflict but does not eliminate it entirely. If a high-ranking individual succeeds in maintaining his status, jealousy and frustration among subordinates can increase and thus contribute to reinforcing their low social position. If a dominant individual is challenged successfully, his ensuing psychological disorganization can be just as profound. The extreme effect is observed in chickens, where a high-ranking hen soon dies if she is deposed.

Despite such possible detrimental consequences for specific single individuals, the inequalities produced by a rank system nevertheless are adaptive for the group as a whole. If food is scarce, for example, the lowest-ranking members (or "omega individuals") well might starve while the dominant ones eat all the available food. But at least some individuals will survive under such conditions. If a rank system were absent, however, the whole group could be wiped out, for after the available food were shared equitably a single share might not suffice for survival. In reproduction, similarly, the dominant animals are quite likely to be among the healthiest and therefore the best able to care for their offspring adequately. Hence it will be advantageous for the group if the genes of these animals are propagated preferentially.

Yet there are built-in safeguards that tend to make too high a status disadvantageous in many circumstances. In chickens, for example, alpha hens sometimes are so aggressive that they dominate and reject even the rooster. More submissive birds then produce most of the offspring. And very dominant male mice often will fight each other so viciously that they make themselves impotent.

Because they are based on aggressive behavior and thus to a major extent on sex-associated endocrine and neural mechanisms, dominance hierarchies become most pronounced—and socially most important—just before and during the breeding season. In year-round breeders, correspondingly, rank systems remain in full force at all times, as in man.

A quite similar relation with the breeding cycle holds for the second main type of social constraint, territoriality, which has a sex-associated endocrine and neural basis, too. Territoriality is exhibited by social groups in which the available geographic space is subdivided into lots, or domains, each occupied by specific individuals—a single animal or a family or even a larger group.

Figure 16.20 Dominance hierarchies. Peck orders are characteristic in social groups such as flocks of hens.

Figure 16.21 Territoriality. A polygamous family of fur seals, consisting of a single male, several females, and their young. Family groups such as this control a particular territory. The animals (especially the males) tend to ward off any fellow members of the species or other potential competitors that might invade their territory accidentally or deliberately.

Figure 16.22 Male cooperation in nesting gulls. Male (in background) signals to female by means of grass in beak that he is ready to assume nesting duties.

In most cases the primary claimant is a male who stakes out a particular area as his personal territory. If the space available to the whole group is extensive enough to make crowding unnecessary, individual domains generally will be just large enough to provide adequate food supplies and nesting materials for the occupants (Fig. 16.21).

The boundaries of a domain are variously identified and protected by scent markers (urine, for example), by intensive patrolling and vocalizing along the borders, and by immediate aggression against intruders. If a male occupant is still without mates, intruding females usually are not attacked, particularly if they behave submissively or display sex recognition signs. On the contrary, males try to attract females by conspicuous displays of their own.

When an occupant leaves his domain and enters another, his behavior becomes wary or even submissive. Being an intruder he prepares to confront the local residents, and even the lowest-ranking of these normally is dominant over any outsider of the same species. Thus even small young animals generally succeed in driving outsiders many times larger out of the territory, and the psychological edge of human property owners against intruders is well known (and also is reflected and reinforced in our many legal safeguards). A given geographic region normally serves as home territory for numerous species simultaneously, but in many cases the occupants of a domain will not tolerate other similar species in the same area. Thus, some species of fishes, birds, and mammals display chronic antagonism toward the members of certain other species and will chase such intruders out of their territory. For example, dogs, which are strongly territorial, invariably chase intruding cats.

Adaptively, a territorial system distributes the individuals of a social group in space, and it thereby probably aids in reducing conflict, notwithstanding occasional border disputes. It also reduces competition over the food supply available to the group. Indeed, by making overutilization of the food sources less likely, territoriality facilitates a long-term occupation of an area. Moreover, a territorial organization contributes significantly to population control. As a population grows numerically, the tendency of its original territory to become subdivided into more but smaller individual domains is likely to be counteracted at some stage by increasing aggression and competition. Thereafter, further growth in numbers will tend to spread a population over a progressively larger area. The regions along the periphery then will eventually abut against unsuitable territory, and at such a stage net population growth is likely to come to a halt.

Territoriality also promotes propagation. Since males are spaced out and more or less bound to their domains, competition over females is lessened. Moreover, mating, nesting, and care of young are facilitated in the comparatively undisturbed sanctuary of a domain. Further, because a territorial system makes all adults effective reproducers, it obviates the psychological castration often observed in dominance hierarchies.

Recognition and familiarity are less important in territorial than in dominance systems. Indeed, territorial organizations tend to be specially characteristic of animals that form comparatively large social units. Also, such animals usually exhibit relatively strong exploratory drives; their coloration patterns are largely inconspicuous and serve in camouflage and protection more than in recognition; and males and females tend to be more nearly alike in coloration, size, and appearance than is generally the case in dominance systems. Moreover, the males typically cooperate with the females in carrying out parental duties. In some fishes and birds, it actually is the male that protects the nest or incubates the eggs (Fig. 16.22).

The fine structure of territorial organizations, like that of dominance systems, varies greatly not only for different species but also for a given species at different times. Also, changed conditions can lead to shifts from territoriality to dominance or the reverse. In many cases, for example, territoriality manifests itself only during the breeding season, while at other times a dominance system is in force. In animals that maintain territories the year round dominance often passes from the male to the female during the breeding season. And a normally territorial group can shift to a dominance hierarchy altogether if its living space becomes too small. This is a common occurrence in domesticated animals, for example, and among caged groups in zoos and laboratories. Finally, a social group can be bound together by complex and simultaneous combinations of territoriality and dominance, as is illustrated particularly well in human society.

Review Questions

1. Describe some of the general characteristics of reactive, active, and cognitive behavior. Give specific examples of each of these behavioral categories. What kinds of learning are common for various categories of behavior?

2. What are the identifying features of exploratory behavior, play, and manipulative behavior? Give some examples of visual and chemical stimuli that can act as releasers in particular animals. What are pheromones? Give some examples of their action.

3. What are the specific characteristics of imprinting? How does imprinting differ from other kinds of learning? What stimuli generated by a mother bird can evoke a following response in her young?

4. Describe several different rhythmic behaviors. What kinds of experiments have suggested that animals have biological clocks? What is meant by entrainment, and to what extent can rhythmic behavior become entrained?

5. How do changing photoperiods affect animal behavior? What are tropisms and kineses? Give specific examples of each. What is meant by sun compass orientation? Name different groups of animals in which this process occurs.

6. Show how, at different times of day or night, honeybees communicate distances and directions of feeding areas to other bees. Is this communication capacity inherited or learned?

7. What kinds of stimuli usually clock the onset of breeding seasons? Of seasonal migrations? What is the adaptive advantage of seasonal migrations? By what mechanisms do homing pigeons and other birds migrate? To what extent is prior experience important in seasonal migrations? What is chemotaxis?

8. Describe some of the characteristics of social populations and some of the probable adaptive advantages of social living. Describe the general organization of insect societies. What forms of polymorphism occur in such societies? How do different polymorphic variants develop?

9. Describe the organization of vertebrate societies and contrast it with that of insect societies. How does the relation of individual to group differ in the two kinds of societies? Does polymorphism occur among vertebrates?

10. What is the role of individual aggression in social groups? What appears to be the behavioral relation between threat display and combat? Describe the nature of threat displays. What are re-direction and displacement responses? What are the functions of such responses?

11. Describe the endocrine and neural mechanisms that regulate aggressive and submissive behavior. To what extent is aggressive behavior inherited and to what extent is it learned?

12. Name and describe as many kinds of behavior as possible that are exhibited under the specific influence of sex-associated neural and endocrine mechanisms.

13. Describe the characteristics and adaptive advantages of dominance hierarchies. What are the usual organizational characteristics of societies in which dominance hierarchies are effective? What is psychological castration? What are the disadvantages of dominance hierarchies?

14. Describe the characteristics and adaptive advantages of territoriality. What are the usual characteristics of societies in which territorial organizations are effective? How does territoriality aid in (a) reducing conflict, (b) promoting reproduction? What are the limiting conditions that would make a territorial organization ineffective?

15. How do dominance and territorial systems become established? Under what conditions can one kind of system change to the other? Describe some aspects of human society in terms of aggression, dominance, and territoriality.

Collateral Readings

Ardrey, R.: "The Territorial Imperative," Atheneum, New York, 1966. A book on territoriality, with emphasis on the "instinctive" basis of both territoriality and aggression. See also the Lorenz entry and the critique by Montagu cited below.

Berlyne, D. E.: Conflict and Arousal, Sci. American, Aug., 1966. The role of both phenomena in learning is examined.

Brown, F. A.: Biological Clocks and the Fiddler Crab, Sci. American, Apr., 1954. One of the early studies on rhythmic behavior.

Carthy, J. D.: "The Behavior of the Invertebrates," G. Allen, London, 1958. Pertinent in the context of this chapter.

DiCara, L. V.: Learning in the Autonomic Nervous System, Sci. American, Jan., 1970. Learning can influence even "involuntary" functions such as heartbeat and intestinal contractions.

Dilger, W. C.: The Behavior of Lovebirds, Sci. American, Jan., 1962. Differences in the nest-building activities of different species shed some light on the evolution of social behavior.

Esch, H.: The Evolution of Bee Language, Sci. American, Apr., 1967. Sound communication may have preceded visual communication. See also the article by Wenner, below.

Etkin, W. (ed.): "Social Behavior and Organization among Vertebrates," University of Chicago Press, Chicago, 1964. A good, popularly written examination of vertebrate societies. Recommended.

Guhl, A. M.: The Social Order of Chickens, Sci. American, Feb., 1956. A close look at a specific dominance hierarchy.

Hess, E. H.: Imprinting in Animals, *Sci. American,* Mar., 1958. An article by a well-known student of the subject.

Hinde, R. A.: ''Animal Behavior,'' McGraw-Hill, New York, 1966. This text places special emphasis on the psychological aspects of behavior. With valuable references at the end of the book.

Holst, E. von, and U. von Saint Paul: Electrically Controlled Behavior, *Sci. American,* Mar., 1962. The nature of drives is examined by experiments with electrodes placed in the brain of chickens.

Jacobson, M., and M. Beroza: Insect Attractants, *Sci. American,* Aug., 1964. Natural and man-made pheromones are discussed.

Johnson, C. G.: The Aerial Migrations of Insects, *Sci. American.* The article examines the seasonal migrations of certain insects.

Lorenz, K.: ''On Aggression,'' Harcourt, Brace, New York, 1966. This book places emphasis on the ''instinctive'' nature of aggressiveness. It is in line with the book by Ardrey (cited above) but is criticized by Montagu (see below).

Montagu, M. F. A.: ''Man and Aggression,'' Oxford University Press, London, 1969. A counterargument to the books by Lorenz and Ardrey (cited above), with emphasis on the importance of learned and cultural factors in aggressiveness.

Mykytowycz, R.: Territorial Marking by Rabbits, *Sci. American,* May, 1968. A case study of territoriality.

Tinbergen, N.: The Curious Behavior of the Stickleback, *Sci. American,* Dec., 1952.

————: ''Social Behavior in Animals,'' Wiley, New York, 1953.

————: The Evolution of Behavior in Gulls, *Sci. American,* Dec., 1960.

    In these three readings a well-known student of animal behavior discusses various expressions of social life.

Van der Kloot, W. G.: ''Behavior,'' Holt, New York, 1968. A highly recommended paperback.

Von Frisch, K.: ''Bees: Their Vision, Chemical Senses, and Language,'' Cornell University Press, Ithaca, N.Y., 1950.

————: Dialects in the Language of Bees, *Sci. American,* Aug., 1962.

————: ''Dance Language and Orientation,'' Harvard, Cambridge, Mass., 1967.

    The discoverer and foremost ''translator'' of bee language discusses the communication system of these animals.

Washburn, S. L., and T. DeVore: The Social Life of Baboons, *Sci. American,* June, 1961. A case study of a complex dominance hierarchy.

Wenner, A. M.: Sound Communication in Honeybees, *Sci. American,* Apr., 1964. Bees are now known to communicate by sound in addition to vision-dependent waggle dances.

Wilson, E. O.: Pheromones, *Sci. American,* May, 1963. A good article on animal communication through chemicals.

Wynne-Edwards, V. C.: Population Control in Animals, *Sci. American,* Aug., 1964. Various social behaviors can limit reproduction and thus contribute to population control.

The following articles from *Scientific American* are adequately described by their titles:

Boycott, B. B.: Learning in the Octopus, Mar., 1965.

Butler, R. A.: Curiosity in Monkeys, Feb., 1954.

Carr, A.: The Navigation of the Green Turtle, May, 1965.

Eibl-Eibesfeldt, I.: The Fighting Behavior of Animals, Dec., 1961.

Emlen, J. T., and R. L. Fenney: The Navigation of Penguins, Oct., 1966.

Ferster, C. B.: Arithmetic Behavior in Chimpanzees, May, 1964.

Flyger, V., and M. R. Townsend: The Migration of Polar Bears, Feb., 1968.

Gilbert, P. W.: The Behavior of Sharks, July, 1962.

Gleitman, H.: Place-learning, Oct., 1963.

Hasler, A. D., and J. A. Larson: The Homing Salmon, Aug., 1955.

Sauer, E. G. F.: Celestial Navigation by Birds, Aug., 1958.

Schneirla, T. C., and G. Piel: The Army Ant, June, 1948.

Shaw, E.: The Schooling of Fishes, June, 1962.

Thorpe, W. H.: The Language of Birds, Oct., 1956.

SPECIES AND ECOSYSTEM

The interrelations among animals, other organisms, and their environments are the special concern of the biological subscience of *ecology*. This chapter deals with two major ecological associations, the *species* and the *ecosystem*. A species consists of like individuals that live together in social or nonsocial *populations,* and groups of populations belonging to several different species form a *community*. Communal populations include free-living organisms and, almost invariably, also parasites and other types that live in so-called *symbiotic* associations. An ecosystem is composed of all the living components of a community and the whole physical environment in which the community exists.

The Species

Structures and Functions

As pointed out in Chap. 7, a species is a taxonomic unit that encompasses animals capable of interbreeding with one another. Thus, all the bullfrogs or all the human beings on earth represent a species. Several types of bonds unify the members of a species as a *natural* grouping of organisms.

First, a species is a *reproductive* unit; by definition the member animals are capable of interbreeding, but members of two different species normally are not cross-fertile. Actually not all species can be defined uniquely in this way. Certain closely related ones, and occasionally even more distantly related ones, are known to be interfertile in some cases, although in nature they often are isolated from one another by various breeding barriers. For example, horses and asses produce *mules.*

In some cases, on the contrary, interbreeding is not possible between two members of the same species. For example, the protozoan taxonomic species *Paramecium aurelia* contains 16 distinct hereditary mating groups called *syngens,* or "biological species." Mating within a syngen can occur, but mating between syngens cannot. Such infertility among the members of a species reaches extreme form in asexual types, in which interbreeding is completely absent (Fig. 17.1). Common reproductive bonds evidently characterize what appears to be more than a single species in some cases and less than a whole species in others. But for the vast majority of animals the definition of a species as an interbreeding unit does hold, and reproduction does represent an important unifying link for species members.

Because it is a reproductive unit, each species also is an *evolutionary* unit; since interbreeding has taken place in a species throughout its history, the member animals are more closely related to one another than to members of any other species. As a result, the members of a species have in common a basic set of structural and functional traits. However, no two animals are exactly alike, and the

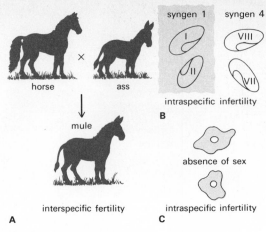

A interspecific fertility

B intraspecific infertility

absence of sex

C intraspecific infertility

Figure 17.1 Species definition: exceptions. A, horses and asses are distinct species, yet matings between them can occur. *B,* mating of types I and II of *Paramecium aurelia* forms a syngen within which mating can occur. Similarly, types VII and VIII form a mating syngen. But cross mating between syngens cannot occur, even though all such syngens belong to the same species. *C,* organisms (such as *Amoeba proteus*) are sexless and thus by definition infertile, yet all individuals nevertheless belong to the same taxonomic species.

Figure 17.2 Individual variation. A and *B* are umbrella birds that belong to the same species, *Cephalopterus ornatus.* But they are members of different populations, and the structural differences between the birds are quite pronounced. Technically these birds are said to belong to different subspecies of the same species.

A

B

members of a species actually differ from one another quite considerably. Superimposed on the common traits, *variations* of structure, function, or both, occur in each species. The range of such variations in one species can be directly continuous with the range in a closely related species.

Therefore, two animals of two different species sometimes might not differ too much more than two animals of the same species (Fig. 17.2).

Variations can be *inheritable* or *noninheritable.* The first are controlled genetically and can be transmitted to offspring. Noninheritable variations are the result of developmental or environmental influences, and they therefore disappear with the death of the individuals that exhibit them. Thus only inheritable variations are significant in determining the lasting traits of a species. If a man is an athlete, his muscular system is likely to be developed much more than in the average person. This is a noninheritable variation; the degree of muscular development depends primarily on exercise, not on heredity. By contrast, blood type or skin color are examples of hereditary variations. They are part of the genetic inheritance from parents and earlier forebears and in turn will influence the traits of future generations (Fig. 17.3).

Some variations appear to be associated with climates and geography. In warm areas, for example, individuals of many animal species tend to have smaller body sizes, darker colors, and longer ears, tails, and other protrusions than fellow members of the same species living in cold climates (Fig. 17.4). Such structural variations usually are adaptive, or advantageous in the different environments. Thus, smaller bodies and longer ears give animals a larger skin surface relative to the body volumes. Surface evaporation then is comparatively rapid, and the resulting cooling effect is of considerable benefit. The converse holds in a cool climate. However, in many instances it is difficult to recognize the adaptive value of a variation, as in most of the human racial differences discussed in Chap. 15.

Because of the variations of their members, all species exhibit a greater or lesser degree of *polymorphism;* the members have "many shapes." For example, a species always includes variants in the form of immature and mature individuals. Often also, as noted in Chap. 16, males and females exhibit a form of polymorphism called *sexual dimorphism* (see Fig. 16.12). Moreover, a species usually encompasses subordinate groups such as subspecies and subsubspecies, and these too are distinguished by particular variations. Polymorphism actually can be so pronounced that the close relation of two individuals of the same species becomes evident only through careful study. Good examples are the social insects (see Chap.

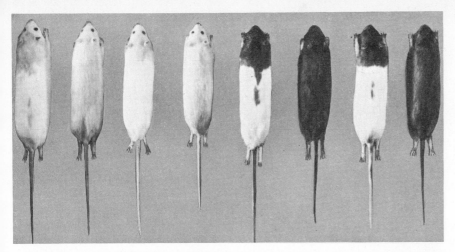

Figure 17.3 Inheritable variations. These are littermate rats produced by the same two parents. Considerable variation in coat color is evident. Such differences are controlled genetically and are superimposed on common basic traits. For example, despite the color differences, all the littermates are distinctly rats and share not only the traits common to the species but also those that are specifically characteristic of their breeding line.

16). Such instances of polymorphism are expressions of structural *specialization*. And where animals exhibit great polymorphic diversity, a high degree of functional interdependence usually follows as well. In a colony of social insects, for example, only the whole colony, with repre-

Figure 17.4 Variations and climate. Evaporation surfaces tend to be larger in warm-climate animals than in corresponding cool-climate ones. For example, the Arabian desert goat shown here (*A*) has external ears very much larger than those of related types in temperate regions (*B*).

A   B

sentatives of all types of polymorphic variants, is functionally self-sufficient.

By virtue of being evolutionary units all species are also *ecological* units. Each is defined by its *ecological niche,* its place in nature: it inhabits a certain geographic region, it uses up particular raw materials in that region, and it produces particular byproducts and endproducts. For example, the environment offers numerous opportunities for carnivorous modes of animal existence, all differing from one another in hundreds of fine details. Thus, a carnivorous species can subsist by being terrestrial, aquatic, or aerial; sessile or motile; a cold-, temperate-, or warm-climate type; a type that hunts in daytime or nighttime; a form that specializes in small prey or large prey. And each such coarse category contains innumerable finer categories of possible carnivorous ways of life. All such different opportunities represent ecological niches, and given species occupy them. Therefore, just as each species is identified by a particular set of structural, functional, and evolutionary characteristics, so it also is identified uniquely by its ecological niche. No two species have precisely identical niches, and a given niche cannot be occupied indefinitely by more than one species (Fig. 17.5).

Such niches are associated intimately with geography; in similar kinds of environments, even if widely separated, species with similar ways of life will be found. For example, widely different localities offering similar conditions of soil and climate will support prairies composed of grass species having similar requirements. Prairies in turn offer opportunities for grazing animals, and each prairie region of the world actually has its own animal species filling available grazing niches—antelopes in Africa, bisons in North America, kangaroos in Australia. In like manner, species on high mountains occupy similar ecological niches. Several similar niches also can be available in a single territory. For example, the Central African plains support not only numerous types of antelopes but also zebras, giraffes, and other grazing species. The ways of life of such species overlap in many respects, but they are not precisely identical in all details; each species normally fills a unique niche.

By being adapted to the same niche, the members of a species are linked together through powerful bonds that have cooperative and competitive aspects. For example, *intraspecific cooperation* often is necessary to execute the way of life of

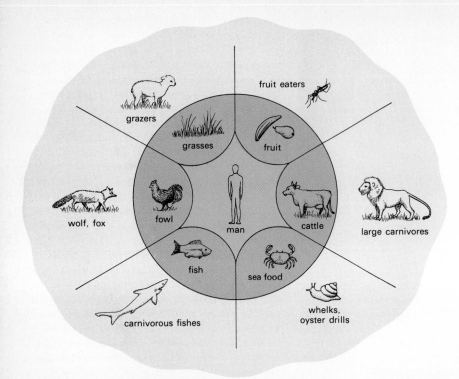

Figure 17.5 The ecological niche of man. It is characterized in part by man's requirement of foods shown inside the circle around man. These food organisms also happen to form parts of the ecological niches of other animals, as indicated outside the circle, but such animals also use foods not used by man. Thus although these niches partly overlap man's, each is nevertheless distinct and characteristic of a particular species. Ecological niches are defined not only by food organisms but also by geographic territory, waste products, structures, and other attributes of species.

exception of occasional direct confrontation, most animal competition occurs only indirectly through the environment, and physical or visual contact is not or need not be involved.

Development and Geography

A species occupies its ecological niche for the duration of its evolutionary life—on the order of a million years on the average. In the course of this time a species typically has a characteristic history. It originates in a small home territory, spreads from there over as wide a geographic range as conditions permit, then dies out in various localities of that range and eventually becomes extinct (Fig. 17.6). Species extinction is not a theoretically inherent necessity, but no past species is known to have survived longer than a few million years at most. What factors govern this pattern of species development?

As shown in Chap. 13, species formation, or *speciation*, occurs when, in a parent species, a population becomes isolated reproductively from neighboring sister populations. The isolation can develop in various ways, but the common result is that the isolated population no longer can interbreed with the rest of the parent species and in time comes to form a new descendant species. In this process of being ''born,'' a new species also acquires its own ecological niche. It can do so, for example, by occupying a niche that had been free in the territory. Or if a change in the physical environment has created a new niche, the new species may come to occupy it. Or the new species can create its own niche by evolving a way of life that overlaps partially with the ways of life of similar species in the area. Or the new species simply might encroach on a niche that another species in the area already occupies.

During its early development a new species is said to be *endemic;* its members form just one or

an animal species; hunting may have to be done in cooperative packs, migrations may have to be undertaken in the comparative safety of herds or flocks, and groups may be required for the construction of complex nests or hives. But inasmuch as the members of a species must share the living space and food sources in their common niche, *intraspecific competition* occurs as well. With the

Figure 17.6 Species: general life history.

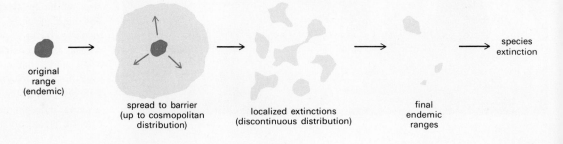

a few populations, all localized in a single small territory. Further development depends on the geographic extent of the ecological niche of the species. In general, a species tends to expand in widening circles around the home territory until the boundaries of its niche are reached—until ecological barriers impose conditions to which the species is not adapted. Such barriers are biological as well as physical. For example, further spreading in a particular direction can be blocked by the presence of competing species or by the absence of appropriate food sources. Physical barriers include water, deserts, or mountains that can stop the expansion of a terrestrial species; land usually prevents an aquatic species from spreading. Among more subtle physical barriers on land are seasonal and daily temperatures, precipitation and soil characteristics, water supply, and numerous other variables. In the ocean, species often are confined to particular temperature zones, salinity zones, pressure zones, and many more. Similarly, dispersal opportunities in fresh water can depend on strengths of currents, amount of silting and pollution, oxygen content, bottom conformation, and a wide variety of other conditions.

Thus if the boundaries of its ecological niche are close to the original home territory, a species cannot expand significantly and then must remain endemic. Such species usually become extinct comparatively rapidly, as soon as physical or biological changes in the geographically limited environment make survival impossible. But most species can find the living conditions they require in more extensive areas and they usually succeed in spreading quite widely. Many cover all or major parts of a continent or an ocean, and many others expand even farther and become *cosmopolitan*, distributed around the world.

But after a species becomes cosmopolitan or at least manages to become distributed widely, its populations in numerous localities cease to exist sooner or later. On a time scale of thousands of years, such local disappearances are brought about in three ways: a population dies out, or it emigrates, or it evolves into a new species. The underlying causes can be physical, biological, or both. For example, if a locality undergoes slow physical changes, a population living there must respond to such changes in one of the three ways mentioned. Biological change is promoted by other species living in the same locality. For example, if two populations of different species have a predator-prey relationship, then disappearance of the prey for any reason will be a biological change to which the predator in many cases must respond in one of the three ways above. Another frequent cause of biological change is the invasion of a region by populations of other species that occupy very similar ecological niches. The result then is *interspecific competition* for the same living space or the same food sources or both. Coexistence might be possible at first, but in time one population usually becomes dominant and those of other species then must emigrate, evolve into new species, or die out.

In widely distributed and cosmopolitan species, therefore, population gaps are likely to appear in various parts of the territorial range. In time such gaps tend to become progressively more numerous, and what may have been a fairly continuously distributed species earlier eventually becomes a discontinuous array of more or less isolated populations. Literally hundreds of species actually consist of widely separated populations in various parts of the globe. They represent remnants of previously contiguous populations. Ultimately a species again becomes endemic and consists of but few localized populations or even just a single one. Total extinction then is usually not far off (Fig. 17.7).

Some 1½ million species are known and identified today, and ten or more times that many probably exist. About 10,000 new ones are described

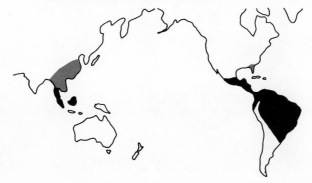

Figure 17.7 Restricted distribution of animals with formerly continuous ranges. *Black areas,* present distribution of tapirs through Malaya, Sumatra, Borneo, and most of central and northern South America. *Color areas,* present distribution of alligators through southern Asia and Florida. At one time alligators ranged throughout Asia and North and Central America.

each year. Apart from cosmopolitan species, characteristic groups of terrestrial species are found on each of the continental land areas of the world. Based especially on the types of birds and mammals present, six major *zoogeographic regions* can be identified, as shown in Fig. 17.8. The figure indicates, for example, that the widely separated faunas of Alaska and northern Mexico by and large resemble each other more than the directly adjacent faunas of northern Mexico and southern Mexico.

If species are considered singly, the pattern of their life histories in most cases accounts adequately for their present global distribution. But the distribution of groups of species is not always explainable so readily. For example, we know that pouched marsupial mammals exist today not only in Australia but also in North America, where the opossum is the only surviving representative. It is also known from fossils that, about 100 million years ago, groups of marsupials lived in southern Asia as well as South America, and that from these regions they later spread to Australia and North America, respectively. To explain this distribution it could be postulated that the first ancestral marsupials evolved either in some Asian or in some

American territory, and that they then expanded from this single original location to both southern Asia and South America. Such a hypothesis cannot be ruled out entirely, but the available evidence makes it rather unlikely. Or it could be postulated that marsupials originated independently in two different regions of the world. This hypothesis, too, is possible but not particularly promising. A third and most likely explanation is based on the idea of *continental drift*.

First proposed in the early part of this century, it postulates that all the land mass of the globe formed a single continent, *Pangaea*, during the Carboniferous, roughly 250 million years ago (Fig. 17.9). Pangaea then is assumed to have split into a northern *Laurasia* (or *Holarctica*) and a southern *Gondwana*, with a *Tethys* sea between them. During the late Mesozoic each of these land masses is considered to have split further, Laurasia into what eventually was to become North America and Eurasia, Gondwana into land now represented by Antarctica, Africa, India, Australia, most Pacific islands, and South America. All these fragments are thought to have drifted to their present locations and to have become joined here and there, forming the familiar land masses of today.

Figure 17.8 The six zoogeographic regions. For each supercommunity, three representative and characteristic mammals are indicated. Within each supercommunity the positioning of the animals is dictated largely by the drawing space available, not by geographic considerations.

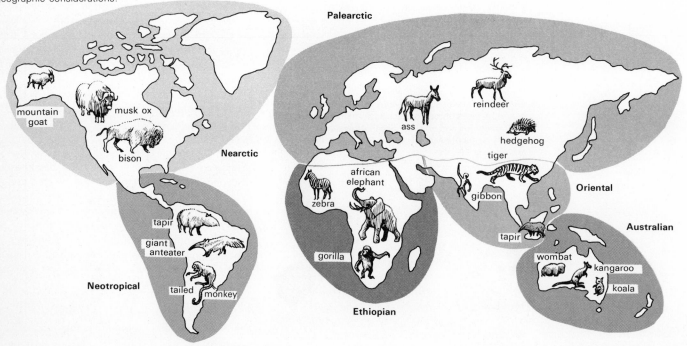

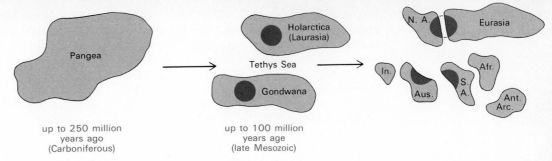

Figure 17.9 Continental drift. Continental areas at right, identified by abbreviations, are assumed to have drifted to their present positions during the last 100 million years. Dark-shaded areas in Holarctica and at right indicate how the distribution of eels might be accounted for. Dark-shaded areas in Gondwana and at right similarly suggest an explanation for the distribution of marsupials.

The hypothesis was suggested originally by the contours of the present continents, some of which seem to fit together like pieces of a jigsaw puzzle (for example, South America and Africa). Moreover, continental drift could account for numerous other geographic and geologic features of the earth. Above all, the drift hypothesis was neatly consistent with the known zoogeographic attributes of the world, as outlined by Fig. 17.8. For example, marsupials could have originated on Gondwana, and when this continent later foundered the animals would have become split into southern Asian and South American groups. A drift hypothesis could also explain curious migration patterns such as those of eels. These animals travel to common North Atlantic spawning grounds from both American and European rivers. Thus if North America and Europe actually were part of a Holarctic land mass as postulated, ancestrial eels could originally have spawned in a common central Holarctic river system. Their present descendants therefore still might travel from both west and east to the same ancestral spawning area, which now, however, happens to lie in the middle of an ocean (see Fig. 17.9).

Geologic evidence for continental drift has recently become quite strong and convincing, and many former puzzles of zoogeography soon may be on the way to being solved.

Population

Every species consists of one or more populations, or *species-populations*. Examples of such relatively stable, geographically localized groups are the minnows in a pond, the earthworms in a plot of ground, or the people in a village. Individual animals multiply and die, emigrate or immigrate, but collectively a population persists. It can split into subpopulations or fuse with adjacent sister populations, yet the basic characteristics of the group as a whole do not thereby change. All members of a population share the same food sources and the same local territory, and the members also interbreed more or less preferentially with one another. In addition, interbreeding with members of sister populations occurs fairly frequently. A population thus is a reproductively cohesive unit, integrated more loosely with other such units.

The significant structural characteristics of populations are their *dispersion* and *growth* patterns. Dispersion is a measure of population density. The growth pattern is influenced by the extent and the resources of the territory, by the balance between reproduction and mortality, by rates of emigration and immigration, and by the effects of population density on growth. Such effects can be of three kinds. In some species population growth occurs the more rapidly the lower the population density. In other species growth rates tend to increase geometrically until the limits of the available living space and food supply are reached. And in still other species population growth is most intense at intermediate densities; both undercrowding and overcrowding here depress growth.

The usual net result of all variables affecting population growth is that, over appreciable periods of time, *natural populations tend to remain constant in size.* The conditions that promote and those that limit growth tend to balance each other, and the numbers of individuals in a population then normally do not change significantly on a long-term basis (see also below). Many human populations are notable exceptions to this rule, a subject discussed in the next chapter.

A significant functional attribute of populations is that they *interact* with all components of their surroundings. The members of a population interact with one another, with sister populations, with populations of other species in the same territory, with the territory itself, and with the physical environment in general. Indirect interactions are extensive and continuous; each member of a population affects every other one through its impact on the environment. Thus even two earthworms some distance apart affect each other by, for example, drawing on the same supply of nutrients in the soil. Direct interactions may not occur or, as among solitary animals, may be limited to reproductive contact. By contrast, the members of certain animal populations interact far more extensively, directly as well as indirectly. The animals in such cases live together as more or less closely knit *social* populations, a topic already examined in Chap. 16.

The Ecosystem

A community is a localized association of several populations of *different* species. Almost always, a community contains representatives of plants, animals, and microorganisms, all being required for group survival. Communities together with the physical environments in which they live represent ecosystems, the largest subunits of the biosphere; examples are a pond, a forest, a meadow, a section of ocean shore, a portion of the open sea, a coral reef, or a village with its soil, grasses, trees, people, bacteria, cats, dogs, and other living and nonliving contents.

Structure and Growth

The living portion of an ecosystem exhibits a characteristic species structure; a few species are represented by large populations and many species are represented by small populations. For example, the animals in a jungle community often comprise a few large populations of just two or three kinds of monkeys but many small populations of numerous kinds of birds, bats, snakes, and other animals. Plants are likely to be represented by several dozen species of trees, but populations of just two

or three species can make up to 70 to 80 percent of the total number of trees. In general also, species diversity in an ecosystem is inversely proportional to the sizes of organisms. Thus, a forest is likely to contain more species of insects than of birds and more species of birds than of large mammals. The reasons for such correlations will soon become apparent.

Like other living units communities grow, develop, pass through mature phases, reproduce, and ultimately die. The time scale may be in hundreds and thousands of years. Such life cycles result from an interplay between the living and the nonliving components of an ecosystem. Being specialized to occupy particular ecological niches, different species must live in different types of environment. The physical characteristics of a given region therefore determine what types of organisms can settle there originally. By its very presence, however, a particular set of organisms gradually alters local conditions. Raw materials are withdrawn from the environment in large quantities, and metabolic wastes are returned. The remains of dead organisms return to the environment, too, but not necessarily in the same place or the same form in which they were obtained. In time, therefore, communities bring about profound redistributions and alterations of vast quantities of the earth's substance.

Later generations of the original community then may find the changed local environment no longer suitable, and the populations must resettle elsewhere or readapt or die out. The result, as noted earlier, is gradual development of population gaps in species. A new community of different plant and animal populations therefore can come to occupy the original territory, and as this community now alters the area according to its own specializations, type replacement, or *ecological succession*, eventually will follow once more.

Continued ecological succession ultimately produces a *climax community:* a set of populations that changes the local environment but repeatedly re-creates the original conditions more or less exactly. Good examples are the North American prairie and forest belts and the communities in large lakes and in the ocean. Climax communities represent ecological steady states that last as long as local physical conditions are not altered drastically by climatic or geologic upheavals. If that happens, communal death usually follows. New communities then might develop by immigration,

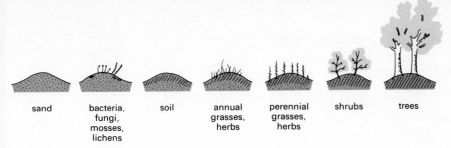

| sand | bacteria, fungi, mosses, lichens | soil | annual grasses, herbs | perennial grasses, herbs | shrubs | trees |

Figure 17.10 Ecological succession. Left to right, a sequence of seres from virgin land (sand) to climax (trees). The total living mass increases during this sequence.

or the remnants of the old community might undergo major evolutionary readjustment.

In an ecological succession leading to a climax community, each developmental stage is a *sere* (Fig. 17.10). A sequence of seres usually is characterized by repeated changes in the types of population present and by an increase in species diversity and the total quantity of living matter. The particular seres that can be expected in a given region often are fully predictable. For example, if the original physical environment is sand or equivalent virgin territory, then the dominant organisms in the seres generally follow the pattern: soil-formers (bacteria, lichens, mosses) → annual grasses and herbs → perennial grasses and herbs → shrubs → trees. Characteristic animal populations are associated with each seral plant population. Development from sand to forest climax can require on the order of 1,000 years. If soil is already present at the start, then a grass

climax can be attained in about 50 to 100 years, a forest climax in about 200 years or more.

Clearly, *turnover* occurs on the level of the ecosystem just as it does on all other levels of living organization. After a climax has been attained, a community exhibits a numerical steady state in all populations present: the numbers of individuals remain relatively constant on a long-term basis. In a large, permanent pond, for example, the numbers of algae, frogs, minnows, and other organisms stay more or less the same from decade to decade. Annual fluctuations or cycles of several years are common, but over longer periods of time constancy of number is characteristic in most natural communities.

Three key factors create and control these striking numerical balances: *nutrients, reproduction,* and *protection.* They are the main links that make the populations of a community interdependent.

Links and Balances

A stable ecosystem generally has a four-part nutritional structure (Fig. 17.11). The *abiotics* are the nonliving physical components of the environment on which the living community, or *biomass,* ultimately depends. The *producers* are the green photosynthetic food-creating autotrophs that live entirely on the abiotic portion of the ecosystem. The *consumers* are heterotrophs, largely animals. Herbivorous animals feed on the producers, and carnivorous types consume one another as well as the herbivores. Lastly, the *decomposers* are saprotrophs, bacteria and fungi, that live on the excretion products and the dead bodies of all producers and consumers. Decomposers bring about decay and a return of raw materials to the abiotic part of the ecosystem. The physical substance of an ecosystem thus *circulates,* from the abiotic part through the biomass back to the abiotic part. This circulation is maintained by the continuous influx of energy from the sun.

However, energy dissipates in the course of such cycles, and as raw materials pass from the environment to plants and from plants to animals, these transfers are not 100 percent efficient. More than a pound of inorganic materials is needed to make a pound of plant matter, and more than a pound of plant food is needed to make a pound of animal matter. Similarly, more than 300 lb of antelope meat or even lion meat is required to produce a 300-lb lion. This inescapable circum-

Figure 17.11 Nutritional structure of an ecosystem. Abiotics represent the physical nonbiological components of the system. Arrows indicate flow of energy and materials.

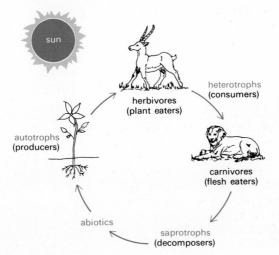

sun

herbivores (plant eaters)

heterotrophs (consumers)

autotrophs (producers)

carnivores (flesh eaters)

abiotics

saprotrophs (decomposers)

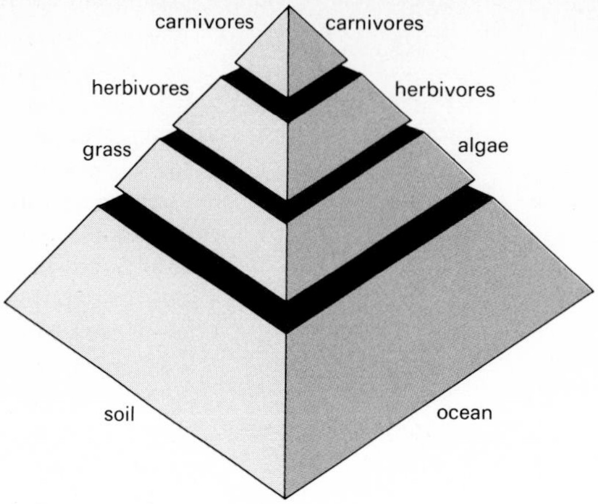

Figure 17.12 Food pyramids. Soil and ocean support plant life; plant-eating animals (herbivores) subsist on the plants; and flesh-eating animals (carnivores) subsist on the herbivores.

stance results in *food pyramids* in the ecosystem (Fig. 17.12). So many tons of soil can support only so many *fewer* tons of grass. Grass in turn supports herbivores that together weigh less than the grass. And only a relatively small weight of carnivores can find sustenance in such a community. Several acres of ground therefore might just suffice to support a 150-lb man. Such a pyramid of productivity and weights also describes a pyramid of number and sizes. Prey generally is smaller than predator, and the balanced community might contain millions of individual ants but only one man.

Food pyramids are among the most potent contributors to communal steady states; significant variations of numbers at any level of a pyramid soon bring about automatic adjustments at every other level. For example, an overpopulation of carnivores can result in a depletion of herbivores, since a greater number of herbivores is eaten. This depletion can lead to starvation of

carnivores, hence to a reduction of their numbers. Underpopulation of carnivores then could result in overpopulation of herbivores, since fewer herbivores are eaten. But these fewer carnivores could be well fed and could reproduce relatively rapidly, which would increase their numbers again. As a result, although the numbers of all kinds of organisms can undergo short-term fluctuations, the total quantities generally remain relatively constant over the long term (Fig. 17.13).

The territory of an ecosystem usually supports more than one food pyramid. Each represents a different *food chain* that usually ends with a different carnivore. For example, a lion would find it extremely expensive in terms of locomotor energy to live on insects, worms, or even lizards and mice. Bigger prey like antelope or zebra is more appropriate. On the other hand, insects and worms are suitable food for small birds, and small birds, lizards, and mice provide adequate diet for larger predatory birds. In this example two food pyramids are based on the same plot of land. The pattern is generally much more complex. Different types of plants in one territory can sustain many different herbivores. These can form the basis of different, intricately interlocking animal food chains. As in the case of elephants, a herbivore can itself represent the peak of a food pyramid.

In a balanced ecosystem the total biomass yields just enough dead matter and other raw materials to replenish the soil or the ocean. This circumstance permits the continued existence of the various food pyramids above ground or in water. In such delicate nutritional interdependencies, minor fluctuations are rebalanced fairly rapidly. But serious interference, by disease, by man, or by physical factors, is likely to topple the whole pyramid. If that happens the entire community often ceases to exist. Man's impact has been particularly detrimental in this respect, a subject examined further in the next chapter.

A second link between the populations of a community is reproductive interdependence. Familiar examples are the pollinating activities of insects and the seed-dispersing activities of a large variety of animals, man included. It is fairly obvious how such activities contribute to population balance: any reduction of the animal populations restricts the reproduction of the plant populations, and vice versa. Many other examples illustrate the same type of population control. Thus, birds like cuckoos lay eggs in nests of other birds. Insects

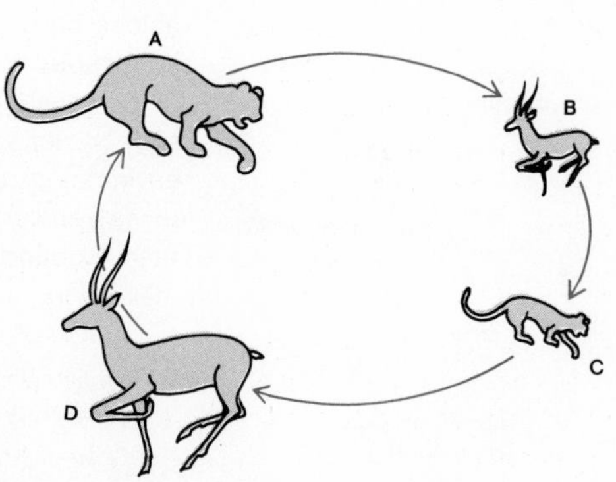

Figure 17.13 Population balance. A large carnivore population reduces the herbivore population by predation (*A, B*). The food supply of carnivores thus is decreased and results in starvation and decrease of the carnivore population (*C*). The herbivore population then can flourish again (*D*), and an increase in the numbers of carnivores can follow once more.

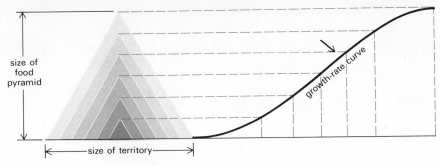

Figure 17.14 *Population growth.* As the number of individuals increases, more territory will be occupied and the food pyramid will become wider and higher (left part of diagram). Rates of population increase are indicated in the curve at right; rates increase to a turning point (arrow), then decrease.

such as gall wasps embed their eggs deep in the tissues of particular plants, where the hatching larvae find food and protection. And many insects deposit eggs on or under the skin of various animals (Color Fig. 8).

Reproduction and nutrition play important roles in the growth and geographic expansion of a community. In new territory, a pioneer community first will form a small food pyramid that still might be too "low" to support any large animals. Abundance of food and absence of competition can promote high reproduction rates and a rapid increase of numbers at all levels of the pyramid. The base of the pyramid therefore will widen and a larger area of the territory will be occupied. Sizable herbivores and even a few larger carnivores gradually might be assimilated into the community. As a result the rate of predation will increase, which in turn will slowly decrease the net reproductive gain. A turning point will be reached eventually. Prior to it the community grows at an increasing rate; after it the community still grows, but at a decreasing rate. Net expansion finally comes to a standstill, and from then on the pyramid retains relatively stable proportions (Fig. 17.14).

Such a growth pattern depends on the absence of external restrictions on the expansion of a community. If geographic or biological barriers do limit expansion, then communal growth is stopped prematurely. The food pyramid in such a limited community might never become high enough to support large animals. One searches in vain for stag in a tiny forest, for large fish in a small pond. But the tiny forest is likely to abound in worms, mice, and small birds, the small pond in algae, protozoa, and frogs.

In a community incapable of further expansion, steady reproduction can produce a centrifugal *population pressure*. If emigration is not possible or if it is not sufficiently effective, numbers will be decimated by starvation or even sooner by epidemic diseases. The latter spread rapidly through an overpopulated, undernourished, spatially restricted community. Even if disease affects only one of the component populations, the whole communal web is likely to be disrupted.

The third main link in a community is protective interdependence. Plants usually protect animals by providing shelter against enemies and adverse weather. If the opportunities for protection are reduced, both the animal and plant populations can suffer. For example, if an overpopulation of insects makes plant shelter inadequate, the insects will become easier prey for birds and bats. But this circumstance can also decimate the plant populations, for their major pollinating agents may no longer be sufficiently effective.

Many animals are protected from others by *camouflage,* involving body color or body shape or both. Probably the most remarkable instance of color camouflage is *mimicry,* widespread particularly among butterflies and moths. In certain of these the pigmentation patterns are virtually indistinguishable from those of other, unrelated species. Such mimicked species usually are strong or fast and have few natural enemies. An animal resembling even superficially another more powerful one then will be protected, too, by scaring off potential predators. Insects also display a variety of structural camouflages. For example, the individuals of certain species are shaped like leaves, branches, or thorns, an adaptive device serving not only defensively but also as a disguise that misleads potential victims (Fig. 17.15 and Color Fig. 9).

Many other protective devices are common. Various birds and some mammals mimic the song and voice of other species, either defensively or as an aggressive lure. The hermit crab protects its soft abdomen in an empty snail shell of appropriate size. Schools of small pilot fish scout ahead of large sharks and lead their protectors to likely prey. Significant protection is afforded also by man, through domestication, game laws, parks, and sanctuaries.

These various examples illustrate how the member populations of a community are specialized nutritionally, reproductively, and protectively. Carnivorous populations cannot sustain themselves on plant food and not even on every kind of animal food. Herbivorous populations are incapable of hunting for animals. The photosynthe-

Figure 17.15 Camouflage. A, dead-leaf butterfly (top of photo), resembling foliage in form and color. *B,* leaf insect, resembling a dried and fallen leaf. *C,* South African mantis, resembling stem and branches of a bush. See also Color Fig. 9.

A   **B**   **C**

sizing populations depend on soil or ocean, and the decomposers cannot do without dead organisms. These are profound specializations in structure and function, and they imply a lack of self-sufficiency of individual populations. Communal associations of populations evidently are a necessity. Indeed they are but extensions, on a higher biological level, of the associations of cells in the form of tissues, organs, and organ systems.

In a large number of cases such specializations of the populations in a community are even more profound; for a community contains not only free-living populations but also individuals of different species that live together in more or less permanent physical contact. These associations are instances of *symbiosis,* an expression of the most intimate form of communal life.

Symbiosis

There is no major group of animals that does not include symbiotic species, and there is probably no individual animal that does not play *host* to at least one *symbiont.*

Symbiosis occurs in two basic patterns. In *facultative* associations, two different animals "have the faculty" of entering a more or less intimate symbiotic relationship. But they need not necessarily do so, being able to survive as free-living

forms. In *obligatory* associations, one animal *must* unite symbiotically with another, usually a specific one, if it is to survive. The ancestors of obligatory symbionts invariably have been free-living animals that in the course of history have lost the power of living on their own. Before becoming obligatory symbionts, they formed facultative associations with animals on which they came to depend more and more.

A symbiont affects its host in different ways. In *mutualism,* both associated partners derive some benefit from living together. *Commensalism* benefits one of the partners and the other is neither helped nor harmed by the association. *Parasitism* is of advantage to the parasite but is detrimental to the host to greater or lesser extent.

Mutualism and Commensalism

An example of mutualism is the tickbird-rhinoceros relationship. Tickbirds feed on skin parasites of rhinoceroses; the latter are relieved of irritation and are warned of danger when the sharp-eyed birds fly off temporarily to the security of the nearest trees. This is an example of facultative mutualism; both tickbirds and rhinoceroses can get along without each other if necessary.

A somewhat greater degree of physical intimacy is exhibited in the mutualism of sea anemones and hermit crabs. Sea anemones attach themselves to empty snail shells, and hermit crabs use these shells as protective housings. The sea anemone, an exceedingly slow mover by itself, thereby is helped in its search for food and in geographic dispersal. The hermit crab in turn benefits from the disguise. This too is a facultative association; sea anemones and hermit crabs can and largely do live on their own.

The most intimate forms of mutualism involve symbionts that live directly inside their hosts. For example, in the gut of termites live flagellate protozoa capable of digesting the cellulose of wood. Termites chew and swallow wood, the intestinal flagellates then digest it, and both kinds of organisms share the resulting carbohydrates. Thus, to the detriment of man, termites can exploit unlimited food opportunities open to very few other animals. And the protozoa receive protection and are assured of a steady food supply.

Virtually every animal with an alimentary canal houses billions of mutualistic bacteria in the lower gut. The bacteria draw freely on substances not digested by the host, and the result of this bacterial activity is the decay of materials that the host later eliminates. The host generally benefits from this auxiliary bacterial digestion and in many instances is dependent also on certain of the bacterial byproducts. For example, man and other mammals obtain many vitamins in the form of "waste" materials released by the bacteria of the gut.

Many mutualistic associations probably have evolved from parasitic ones. For example, mutualistic intestinal bacteria might be descendants of originally parasitic ancestors. It is advantageous to a parasite not to jeopardize its own survival, and in the course of evolution this circumstance well might have tended to change a parasitic relation to one that might be beneficial or at least harmless for the host.

Commensalism is illustrated, for example, by a species of small tropical fish. Individuals of this species find shelter in the cloacas of sea cucumbers. The fish darts out for food and returns, to the utter indifference of the host. The so-called shark sucker, or *remora,* provides another example (Fig. 17.16). In this fish a dorsal fin is modified as a holdfast device. By means of it the fish attaches to the underside of sharks and thereby secures scraps of food, wide geographic dispersal, and protection. The shark neither benefits nor suffers significantly. In yet another example, barnacles can attach to the skin of whales and can thereby secure geographic distribution and wide feeding opportunities. In this instance a trend toward parasitism is in evidence; in some cases the barnacles send rootlike outgrowths into the whale that eat away bits of host tissue.

Parasitism

Ways of Life Symbiosis in general and parasitism in particular tend to be most prevalent among heterotrophs, organisms that must obtain nutrients from others and among which competition for food therefore is most intense—viruses, bacteria, fungi, and animals. Many animal groups are wholly

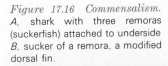

Figure 17.16 Commensalism. A, shark with three remoras (suckerfish) attached to underside *B,* sucker of a remora, a modified dorsal fin.

A

B

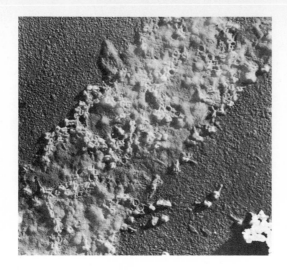

Figure 17.17 Bacteriophages. The photo shows the remnants of a bacterium after attack by bacteriophage viruses. The virus parasites are the small rodlets with knobby ends.

parasitic, and practically all others include at least some parasitic subgroups.

As a way of life parasitism is almost as ancient as life itself (see Chap. 14). So advantageous and economical is the parasitic mode of existence that many parasites are infested with smaller parasites of their own. For example, a mammal might harbor

parasitic worms, these can be invaded by parasitic bacteria, and the bacteria often are infected by *bacteriophages,* or viruses that parasitize bacteria (Fig. 17.17). The presence of one parasite inside another is called *hyperparasitism.* This common condition represents a natural consequence of the very principle of parasitism. Hyperparasitic relationships form inverted food pyramids in the pyramids of the larger community.

In most cases a host cannot readily prevent a parasite from becoming attached to its body surface. Numerous *ectoparasites* exploit this possibility. Equipped with suckers, clamps, or adhesive surfaces, they hold on to the skin of a host and feed on the internal body fluids with the aid of cutting, biting, or sucking mouth parts. Examples are leeches, lice, ticks, mites, and lampreys. Inside the body of a host live *endoparasites,* which must breach more formidable defenses. Cellular enzymes of a host, digestive juices and strong acids in the alimentary tract, antibodies in the blood, and phagocytic cells that engulf foreign bodies are among the defensive agents guarding against the invader. Overcoming such defenses means *specialization:* formation of resistant outer cuticles or

Figure 17.18 Parasite attachment. A, an ectoparasitic wood tick attached to the fur of a deer. *B,* section through the anterior part of an endoparasitic hookworm (a nematode) clamped to the mucosal lining in the gut of a host.

A

B

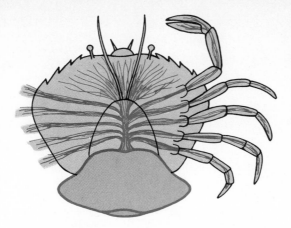

Figure 17.19 Parasitic simplification. The crustacean parasite *Sacculina* (colored) is attached to the abdomen of a crab. The exteriorly formless adult parasite, related to barnacles, spreads tissue outgrowths into the host. (*Adapted from Calman.*)

cyst walls, development of hooks or clamps that hold on to the gut wall, or secretion of enzymes that erode a path through host tissues (Fig. 17.18).

Specialization also involves selection of *specific* hosts. Parasites can survive only in hosts that, for example, contain particular types of nutrients and that offer appropriate invasion routes by which such nutrients actually can be reached. Once an infection is achieved, it is to the obvious advantage of a parasite to keep its host alive; for when a parasite kills a host, the killer generally is killed as well. Indeed, the virulence of a parasitic species often decreases with time. When a parasite-host relationship is first established, the invader is likely to be disease-causing, or *pathogenic*. Two parallel evolutionary trends then tend to reduce this pathogenicity. One is natural discrimination against infected hosts—the least resistant will have less chance of surviving. At the same time, the less harmful a parasite, the better will be its own chance of surviving. Many parasites are only mildly pathogenic, or not at all, often indicating long association with a particular host. In time the parasitic relation actually can become commensalistic or mutualistic.

Once established in a host, a parasite is more or less embedded in nutrients and does not require much locomotor equipment, many sense organs, or fast nervous reflexes. Indeed, structural and functional *simplification* is a nearly universal characteristic of parasites. Here is the ultimate expression of the principle that loss of self-sufficiency tends to be proportional to the degree of interdependence of animals.

Structural simplification is exhibited by, for ex-

ample, tapeworms. They have only a reduced nervous system, a greatly reduced muscular system, and not even a vestige of a digestive system (see Fig. 21.12). Almost like blotting paper, the worms soak up through their body walls the food juices in the host gut. Even more simplified is *Sacculina,* a relative of barnacles that parasitizes crabs. The parasitic adult is little more than a formless, semifluid mass of cells that spreads through a crab like a malignant tumor. The invader later gives rise to recognizably typical, free-swimming crustacean larvae, which attach to crabs, enter them, and then metamorphose to the simplified adults (Fig. 17.19).

Simplification also affects metabolic activities. For example, the synthetic capacities of a parasite are almost invariably restricted. The animal has become dependent on its host to supply it, in prefabricated form, with most of the components of its living substance. Simplification is probably an adaptive advantage, for the reduced condition usually is more economical than the fully developed condition of the free-living ancestor. A tapeworm, for example, need not divert energy and materials toward maintenance of elaborate nervous, muscular, or digestive systems, which are unnecessary anyway in this parasitic way of life.

Reproduction Parasites typically are exceedingly prolific. The enormous reproductive potential represents a solution of a major problem confronting the parasite, particularly the endoparasite: how to get from one host to another of the same species.

Parasites succeed by *active transfer* and *passive transfer,* both of which involve reproduction. In active transfer, one stage of the life cycle is free-living and motile, and this stage transfers from one host to another through its own powers of locomotion. For example, the adult may be parasitic but the embryo or larva is free-living and capable of locomotion, as in *Sacculina.* Or the immature phase may be the parasite, the adult then being free-living and motile, as in a number of parasitic insects (see Color Fig. 8).

Passive transfer is encountered where a parasite is not motile at any life-cycle stage. Propagation here is accomplished by wind, by water, or by *intermediate hosts.* The last offer a means of transfer not quite as chancy as distribution by wind or water. For example, tapeworms that infect man use one of the easiest routes into and out of their hosts, the alimentary tract. Entering through the

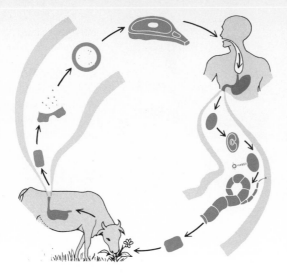

Figure 17.20 Beef tapeworm life cycle. Ripe sections of the worm pass with the feces from the human gut. Eggs are released from these sections in the gut of cattle. Walled, hook-bearing tapeworm embryos then encapsulate in beef muscle, and the embryos become adults in the intestine of man. The head (scolex) of the worm is invaginated at first but it soon everts, and with the hooks and newly developed suckers it attaches to intestinal tissues. See Fig. 21.12 for illustration of adult worm.

Figure 17.21 Chinese liver fluke life cycle. A, adult in liver of man. *B,* egg released with feces and eaten by snail. *C,* miracidium larva in snail. *D,* sporocyst. *E,* one of many redias formed from a sporocyst. *F,* cercaria. *G,* cercarias escape from snail and encapsulate in fish muscle. Man then eats undercooked fish, and an adult fluke develops in the host alimentary system. See also Figs. 21.9 and 21.10.

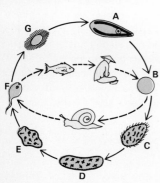

host's mouth by way of eaten food, tapeworms spend their life directly in the gut cavity of man, and they release their eggs through the anus of the host, along with feces. The problem now is to transfer by passive means to another human host.

Since man does not eat feces, the eggs cannot reach new human hosts directly. However, tapeworms take advantage of the food pyramids of which man is a member; man eats beef, and cattle eat grass. A ready-made pathway from grass to man thus exists, and transfer can be achieved if, as happens on occasion, human feces are deposited on grass. Tapeworm eggs clinging to such vegetation then are often eaten by cattle (Fig. 17.20).

In the intestine of a cow, a tapeworm egg develops into an embryo that bores a path through the gut wall into the cow's blood stream. From there the embryo is carried to beef muscle, where it encapsulates and matures. If man later eats raw or partially cooked beef, the capsule surrounding the young tapeworm is digested in the human gut and the free worm hooks on to the intestinal wall of its new host.

This history is by no means unusual, for many kinds of parasites use well-established food chains in transferring to new hosts. Often there is more than one intermediate host, as in the life cycle of

the Chinese liver fluke (Fig. 17.21). The adults of this parasitic flatworm infest the liver of man. Fertilized eggs are released through the bile duct into the gut of the host and pass to the outside with the feces. If the feces get into ponds or rivers, as happens frequently, the eggs develop into so-called *miracidium* larvae.

Such a larva then must enter a snail and there the miracidium develops into another larval type, called a *sporocyst*. Inside it develop many *redia* larvae, which escape and feed on snail tissue. Then each of the redias produces yet another set of many larvae, called *cercarias*. These fourth-generation larvae escape from the snail, and if within a short time they happen to encounter a fish, they bore into it and encapsulate in muscular tissue. If man later eats raw or incompletely cooked fish, the young adult flukes find their way from the human gut to the liver.

This cycle involves two intermediate hosts, the snail and the fish. Transfer is partly passive (man to snail, fish to man), partly active (snail to fish). Note particularly the multistage, larva-within-a-larva type of development. Characteristic of flukes generally, larval polymorphism of this sort represents a highly efficient method of enormously increasing the number of reproductive units. A single fluke egg is estimated to yield a final total of some 10,000 cercarias—and a single adult fluke can produce many tens of thousands of eggs. Hence the chances become fairly good that at least some of the millions or billions of larvae will reach final hosts.

Through active locomotion, through physical agents such as air and water, and through routes involving food pyramids and intermediate hosts, parasites have solved their transfer problems most successfully—so successfully, indeed, that there are many more individual parasites in existence than free-living animals.

A community consists of various kinds of free-living and various kinds of symbiotic populations. Which particular ones of each type actually compose a given community is determined largely by the nature of the physical, abiotic portions of ecosystems, as the following chapter will show.

Review Questions

1. Define ecosystem, biosphere, species, polymorphism, dimorphism, community, and give examples. What characteristics does a species exhibit as an evolutionary unit, a reproductive unit, and an ecological unit? What is an ecological niche?

2. Distinguish between interspecific fertility and intraspecific infertility. What is a syngen? Is the usual definition of "species" universally applicable? Distinguish between intraspecific cooperation and intraspecific competition and give examples.

3. Describe the typical life cycle of a species. What are endemic and cosmopolitan species? What effect does interspecific competition have on species survival?

4. Describe the continental drift hypothesis and show how it helps to explain the geographic distributions of certain animals. Describe the geographic and zoological characteristics of the six zoogeographic zones.

5. What is a species-population and what is its relation to a species? What are the structural characteristics of a population with reference to dispersion and growth?

6. What are some of the physical and biological limiting factors of population growth? In what ways does a population interact with its physical environment? Its biological environment?

7. Describe the species structure of a community. What factors produce and maintain such a structure? Describe the life cycle of a community.

8. What are ecological succession, climax communities, seres? Describe the nutritional structure of a community. What factors maintain such a structure?

9. What are food pyramids and what conditions produce and maintain them? Show how nutritional factors contribute to the long-range numerical constancies in communities.

10. Describe reproductive links that make the population of communities interdependent. Show how such links contribute to numerical population balances.

11. Give examples of mimicry and other forms of camouflage in animals. What other kinds of protective links unite the members of communities? How do all such links contribute to numerical population balances?

12. What are the various forms of symbiosis and how are they defined? Give specific examples of each. Distinguish between obligatory and facultative symbiosis.

13. What general structural and functional characteristics distinguish parasites from free-living forms? What is hyperparasitism? What is the adaptive advantage of parasitic simplification?

14. Distinguish between active and passive transfers in parasite life cycles. What is the role of food pyramids in parasite transfers? What are intermediate hosts?

15. Review the life cycles of tapeworms and liver flukes and show what general modes of parasite transfer are illustrated by these cycles.

Collateral Readings

Cole, L. C.: The Ecosphere, *Sci. American,* Apr., 1958. The amount of life sustainable on earth is assessed from the point of view of ecological interrelations of organisms and their environment.

Dodson, E. O.: "Evolution: Process and Product," Reinhold, New York, 1960. This book contains comprehensive sections on biogeography and species distribution.

Dunbar, M. J.: The Evolution of Stability in Marine Environments: Natural Selection at the Level of the Ecosystem, *Am. Naturalist,* vol. 94, 1960. A case study of the development of stable communities.

Harzen, W. E.: "Readings in Population and Community Ecology," Saunders, Philadelphia, 1964. A compilation containing discussions of many topics covered in this chapter.

Hurley, P. M.: The Confirmation of Continental Drift, *Sci. American,* Apr., 1968. An account on some of the most recent evidence in support of the hypothesis.

Kendeigh, S. C.: "Animal Ecology," Prentice-Hall, Englewood Cliffs, N.J., 1961. Useful for additional data and review.

Kurtén, B.: Continental Drift and Evolution, *Sci. American*, Mar., 1969. On the zoogeographic effects of continental drift on reptilian and mammalian evolution.

Limbaugh, C.: Cleaning Symbiosis, *Sci. American*, Aug., 1961. This article describes mutualistic and commensalistic relationships among marine fishes.

Odum, E. P.: "Ecology," Holt, New York, 1963. This is a useful paperback for general background on topics covered in this and the following chapter.

Rogers, W. P.: "The Nature of Parasitism," Academic, New York, 1962. Well worth consulting for a more thorough study of this aspect of symbiosis.

Smith, R. L.: "Ecology and Field Biology," Harper & Row, New York, 1966. The first sections of this book provide a good discussion of basic ecological principles.

Solomon, M. E.: "Population Dynamics," St. Martin's, New York, 1969. A highly recommended paperback on various aspects of population ecology.

Wilson, J. R.: Continental Drift, *Sci. American,* Apr., 1963. A historical outline of the hypothesis and a description of some of the recent supporting evidence. See also the more recent articles by Hurley and Kurtén, above.

18

HABITAT
AND BIOSPHERE

The sum of all ecosystems on earth represents the *biosphere,* and the physical portion of this largest ecological unit is the global environment. Everything living ultimately depends on this environment; it sustains life, orients evolution, and provides the specific homes, or *habitats,* in which communities actually exist.

Habitats

With the possible exception of the most arid deserts, the high, frozen mountain peaks, and the perpetually icebound polar regions, probably no place on earth is devoid of life. The two main types of habitat are the *aquatic* and *terrestrial.* Both range from equator to pole and from a few thousand feet below to a few thousand feet above sea level. *Ocean* and *fresh water* are the main subdivisions of the aquatic habitat, and *air* and *soil* of the terrestrial.

The Sea

The most familiar attribute of sea water is its high mineral content. The proportions of the different types of salt are almost the same all over the globe, as a result of thorough mixing of all waters by currents. Fifty-five percent of all ions present are chlorine, thirty percent are sodium. Thus, more than four-fifths of the total mineral content consists of table salt.

However, the total salt concentration, or *salinity,* varies considerably from region to region. The highest salinities occur in tropic waters, where high temperatures and extensive evaporation concentrate the oceanic salts. In the Red Sea the salt concentration is 4 percent, one of the highest known. At higher latitudes, by contrast, sea water evaporates less and therefore is less salty. Moreover, salinities are lower for often several hundred miles around the mouths of great rivers. The lowest known salinity is that of the Baltic Sea, where it is 0.7 percent (that of human blood is 0.9 percent). Salinity determines the density, or buoyancy, of ocean water, density being the greater the higher the salinity. Both salinity and buoyancy are of considerable significance to all marine life. Salinity affects, for example, the mineral and osmotic balances of living matter, and buoyancy aids in counteracting the pull of gravity.

An ocean basin has the general form of an inverted hat (Fig. 18.1). A gently sloping *continental shelf* stretches away from the coastline for about 100 miles (discounting often extreme deviations from this average). The angle of descent then changes fairly abruptly and the shelf grades over to a steep *continental slope.* This slope eventually levels off and becomes the ocean floor, a more or less horizontal expanse called the *abyssal plain.* Mountains rise from it in places, with peaks sometimes so high that they rear up above sea level as islands. Elsewhere the plain is scarred by rifts, the deepest being the approximately 7-mile

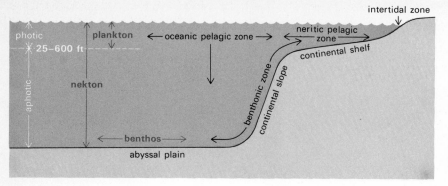

Figure 18.1 *The structure of an ocean basin.* The littoral zone (not labeled) is the part of the benthonic zone that forms the floor of the continental shelf.

deep Japan and Philippine Trenches along the western edge of the Pacific.

Three major habitats can be distinguished in such a basin. The sea floor from the shore out to the edge of the continental shelf forms a *littoral zone.* Beyond it, the sea floor along the continental slope and the abyssal plain represents the *benthonic zone.* And the water that fills the ocean basin is the *pelagic zone.* Sunlight penetrates water only to an average depth of about 250 ft and to at most 600 ft in certain seas. Within this sunlit *photic zone* light dims progressively to zero. As a consequence, photosynthesizing organisms can exist only in the uppermost layers of the sea. Animal life directly dependent on this vegetation therefore must remain near the surface as well. In sharp contrast, the dark region below the sunlit water, called the *aphotic zone,* is completely free of photosynthetic types and contains only animals, bacteria, and possibly fungi.

On the basis of its relationship to these habitats, marine life can be classified generally as *plankton, nekton,* and *benthos.* Plankton includes all drifting or floating types. Most of them are microscopic and are found largely in the photic zone. Although many of these forms have locomotor systems, they are nevertheless too weak or too small to counteract water currents (Color Fig. 10). Nekton comprises the strong swimmers, capable of changing stations at will. All nektonic types thus are animals, and they are found along the surface as well as in the sea depths. The benthos consist of crawling, creeping, and attached animals along the sides and the bottom of the ocean basin.

The planktonic organisms in the surface waters include teeming trillions of algae that as a group probably photosynthesize more than twice as much food as all land plants combined. Collectively called *phytoplankton,* this oceanic vegetation represents the richest pasture on earth; directly or indirectly it forms the nutritional basis of all marine life. Most of the algal types in this "grass of the sea" are microscopic, and among them the single-celled, yellowish- or brownish-green *diatoms* are probably the most abundant (Color Fig. 11).

Living side by side with the phytoplankton in open surface water are the small nonphotosynthetic forms. These include bacteria, protozoa, and members of the *zooplankton:* eggs, larvae, tiny shrimp (krill) and copepods, and other small animals carried along by surface drift. They feed directly on the microscopic vegetation. Most of the nekton, largely fishes and marine mammals, comes to these waters and feeds either on zooplankton or on photoplankton directly.

Through its influence on water temperature the sun has important effects on the distribution and abundance of plankton, hence on marine life in general. When the temperature of surface water is high, as in tropical seas throughout the year and in northern and southern seas in summer, the surface water cannot mix readily with the colder water below; a warm water layer is less dense and therefore lighter than a colder layer, and it "swims" on top of the colder layer without mixing. The boundary between the two layers is a *thermocline.* Organisms above such a thermocline deplete the water of mineral raw materials, and after the organisms die these materials sink down without being returned to the surface by vertical mixing. As a result the amount of surface life is limited, and warm seas actually are relatively barren. In cold seas, by contrast, surface and deeper waters have roughly the same low temperature, thermoclines tend to be less pronounced or even absent, and vertical mixing of water can occur more readily. Minerals then are recirculated more rapidly and surface life can be correspondingly more abundant. The perennially cold artic, antarctic, and subpolar waters actually support huge permanent plankton populations. And as is well known, the best commercial fishing grounds are in the high north and south, not in the tropics, and the best fishing seasons are spring and fall, not summer.

Warm and cold seas differ not only in the total amount but also in the diversity of life: warm oceans typically sustain small populations of many species, and cold oceans harbor large populations of comparatively few species. The reason is that

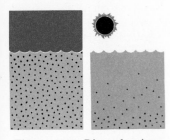

Figure 18.2 *Diurnal migrations* of plankton. At night (*left*) plankton is distributed throughout the dark surface water. In daytime (*right*) plankton migrates away from the surface to darker, deeper waters.

higher temperatures promote all reactions, including those leading to evolution. Warm-climate life therefore will tend to become more diverse than cold-climate life. However, the limited mineral content of warm seas tends to keep the numbers of individuals low.

The sun also produces *diurnal migrations* of most planktonic organisms. During the night the organisms are distributed vertically throughout the surface waters, but during the day most of the plankton shuns the bright light and moves down to the dim layers of the photic zone. Larger animals feeding on plankton migrate up and down correspondingly. As a result even richly populated seas are quite barren on the surface during the daytime, and it is well known that surface fishing is most successful at night (Fig. 18.2).

A

B

Figure 18.3 *Animal coastal life*. A, *Caryophyllia*, a coral-forming cnidarian of warm shallow water. B, *Eudistylia*, a feather-duster worm (distantly related to clamworms and earthworms) that forms and lives in secreted tubes attached along shallow coastal bottoms. The feathery tentacles trap minute food organisms. See also Color Fig. 1.

Nearer to shore, in the water above the littoral zone, even a bottom dweller is likely to be within the range of sunlight. The main problem here is to remain attached to solid ground, for the force of waves and ground swells is considerable. In the tide zone, moreover, a still greater problem is the rhythmic alteration between aquatic and essentially terrestrial conditions twice every 24 hours. As the meeting ground of water, land, and air, the tidal region actually is among the most violently changing habitats on earth.

Vegetation in this coastal region is again largely algal. In addition to the single-celled and small planktonic types, multicellular algae abound. Most of these seaweeds are equipped with specialized holdfasts that anchor the organisms to underlying ground. Animals in these waters include representatives of almost all major groups. In addition to planktonic types the region contains numerous nektonic animals, largely fish, as well as attached, burrowing, and creeping forms adapted to rocky, muddy, or sandy bottoms (Fig. 18.3).

The contrast between the surface habitat within reach of the sun and the aphotic zone below it is dramatic. As the one is forever fluctuating, so the other is perennially steady and relatively unchanging. Several unique physical conditions characterize this world of the sea depths.

First, in the total absence of sunlight, the region is pervaded with an eternal blackness of a kind found nowhere else on earth. Second, seasons and changing weather are virtually absent. Localized climatic changes do occur, either as a result of occasional submarine volcanic activity, or, more regularly, through deep-sea currents. These produce large-scale shifts of water masses and, incidentally, bring oxygen to even the deepest parts of the ocean. The deep waters also are perpetually cold. Temperatures range from about 10°C at the top of the dark zone to about 1°C along the abyssal plain.

Third, water pressure increases steadily from the surface down, 1 atmosphere (atm) for every 33 ft of descent. In the deepest trenches of the ocean the pressure therefore is about a thousand times as great as at sea level. And last, a continuous slow rain of the remains of dead surface organisms drifts down toward the sea bottom. A good deal of this material dissolves completely during the descent. But much microscopic mineral matter reaches the abyssal plain, where it forms ever-thickening layers of ooze. Accumulating over the

A

B

Figure 18.4 Deep-sea life. A, an angler fish with a stalked, luminescent "lantern" over the mouth. Note the vertical position of the mouth, which facilitates catching prey lured to the light of the lantern. *B,* the angler fish shown here is a female. The structure above the eye is a parasitic male, which is carried about permanently attached. This neatly solves the problem of finding a mating partner in the dark. Many of these large-mouthed, dagger-toothed fishes are surprisingly small; for example, the fish shown here fits comfortably into a person's palm. *C,* another "lantern" fish, in which not only the organ above the mouth but also the "beard" is probably biolumines-cent.

C

millennia, the older layers eventually are com-pressed to rock. Vertical bore samples of such rock have revealed a great deal about the past history of the oceans and their former surface inhabitants.

Contrary to early beliefs that life should be

impossible in such an environment, a surprisingly rich diversity of organisms has been found to exist in it virtually everywhere. Apart from bacteria and perhaps fungi, the community is characteristically animal; photosynthetic organisms cannot live in perpetual darkness. Nearly all animal groups are represented, many by strange and bizarre types uniquely adapted to the locale (Fig. 18.4). With the exception of animals such as toothed whales, which can traverse the whole ocean from bottom to surface, most of the nektonic deep-sea forms are adapted to particular water pressures. Such animals therefore are rigidly confined to limited pressure zones at fixed depths. Food must be obtained either from the dead matter drifting down from the surface—a meager source, particularly in deeper water—or from other nektonic types.

This last condition makes the deep sea the most fiercely competitive habitat on earth. The very structure of the animals underscores their violently carnivorous, "eat-or-be-eaten" mode of existence. For example, most of the fishes have enormous mouths equipped with long, razor-sharp teeth, and many can swallow fish larger than themselves. Since the environment is pitch-black, one of the critical problems for these animals is, to begin with, *finding* food. A highly developed pressure sense provides one solution. Turbulence in the water created by nearby animals can be recognized and, depending on the nature of the disturbance, can be responded to by flight or approach.

Another important adaptation to the dark is bioluminescence. Many of the deep-sea animals have light-producing organs of different shapes, sizes, and distributions. The light patterns emitted probably serve partly in species recognition. Iden-tification of a suitable mate, for example, must be a serious problem in an environment where every-thing appears equally black. Another function of the light undoubtedly is to warn or to lure. Certain fish, for example, carry a "lantern" on a stalk protruding from the snout. An inquisitive animal attracted to the light of the lantern will discover too late that it has headed straight into powerful jaws (see Fig. 18.4).

The Fresh Water

Physically and biologically the link between ocean and land is the fresh water. Among descendants of ancestral marine animals that invaded the rivers, some adapted to the brackish water in river

A

C

B

Figure 18.5 Freshwater life. A, a swamp region, characterized by many plants adapted secondarily to aquatic existence. These plants can resist displacement by flowing water. *B,* freshwater plankton. The algal components here are largely *Volvox,* and the animals are crustacean water fleas (*Daphnia*) and copepods (*Cyclops*). *C,* amphibian eggs enveloped by sticky jelly coats. Attachment of such eggs to aquatic vegetation is an animal adaptation to flowing fresh water.

could leave the ocean entirely. Certain of such freshwater types then managed to gain a foothold among the terrestrial forms, some later returned to water and adapted secondarily to an aquatic existence (for example, some snails, many insects). Thus, animals that inhabit the fresh water today represent a rich and major subdivision of the living world.

In addition to rivers, freshwater habitats also include lakes, ponds, and marshes. Three main conditions distinguish all such environments from the ocean. First, the salinity is substantially lower—indeed, lower than the salt concentrations inside animals. The osmotic consequences of this difference and the excretory adaptations evolved by freshwater animals have been discussed in Chap. 9. Second, much of the fresh water flows in strong, swift currents. Where these occur, passively floating life so typical of the ocean surface is not likely to be encountered. On the contrary, the premium will be either on maintaining firm anchorage along the shores and bottoms of rivers or on ability to resist and overcome the force of currents by muscle power.

Indeed, the vegetation found in rivers consists almost entirely of plants with rootlike holdfasts or actual roots—reed grasses, water foxtails, wild rice, and watercress. But where fresh water is not flowing strongly, not only rooted by also floating planktonic vegetation can be exceedingly abundant. In stagnant or near-stagnant water, algal communities forming continuous layers of green surface scum are particularly conspicuous (see Color Fig. 11). Among animals, similarly, those in quiet fresh waters include both planktonic and nektonic types, but those in swiftly flowing water are either attached or nektonic. The eggs of nektonic animals are enveloped by sticky jelly coats that adhere firmly to plants or other objects in the water, and the young are strongly muscled from the moment they hatch (for example, fishes and freshwater vertebrates in general; Fig. 18.5).

A third major distinction of the fresh water is that, with the exception of the very large lakes, it is affected much more by climate and weather than the ocean. Bodies of fresh water often freeze over in winter and can dry up completely in summer. Water temperatures change not only seasonally but also daily. Gales or floods bring bottom mud and silt to the surface. A large number of factors can alter flow conditions and produce, for example, stagnant water, or significantly altered

mouths or to a life spent partly in the ocean, partly in fresh water (for example, salmon, eels). Most on land. Of these, some continued to spend part of their lives in or near fresh water (for example, frogs), but more became wholly terrestrial. And

A

B

C

The Land

All land life is sustained by air and, directly or indirectly, also by soil. Air and soil are to the terrestrial habitats what the surface waters of the ocean are to the marine.

Like air, soil is itself a terrestrial home, providing a habitat for a vast array of subsurface organisms. And by creating the conditions necessary for the survival of all other terrestrial organisms, soil transforms the land surface into actually usable living spaces. Two other factors play a vital role, annual *temperature* and *precipitation*. As these vary with geographic latitude and altitude, they divide soil-covered land into a number of distinct habitat zones, or *biomes*. The six main biomes are *desert, grassland, rain forest, deciduous forest, taiga,* and *tundra.*

The first three are common in but not confined to the tropics. These biomes are characterized by comparatively high annual mean temperatures and by daily temperature variations that are greater than the seasonal variations; day and night temperatures tend to differ more than the average summer and winter temperatures. Variations in the amount of precipitation account for many of the different characteristics of these habitats.

A *desert* usually has less than 10 in. of rain per year, concentrated mainly in a few heavy cloudbursts. Desert life is well adapted to such conditions. Most plants grow, bloom, and produce seeds all in a matter of days after a rain. Since the growing season therefore is restricted greatly, such plants stay relatively small. Leaf surfaces are often reduced to spines and thorns (as in cacti), which minimize water loss by evaporation. Desert animals too are generally small, and they include many burrowing forms that can escape the direct rays of the sun under the ground surface. In most deserts, mammals and birds are comparatively rare or absent altogether; maintenance of constant body temperature is difficult or impossible under conditions of great heat and virtually no water. Cold-blooded animals can get by much more easily (Fig. 18.6).

The *grassland* extends from the tropics to much of the temperate zone. The more or less synonymous terms ''prairie,'' ''pampas,'' ''steppe,'' ''puszta,'' and many other regional designations underscore the wide distribution of this biome. A common feature of all grasslands is intermittent, erratic precipitation, usually amounting to about

Figure 18.6 The habitat of the desert. A, a desert landscape with tall saguaro cacti, barrel cacti (center foreground), cholla cacti (foreground), and palo verde trees. *B,* bobcat, and *C,* kangaroo rat, two mammals characteristic of many deserts.

chemical content, or situations facilitating infectious epidemics. In effect, the fresh water shares the environmental inconstancies of the land in very large measure.

Figure 18.7 The habitat of the grassland. *A*, a typical grassy plain. *B*, oryx, and *C*, wild buffalo, two mammals characteristic of African grasslands.

10 to 40 in. annually. Grasses of various kinds, from short buffalo grass to tall elephant grass and thickets of bamboo, are particularly adapted to irregularly alternating periods of precipitation and dryness. Grassland probably supports more animal species than any other terrestrial habitat. Different kinds of mammals are particularly conspicuous (Fig. 18.7).

In those tropical and subtropical regions where torrential rains fall nearly every day and where a well-defined rainy season occurs in winter, plant growth continues the year round. Such areas support *rain forests,* composed of populations of up to several hundred different species of trees. Rain forests cover much of central Africa, southern and southeastern Asia, Central America, and the Amazon basin of South America. Trees in such forests normally are so crowded together that they form

A

B

Figure 18.8 The habitat of the rain forest. A, Caribbean rain forest, with palms the predominant (but not only) trees. Note dense population of plants. *B,* an African rain forest. Note moss plants hanging from tree branches. The high moisture content of the atmosphere (resulting from release of much vapor by flora) permits growth of the mosses in such forests.

a continuous overhead canopy of branches and foliage. This cover cuts off virtually all the sunlight, much of the rain water, and a good deal of the wind. As a result the forest floor is exceedingly humid and quite dark, and it is populated by plants that require only a minimum of light. Animal communities too are stratified vertically, according to the several very different habitats offered between canopy and ground (Fig. 18.8).

Apart from extensive grasslands and occasional deserts, the most characteristic biome of the temperate zone is the *deciduous forest.* The basic climatic conditions here are cold winters, warm summers, and well-spaced rains that bring some 30 to 40 in. of precipitation per year. Moreover, day and night temperatures tend to differ less than the average summer and winter temperatures. Winters make the growing season discontinuous, and the flora is adapted to this condition. Trees are largely deciduous—they shed their leaves and hibernate; and although annual plants die in winter, they produce seeds that withstand the cold weather. A deciduous forest differs from a rain forest in that trees are spaced farther apart and in that far fewer species are present, perhaps no more than 10 to 20. Maple, beech, oak, elm, ash,

and sycamore are among the common trees of a deciduous forest. The many familiar animal types in this biome include deer, bears, raccoons, foxes, squirrels, and, characteristically, woodpeckers (Fig. 18.9).

North of the deciduous forests and the grasslands, across Canada, northern Europe, and Siberia, stretches the *taiga.* This is a biome of long, severe winters and of growing seasons limited to the few months of summer. Moose, wolves, and bears are most representative of the fauna, and hardy conifers of the forest flora. These forests differ from others in that they usually consist of a single species of tree. For example, spruce may be the only kind of tree present over a very large area. Another conifer species might be found in an adjacent, equally large area. The taiga occurs mainly in the northern hemisphere—little land exists in corresponding southern latitudes (Fig. 18.10).

The same circumstance makes the *tundra* a predominantly northern biome. Much of the tundra lies inside the Arctic Circle, hence there can be continuous night during the winter season and continuous daylight of comparatively low intensity during the summer. Some distance below the

A

B

C

Figure 18.9 The habitat of the deciduous forest. A, a typical deciduous forest, with maples the predominant trees. *B,* mountain lion, and *C,* Virginia deer, two mammals characteristic of many deciduous forests.

surface ground is frozen permanently (*permafrost*). Above ground frost can form even during the summer—plants often freeze solid and remain dormant until they thaw out again. The effective growing season is very brief, as in the desert, and indirectly the limiting factor again is water supply: frozen water is functionally equivalent to absence of water. Trees are completely absent, and the vegetation consists largely of lichens, mosses, and low shrubby plants. Also present are herbs with brilliantly colored flowers, many blooming simultaneously during the growing season. Conspicuous among the animals are hordes of insects, particu-

larly flies, and a considerable variety of mammals: caribou, arctic hares, lemmings, musk oxen, and polar bears. Birds are largely migratory, leaving for more southern latitudes with the coming of winter (Fig. 18.11).

Life does not end at the northern margin of the tundra but extends farther into the ice and bleak rock of the soilless polar region. Polar life is almost exclusively animal and includes types such as walrus, seals, and penguins. This fauna is not really terrestrial, however, but is based on the sea (Fig. 18.12).

The horizontal sequence of biomes between

A

B

C

Figure 18.10 The habitat of the taiga. A, a typical taiga landscape, with single (coniferous) tree species covering a large area. B, moose, and C, wolverine, two mammals characteristic of the taiga.

equator and pole is repeated more or less exactly in a vertical direction, along the slopes of mountains (Fig. 18.13). Here too temperature and precipitation are the decisive variables. On a high mountain in the tropics, for example, the succession of biomes from mountain base to snow line is tropical rain forest, deciduous forest, coniferous forest, and lastly low shrubby growths and lichens.

The farther north a mountain is situated, the more northern a biome covers its base and the fewer biomes cover its slopes. In the taiga, for example, the foot of a mountain is coniferous forest and the only other biome higher up is the zone of low shrubby plants. Thus, habitats spread over thousands of miles of latitude are telescoped into a few thousand feet of altitude.

A

B

C

Figure 18.11 The habitat of the tundra. A, Alaskan tundra landscape. Note complete absence of trees. B, musk ox, and C, ptarmigan, two animals characteristic of the tundra.

The nature of any kind of communal home, terrestrial, freshwater, or marine, evidently is determined by a few recurring physical variables. Among them are solar light, solar heat, geographic latitude, vertical depth and altitude, precipitation, wind and water currents, and the chemical composition of the locale. Variables like these are global in scope, and together they add up to large-scale "environment." With this we reach the most comprehensive ecological unit on earth, the biosphere as a whole, in which a thin surface layer of the planet maintains a carpet of life.

The Biosphere

The most important general property of the environment is that it is forever changing, at every level from the submicroscopic to the global. Astrophysical, meteorologic, geologic, geochemical, and biological forces alter every component of the earth sooner or later, very rapidly in some cases, slowly in others. A basic reason for this unceasing change is that the earth as a whole, living matter included, is an *open system*. Such systems ex-

A   B

Figure 18.12   Animals of the polar and subpolar habitats. A, seals, B, king penguin (molting).

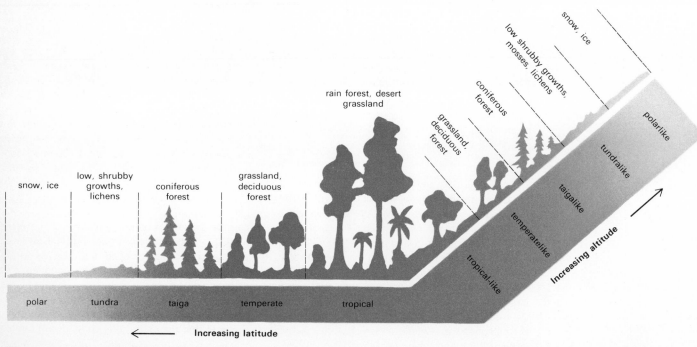

Figure 18.13   Mountain habitats. The sequence of habitat zones between equator and pole is repeated altitudinally between the base and the top of a mountain.

change materials, energy, or both with their sur-roundings. By contrast, a *closed system* exchanges nothing with its surroundings (see also Chap. 3). To be sure, the amounts of material exchanged between earth and space are negligible. But *energy* enters and leaves all the time: heat, light, X rays, ultraviolet rays, electric waves, and other forms of solar energy beam to earth uninterrupt-edly, and enormous amounts of energy radiate away, mainly in the form of heat. So long as the sun shines and the earth spins, therefore, the energy flow creates balance-upsetting disturb-ances. Every imbalance then creates new imbal-ances of its own, and as a general consequence the earth's environment is forever changing.

Being produced primarily by the sun and the motion of the earth, such changes tend to occur in rhythmic *cycles*. Daily and seasonal climatic cycles are familiar examples. Other cycles are less readily discernible, for their scale often is too vast or too minute or they occur too fast or too slowly for direct observation. Living matter is interposed in these cycles; and as the earth's components circulate, some become raw materials in living processes.

Living organisms therefore can be envisaged as temporary constructions built out of materials "borrowed" from the environment for a short time. Billions of tons of materials are withdrawn from the environment by billions of organisms all over the world, are made components of living matter, are redistributed among and between or-ganisms, and finally are returned to the environ-ment. Through such *nutrient cycling* organisms contribute actively to the large-scale movement of earth substances. And because of such cyclic

movements the physical earth *conserves* all its raw materials on a long-term basis. An indefinitely continued, repeated re-creation of living matter thereby becomes possible, and the continuity of life clearly depends on the parallel continuity of death.

The physical portion of the biosphere consists of three global subdivisions. The *hydrosphere* in-cludes all liquid components—the water in oceans, lakes, and rivers, and on land. The *litho-sphere* comprises the solid components, the rocky substance of the continents. And the *atmosphere* is the gaseous mantle around the hydrosphere and the lithosphere. Living organisms depend on raw materials from each of these subdivisions. The hydrosphere supplies *water;* the lithosphere con-tributes all other *minerals;* and the atmosphere is the source of *oxygen, nitrogen,* and *carbon di-oxide.* Together, these inorganic materials provide all chemical elements for construction and main-tenance of living matter (Fig. 18.14).

The Hydrosphere

Representing the most abundant mineral of the planet, water covers some 73 percent of the earth's surface entirely and is a major constituent of the lithosphere and the atmosphere. Water also is the most abundant component of living matter (see Chap. 4).

The basic cycle that moves and conserves water in the environment is quite familiar. Solar energy evaporates water from the hydrosphere to the atmosphere. Later cooling and condensation of the vapor at higher altitudes produces clouds, and precipitation as rain or snow then returns the water to the hydrosphere. This is the most massive process of any kind on the earth's surface, con-suming more energy and moving more material than any other.

Aquatic animals obtain water directly from their liquid environment, and they excrete some of it back while they live. After death the remainder is returned through decay. Terrestrial forms draw water from the reservoir in soil and in bodies of fresh water. The animals retain some of the water in their bodies and the rest is excreted, partly as liquid water back to the hydrosphere, partly as water vapor to the atmosphere. Indeed, water moves from hydrosphere to atmosphere far faster through the "pump" of living organisms than if it were simply allowed to evaporate directly from

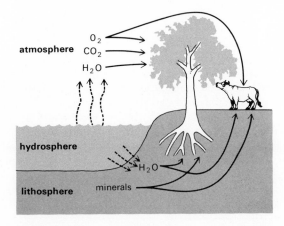

Figure 18.14 Environment and living matter. An illustration of the material contributions of each of the three subdivisions of the environment to the mainte-nance of living matter.

atmosphere

O_2
CO_2
H_2O

hydrosphere

lithosphere   minerals

H_2O

the hydrosphere. Terrestrial types therefore actively accelerate the global water cycle, and sometimes this can have an effect on climate. For example, the trees in jungles release so much water vapor that the air over vast areas remains permanently saturated with moisture, cloudbursts then occurring almost every evening. After terrestrial organisms die, any liquid water in their bodies returns again to the hydrosphere through decay.

Apart from being a prime nutrient water also influences animals through its effect on almost all aspects of climate and weather. In the ocean, water warmed in the tropics becomes light and rises to the surface, whereas cool polar water sinks. These up-down displacements bring about massive horizontal shifts of water between equator and pole. The rotation of the earth introduces east-west displacements. These effects, together with the driving force of similarly patterned winds, result in *ocean currents*. Such currents influence climatic conditions not only in the sea but also in air and on land.

Another climatic effect is a consequence of the thermal properties of water. Of all liquids water is one of the slowest to heat or cool, and it stores a very large amount of thermal energy. The oceans thus become huge reservoirs of solar heat. The result is that sea air chilled at night becomes less cold because of *heat radiation* from water warmed by day. Conversely, sea air warmed during the day becomes less hot because of *heat absorption* by water cooled at night. Warm or cool onshore winds then moderate the inland climate in daily patterns. Similar but more profound effects produce seasonal summer and winter patterns.

Third, global climates are influenced by the relative amount of water locked in *polar ice*. Temperature variations averaging only a few degrees over the years, produced by still poorly understood geophysical changes, suffice for major advance or retreat of polar ice. Ice ages have developed and waned during the last million years, and warm *interglacial* periods, characterized by ice-free poles, have intervened between successive advances of ice. At present the earth is slowly emerging from the last ice age (see Chap. 15), and as polar ice is melting water levels are now rising and coast lines are gradually being submerged. Moreover, deserts are presently expanding; snow lines on mountains are receding to higher altitudes; in given localities more days of the year are snow-free; and the flora and fauna native to given latitudes are slowly spreading poleward.

Evidently, by influencing temperature, humidity, amount of precipitation, winds, waves, and ocean and river currents, the hydrosphere plays a major role in determining what kinds and amounts of animals can live in particular regions.

The Lithosphere

This subdivision of the environment supplies most *mineral* nutrients to animals. Also, it forms the inorganic base of *soil,* required specifically by numerous subterranean animals.

Like the world's water, the rocky substance of the earth surface moves in a gigantic cycle. But here the rate of circulation is measured in thousands and millions of years. One segment of this global mineral cycle is *diastrophism,* the vertical uplifting of large tracts of the earth's crust through a variety of geologic forces. Major parts of continents or indeed whole continents can undergo such slow diastrophic movements. The most striking example is *mountain building*. Presently the youngest and highest mountains are the Himalayas, the Rockies, the Andes, and the Alps. All of them were thrown up some 70 million years ago, and the earth surface in these regions is not completely settled even now.

Quite apart from the tremendous upheaval caused by mountain formation itself, such an event has long-lasting effects on climate, hence on animals. The appearance of a high mountain barrier is likely to interfere drastically with continental air circulation, and moisture-laden ocean winds no longer may be able to pass across the barrier. Rain then will continue to fall in the region between ocean and mountain and this region can become lush and fertile. But the region on the other side of the mountain will be arid and desert conditions are likely to develop (Fig. 18.15). Along the Himalayas, for example, India on the oceanside is fertile but a belt of deserts lies north of the mountains. Similarly, California on the oceanside of the Sierras is fertile, but deserts lie east of the Rockies, in Arizona and New Mexico. Animals living on either side of a newly formed mountain range must adapt to the new environmental conditions by evolution. Indeed, as pointed out in Chap. 15, periods of extensive mountain building have always been followed by major evolutionary turnover among animals.

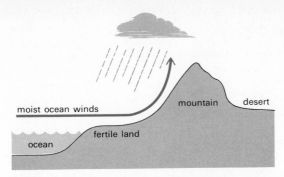

Figure 18.15 Mountains and climate. A mountain deflects moisture-rich ocean winds upward and causes rain to remain confined to the slope facing the ocean. That slope therefore will be fertile, but the far slope will become a desert.

The second segment of the global lithospheric cycle involves *gradation,* the lowering of high land and the leveling of mountains. These changes occur by actual geologic sinking of land and by *erosion.* Physical erosion through water, wind, and changes in temperature results in a breaking of rock. For example, the shearing and pulverizing effects of water and ice on rock are well known (Fig. 18.16). Chemical erosion, brought about by the often acid products of decay, leads to a dissolution of rock. All such erosive forces contribute to converting large stones to smaller ones and small pebbles to tiny sand grains and microscopic rock fragments. Erosion thereby plays a major role in forming the mineral components of soil.

Figure 18.16 Erosion. This canyon was cut by the erosive action of the stream seen in the photo.

Through chemical erosion, moreover, water dissolves small quantities of rock and acquires a mineral content as a result. Accordingly, as rain water runs off high land it becomes progressively laden with minerals. Dissolved in streams, rivers, and soil water, such minerals serve as raw materials for terrestrial and freshwater animals. After these die and decay the minerals return to the hydrosphere and eventually are carried to the sea. Therefore, as the lithosphere is slowly being denuded of mineral compounds the marine portion of the hydrosphere accumulates them. It was partly by this means that the early seas on earth acquired their original saltiness, and as the global water cycle now continues it makes the oceans even saltier. Animals in the sea freely use the mineral ions as nutrients.

The death of marine organisms then helps to complete the global mineral cycle. Many marine forms use mineral nutrients in the construction of protective shells and supporting bones. After these organisms die their mineral components rain down to the sea floor where, as already pointed out, the accumulating layers of ooze ultimately compress to rock. The global lithospheric cycle becomes complete when such a section of sea bottom or low-lying land generally is subjected to new diastrophic forces. High ground or mountains are thereby regenerated, and parts that were sea floor originally are thrust up as new land in the process (Fig. 18.17).

The Atmosphere

Like the hydrosphere, the atmosphere is subjected to physical cycles by the sun and the spin of the earth. Warmed equatorial air rises and cooled polar air sinks, and the axial rotation of the earth shifts air masses laterally. The resulting global air currents have basically the same general pattern as the ocean currents, and indeed these winds are the main forces that produce the ocean currents. Like the water circulation, moreover, the global air circulation influences climatic conditions substantially.

Equally significant for animals are the chemical cycles of the atmosphere. Air consists mainly of oxygen (O_2, about 20 percent); carbon dioxide (CO_2, about 0.03 percent); nitrogen (N_2, about 79 percent); water (in varying amounts, depending on conditions); and minute traces of inert gases. Except for the inert gases, all these components

Figure 18.17 The global mineral cycle. Minerals absorbed by terrestrial plants and animals return to soil by excretion and death. Rivers carry soil minerals to the ocean, where some of them are deposited at the bottom. Portions of the sea bottom then are occasionally uplifted by geologic forces, which reintroduce minerals to a global cycle.

of air serve as raw materials, and each circulates in a global cycle in which organisms play a conspicuous role. Also, all the gases are dissolved in natural waters; hence aquatic as well as terrestrial organisms have access to the aerial raw materials.

Atmospheric oxygen enters the living world as a respiratory gas, and in the course of respiration it combines with hydrogen and forms water (see Chap. 5). This water becomes part of the general water content of living matter, and as such it undergoes three possible fates. First, it can be excreted immediately to the hydrosphere or at-

mosphere. Second, it can be used in the construction of more living matter as a source of the elements hydrogen and oxygen. Structural oxygen of this sort remains in an animal until decay after death releases most of it in the form of H_2O or CO_2, which return to the environment. Third, water is a raw material in photosynthesis, where the hydrogen participates in food manufacture and the oxygen returns to the environment as a byproduct (Fig. 18.18).

In effect, oxygen enters organisms only through respiration and leaves only through photosynthesis. In intervening steps the oxygen is incorporated in water, and in this form it can interlink with the water cycle or indirectly with the carbon cycle. Atmospheric oxygen is the source of an ozone (O_3) layer that envelops the earth at an altitude of some 10 miles. This layer shields organisms by preventing most of the solar ultraviolet and X rays from reaching the earth's surface.

Atmospheric carbon dioxide is the exclusive carbon source of living matter. The gas enters the living world through photosynthesis, in which it is a fundamental raw material (Fig. 18.19). In this manner CO_2 contributes to the formation of food substances, and these are partly respired, partly used in construction of more living matter. In the first case CO_2 reappears as a byproduct that is returned to the environment immediately, and in the second it is a decay product returned only after death. The interrelations of the CO_2, O_2, and H_2O cycles are outlined in Fig. 18.20.

The carbon dioxide content of the atmosphere is replenished also through forest and other fires and through burning of industrial fuels. Such processes actually represent a long-delayed completion of the carbon cycle. The combustible substances in wood, coal, oil, and natural gas all are organic compounds that were manufactured through photosynthesis, in many cases millions of years ago. Aerial CO_2 then was used up, and the gas is returned to the atmosphere only now. Atmospheric CO_2 acts as a heat screen. Solar energy that reaches the earth's surface is largely converted to heat, but CO_2 retards radiation of earth heat into space. The gas therefore has a "greenhouse" effect that probably contributes to the present warming up of global climates (see also below).

The nitrogen of the atmosphere is the ultimate source of this element in living organisms. However, aerial nitrogen is chemically relatively inert,

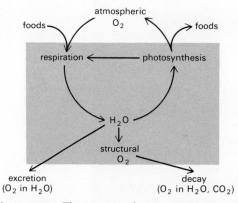

Figure 18.18 The oxygen cycle. The parts of the cycle that occur inside organisms are shown in the colored rectangle.

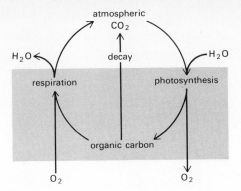

Figure 18.19   The carbon cycle. The parts of the cycle that occur inside organisms are shown in the colored rectangle.

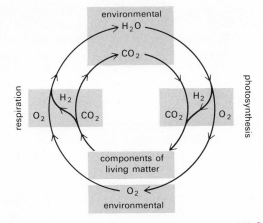

Figure 18.20   Interrelations of the O_2, CO_2, and H_2O cycles.

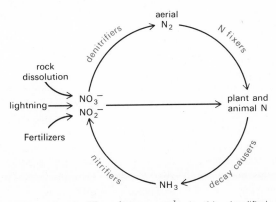

Figure 18.21   The nitrogen cycle. In this simplified representation the participating bacterial types are indicated in color over the curved arrows.

and the majority of organisms cannot use it directly. For example, although terrestrial animals obtain an abundance of nitrogen with every breath, all of it is again exhaled unused. Aerial N_2 can be used directly only by so-called *nitrogen-fixing* organisms, mainly certain bacteria and blue-green algae living in soil and water. These absorb N_2 as a nutrient and are able to make it part of organic substances.

Such usable fixed nitrogen then becomes available to plants and through plant foods to animals (Fig. 18.21). Another important source of usable nitrogen is the nitrate ion, NO_3^-, present in water and soil. Plants (but not animals) can convert nitrates to usable organic substances, and these again become available to animals via plant foods. Through decay after excretion and death, nitrogenous compounds in plants and animals then return to water and soil in the form of ammonia, NH_3. This substance is a nutrient for so-called *nitrifying bacteria*. One type of these converts ammonia to nitrite ions, NO_2^-; a second type absorbs the nitrite and converts it to *nitrate* ions, NO_3^-. This combined action of the nitrifying bacteria thus replenishes the supply of environmental nitrates.

Replenishment is achieved also in at least three other ways, through dissolution of rock by water, through addition of fertilizers to soil by man, and to a small extent through formation of nitrates in air, when lightning combines aerial nitrogen and oxygen. The nitrate supply in the environment then is drawn on not only by plants but also by so-called *denitrifying bacteria*. They use nitrates as nutrients and ultimately convert them to molecular nitrogen, N_2. This gas escapes to the atmosphere, and the global nitrogen cycle is thereby completed.

The cycle evidently depends on four different sets of bacteria: nitrogen fixers, decay causers, nitrifiers, and denitrifiers. These bacteria act as they do, not because they are aware of the grand plan of the global nitrogen cycle, but because they derive immediate metabolic benefits from their action; the nitrogenous compounds they absorb are nutrients, and the altered compounds they eliminate are their waste products. The nitrogen cycle also illustrates, perhaps as sharply as any one process can, how thoroughly the living and nonliving components of the biosphere are interdependent and interconnected. Indeed such a total ecological integration is absolutely essential for the continued existence of the biosphere and all its subordinate parts.

A   B   C

Figure 18.22   Pollution of land, water, and air. *A,* country road in Florida. *B,* detergent foam on the Mississippi river near Alton. *C,* smog over New York City.

The Ecological Crisis

The interrelation of all parts of the biosphere is probably the single most important factor that man today must be concerned with if he and the animal world as a whole are to survive. Of all biospheric components man is by far the most influential, and his net ecological impact to date has been to bring about a rapidly accelerating disruption of the links and balances on which the existence of animals and all other living things depends. Indeed, human survival itself now is seriously threatened by this "environmental decay." Broadly speaking the man-made ecological crisis manifests itself in three parts of the biosphere: the physical environment, in which we have created a crisis of resources for all life, our own included; the biological environment, in which we must cope with the twin crises of overpopulation and undernourishment; and the cultural environment, which has generated a crisis of technology and science.

The physical resources of the globe include all environmental components on which life depends—land, water, and air, the providers of living space and basic raw materials. A crisis arises because our withdrawals from these finite resources and our additions to them have been excessive and have seriously impoverished both space and supplies. On land, for example, we

increasingly withdraw usable territory for animals and plants by building houses, roads, and other installations incompatible with nonhuman life. One result is a curtailment of photosynthesis, with a corresponding reduction in the amount of oxygen liberated by plants to the air. At the same time we withdraw and burn enormous quantities of oil, coal, and gas, which further depletes the oxygen of the atmosphere.

Moreover, we add so much nitrogen- and phosphorus-containing waste to water resources that many of them become overfertilized. The consequence is overgrowth of aquatic microorganisms and depletion of oxygen in water, often followed by death of the microorganisms and all other aquatic life as well. Lake Erie is a good example of a body of water that is virtually dead biologically—and such results tend to be irreversible for many decades. A continuing oxygen depletion ultimately can be lethal for man, too, for once the rate of depletion through fuel combustion and pollution exceeds the rate of global photosynthesis, the atmosphere will begin to lose oxygen. Although some 70 percent of all photosynthesis on earth occurs in the ocean, even there, especially in coastal regions, man now adds pollutants to such an extent that the life of plankton—hence oxygen liberation—is affected adversely (Fig. 18.22).

Furthermore, our burning of fossil fuels (particularly in automobiles and planes) adds so much CO_2 to the atmosphere that the "greenhouse" effect referred to earlier becomes immediately worrisome. It has been estimated that at present combustion rates the amount of excess CO_2 accumulated by the end of this century will be enough to cause polar ice to melt in from 400 to 4,000 years. Assuming a melting time of 1,000 years for the Antarctic cap alone, the sea level would rise 4 feet every 10 years, enough to inundate most major cities and all coastal areas around the globe. Moreover, the effect of excess CO_2 and other pollutants on air quality and local climates is appreciable even now, especially in urban areas. In a city like Los Angeles, for example, the *daily* outpouring of aerial pollutants includes some 150 tons of volatile acids, 400 tons of sulfurous compounds, 500 tons of nitrogenous substances, and 1,500 tons of organic materials (among which the hydrocarbons become irritants in smog after exposure to sunlight and oxygen). To that should be added some 2 tons of dust settling on every square mile every year. There can be little question that the combined deleterious effects on plant crops and animals, and the clinical and subclinical effects on people, are substantial. Indeed such effects are usually so pervasive and diverse that they defy precise identification and measurement.

Aerial pollutants absorbed by water and wastes added to water directly affect life quite as decisively. Many pollutants decay too slowly or not at all, and then they pass through aquatic food chains to fish and other water animals eaten by man, or from water animals to land forms that in turn are eaten by man. An important *amplification* phenomenon is usually observed in such cases: both in water and on land the concentration of a pollutant increases at each step during its transfer through a food chain. For example, it has been shown that for any given concentration of insecticide in soil, earthworms in that soil acquire concentrations up to 40 times higher; and in birds feeding on such earthworms the concentration rises up to 200 times. Similar amplifications have been demonstrated for many other pesticides, for synthetic detergents, and for the components of radioactive fallout.

Since 1965 the use of undegradable detergents has been banned by law, but slowly decaying pesticides such as DDT still persist virtually everywhere, some 20 years after their first use—and long after many kinds of insects have become resistant to the compound. Similarly, fallout products such as strontium 90 will persist for decades or centuries. An additional problem is that the despoilation of resources often is irreversible or nearly so and that such effects are cumulative. The overall result is a continuing impoverishment of the physical environment and a steadily deepening resource crisis.

The sheer massiveness of the human presence aggravates this crisis enormously. Human protoplasm well might outweigh the amount of living matter of any other land animal, and our increasing numbers actually create or seriously contribute to all other ecological crises. Indeed, overpopulation is generally considered the single most critical of all the critical problems we face. Mankind now doubles in numbers every 30 to 35 years, or once every generation, so that by the year 2000 there will be 7½ billion people, roughly twice as many as at present. The annual growth rate currently is about 2 percent, but this rate itself is increasing. Our net global population gain amounts to 2 persons every second, which means 200,000 more people, or a new Peoria, every day; 6 million more, or a new Moscow, every month; and 70 million more, or a new Brazil, every year. The basic problem—though by no means the only one—is food supply, which is inadequate now and is bound to become more so. According to expert consensus global food production is not likely to keep pace with the inexorable population increase, and widespread famines are believed to be virtually unavoidable in the years ahead (Fig. 18.23).

New food technologies conceivably might help to forestall catastrophe. Indeed, laboratory methods are rapidly becoming available for a "made-to-order" creation of improved kinds and even new kinds of food plants. However, accelerated and ever more ingenious research by itself cannot solve the global food problem. Technical improvements must be incorporated in routine agricultural practices, yet for various reasons this is often difficult or impossible. For example, many new strains and varieties of food plants developed in the United States do not grow well in tropical regions, the main centers of the food crisis. And when an attempt is made to develop new strains in tropical and subtropical areas directly, success is hardly ever rapid. A project to create the right kinds of high-yield corn in Mexico took 25 years to complete (quite apart from concentrated re-

A

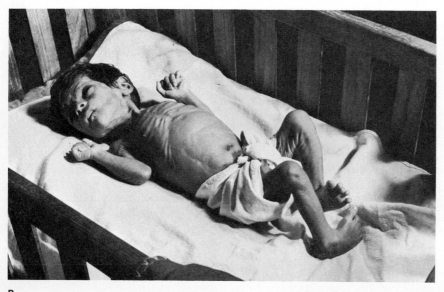

B

Figure 18.23   Overpopulation and undernourishment.

countries, but by now the extensive reserves are long gone. In many cases, moreover, receiver nations that have come to rely on donated food have delayed improving their own food production, and unchecked population increases in such nations have more than nullified the very temporary effects of food shipments. Thus, the two-thirds of mankind that was undernourished 20 years ago is still undernourished today, except that now this two-thirds comprises not 1 billion but 2 billion people. And the degree of malnutrition is still increasing.

A more direct way of solving the population problem would be birth control. Widely regarded as a more hopeful and more effective procedure than attempts to increase the food supply, birth control has indeed been successful in technically advanced nations. But success has been far less impressive in countries where the population and food crises are at their worst. In many places large populations still are considered evidence of national strength, and in others various religious and economic beliefs continue to foster excessively large families.

Some ecologists now believe that a solution of the population crisis and of the ecological crisis as a whole will require a new world-wide morality, a kind of eleventh commandment: at your death leave the ecosystem in a state no worse than it was at your birth; or, in specific relation to the population problem, make it part of your personal philosophy that two people should leave no more than two children. Such an ethic would lead to a voluntary but nevertheless absolute restriction of human numbers, and like other moral precepts it could have social and even legal reinforcement. Thus, exceeding the birth limitation could come to be regarded morally just as reprehensible, and just as punishable, as theft or murder; in a real sense excessive reproduction does rob and murder the life of future generations. Similarly, while any population would retain the freedom to breed itself into starvation, it would be immoral for outsiders to donate food since such a course would aggravate the global problem and indeed the local problem itself.

To be effective, a morality of this sort clearly would have to be truly world-wide; and in view of a lack of global agreement on far lesser issues, one has reason to doubt its rapid adoption. Yet, without it only two other solutions appear to re-

search aid by the United States and an expenditure of millions of dollars). That much time is simply no longer available today in most food-poor regions.

Besides, successful introduction of modern agricultural practices and new food technologies depends on full support of local farming populations; yet in many places such support is lacking or takes too long to foster. Until recently food-rich nations have shipped their surpluses to needy

main: reduction of numbers through nuclear holocaust, conventional wars no longer being sufficiently destructive to affect global population growth; and the "natural" solution through which the ecosystem always balances itself on its own accord—large-scale periodic waves of famine and death whenever the food supply falls below minimum sustenance levels.

While the food crisis is the most serious consequence of overpopulation in developing nations, the main consequence in advanced nations has been a cultural and technological crisis. Excessive numbers are a primary cause of crowded communities, sprawling slums, noise and congestion, foul air and environmental ugliness, and all the other physical manifestations of the "urban crisis." Biologically, moreover, life is often stunted, too tense, or simply dissatisfying, and in many cases life spans are shortened unduly by the excessive stresses generated by crowded conditions. Above all, overpopulation produces an increasingly automated, computerized, dehumanized culture, in which impersonal institutions frequently overwhelm many of the aspirations of the individual. Even now machines already control us as much as we control them.

Many people therefore have been prompted to wonder if science and technology have not gone "out of control," and thus if it might not be better to abolish these enterprises. It should be quite apparent, however, that suggestions of this kind are merely superficial responses to the symptoms, not the causes, of our cultural crisis. Our basic problem is not science and technology as such, but knowing—or rather not knowing—what to use them for. Too often science and technology have been employed simply to abet human greed or nationalistic ends, and too often has there been little concern about the consequences of their employment. For example, detergents and gasoline engines have proved to be exceedingly useful products, but it was not foreseen that they also polluted; and since their pollutant effects have become known it has taken too long, with too much opposition, to institute remedies. Similarly, nuclear explosions have served primarily as experiments in engineering and as instruments of national policy, and even now they are not treated as they should be, as major ecological interventions that affect *all* parts of our biosphere. Each technological advance always introduces risks as well as benefits, and deciding on a particular balance between risk and benefit is fundamentally *not* a scientific matter; as pointed out in Chap. 1 it is a social and cultural one, involving value judgments by all.

The answer in effect is not less science but more, for only through science are we at all likely to have some hope of coming out of the ecological crisis as a still viable species. Realization of this hope actually would not even require major scientific advances or new technologies; we already have enough information and know-how to make a significant start at improving our ecosystem now. Instead, the essential requirement is a collective willingness to put our knowledge to work. However, development of such a willingness demands certain basic changes in our attitudes. First, human action would have to be guided by the recognition that man and his environment form a single, closed, global system of integrated components. In this system every man is bound to every other not simply through common biological ancestry or through philosophic dreams of oneness, but, far more directly, through the same atmosphere, the same water, the same food-producing soil. The ancient query, "Am I my brother's keeper?" demands a very *practical* answer today.

Second, mankind would have to moderate its arrogant habit of trying to dominate the environment and would have to begin living with it in true *partnership*. We would have to cultivate the idea of being stewards only, not self-appointed masters, of what one observer has called our common "spaceship Earth." And as watchful stewards we would have to recognize that our technology by its very nature tends to be parasitic on the environment, and that destructive forces nearly always work more rapidly than those of rebirth; the animal we shoot in a second may have taken 10 years to develop. If we started the environmental restoration right now, it would take us from one to three generations, or on the order of 100 years at the very least, to resolve the present ecological crisis. We would need to be patient, therefore, and we would need to launch an immediate, permanent program of *planned ecological management*. Our basic aim would have to be an improvement in the quality of life—*all* life—with considerations of quantity distinctly secondary.

And third, in our role of environmental stewards, we no longer could afford to have our technological innovations produce unexpected ecological side effects; we would have to foresee and prepare for such effects. The necessary foresightedness can be generated today through ''systems analysis,'' a computer-based management technique. Mathematical models of our ecosystem and its subordinate parts could be constructed, and any change contemplated by us could be fed into these models. Through computer analysis we would then be able to find out ahead of time if the effects of the planned changes were consistent with our goals or, if not, how contemplated changes would have to be modified to make the effects acceptable.

Ecological analyses of this sort have not yet begun; they are exceedingly laborious and would require all our ingenuity and dedication, with contributions from all fields of knowledge. Yet we know *how* to begin them. And we also know that, unless we do begin them, man who has uncaringly tried to overwhelm the earth is about to be overwhelmed by his own hand.

Review Questions

1. What is the structure of an ocean basin? What are the major habitats of such a basin? What role does the sun play in creating subdivisions in these habitats? What physical conditions characterize the various subdivisions?

2. Define plankton, nekton, and benthos. Give specific examples of each. Where in the ocean are each of these types of organism found?

3. How do different oceans vary in water density and salinity? What are the proportions of oceanic salts? What factors make life in tropic waters generally less abundant but more diverse than in temperate and subpolar waters?

4. What are thermoclines, diurnal migrations? What physical and biological conditions characterize the sea depths?

5. Review the physical characteristics of freshwater habitats. What major groups of animals occur in fresh water, and in what general ways are they adapted to this habitat?

6. What are the main terrestrial habitats, and what physical and biological conditions characterize each of them? In what way are the terrestrial habitats at different latitudes related to those at different altitudes?

7. What factors maintain the global environment in a state of continuing change? What are open and closed physical systems? How do animals contribute to cyclic environmental change?

8. Describe the global water cycle What forces maintain it? How do animals participate in this cycle? In what different ways does the world's water influence climates?

9. What are diastrophism and gradation? How does the formation of mountains influence climates? Cite examples. In what ways is the global lithospheric cycle of nutritional importance?

10. Describe the general pattern of mineral cycles. On the basis of this, construct diagrams showing the pattern of global phosphate and calcium cycles.

11. What is the chemical composition of the atmosphere? Which of these components do not play a role in the maintenance of animals?

12. Describe the oxygen and carbon cycles and show how they are interlinked.

13. Outline the global nitrogen cycle. How many different groups of bacteria aid in the maintenance of this cycle? What is the role of decay in the atmospheric, lithospheric, and hydrospheric cycles?

14. Identify the main contributing factors to our present ecological crisis. Describe the amplification effect of pollutants. What specific conditions jeopardize the supply of atmospheric oxygen? The maintenance of aquatic life?

15. What are the ecological consequences of human overpopulation? Outline your own solutions to our various environmental crises, and systematically examine the ecological consequences of these suggestions.

Collateral Readings

Berrill, N. J.: "The Living Tide," Dodd, Mead, New York, 1951. A beautifully written book on the communities of the seashore.

Calhoun, J. B.: Population Density and Social Pathology, *Sci. American*, Feb., 1961. Experiments on rats reveal some of the drastic effects of overcrowding.

Carson, R.: "The Sea Around Us," Oxford University Press, Fair Lawn, N.J., 1951. A classic nontechnical book on the physics, chemistry, and biology of the sea, written for popular consumption. Strongly recommended.

————: "The Silent Spring," Houghton Mifflin, Boston, 1962. This book (which caused considerable controversy when it first appeared) examines the disruptive effect of man's use of chemicals and other agents on the ecological balance of nature.

Clark, J. R.: Thermal Pollution and Aquatic Life, *Sci. American*, Mar., 1969. On the importance of eliminating waste heat now added to natural waters.

Commoner, B.: "Science and Survival," Viking, New York, 1966. A penetrating examination of our resource crisis and the cultural crisis generated by our advanced technology. Strongly recommended.

Deevey, E. S.: The Human Population, *Sci. American*, Sept., 1960. An examination of the total amount of human life the earth may be able to support.

Dubos, R.: "So Human an Animal," Scribner, New York, 1968. An exceedingly well-written, nontechnical analysis of all the undesirable ecological and cultural aspects of the contemporary human scene.

Fairbridge, R. W.: The Changing Level of the Sea, *Sci. American*, May, 1960. This article discusses the effects of ice ages and of diastrophic movements of the ocean floor.

Fremlin, J. H.: How Many People Can the World Support? *New Scientist*, vol. 24, p. 285, 1964. A physicist estimates the possible upper limit of human numbers on the basis of the total body heat produced, and he indicated the changes in way of life that would be necessary to maintain such a population. Extremely thought-provoking.

Isaacs, J. D.: The Nature of Oceanic Life, *Sci. American*, Sept., 1969. On the food chains and food pyramids in the ocean. The whole Sept., 1969, issue deals with the ocean and thus provides valuable background material.

McDermott, W.: Air Pollution and Public Health, *Sci. American*, Oct., 1961. An examination of the subtle effects of aerial pollutants on human health.

Miner, R. W.: "Fieldbook of Seashore Life," Putnam, New York, 1950. The emphasis is on invertebrates of the Atlantic Coast.

Nichol, J. A. C.: "The Biology of Marine Animals," Interscience, New York, 1960. The subject is treated on a functional basis.

Opik, E. J.: Climate and the Changing Sun, *Sci. American*, June, 1958. Changes in the radiation rate of the sun may have contributed to the development of ice ages.

Peakall, D. B.: Pesticides and the Reproduction of Birds, *Sci. American*, Apr., 1970. Chlorinated hydrocarbons upset breeding behavior and egg production in carnivorous types such as hawks and pelicans.

Pirie, N. W.: Orthodox and Unorthodox Methods of Meeting World Food Needs, *Sci. American*, Feb., 1967. An examination of required changes in cultural attitudes in attempts to solve the food crisis.

Plass, G. C.: Carbon Dioxide and Climate, *Sci. American*, July, 1959. CO_2 contributes to the regulation of global temperature, and the gas therefore plays an important ecological role.

Woodwell, G. M.: The Ecological Effects of Radiation, *Sci. American,* June, 1963.

———: Toxic Substances and Ecological Cycles, *Sci. American,* Mar., 1967. The ecological amplifications and other effects of pollutants are discussed in both articles.

The following readings from *Scientific American* have self-explanatory titles; they deal with the relation of the living to the nonliving components of the biosphere and with the nature of various habitats:

Deevey, E. D.: Life in the Depth of a Pond, Apr., 1951.
Ingle, R. M.: The Life of an Estuary, May, 1954.
MacInnis, J. B.: Living Under the Sea, Mar., 1966.
Munk, W.: The Circulation of the Oceans, Sept., 1955.
Nicholas, G.: Life in Caves, May, 1955.
Pequegnat, W. E.: Whales, Plankton, and Man, Jan., 1958.
Powers, C. F., and A. Robertson: The Aging Great Lakes, Nov., 1966.
Ryther, F. H.: The Sargasso Sea, Jan., 1956.
Stetson, H. C.: The Continental Shelf, March, 1955.
Vevers, H. G.: Animals of the Bottom, July, 1952.
Walford, L. A.: The Deep-sea Layers of Life, Aug., 1951.
Wexler, H.: Volcanoes and World Climate, Apr., 1952.

Up to this point the emphasis has been on the animal heritage, the traits and attributes of animals as a whole—patterns of body construction, functions at different levels of organization, developmental and evolutionary histories, and ecological associations. From now on all such attributes are considered jointly and the emphasis shifts to particular kinds of animals—the forms and combinations in which specific animal types actually exhibit the components of their general heritage.

The chapters of this unit are organized along taxonomic and evolutionary lines. The phylum is the basis of discussion; superphyla or related groups of phyla form broader frames of reference; and the general objective is to identify and characterize the major kinds of animals, in most cases at least down to the level of taxonomic orders. Pursuit of this objective provides insight not only into the vast diversity of the animal creatures on earth, but also into the innumerable ways in which the problems of living have found "animal" solutions.

UNIT 2
THE KINDS
OF ANIMALS

PROTOZOA
a Linnean subkingdom; animal-like protists. *Protozoa*

METAZOA
a Linnean subkingdom; multicellular animals.

Branch MESOZOA
adults blastulalike. *Mesozoa*

Branch PARAZOA
adults more complex than blastula. *Porifera*

Branch EUMETAZOA
adults above tissue level.

Grade RADIATA
radial animals with organs. *Cnidaria, Ctenophora*

Grade BILATERIA
bilateral animals with organ systems

Subgrade ACOELOMATA
animals without body cavities. *Platyhelminthes, Nemertina*

Subgrade PSEUDOCOELOMATA
animals with pseudocoelic body cavities. *Aschelminthes, Acanthocephala, Entoprocta*

Subgrade COELOMATA
all other animals.

The phyla examined in this part represent an evolutionary radiation of at least four main lines of descent; protozoa exemplify one, sponges another, the radiate group a third, and the bilateral animals a fourth. The mesozoans might or might not constitute a fifth original line of descent. These various phyla have close affinity to the most ancient animals and animal-like forms on earth. As such they presumably offer clues about what the base of the bush of animal evolution might have been like.

PART 7
NONCOELOMATE GROUPS

19

PROTOZOA

Phylum Protozoa

protists at cellular grade of construction; heterotrophic; predominantly motile and unicellular; cells in colonies are alike and can survive as isolated individuals; flagellate-ciliate, amoeboid, and coccine states of existence.

Subphylum Mastigophora

flagellate protozoa.

Subphylum Sarcodina

amoeboid protozoa.

Subphylum Sporozoa

spore-forming (coccine) protozoa.

Subphylum Cilophora

ciliate protozoa.

Protozoa are *protists,* with structural and evolutionary affinities to slime molds, fungi, and algae. Protozoa therefore are not "animals" (see discussion in Chap. 14). However, they are the most animal-*like* protists. As a group they have exploited the unicellular way of life more diversely than any other organisms, and by any criterion they are among the most successful of all living creatures. Major segments of the biosphere owe their development and present organization to the protozoa.

General Characteristics

Some zoologists regard protozoa as "noncellular" or "acellular" types, or organisms "not subdivided into cells." Such a view implies, first, that the cell theory is invalid, since "true" cells would exist only among multicellular forms; second, that *all* organisms other than multicellular ones are acellu-

lar, and that real cells therefore could have originated only *after* acellular organisms had evolved; and third, that the cells of *all* multicellular organisms could have originated only by septation, or internal partitioning (see Chap. 14), for if they had originated by aggregation the units that aggregated would have been "cells" already.

These implications make the acellular conception untenable. For example, an egg is universally admitted to be a "true" cell; yet numerous protists, postulated to be "acellular," can function as eggs directly. How then could an "acellular" unit be a "cellular" unit at the same time; and how could eggs have existed before cellular units are supposed to have existed? Also, even ardent septationists do not maintain that multicellularity in *all* cases could have arisen *only* by septation, yet one is forced to such a position by the "acellular" view. Most advocates of the "acellular" idea are zoologists who champion the ciliate protozoa as the ancestors of the Eumetazoa. Others, impressed by the great complexity and diversity of protozoa, assume that by attaching the label "acellular" they somehow elevate the organisms to some status more elaborate than "just" cells.

Actually the cell concept encompasses even the most complex protozoa very adequately. The essence of that concept is that the functional units of living matter are differentiated as nuclear, predominantly genetic structures and cytoplasmic, predominantly nongenetic ones. Consequently, "acellular" or "noncellular" really means that such a differentiation does not exist. But protozoa display a very obvious nucleocytoplasmic differentiation and they therefore are clearly "cellular." Any other designation not only denies their membership in the living fabric of the earth but also introduces more problems, unnecessarily so, than it intends to solve. Accordingly, protozoan units here are regarded to be *unicellular* and homologous to individual cells of multicellular types. Such a view does not prejudge the methods through which multicellularity arises, and it is semantically consistent with the observation that a single cell is "not subdivided into cells."

It is a frequent custom to include among protozoa numerous photosynthetic organisms that botanists claim to be algae. The issue relates directly to the inadequate Linnaean practice of rigidly labeling all organisms as either "plants" or "animals." In line with the discussion of this point in Chap. 14, protozoa here are considered to contain nonphotosynthetic, heterotrophic types *only;* photosynthetic forms such as *Euglena* or *Volvox* are algal protists, not protozoan ones.

Protozoa form four main groups: flagellate, amoeboid, spore-forming, and ciliate types. The consensus is that the flagellate and amoeboid groups are closely related, that flagellate ancestors probably have given rise to the ciliate group, and that the origin of the spore-forming group is obscure. Opinion is less uniform as to how the four groups should be classified. Traditionally each has been considered as a taxonomic class and in more recent practice a subphylum. All four together have been and in general still are regarded as a single phylum. Cogent arguments also can be made for a subdivision into several phyla. In any event the whole group represents a subkingdom in a Linnaean sense, and a similarly high ranking is implied if protozoa are regarded as a main subgroup of a category Protista.

Protozoa exhibit all known forms of heterotrophism. Most species are free-living holotrophs that feed on microscopic animals and other protozoa, including fellow members of the same species in cannibalistic forms (see Color Fig. 15). "Herbivorous" types depend on bacteria and microscopic algae; omnivorous types subsist on any microscopic food. Protozoa in turn are a food source for many animals. Some protozoan species are saprotrophic, and many are symbiotic, particularly parasitic. The spore-formers are exclusively parasitic and all other groups include some parasitic subgroups. Protozoa are themselves hosts to various parasites, mainly bacteria, fungi, and other protozoa. Also, many are hosts to photosynthetic algae (*zoochlorellae* and *zooxanthellae*), which live mutualistically in protozoan cytoplasm.

The organisms are components of all ecosystems in all aquatic environments, in soils, and generally in any environment that contains some moisture. Indeed, protozoa occupy almost as many different ecological niches in the microscopic world as Metazoa do elsewhere, and they exist in numerous niches not open to larger organisms. This tremendous ecological diversity is reflected in the number of protozoan species. Figures often quoted are on the order of 15,000, but there are known to be more than that many foraminiferan species alone. Moreover, very many animals (and some plants) are hosts to at least one unique type of protozoan parasite, which means that protozoan species well could number in the hundreds of thousands. Conservatively, at least 100,000 species of protozoa can probably be assumed to exist at present; and some 10,000 extinct species have been described to date as well.

Free-living protozoa are primarily flagellate and amoeboid. Coccine potentials are exhibited by the spore-formers, but sporine types are absent (see also Chap. 14). The protozoan cell (Fig. 19.1) is either naked or surrounded by a flexible to rigid *pellicle,* a cuticle composed of a variety of organic and horny substances. Many species secrete shells (*tests*) as permanent external covers, and some manufacture *loricae,* secreted housings within which the organisms can move. Numerous protozoa also secrete temporary shells, or *cysts.* In a *protective* cyst, viable for up to 5 years in some cases, a protozoon can remain dormant and withstand unfavorable environmental conditions; in a *reproductive* cyst, division or sexual processes take place. Encysted states are expressions of proto-

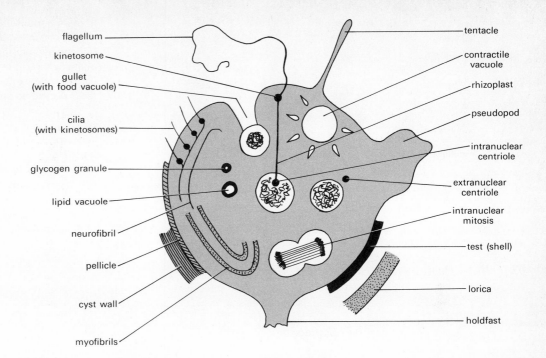

flagellum
kinetosome
gullet
(with food vacuole)
cilia
(with kinetosomes)
glycogen granule
lipid vacuole
neurofibril
pellicle
cyst wall
myofibrils

tentacle
contractile
vacuole
rhizoplast
pseudopod
intranuclear
centriole
extranuclear
centriole
intranuclear
mitosis
test (shell)
lorica
holdfast

Figure 19.1 Protozoan organelles. No single protozoan has all the organelles shown in this composite sketch, and most protozoa have others, not shown here.

zoan polymorphism. Other such expressions are common: sexual partners or gametes can differ structurally; young and adult stages can differ; and flagellate states often become nonflagellate or amoeboid and vice versa. Cell sizes among protozoa range from about 2 or 3 μ to several millimeters and more. Some of the foraminifera attain diameters of up to 10 cm or more.

Under the cell surface in a number of protozoan groups are conductile neurofibrils, contractile myofibrils, and contractile vacuoles. The latter occur in virtually all freshwater forms as well as in a few marine and parasitic types. Ingestive structures in holotrophic protozoa are pseudopodia, gullets, and in certain cases tentacles. As in animals, the diet of protozoa must contain organic carbon and nitrogen, some 10 or more amino acids, and several vitamins. Digestion takes place inside food vacuoles, and reserve foods occur mainly as polysaccharide granules (predominantly glycogen) and as lipid droplets or vacuoles. Internal distribution of nutrients is achieved by diffusion and cyclosis, and gas exchange and excretion take place directly through the cell surface. If contractile vacuoles are present they are the main osmoregulating and water-balancing organelles.

Although in most cases definite sensory structures cannot be identified, protozoa nevertheless are exquisitely sensitive to their surroundings. They can "taste" food and refuse to ingest unsuitable materials; they give distinct avoidance responses to unsuitable temperatures, light, electric charges, pH, mechanical stimuli, and dissolved chemicals; they seek out optimum environments by trial-and-error behavior; and many of them have been trained through conditioning to give "learned" responses to various stimuli.

Protozoa are uninucleate or multinucleate. Mitotic division is *intranuclear* in the vast majority of species; two sets of chromosomes are formed inside the original nucleus, which then constricts into two. Centrioles are not present in some protozoa; in other cases such granules either are intranuclear or, less often, extranuclear, in which case a rhizoplast frequently connects with a kinetosome near the cell surface.

The basic forms of protozoan reproduction are equal binary fission, unequal binary fission (*budding*), and multiple fission (*sporulation*, or spore formation). Equal binary fission is by far the most common. A few multinucleate types reproduce vegetatively by fragmentation (*plasmotomy*), a process in which the organism divides into two

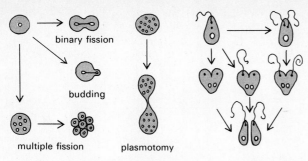

Figure 19.2 *Protozoan reproductive patterns.* Diagram at right shows that, during division, neither offspring or one offspring or each offspring can inherit flagella (or other formed organelles) directly from the parental cell.

binary fission

budding

multiple fission

plasmotomy

or more parts without nuclear divisions, the nuclei merely becoming distributed among the offspring. Regeneration potentials are high in general. Many multinucleate forms can be cut into virtually as many pieces as there are nuclei, each piece then redeveloping as a complete organism (Fig. 19.2).

Sex occurs through syngamy or conjugation (see Fig. 10.4). Hermaphroditism is common, and such species either are cross-fertilizing, self-fertilizing (autogamous), or both. With some exceptions to be referred to, protozoan life cycles are diplontic; meiosis takes place during gamete formation. In parasitic types the life cycles usually are exceedingly complex and geared to the ways of life of specific hosts.

Flagellate Protozoa   Subphylum Mastigophora

one or more flagella present throughout life or at given stages; predominantly uninucleate; reproduction by longitudinal binary fission; sexuality by syngamy.

Class Zooflagellata

Order Rhizomastigida   one to many flagella; permanently amoeboid. *Mastigamoeba, Mastigina, Mastigella, Multicilia*

Order Protomastigida   one to three flagella; often amoeboid; cells naked, without gullet. *Bodo, Codosiga, Trypanosoma*

Order Polymastigida   two to many flagella; organelles often in duplicate or multiple sets. *Giardia*

Order Trichomonadida   three to five flagella, plus one forming edge of undulating membrane. *Trichomonas*

Order Hypermastigida   numerous flagella; body complex. *Trichonympha*

Order Opalinida   numerous flagella, shortened to cilia, arranged in rows. *Opalina*

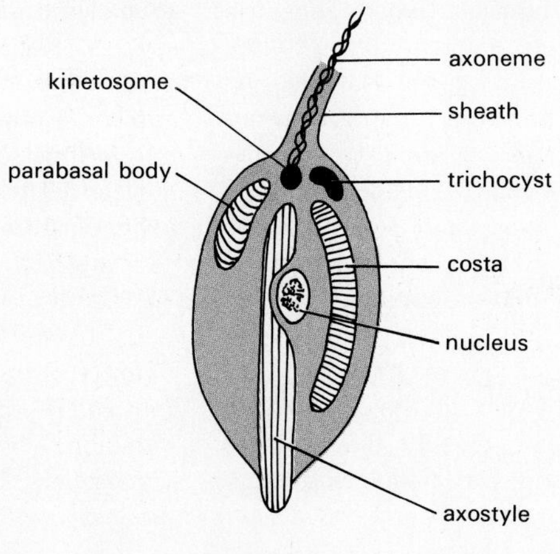

Figure 19.3 Kinetosome derivatives. This composite diagram shows some of the mastigophoran organelles associated with the kinetosome. The size of the flagellum base is exaggerated in relation to the cell as a whole. The outer flagellum sheath covers the 11 fibrils that form the axoneme attached to the kinetosome.

kinetosome

parabasal body

axoneme

sheath

trichocyst

costa

nucleus

axostyle

With the exception of the order listed last, the above orders represent a series that ranges from predominantly free-living, simply constructed types to predominantly symbiotic, complexly constructed types. The former appear to have given rise to the latter. The body of the organisms is elongated in the direction of motion, and where flagella are few in number they originate anteriorly (Fig. 19.3). A flagellum is believed to propel an organism on the principle of the screw; flagellate protozoa typically spiral through the water. Each flagellum consists of an outer sheath continuous with the cell surface and, as noted in Chap. 4, an inner bundle of 11 fibrils. Referred to collectively as the *axoneme*, these fibrils are produced by and attached to a kinetosome. In most species with multiple flagella,

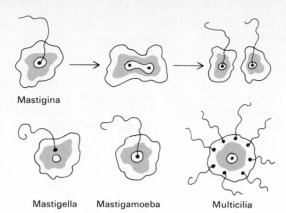

Figure 19.4 Primitive Mastigophora. These organisms are both amoeboid and flagellate permanently. In *Mastigina* and *Mastigella,* the basal granule serves jointly as centriole and kinetosome; the division pattern in *Mastigina* is sketched at top. In the other two genera shown, separate granules perform the functions of centrioles and kinetosomes. Note the rhizoplast in *Mastigamoeba.*

Mastigina

Mastigella   Mastigamoeba   Multicilia

Figure 19.5 Zooflagellates. In *Bodo* a trichocyst is indicated. The electron microscope shows that the collar of collar flagellates is a circlet of long cytoplasmic extensions, as indicated here. In *Trypanosoma,* note the posterior kinetosome and the undulating membrane edged by the flagellum. In the polymastigid *Giardia,* note the two nuclei (with centrioles) and the symmetrically paired kinetosomes, rhizoplasts, and flagella. *Trichomonas* has an axostyle, a parabasal body, a gullet, an undulating membrane, a nucleus with intranuclear centriole, and a kinetosome with additional flagella. The hypermastigid *Macrospironympha* has a spiral, ribbonlike kinetosome and hundreds of flagella emanating from it. *Trichonympha* is similarly hyperflagellate. *Opalina* is characterized by diagonally spiraling rows of kinetosomes, each such granule bearing a short flagellum. The nucleus is without centriole, as in ciliate protozoa. (*Adapted from Minchin, Kudo, Lapage.*)

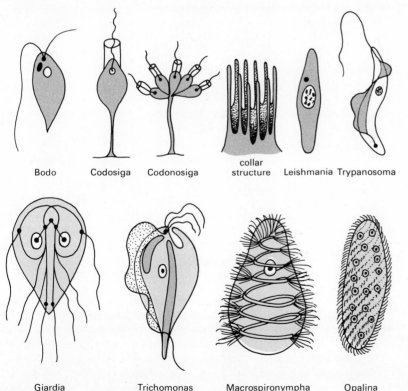

Bodo   Codosiga   Codonosiga   collar structure   Leishmania   Trypanosoma

Giardia   Trichomonas   Macrospironympha   Opalina

each flagellum originates from a separate kinetosome. But in some forms, notably the Hypermastigida, hundreds of flagella can be anchored on a single, greatly enlarged, ribbon-shaped kinetosome (see Fig. 19.5).

In Rhizomastigida and Opalinida the flagellum is the main product of the kinetosome, but in the other orders the basal granule also gives rise to a variety of other organelles (see Fig. 19.3). Thus, most groups have a *parabasal body,* which exhibits a Golgi-like structure under the electron microscope. An *axostyle* is often present, composed of bundles of fibrils attached to the kinetosome and possibly serving a supporting function. Similarly supporting appears to be the *costa,* which under the electron microscope displays a crossbanded structure like that of collagen fibers in animal connective tissues. Other kinetosomal products include dischargeable *trichocysts* that in some instances are believed to paralyze potential prey and in others probably aid in anchoring a cell during feeding.

The Rhizomastigida are excellent examples of organisms with joint flagellate and amoeboid traits (Fig. 19.4). In some of these amoeboid flagellates the flagella attach to granules that are either intranuclear (as in *Mastigina*) or extranuclear (as in *Mastigella*) and that serve *both* as kinetosomes and centrioles. Such a dual role of the granules unquestionably represents a primitive condition. At cell division the flagellum usually is resorbed, the kinetosome-centriole divides, and each daughter granule produces a new flagellum in each offspring cell. Fission occurs longitudinally, as in flagellates generally. In most members of the other flagellate orders, centrioles and kinetosomes typically are separate organelles. The centrioles again are intranuclear or extranuclear, the kinetosomes always extranuclear.

The Protomastigida include free-living saprotrophic types like *Bodo* and the holotrophic *choanoflagellates,* or collar flagellates, so named after the membranous sleeve around the single flagellum (Fig. 19.5). Under the electron microscope such a collar is seen to consist of a circlet of fingerlike cytoplasmic extensions from the main part of the cell. *Codosiga* is a stalked, sessile, solitary collar flagellate; other genera, among them *Codonosiga,* form stalked sessile colonies. *Proterospongia* was long thought to belong to this group of colonial flagellates, but these "organisms" have since been shown to be fragments of sponges. Among numerous symbiotic and parasitic Protomastigida, the trypanosomes are intracellular parasites in lymph and blood cells of various animals, vertebrates and mammals in particular. For example, a species of *Leishmania* is the causative agent of kala azar, a serious oriental disease that affects lymphoid tissues; and various species of *Trypanosoma* are blood parasites that produce sleeping sickness in man and other mammals.

Polymastigida are mostly intestional symbionts in insects and vertebrates. Many, like *Giardia,* are interestingly bilateral, with symmetric duplicate sets of nuclei, flagella, axostyles, and other organelles (see Fig. 19.5). Certain species of both *Giardia* and *Trichomonas* occur in man as free-swimming holotrophic or saprotrophic commensals. They are transmitted in encysted condition via feces. The Hypermastigida are the most complex flagellates. All are symbionts in cockroach and termite intestines, where the flagellates digest the wood eaten by their hosts for the benefit of both hosts and symbionts. It has been shown that the timing of the sexual process of the flagellates is geared to the molting time of the hosts. Undoubtedly, the numerous flagella of these flagellates are advantageous adaptively in the thick, viscous contents of the host gut.

The Opalinida live as saprotrophic commensals in the gut of amphibia. These flagellates are of considerable interest from an evolutionary standpoint, for their body is covered uniformly with neat rows of shortened flagella that are indistinguishable from cilia (see Fig. 19.5). Each cilium originates at a separate kinetosome. Because of this ciliation pattern and the absence of centrioles, opalinids have been classified until recently as a primitive group of ciliate protozoa. However, fission is longitudinal, a single type of nucleus is present, and mating occurs by syngamy, traits that are distinctly flagellate, not ciliate. Conceivably, the opalinids represent still living descendants of an evolutionary transition stage between flagellates and ciliates, a possibility strengthened further by the resemblance of opalinids to flagellates such as *Multicilia* (see Fig. 19.4). Ancestral flagellates thus might have given rise to two main evolutionary lines, one leading via Rhizomastigida and Protomastigida to the Hypermastigida, the other via opalinidlike stocks to the ciliate protozoa. In some classifications the Opalinida are ranked as a superorder of flagellates.

Amoeboid Protozoa   *Subphylum Sarcodina*

Feeding and locomotion by pseudopodia; uni- and multinucleate; reproduction by binary and multiple fission and plasmotomy; sexuality by syngamy.

Class Actinopodea   pseudopodia are axopodia.

Order Helioflagellida   with flagella. *Dimorpha*

Order Heliozoida   radial to spherical, with vacuolated outer cytoplasm. *Actinophrys, Actinosphaerium*

Order Radiolarida   radial to spherical, with central capsule and skeletal shell predominantly of silica. *Heliosphaera*

Class Rhizopodea   pseudopodia are not axopodia.

Order Amoebida   without shells. *Naegleria, Amoeba, Pelomyxa*

Order Testacida   with noncalcareous shells. *Arcella, Difflugia*

Order Foraminiferida   with calcareous shells. *Lagena, Globigerina*

The subphylum is divided into two classes according to the nature of the pseudopodia (Fig. 19.6). In the Actinopodea, an *axopodium* is a cytoplasmic extension supported internally by a stiff, nonmotile axoneme, a fibril bundle of the kind present in a flagellum. This axoneme originates either at a kinetosome or at an extranuclear or intranuclear centriole. In the Rhizopodea, the pseudopods without axonemes are of three main types: slender, terminally branching pseudopods are *filopodia;* broad, rounded ones are *lobopodia;* and filamentous, netlike ones are *rhizopodia.*

The Helioflagellida undoubtedly are a primitive group with close affinities to the flagellates. In

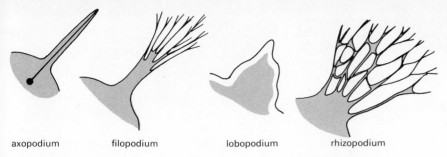

axopodium   filopodium   lobopodium   rhizopodium

Figure 19.6 Pseudopodia. In an axopodium the axoneme is a stiff, internal supporting rod, structured like a flagellum and emanating from an extranuclear or intranuclear kinetosome-centriole. Axonemes are absent in the other pseudopodial types.

Dimorpha, for example, a single extranuclear kinetosome-centriole gives rise to two anterior motile flagella and also to numerous axopodia (Fig. 19.7). As in Actinopodea generally, the naked exterior cytoplasm of an axopodium is sticky and traps food organisms that come into contact with it. A food vacuole then forms at the point of contact and this vacuole later flows into the main portion of the protozoon.

Heliozoida, the ''sun animalcules'' or Heliozoa, lack motile flagella. *Actinophrys* is uninucleate, with an intranuclear kinetosome-centriole. *Camptonema* is multinucleate, each nucleus having its

Figure 19.7 Actinopod Sarcodina. A, the axopodium shows the stiff axoneme and a food vacuole in the axopodial cytoplasm. In *Dimorpha* (a helio-flagellid), note the motile anterior flagella and the stiff axopodia, all radiating from an extranuclear centriole. In *Actinophrys* (a heliozoid), note the foamy cortex with contractile and food vacuoles, and the denser medulla with a nucleus. Axonemes radiate out from an intranuclear centriole. The heliozoids *Camptonema* and *Actinosphaerium* are multinucleate, the former with intranuclear and the latter with extranuclear centrioles. The general structure of a radiolarian shows the vacuolated cortex with free spicules, the perforated capsule, and the interior medulla with nuclei. In the silica skeleton of a radiolarian note three spherical lattices one inside the other, the outermost with projecting spines. The three lattices are kept in position by interconnecting struts. *B,* skeletal shells of radiolarians. Note latticelike construction (approx. ✕50). See also Color Fig. 12. (*A, adapted from Bütschli.*)

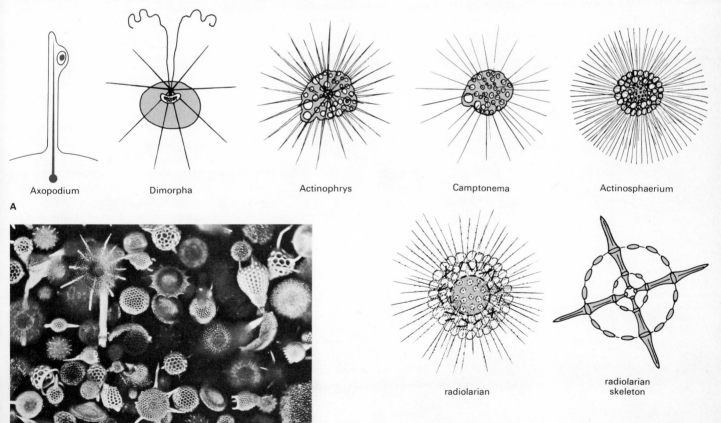

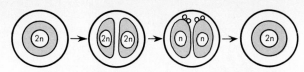

binary fission | 2 meiotic divisions | fertilization

Figure 19.8 Sex in Actinophrys, by autogamous self-fertilization. Inside a cyst wall the cell divides, each daughter undergoes meiosis (with polar-body formation), and the resulting single gamete cells then fuse and form a zygotic diploid adult.

own internal kinetosome-centriole *Actinosphaerium* is multinucleate, too, but here the kinetosome-centriole granules are extranuclear. In all these forms the axopodia can be shortened temporarily or resorbed completely, as during reproductive and sexual stages and during a rolling locomotion on solid surfaces.

The main part of the body consists of a highly vacuolated *cortex* on the outside and a denser, nucleus-containing *medulla* in the center. Heliozoa are largely freshwater types, with contractile vacuoles in the cortex. Some Heliozoa have latticelike silica skeletons around the surface and the axopodia then protrude through the spaces in the lattices. Reproduction occurs by binary fission, budding, and plasmotomy. Sex by self-fertilization is known for *Actinophrys* and *Actinosphaerium* (Fig. 19.8).

The Radiolarida, or Radiolaria, are marine and planktonic. Many species have diameters of several centimeters. The body of these organisms is subdivided into an inner medulla and an outer cortex by a *capsule,* a membrane with numerous perforations and composed of chitinlike material (Color Fig. 12 and see Fig. 19.7). The medulla contains one to many nuclei and the cortex is highly vacuolated. Radiolaria are noted for their exceedingly complex, beautifully sculptured skeletons. Most of these are made of silica; in one group they consist of strontium sulfate. In many cases several skeletal spheres are present, one inside the other, and the outermost sphere usually is ornamented with spines, needles, and hooklike extensions. Moreover, needle-shaped *spicules* lie free in the cytoplasm of many species. Protruding through the perforated skeletons are the pseudopodia, which are axopodia in most cases but filopodia in some.

In certain forms, reproduction by binary fission includes a division of the skeleton into two halves, each daughter cell then regenerating the missing half. In most species one daughter inherits the parental skeleton and the other escapes and develops a new skeleton of its own. Some Radiolaria are or become multinucleate and then subdivide by multiple fission into biflagellate cells that can function as gametes.

Buoyed up by the vacuolated cytoplasm and the floating surfaces formed by pseudopodia and skeleton, the Radiolaria are part of the plankton in all oceans. After death, the silica skeletons resist dissolution well during their slow fall to the depths. Such skeletons become a major component of the bottom deposit of deep oceans; some 5 percent of the world's ocean floor consists of "radiolarian ooze." Compressed to rock, these silica deposits form *flint* and *chert.* Radiolaria are among the few fossil organisms known from Precambrian times.

In the class Rhizopodea, the order Amoebida includes flagellate amoebae that correspond to the helioflagellates among the Actinopodea. Thus, *Naegleria* is permanently amoeboid but under certain conditions it can develop from one to three flagella (Fig. 19.9). These are attached to a kinetosome that connects by a rhizoplast to an intranuclear centriole. *Naegleria* does not have axopodia, however, but lobopodia. The other members of the order are largely lobopodial, too.

Figure 19.9 Rhizopod Sarcodina. A, Naegleria is an amoeboid type that also has flagella. *Entamoeba histolytica* is shown with ingested blood corpuscles in food vacuoles. The shell of *Difflugia* is made of sand grains, that of *Arcella,* of chitinous substances. *B, Pelomyxa,* one of the naked, multinucleate amoeboid types (with paramecia around it and *Amoeba proteus* near top, for comparison; approx. ×70). An amoeba is also shown in Fig. 2.7. See also Color Fig. 13. (*B, courtesy Carolina Biological Supply Company.*)

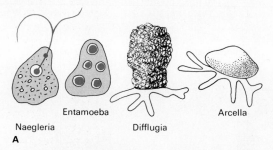

Naegleria | Entamoeba | Difflugia | Arcella

A

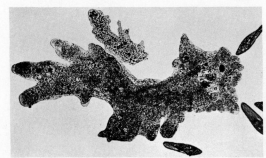

B

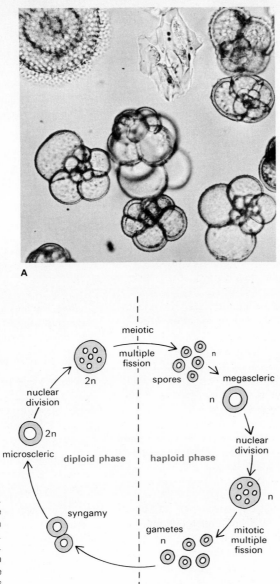

Figure 19.10 Foraminifera. A, Globigerina, a common genus. The helically coiled shells have minute openings through which the pseudopodia project (approx. ×100). See also Color Fig. 14. *B,* the foraminiferan life cycle. In this alternation of generations, the diploid microscleric adult is small-shelled, the haploid megascleric adult, large-shelled.

multinucleate genus that reproduces by plasmotomy.

Most members of the Amoebida are freshwater and marine bottom creepers, largely omnivorous and subsisting on bacteria and other protozoa. For example, amoebae feed on ciliates such as paramecia (Color Fig. 13). The Testacida occupy similar ecological niches. The shells of these protozoa usually have one opening from which lobopodia of filopodia are protruded. *Arcella* secretes a chitinlike test. *Difflugia* manufactures organic secretions to which it cements sand grains as a protective cover.

The Foraminiferida, or Foraminifera, correspond to the Radiolaria among the Actinopodea. The organisms are marine, mostly deep-water bottom creepers, often very large, and exceedingly abundant. *Globigerina* is among several planktonic surface floaters (Fig. 19.10 and Color Fig. 14). Some of the foraminiferan tests, or shells, are single-chambered, as in *Lagena.* Such tests usually are composed of organic (often chitinlike) materials to which sand grains or other foreign bodies can become attached. Other kinds of shells have many interconnected chambers and are impregnated with secreted calcium carbonate. Exterior perforations permit extrusion of pseudopods, which are sticky rhizopodial networks. The skeletal perforations give the group the name Foraminifera, or "hole bearers."

The shells are formed in a large variety of species-specific patterns. Shells of dead organisms sink toward the sea floor, but they dissolve before they reach very great depths. At intermediate depths, however, they accumulate in fantastically large quantities. Some 30 to 40 percent of the world's ocean floor is covered with foraminiferan ooze (and with *Globigerina* ooze specifically in numerous localities). *Limestone* and *chalk* are the rocky products of such bottom deposits. Uplifted geologically, they form limestone mountain ranges and chalk cliffs such as those along the English Channel coasts. Foraminifera, too, are known from at least early Paleozoic times.

Foraminifera with many-chambered shells have life cycles that include two generations of adults (see Fig. 19.10). A young, small-shelled (*microscleric*) organism is uninucleate at first but later becomes multinucleate by mitotic nuclear divisions. A process of multiple fission then takes place that includes meiosis. The resulting haploid offspring cells are *spores.* Escaping from the par-

The "common" amoeba is *Amoeba,* and the most familiar species, one of the best known organisms of all kinds, is *Amoeba proteus.* In this highly specialized (note: *specialized*) form, kinetosomes are absent and the flagellary potential has been lost (see Fig. 2.7). The related genus *Entamoeba* is symbiotic in the intestine of many animals. One parasitic species, *E. histolytica,* causes amoebic dysentery in man. *Pelomyxa* is a large,

ent shell, they later develop into large-shelled (*megascleric*) haploid individuals. These too become multinucleate eventually and then undergo *mitotic* multiple fission. The products so formed are haploid gametes. After syngamy, the diploid zygotes mature as small-shelled individuals that start a new life cycle. Such a cycle, characterized by meiosis during spore-formation, thus involves an alternation of two dimorphic generations of adults. It is a *diplohaplontic* cycle, of a kind also encountered in many algal groups and in all Metaphyta.

The Sarcodina as a whole undoubtedly represent a highly polyphyletic group. Various algal flagellate stocks may have given rise to the amoeboid flagellates and to the flagellate amoebae. Moreover, the nonflagellate amoebae could have evolved by loss of chlorophyll from numerous algal amoeboid types (see also Chap. 14). Furthermore, the Radiolaria and Foraminifera are similar in many respects to certain shelled flagellate groups of golden-brown algae, and the life cycles of these protozoan orders likewise suggests algal affinities. Other protozoa—zooflagellates, for example—could have contributed to sarcodine evolution, too. Little can therefore be said about the relations of the groups within the subphylum or in the phylum as a whole.

Spore-Forming Protozoa

Subphylum Sporozoa

exclusively parasitic; cells without contractile vacuoles; coccine, uni- and multinucleate; reproduction by multiple fission; sexuality by syngamy; life cycle typically haplontic.

Class Telosporidia   spores naked or encapsulated; schizogony or gamogony often absent; polar capsules absent.

Subclass Gregarinida   extracellular parasites, often with myonemes; spores encapsulated; schizogony absent (*Gregarina, Monocystis*) or present (*Schizocystis, Ophryocystis*).

Subclass Coccidia   intracellular epithelial parasites; spores encapsulated; schizogony present, gamogony absent. *Eimeria*

Subclass Hemosporidia   intracellular blood parasites; spores naked; schizogony present, gamogony absent. *Plasmodium*

Class Cnidosporidia   sporogony absent; polar capsules present.

Class Acnidosporidia   sporogony absent; polar capsules absent.

The life cycles of these protozoa include spore-forming stages. A single cell undergoes multiple fission, and each of the spore cells so formed is uninucleate. After a spore has become established in a host as a mature parasite, it eventually becomes multinucleate before the next multiple fission. As many as three successive spore-forming generations can occur in a single life cycle. Also, a cycle can require one or more intermediate hosts in addition to a main host. Sporozoa thus represent coccine forms; cell division takes place primarily during reproduction.

A complete life cycle that incorporates all stages has the following general pattern (Fig. 19.11). A vegetative haploid sporozoan in a particular host is a *trophozoite*. Uninucleate at first, it later develops into a multinucleate reproductive individual called a *schizont* (or *agamont*). This cell then undergoes multiple fission, or *schizogony* ($=$ *agamogony*). The resulting spore cells, *merozoites*, either develop as new trophozoites that reinfect other body parts of the host, or they become sexual individuals called *gamonts*. These undergo *gamogony*, a process of multiple fission resulting in gametes. The latter fuse pairwise and the diploid zygotes so formed then undergo still another process of multiple fission. This division, *sporogony*, now includes meiosis. The cells produced are all haploid, and they become *sporozoites* either directly or via encapsulated stages. Free sporozoites finally become vegetative trophozoites, and a new cycle is thereby initiated. In effect, the only diploid stage is the zygote. Life cycles with haploid adults and meiosis in the zygote are said to be *haplontic*; they are common also among algae, slime molds, and fungi.

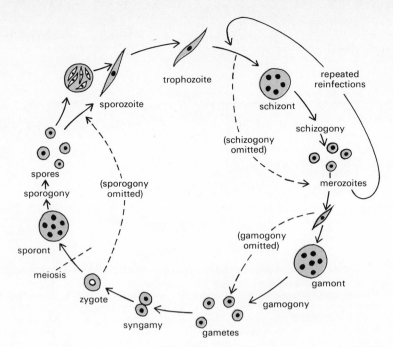

Figure 19.11 *The basic sporozoan life cycle.* In it, the zygote is the only diploid stage. Three haploid generations succeed each other, each characterized by a process of multiple fission (schizogony, gamogony, sporogony). The first is omitted in many telosporidians, the second in various different sporozoan groups, and the third in cnidosporidians and acnidosporidians. After sporogony, spores become free sporozoites directly or encapsulated sporozoites first.

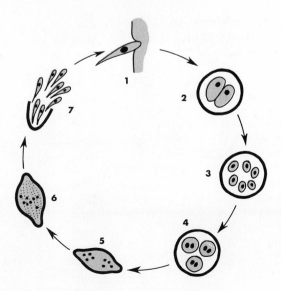

Figure 19.12 *Life cycle of Monocystis.* 1, adult trophozoite in seminal vesicle of earthworm host. 2, a syzygy chain of two gamonts encapsulated in gamocyst. 3, gamogony, resulting in gametes. 4, pairwise syngamy, resulting in zygotes. 5, individual zygote, encysted, nucleus undergoing two meiotic and one mitotic divisions. 6, sporogony, resulting in formation of eight sporozoites. 7, escape of sporozoites from original zygote cyst (and gamocyst) in new host. (*After Borradaile.*)

Variations of the basic sporozoan pattern result from an absence of any of the three phases. If schizogony is omitted, the trophozoites become gamonts directly. If gamogony is omitted, the gamonts become gametes directly. And if sporogony is omitted, the zygotes become sporozoites and thus trophozoites directly. Further variations result from the presence or absence of intermediate hosts in given life cycles.

The Sporozoa parasitize all phyla, including other protozoa and even other Sporozoa. Most require specific hosts. In the class Telosporidia, the gregarines are found in numerous invertebrate phyla but apparently not in vertebrates. Trophozoites live extracellularly in the intestine and in body cavities, where they adhere to tissues by means of various holdfast devices. For example, *Monocystis* is parasitic in the seminal vesicles and sperm ducts of earthworms. *Gregarina* has a similar life history in the intestine of insects (Fig. 19.12). These gregarines are relatively harmless, for in the absence of schizogony their numbers in any one host stay low. However, schizogony does occur in other gregarines, and repeated vegetative reinfection cycles here produce numerous parasitic individuals that cause considerably more damage. Examples are *Schizocystis,* an intestinal parasite of larval flies, and *Ophryocystis,* a parasite in the Malpighian excretory tubules of beetles.

Telosporidians of the subclass Coccidia occur in most coelomate animals, vertebrates included. The parasites live in the epithelial cells of the intestine, liver, spleen, and other organs. Coccidia are relatively damaging, again because of schizogony. For example, *Eimeria* is the causative agent of coccidiosis, a disease common in domesticated and other mammals and characterized by severe digestive disturbances.

The subclass Hemosporidia includes the best-known and most-studied sporozoan genus, *Plasmodium,* various species of which cause malaria in mammals, birds, and occasionally lizards. Man is subject to infection by four species, each responsible for a different type of malaria. Repeated cycles of schizogony and release of merozoites from red blood corpuscles result in successive attacks of fever, the time interval between attacks being a main diagnostic feature of each of the four kinds of malaria. The plasmodial life cycle requires a bloodsucking intermediate host, the *Anopheles* mosquito in man, the *Culex* mosquito in birds. In the intestinal tissues of such insects gamete

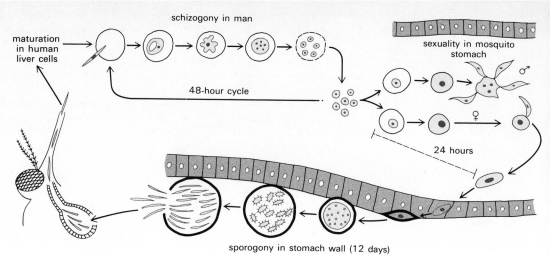

schizogony in man

maturation
in human
liver cells

48-hour cycle

sexuality in mosquito
stomach

♂

♀

24 hours

sporogony in stomach wall (12 days)

Figure 19.13 Malaria: life cycle of the sporozoan parasite *Plasmodium vivax.* Mature parasitic cells (trophozoites) invading human red blood corpuscles (top left) undergo multiple fission (schizogony) either inside the corpuscles, as shown, or outside them. Offspring cells (merozoites) destroy corpuscles and cause an attack of fever. They also reinfect new corpuscles and then lead to repetition of a 48-hr fever cycle. Merozoites entering red corpuscles also function as gamete producers (gamonts), and if human blood is sucked by an *Anopheles* mosquito, the gamonts break free in the mosquito stomach, transform to male and female sex cells, and bring about fertilization. The fertilization product then encysts in the stomach wall and undergoes meiotic multiple fission (sporogony). The resulting free spore cells (sporozoites) later migrate through the body cavities and organs of the insect, including the salivary glands. From there the sporozoites are injected into the human circulation when the mosquito bites a man. Sporozoites mature in human liver cells, and fully formed trophozoites then begin a new life cycle.

formation and sporogony take place, without apparent harm to the hosts. Sporozoites then invade the salivary glands of the insects, and from there the parasites are injected into the blood streams of new main hosts through mosquito bites (Fig. 19.13).

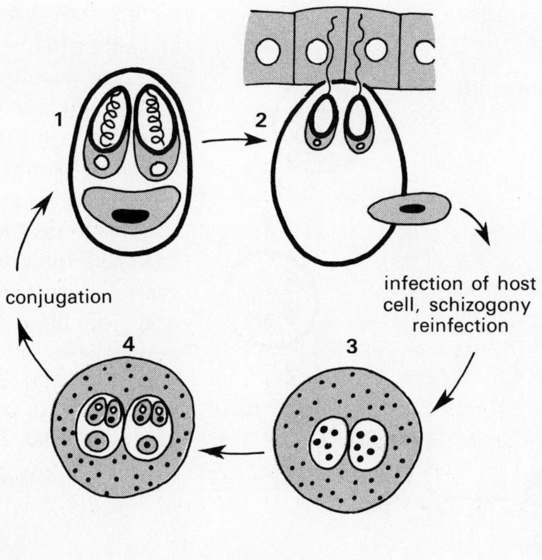

Figure 19.14 Life cycle of a cnidosporidian. 1, mature spore after conjugation, with zygote (bottom) and two polar-capsule-containing cells (top). 2, discharge of polar capsules attaches the spore to the intestinal lining of host, and zygote (trophozoite) is set free. 3, after schizogony, merozoites grow as syncytial complexes in which multinucleate spore-forming cells appear. 4, the spore-forming cells develop polar capsules and spore cells; conjugation of latter completes the cycle.

1

2

conjugation

infection of host cell, schizogony reinfection

4

3

The class Cnidosporidia is characterized by walled spores that contain *polar capsules* with coiled, hollow filaments (Fig. 19.14). In the intestine of a host the spore walls are digested and the threads of the polar capsules are discharged. Everted like the finger of a glove, such a thread can pierce the intestinal lining and fasten the spore cell to the gut wall. Trophozoites are amoeboid. They become multinucleate, eventually undergo schizogony, and the amoeboid merozoites then mature as multinucleate and in some cases multicellular gamonts. The latter give rise to sporelike gametes with polar capsules, and after syngamy a new generation of amoeboid trophozoites is formed. These life cycles are exceedingly complex and many details are not yet known. The members of the class are parasitic in annelids, arthropods, and fishes.

The Acnidosporidia are similarly amoeboid, but their spores lack polar capsules. Some of these Sporozoa inhabit the muscles and connective tissues of various vertebrates, including particularly mammals and occasionally man. Others parasitize various tissues of aquatic invertebrates such as annelids. As in the preceding class, many details of the life histories are still unknown.

Like Sarcodina, Sporozoa probably are a polyphyletic assemblage. The Telosporidia have possible affinities to free-living flagellate-amoeboid protists, but spore formation and haplontic life cycles are reminiscent of coccine protists, both algal and fungal. The other two sporozoan classes do not appear to be related too closely to the telosporidians. Polar capsules appear to be originally evolved adaptations to parasitism, and it is possible that Acnidosporidia might have arisen from Cnidosporidia by loss of the polar capsules. Because sporozoa are parasitic, one cannot be sure whether any of their traits is original and primitive or a secondary adaptation to parasitism. On balance, the Sporozoa probably constitute at least three and possibly more separate groups of protists, all or some of which may or may not have any direct evolutionary relation to other protozoa.

Ciliate Protozoa   *Subphylum Cilophora*

cilia present throughout life or at young stages, generally in rows; with micronuclei and macronuclei, often numerous; centrioles absent; reproduction by binary fission, largely transverse; sexuality by conjugation.

Class Ciliata

Subclass Holotrichida   ciliation uniform over entire body.

Order Gymnostomatida   gullet simple, with trichites. *Prorodon, Didinium*

Order Trichostomatida   gullet with vestibule.

Order Chonotrichida   ciliation reduced; vase-shaped body; ectocommensalistic on crustacea.

Order Apostomatida   rosette around gullet; ectocommensalistic on marine crustacea.

Order Astomatida   gullet absent; parasitic in annelid gut.

Order Suctorida   stalked; feeding by tentacles; reproduction by budding.

Order Hymenostomatida   gullet ciliation elaborate. *Paramecium, Tetrahymena*

Order Thigmotrichida   gullet absent or located posteriorly.

Order Peritrichida   ciliation transverse, fission longitudinal; many forms stalked, also colonial. *Vorticella, Epistylis*

Subclass Spirotrichida   ciliation nonuniform, largely reduced, compound ciliary organelles conspicuous.

Order Heterotrichida   adoral area elaborate, body ciliation uniform. *Stentor, Blepharisma*

Order Oligotrichida   adoral area elaborate, body ciliation absent.

Order Tintinnida   adoral area elaborate; body in shell or lorica.

Order Odontostomatida   little ciliation, body compressed laterally.

Order Entodiniomorphida   body ciliation absent, internal structure complex; endocommensalistic in herbivorous mammals. *Cycloposthium*

Order Hypotrichida   body ciliation absent, cirri present. *Euplotes, Stylonychia*

The ciliates form a more nearly homogeneous evolutionary group than the other protozoan subphyla. They also are the most complexly elaborated protozoa and the most diversely specialized of all known cell types. The organisms probably are monophyletic, with an ancestry among early zooflagellates, possibly opalinidlike stocks. Ciliates include some five thousand described species. Most of them are free-living in all aquatic environments. Some are commensalistic in various animals and comparatively few are parasitic. The majority are solitary motile organisms; some are

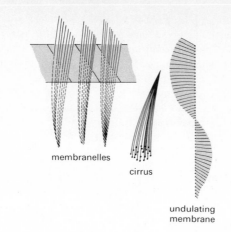

membranelles

cirrus

undulating
membrane

Figure 19.15 Surface organelles in ciliates: locomotor structures formed from variously fused cilia. Each membranelle has a triangular basal portion that forms an anchor in the cytoplasm; in the diagrams of a cirrus and an undulating membrane only the parts that project beyond the cell surface are shown.

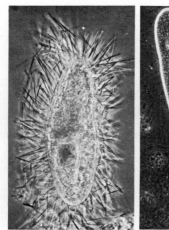

A   B

Figure 19.16 Trichocysts and vacuoles. A, Paramecium discharging its trichocysts. *B,* the rosette-shaped system of contractile vacuoles in *Paramecium.* One rosette is near each end of the organism.

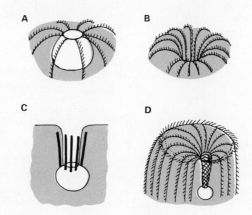

A   B

C   D

Figure 19.17 Mouth structure in ciliates. A, simple mouth opening on cell surface, food vacuole formed directly underneath (as in *Prorodon*). *B,* mouth located at bottom of vestibular pit. *C,* like *B* but vestibule supported by circlet of rodlike trichites (as in *Didinium*). *D,* vestibular depression and peristome area (adoral zone) circled by ring of adoral membranelles; vestibule leads into funnel-shaped cytopharynx, with mouth and food vacuole at bottom (as in *Stentor*).

sessile, a few forming branching, treelike colonies.

The ciliate body usually is plastic but has a species-specific shape and in most cases also a distinct longitudinal axis. A pellicle covers the outside, and protruding through are the cilia (Fig. 19.15). These typically are arranged in orderly rows and they beat in *metachronous* rhythm, a coordinated, wavelike sequence. Cilia are attached internally to kinetosomes that are joined in rows by complex systems of fibrils (*kinetodesmas*). Also present under the pellicle is a system of conductile neurofibrils that presumably controls the ciliary beat. In many species contractile myofibrils (*myonemes*) parallel the rows of kinetosomes. Rows of dischargeable *trichocysts* often are present as well (Fig. 19.16).

Cilia frequently form compound organelles. For example, cilia in a row can be fused as a sheetlike *undulating membrane* that can serve in locomotion or produce food-bearing currents. Tapered tufts of fused cilia are *cirri*, strong bristlelike organelles that function as locomotor legs. Fused cilia from several rows form *membranelles*, tiny paddles that create a strong beat (see Fig. 19.15).

Most ciliates have a permanent gullet, or *cytostome*. The simplest type is a *mouth* opening on the body surface (Fig. 19.17). In more complex gullets the mouth is at the bottom of a *vestibule*, a funnel-shaped depression in the body surface. The most complex cytostomes have a vestibule that leads to a short canal, the *cytopharynx*, at the bottom of which is the ingestive area proper. Ciliates feed either by using the cytostome to catch prey, often larger than the ciliate itself, or by creating water currents that bring microscopic food to the gullet. In the first case the mouth area or vestibule often is strengthened by a circular set of stiff rods called *trichites*. In the second case a specialized *adoral zone*, or *peristome*, surrounds the gullet area and produces the food-bearing currents. The adoral zone usually is composed of membranelles, and additional membranelles as well as undulating membranes frequently are in the cytopharynx.

Food vacuoles migrate over a more or less definite path inside the ciliate body, and digestive remains are egested at a *cytopyge*, a fixed point often located in or near the cytopharynx. Contractile vacuoles occur at fixed positions near the body surface, and in many cases definite cytoplasmic channels form an internal drainage system that

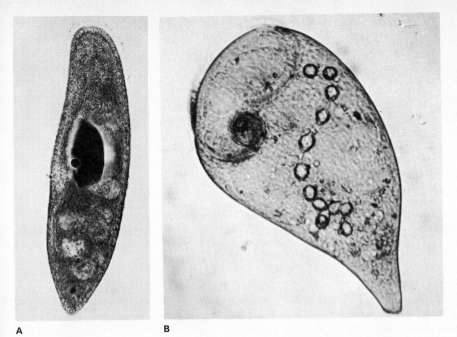

A **B**

Figure 19.18 Ciliate nuclei. A, *Paramecium caudatum* stained to reveal the macronucleus (large dark central body) and the smaller micronucleus, which partly overlaps it on one side (approx. ×600). B, *Stentor coeruleus,* from life, Note region of vestibule, ring of membranelles, gullet, macronuclear chain, and holdfast (partly contracted at end opposite gullet). Faint rows of body cilia are also visible (approx. ×200). (A, courtesy *Carolina Biological Supply Company.*)

Figure 19.19 Ciliate reproduction. Top, fission in *Paramecium.* The micronucleus undergoes mitosis and the macronucleus elongates and constricts into two parts. Also, the original mouth organelles are resorbed and each offspring develops a new set. *Bottom,* fission in *Stentor.* The nodulated macronucleus (as in Fig. 19.18) condenses to a compact mass and then reelongates as each micronucleus divides mitotically. Concurrently the future anterior offspring inherits the original set of mouth organelles, while the future posterior one develops a new set. After the two offspring are constricted apart, the macronuclear portion inherited by each renodulates.

leads to the contractile vacuoles (see Figs. 19.16 and 9.26).

Ciliates are multinucleate. They have at least one and often many (up to several hundred) *micronuclei,* and one or many (up to several dozen) *macronuclei* (Fig. 19.18). The diploid micronuclei contain typical chromosomes but the macronuclei do not, at least not in the usual identifiable form.

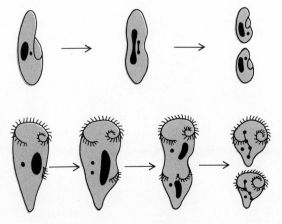

Instead, a macronucleus is believed to consist of numerous "subnuclei," each with a diploid set of genes but without the typical chromosomal organization; the genetic substance appears as a homogeneous material under the light microscope.

Micronuclei produce and exert long-range control over the macronuclei, and they control sexual processes. Macronuclei govern all metabolic and developmental functions and they maintain the visible traits of the organism. Micronuclei can be lost or removed and a ciliate still survives as a vegetative individual that can continue to reproduce normally by fission. But if the macronuclei are removed or lost, structures such as gullet, membranelles, cirri, contractile vacuoles, and body cilia all degenerate and the organism dies—even if the micronuclei still are present. Macronuclear functions ultimately are determined by the micronuclear chromosomes.

This unique specialization of the genetic material in two kinds of nuclei relates directly with the high degree of cytoplasmic specialization. The whole complex surface apparatus of a ciliate is controlled by the macronuclei; and it appears that such a level of cell specialization could evolve only by a parallel evolution of a separate nuclear machinery for it. Complex ciliation and macronuclei go together, and apparently there cannot be a ciliate level of structure without a macronucleus.

This circumstance provides another major argument against a ciliate hypothesis of eumetazoan origins. Proponents usually postulate a ciliate ancestor *without* macronucleus—a conception completely at variance with actual ciliates. And if a ciliate ancestor *with* macronucleus were postulated, what would have been the fate of this organelle in the supposed metazoan descendants? These are totally without any signs or remnants of it, and the single type of nucleus of all their cells controls *all* activities. What the ciliate nuclei actually do suggest is that the subphylum represents a specialized, separate line of cellular evolution, a line that has developed a high level of cytoplasmic specialization based on a unique nuclear machinery. Ciliates presumably have given rise to nothing but more ciliates.

A capacity of encystment is in evidence throughout the ciliate subphylum. Reproduction occurs by transverse binary fission in virtually all cases, though longitudinal fission is the rule in one group (order Peritrichida) and budding in another (order Suctorida). In fission (Fig. 19.19), the

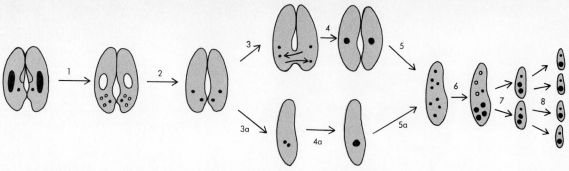

Figure 19.20   Sexuality in ciliates. The sexual process in *Paramecium* is illustrated (see also Fig. 10.4). 1, meiosis in each conjugating partner, followed by degeneration of macronucleus and of three haploid micronuclei resulting from meiosis. 2, mitotic division of remaining haploid nucleus, yielding one stationary and one migratory gamete nucleus (the latter situated near surface in paroral cone). 3 and 4, nuclear exchange and formation of synkaryon (diploid zygote nucleus). 3a and 4a, autogamy, or self-fertilization (fusion of haploid gamete nuclei within same individual). 5 and 5a, three mitotic nuclear divisions, resulting in eight diploid nuclei. 6, four of the eight become macronuclei, three degenerate, and one becomes a micronucleus. 7 and 8, cell and micronuclear divisions and parceling out of macronuclei, till each exconjugant has normal nuclear complement.

micronuclei divide mitotically and intranuclearly; centrioles are absent. Macronuclei divide *amitotically* by splitting into approximately equal halves. Each daughter cell thereby receives a roughly equal number of macronuclear gene sets. In many species the gullet, membranelles, cirri, and other surface structures are resorbed at the onset of fission, and each daughter cell then develops a new set. In other species the anterior daughter cell inherits the parental organelles and the posterior one develops a new set. Ciliates have a highly developed regeneration capacity. In numerous instances almost any cell fragment with at least one macronucleus can regenerate all missing structures.

Virtually all ciliates are hermaphroditic and the sexual process is conjugative. Many species include two or more mating types, conjugation oc-

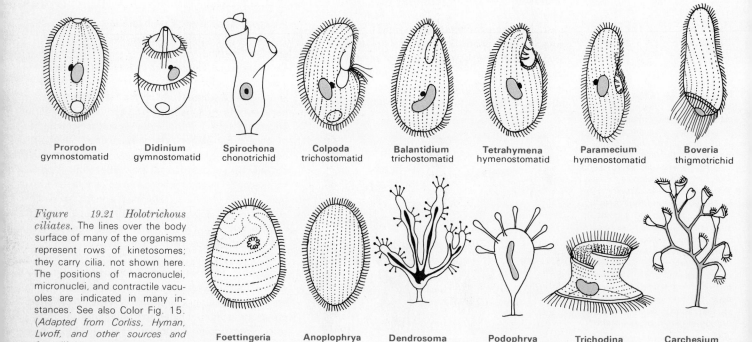

Prorodon gymnostomatid   **Didinium** gymnostomatid   **Spirochona** chonotrichid   **Colpoda** trichostomatid   **Balantidium** trichostomatid   **Tetrahymena** hymenostomatid   **Paramecium** hymenostomatid   **Boveria** thigmotrichid

Figure 19.21 Holotrichous ciliates. The lines over the body surface of many of the organisms represent rows of kinetosomes; they carry cilia, not shown here. The positions of macronuclei, micronuclei, and contractile vacuoles are indicated in many instances. See also Color Fig. 15. (*Adapted from Corliss, Hyman, Lwoff, and other sources and from life.*)

Foettingeria apostomatid   **Anoplophrya** astomatid   **Dendrosoma** suctorian   **Podophrya** suctorian   **Trichodina** peritrichid   **Carchesium** peritrichid

curring only between such types (see also discussion of *syngens*, Chap. 17). Conjugation involves a temporary partial fusion of two individuals, usually near their gullet areas. Concurrently the macronuclei degenerate and the micronuclei undergo meiotic division. Nuclear exchange between mating partners follows and a zygote nucleus (*synkaryon*) forms in each conjugant. Mitotic divisions of the synkaryon then give rise to new micronuclei and macronuclei (Fig. 19.20).

Some ciliates (the peritrich *Vorticella*, for example) exhibit mating dimorphism. One mating partner is a large *macroconjugant,* the other a smaller *macroconjugant.* After conjugation only the macroconjugant survives and among its later offspring new microconjugants develop. In *Paramecium* and several other genera, the successive generations descended from a single exconjugant pass through a kind of "super life cycle," also called a *Maupasian* or "physiological" cycle. Early vegetative generations after conjugation are juvenile and sexually immature; they can reproduce but not conjugate. A mature phase then follows, during which conjugation can occur. If sex actually takes place the participants become "rejuvenated" and their immediate descendants again are juvenile. But if conjugation is prevented (by isolation, for example), a senile phase eventually follows. It is characterized by nuclear and genetic abnormalities and other signs of aging. Such a line sooner or later loses conjugation capacity and ultimately dies out.

In many cases, genetic or actual death of a line is avoided even in the absence of conjugation by self-fertilization, or autogamy (see Fig. 19.20). In this process two sister gamete nuclei fuse in a single individual, which has the same rejuvenating effect as conjugation; the descendants after autogamy are juveniles.

The single class in the ciliate subphylum consists of two subclasses (Figs. 19.21 and 19.22).

Figure 19.22 Spirotrichous ciliates. Stylonchia is shown in side view. Otherwise as in Fig. 19.12. See also Color Fig. 16. (*Adapted from Corliss, Bütschli, and other sources and from life.*)

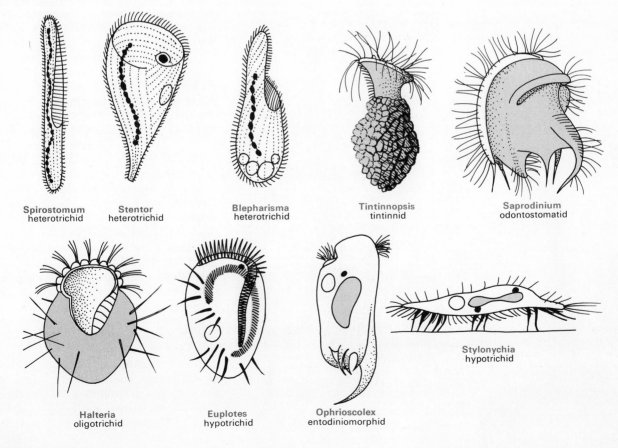

Spirostomum
heterotrichid

Stentor
heterotrichid

Blepharisma
heterotrichid

Tintinnopsis
tintinnid

Saprodinium
odontostomatid

Halteria
oligotrichid

Euplotes
hypotrichid

Ophrioscolex
entodiniomorphid

Stylonychia
hypotrichid

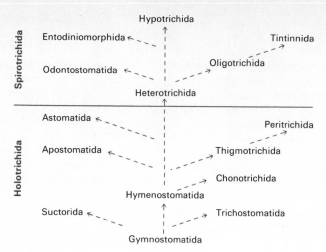

Figure 19.23 *Ciliate evolution.* All connecting paths should be regarded as tentative. (*Adapted from Corliss.*)

Figure 19.24 *Suctoria. A,* internal (left) and external (right) budding. *B,* a suctorian. The tentacles radiating out from the attached cell body suck up the juices of trapped prey (×400).

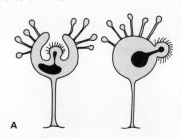

A

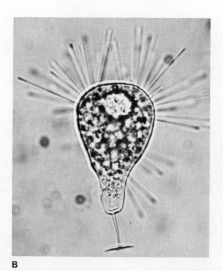

B

The main evolutionary path is believed to have led from Gymnostomatida to Hymenostomatida to Heterotrichida to Hypotrichida, with side branches at each step (Fig. 19.23). This main path is characterized by a progressive increase in the complexity of the ciliary apparatus and the gullet structures. Thus, primitive gymnostomes (*Prorodon*) have only a simple mouth and lack compound ciliation. Hymenostomes (*Paramecium, Tetrahymena*) have a gullet with vestibule and cytopharynx, the latter equipped with a four-part "tetrahymenal" arrangement of three membranelles and one undulating membrane. Heterotrichs (*Stentor*) in addition have a complex peristome ringed by sets of adoral membranelles. And hypotrichs (*Euplotes*) have the most complex sets of membranelles, cirri, and undulating membranes.

Gymnostomes are saltwater and freshwater forms. The order includes *Didinium*, a carnivore with a protrusible proboscis by means of which it can engulf a whole paramecium. The four orders Trichostomatida, Chonotrichida, Apostomatida, and Astomatida consist largely of commensalistic and parasitic types; *Balantidium* is the only known ciliate parasite of man. Suctorida are stalked, sessile types that trap prey with tentacles. By some still poorly understood process the organisms suck the contents of the prey through the tentacles into their bodies. Adult suctorians lack body cilia (but do have kinetosomes). Offspring arise by budding. A bud develops cilia, escapes from the parent, and after swimming about for some time settles and becomes a sessile adult without cilia (Fig. 19.24).

In the peritrichs, the body surface (but not the peristome) lacks cilia. Kinetosomes are arranged in transverse rows and fission is longitudinal. Sessile adults include stalked solitary forms such as *Vorticella* and branched colonial types such as *Epistylis* (Fig. 19.25).

In the subclass Spirotrichida, the order of heterotrichs has a uniform body ciliation but, as noted, the peristome is quite elaborate. Many heterotrichs are beautifully pigmented in green, blue, pink, orange, and other colors. Most are free-living, some are commensals. Among freshwater types, *Spirostomum* and some species of *Stentor* are large enough to be visible with the naked eye.

The oligotrichs include *Halteria*, a type without general ciliation but with bristles that are used as tiny stilts in a jumping form for locomotion. The tintinnids are marine and pelagic, without cilia

Figure 19.25 Vorticella, a peritrich. A, daughter individuals just after fission. One daughter inherits the stalk of the parent (with spiral myoneme); the other develops a posterior ciliary girdle and will migrate away and settle elsewhere via a newly formed stalk. B, expanded individual.

A

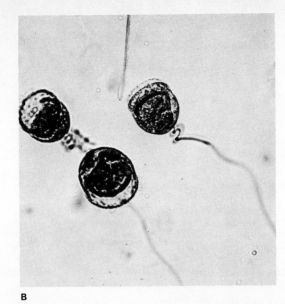

B

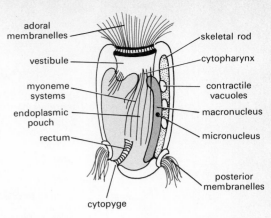

Figure 19.26 Cycloposthium, one of the most complex ciliates. The whole structural complexity of this ciliate cannot become apparent, however, in a sketch like this. (After Dogiel.)

except in the peristome area. Their secreted housings have been preserved as fossils, the only ciliate fossils known. The odontostomes live in sewage and in natural waters in which the oxygen content is low. Their bizarrely shaped bodies are compressed laterally and the ciliation is highly reduced.

The remaining two orders, Entodiniomorphida and Hypotrichida, include the most complex of all ciliates (and therefore probably the most complex cells). For example, Cycloposthium, an intestinal commensal of herbivorous mammals, contains skeletal supports, virtually a complete alimentary "system" with a permanent rectal tube, a "muscular" system composed of numerous sets of myonemes, and also neurofibrils, systems of con-

tractile vacuoles, and complex groups of membranelles in various regions (Fig. 19.26). General ciliation is absent. This is true also in the hypotrichs, ciliates that are flattened dorsoventrally and in which ventral cirri serve as walking legs. Hypotrichs are largely free-living in both salt and fresh water (Color Fig. 16).

Even a brief account thus shows that protozoa are far from "simple." They are small, to be sure, but in this very smallness lies perhaps their most remarkable property: despite being limited to unicellular bodies they can be as diverse, varied, and complex in the microsphere of life as only very few groups of far larger organisms can be in the macrosphere.

Review Questions

1. Give taxonomic definitions for protozoa and each of the main subgroups. Distinguish protozoa from (a) other protists, (b) Metazoa. What are the conceptual consequences of regarding protozoa as "acellular" forms?

2. Review the possible evolutionary relations of protozoan groups to one another and to other organisms. Describe the general structure of protozoan cells. How does reproduction occur? Describe the life cycle of the majority of protozoa.

3. State the characteristics of Mastigophora. How are the zooflagellate orders probably related? Describe the structure of a zooflagellate. What are axonemes, parabasal bodies, axostyles, costae, trichocysts?

4. In what way do the Rhizomastigida suggest a close interrelation of Mastigophora and Sarcodina? Name a representative genus of several mastigophoran orders and describe its structural and ecological characteristics.

5. What do *Mastigamoeba* and *Naegleria* have in common? Why are they not both classified as members of the same subphylum or class? Name the classes and orders of Sarcodina and a representative genus of each. Describe the structure of various types of pseudopodia and indicate in which group each occurs.

6. What is the structure of Heliozoa? How do these organisms feed? Reproduce? Undergo sexual processes? Describe the structure of Radiolaria. Where do these organisms live?

7. What are Foraminifera and where do they live? Describe their structure. Review their life cycles and indicate what makes such cycles diplohaplontic.

8. Describe the basic life cycle of Sporozoa and distinguish between schizogony, sporogony, and gamogony. Review the classification of the subphylum and indicate the type of life cycle encountered in each major group. What is a polar capsule? Describe the life cycle of the malarial parasite *Plasmodium*.

9. What is the general structure of a ciliate? What are cirri, membranelles, and undulating membranes? How does gullet structure vary in different ciliate groups? Distinguish between micronuclei and macronuclei, both structurally and functionally.

10. Show how a ciliate divides. What are the detailed events in ciliate conjugation? What is autogamy? A Maupasian life cycle? Name various ciliate orders, indicate their characteristics, and review their postulated evolutionary interrelations.

Collateral Readings

Calkins, G. N., and F. M. Summers (eds.): "Protozoa in Biological Research," Columbia, New York, 1941, or Hafner, New York, 1964 (reprinted edition).

Hawking, F.: The Clock of the Malaria Parasite, *Sci. American,* June, 1970. Mosquitoes most commonly feed at night, and the life cycle of *Plasmodium* is timed so that infective stages are ready for transfer from insect to man at just the feeding periods of the insects.

Hunter, S., and A. Lwoff (eds.): "Biochemistry and Physiology of the Protozoa," vols. 1 and 2, Academic, New York, 1951, 1955. Advanced, excellent for supplementary data.

Hyman, L.: "The Invertebrates," vol. 1, "Protozoa through Ctenophora," McGraw-Hill, New York, 1940. A detailed account on Protozoa is given in the third chapter of this important treatise.

Jahn, T. L., and F. F. Jahn: "How to Know the Protozoa," Brown, Dubuque, Iowa, 1949. This book greatly facilitates the identification and characterization of these organisms.

Kudo, R. R.: "Protozoology," 5th ed., Charles C Thomas, Springfield, Ill., 1966. A large standard text.

Rivera, J. A.: "Cilia, Ciliated Epithelium, and Ciliary Activity," Pergamon, New York, 1962. The title is sufficiently descriptive.

Russell, P. F.: The Eradication of Malaria, *Sci. American,* June, 1952. Problems in the control of this widespread disease are examined.

Sonneborn, T. M.: Breeding Systems, Reproductive Methods, and Species Problems in Protozoa, in "The Species Problem," American Association for the Advancement of Science, Washington, D.C., 1957. An important paper on protozoan biology, with special emphasis on *Paramecium* and its mating types.

Tartar, V.: "The Biology of *Stentor*," Pergamon, New York, 1961. A review, with special attention to research on the development, regeneration, and morphogenesis of this ciliate.

Wichterman, R.: "The Biology of *Paramecium*," McGraw-Hill, New York, 1953. Highly recommended not only for further data on this ciliate but also for additional background on Protozoa generally.

20

MESOZOANS, SPONGES, RADIATES

METAZOA

multicellular animals; oögamous, sexuality by syngamy; life cycle diplontic; development includes embryos and, typically, larvae.

BRANCH MESOZOA

animals at tissue grade of construction; adult body a stereoblastula. *Mesozoans*.

BRANCH PARAZOA

animals at tissue grade of construction; adult body more complex than a blastula. *Sponges*

BRANCH EUMETAZOA

animals above tissue grade of construction; embryonic germ layers homologous (but not always analogous).

Grade RADIATA

animals attaining organ level of complexity; with primary and (typically) secondary radial symmetry. *Cnidarians, ctenophores.*

Mesozoans   *Phylum Mesozoa*

parasitic in squids and various other invertebrates; body composed of single surface layer of ciliated digestive and locomotor cells and one to several interior reproductive cells; cell numbers and arrangements constant for a species; life cycles complex, incompletely known, with vegetative and sexual generations.

Many zoologists regard this small group of minute animals as degenerate offshoots of flatworms, specifically flukes. The main basis for such a view is the mesozoan life cycle which, like that of some of the flukes, includes vegetative and sexual generations. The first take place in some still unidentified intermediate hosts, the last in molluscan main hosts.

Other investigators regard mesozoan life cycles as adaptations to parasitism, and they point out that Mesozoa have so simple a structure that a flatworm affinity is unlikely. Instead, the animals are postulated to be on a level between protozoa and cnidarians, and a protozoan derivation is therefore suggested. The issue remains obscure, not only because of the parasitism of the animals, but also because the body layers of Mesozoa do not correspond to ectoderm and endoderm (contrary to what might be expected in flatworm derivatives). The outer tissue is digestive and the inner cells are purely reproductive, never digestive (Fig. 20.1). Thus the evolutionary status of the Mesozoa remains an enigma at present.

Sponges

Phylum Porifera

"pore-bearing" animals; marine and freshwater; adults sessile, often in colonies; body radial to asymmetric; alimentation in channel system with choanocytes; development regulative; larvae flagellated and free-swimming.

Class Calcarea   chalk sponges; spicules calcareous, with one, three, or four rays; larvae amphiblastulae or stereoblastulae. *Leucosolenia, Sycon, Leuconia*

Class Hexactinellida   glass sponges; spicules silicaceous, with six rays, often fused in continuous networks; surface epithelium absent; choanocyte chambers in layer. *Hexactinella, Euplectella*

Class Demospongiae   horny sponges; spicules of spongin, silica, or both, sometimes none; silica spicules with one to four rays; body of leuconoid type; larvae amphiblastulae or stereoblastulae. *Spongia*

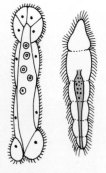

Figure 20.1 Mesozoa. Left, section through nematogen of a dicyemid, the mature parasitic stage in cephalopod hosts. By nuclear divisions in the single elongated interior cell, more cells are formed that escape and produce new generations of nematogens. Upon sexual maturity of the host, interior cells of a nematogen give rise to a rhombogen stage in which the internal cell produces larva-forming cells. Larvae leave the host and presumably enter intermediate hosts in which the sexual process takes place. *Right,* mature male of an orthonectid, with testis (color). The exterior tissue of these animals consists of numerous small cells. Fertilized eggs develop within the female, and the larvae then escape and infect hosts where orthonectids have the form of multinucleate amoeboid masses. From the latter, males and females develop and escape from the hosts. (*After Hyman.*)

Some 150 species of horny sponges live in fresh water; all others of the about 5,000 known species in the phylum are marine. Chalk sponges are largely shallow-water forms, glass sponges mostly inhabit deep water, and horny sponges are found in all regions, from the tide zone to depths of 3 to 4 miles. Many of the glass sponges are anchored to the sea floor by means of ropelike tufts of skeletal materials (Fig. 20.2).

Sponges provide a home for numerous other animals. Shrimps, crabs, copepods, sea anemones, octopuses, barnacles, and brittle stars are among many types that shelter in the spaces of sponge colonies. Such colonies often must compete for living space with corals, ascidians, ectoprocts, and other sessile colonial forms. Snails and chitons prey on sponges, and some crabs give themselves a protective disguise by attaching sponge pieces to their backs. Horny sponges often occur as encrustations in objects in water. Many kinds of sponges are pigmented red, yellow, blue, or black.

The body of a chalk sponge is made up of an external epithelium, an internal epithelium, and an intermediate *mesogloea* (Fig. 20.3). The first is composed of *pinacocytes,* flat polygonal cells that form an epidermal surface. Many of the pinacocytes are *myocytes* that contain contractile myofibrils. The interior epithelium is formed from *choanocytes,* flagellated collar cells that maintain a flow of food-bearing water through the sponge.

The mesogloea consists of jelly and embedded amoeboid mesenchyme cells, or *amoebocytes.* These function variously as digestive cells, skeleton-forming cells, food-storing cells, gamete-forming cells (*archeocytes*), nurse cells for eggs, pigment cells, cells that harbor symbiotic algae (*zoochlorellae*), and jelly-secreting cells. Some or most of the amoebocytes can transform from one variety to others.

Chalk sponges exhibit three degrees of architectural complexity (Fig. 20.4). The simplest types, exemplified by *Leucosolenia,* have an *asconoid* structure. Such a sponge is essentially a straight-walled sac, the wall being formed by an exterior pinacocyte layer, mesogloea underneath, and an interior choanocyte layer. The cavity of the sac is a *spongocoel.* Water enters it through tubular pore cells (*porocytes*) scattered in the body wall. Probably derived from amoebocytes, porocytes are contractile and can close their water-admitting canals. Water leaves a sponge at an *osculum,* a wide exit from the spongocoel.

More complexly constructed are *syconoid* sponges such as *Sycon.* In these the body wall is deeply pitted or folded along its interior surface. The spaces between the folds, called *radial canals,* are lateral extensions of the spongocoel. Choanocytes line the radial canals only, and the rest of the spongocoel is covered by pinacocytes. Water enters the sponge at numerous *ostia,* porelike spaces between the external pinacocytes; poro-

A   B   C

Figure 20.2 Sponge types. A, a calcareous type. *B,* Venus flower basket, one of the glass sponges. See also Fig. 2.11. *C,* a horny sponge. To this group also belong the bath sponges.

cytes are absent. Intercellular spaces in the meso-gloea form *incurrent canals* from the ostia to the radial canals. Water again leaves via spongocoel and osculum.

The *leuconoid* sponges, exemplified by *Leuco-*

Figure 20.3 Tissues and cells of a sponge. Left, the three basic tissue layers. In the mesogloea, free skeletal spicules and amoebocytes are indicated. A porocyte is a tubular cell with a water canal into the sponge. *Right,* some of the cell types. Myocytes are modified pinacocytes with contractile myofibrils. The choanocyte collar consists of a circlet of cytoplasmic extensions, as shown (see also Fig. 19.5). Variants of amoebocytes form numerous other cell types with different functions. Sponge spicules are shown in Fig. 20.6.

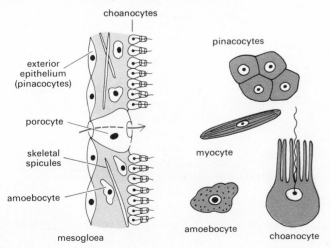

choanocytes

exterior
epithelium
(pinacocytes)

porocyte

skeletal
spicules

amoebocyte

mesogloea

pinacocytes

myocyte

amoebocyte   choanocyte

nia and architecturally the most complex, have their choanocytes in small interconnected chambers. Incurrent canals lead to such a chamber through several *prosopyles,* tiny spaces between the choanocytes. Water exits from a chamber at an *apopyle,* an opening to an *excurrent canal.* Such canals from adjoining chambers form progressively larger channels, and the largest ultimately lead to the exterior at one or more oscula. The structural series from asconoid to syconoid to leuconoid types thus exhibits a progressively greater subdivision of the choanocyte-lined spaces and an increasing complication of the channel system to and from these spaces.

In any sponge, the diameter of the water exits is greater than the combined diameters of all the incurrent openings. Water therefore enters a sponge at greater velocity and lower pressure than it leaves. Thus the water is actively sucked into the sponge and actively pushed out of it by the choanocytes. Food in the water consists largely of bacteria, microscopic algae, and organic debris. The food-bearing water in effect passes through a succession of progressively finer screens; ostia average 50 μ in diameter, prosopyles are about 5 μ wide, and the slits between the cytoplasmic extensions in the choanocyte collars have widths of about 0.1 μ. Hence only the smallest food

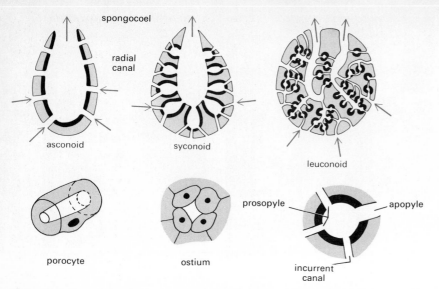

Figure 20.4 Sponge structure. In these three levels of complexity, black layers indicate position of choanocytes. Lower figures outline the structure of the incurrent water pores.

particles can be engulfed by choanocytes. Larger food is taken up by archeocytes along the incurrent canals. After these cells digest food intracellularly in protozoan fashion, they migrate to the body surface or to the excurrent canals and discharge indigestible matter there.

Virtually all horny sponges are of the leuconoid type. Glass sponges have a different architecture

(Fig. 20.5). In the basically syncytial body of these animals, mesenchyme cells are joined together as wide-meshed networks. Individual cells can detach themselves, migrate about, then rejoin the syncytium. An epidermal epithelium is absent. The choanocytes are in chambers with wide apopyles, and the basal portions of these cells are fused in a netlike arrangement that leaves tiny prosopylar spaces. The chambers themselves are arranged as a layer in the mesenchyme between the body surface and the spongocoel.

Although claims to the contrary have been made, sponges appear to lack nerve cells. The animals respond to external stimuli nevertheless. For example, bright light or mechanical irritation can initiate a slow contraction that gradually spreads from the point of stimulation over the rest of the sponge. Such responses are carried out by the myocytes. Also, myocytes around ostia and oscula can widen or narrow these openings and thereby regulate the water flow. To be sure, behavioral reactions are neither rapid nor versatile.

Contraction responses could not be too varied in any event, for the rigid skeletal elements of sponges, the *spicules*, preclude any substantial amount of body deformation. A spicule consists of up to six needlelike rays and is said to be monactinic, diactinic, triactinic, tetractinic, pentactinic, or hexactinic. The amoebocytes that manufacture spicules are, according to the class of sponges, *calcoblasts, silicoblasts,* and *spongioblasts.* Their skeletal products are composed of, respectively, calcium carbonate, silica, and spongin, a horny scleroprotein (Fig. 20.6).

In chalk sponges, a calcoblasts deposits calcium carbonate around an organic fibril in the cytoplasm. As this ray grows the cell divides and leaves most of the ray exposed. One daughter cell (''founder'') continues to shape the ray, and the other cell (''thickener'') secretes more skeletal material around it. Groups of calcoblasts cooperate in producing complex spicules. The cells of a group align in such a way that, after each manufactures a single ray, the several rays can be fused in a particular pattern.

Spicule formation occurs in similar fashion in the glass sponges. In many of these the spicules project out from the body surface, and in forms like *Euplectella,* the Venus flower basket, the spicules are fused as a continuous glassy latticework (see Fig. 20.2). A sieve of interlaced spicules also covers the osculum. In the horny sponges spicules

Figure 20.5 The syncytial structure of glass sponges. Left, section through such a sponge, showing absence of definite epidermal layer, the mesenchymal net, and the layer of thimble-shaped choanocyte chambers. (*From Hyman after Schulze.*) A single chamber is sketched at right (top), the wide opening forming the apopyle. The bases of adjacent choanocytes are fused syncytially, leaving prosopylar openings between the nucleated portions (bottom). (*After Ijima.*)

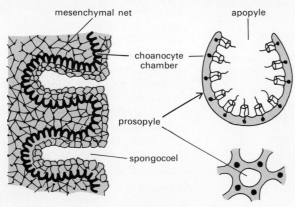

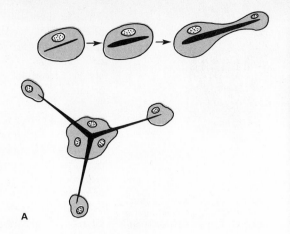

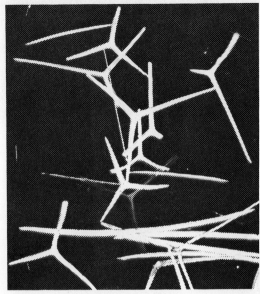

A

B

Figure 20.6 Spicules. A, top, formation of a spicule. A single cell (far left) manufactures a monactinic spicule around an organic fibril. Division of the cell results in a "founder" at the base of the developing spicule and a "thickener" at the tip. The founder shapes the spicule. *Bottom,* groups of cells cooperate in the formation of many-rayed spicules. *B,* isolated spicules of a calcareous sponge (*B,* courtesy *Carolina Biological Supply Company.*)

are produced by linear rows of spongioblasts. In all sponges, the producing cells usually migrate away or degenerate after a spicule has been formed. Some primitive horny sponges lack skeletal components.

Vegetative reproduction occurs extensively in sponges. The animals can reproduce by *budding,* a process in which groups of amoebocytes and archeocytes collect at some surface region of a parent sponge (Fig. 20.7). Such buds then either fall off and develop on their own or remain on the parents and form colonies. Sponges can also be fragmented and each piece usually reconstitutes a whole individual. Indeed, if a sponge is dissociated into a suspension of loose cells, the separated cells migrate together in amoeboid fashion and reconstitute an intact, normally structured sponge. If cell suspensions of two different species are mixed together, the cells "recognize" which species they belong to and form two distinct wholes.

Under unfavorable conditions sponges often form *reduction bodies,* small globules composed of outer pinacocyte layers and inner amoebocytes. Such bodies can develop into new individuals. All freshwater and some marine sponges produce *gemmules* under conditions of drought or low temperature. A gemmule is a reduction body covered by an external layer of columnar cells with spicules. The internal cells again include amoebocytes and archeocytes. Note that none of these vegetative reproductive units normally contains choanocytes; such cells develop later from some of the amoebocytes. In a normal sponge, also, a choanocyte can resorb its flagellum and collar and become a migratory amoebocyte.

Most sponges are hermaphroditic. Self-fertilization is rare, for sperm and eggs usually are not formed at the same time. The gametes arise from archeocytes, and after meiosis the eggs are amoeboid, the sperms flagellate. Eggs come to lie under the choanocyte layer, where they often receive food from amoebocytic nurse cells (Fig. 20.8). Sperms leave the parent and enter another sponge with the incoming water current. A sperm becomes trapped either by an amoebocyte or by a

Figure 20.7 Vegetative reproduction in sponges. In budding, a bud can form either a separate (left) or an attached (right) individual. In a gemmule, elongated spicules and pinacocytes form exterior layers and amoebocytes fill the interior. The structure of a reduction body is similar, except that spicules have different forms and arrangements.

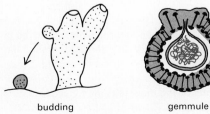

budding gemmule reduction body

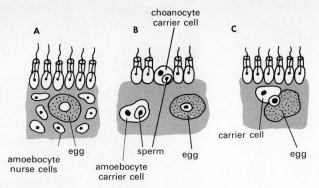

choanocyte
carrier cell

amoebocyte
nurse cells

egg

sperm

amoebocyte
carrier cell

egg

carrier cell

egg

A   B   C

Figure 20.8 *Sponge gametes and fertilization.* *A,* egg surrounded by amoebocytic nurse cells underneath the choanocyte layer of the parent. *B,* formation of sperm-carrying cells, either from choanocytes or from amoebocytes. *C,* migration of a carrier cell and transfer of sperm to egg, resulting in fertilization.

choanocyte that then transforms to an amoebocyte. Such a cell functions as an amoeboid "carrier" cell; it migrates to an egg and transfers the sperm into it.

Chalk sponges develop according to two main patterns. In one, exemplified by *Sycon,* cleavage of the zygote produces a hollow blastula in which the cells of one half have flagella directed into the blastocoel (Fig. 20.9). During later cleavages an opening forms in the nonflagellated half, and the whole blastula then turns inside out through this opening. The flagella of the flagellated cells thus

Figure 20.9 *Sponge development,* Sycon pattern. 1, blastulalike embryo underneath maternal collar cell layer. 2, inversion (embryo turns inside out through opening in nonflagellated half, resulting in exteriorized flagella). 3, amphiblastula larva. 4, invagination of flagellated half into nonflagellated half. 5, larva settles. 6, open end of embryo closes over and osculum breaks through at the free side, establishing the basic (asconoid) structure. (*Adapted from Hartmann and Brien.*)

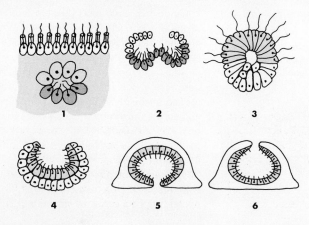

1   2   3

4   5   6

come to point outward. Such a curious process of embryonic *inversion* is encountered not only in sponges but also in the development of spherical colonial algae such as *Volvox*—an interesting instance of parallel evolution in similar embryonic situations.

After inversion, the sponge embryo escapes from the adult as a hollow *amphiblastula* larva. It swims about for a time, eventually settles, and the flagellated hemisphere then invaginates into the nonflagellated hemisphere. A basic asconoid structure is thereby established, with the nonflagellated cells on the outside and the flagellated cells on the inside. The outer cells give rise to pinacocytes, porocytes, and calcoblasts. The inner cells form choanocytes and all other amoebocytes.

In a second pattern of development, exemplified by *Leucosolenia,* the zygote first produces a hollow coeloblastula and it then becomes a solid, externally flagellate *stereoblastula* larva (Fig. 20.10). After this larva settles inversion takes place; the exterior cells lose their flagella and migrate to the interior, where they form choanocytes. Concurrently the interior cells migrate to the exterior, where they develop into pinacocytes and amoebocytes. A spongocoel becomes hollowed out later and an asconoid structure results in this manner. *Leucosolenia* remains asconoid, but in other chalk sponges further structural transformations to syconoid or leuconoid adults can occur.

In glass sponges, the few genera studied develop via stereoblastula larvae, as above, and in horny sponges either stereoblastula or amphiblastula larvae are formed. Adult leuconoid structure is established directly in some cases, indirectly via asconoid and syconoid stages in others (Fig. 20.11).

The developmental patterns of sponges all have one characteristic in common; they do *not* correspond to the patterns encountered in Eumetazoa. Thus the exterior cells of sponge embryos are not "ectoderm," for they eventually form some of the mesenchyme as well as the *interior* choanocyte layer; and the interior cells of sponge embryos are not "endoderm," for they later produce the remainder of the mesenchyme as well as the *exterior* epidermal layer. In effect, the embryonic layers of sponges are not homologous to those of other animals. This absence of embryonic homology is one of the main reasons for regarding sponges as an independent branch of metazoan evolution.

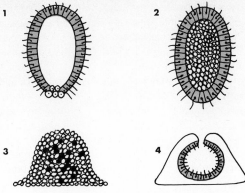

Figure 20.10 Sponge development, Leucosolenia *pattern.* 1, coeloblastula with posterior nonflagellated cells. 2, stereoblastula larva formed by ingrowth and proliferation of nonflagellated cells. 3, settling of larva and inversion—migration of flagellated cells (black) to center, nonflagellated cells (white) to outside. 4, arrangement of flagellated cells as choanocyte layer and formation of osculum, establishing asconoid structure. This stage later can develop further and acquire a syconoid or leuconoid architecture, depending on the genus. (*Adapted from Hartmann, Metschnikoff.*)

Other reasons include the presence of choanocytes (and, indeed, of unique porocytes); the pattern of alimentation based on the water-channel system; and the construction of the body generally, which represents a loosely integrated grouping of cells at an advanced colony or primitive tissue level of organization.

Because of their choanocytes, sponges are traditionally considered to have evolved from zooflagellates, particularly stocks related to the collar flagellates. However, sponge choanocytes are secondarily developed adult cells not present in embryos; and choanocytes are not the predominant sponge cells in any case, since the majority of cells exhibit an *amoeboid* state of existence. It is therefore quite possible that choanocytes in protozoa and sponges are independent evolutionary developments. If so, sponges might trace their ancestry back to stocks that were primarily amoeboid but also had flagellate potentials. Such stocks need not have been protozoan, moreover, but could have been more broadly protistan in nature.

Within the sponge phylum, the three classes presumably represent separate evolutionary lines descended from a common ancestor. In each line asconoid types without skeletons might have been primitive, and in many cases syconoid and leuconoid types with increasingly elaborate skeletons could have evolved later. Fossil sponges with complex skeletons are known from earliest Cambrian times.

Cnidarians

Phylum Cnidaria

coelenterates; largely marine; tentacles with *cnidoblasts* containing *nematocysts;* radial body symmetry; alimentary cavity with single opening; polymorphic, with medusae and/or polyps; solitary or colonial; development regulative, via *planula* larvae.

Figure 20.11 Development of horny sponges (Halisarca). 1, stereoblastula larva. 2, settling, inversion, and hollowing out of spongocoel. 3, osculum formation, asconoid stage. 4, proliferation and folding of choanocyte layer, syconoid stage. 5, pinching off of choanocyte chambers, young adult leuconoid stage. (*Adapted from Lévi.*)

Class Hydrozoa	hydrozoans; medusa and polyp phases alternating or either phase reduced or omitted.
Class Scyphozoa	jellyfishes; medusae dominant.
Class Anthozoa	corals, sea anemones, sea fans, sea pens; polyps only.

General Characteristics

The approximately ten thousand species of cnidarians form a phylum of first importance both historically and ecologically. Generally believed to represent the most primitive living Eumetazoa, the phylum is of special interest as a group at or close to the base of the animal evolutionary bush. Today

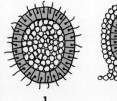

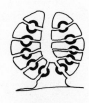

1   2   3   4   5

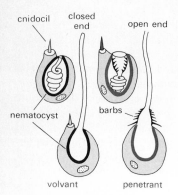

Figure 20.12 Cnidoblasts, undischarged and discharged. Numerous variants of each of the two basic types shown are known.

Figure 20.13 Cnidarian structure. A, the oral-aboral axis is long compared with the body diameter in a polyp, whereas the reverse holds true in a medusa. Both polyp and medusa are variants of a single common body pattern. *B*, some cell types in the three body layers, as in jellyfishes and sea anemones. In these animals the inner ends of the epidermal and gastrodermal cells are pseudopodial, as shown, but in Hydrozoa the bases of these cells are T-shaped and contain contractile myofibrils (see Fig. 20.14). Correspondingly, Hydrozoa lack separate muscle cells. Cells in the mesogloea secrete jelly, considerable amounts of it in medusas. Nerve cells are part of the epidermis, gastrodermis, or both (and thus they lie closer to either layer than shown here).

cnidarians are possibly the most common macroscopic animals of the oceans, and they have shaped substantial parts of the earth surface as the main builders of coral reefs and coral islands. Some groups are benthonic in deep waters, but most inhabit the shores and the shallow sunlit zones of tropical and subtropical seas. Dead, discolored laboratory specimens usually reveal very little of the delicate watery transparency of most of these animals in the living state; nor do such specimens indicate their myriads of living colors and often flowerlike shapes or their eerie bioluminescent beauty at night.

Cnidaria occur as sessile or motile forms and as solitary animals or colonies. Both individuals and colonies can be near-microscopic or exceedingly large. The largest solitary form is a species of jellyfish that can become up to 12 ft across and have trailing tentacles up to 100 ft long. Cnidaria are exclusively carnivorous. They eat any animals their tentacles can hold and paralyze, and their highly distensible bodies can admit prey far larger than themselves. Only very few species are parasitic, but many animals of other phyla live symbiotically with cnidarians and even more make their home among cnidarian growths.

Although the term ''coelenterates'' still is widely used, the phylum now is named after the most common trait of the animals, the presence of *cnidoblasts*. These are stinging cells found singly or in groups around the mouth, on the tentacles, and elsewhere on and in the body. A cnidoblast contains a horny stinging capsule, or *nematocyst*, similar in some respects to the polar capsules of sporozoan groups (Fig. 20.12). In a nematocyst is a coiled, hollow thread with a closed or open free end. From the cnidoblast projects a pointed spike, a *cnidocil*. Appropriate stimulation of it leads to explosive discharge of the nematocyst thread, a process in which the thread turns inside out. Threads with closed ends (*volvants*) function in trapping and holding prey. Those with open ends (*penetrants*) pierce through the body surface of prey and secrete a paralyzing toxin. A massive injection of cnidarian toxin is powerful enough to kill a man.

Discharge of nematocysts appears to be brought about by mechanical, chemical, and probably also nervous stimuli. Well-fed cnidarians do not discharge their nematocysts. Similarly, commensals and animals that feed on cnidarians do not cause nematocyst discharge. In some animals, indeed, cnidoblasts of eaten cnidarians migrate from the gut to the body surface of the eaters, and these cells become part of the skin and serve as in cnidarians.

Cnidaria are basically dimorphic, the two structural forms being the *polyp* and the *medusa* (Fig. 20.13). Both forms are organized radially around an *oral-aboral* main axis from the alimentary opening to the opposite end. If this axis is long in relation to the diameter of the body, the animal is a cylindrical polyp. If the axis is short in relation to the diameter, the animal is an umbrella- or bell-shaped medusa. Polyps are sessile vegetative individuals, medusas are free-swimming, sexual individuals. Both types bear tentacles, and both or either one only occurs in a given life cycle.

Any cnidarian, polyp or medusa, contains a central *coelenteron*, or *gastrovascular cavity*, with a single alimentary opening that serves both as mouth and anus. The body around this cavity

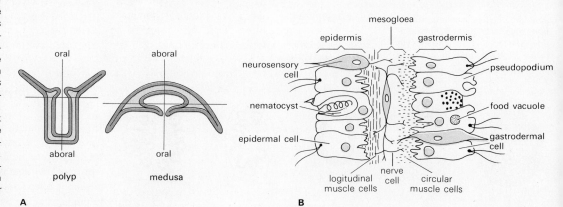

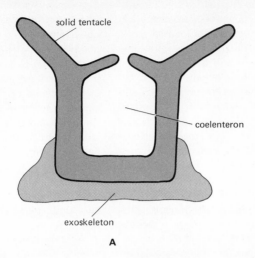

solid tentacle

coelenteron

exoskeleton

A

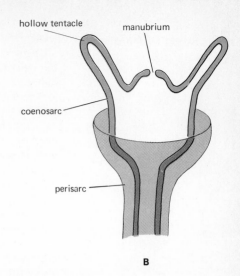

hollow tentacle

manubrium

coenosarc

perisarc

B

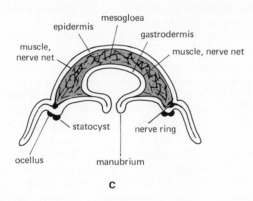

mesogloea

epidermis

gastrodermis

muscle,
nerve net

muscle, nerve net

statocyst

nerve ring

ocellus

manubrium

C

D

Figure 20.14 Cnidarian structure. A, B, sectional diagrams of polyps with solid and hollow tentacles. The coenosarc consists of all living body layers; the perisarc, of the exoskeletal layer. A is a solitary polyp; B is a hydranth, one of many polyps in a branching colony. C, sectional diagram of a medusa showing the three layers of the body. D, a T-shaped epitheliomuscular cell, as in Hydrozoa.

consists fundamentally of three tissue layers derived from the three embryonic germ layers (see Fig. 20.13). The main derivative of the ectoderm is an epithelial (occasionally syncytial) *epidermis.* The main endodermal derivative is an epithelial *gastrodermis.* Between these two is a mesodermal *mesogloea,* a poorly to well-developed connective tissue. The mouth areas of polyps and medusas and also the bell margins of many medusas bear *tentacles.* Hollow tentacles are formed from all three body layers and the internal spaces are extensions of the coelenteron; solid tentacles lack gastrodermal tissue.

The epidermis usually forms an exoskeleton in polyps (Fig. 20.14). Most hydrozoan polyps secrete a *perisarc,* a tube of transparent chitin. Numerous corals embed their bases and sides in massive calcareous secretions, which in many

cases accumulate as the main constituents of coral reefs. Where an exoskeletal cover is absent, as in medusas and in parts of polyps, the epidermis can be smooth (Hydrozoa), flagellated (Scyphozoa), or ciliated (Anthozoa). In the Hydrozoa such epidermal cells are *epitheliomuscular;* the expanded bases of the cells contain longitudinally oriented myofibrils. In the two other classes the inner ends of the epidermal cells are narrow and pseudopodial. These animals have a layer of distinct muscle cells under the epidermis.

Scattered in the epidermis of all classes are neurosensory cells that probably serve as receptors for thermal, mechanical, and chemical stimuli; and cnidoblasts, concentrated particularly in the oral region and the tentacles. The epidermis of medusas usually also contains sense organs for vision and balance. The former are *ocelli,* either eyecups with or without chitinous lenses or simple eye-spots. The balance organs are *statocysts* with statoliths. Such organs usually are located along the bell margins, particularly at the bases of the tentacles (see Figs. 20.14 and 8.37).

The gastrodermis consists largely of flagellate-amoeboid cells that engulf small food particles and digest them intracellularly. Each such cell normally bears two flagella that are resorbed during amoeboid activity and then regrown from the single kinetosome. In Hydrozoa the expanded bases of these cells again contain myofibrils, here aligned circularly. Scyphozoa and Anthozoa usually have a muscle layer under the gastrodermis. The gastrodermis also contains large numbers of gland cells. Most of them secrete digestive enzymes into the gastrovascular cavity and participate in extracellular digestion. Present in addition are neurosensory cells and, in certain gastrodermal regions, also cnidoblasts. In many species the gastrodermal cells harbor symbiotic algae (zoochorellae or zooxanthellae).

Underlying the epidermis and often also the gastrodermis are *nerve nets*. Neuron fibers from these nets connect with myofibril-containing cells, muscle cells, and neurosensory cells. The greatest concentration of neurons occurs in the mouth and tentacle regions, and in a medusa the net also includes a *nerve ring* around the bell margin. This ring receives impulses from the sense organs and emits impulses to the myofibrils or muscles that produce contraction of the bell.

Nerve nets, muscles, and the epidermal and gastrodermal epithelia cover the mesogloea. In its most primitive form, as among hydrozoan medusas, this middle layer consists largely of jelly and some embedded mesenchyme cells and connective tissue fibers (Fig. 20.15). The jelly, about 95 percent or more water, makes up the bulk of such

a medusa. From this condition one line of evolution apparently has led to a less elaborate mesogloea, another line to a more elaborate one. A less complex mesogloea is encountered in the majority of hydrozoan polyps, where the jelly forms a thin layer that contains neither cells nor fibers. The extreme stage of simplification is found in types such as *Hydra,* in which a mesogloea is absent. A more complex level of mesogloeal development is exhibited by the Scyphozoa, in which the jelly contains numerous mesenchyme cells and many fibers. The highest degree of elaboration is reached in the Anthozoa, where the mesogloea is a true connective tissue with large numbers of cells and fibers embedded in small amounts of jelly.

A mesogloea is the more resilient the more fibers it contains, and despite its high water content it actually is quite firm and elastic. This elasticity plays a role in the locomotion of a medusa. The bell margin contracts muscularly and pushes water out from under the bell, which leads to propulsion in the opposite direction. Muscles then relax, followed by elastic recoil of the mesogloea. Repeated contraction-recoil cycles produce a rhythmic, spurting locomotion. The fibers of the mesogloea presumably are secreted by the mesenchyme cells.

Most of the mesenchyme cells are amoeboid, and some of them migrate and become part of the epidermis and gastrodermis. These so-called *interstitial cells* are believed to be equivalent to the archeocytes of sponges and the source of the sex cells. Either interstitial cells or other mesenchyme cells probably give rise also to muscle cells and cnidoblasts. The mesogloea is largely of ectodermal origin and cnidarians therefore can be considered to have an ectomesoderm.

The main tissue types of cnidarians thus are of

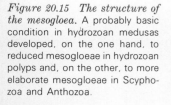

Figure 20.15 *The structure of the mesogloea.* A probably basic condition in hydrozoan medusas developed, on the one hand, to reduced mesogloeae in hydrozoan polyps and, on the other, to more elaborate mesogloeae in Scyphozoa and Anthozoa.

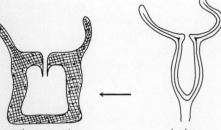
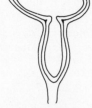

anthozoan polyps
(true connective tissue)

hydras
(no mesogloea)

hydrozoan medusae
(some cells, fibers)

most hydrozoan polyps
(no cells, fibers)

scyphozoan medusae
(many cells, fibers)

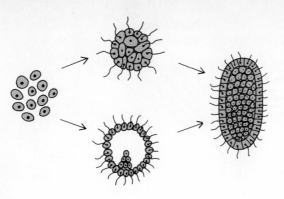

Figure 20.16 Early development. Irregularly cleaved blastomeres (left) become a morula or stereoblastula in most cases (top center) and a coeloblastula in some cases (bottom center). In all cases a stereogastrula (right) forms ultimately. This stereogastrula represents the common planula stage of cnidarian development.

four kinds, epithelial, nervous, muscular, and mesenchymal. In some cases a fair amount of tissue plasticity is in evidence. For example, if *Hydra* is turned inside out experimentally, the cells can migrate back to normal positions and form a normal polyp. And, as in sponges, if hydras are dissociated into cell suspensions the loose cells migrate together and reconstitute whole intact animals.

Cnidaria reproduce by budding, by fragmentation, and by gametes. Budding capacity is particularly extensive and in many cases even the larvae can bud. A polyp bud arises as a fingerlike outgrowth of usually all layers of the body wall, and continued budding can produce a polyp colony in which all individuals share a common, continuous gastrovascular cavity (see Fig. 20.17). Polyps can also form buds that develop as medusas, and these can bud off more medusas in turn, usually in the mouth region or at the bell margin. Reproduction by fragmentation can occur by spontaneous splitting, as in sea anemones (see Fig. 10.2), or as regenerative reproduction following injury. The regeneration potential is particularly high in polyps.

Most cnidarians are separately sexed. Gametes arise seasonally, probably from interstitial cells, and accumulate in given regions as testes or ovaries (see Fig. 20.21). Mesenchyme nurse cells often supply growing eggs with yolk and in some cases an egg is formed by fusion of several cells. In many cases the gametes are shed; in others the eggs are fertilized and undergo part or all of their early development while attached to the parental body.

Cleavage is superficial in some cnidarians, radial and holoblastic in most. The result is a morula or stereoblastula, rarely a coeloblastula. Gastrulation by ingression or delamination then produces a stereogastrula, the *planula* stage characteristic of the phylum (Fig. 20.16). Depending on how long a parent retains its eggs, if at all, the embryo can be set free as a larva at any time before or not until the planula stage has developed. The further fate of the planula differs for the three classes.

Hydrozoans   **Class Hydrozoa**   polyp and medusa phases alternating or either phase reduced or omitted; medusa usually with *velum* and with ocelli and/or statocysts; alimentary cavity not partitioned; gonads usually epidermal; solitary and colonial; mostly marine, some in fresh water.

Order Hydroida   hydroids; polyp and medusa phases usually alternating. *Obelia, Tubularia, Hydra*

Order Milleporina   millepore corals; colonies on extensive calcareous exoskeletons; polyps dimorphic, medusas reduced.

Order Stylasterina   like millepores in most respects.

Order Trachylina   sessile polyp phase absent. *Gonionemus*

Order Siphonophora   floating composite colonies of polyps and medusas. *Physalia*

The planula of a hydroid typically develops into a polyp colony. A free-swimming larva settles and elongates as a polyp *stem*, and at the free end then forms a *hydranth*, a bell-shaped terminal part ringed by whorls of tentacles (Fig. 20.17). In the center of this tentacle wreath is the mouth, located at the tip of a conical *manubrium*. Many hydroids produce *stolons*, stalks along the ground from which upright stems can bud. Branch stems also grow out laterally from an upright stalk and develop hydranths at the free ends. The living tissues of such a colony collectively represent the *coeno-*

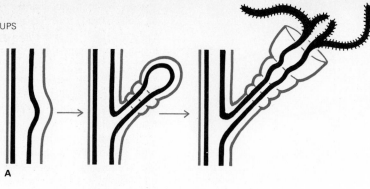

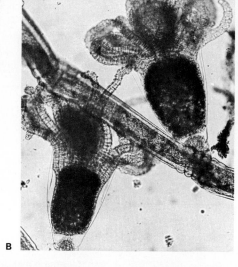

B

Figure 20.17 Hydranths. *A*, feeding polyp forms as shown by budding from upright stem or horizontal stolon of parent colony. The exterior cover indicated is a cup perisarc (as in *Obelia*), and all body tissues (coenosarc) are represented in section as a single colored layer. Note manubrium and tentacles on mature hydranth (right). *B*, close-up view of two hydranths on branching *Obelia* colony.

sarc, in distinction to the secreted perisarc, a thin chitinous exoskeleton. In forms such as *Obelia* the perisarc widens as a cup around each hydranth; in others, *Tubularia* among them, the perisarc terminates below a hydranth, which remains exposed.

Hydranths are feeding individuals, or *gastrozooids*. Polyp colonies also produce polymorphic variants of hydranths (Fig. 20.18). One type is a *dactylozooid*, a protective individual without mouth, tentacles, or alimentary cavity but richly studded with batteries of sting cells. Dactylozooids

Figure 20.18 Polymorphic variants. *A*, dactylozooid, with batteries of stinging cells and without mouth or tentacles. *B*, gonozooid, studded with medusa-producing gonophores. In types such as *Obelia,* a perisarc envelops the gonozooid as shown. *C*, detail of a polyp colony with gastrozooids (hydranths) and gonozooids. The exterior layer is a transparent chitinous exoskeleton. In gastrozooids note the tentacles, and in gonozooids, the medusas at various developmental stages under the chitinous perisarc.

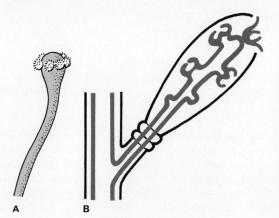

A   B

C

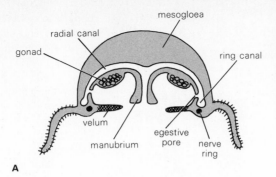

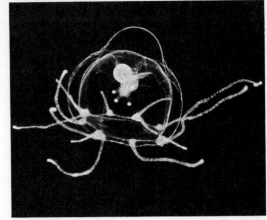

Figure 20.19 *Hydrozoan medusae*. *A*, sectional diagram. *B*, a medusa from life. Note mouth containing manubrium in center of deep bell, smooth bell margin, and tentacles. See also Color Fig. 17.

may or may not form in given species, but nearly all hydroids produce reproductive medusa-forming polyps. In many cases such polyps are *gonozooids*, stalked, perisarc-enclosed individuals as in *Obelia*. A gonozooid buds off *gonophores* in anteroposterior succession, and each gonophore develops as a medusa. Mature medusas detach from gonozooids and become free-swimming.

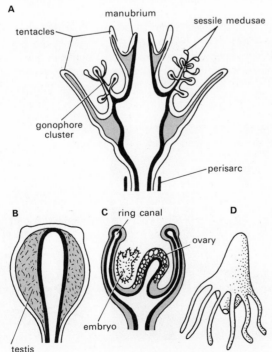

Figure 20.20 *The life cycle of Tubularia*. *A*, section through a hydranth, with gonophore-bearing gonozooids on manubrium. Epidermis white, gastrodermis black, gastrodermal tentacle-supporting cells in color. *B*, immature male gonophore (which eventually acquires a medusoid structure resembling *C*). *C*, mature female gonophore, representing a sessile medusa as indicated in *A*. The embryo is a late planula, just developing tentacles. *D*, young actinula larva, developed from planula and set free by parent medusoid. The larva eventually settles at the aboral end and grows as a polyp colony with hydranths, as in *A* and Color Fig. 18.

On the underside of a hydrozoan medusa is a four-cornered manubrium, often equipped with long trailing arms or tentacles (Fig. 20.19 and Color Fig. 17). The manubrium leads to a gastrovascular cavity from which four *radial canals,* 90° apart, pass toward the bell margin. There each canal opens to a common circular *ring canal*. Near this juncture a small egestive pore in the radial canal leads to the outside. A nerve ring parallels the ring canal. The bell margin of most hydrozoan medusas is extended as a *velum,* a contractile, shelflike membrane on the underside. This structure aids in locomotion by increasing the force of bell contraction and by reducing the opening through which water is expelled from the underside of the bell. *Obelia* and a few other hydrozoan medusas lack a velum. The gonads develop as epidermal folds, usually along the side of the manubrium or underneath the radial canals.

In many hydrozoans the medusa phase is suppressed to a greater or lesser extent. Polyp colonies in such cases typically give rise to polymorphic variants called *medusoids*. These are stalked buds formed almost anywhere on a colony. In *Tubularia,* for example, medusoids arise on the manubria of the hydranths (Fig. 20.20 and Color Fig. 18). A medusoid produces gonophores, and these generally develop as *sessile* medusas, structurally simplified individuals that are not set free. Thus, fertilized eggs of *Tubularia* are retained in the sessile medusas. After a planula stage has developed the embryo acquires tentacles, a mouth, and a rudimentary alimentary cavity. When such a polyplike stage has been reached, the em-

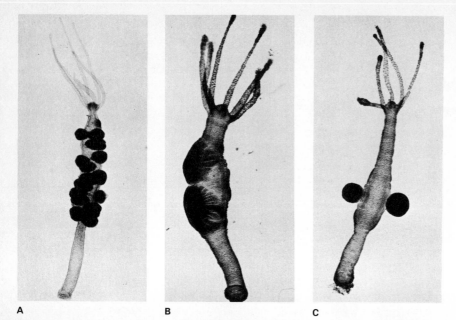

A   **B**   **C**

Figure 20.21 Gonads and embryos in **Hydra.** *A and B,* adults showing position of testes and ovaries, respectively. *C,* embryos develop in cysts that protrude from the body wall of the parent. Such cysts eventually drop off and independent offspring hydras emerge directly.

body levels. Mature eggs protrude through the epidermis and after fertilization they are retained there. An encysted stereogastrula forms, which then falls off the parent and remains dormant for some weeks or months, depending on environmental conditions. When development is resumed a new polyp arises directly. None of the freshwater cnidarians has free-swimming larvae, undoubtedly an adaptation to their particular environment.

The other orders of the Hydrozoa likewise exhibit greater or lesser suppression of either the polyp or the medusa phase. A greatly abbreviated medusa phase is characteristic in the Milleporina, the millepore corals (Fig. 20.22). These reef formers live in shallow tropical seas, where they produce colonies up to 2 ft in size. The living portion of a colony is a thin film on the surface of a massive calcareous exoskeleton. Individual polyps sit in small pits that are interconnected by a network of tunnels in the exoskeleton. Polyps include gastrozooids and dactylozooids. Gonophores are budded into some of the tunnels. Reduced medusas form there, and after they are set free they live only a few hours, just long enough to produce gametes. Stylasters are similar to the millepores; a pit in which a polyp sits here has a calcareous spike at the bottom.

In contrast to these animals, the sessile polyp phase is absent in the Trachylina (Fig. 20.23). This order is of considerable theoretical interest, for the animals are widely believed to represent the most primitive cnidarians and thus the most primitive living Eumetazoa. In the trachylines a planula develops into a motile actinula larva, which then

bryo is liberated as an *actinula* larva. It leaves the sessile medusa, creeps about with the aid of its tentacles, and later settles and produces a new polyp colony.

A medusa phase is suppressed completely in forms such as *Hydra,* one of the few freshwater cnidarians. In these animals gonads develop in the epidermis of the solitary polyp (Fig. 20.21). Ovaries typically form near the base, testes at higher

Figure 20.22 Millepores. A, calcareous deposit of a millipore coral, a hydrozoan in which the polyp phase is dominant. *B,* section through surface region of *Millepora,* showing continuous surface epidermis (color), the subsurface canal system, and the pits for the dimorphic polyps. In the reduced medusa, note the eggs (two are shown).

A

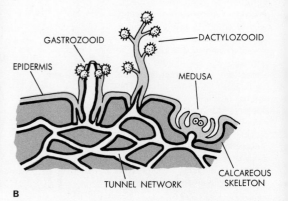

B

A **B**

Figure 20.23 Trachylines. A, actinula larva of *Gonionemus,* budding off another actinula. This stage has arisen from a planula. *B,* adult medusa of *Gonionemus* (from life), with gonads along radial canals. Note velum. Fertilized eggs become planulae which form actinulae directly, without sessile polyp stages.

phase of the life cycle. Thus, the polyps of the present cnidarians would represent long-lived, neotenous larvae. And once both polyps and medusas could be formed in one life cycle, either phase then could be emphasized or suppressed in adaptation to particular modes of existence. The trachylines thus could have been ancestors of all three modern classes of Cnidaria. Such a view therefore postulates that radial free-swimming medusas were the first cnidarians and that the bilateral sessile polyps of Anthozoa were the last and most specialized (see also further discussion below).

If trachylines are primitive, the Siphonophora (now often classified as more than one order) certainly are advanced. These pelagic, predatory, social colonies are made up of various types of polymorphic variants of polyps and medusas not encountered in other Hydrozoa. Polyplike polymorphs include gastrozooids, two or more kinds of dacylozooids, and gonozooids. Medusalike polymorphs include attached medusa bells, gonophores, and *pneumatophores,* or floats that buoy up the whole colony. The Portuguese man-of-war *Physalia* has an air-filled baglike float (Color Fig. 19 and see Fig. 16.12); the float of *Velella* is disk-shaped, with an upright "sail" that catches ocean winds. The poisons released by their stinging batteries make the members of this order exceedingly dangerous even out of water.

becomes an adult medusa directly. The original cnidarian stock is postulated to have been similarly medusoid and free-swimming, with a life cycle that included planula and actinula stages. Further, it is believed that some of the early actinula larvae later became established as a prolonged, sessile

Scyphozoans **Class Scyphozoa** Jellyfishes; all marine; pelagic medusa stage dominant, without velum; epidermis flagellate, with cnidoblasts in all body regions; subepidermal muscle cells; mesogloea with abundant cells and fibers; alimentary cavity with gastric filaments, typically partitioned by four gastric septa; gonads endodermal; development via planula, *scyphistoma,* and *ephyra* larvae.

Scyphomedusas can be distinguished readily from hydromedusas by the absence of a velum and the presence of scalloped *lappets* along the bell margin. The number of lappets usually is divisible by four and the body as a whole is organized in quadrants (Fig. 20.24 and Color Fig. 20).

In jellyfishes such as the common *Aurelia,* the manubrium leads to a small gastrovascular cavity that branches out into four radial canals, each with numerous branches. These canals continue into the tentacles along the bell margin. Ring canals generally are absent, though not in *Aurelia.*

In the majority of scyphomedusas, however, canal systems are not formed. Instead, the large

alimentary cavity is partitioned into four *gastric pouches* formed by outfoldings of the gastrodermis and underlying mesogloea. Each outfolding (*gastric septum*) is pierced by a perforation that makes adjacent gastric pouches continuous. The free edges of the septa bear fringes of *gastric filaments,* short enzyme-secreting tentacles. Such filaments occur in all Scyphozoa, even those without gastric pouches.

Each of the four gastric septa contains a mesogloeal cavity that opens to the outside through a pore on the underside of the umbrella. Called *subumbrellar funnels,* these cavities fill with sea water and presumably facilitate breathing. The

Figure 20.24 Scyphozoa. A, the common jellyfish *Aurelia*. Note the tentacle-bearing lappets around the bell margin, the canal system in the bell, the four horseshoe-shaped gonads near the center, and the mouth-bearing manubrium curving from center to lower left. See also Color Fig. 20. *B*, the canal system and symmetry in *Aurelia*. Interradial and periradial canals are branched, adradials are unbranched. Heavy dashed lines mark the positions of the tentacular arms of the manubrium. A ring canal, present in *Aurelia*, is not found in most other scyphomedusas. *C*, horizontal section indicating the structure of most types of medusa in this class (*Aurelia* does not have gastric pouches and septa as shown here). (*A*, courtesy *Carolina Biological Supply Company*.)

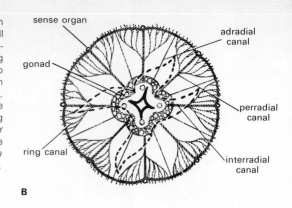

B

A

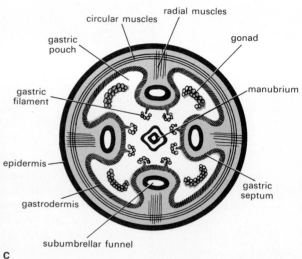

C

Figure 20.25 Scyphozoan development. A, scyphistoma larvae of *Aurelia* hanging from underside of rock, young larva at right, strobilating older larvae in middle and at left. Note ephyras being cut off at left. *B*, individual free ephyra larva of *Aurelia*, seen from underside.

A

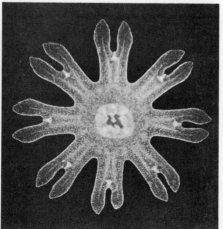

B

distinct muscle cells of Scyphozoa are ectomeso-dermal and form a layer under the epidermis. Particularly prominent are a circular muscle band around the bell margin, radial bands to the bell margin, and a longitudinal band that traverses the base of each gastric septum. Muscles under the gastrodermis are not well developed. Along the margin of the umbrella are sense organs and short, often structurally reduced tentacles. With some exceptions, most scyphomedusas lack marginal nerve rings.

The animals usually have four gonads, located in the gastrodermis along the floor of the gastric pouches. The sexes generally are separate. A planula settles and develops into a *scyphistoma*, a solitary polyplike larva that often lives for years (Fig. 20.25). Ultimately it transforms either to a single free-swimming larval medusa, the *ephyra*, or to several ephyras. A single ephyra is formed in *Pelagia*, for example. In *Aurelia*, by contrast, numerous ephyras arise through a budding process called *strobilation*. Buds here develop like body segments in oral-aboral succession, and when the topmost bud is mature it is cut off as a free ephyra.

Strobilation is the chief form of vegetative reproduction among Scyphozoa. Adult medusas, formed from ephyras, reproduce exclusively through gametes.

Anthozoans

Class Anthozoa   polyps only, all marine; externally radial but internally bilateral; with oral disk, pharynx, and siphonoglyphs; epidermis ciliated; subgastrodermal muscles well developed; mesogloea a fibrous connective tissue; alimentary cavity partitioned; gonads endodermal; development via planula.

Subclass Octocorallia   body sections eight or multiples; one ventral siphonoglyph; always colonial; usually with endoskeleton.

Subclass Hexacorallia   body sections six or multiples; two or more siphonoglyphs; solitary or colonial.

The oral end of anthozoan polyps is a flat *oral disk,* ringed by tentacles (Fig. 20.26). In the center of the disk the epidermis continues past the mouth opening as a short internal tube, the *pharynx.* The mouth is oval or a slit, one expression of anthozoan bilaterality: a plane through the slit marks right and left body halves. In Octocorallia one end of the slit has a pharyngeal band of flagellated cells, the *siphonoglyph.* This end is said to be ventral, the other dorsal. More than one siphonoglyph is usually present in the Hexacorallia. Siphonoglyphs maintain a flow of water to and from the alimentary cavity and thereby aid in breathing.

The alimentary cavity is partitioned by *radial septa,* outfoldings of the gastrodermis and mesogloea. At the oral end the septa connect to the oral disk and the pharynx, and they become narrower toward the base of the animal. Eight septa occur in the Octocorallia, six or multiples of six in the Hexacorallia. At the free edge of each sep-

tum is a thickened *septal filament,* a glandular band studded with cnidoblasts. A strong longitudinal muscle lies under the gastrodermis along the ventral surface of each septum, another expression of internal bilaterality: the muscles on the two most ventral septa face each other; those on the two most dorsal ones face away from each other. Sets of circular muscles ring the body of the polyp. The gastric septa also bear the gonads under the gastrodermis. Sexes most often are separate. Eggs are shed in some forms, retained up to the planula stage in others. Planulae settle and develop directly into polyps.

Anthozoa comprise some two-thirds of all cnidarian species. Octocorallia are exclusively colonial, and their branching, often delicately plantlike and multicolored growths, are among the most beautiful objects in the sea (Color Fig. 21). Colonies are supported by endoskeletons that are either calcareous or horny, or both. Such skeletons arise

Figure 20.26 Anthozoan structure. A, cutaway section through a sea anemone, to show the gastric partitions. *B* and *C,* the symmetries and ground plans of Octocorallia and Hexacorallia, with longitudinal muscles on the ventral surfaces of the gastric septa and with one or two siphonoglyphs. See also Color Figs. 21 and 22.

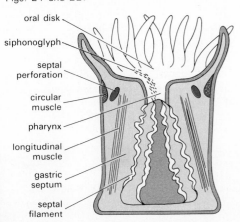

oral disk
siphonoglyph
septal perforation
circular muscle
pharynx
longitudinal muscle
gastric septum
septal filament

A

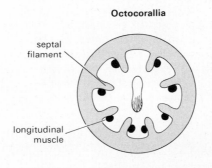

Octocorallia

septal filament

longitudinal muscle

B

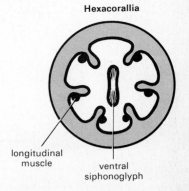

Hexacorallia

longitudinal muscle

ventral siphonoglyph

C

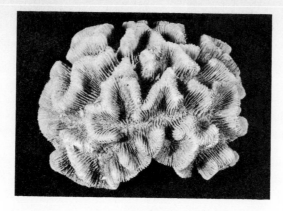

Figure 20.27 Hexacorallia: corals and reef formers. Skeleton of a brain coral, a stony coral of the order Madreporia. Another madreporian is shown in Fig. 18.3. See also Color Figs. 21 and 22. (Courtesy *Carolina Biological Supply Company.*)

as spicules, formed as in sponges by mesenchyme cells. In many cases the spicules are fused as a continuous supporting framework. Horny skeletons most often consist of rods that pass through the cores of stems and branches. Polyps are arranged around these rods.

Animals in this subclass include the purple organ-pipe corals (*Tubipora*), the blue corals (*Heliopora*), the soft corals (*Alcyonium*), the sea pens (*Pennatula*), and the ''gorgonian'' corals, among which are sea whips, sea feathers, and the common sea fans (*Gorgonia*). These gorgonians have horny skeletons, but a calcareous one is found in the gorgonian *Corallium*, the red, or precious, coral.

In the subclass Hexacorallia are the sea anemones and the stony corals (Color Fig. 22). Sea anemones are solitary and attached to rock like suction cups: the animals lift the central portions of their basal disks away from the stone surface. Occasionally an anemone will relinquish its hold and creep slowly to another location. Anemones such as *Adamsia* live mutualistically with hermit crabs, attached to the snail shells the crabs inhabit. The stony corals secrete massive globular exoskeletons up to several yards in width, and individual polyps live in tiny pits along the surface. The various genera are identified by the characteristic surface configurations of their exoskeletons. For example, brain corals (*Meandra*) suggest

cerebral hemispheres (Fig. 20.27); fungus corals (*Fungia*) are reminiscent of the gill plates on the underside of a mushroom.

Stony corals contribute most to reef formation. Other reef-forming organisms include, for example, the coralline (red) algae, the foraminiferan protozoa, and corals such as *Millipora* and *Tubipora*. Organisms of this type largely require warm, shallow seas, and reefs today occur only in a geographic belt bounded approximately by the Tropic of Cancer and the Tropic of Capricorn. Anthozoa most particularly, but also members of the other classes, are known as fossil forms from early Paleozoic times on. The earliest fossil is a jellyfish from the very late Precambrian.

As noted in Chap. 14, a minority opinion holds that sessile bilateral Anthozoa are primitive cnidarians and that the Hydrozoa are the most advanced class. The issue has important implications for the problem of the ancestors of the Bilateria. If Hydrozoa are primitive, then radial animals have given rise to bilateral ones; but if Anthozoa are primitive, then Eumetazoa could have been bilateral from the start, a basic postulate of the ciliate-flatworm hypothesis of eumetazoan origins. Cnidarian structure provides a further argument against this latter hypothesis.

The assumption of anthozoan primitiveness would imply that true muscle cells must have evolved first, and that such cells later degenerated or somehow fused with epithelial cells to form the epitheliomuscular cells of Hydrozoa. Such a possibility is hardly credible, especially since the reverse hypothesis is entirely plausible. Epitheliomuscular cells (reminiscent of sponge myocytes and myoneme-containing protozoan cells) later could readily have specialized in two different directions, one leading to primarily epithelial cells, the other to primarily contractile cells. On this view, Hydrozoa well could have been ancestral to Anthozoa. The bilaterality of Anthozoa then would represent an independent adaptation within the class, and the bilaterality of the later Eumetazoa would have been derived separately from radial Hydrozoa.

Ctenophorans   ## Phylum Ctenophora

comb jellies; all marine and hermaphroditic; body organized in quadrants; mesogloea with muscle cells; locomotion by eight meridional comb plates; gonads endodermal; development mosaic, via *cydippid* larvae.

Class Tentaculata   with tentacles and colloblasts.

Class Nuda   without tentacles; alimentary cavity with large pharynx.

The 80 species of this phylum have world-wide distribution in all oceans. Mostly small, pelagic, and bioluminescent, the animals form part of the zooplankton and they feed on plankton themselves (Fig. 20.28). Polymorphic or colonial forms do not occur. Most comb jellies are glassily transparent and globular in shape, but some are flattened. For example, the 1- to 2-yd-long Venus girdle (*Cestum*) has the form of a laterally compressed band; *Coeloplana* is flattened in an oral-aboral direction and is adapted to a creeping mode of life.

The body consists of epidermis, subepidermal nerve net, bulky mesogloea with mesenchyme and muscle cells, gastrodermis, and a gastrovascular cavity with one main opening (Fig. 20.29). The epidermis bears eight meridional rows of locomotion-producing *comb plates*, two rows per body quadrant. A comb plate is a short band of cilia, essentially like a protozoan membranelle.

At the aboral pole of the animal is a statocyst, more complexly organized here than the equivalent sense organ of Cnidaria. In a ctenophoran statocyst a statolith is supported by four stiff bristles, or *balancers*. Fused cilia form a dome over the statolith. Two ciliary tracts lead from each balancer over the epidermal surface to the two comb rows of that body quadrant. On opposite sides of the statocyst are two ciliated *polar fields*, areas believed to be chemoreceptive.

Comb jellies of the class Tentaculata typically have two tentacles, one on each side of the body between two comb rows. A tentacle is anchored at the bottom of a deep epidermis-lined pit in the body. The tentacles contain longitudinal muscles and can be withdrawn completely. Numerous *lateral filaments* give tentacles a feathery appearance. On these filaments are *colloblasts*, specialized epidermal adhesive cells. They serve in trapping planktonic food organisms, which the tentacles then convey to the mouth (see Fig. 20.29).

The mouth leads to a tubular, epidermis-lined *pharynx*, a ciliated canal serving largely in extracellular digestion. From this tube food passes to a gastrodermis-lined *stomach*, a central gastrovascular cavity. Series of canals that lead away from the stomach have a flagellated or ciliated gastrodermal lining, and the cells carry out intracellular digestion. Two of the four small aboral anal canals are blind-ended and the other two conduct indigestible remains to the outside through *anal pores*.

Vegetative reproduction is unknown in ctenophores. The endodermal gonads of these hermaphroditic animals are located along the meridional

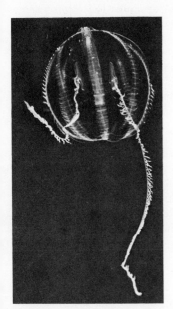

Figure 20.28 Ctenophores. Pleurobrachia, photographed from life. Note locomotor comb plates in rows on globular body, fringed (adhesive-cell-containing) tentacles, and position of sensory balancing organ at top of animal.

Figure 20.29 Ctenophore structure. A, arrangement of gastric canals and comb rows on one side. *B,* structure of aboral sense organ, top and side views. *C, left,* cross section through a lateral filament of a tentacle, with colloblasts forming an external layer; *right,* single colloblast cell with adhesive vesicles on outside and a spiral and a straight filament extending from the nucleated portion.

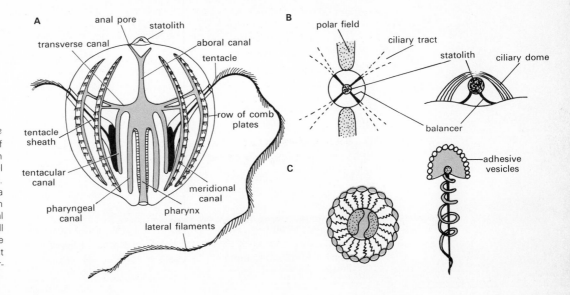

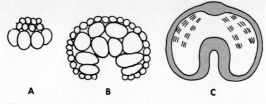

Figure 20.30 Ctenophore development. *A,* 16-cell stage, with a tier of eight macromeres and one of eight micromeres; cleavage here has a biradial (disymmetric) symmetry. *B,* section showing epibolic gastrulation. *C,* a young cydippid larva, with invagination of pharynx and early gastrocoel; early comb rows have formed, and the aboral sense organ is beginning to differentiate.

canals, and gametes are shed through the gastro-vascular channels. In sharp contrast to the regulative development of cnidarians, ctenophore development is mosaic (Fig. 20.30). At the 16-cell stage an embryo consists of a tier of eight micromeres and one of eight macromeres. During later gastrulation the micromeres grow epibolically over and around the macromeres. The middle body layer arises by ingrowth of cells from the ectoderm. The larva characteristic of the whole phylum is the *cydippid,* a globular stage that develops comb rows and tentacles. In forms like *Pleurobrachia,* larva and adult are quite similar and in such cases only few metamorphic changes take place. In other instances, as in *Cestum* or *Coeloplana,* formation of the adult requires considerable additional development.

As a group, Ctenophora display fundamental affinities to the Cnidaria, although many of their features are more specialized. It is now generally believed that ctenophores arose from primitive trachyline medusalike stocks as in independent line. Based mainly on exceptional creeping types such as *Coeloplana,* earlier hypotheses have postulated a direct evolutionary connection between ctenophores and flatworms. Such concepts have proved to be untenable and are now discounted.

Review Questions

1. Review the status of the Mesozoa. What are the structural, developmental, and ecological characteristics of these animals?

2. Define the phylum Porifera taxonomically. Give definitions for each class and name representative genera. What is the general ecology of sponges?

3. Describe the cell types and tissues of sponges. What structures does each cell type produce? Describe the three architectural variants of sponges and indicate groups and genera in which each variant is encountered.

4. What are porocytes, prosopyles, apopyles, spongocoels, ostia, oscula, radial canals, and incurrent and excurrent canals?

5. What is the course of water through a sponge in each of the classes? How is the water flow maintained and adjusted and what are the hydrodynamic characteristics of this flow?

6. Show how a sponge feeds. Which cell types participate in ingestion, digestion, and egestion? What responses to external stimuli can a sponge make?

7. Describe the structure of spicules and show how such supports are manufactured. What cell types are spicule-forming cells derived from?

8. Describe the various forms of vegetative reproduction in sponges. What adaptive advantages can you suggest for the occurrence of special reproductive methods in freshwater species? Can sponges regenerate?

9. Describe the process of fertilization in sponges. What cell types give rise to the gametes? Show how embryonic and larval development take place in (*a*) *Sycon,* (*b*) *Leucosolenia.* What is embryonic inversion and in what variant forms does it occur in (*a*) and (*b*)?

10. Review the evidence suggesting that Mesozoa, Parazoa, and Eumetazoa represent independent branches of animal evolution. Review arguments for and against the traditional hypothesis that sponges have evolved from flagellate protozoa.

11. Give taxonomic definitions of Eumetazoa, Radiata, Cnidaria, and Ctenophora. Describe the ecology of Cnidaria and the body symmetry of these animals.

12. What general structural features distinguish polyps and medusas? Describe the structural and functional characteristics of cnidoblasts.

13. What cell types occur in the cnidarian epidermis, gastrodermis, and mesogloea? How does the organization of these tissues vary in the three cnidarian classes? What is the structure of a tentacle?

14. Review the characteristics of the skeletal elements of Cnidaria. How are nervous and muscular elements organized?

15. How does a cnidarian ingest, digest, and egest? What is the role of the mesogloea in locomotion? How does medusa locomotion take place? What is the embryonic origin of the mesogloea and the gametes?

16. Describe the reproductive process of Cnidaria. How and where does budding occur and where do gonads form? Describe cnidarian development up to the planula stage.

17. Define Hydrozoa taxonomically and name representative genera of each order. Describe the structure of a hydrozoan polyp colony. What are perisarc, coenosarc, stolon, hydranth, manubrium? What are the characteristics of polymorphic variants?

18. How do polyps of *Obelia* and *Tubularia* form medusas? Describe the structure of a hydrozoan medusa. What is an actinula? In what respects is *Hydra* an atypical hydrozoan?

19. Review the organization of millepore corals, siphonophorans, and trachyline medusas. What is the possible evolutionary significance of the last-named? What is a pneumatophore?

20. Define Scyphozoa taxonomically and describe the structure of a medusa in that class. How does *Aurelia* differ in organization from related medusas? What are gastric filaments, subumbrellar funnels? Describe the scyphozoan life cycle. What are the characteristics of the larvae?

21. Define Anthozoa taxonomically and distinguish between the subclasses. Describe the structure of a sea anemone. In what ways do its tissues differ from equivalent ones of other classes?

22. Define septal filament, siphonoglyph. In what respects are Anthozoa bilateral? Describe the life cycle of an anthozoan. What kinds of animals belong to the Octocorallia and the Hexacorallia? Which cnidarians contribute to formation of coral reefs?

23. Review the possible evolutionary interrelations among the cnidarian classes and of Cnidaria as a whole to Ctenophora and to Bilateria. Review hypotheses on the possible ancestors of Cnidaria and discuss the relative merits of various views.

24. Characterize each of the ctenophore classes taxonomically and name representative genera of each class. Describe the exterior structure of ctenophores, with particular attention to locomotor, sensory, and ingestive organs.

25. Describe the gastrovascular system of ctenophores. Review ctenophore development and show how it differs from cnidarian development. How many described species are known in ctenophores, cnidarians, sponges, and protozoa?

Collateral Readings

Berrill, N. J.: "Growth, Development, and Pattern," Freeman, San Francisco, 1961. Several chapters of this valuable book (particularly chaps. 8 and 9) deal specifically with the growth and development of cnidarians (including the formation of polymorphic variants). Other sections analyze growth processes in various invertebrate groups.

Brien, P.: The Fresh-water Hydra, *Am. Scientist,* vol. 48, 1960. A detailed examination of this common but atypical cnidarian.

Hyman, L.: "The Invertebrates," vol. 1, McGraw-Hill, New York, 1940. Chaps. 4 and 6 through 8 contain detailed accounts on the phyla discussed in this chapter.

Lane, C. E.: Man-of-war, the Deadly Fisher, *Nat. Geographic,* Mar., 1963. A well-illustrated article on *Physalia.*

Lenhoff, H. M., and W. F. Loomis (eds.): "Symposium on the Physiology and Ultrastructure of *Hydra* and Some Other Coelenterates," University of Miami Press, Coral Gables, Fla., 1961. The contents are described adequately by the title.

Mackie, G. O.: The Evolution of the Chondrophora (Siphonophora), *Transactions of the Royal Society of Canada,* vol. 53(V), p. 7, 1959. A paper on the behavior of *Velella* and related forms, suggesting a tubularian affinity of these animals.

Wiens, H. J.: "Atoll Environment and Ecology," Yale, New Haven, Conn., 1962. A book on coral reefs and reef formers.

Wilson, H. V., and J. T. Penney: Regeneration of Sponges from Dissociated Cells, *J. Exp. Zool.,* vol. 56, 1930. An account on the original experiments showing that loose sponge cells can migrate together and reconstitute a whole sponge.

21

ACOELOMATES, PSEUDOCOELOMATES

Grade BILATERIA   animals at organ-system level of complexity; with primary and typically also secondary bilateral symmetry.

Subgrade ACOELOMATA   development mosaic, cleavage typically spiral; protostomial, adult mouth formed at or near embryonic blastopore; mesoderm fills space between body surface and alimentary system; skeletal or breathing systems absent; nervous system basically a nerve net with localized ganglionic thickenings; excretion by protoephridial flame-bulb system; with and without larvae. *Flatworms, Ribbon worms*

Subgrade PSEUDOCOELOMATA   development mosaic, cleavage typically in spiral or bilaterally modified pattern; protostomial, adult mouth formed at or near embryonic blastopore; mesoderm leaves pseudocoelic, blastocoel-derived body cavity lined by ectoderm and endoderm; nervous system not a nerve net; skeletal, circulatory, or breathing systems absent; excretion largely by protonephridial flame-bulb system; with and without larvae. *Sac worms, Spiny-headed worms, Entoprocts*

Flatworms   **Phylum Platyhelminthes**
body flattened dorsoventrally, alimentary system with single opening; without circulatory system; largely hermaphroditic. *Turbellaria*, free-living flatworms; *Trematoda*, flukes, parasitic; *Cestoidea*, tapeworms, parasitic.

Of some ten thousand species of flatworms, two-thirds are parasitic. The Turbellaria include a few symbiotic types, but most members of the class live free in fresh water, in the littoral regions of the oceans, and in some cases in moist terrestrial environments. These free-living forms are carnivorous (and occasionally cannibalistic). In general flatworms are small animals, ranging from microscopic sizes to lengths of 2 to 3 in. However, some of the tapeworms can be up to 15 to 20 yd long. Most flatworms are elongated, conspicuously flattened, and have a head region that contains the chief nerve centers and sense organs. The mouth is on the underside of the body, commonly in a midventral or more anterior location. Free-living flatworms undoubtedly are the primitive members of the phylum and the parasitic members are evolutionary derivatives.

Turbellarians   **Class Turbellaria**   free-living flatworms; epidermis cellular or syncytial, without cuticle.

Order Acoela   marine; without intestine.

Order Rhabdocoela   marine and freshwater; intestine straight and saclike.

Order Alloeocoela   marine, freshwater, and terrestrial; intestine straight, often with lateral pouches.

Order Tricladida   planarians; marine, freshwater, and terrestrial; intestine with three branches.

Order Polycladida   marine; intestine with many branches.

The Acoela are the primitive members of the class and thus the most primitive Bilateria. The animals on the whole resemble planula larvae (Fig. 21.1). They frequently are microscopic, sometimes up to $\frac{1}{2}$ in. long. A ciliated (often syncytial) epidermis covers the outside, and the interior is filled with digestive-mesenchyme cells. A simple mouth opening lies midventrally, but other formed alimentary structures are lacking; food passes through the mouth directly to the interior cells. In some acoels the epidermis is epitheliomuscular as in Hydrozoa. In other forms separate muscle cells usually lie under the epidermis. The nervous system is a nerve net, concentrated anteriorly as a poorly defined nerve ring or brain ganglion and elsewhere as longitudinal neuron strands or nerve cords. A statocyst lies anteriorly, and some acoels also have ocelli. Excretory systems are absent. Gametes of both sexes form in clusters directly from mesenchyme cells. Primitive sperm ducts develop seasonally; eggs are shed without ducts through the mouth or ruptures in the body surface. Zygotes cleave spirally (see also below), and the stereogastrula is substantially an adult worm.

As pointed out in Chaps. 14 and 20, acoels are thought to have evolved from hydrozoan planulalike stocks (Fig. 21.2). Some of these would have become actinulae with tentacles and thus cnidarians. Others would have adopted a creeping mode of life and developed bilaterality; the planula would have flattened down and become elongated in the direction of motion. This process would have left the blastopore mouth in a ventral position but shifted the main neural concentration to the anterior end. The resulting primitive acoels would have inherited epitheliomuscular cells and a potential of developing a saclike gastrovascular cavity. All other major flatworm characteristics would have been independent inventions of the turbellarian descendants of acoels.

In these descendants the alimentary system consists of a mouth, a well-developed and usually eversible pharynx, and a saclike intestine (Fig. 21.3). Intestinal cells are ciliated and phagocytic; digestion occurs both extracellularly and intracellularly, as in cnidarians. The middle body layer, often called *parenchyma* in acoelomate animals, is a connective tissue composed largely of a mesenchymal syncytium and free amoeboid cells.

The ciliated epidermis is syncytial in the primitive orders, cellular in more advanced ones. It rests on a basement membrane, and in many cases the nucleus-containing portions of epidermal cells are drawn through the membrane and lie deep in the mesenchyme. Among such sunken cells are mucus-secreting gland cells that secrete a slime track on which the animals move by ciliary action. Certain gland cells manufacture dischargeable *rhabdites*, rodlike bodies containing a viscous substance that may play a role in mucus formation; their exact function is unknown. Glandular cells also form multicellular glands that function as adhesive organs or in trapping prey. Many flatworms feed on hydras and other cnidarians, and

Figure 21.1   Acoel flatworms. A, frontal view. B, side view. The epidermis is drawn as a syncytium although in many cases it is epithelial and ciliated. The nervous system is part of the epidermal tissue layer.

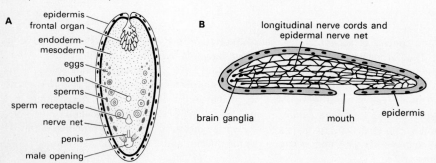

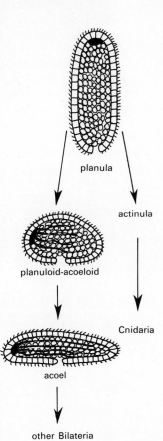

Figure 21.2 *Flatworm derivation*, as from a planuloid ancestor. In a radial, pelagic planula, sensory structures are situated aborally, opposite the blastopore mouth. If the oral-aboral axis became shortened and if, in conjunction with a creeping mode of life, the sensory center also shifted in the direction of forward propulsion, then a bilateral acoel structure could have been derived. (*Adapted from Hyman.*)

in such cases the epidermis usually contains functional cnidoblasts that have migrated from the intestine to the surface.

Under the epidermis lie two or more layers of muscle, and muscles also traverse the parenchyma. This body musculature permits flatworms to swim or creep by sinuous movements. Other muscles operate the pharynx and the copulatory organs. The nervous system fundamentally is a nerve net, with ganglia or rings anteriorly and sets of longitudinal cords passing posteriorly. In advanced flatworms ventral nerve cords and their cross connections are developed more than the others, resulting in a basic ladder-type system (Fig. 21.4). Also, the whole net with its cords and ganglia has sunk deeper into the body, coming to lie in or under the muscle layers or even in the mesenchyme. However, numerous neurosensory cells are "left behind" in the epidermis: mechanoreceptors and chemoreceptors occur abundantly all over the body surface, and more elaborate sensory areas are found in the head region.

Statocysts are common, particularly in marine types, and, the cave-dwelling flatworms excepted, virtually all others have one or more pairs of ocelli, largely of the eyecup type (see Fig. 8.37).

Excretion occurs in a protonephridial flame-bulb system. It is embedded in mesenchyme and opens in various body regions at several nephridiopores. Also in the mesenchyme is the reproductive system, the most complex component of the flatworm body (see Fig. 21.4). Both male and female systems develop seasonally in each individual. The male system typically contains numerous testes arranged as anterposterior pairs. Sperms pass through a pair of sperm ducts that terminate in a single, muscular, mid-ventrally located tube, the copulatory organ. Often equipped with hooks, this organ projects into a genital chamber that opens to the outside behind the mouth. In many worms the gonopore opens into the pharynx.

The female system contains one to many pairs of ovaries, a pair of oviducts, and numerous yolk glands alongside the oviducts or part of the ova-

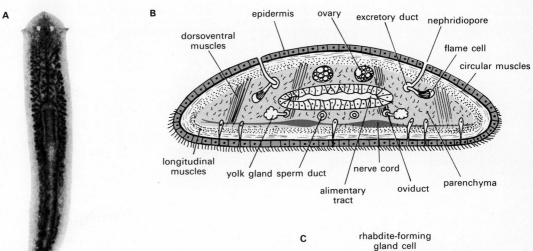

Figure 21.3 *Planarian structure*. A, the free-living planarian *Dugesia*. Note eyes, pointed lateral lobes (auricles) at level of eyes, pharynx in middle of underside, and darkly stained parts of the alimentary system. The pharynx can be protruded and retracted. B, cross section through the anterior portion of a planarian. The epidermis is wholly ciliated or only partly, as shown here. Mesenchyme makes up the parenchyma. C, detail of a section of ventral epidermis, to show the sunken nucleated portions of the epidermal cells.

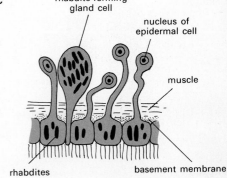

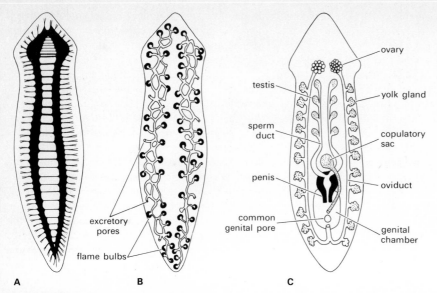

Figure 21.4 *Planarian structure. A,* scheme of the ladder-type nervous system. *B,* the protonephridial excretory system. *C,* the reproductive system. Both male and female systems are present in these hermaphrodites.

there. Many flatworms can be self-fertilizing, but mutual internal cross-fertilization takes place in most cases. In some worms mating occurs by hypodermic impregnation, a process in which the copulatory organ of one worm hooks anywhere into the epidermis of another. Ejected sperms then migrate to the eggs in the oviducts.

In polyclads and acoels, zygotes are shed into the sea. Cleavage is spiral and gastrulation is epibolic. Mesoderm arises in part from the 4d blastomere formed at earlier stages, in part from ectoderm cells that migrate to the interior (Fig. 21.5). This middle body layer later forms the parenchyma and the muscular, excretory, and reproductive systems. Polyclads form a *Mueller's larva,* a ciliated swimming form with eight epidermal lobes. Metamorphosis involves mainly a resorption of these lobes.

In all other orders the eggs are *ectolecithal,* with yolk in separate *yolk cells* formed in either the ovaries or the yolk glands. The oviduct then secretes a cyst wall around each fertilized egg and its surrounding yolk cells. Owing to the presence of yolk cells, embryonic processes in such cysts are greatly modified. Blastomeres at first lie free in the cysts as separate cells (see Fig. 21.5). Later they aggregate more or less directly as body parts of a future worm. Larvae are not formed, and young adults emerge directly from the egg cysts after these have been released by the worms.

ries. The oviducts lead to the genital chamber where the male system terminates or to a separate genital chamber. A *copulatory sac* that projects into the genital chamber receives the copulatory organ of the mating partner. Sex cells arise in the mesenchyme, migrate to the gonads, and mature

Figure 21.5 *Flatworm development. A,* development of polyclad turbellarians. 1, postgastrula stage, attained after regular spiral cleavage and epibolic gastrulation; the mouth opening has formed, and the alimentary pouch is organizing. 2, later stage, larval alimentary system fully formed. 3, Mueller's larva, side view, showing the eight ectodermal lobes and the position of internal organs. *B,* egg cyst with ectolecithal eggs of directly developing turbellarians (such as triclad planarians).

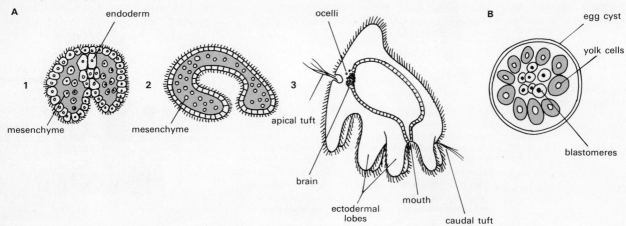

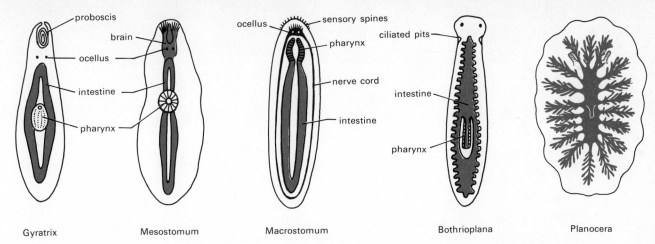

Figure 21.6 Free-living flatworms. First three diagrams on left illustrate rhabdocoel genera. Note the straight digestive pouches. *Bothrioplana* is an alloeocoel. Note lateral diverticula along digestive pouch. *Planocera* is a common polyclad. Note anterior tentacles, many-branched digestive sac. (*Adapted from Hyman.*)

Some groups (the freshwater rhabdocoel *Mesostoma,* for example) are ovoviviparous; young adults form inside the female system.

Flatworms are noted for their extensive vegetative reproduction. Many of the animals can fragment spontaneously into two or more anteroposterior parts, each forming a whole worm. Triclad planarians such as *Dugesia* have long been among the most widely used animals in regeneration research. If these animals are subjected to prolonged starvation, moreover, they undergo ''degrowth,'' or partial self-destruction. They then use the raw materials so gained to maintain themselves as intact worms, though on a progressively smaller scale. Planarians also can be trained by conditioning. The worms then can be cut into several pieces, and after the pieces have regenerated to whole animals each still ''remembers'' the training of the original worm. RNA has been implicated in these knowledge-storage and knowledge-transfer processes.

One group of early acoels presumably gave rise to the Rhabdocoela (Fig. 21.6). The original rhabdocoels must have evolved all the characteristic features of flatworms and become highly diversified in the process; modern rhabdocoels actually are often classified as three separate orders. The primitive members still contain four pairs of longitudinal nerve cords, but in more advanced types a ventral pair already forms a well-developed ladder-type system. Rhabdocoels such as *Gyratrix* have a highly differentiated protrusible proboscis, of a type very similar to that in the Nemertina. Rhabdocoels actually are believed to have been ancestral to the nemertines and indeed also to all the flukes and tapeworms.

A second group of early acoels probably gave rise to the Alloeocoela. These worms have a straight intestine with numerous lateral pouches (*diverticula*) that increase the digestive and absorptive area. Intestinal diverticula also occur in the Tricladida, an order presumably derived directly from the Alloeocoela. The most familiar triclads—and most familiar flatworms generally—are the freshwater *planarians*, of which *Dugesia* (= *Euplanaria*) is the best known genus (see Fig. 21.3). The many species of these animals have well-developed eyes and, laterally at the level of the eyes, pointed sensory lobes called *auricles*. Triclads also include marine types such as *Bdelloura*, a commensal on horseshoe crabs (Fig. 21.7), and terrestrial types such as *Bipalium*, which becomes up to 1 ft long.

Acoels most probably gave rise directly to the marine Polycladida. These are comparatively large, oval, platelike animals in which the intestine is divided into numerous branches. The worms usually have a pair of short tentacles equipped with ocelli, and additional groups of ocelli frequently occur in other body regions as well. Posterior and ventral adhesive organs are common in the group.

Flukes   *Class Trematoda*   holotrophic parasites; epidermis absent, body covering a cuticle; one or more suckers.

Subclass Digenea   liver and blood flukes; endoparasitic, with up to three intermediate hosts.

Subclass Aspidogastrea   mainly endoparasitic, with and without intermediate hosts.

Subclass Monogenea   mainly ectoparasitic, without intermediate hosts.

The most notable structural differences between flukes and rhabdocoels are epidermal. First, flukes lack an epidermis and are covered instead by a tough *cuticle* (Fig. 21.8). Secreted in the mesenchyme, this cuticle is a horny scleroprotein. Second, sense organs and sensory cells are greatly reduced in number and eyes or statocysts are largely absent. Third, flukes have one or more well-developed muscular *suckers* or equivalent adhesive organs. In the Digenea one sucker usually lies anteriorly, another mid-ventrally. In the Monogenea the main sucker is posterior and is often equipped with hooks or claws. Anteriorly, adhesive disks or pits are present but not usually a sucker. The Aspidogastrea have a ventral plate subdivided into rows of suckers. Cuticle, lack of sensory receptors, and adhesive devices all are adaptations to the parasitic mode of life.

Internally the nervous, muscular, and excretory systems are essentially rhabdocoel-like. The intestine usually divides into two blind-ended branches (though in some forms each branch opens at a posterior anal pore). Many flukes have networks of *lymphatic channels* in the parenchyma that are thought to facilitate circulation of the body fluids. With some exceptions in which the sexes are separate (notably blood flukes of the genus *Schistosoma*), flukes are hermaphroditic. Self-fertilization can occur on occasion but cross-fertilization is the general rule. Copulation takes place as in Turbellaria, and fertilized eggs surrounded by yolk cells encyst in the oviduct, here a wide, coiled tube that functions as a uterus; egg cysts can be retained in it for considerable periods (Fig. 21.9).

In the Digenea, or "digenetic" trematodes, the life cycle typically includes three or four successive larval stages and up to four different hosts. The main or final hosts are vertebrates, and the parasites live in the intestine, blood, and body cavities. For example, *Schistosoma japonicum* (see Fig. 21.11) inhabits the intestinal blood vessels of man. Released egg cysts accumulate till the blood vessels rupture, and the cysts then pass to the gut cavity from which they are discharged with the feces. Egg cysts of *Clonorchis sinensis*, the Chinese liver fluke, reach the outside via bile duct and intestine.

The succession of intermediate hosts and larval stages (*miracidium* → *sporocyst* → *redia* → *cercaria*) is diagramed in Fig. 21.10 (and see also Fig. 17.21). These multiple larval stages are instances of vegetative reproduction by *internal bud-*

Figure 21.7   Triclads and polyclads. A, Bdelloura, a commensalistic marine triclad. Note nerve cords (light), pharynx, and faintly visible digestive tract with diverticula. The posterior part of the worm is an adhesive pad. *B, Prostecereaus,* a polyclad. The anterior end is near top.

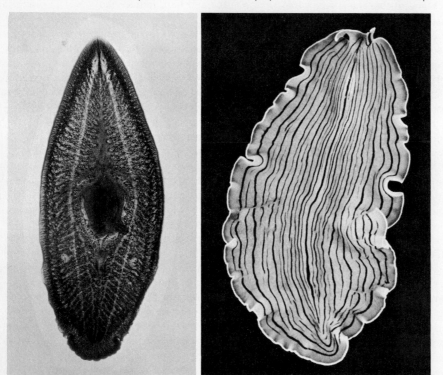

A

B

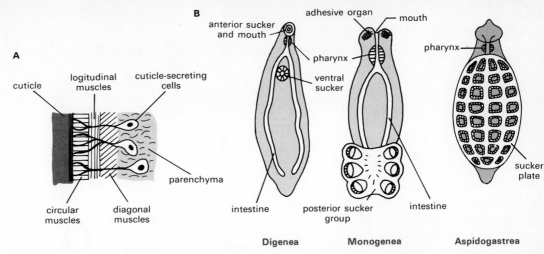

Figure 21.8 *Flukes.* *A,* diagram of a surface section of a fluke, showing the tissues and the absence of an epidermal layer. *B,* outline representations of the basic structure of each of the three orders of trematodes; Digenea represented by *Dicrocoelium,* Monogenea by *Polystomoides,* Aspidogastrea by *Aspidogaster* (ventral view; note ventral sucker plate subdivided into four longitudinal columns and several transverse rows of individual suckers).

Figure 21.9 *Trematodes.* The structure of *Clonorchis sinensis,* the Chinese liver fluke, is shown. In many flukes a shell gland is not present, and the testes can be large and branched as in the photo or globular as in the diagram.

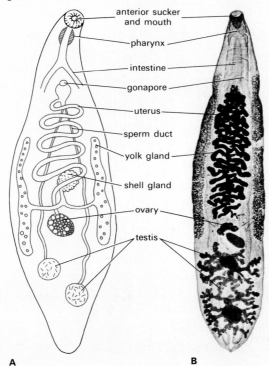

ding. Thus, certain of the stages produce internal *germ balls;* they contain *propagatory cells* that give rise to the next larval stage. Cercariae of *Schistosoma* enter their final host, man, through his skin, migrate in the circulation to intestinal blood vessels, and become adults there. *Clonorchis* cercariae enter fish intermediate hosts and encyst in fish muscle. Raw or partially cooked fish then must be eaten by man before the encysted flukes become adults.

Aspidogastrea parasitize clams and snails. Egg cysts develop there into larvae and then adults. If infected mollusks are eaten by fish, however, the flukes become adults there. The Monogenea, or ''monogenetic'' trematodes, are largely ectoparasites on fish and amphibia. For example, *Gyrodactylus* lives on fish gills. An interesting phenomenon in this genus is the continuous development of embryos: an early embryo forms a second embryo inside it, the second one forms a third inside it, and often a fourth generation embryo forms inside the third (Fig. 21.11).

Because of their harmfulness, their endoparasitic nature, and their complicated life cycles, the Digenea usually are regarded as the most primitive flukes. Recall here also the often presumed evolutionary connection between trematodes, Digenea particularly, and the Mesozoa (see Chap. 20).

Figure 21.10 Fluke life cycles: Clonorchis sinensis (outer pathway) and *Schistosoma japonicum* (inner pathway). Broken lines mark the extent of developmental phases passed in given environments, as indicated.

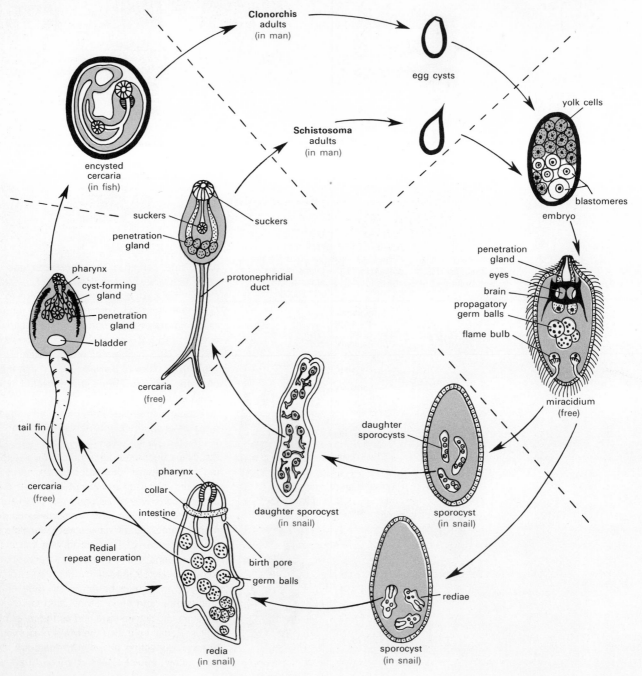

A

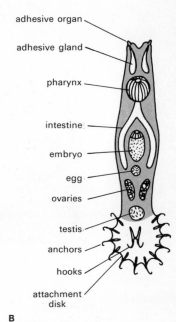

B

Figure 21.11 Monogenetic trematodes. A, an adult male blood fluke. Schistosoma japonicum. B, Gyrodactylus. Note the developing embryo inside the adult. (B, Adapted from Mueller, Van Cleave.)

Tapeworms **Class Cestoidea** fluid-feeding parasites; epidermis absent, body covering a cuticle; without alimentary system.

Subclass Eucestoda body with segmental proglottids.

Subclass Cestodaria body nonsegmented.

Tapeworms (cestodes) are believed to have evolved from early rhabdocoels as a specialized line of fish parasites. Later the group came to infect all other vertebrates as well. As pointed out in Chap. 17, tapeworms absorb molecular nutrients directly through the body surface. This surface again is a secreted cuticle. Muscles lie under it, and the interior is filled solidly with parenchyma. At the head end are reduced brain ganglia, and a longitudinal nerve cord runs along each edge of the body. A simplified protonephridial system is present, with longitudinal ducts paralleling the nerve cords (Fig. 21.12).

The majority of tapeworms belong to the subclass Eucestoda. In these animals the body con-

sists of a head (*scolex*) equipped with six attaching hooks and several suckers; a short *neck;* and a series of segmental *proglottids.* These are formed continuously behind the neck—a form of vegetative reproduction similar to the strobilation process of the scyphistoma larvae of jellyfish. Young proglottids are broader than long; older ones gradually become longer than broad. In each, the lateral nerve cords and excretory ducts are joined by their own cross connections.

Each proglottid contains a complete male and female reproductive system (see Fig. 21.12). The male system matures earlier, hence proglottids near the middle of a worm contain ripe sperms, those closer to the posterior end contain ripe eggs.

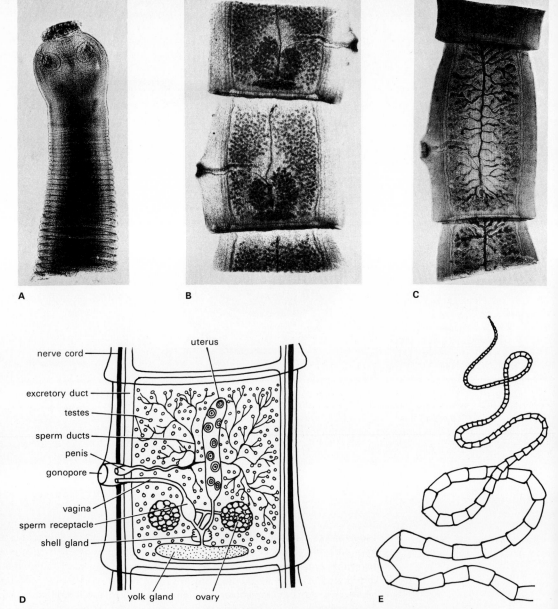

Figure 21.12 Tapeworms. A, head, or scolex, with hooks and suckers. *B,* segmental sections, or proglottids, near middle of body. *C,* proglottids near hind end of body. Tree-shaped structures in *B* and *C* are reproductive organs. Note testes filling proglottids in *B* and genital pores opening on the sides. In *C,* the uterus filled with eggs is conspicuous. *D,* structure of a proglottid. *E,* whole worm, showing change in length and breadth of proglottids with increasing distance from scolex.

Fertilization is internal and can occur in four different ways: by hypodermic impregnation (in some cases); by self-fertilization in a single proglottid; by cross-fertilization between two proglottids of the same worm (the most common process); or by cross-fertilization between two different worms. Zygotes together with yolk cells become encysted and accumulate in the enlarging uterus. All other parts of the reproductive systems then degenerate. Ripe proglottids of this type break off at the hind end of a worm and are discharged with the feces of the host.

The subsequent life history of one of the tapeworms of man, the beef tapeworm *Taenia saginata,* has been outlined in Chap. 17 (Fig. 17.20). Numerous other tapeworms similarly have only one intermediate host, but many have two. For example, the fish tapeworm *Diphyllobothrium latum* (over 50 ft long and by far the most harmful of the tapeworms of man) uses copepods and fish as intermediate hosts. On occasion man eats the undercooked meat of an animal that serves as intermediate host for a tapeworm in which the final host is not man but another animal. These tapeworms will never complete their life cycle, for man is not normally eaten by the usual final host. If he were, man would long since have become a regular intermediate host. Life cycles with multiple intermediate hosts might have evolved from such originally accidental infections of given animals.

Cestodaria are believed to be neotenous offshoots of Eucestoda: the animals retain an unsegmented body, like eucestode larvae. These worms infect the intestines and coelomic cavities of primitive fishes, attaching themselves by means of posterior holdfast organs. Larvae mature in invertebrate intermediate hosts, and if these are eaten by fishes the larvae become adults there.

Ribbon Worms Phylum Nemertina (*Rhynchocoela*)

proboscis worms; alimentary system with separate mouth and anus; circulatory system present; eversible proboscis in rhynchocoel; sexes separate.

Class Anopla   mouth posterior to brain; nervous system subepidermal.

Class Enopla   mouth anterior to brain; nervous system submuscular.

The ribbon worms encompass some six hundred species of largely marine animals. A few live in fresh water and some are terrestrial. Most marine forms are bottom dwellers in sand, mud, or under stones along the coasts of North Temperate regions. The worms range in length from less than an inch to under 2 ft, but one species of *Lineus* can be up to 100 ft long. Many types are brightly pigmented, red, green, brown, and other colors often forming striped patterns (Fig. 21.13).

The body of the worms is covered with a ciliated, glandular epidermis. As in flatworms, many of the epidermal cells contain rhabdites. Some nemertines secrete mucus tubes in which they live or copulate. Under the epidermis is a connective tissue dermis, the two layers forming a well-

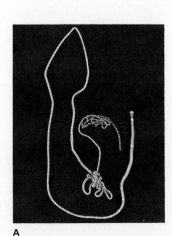

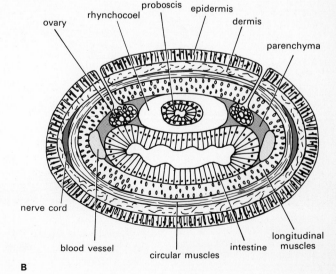

Figure 21.13 Nemertines. A, the proboscis worm *Tubulanus.* B, cross section through the nemertine body. Note the sparseness of parenchymal spaces and the glandular epidermis. The proboscis is lined by an epithelium that overlies several radial and longitudinal muscle layers.

A

B

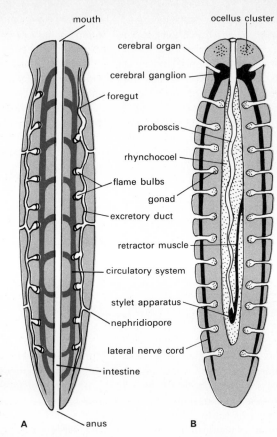

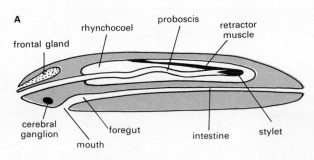

Figure 21.14 Nemertine structure. A, alimentary, circulatory, and excretory systems. *B,* nervous, sensory, reproductive, and proboscis systems.

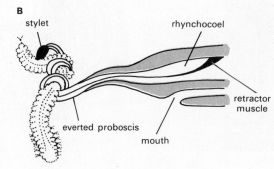

Figure 21.15 The proboscis apparatus. A, position of rhynchocoel and proboscis in rest condition. *B,* everted proboscis entangling and capturing a clamworm.

defined skin. The body-wall musculature underneath consists of two or more layers. The mesenchyme is a connective tissue that fills all spaces not occupied by other tissues and organs. Such spaces are not very extensive; nemertines are compactly built animals.

The alimentary system is a one-way tube; mouth, foregut, intestine, and anus are the main components, and the wall is a single layer of ciliated cells (Fig. 21.14). Digestion occurs extracellularly and food is moved through the tract by contractions of the body wall. The nemertines also have a closed circulatory system. It lies in the mesenchyme and consists mainly of two longitudinal lateral vessels joined anteriorly and posteriorly. A longitudinal dorsal vessel or cross-connecting transverse vessels often are present as well. In a few nemertines the blood is pigmented (red, orange, yellow, or green); the red pigment is hemoglobin in blood cells.

Excretion is accomplished by a protonephridial flame-bulb system (see Fig. 21.14). The nervous system is a nerve net with a pair of brain ganglia and a pair of lateral nerve cords. Numerous accessory ganglia and secondary cords connect with the main cords. Many nemertines have paired anterior *frontal organs,* pits or canals lined with ciliated cells and probably serving as chemoreceptors. Up to two hundred or more eyecup ocelli can be clustered over the front end of the body. Some nemertines have paired statocysts over the brain ganglia. A notable phylum characteristic is a pair of *cerebral organs,* anterolateral pits lined with glandular cells. These organs are believed to serve chemoreceptor and possibly also endocrine functions.

The most distinctive trait of a ribbon worm is its *proboscis,* an anterior muscular tube that in the rest state lies in a closed, fluid-filled cavity dorsal to the intestine, the *rhynchocoel* (Fig. 21.15). The proboscis often is looped and therefore can be longer than the worm itself. The blind-ended tip of the proboscis is attached by a retractor muscle to the wall of the rhynchocoel. Muscular contraction of this wall results in an explosive eversion of the proboscis to the outside.

The animal uses this organ in locomotion, in burrowing, and particularly in trapping annelids and other worms, prey that becomes encoiled by the proboscis. In many cases the tip of the proboscis is armed with a sharp-pointed *stylet* and with glands that secrete poison into a wound made by the stylet. In one nemertine the proboscis branches in bushlike fashion from the base forward, ending in up to 32 separate tips. The proboscis apparatus and nemertine structure as a whole give evidence of a fairly close evolutionary relation between ribbon worms and rhabdocoel flatworms.

Like flatworms, nemertines can reproduce vegetatively by fragmentation. When irritated unduly, forms like *Lineus* are known to fragment spontaneously into a dozen or more pieces. Each piece then regenerates as a smaller worm. In many worms, also, a new proboscis can grow if the original one breaks off accidentally.

Most ribbon worms are separately sexed, although hermaphroditism is common in freshwater and terrestrial groups. Gonads arise in mesenchyme as anteroposterior pairs, and each acquires a separate duct and gonopore (see Fig. 21.14). In a few groups (exemplified by *Prostoma*) development occurs internally and directly; larvae are absent and young adults form ovoviviparously. Most species are externally fertilizing, however. Cleavage is mosaic and spiral, a coeloblastula

forms, and gastrulation takes place by embolic invagination in most cases, by ingression or epiboly in the others. (Being mosaic, the embryo cannot regenerate, but the adult regenerates readily. A similar disparity occurs in entoprocts and tunicates.)

Larvae are of two types (Fig. 21.16). The more common *pilidium* larva (as in *Cerebratulus*) is a helmet-shaped ciliated form, with an aboral *apical tuft* of long cilia and two epidermal *oral lobes* on each side of the mouth. The alimentary system terminates interiorly as a blind *stomach*. *Desor's larva* (as in *Lineus*) remains in its egg membrane and does not develop an apical tuft or oral lobes. In other respects this larval type is similar to the pilidium and also develops similarly.

Metamorphosis is a drastic transformation. The larval ectoderm invaginates in a number of places and buds off small *imaginal disks* to the interior. These eventually grow as a continuous layer of ectoderm around the larval stomach. The layer and its contents then elongate and develop into the future worm. All external parts—the bulk of the original larva—degenerate and are cast off

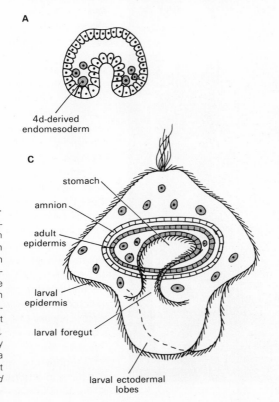

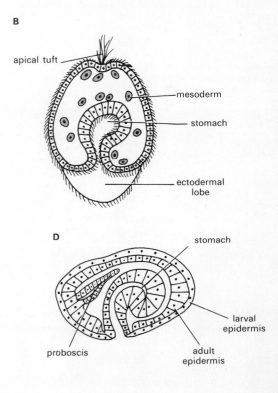

Figure 21.16 Nemertine development. *A,* embolic gastrulation, endomesoderm forming from 4d teloblast cell. *B,* section through early pilidium larva, as in *Cerebratulus.* *C,* late pilidium, beginning of metamorphosis; the amnion is complete and has given rise to the adult epidermis. Structures outside the amnion will not become part of the adult. *D,* Desor's larva, as in *Lineus.* In early stages, the structure of the larva corresponds substantially to that of an early pilidium. (*B,* adapted from Coe; *D,* after Arnold.)

Of the three pseudocoelomate phyla, sac worms, spiny-headed worms, and entoprocts, the first two are not yet stabilized taxonomically; the spiny-headed worms well could be regarded as one of the sac worm classes. It also has been suggested that sac worms and spiny-headed worms be regrouped as two differently constituted and differently named phyla. Alternatively, the various sac worm classes could be—and indeed often are—treated as separate phyla. Whatever taxonomic system finally proves to be most appropriate, there is little doubt that the third pseudocoelomate group, the entoprocts, do form a clearly separate phylum unit.

A fourth phylum group, the priapulids, is discussed after the pseudocoelomates. Until recently these exceedingly enigmatic animals were regarded as a class of sac worms, but now it appears that the group is not even pseudocoelomate.

Sac Worms *Phylum Aschelminthes*

body with cuticle, often segmented superficially; number and arrangement of cells or nuclei constant for each species; musculature typically not in layers; pharynx usually highly differentiated; sexes largely separate. *Rotifera, Gastrotricha, Kinorhyncha, Nematoda, Nematomorpha*

Rotifers *Class Rotifera*

microscopic aquatic animals; with anterior wheel organ; pharynx with jaws; flame-bulb protonephridia.

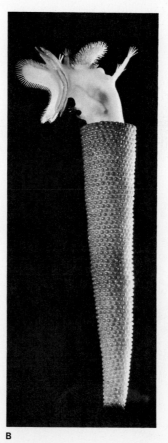

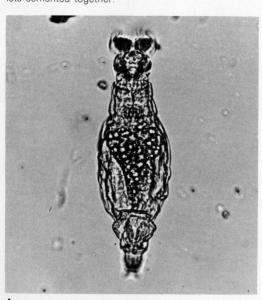

Figure 21.17   Rotifers. A, Philodina, from life. B, glass model of Melicerta, a sessile type that constructs its housing from fecal pellets cemented together.

A

B

Most of the approximately fifteen hundred species of rotifers are freshwater types, the rest are marine. The animals live as swimmers, creepers, floaters, and sessile forms, and their shapes reflect their modes of life. Floaters tend to be globular and saclike; sessile species are vaselike and usually enveloped by a lorica or a cuticular envelope; and creepers and swimmers are elongate and roughly worm-shaped (Fig. 21.17). The animals are about as large as ciliate protozoa. The major distinguishing feature of the class is the *corona,* or wheel organ, an anterior wreath of cilia used in swimming locomotion and in creating food-bearing water currents.

Each member of a species is constructed from exactly the same number of embryonic cells. After these become syncytial in the adult, each individual still retains the same number of nuclei at exactly fixed locations. Thus, any two members of a species are structurally identical and the architecture of each species can be mapped out precisely, cell for cell and nucleus for nucleus. Such cell and nuclear constancies are exhibited also by most other sac worm classes. By actual count a rotifer consists of 1,000 to 2,000 cells or nuclei, a particular number in that range identifying each species.

The body of a rotifer usually contains a *head,* a *trunk,* and a tapered *foot* (Fig. 21.18). The end of the foot frequently bears two *toes,* each with

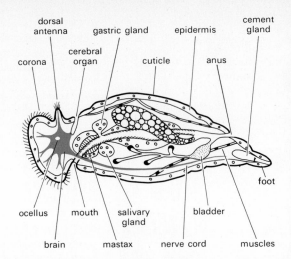

Figure 21.18 Rotifer structure, side view.

a *cement organ*. The toes are used for temporary attachment and a creeping, caterpillarlike locomotion. The body is covered with a tough scleroprotein cuticle secreted by the thin epidermis underneath. Transverse ringlike grooves in the cuticle

Figure 21.19 Rotifer development. A, the life cycle. Diploid amictic eggs (produced by a single meiotic division) develop parthenogenetically into new generations of females. Mictic eggs (produced by two normal meiotic divisions) develop parthenogenetically into simplified haploid males, which produce sperms. If sperms then fertilize mictic eggs, encysted diploid winter eggs are formed. These produce new generations of diploid females the following spring. B, stereoblastula, beginning of gastrulation. C, sagittal section through later embryo. The large invaginated cell gives rise to the ovary and the reproductive system. Mesoderm arises from ectoderm. (*Adapted from Tannreuther, Nachtwey.*)

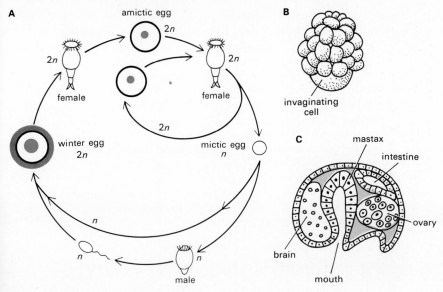

give the animal a segmented appearance and make the body highly flexible.

In the interior the lymph-filled pseudocoel contains amoeboid mesenchyme cells, Muscle cells are not arranged as distinct layers but crisscross in specific patterns through the pseudocoel. The alimentary tract is an internally ciliated, syncytial tube. Ventral to the wheel organ is the mouth, and behind it the pharynx (*mastax*) is a complex muscular chewing organ that contains cuticular jaws studded with teeth and receives secretions from salivary glands. The detailed construction of the mastax differs for different species.

From a brain ganglion dorsal to the pharynx lead two main ventral nerve cords. Just behind the brain ganglion lies a glandular *cerebral organ,* believed to be homologous to the frontal organs of acoelomates. Most rotifers have an *ocellus* directly over the brain ganglion. A characteristic feature is a *dorsal antenna,* a small surface projection dorsal to the brain ganglion. The precise function of this innervated and ciliated organ is obscure. The excretory system consists of a single pair of flame-bulb clusters in the pseudocoel. At their posterior ends the ducts from the flame bulbs form a urinary bladder that opens in the hind part of the intestine.

In one group of rotifers, exemplified by the common freshwater genus *Philodina,* males are unknown and propagation occurs exclusively through parthenogenetically developing eggs. In other types males are small, structurally simplified individuals without alimentary systems and capable only of producing sperms (Fig. 21.19). Mating takes place most often by hypodermic impregnation, the injected sperms migrating through the pseudocoel to the ovaries. A female can produce only as many eggs as there are nuclei in the syncytial ovary (from 10 to 50).

In bisexual forms the eggs are of two types. One type undergoes but a single meiotic division during maturation. Such *amictic* eggs are diploid; they cannot be fertilized and develop parthenogenetically into females (see Fig. 21.19). Successive generations of these females increase the rotifer population in a pond during spring and summer, when conditions are favorable. As temperatures fall at the approach of winter, the eggs formed pass through two normal meiotic divisions. Such *mictic* eggs are haploid. If they are not fertilized, as would necessarily be the case in the first mictic eggs of the population, they develop par-

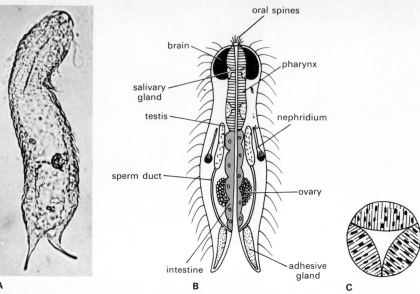

oral spines
brain
salivary gland
testis
sperm duct
intestine
pharynx
nephridium
ovary
adhesive gland

A
B
C

Figure 21.20 Gastrotrichs. A, Lepidodermella, from life. B, internal organs of Chaetonotus, top view (uterus not shown). C, cross section of pharynx, showing radially aligned muscles and triangular outline of pharyngeal canal. (A, Carolina Biological Supply Company; B, C, adapted from Zelinka.)

Cleavage is spiral in some cases, bilateral in others. The resulting stereoblastula gastrulates by emboly; only a single cell invaginates, the source of the future gonads (see Fig. 21.19). The mesoderm is almost entirely ectomesodermal. Development is mosaic and even adult rotifers cannot regenerate. However, freshwater rotifers are remarkable in being able to withstand drying out. Slow desiccation shrivels such animals to cystlike bodies and in this condition they can remain dormant for years. When water becomes available again the animals "inflate" back to normal and resume active life.

Experiments have shown that the first egg formed by a rotifer develops into a longer-lived offspring than the second, third, or any later egg. When a longer-lived offspring produces eggs in turn, the first of these becomes a still longer-lived offspring. Thus, by continued selection of first eggs in successive generations, a line of progressively longer-lived generations can be raised. This phenomenon well could be an adaptation to the seasonal life of these animals. Inasmuch as first-formed individuals live longer they may have a better chance of surviving till fall, when production of winter eggs will ensure species survival into the following year.

thenogenetically into males. But if they are fertilized by some of the males that now appear in the population, the mictic eggs acquire a cyst wall in the oviduct and are shed as resistant *winter eggs*. These develop into amictic females in the following spring.

Gastrotrichs   **Class Gastrotricha**   microscopic aquatic animals; with cilia and cuticular spines.

The two hundred or so species in this class have characteristics that in part resemble those of rotifers, in part those of nematodes (Fig. 21.20). A rounded head usually is set off from the trunk by a slight constriction. The ciliated ventral surface produces a smooth, gliding type of locomotion. Dorsally the body bears spines, bristles, or scales. Epidermis, musculature, and nervous system are essentially as in rotifers. Cell-free lymph fills a not very extensive pseudocoel. Some gastrotrichs have epidermis-derived noncellular membranes that divide the pseudocoel partially into compartments.

The muscular pharynx resembles that of nematodes greatly. It extends over roughly one-third of the body, it has one or more bulbous enlarge-

ments, and its interior canal has a triangular cross-sectional shape. An excretory system (of protonephridial flame bulbs) occurs in only one group (exemplified by *Chaetonotus*). Gastrotrichs are basically hermaphroditic. In some species the male reproductive system has degenerated and these animals propagate exclusively as parthenogenetic females. Development is mosaic. The first three cleavages are spiral, later ones, bilateral. A coeloblastula results that gastrulates embolically.

Because gastrotrichs resemble nematodes in cuticular specializations and in pharynx structure, they are believed to represent an evolutionary link between rotifers and nematodes.

Kinorhynchs   **Class Kinorhyncha**   microscopic, marine; without cilia; body segmented superficially

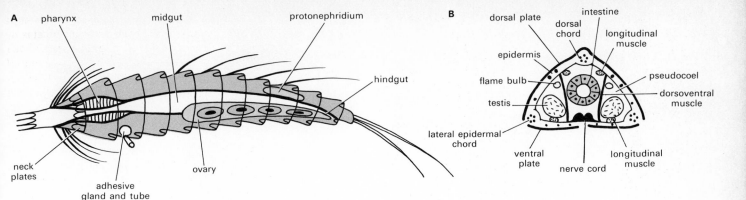

A
pharynx
midgut
protonephridium
hindgut
neck plates
ovary
adhesive gland and tube

B
dorsal plate
intestine
dorsal chord
epidermis
longitudinal muscle
flame bulb
pseudocoel
dorsoventral muscle
testis
lateral epidermal chord
ventral plate
nerve cord
longitudinal muscle

Figure 21.21 Kinorhynchs. A, side view of Echinoderes. B, cross section through tenth segment of Pycnophyes. (Adapted from Zelinka.)

The approximately sixty species of this class are abundant in oceanic mud bottoms. The body of the animals has 13 conspicuous body divisions (Fig. 21.21). The first is a *head*, with mouth and a circlet of *oral spines*, that can be retracted into the second and third "segments." The *trunk* is covered by one dorsal and two ventral chitinous plates that lap over the segment behind. Spines and bristles stud the cuticle.

The syncytial epidermis is thickened dorsally and laterally as a series of *longitudinal chords*, a feature that relates the animals to nematodes. Indeed, the structure of the pharynx is similar to that of nematodes and gastrotrichs. The body musculature has a segmental arrangement. Simi-

larly segmental is the nervous system, which is so intimately joined to the epidermal layer that, with the exception of the double ventral cords, the two tissues appear to be one. The pseudocoel is spacious and contains lymph with cells. One pair of modified flame bulbs is situated in the tenth and eleventh body divisions. Sexes are separate and the pattern of embryonic development is unknown. Juvenile stages are quite unlike the adults and at the end of each stage the cuticle is molted and a larger one forms. It is generally agreed that, in view of their pseudocoelomate structure and their specific particular traits, kinorhynchs are related to nematodes and gastrotrichs.

Roundworms   ***Class Nematoda***   cylindrical, elongated worms without cilia; excretory system not protonephridial; free-living and parasitic.

After the insects, roundworms are probably the most abundant of all animals. Named species total only about twelve thousand at present, but new ones are being described at an average rate of one per day. Virtually every species harbors at least one type of parasitic nematode, and roundworms also parasitize plants. Further, the free-living species in fresh water, ocean, and soil probably are even more numerous than the parasitic ones. Informed guesses therefore place the number of existing nematode species at about $\frac{1}{2}$ million. Most nematodes are under 2 in. long, although some parasitic types have lengths of well over a yard.

The worms are cylindrical in cross section and at the tapered ends are the mouth and anus,

respectively (Fig. 21.22). A three-layered cuticle covers the body. From the outside in, the layers are protein (similar to keratin), a net of spongy fibers, and a meshwork of collagen fibers on a basement membrane. The epidermis, cellular or syncytial, is thickened on the inner surface as four *longitudinal chords*, one dorsal, one ventral, and one on each side. The lateral chords are visible externally as faint lines. If the epidermis is syncytial its nuclei are situated only in the chords. In each body quadrant between the chords is a layer of body wall muscles, with cells aligned longitudinally only; the worm can bend but not lengthen (nor produce peristaltic contractions).

The nervous system consists of a nerve ring

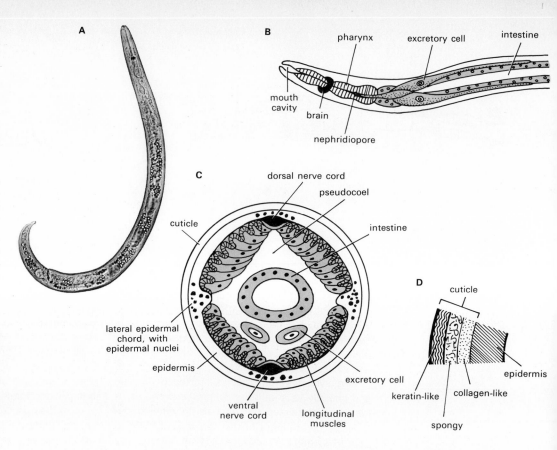

A

B

pharynx   excretory cell   intestine

mouth cavity

brain

nephridiopore

C

dorsal nerve cord

pseudocoel

cuticle

intestine

lateral epidermal chord, with epidermal nuclei

epidermis

ventral nerve cord

longitudinal muscles

excretory cell

D

cuticle

keratin-like   collagen-like

spongy

epidermis

Figure 21.22 Nematode structure. A, whole view of a mature female of the nematode *Paratylenchus,* which causes disease in plants. *B,* ventral view of organs in anterior portion of a nematode. *C,* cross section at level of excretory glands. The contractile portions of the longitudinal body-wall muscles (cross-barred) are adjacent to the epidermis, the noncontractile portions curve toward the nearest longitudinal nerve cord. In some nematodes a longitudinal excretory canal runs within each lateral epidermal chord and connects with the excretory glands. The lateral chords mark the position of the externally visible lateral lines. *D,* the layers of the cuticle secreted by the epidermis. *(B, adapted from Chitwood.)*

Figure 21.23 Nematode structure: cross section through the pharyngeal region of *Ascaris.* Note exterior cuticle, longitudinal chords, muscle quadrants, and thick central pharynx with triangular interior canal. *(Courtesy Carolina Biological Supply Company.)*

around the pharynx, a dorsal and ventral nerve cord, from one to three pairs of lateral cords, and various associated ganglia. In the alimentary system, cuticular jaws or teeth often occur around the mouth. As in gastrotrichs, the well-developed pharynx is a tube of epithelial, glandular, and muscular tissue, with a canal that has a three-cornered cross section (Fig. 21.23). The excretory system is unique. It consists of a pair of *ventral glands,* large cells under the pharynx that lead to anterior nephridiopores. Such a system is not protonephridial and it does not resemble any other excretory system of animals (see Fig. 21.22).

Many species are hermaphroditic, and in some worms the same gonad produces sperms first, which are stored, then eggs that are fertilized by the stored sperms. Certain species are exclusively parthenogenetic. In the vast majority of nematodes the sexes are separate. Males are smaller than females and usually are identifiable by their curled posterior ends (Fig. 21.24). The tubular gonads are single or paired. The sperms of nematodes are

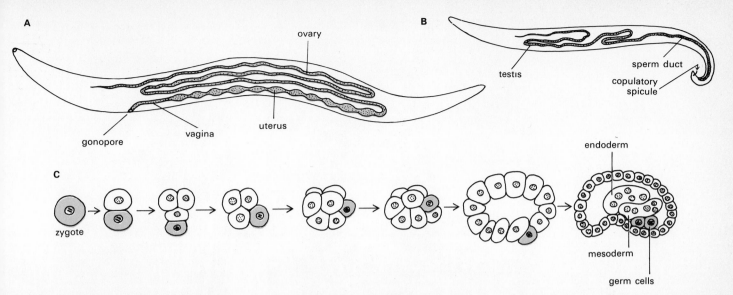

Figure 21.24 *Reproduction and development* in *Ascaris*. *A*, female worm, and *B*, male worm, side views, each showing position of reproductive system. Only the left gonads are shown; the terminal portions of the systems are unpaired and are situated along the ventral midline. Note the smaller size and the curled posterior end of the male. In nematodes other than *Ascaris*, the female gonopore can be located in the middle or the posterior portion of the body. *C*, early development. The anterior end is toward the left. Blastomeres in which chromosome diminution has not occurred are shown in color. The next-to-last stage is a sagittal section through the blastula; the last stage, a section through a postgastrula. A mouth invagination is beginning to form in the blastopore region, and cells in the interior are becoming syncytial. (*C, adapted from Boveri.*)

unique in that they are amoeboid, not flagellate (and nematodes as a whole are unique in that flagella or cilia are totally absent). Fertilization occurs internally. The males of many species have a copulatory spicule near the anus, used to widen the female gonopore during introduction of sperms. These enter eggs near the ovary, and zygotes become encysted and begin to develop during their passage through uterus and vagina.

Cleavage follows a modified spiral pattern (see Fig. 21.24). Development is mosaic and is particularly noteworthy for the very early differentiation of the cellular source of the later reproductive cells. At the two-cell stage in *Ascaris*, for example, only one of the two cells retains normal chromosomes. Those of the other cell become reduced to small granular bodies, a process known as *chromosome diminution*. In all following divisions of the first cell, similarly, only one daughter retains normal chromosomes and diminution occurs in the other. The cell line with normal chromosomes will eventually form gametes and will continue to divide, but all cells with diminished chromosomes soon stop dividing and then represent the final cell complement of the body of the worm. Thus, adult traits of all but the reproductive cells become fixed

permanently and very early by *chromosomal* specializations; and the source of the gametes is marked precisely from the zygote stage on, by cells with normal chromosomes.

During their larval development nematodes typically molt their cuticles, usually four times in succession, and enlargement of the larvae takes place at these molting stages. The life cycles of the parasitic roundworms are more varied than those of almost any other animals. An infective stage is reached at a given point in development or during the adult phase. Up to that point the worms are free-living, and when they become infective they must enter a specific plant or animal host within a short time or perish. Numerous nematodes require one or more intermediate hosts.

Man harbors some fifty species of nematodes. Most of these are relatively harmless, but some cause serious diseases (Fig. 21.25). Among them are the *trichina worms* (*Trichinella spiralis*), introduced into the human body from insufficiently cooked pork; the *hookworms* (such as *Necator americanus*), which live in soil and infect man by boring through his skin; the *guinea worms* (such as *Dracunculus medinensis*), which develop as larvae in copepods, enter man via copepod-

A

B

Figure 21.25   Parasitic round-worms. A, trichina worm larvae, encapsulated in pig muscle. If infected pork is cooked improperly, the larvae are digested out in the intestine of the host, and the worms then invade the host tissues. *B,* elephantiasis of legs and feet, caused by filaria worms.

containing drinking water, and form ulcerating skin blisters that release the larvae of the next generation; and the *filaria worms* (such as *Wuchereria bancrofti*), transmitted by mosquitoes and causing blocks in human lymph vessels. The resulting disease, *elephantiasis,* is characterized by immense swellings. One of the relatively less dangerous roundworms is *Ascaris lumbricoides,* a species that lives in the intestine of man and other mammals. Intestinal nematodes are usually implicated when an animal is said to suffer from "worms."

All nematodes, parasitic or free-living, probably have evolved from marine nematode ancestors. These in turn undoubtedly have been derived from the same stocks that also have given rise to the kinorhynchs and gastrotrichs.

Hairworms   **Class Nematomorpha**   without epidermal chords or excretory systems; alimentary system reduced; parasitic in arthropods during young stages.

The eighty species of these worms are readily identified by their very great lengths relative to their diameters (Fig. 21.26). Ranging in size from less than ¼ in. to nearly 2 yd, the animals resemble roundworms in general structure. However, epidermal chords and excretory systems are lacking, and the alimentary tract remains undeveloped and nonfunctional at either the mouth, the anus, or both openings. The pseudocoel is partitioned by membranes as in gastrotrichs.

The sexes are separate. Cleavage follows the nematode pattern and gastrulation is embolic. Larvae can remain free-living for only a short time and must soon enter insect hosts such as grass-

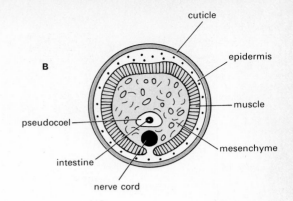

Figure 21.26 Hairworms. A, outline sketch of Paragordius. B, cross section through a nematomorph. Note absence of epidermal chords and excretory organs. (Adapted from Montgomery.)

hoppers and beetles. The larvae have a retractible proboscis apparatus that is lost at metamorphosis, when the worms leave their hosts and become free-living in bodies of water.

Hairworms clearly are related to nematodes.

Among Aschelminthes as a whole, therefore, rotifers appear to provide a link between flatworms on the one hand and all other aschelminth classes on the other.

Spiny-headed Worms

Phylum Acanthocephala

parasitic, with arthropod intermediate hosts and vertebrate final hosts; nuclear constancy in adult; with armed proboscis; alimentary system absent; excretory system protonephridial if present.

This group comprises some six hundred species of intestinal parasites. Their structure relates them to the sac worms, and indeed the animals sometimes are regarded as a class of Aschelminthes. Most of the acanthocephalans are quite small, though some attain lengths of about 1 ft. The worms are identified by a short anterior *proboscis*, armed with numerous recurved spines or hooks (Fig. 21.27). This organ of attachment can be withdrawn into the anterior portion of the spiny or warty *trunk*. Like most of the aschelminths, the Acanthocephala exhibit a syncytial structure as adults and the nuclear number is constant for each species.

In many cases the external cuticle is superficially segmented. In the epidermis, a unique *lacunar system* consisting of fluid-filled channels is formed by unlined spaces in the epidermal cytoplasm. This system does not communicate with any other parts or cavities of the body and is believed to function in food transport; the worms are fluid feeders without alimentary systems, and food absorbed through the body surface is thought to be distributed by the lacunar system. At the juncture of proboscis and trunk, the epidermis extends to the pseudocoel as *lemnisci,* two elon-

gated bodies thought to serve as reservoirs for the lacunar fluid when the proboscis is retracted.

Under the epidermis lie two layers of syncytial muscle fibers. The nervous system consists of an anterior brain ganglion and two longitudinal lateral cords. Sensory receptors are greatly reduced. Most acanthocephalans lack excretory systems. Where such a system is present it is a cluster of flame bulbs without nuclei. They empty into a nucleus-containing excretory duct that opens into the reproductive duct.

The sexes are separate and fertilization is internal. A pair of reproductive organs lies in *ligament sacs,* longitudinal membranous (but noncellular) chambers in the pseudocoel. Zygotes are retained in these sacs until larvae have formed. Development is mosaic. The cyst wall around a zygote prevents formation of a typical spiral cleavage pattern. The blastomeres soon fuse and the embryo becomes syncytial very early, without the usual gastrulation process. The larva is solid at first but later acquires a pseudocoel around the developing ligament sacs. Released larvae must enter the body cavity of intermediate insect hosts. If an infected insect is then eaten by a final vertebrate host, the larvae attach to the intestinal lining of the vertebrate and mature into adults.

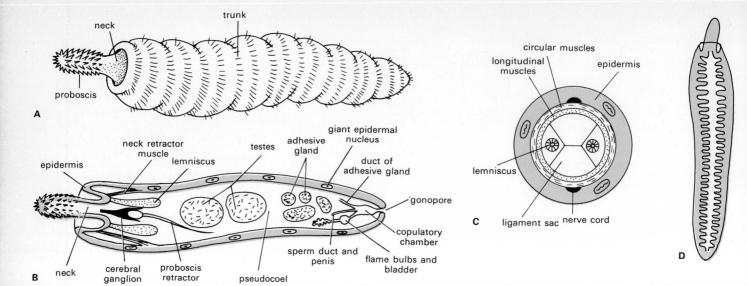

Figure 21.27 Acanthocephala. A, external appearance of acanthocephalan, proboscis extended. *B,* longitudinal section through *A,* proboscis retracted, body-wall muscles and female reproductive system not shown. Note the giant nuclei of the epidermis. *C,* cross section through lemniscal region, showing ligament sacs. More posteriorly these sacs envelop the gonads. In some acanthocephalans the nerve cords are lateral and the ligament sacs are reduced and incomplete. *D,* the lacunar system in the epidermis, dorsal view. Transverse branch channels lead from a longitudinal main channel. (*B, adapted from Yamaguti; C, D, after Meyer.*)

Entoprocts *Phylum Entoprocta*

adults stalked and sessile, solitary and colonial; with mouth and anus inside a circlet of ciliated tentacles; development via unique larvae.

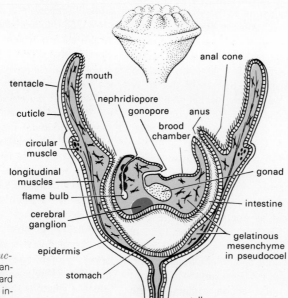

Figure 21.28 Entoproct structure. In this sagittal section, anterior is toward left, dorsal toward bottom. Inset diagram shows individual with tentacles tucked in.

The sixty species of entoprocts are the only pseudocoelomates that are not wormlike. All but the freshwater genus *Urnatella* are marine. Types like *Loxosoma* are solitary, and others, such as *Pedicellina,* form encrusting colonies on rocks, shells, and algae in shallow water. The individual animals largely are microscopic; none exceeds $\frac{1}{4}$ in. in size.

An entoproct is attached to a solid surface by a stalk, a continuation of the body wall (Fig. 21.28). The remainder of the body is organized around a U-shaped alimentary system, mouth and anus opening away from the attachment surface. This oral (ventral) side is ringed by ciliated tentacles. Entoprocts are *filter feeders;* microscopic food organisms are strained out by the tentacles from water currents created by the cilia.

A cuticle covers the body and the stalk but not the tentacles. The cellular epidermis is underlain by muscles that continue into the tentacles and allow these to be curled in. A muscle ring circling

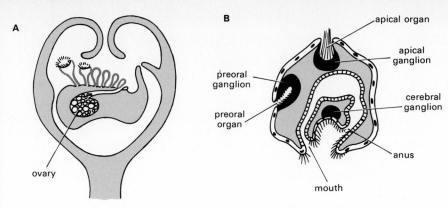

Figure 21.29 Entoproct development. A, section through adult (with incurled tentacles), showing series of embryos at different stages (color) in brood pouch. B, section through a free larva.

around the bases of the tentacles also enables the animal to contract its ventral side with the tentacles tucked in. Tentacles are hollow and contain extensions of a gelatinous, mesenchymal pseudocoel. In the curve of the alimentary U lies the main nervous ganglion, and near it is a pair of sex organs. Nerves from the main ganglion lead to subsidiary ones near the tentacle crown. The excretory system consists of a single pair of flame

Figure 21.30 Entoproct development. A, metamorphosis. The larva settles at the oral side, the internal organs rotate through 180°, and growth of stalk and tentacles then establish the adult organization. B, budding from horizontal stolon. The ectodermal alimentary invagination later will differentiate as an alimentary system, which thus arises from parental ectoderm. Parental mesenchyme gives rise to the mesodermal components of the budded individual.

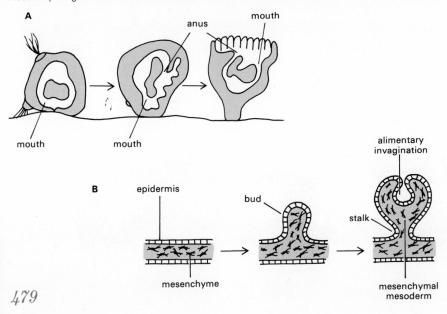

bulbs and ducts that open between mouth and anus.

In most entoprocts the sexes are separate and fertilization is internal. Eggs develop in a "brood chamber," a depression between gonopore and anus (Fig. 21.29). Early embryos just entering the chamber push aside the older embryos already there; hence a brood chamber usually contains a series of embryos at all stages of development. Cleavage occurs in a modified spiral pattern similar in some respects to that in rotifers. The mosaic blastomeres form a coeloblastula that gastrulates by emboly.

The free-swimming conical larva often is said to be "trochophorelike" but is actually very unlike a trochophore and is unique to entoprocts. It has an apical tuft, a girdle of cilia around the blastopore-mouth region, and a sensory *preoral organ* near the mouth. Later larvae develop a complete U-shaped alimentary tract. A remarkable metamorphosis eventually follows (Fig. 21.30). The larva settles with its mouth-anus side directed downward, and then its whole U-shaped alimentary system undergoes an internal rotation of 180°, which brings the mouth and anus away from the attached side to the adult position. A crown of tentacles develops later.

The adults (but not the mosaic embryos) regenerate readily. Entoprocts also reproduce abundantly by budding and form colonies in this manner. Individuals arise from horizontal stolons or from upright stalks. A bud starts as a localized epidermal outgrowth that contains some mesenchyme. The epidermal layer then invaginates and nips off an internal vesicle that will become the alimentary tract. Thus, although the alimentary system usually is endodermal, it arises here from ectoderm. This development of a body part from a "wrong" germ layer is common in budding and is encountered also in other phyla. It indicates clearly that although germ layers are homologous they are not always analogous (see Chap. 14); animals can develop their parts from any parental structures that happen to be available, regardless of the original germ-layer source of these structures.

The evolutionary affinities of entoprocts are obscure. The phylum probably represents an independent branch of pseudocoelomate animals, presumably derived from very distant acoelomate stocks that also produced the coelomates (Fig. 21.31).

Phylum Priapulida

marine; body up to about 6 in. long, with invertible prosoma and superficially segmented trunk; with solenocytic protonephridia.

The taxonomic affinities of this small phylum (five or six species) are still obscure. Until recently the animals were believed to have a pseudocoel lined by a noncellular membrane, as in many of the sac worms and in acanthocephalans. On this basis and in view of a number of anatomic correspondences with kinorhynchs, the priapulids generally were classified as a class of Aschelminthes. New studies indicate, however, that the lining membrane of the priapulid body cavity is cellular and therefore probably a peritoneum. The body cavity then would be a true coelom—if embryologic studies confirm that a coelom is actually formed. If so, the animals must be placed among the Coelomata as a separate phylum. At present, unfortunately, priapulid embryology is not yet fully known and the status of the group thus remains unsettled. For the moment it is probably best to regard the group as a separate phylum in any case, and to classify it with the coelomate animals whenever embryologic confirmation becomes available.

Priapulids live in the mud bottoms of shallow shore regions. The anterior part of the body is a bulbous *prosoma* ridged with wartlike sensory and glandular papillae (Fig. 21.32). It can be retracted into the cylindrical *trunk*. The body wall consists of cuticle, cellular epidermis, nervous system, muscle layers, and the presumably peritoneal inner lining. This lining is continuous with the covering membrane of the alimentary tract. Circular and radial muscles operate the pharynx, which is lined with a toothed interior continuation of the cuticle that covers the outside. In the body cavity lie extensive clusters of solenocytic nephridia (and the presence of such organs reinforces the probability that the animals are coelomate). On each side of the body the nephridia open together with a gonoduct at a pore near the anus. Sexes are separate and the larvae are similar to the adults.

Figure 21.31 Noncoelomate Bilateria. The likely evolutionary interrelations of acoelomate and pseudocoelomate animals are shown. Read from bottom up.

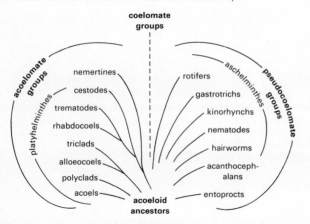

Figure 21.32 Priapulids. A, external view. *B,* sectional view. The retractors in the proboscis (shown only in part) attach to the inner surface of the trunk body wall. (*Adapted from Theel.*)

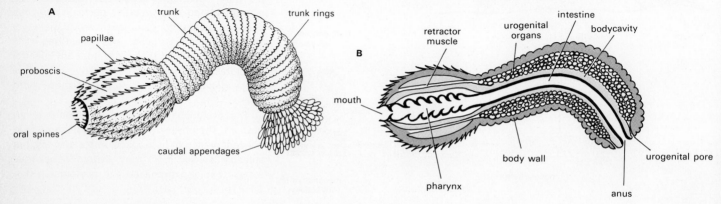

Review Questions

1. Give taxonomic definitions of Bilateria, Acoelomata, Platyhelminthes, and Nemertina. Similarly define the classes of flatworms and characterize each turbellarian order.

2. Describe the structure of the Acoela and contrast it with that of a planula. How do Acoela develop?

3. Describe the tissue and system organization of turbellarians. How is the reproductive system of planarians organized and how does fertilization take place?

4. Describe the development of turbellarians with ectolecithal eggs and of polyclads. What larval types occur in polyclads? What are the probable evolutionary interrelations of the turbellarian orders and the other flatworm classes? What is the basis for such views?

5. In what structural respects do flukes and turbellarians differ? What are Digenea? What are the life cycles of *Schistosoma* and *Clonorchis?* Describe the larval stages of these worms.

6. Distinguish the subclasses of tapeworms and describe the structure of *Taenia*. What is the organization of the reproductive system in a proglottid? In which different ways can fertilization occur? Review the life cycle of tapeworms.

7. Describe the structure and function of the nemertine proboscis apparatus. Review the sequence and structure of the body tissues from the epidermis inward.

8. By what means is food moved through the nemertine gut and digested? What is the organization of the nemertine nervous, circulatory, and excretory systems?

9. Describe the development of nemertines. Describe the structure of a pilidium and the events of metamorphosis. Are acoelomate animals derived from mosaic or regulative eggs, and are they protostomial or deuterostomial?

10. What are the environments of flatworms and nemertines? Be specific and name the actual groups or genera. How abundant are both phyla in numbers of species? Can these animals regenerate and do they reproduce vegetatively?

11. Give taxonomic definitions of pseudocoelomates, aschelminths, acanthocephalans, and entoprocts. Name the classes of aschelminths and indicate the characteristics that justify inclusion of these animals in a single phylum. On what grounds could Acanthocephala be regarded as an aschelminth class?

12. Describe the structure of a rotifer. What is cell constancy and which animal groups exhibit it? Distinguish between amictic and mictic eggs and show by what processes each kind is formed. Describe the life cycle of bisexual rotifers.

13. What is the structure of a gastrotrich? In what respects is the pharynx similar to that of nematodes? What is the structure of a kinorhynch? Is the body segmentation superficial or metameric? Describe the excretory system of all aschelminth classes. Which aschelminth groups are hermaphroditic?

14. In what environments can members of the aschelminth classes be found? Describe the organization of the nervous system in all aschelminth classes.

15. What is the structure of a nematode? In what external traits are males and females different? What are longitudinal chords? Describe the development of a nematode. What is chromosome diminution and what is its significance? Name nematodes found in man.

16. How many species of each aschelminth class have been described? In what structural respects does a hairworm differ from a nematode? Describe the life cycle of a hairworm.

17. Describe the structure of a spiny-headed worm. What are lemnisci, the lacunar system, ligament sacs? How does mating take place in all worm-shaped pseudocoelomates?

18. What is the structure of an entoproct. How does such an animal feed? Mate? Where does development take place? Describe the organization of the larva and the process of metamorphosis. Outline the structure of a priapulid.

19. Which pseudocoelomate groups exhibit vegetative reproduction and in what forms? Make a diagram of possible evolutionary interrelations among pseudocoelomate and acoelomate groups. On what morphological evidence can you base such a diagram?

Collateral Readings

Berrill, N. J.: "Growth, Development, and Pattern," Freeman, San Francisco, 1961. Chapter 7 of this book examines budding processes.

Best, J. B.: Protopsychology, *Sci. American,* Feb., 1963. A review of the interesting research on learning and memory in whole and regenerating planarians.

Bronstedt, H. V.: Planarian Regeneration, *Biol. Rev.,* vol. 30, 1955. A thorough review of the ample research on this subject.

Cameron, T. W. M.: "Parasites and Parasitism," Wiley, New York, 1956. Among many other parasitic types discussed in this book, flukes, tapeworms, and roundworms receive prominent attention.

Chitwood, B. G.: Nematoda, in "McGraw-Hill Encyclopedia of Science and Technology," rev. ed., vol. 9, 1966. A concise, informative article by one of the foremost students of roundworms.

Hyman, L.: "The Invertebrates," vol. 2., chaps. 10 and 11; vol. 3, chaps. 12–14, McGraw-Hill, New York, 1951. The indicated chapters give detailed accounts on the phyla dealt with in this chapter.

Jenkins, M. M., and H. P. Brown: Sexual Activities and Behavior in the Planarian *Dugesia, Am. Zoologist,* vol. 2, 1963. The contents are described adequately by the title.

Lansing, A.: Experiments in Aging, *Sci. American,* Apr., 1953. A review of experiments showing that the longevity of a rotifer is determined by how old its parent was when it produced the egg that gave rise to the animal.

Moore, D. V.: Acanthocephala, in "McGraw-Hill Encyclopedia of Science and Technology," rev. ed., vol. 1, 1966. A good, short description of the biology of these worms.

Shapeero, W. L.: Phylogeny of Priapulida, *Science,* vol. 133, p. 879, 1961. Evidence for the apparently coelomate nature of the animals is presented.

Grade **BILATERIA**

Subgrade **COELOMATA** with true coelom, formed in various ways as a body cavity lined entirely by mesoderm.

LOPHOPHORATES coelom formed in various unique ways; development protostomial; adults with food-catching lophophore; *Phoronida, Ectoprocta, Brachiopoda*

SCHIZOCOELOMATES coelom formed by splitting of mesoderm; development protostomial; with trochophore larvae. *Mollusca, Sipunculida, Annelida, Echiuroida, Oncopoda, Arthropoda*

ENTEROCOELOMATES all remaining phyla.]

As noted in Chap. 14, unspecifiable acoelomate ancestors are generally assumed to have given rise to basic stocks of coelomate animals. These in turn appear to have produced the three major groups listed above. Lophophorates share some traits with the other two groups, but on the whole they have more in common with the schizocoelomates. For example, both lophophorates and schizocoelomates exhibit mainly spiral cleavage patterns, their development is mosaic, and the embryonic blastopore becomes the adult mouth in protostomial fashion. Yet lophophorates do not appear to be allied obviously or closely to the other coelomates. They probably represent remnants of some of the earliest, most primitive coelomate animals, and presumably they have evolved independently from the schizocoelomate and enterocoelomate lines of descent.

PART 8
COELOMATES: PROTOSTOMES

22

LOPHOPHORATES

LOPHOPHORATES adults sessile or sedentary, filter feeding with ciliated *lophophore;* alimentary tract U-shaped, anus outside lophophore; coelom formed variously, divided into *mesocoel* in lophophore and *metacoel* in trunk; without breathing systems; development mosaic and protostomial, cleavage spiral to radial, larvae unique (not trochophores). *Phoronida, Ectoprocta, Brachiopoda*

The animals in these phyla are *filter feeders;* they strain microscopic food organisms from their aquatic environment by means of ciliated tentacles. These form a *lophophore* in which the mouth is located. The lophophore region represents the forepart of the body, the trunk, the hind part. Each of these two body parts encloses a portion of the coelom, the *mesocoel* anteriorly, the *metacoel* posteriorly. A transverse *peritoneal septum* separates these two cavities partially or completely. All lophophorates form exoskeletons, of different kinds in the three phyla. The larvae of the animals frequently are described as ''trochophorelike'' or ''modified trochophores,'' but they differ from trochophores and actually are unique.

Phoronids   *Phylum Phoronida*

marine, wormlike, tube-dwelling; with horseshoe-shaped lophophore; adult excretion through metanephridia; with closed circulatory system; development mosaic, cleavage spiral to irregular; actinotroch larvae with solenocytic protonephridia.

The only two genera of this phylum (*Phoronis, Phoronopsis*) include 16 described species. The $\frac{1}{2}$- to 6-in.-long animals live in shallow water along sandy or muddy shores, in solitary upright tubes or in tangled masses of many tubes. Representing the exoskeleton of phoronids, the tubes are chitinous parchmentlike secretions in which the animals can move freely (Fig. 22.1).

From the upper end of its tube a phoronid can project the lophophore, a double row of ciliated tentacles set on a double ridge of the body wall. The mouth is a slitlike, crescent-shaped funnel between the two ridges, centered in the middle between the left and right *arms* of the lophophore. A flap of tissue, the *epistome,* overhangs and can close off the mouth. This midregion also is the place where new tentacles are developed; the oldest tentacles are at the outer extremities of the lophophore arms.

The body wall of a phoronid contains a cuticle-

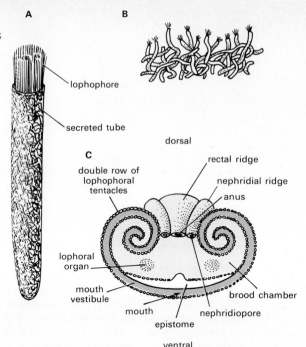

Figure 22.1 *Phoronids.* A, general appearance of a phoronid in straight tube. B, tangled group of phoronids. C, the anterior, lophophoral end of a phoronid. (B, adapted from Shipley; C, adapted from Benham.)

covered glandular epidermis underlain by muscle layers. In the mid-trunk region the longitudinal muscles usually form radial, ribbonlike bundles that project into the coelom (Fig. 22.2). The coelom lining is a syncytial peritoneum that also

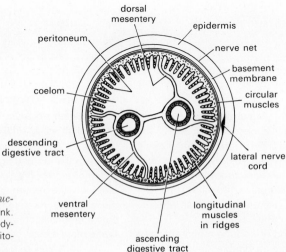

Figure 22.2 *Phoronid structure:* cross section through trunk. Note ridges of longitudinal body-wall muscles, overlain by peritoneum.

traverses the coelom as mesenteries and holds the U-shaped alimentary tract in place. The mesenteries usually divide the trunk coelom into four longitudinal compartments, and the trunk coelom itself is divided off from the lophophore coelom by a transverse *peritoneal septum.* The lophophore coelom extends into all the tentacles and is lined by a peritoneal membrane of its own.

The nervous system is an integral part of the epidermis; neural elements are *intraepidermal,* a primitive trait reminiscent of flatworms. An epidermal nerve ring lies along the outer edge of the lophophore, and a median ganglionic enlargement in this ring represents the main neural center (Fig. 22.3). Branch nerves from the ring pass to the tentacles, and a single longitudinal cord runs through the trunk epidermis along the left side of the body. In some cases a smaller right lateral cord is present, too. The remainder of the body is innervated by branch nerves from the left cord.

The alimentary tract consists of esophagus, stomach, and intestine, and the inner muscosal lining is ciliated throughout. Muscle and connective tissue layers form the wall of the tract between mucosa and peritoneum. The anus opens anteriorly between the lophophore arms. Food strained out by the lophophore is planktonic, diatoms mainly, and digestion is largely intracellular, another primitive trait. The circulatory system consists of two longitudinal trunk vessels interconnected posteriorly and opening anteriorly into a pair of ring vessels in the lophophore base. From these ring vessels projects a loop to each tentacle. An extensive blood sinus in the stomach wall is part of the trunk circulation. Blood contains red cells with hemoglobin. Such cells occur also in the coelomic fluid, which carries amoeboid white cells as well. A pair of metanephridial tubules is located anteriorly in the trunk and is held in place by the mesenteries. These tubules filter the coelomic fluid at ciliated, funnel-like nephrostomes, and they open to the outside at nephridiopores next to the anus (see Fig. 22.1).

Most phoronids are hermaphrodites. Gonads are loose masses along the peritoneum and gametes are shed to the coelom. From there they reach the outside through the metanephridia, which provide the only exits. In some phoronids fertilization takes place right inside the coelom; in other cases the zygotes form externally. Development occurs in the open sea or, more usually, in a *brood space* between the lophophore arms. Each

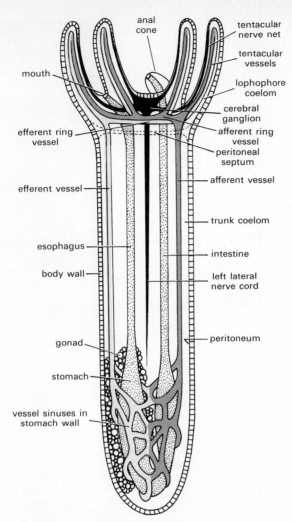

Figure 22.3 Phoronid structure. The alimentary, nervous, circulatory, and reproductive systems. Peritoneal membranes enveloping alimentary tract and blood vessels not shown.

arm embraces a *lophophoral organ,* a glandular depression that produces adhesive secretions believed to assist in retaining embryos in the brood space (see Fig. 22.1).

Cleavage patterns are variously spiral, radial, or irregular. A coeloblastula forms and gastrulation occurs by embolic invagination. The mesoderm arises predominantly by ingrowth of cells from the endoderm. After the embryonic alimentary system has become established, a ring of long cilia, the *telotroch,* develops around the anal region, and another ciliary band, the *postoral girdle,* forms in an oblique plane between mouth and anus (Fig. 22.4). From this girdle grows a wreath of larval tentacles. The ectoderm also gives rise to a *preoral lobe,* a prominent hood over the mouth. The coelom arises near the anus from mesenchyme cells, which migrate and arrange themselves as a hollow dorsal sac. This sac later grows around the intestine and forms mesenteries and an anterior peritoneal septum. At this stage the embryo usually escapes from the brood chamber as an *actinotroch* larva.

This larva is about $\frac{1}{4}$ in. long and leads a free-swimming existence for several weeks. It develops nervous, muscular, and circulatory systems as well as solenocytic protonephridia. These form as a branching ingrowth from a *nephridial pit* near the anus. Also, a progressively deepening invagination of the body wall develops in the midregion of the larva, between mouth and anus. This *metasome pouch* plays a key role during metamorphosis, the main events of which are completed within a few minutes. The larva undergoes spasmic contractions that result in sudden eversion of the metasome pouch. Body-wall muscles and a loop of the alimentary tract are carried along into the everted part, which now represents the adult trunk (see Fig. 22.4). Most other larval components then degenerate. The adult lophophore grows in the region of the larval tentacles and later acquires its own mesocoel from mesenchyme cells. Solenocytes degenerate, but the protonephridial tubules of the larva become part of the adult metanephridia.

Phoronids regenerate well as adults. Cut sections of the trunk can develop into whole individuals and damaged or lost tentacle crowns can be regenerated readily. Like flatworms, moreover, phoronids reproduce vegetatively by spontaneous fragmentation, a normal process in species that live as intertwined groups of individuals.

Ectoprocts   *Phylum Ectoprocta*

moss animals (also *Bryozoa* or *Polyzoa*); always sessile in colonies, formed by budding from single microscopic zooids; without circulatory or excretory systems; development mosaic, cleavage largely radial; with *cyphonautes* or other larvae.

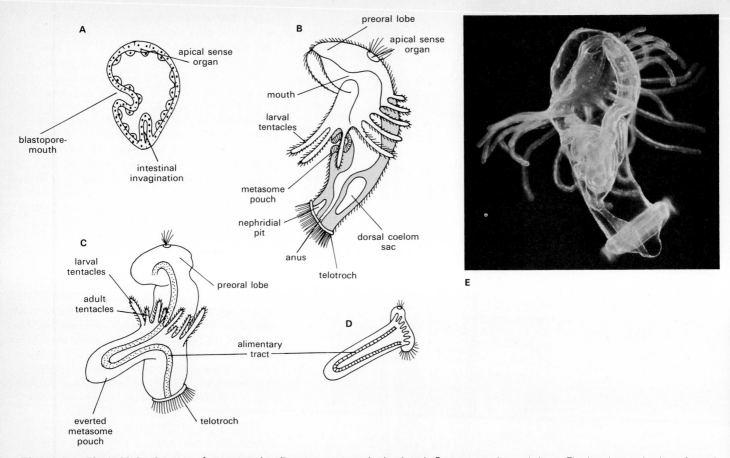

Figure 22.4 Phoronid development. A, postgastrula, alimentary tract nearly developed. *B,* mature actinotroch larva. The larval tentacles have formed. Note invagination of metasome pouch, and coelom sac. The nephridial-pit tissue will extend interiorly and bifurcate into two groups of solenocytes. *C,* metamorphosis. The metasome pouch has everted, carrying a loop of the alimentary tract with it. Adult lophophoral tentacles already have begun to form along the band of degenerating larval tentacles. *D,* establishment of adult by continued elongation of everted metasome pouch and degeneration of other larval parts. *E,* actinotroch larva, from life. Note larval tentacles, digestive tract, metasome pouch at mid-body, and posterior ciliary girdle.

Class Gymnolaemata   marine; lophophore circular, without epistome; body-wall musculature absent; without coelomic connection between zooids; exoskeleton calcified or membranous; often polymorphic.

Class Phylactolaemata   fresh water; lophophore horseshoe-shaped, with epistome; body-wall musculature present; with coelomic connection between zooids; exoskeleton gelatinous; not polymorphic.

Ectoproct fossils date back to Cambrian and possibly Precambrian times. Some 15,000 extinct species have been recognized, and living species are variously stated to number between 5,000 and 10,000. The animals have a remarkably close though only superficial resemblance to the entoprocts.

The individual ectoproct is a microscopic zooid encased in a secreted exoskeleton that is part of its body wall. In the marine forms this exoskeleton is essentially a box open at the unattached side, where the lophophore is protruded. In simply constructed types (*Vesicularia,* for example) the exoskeleton is a chitinous membrane. In most other genera calcareous deposits lie between the chitin membrane and the epidermis (Fig. 22.5), Directly

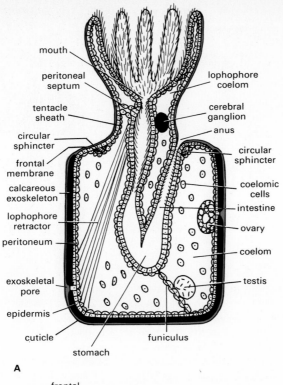

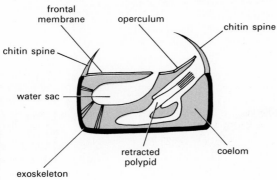

Figure 22.5 Marine ectoprocts. A, section through individual with extended lophophore. *B,* sectional side view of an individual showing operculum, protecting spines, and water sac. (*A, adapted from Marcus.*)

lophophore is a single row of tentacles set on a *tentacle sheath,* which inverts when the lophophore is withdrawn by retractor muscles and then forms a tentacle-enclosing tube. The lophophore is protruded when the coelomic fluid of the animal is put under pressure. In some cases such pressure is generated by the frontal membrane, which can be depressed by muscles. In other cases a sac with a separate opening to the outside can be pumped full with sea water; the sac then squeezes the rest of the interior and forces the lophophore out.

The alimentary tract consists of mouth, ciliated pharynx, esophagus, stomach, intestine, and anus outside the lophophore on the tentacle sheath. The esophagus is lined with epitheliomuscular cells, of a type similar to those of Hydrozoa. The whole alimentary tract is covered externally by the peritoneum. As in phoronids, the coelom is subdivided into mesocoel and metacoel. Neither cavity opens to the outside; reproductive ducts or nephridia are absent. The nervous system consists of a main ganglion between mouth and anus, a nerve ring around the lophophore base, and a nerve net in the body wall. The coelomic fluid contains free amoeboid cells and connective tissue fibers.

Gonads are attached to the peritoneum. Most ectoprocts are hermaphroditic. Fertilization occurs in the coelom and self-fertilization may be the rule. Reproductive cells or early embryos probably get out of the coelom through temporary openings formed in the lophophore region. Some species exhibit *polyembryony,* equivalent to identical twinning; a zygote divides several times and each resulting cell gives rise to a complete, separate individual. Cleavage is radial and a coeloblastula gastrulates by delamination at the vegetal pole. Mesoderm arises from endoderm.

Most marine ectoprocts develop free-swimming *cyphonautes* larvae (Fig. 22.6). These conical ciliated forms have a U-shaped alimentary tract, a basal ciliary girdle ringing the mouth and anus, an ectodermal *adhesive sac* between mouth and anus, and a glandular *pyriform organ* near the mouth. A cyphonautes lives for about 2 months, during which time it also secretes two shell plates, or *valves,* on its surface.

A very drastic reorganization takes place at metamorphosis. The adhesive sac everts and attaches the larva to a solid surface. The whole larva then disintegrates and becomes little more than a mound of loose cells covered by larval ectoderm. This layer invaginates and forms a deepening vesi-

under the epidermis lies the peritoneum; marine forms lack a body-wall musculature. The fixed body wall is customarily called the *cystid* and the movable interior parts, the *polypid.*

The unattached side of the exoskeleton often is covered partially by a *frontal membrane,* a chitinous shield continuous with the lateral body wall. The membrane leaves an opening just large enough for protrusion of the lophophore. In many cases this opening is protected by an *operculum,* a movable cover, and often also by spines that arch over the operculum (see Fig. 22.5). The

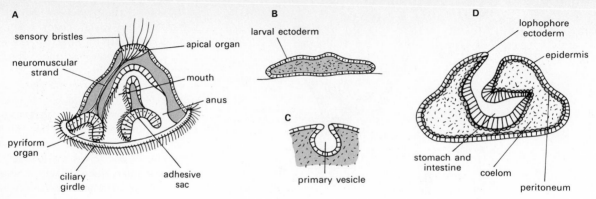

Figure 22.6 Metamorphosis in marine ectoprocts. A, sectional view of cyphonautes larva. *B,* attached larva with disintegrating interior tissues. *C,* formation of primary vesicle from larval ectoderm. *D,* growth of the primary vesicle leads to establishment of basic adult organization. Mesodermal mesenchyme cells (color) are arranged as a peritoneum.

cle, the tissue source of the later lophophore and alimentary system. Around this vesicle develops a mesenchymal sheet, the source of the adult mesoderm and coelomic lining. Evidently, the structure of the larva is totally unrelated to the structure of the later adult. Moreover, *all* adult body parts arise from larval ectoderm, alimentary system included— another instance of the diverse functional potentials of the embryonic germ layers.

The resulting ectoproct is an *ancestrula,* the parent of an adult colony formed by budding. Buds arise when a parent zooid becomes partitioned by an epidermal ingrowth from the body wall (Fig. 22.7). In the budded compartment all structures then develop directly from the epidermis, as in the original formation of the ancestrula from larval ectoderm. Adjacent zooids in a colony have exoskeletal pores in places, covered by the epidermis of each zooid; coeloms therefore do not interconnect.

Many marine ectoprocts are polymorphic. In addition to the feeding *autozooids* various *heterozooids* are formed, modified autozooids in which the cystid portions are overdeveloped and the polypid portions remain underdeveloped. A common type of heterozooid is an *avicularium,* in which the enlarged operculum has the form of a bird's beak. The polypid is absent or reduced to a tiny *setiferous organ.* Avicularia snap their beaks and prevent other organisms from settling on an ectoproct colony (Fig. 22.8). Another type of heterozooid is a *vibraculum,* in which the operculum has become a long bristle that sweeps back and forth over the colony surface. Stolons, stalks, and rootlike anchors in a colony represent heterozooids called *kenozooids.* Some species develop individuals specialized as brood carriers.

In most marine ectoprocts the polypids regularly degenerate into compact *brown bodies,* a process believed to be a substitute for excretion (see Fig. 22.7). The remaining cystids regenerate new polypids and the brown bodies then are digested in the newly developed stomachs. Colonies often are zoned according to these degeneration-regeneration cycles. Active young zooids usually

Figure 22.7 Ectoproct budding and regeneration. A, formation of bud by partitioning of parent cystid, cross-sectional view. *B,* sectional longitudinal view of cystid, with old polypid degenerated to brown body and new polypid beginning to regenerate.

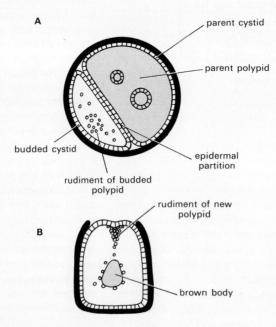

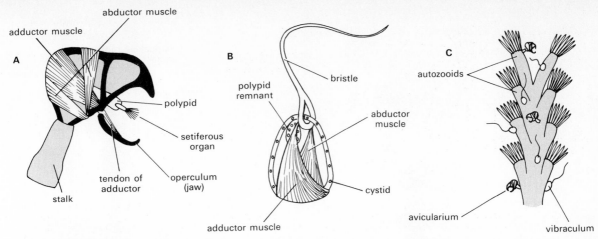

Figure 22.8 Ectoproct heterozooids. A, avicularium of *Bugula.* B, a vibraculum. C, portion of ectoproct colony showing position and comparative sizes of heterozooids. (*B. adapted from Marcus.*)

A   **B**

Figure 22.9 Marine ectoprocts. A, colonies of *Bugula,* suspended from rock overhang. B, colony of *Electra,* on red algae.

are present along the peripheral zone of a colony. Farther inward is a zone of degeneration and brown-body formation, and still farther inward lies a regeneration zone, where active individuals again are present. The oldest zooid, the ancestrula, never develops any gonads. Sexual activity occurs only in zooids farther out, after colonies have attained appreciable size and age (Fig. 22.9).

Freshwater ectoprocts number only about fifty species. Their exoskeletons form large gelatinous masses that often clog drain pipes and spillways. All zooids of a colony are interconnected and share a common, continuous body wall (Fig. 22.10). The lophophore is horseshoe-shaped, not circular as in marine types. The stomach is connected to the base of the body wall by the *funiculus,* a peritoneum-covered muscle-containing tissue cord.

The developmental pattern is adapted to the freshwater environment. The animals are hermaphroditic and all brood their eggs in sacs along the body wall. After cleavage the coeloblastula becomes two-layered, the inner layer representing the peritoneum (Fig. 22.11). The embryo now is the cystid of an ancestrula that produces several further cystid buds at one end. Such a larval colony escapes from its brood sac, settles, and continues to bud. Alimentary systems later develop in each cystid from invaginated ectoderm vesicles, as in the marine types.

Freshwater ectoprocts also reproduce vegetatively through *statoblasts,* formed in the funiculus of an adult. A statoblast consists of an internal cell mass and a protective shell of species-specific construction. Reminiscent of the gemmules of freshwater sponges, statoblasts usually are formed in fall, when the adult colonies disintegrate. Withstanding dry conditions and low winter temperatures, statoblasts germinate in the spring by forming two-layered vesicles as in budding.

These animals are generally considered to be more primitive than their marine relatives. This

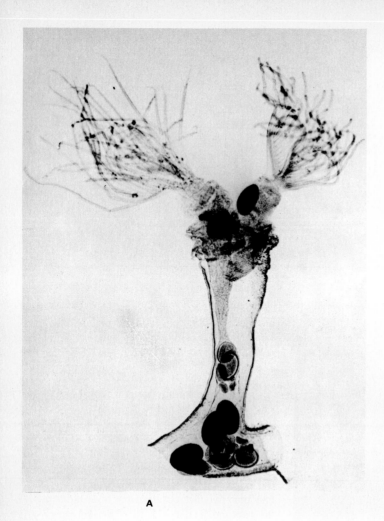

A

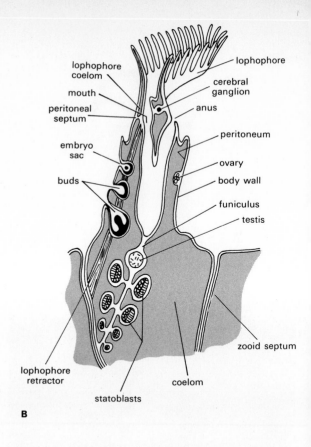

lophophore
coelom

mouth

peritoneal
septum

embryo
sac

buds

lophophore
retractor

statoblasts

lophophore

cerebral
ganglion

anus

peritoneum

ovary

body wall

funiculus

testis

zooid septum

coelom

B

Figure 22.10 Freshwater ectoprocts. A, portion of a colony of *Plumatella.* A statoblast-bearing funiculus is well visible. *B,* general structure. Note horseshoe-shaped lophophore and statoblasts in funiculus. (*Adapted from Brien.*)

Figure 22.11 Development of freshwater ectoprocts. A, embryo forms in sac that has invaginated from maternal epidermis, and additional cystid buds develop at one end of embryo, forming embryo colony. *B,* external top and side views of shells of statoblasts. Hooked type is of *Pectinatella;* caplike type, of *Hyalinella.*

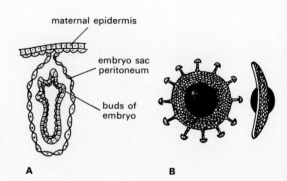

maternal epidermis

embryo sac
peritoneum

buds of
embryo

A

B

assumption is based on the phoronidlike shape of the lophophore, the presence of a body-wall musculature, and the absence of specialized features such as complex exoskeletons, opercula, and polymorphism.

Lamp Shells Phylum Brachiopoda

marine; with dorsal and ventral bivalve shells, often with stalk; shells lined by mantle lobes of body wall; with open circulatory system and metanephridia; development mosaic, cleavage radial; coelom formed as schizocoel or enterocoel; with free-swimming unique larvae.

Class Inarticulata (Ecardines) valves held by muscles only; lophophore without skeletal support;
anus present. *Lingula, Crania*

Class Articulata (Testicardines) valves interlocked; lophophore with skeletal support; anus absent.
Terebratulina

As fossil types the lamp shells are even older and more abundant than the ectoprocts; some thirty thousand extinct species are known. *Lingula* and *Crania* are the most ancient of all fossil genera, going back to at least Ordovician times (although the present *species* are not particularly ancient). Today only some three hundred species exist. Extinct lamp shells attained sizes of more than 1 ft, but living species generally do not exceed lengths of about 3 in.

By virtue of their two-shelled or *bivalve* exoskeletons, brachiopods resemble clams superficially (Fig. 22.12). But whereas the valves are lateral in clams, they are dorsal and ventral in brachiopods. The ventral valve (often borne uppermost in the living animal) usually is the larger one. Leading away from its inner surface is a stalk, or *peduncle,* that anchors many brachiopods to a solid surface. The peduncle passes through a perforation in the ventral valve (hence the name "lamp shell," from the resemblance of the skeleton to an ancient oil lamp). Most stalked brachiopods are attached permanently, but *Lingula* is not. Its peduncle merely sticks in the bottom of a vertical mud burrow along tidal flats, and by using the stalk as a locomotor organ the animal can change its location. *Crania* and some other brachiopods lack peduncles, their ventral valves being cemented directly to a solid surface.

In Articulata the valves are hinged posteriorly by teeth in one valve and corresponding pits in the other. A valve consists of a chitinous cuticle and a layer of calcium carbonate or calcium phosphate underneath. The inner lining is a double-layered flap of body wall, or *mantle* lobe (Fig.

22.13). In the body wall the tissues are epidermis, mesenchymal (and often cartilagelike) connective tissue, and peritoneum. Between the last two layers muscles are present laterally, where the body wall is not in contact with the valves. The peduncle is an extension of the ventral body wall and has the same tissue construction. The coelom extends into the peduncle in forms such as *Lingula,* but in others the stalk is filled secondarily with muscles. Attached to the inner surfaces of the valves are muscles that close and open the valves (see Fig. 22.12).

A brachiopod occupies only the posterior space between the valves, the anterior part, or *mantle cavity,* being filled by the large lophophore. The long tentacles are in a single row and can be protruded beyond the shell margin. In the Articulata the coiled lophophore arms are supported internally by a pair of skeletal prongs that grow out from the dorsal valve (see Fig. 22.12). The mouth, a transverse slit in the center of the lophophore base, leads to an internally ciliated alimentary system. This system is complete only in the Inarticulata, where it is a U-shaped tract with an anus opening to the mantle cavity. In the Articulata the intestine is a blind-ended tube. In both groups the stomach connects with a conspicuous digestive gland ("liver").

A nerve ring around the esophagus is the neural center, and nerves radiate away from it to the lophophore and the posterior body regions. The alimentary system is supported by a dorsal and a ventral vertical mesentery, and over the stomach in the dorsal mesentery lies a simply constructed *heart* (see Fig. 22.13). Leading away from it in

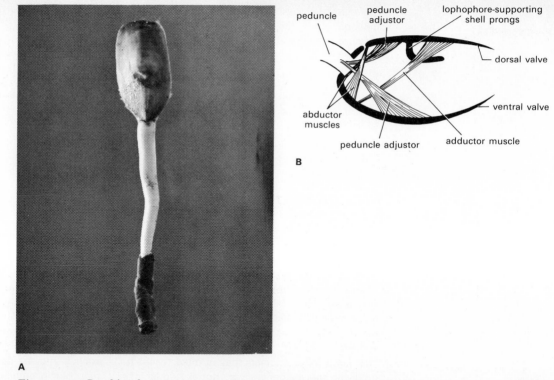

A

B

Figure 22.12 Brachiopods. A, Lingula. The stalk is normally buried vertically in sandy sea bottoms. The lophophore is under the shell (valve). *B,* sectional view showing the valves and some of the valve muscles. The penduncle is the stalk by which the animal is attached.

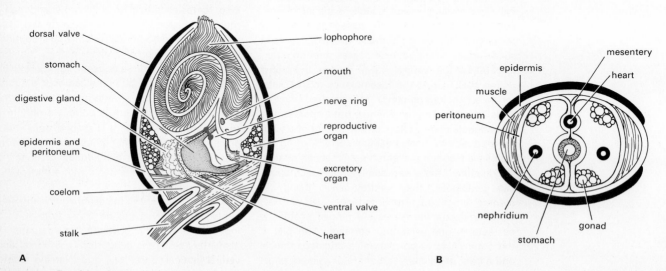

A

B

Figure 22.13 Brachiopod structure. A, longitudinal section. Note the nerve ring around the esophagus and the blind-ended intestine. *B,* cross section. Note that body-wall muscles are present only in the lateral parts of the body wall.

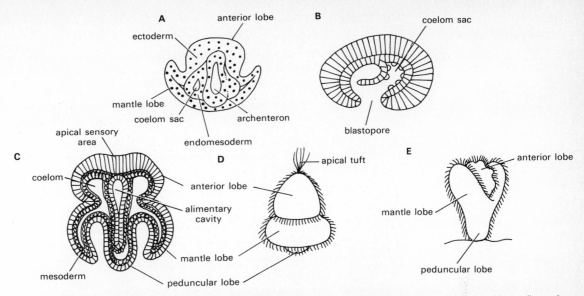

Figure 22.14 Brachiopod development. A, transverse section through *Lingula* larva, showing schizocoelic coelom formation by development of cavity in endomesodermal mass on each side of endoderm; the mantle fold has adult (forward) orientation. *B*, sagittal section through *Terebratulina* embryo, showing development of dorsal endodermal partition cutting off coelomic sac from posterior portion of archenteron. *C*, longitudinal section through early *Terebratulina* larva; coelom has expanded greatly and mantle fold has developed over posterior end. *D*, external view of mature larva. *E*, metamorphosis; after larva settles, the mantle fold turns forward over anterior lobe and the latter gives rise to lophophore. (*Adapted from Conklin, Yatsu, and other sources.*)

both directions are branched, open-ended vessels. *Crania* has a transverse peritoneal septum between mesocoel and metacoel, but in other genera the septum often is developed incompletely. Free cells occur in the coelomic fluid. From the trunk coelom funnel-like nephrostomes lead to metanephridia that open into the mantle cavity and also serve as gonoducts. Gonads (usually four) lie along the peritoneum or the mesenteries.

Sexes are separate and fertilization occurs in the coelom or when the gametes exit through the metanephridia. Zygotes generally are brooded in the mantle cavity. Radial cleavage produces a coeloblastula that gastrulates by emboly in the majority of cases. In *Lingula*, mesoderm arises as a solid cell mass on each side of the invaginated endoderm. These masses then split internally and the cavities so formed represent the coelom (Fig. 22.14). In *Terebratella* and *Terebratulina*, by contrast, a partition grows down from the roof of the archenteron and separates a single dorsal coelomic sac from the endoderm. Thus, the Inarticulata technically appear to have schizocoels, and the Articulata, enterocoels. However, the patterns of formation here differ from those in the schizo-

coelomate and enterocoelomate animals yet to be discussed. Even so, the very occurrence of both schizocoels and enterocoels tends to support the possibility that ancient brachiopods might have been an evolutionary link between, or precursor to, schizocoelomate and enterocoelomate animals proper.

The ciliated, globular larva of brachiopods has an equatorial ectodermal *mantle fold* that grows forward over the oral hemisphere of the larva. From this fold later develop the adult mantle lobes. Valves begin to be secreted even during the larval stages. A lophophore then becomes elaborated around the mouth, and the adult emerges through a very gradual metamorphosis.

Inarticulata probably are the primitive brachiopods—very ancient forms such as *Crania* and *Lingula* lack hinged valves but have complete, U-shaped alimentary tracts. Notwithstanding vague indications of possible affinities to schizocoelomates and enterocoelomates, lophophorates as a whole appear to be a distinct assemblage not obviously or closely allied to other existing coelomates. Indeed, each lophophorate phylum appears to be a distinct line of descent in its own right.

Review Questions

1. Give taxonomic definitions of coelomates, protostomial coelomates, and lophophorates. Which phyla are included in each of these groups? What is a lophophore, a mesocoel, a metacoel, and a peritoneal septum? How does a lophophorate animal feed?

2. In what environments does each of the lophophorate groups occur? How many species are known in each group? What is their possible evolutionary relation to one another and to other coelomates?

3. Describe the general structure of a phoronid. What is an epistome? Describe the organization of the body wall and the subdivisions of the coelom. What is the organization of the nervous, circulatory, and excretory systems?

4. Show how fertilization takes place in phoronids and describe the pattern of development. How does an actinotroch larva form and how does it metamorphose? What is the metasome pouch and what is its role in metamorphosis?

5. Name and define the classes of ectoprocts. Describe the structure of an ectoproct and distinguish between cystids and polypids. How does an ectoproct withdraw and extend its lophophore?

6. Describe the structure of the nervous and alimentary systems of ectoprocts. How do ectoprocts differ from entoprocts and why are these two groups classified as separate phyla? Through what developmental processes does a cyphonautes form? How does the coelom arise?

7. Describe the metamorphosis of ectoprocts and the processes by which adults arise. How do budding and brown-body formation take place? Describe the structure of an ectoproct colony and show how it grows.

8. What kinds of polymorphs occur among ectoprocts and what is their structure and function? How do marine and freshwater ectoprocts differ structurally and reproductively?

9. Define the classes of brachiopods. Describe the structure of a brachiopod and show how this structure differs in the different classes. How does brachiopod symmetry differ from that of a clam? What is the organization of the nervous, circulatory, alimentary, and excretory systems?

10. Describe the development and metamorphosis of brachiopods. What does the developmental pattern suggest with regard to the evolutionary position of brachiopods? Which lophophorate groups are hermaphroditic and which are separatetly sexed? Which groups brood their young and where?

Collateral Readings

Hyman, L: The Occurrence of Chitin in Lophophorates, *Biol. Bulletin,* vol. 114, 1958.

———: "The Invertebrates," vol. 5, McGraw-Hill, New York, 1959. Chapters 19 through 21 are detailed treatments of the lophophorate phyla.

Lynch, W.: Factors Influencing Metamorphosis of *Bugula* Larvae, *Biol. Bulletin,* vol. 103, 1952.

Marsden, J.: Regeneration in *Phoronis, J. Morphol.,* vol. 101, 1957.

Rattenbury, J.: The Embryology of *Phoronopsis viridis, J. Morphol.,* vol. 95, 1954.

The subject matter of these three papers is described adequately by the titles.

23

MOLLUSKS

Subgrade *COELOMATA*

SCHIZOCOELOMATES   development mosaic; cleavage spiral; mesoderm from 4d-derived teloblast cells; coelom formed schizocoelously, by splitting of embryonic mesoderm into outer and inner layers; protostomial; embryonic trochophore stages or free trochophore larvae. *Mollusca, Sipunculida, Annelida, Echiuroida, Oncopoda, Arthropoda*

Although adult schizocoelomates can be as different as clams, earthworms, and mosquitoes, the fundamental embryonic histories of all these animals are very similar and the trochophore larvae of the marine types are virtually identical (see Fig. 11.18). Because of this it is generally believed that a common ancestral stock could have given rise to all the schizocoelomate groups. This stock itself would have evolved from some of the original coelomates, and it eventually produced two broad lines of descent. In one the animals remained unsegmented like the ancestral types, but in the second line metameric segmentation became a more or less conspicuous trait. The unsegmented branch is represented today by the mollusks and sipunculids; the segmented branch, by the four other phyla listed above (see Fig. 14.17).

General Characteristics

Phylum Mollusca

soft-bodied animals; body bilateral, usually with *head,* ventral *foot,* and dorsal *visceral hump,* the latter covered by *mantle* and secreted *exoskeleton;* alimentary system with *radula* and *hepatopancreas;* breathing with gills; circulatory system usually open, with chambered heart; excretion through renal organs; sexes separate or hermaphroditic; development typically schizocoelomate; trochophores followed in many cases by veliger larvae. *Amphineura, Gastropoda, Scaphopoda, Pelecypoda, Cephalopoda*

With 100,000 described species, the mollusks represent the second largest animal phylum; only the arthropods include more known species (and nematodes and protozoa probably include more existing but so far undescribed species). Like large phyla generally, mollusks are adapted to virtually

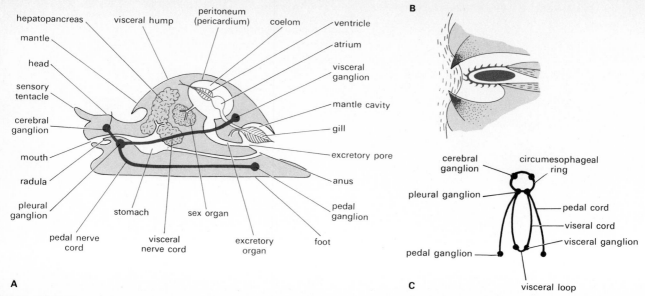

A

B

C

Figure 23.1 Mollusks: structural characteristics. *A,* presumed structure of hypothetical ancestral mollusk in sagittal section. Of paired organs, only those on one side are indicated. Arteries from the ventricle lead to the blood sinuses in all parts of the body, and blood from there returns to the heart via the two excretory organs, the two gills, and the two atria of the heart. *B,* the molluscan radula, as in a squid. The radula is a horny band with recurved teeth and is moved back and forth by muscles around a cartilaginous supporting prop (dark color). Unlike most mollusks, a squid also has horny jaws, as shown here. *C,* the basic molluscan nervous system (top view). Lateral branches and transverse connectives between the two visceral cords and the two pedal cords give the system a somewhat ladderlike appearance.

all available living conditions. The animals occur at all ocean depths, in all kinds of fresh waters, and on land to altitudes as high as the snow line. Most mollusks have lengths on the order of an inch or two, but the phylum also includes the squids, the largest and most highly elaborated of all invertebrates, some of which are 50 to 60 ft long.

Mollusks are known from Cambrian times on. The nautiloids were prominent inhabitants of Paleozoic seas, and ancient snails were among the first of all terrestrial animals. Today mollusks are even more abundant than ever before, and man also uses them as food more widely and in greater quantities than any other invertebrates. Comparatively few members of the phylum are symbiotic, but the animals are themselves hosts to a wide variety of commensals and parasites.

Although the phylum includes animals as different as snails, clams, and squids, the construction of all of them nevertheless can be derived from a single ancestral pattern. This original structural pattern can be hypothesized to have had the following characteristics (Fig. 23.1).

The body consisted of a *head,* which may have borne a pair of sensory tentacles; a broad, ventral

muscular *foot* that served in a creeping form of locomotion; and a dome-shaped dorsal *visceral hump,* which contained the main organ systems. The body wall of the dome was a *mantle* that secreted calcareous spicules in the epidermal layer and was extended as an overhanging rim around the sides of the body, particularly at the posterior end. The space under this posterior rim represented a *mantle cavity.* Projecting into it were paired feathery or leaflike gills, the *ctenidia.* The mouth led to a pharynx with a *radula,* a horny band studded with recurved teeth arranged in various patterns. Muscles moved this band back and forth over a cartilaginous supporting rod in the ventral wall of the pharynx. Protruded through the mouth, the radular apparatus served in rasping pieces of tissue from plant or animal food organisms. From the pharynx food passed through an esophagus to a stomach, which connected with a conspicuous *hepatopancreas* ("liver"). In this organ a substantial amount of digestion occurred intracellularly. Extracellular digestion took place in the intestine, which opened posteriorly to the mantle cavity.

The nervous system consisted of a nerve ring

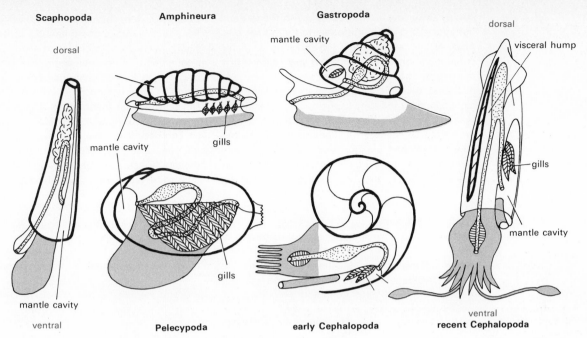

Scaphopoda   **Amphineura**   **Gastropoda**

dorsal

mantle cavity

dorsal

visceral hump

mantle cavity

gills

gills

mantle cavity

gills

mantle cavity

mantle cavity

ventral

ventral

Pelecypoda   **early Cephalopoda**   **recent Cephalopoda**

Figure 23.2 Molluscan body construction. Shells drawn in heavy lines; alimentary systems, stippled; foot in color. The squid is drawn in a position comparable to that of other groups: the tentacle-bearing head-foot end is ventral, the mantle cavity is posterior. The chambered nautilus is in swimming position, with the buoyant shell uppermost.

around the esophagus, thickened dorsally as a pair of *brain ganglia* and ventrally as a pair of *pleural ganglia.* From these ventral ganglia emanated two pairs of longitudinal nerve cords. The *pedal cords* traveled to the foot, where they terminated as a pair of *pedal ganglia,* and the *visceral cords* passed to the dorsal hump, where they formed a pair of *visceral ganglia.* A transverse connection between the visceral ganglia established a *visceral loop.* An open circulatory system contained a heart with one ventricle and two posterior atria, as well as systems of arterial and venous vessels from and to the gills. Additional vessels passed to all other body regions and opened to extensive blood sinuses around all organs.

The coelom was comparatively reduced. Its main component was a *pericardial cavity* around the heart. Leading into this coelomic space anteriorly were the ducts of the gonads, and passing posteriorly were a pair of excretory tubules. These were essentially metanephridial; they connected the pericardial coelom with the exterior, through nephridiopores in the mantle cavity near the anus. Gametes shed to the pericardial coelom left there through the excretory ducts. Development fol-

lowed the characteristic schizocoelomate pattern and included free-swimming trochophores.

This presumed ancestral organization still is preserved to varying degrees in the mollusks now living (Fig. 23.2). In amphineurans the head is reduced and the exoskeleton is more specialized, but in most other respects the ancestral structure is basically unchanged. In gastropods the most conspicuous new feature is a pronounced dorsal growth of the visceral hump, unequally on the left and right sides, resulting in a spiral coiling of the hump and the covering shell. Also, the coiled hump is rotated 180° in relation to the head and the foot, which brings the mantle cavity to an anterior position and shifts the internal organs accordingly. The scaphopods are similarly elongated in a dorsal direction, but coiling does not take place. Pelecypods have adapted to a sedentary filter-feeding existence, and here the body is flattened from side to side, the head has disappeared, and the gills are expanded as ciliary food-collecting organs.

Cephalopods exhibit the most pronounced departures from the ancestral pattern. Early cephalopods evolved a dorsally elongated visceral hump, but growth remained equal on the left and right sides and the hump came to form a flat coil. Also, as the animal grew the covering shell became partitioned into progressively larger compartments and only the last compartment was occupied. All earlier compartments were filled with air, which gave the early cephalopods considerable buoyancy and permitted them to adopt a free-swimming existence. In conjunction with this newly developed pelagic mode of life, the foot became modified partly to muscular prehensile tentacles, which also equipped the animals for predatory, carnivorous activity. The extinct nautiloids and the chambered nautilus today exemplify this early stage of cephalopod evolution. In other cephalopod lines the shells became reduced greatly and the nervous and sensory systems became highly developed in conjunction with the rapid swimming locomotion. The result was the emergence of modern squids and octopuses.

The basic ancestral body evidently made possible numerous and highly varied modifications; the present success of mollusks unquestionably is a consequence of this original adaptive potential.

Chitons   **Class *Amphineura***   marine; head reduced, without eyes or tentacles; nervous system without ganglia; hermaphroditic or separately sexed.

Order Aplacophora   solenogasters; wormlike; with calcareous spicules but shell absent; foot reduced; mostly hermaphroditic.

Order Polyplacophora   chitons; shell consisting of eight overlapping plates; gonoducts with separate openings.

Order Monoplacophora   with caplike shell; most internal organs in serial pairs; sexes separate.

The three amphineuran groups now are often also regarded as separate molluscan classes. The shell-less types, Aplacophora, appear to be primitive and thus perhaps the most primitive living mollusks. The animals have a rudimentary foot in the form of a muscular ridge that projects from a longitudinal groove on the ventral side (Fig. 23.3). The remainder of the body surface is covered by a mantle that contains calcareous spicules. A radula is absent in some cases, well developed in others, and gills too are or are not present in given species. The most certain indication of the primitiveness of these animals is the ancestral nature of the coelom: gonads open into the pericardial space, which communicates with the outside through metanephridial ducts. Other aplacophoran features largely are as in chitons.

Represented by the single living genus *Neopilina*, the Monoplacophora are covered dorsally by a single shell plate, formed by an accumulation of calcareous spicules secreted by the underlying mantle (see Fig. 23.3). Under the rim of the shell is a mantle cavity that circles the animal as in chitons. The reduced head bears a pair of sensory tufts and a pair of tentacles, and a pair of ciliated tissue flaps, the *labial palps*, direct microscopic food toward the mouth. Five pairs of gills project into the posterior part of the mantle cavity. Internal organs similarly occur as multiple pairs: two pairs of atria, five pairs of metanephridial tubules, two pairs of gonads, 10 pairs of transverse neural connections.

These body parts have a regularly spaced anteriorposterior arrangement, justifying the designation "segmented." The segmentation is not metameric, however, since the coelom is unsegmented. Besides, the embryonic development of the animals is still unknown. On present evidence the segmental adult construction appears to be even less distinctive than that of tapeworms, and comparable perhaps to that of certain free-living flatworms in which many internal organs likewise form serial pairs. *Neopilina* was discovered only in the 1950's; before then the Monoplacophora were believed to have been extinct since the Devonian.

Chitons with eight shell plates are known from Ordovician times on. Each shell plate consists of two calcareous layers. These plates do not cover the mantle entirely but leave a *mantle girdle* around the edge of the animal (Fig. 23.4). The mantle cavity is a narrow groove that circles the entire animal around the flat, oval foot. The anus opens posteriorly into the mantle cavity, and on either side of this opening is an excretory pore and

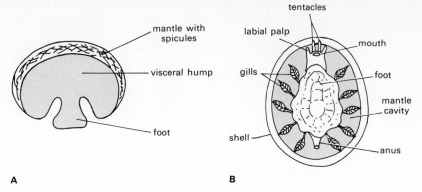

mantle with
spicules

visceral hump

foot

A

tentacles

labial palp

mouth

gills

foot

mantle
cavity

shell

anus

B

Figure 23.3 Amphineura. A, cross section through aplacophoran, to show foot and mantle. *B,* ventral view of monoplacophoran (*Neopilina*). Note one pair of tentacles and labial palps on head and five pairs of ctenidia in mantle cavity. (*Adapted from Moore.*)

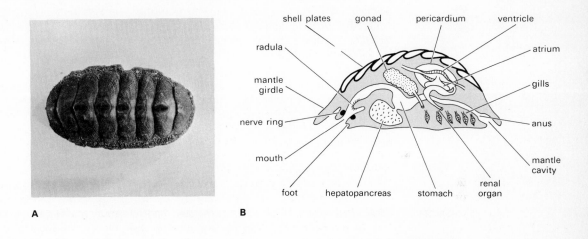

shell plates   gonad   pericardium   ventricle

radula

mantle
girdle

nerve ring

mouth

atrium

gills

anus

mantle
cavity

foot   hepatopancreas   stomach   renal
organ

A

B

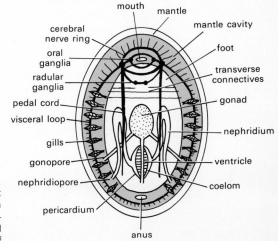

mouth   mantle

cerebral
nerve ring

oral
ganglia

radular
ganglia

pedal cord

visceral loop

gills

gonopore

nephridiopore

pericardium

anus

mantle cavity

foot

transverse
connectives

gonad

nephridium

ventricle

coelom

C

Figure 23.4 Chitons. A, dorsal view. Note the eight shell plates in the mantle. *B,* sagittal section. *C,* organ systems from ventral side. The transverse nerve connections between the pedal cords occur in mid-body and posterior regions as well, but to preserve the clarity of the other structures they are not drawn in.

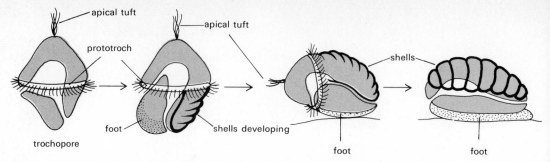

Figure 23.5 Chiton development. Left to right, the trochophore settles on the ventral side and develops shell plates (heavy black lines) and a foot (stippled area). *(Adapted from Heath.)*

a separate gonopore. Some 6 to 80 pairs of small gills project into the mantle cavity laterally. The mouth is anterior and a well-developed radula can be protruded through it. The head as a whole is reduced and not marked off in any way from the trunk. Also, tentacles and eyes are absent. In these respects chitons are specialized and divergent from the probable ancestral mollusks.

In conjunction with the reduction of the head the nervous system has become simplified. Definite ganglionic thickenings usually are absent, and a nerve ring around the esophagus merely extends posteriorly as two pairs of cords. The visceral cords typically join posteriorly and form a loop. The only other significant deviation from the ancestral molluscan pattern is the presence of separate gono-

ducts to the outside, a condition characteristic also of all other mollusks. A pair of tubular excretory ducts still emanates from the pericardial coelom, but they have become complex *renal organs* with distinct filtering and reabsorbing portions, again as in all other mollusks.

Chiton larvae are free-swimming trochophores with solenocytic protonephridia. They eventually develop a foot in the region between mouth and anus and shell plates at the opposite side. A larva then settles with the foot directed downward (Fig. 23.5). Chitons inhabit the rocky sea shores, where they either creep slowly or cling to rocks so tightly that they cannot be pried off without serious injury to the animals. Algal vegetation is the principal food.

Snails **Class Gastropoda**   marine, freshwater, and terrestrial; visceral hump typically coiled, with *torsion* or various degrees of *detorsion;* usually with shell; head with eyes and one or two pairs of tentacles; trochophore and veliger larvae or direct development.

Subclass Prosobranchia   mostly marine, some freshwater and terrestrial; 180° torsion, visceral loop in figure of eight; shell caplike or coiled; one pair of tentacles; foot with or without operculum; sexes usually separate.

Order Archaeogastropoda   aspidobranchs; limpets, abalones; shell caplike or moderately coiled; operculum absent; nervous system not concentrated, internal organs paired.

Order Mesogastropoda   pectinibranchs; periwinkles, cowries; shell well humped, usually coiled; operculum present; nervous system not concentrated; internal organs single.

Order Neogastropoda   whelks, conches; shell

coiled; operculum present; nervous system concentrated; internal organs single.

Subclass Opisthobranchia   sea hares, sea butterflies (pteropods), nudibranchs; mostly marine; 90 or 180° detorsion, visceral loop partly or wholly uncoiled; shells reduced or absent; two pairs of tentacles; foot usually without operculum; almost all hermaphroditic.

Subclass Pulmonata   terrestrial and freshwater; modified torsion; with or without shells; two pairs of tentacles; without operculum; gills absent, air breathing through pulmonary sac; all hermaphroditic; development direct, without larvae.

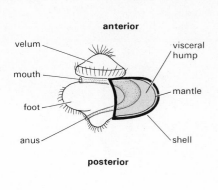

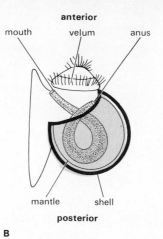

A   B   C

Figure 23.6 Snail development. A, mature veliger, before torsion. The dorsal visceral hump has enlarged greatly, producing a U loop in the alimentary tract. B, veliger after torsion. The alimentary tract now is coiled, with the anus shifted to same (anterior) side as the mouth. C, veliger larva of the marine snail *Rissoa*, seen from the ventral side. The ciliated velum is nearest the viewer; the visceral hump with shell faces away from the viewer.

Order Basommatophora   water snails; eyes at base of posterior tentacles.

In this largest molluscan class (80,000 species), the most characteristic group-identifying events take place during the larval stages. A trochophore develops in the usual manner. This larva then becomes a *veliger,* a form with an elaborate ciliary girdle (*velum*) used as a swimming organ (Fig. 23.6). In the veliger larva a foot develops as in chitons, between mouth and anus. But then the developing visceral hump enlarges greatly and pushes up in a dorsal direction. This upward growth occurs more or less unequally on the left and right sides, resulting in a spiral coiling of the visceral hump and the shell covering it. The upward growth also has the effect of pulling the alimentary tract to a U shape, until the anus comes to lie quite close to the mouth.

A distinct separate event then takes place. Referred to as *torsion,* it is achieved in a few minutes: the whole visceral hump of the veliger rotates 180° in relation to the rest of the body, usually in a counterclockwise direction (Fig. 23.7A, B and see Fig. 23.6). As a result, the mantle cavity comes to lie anteriorly, above the head; the gills are anterior; the anus, excretory pores, and gonopores all are anterior, in the mantle cavity; the alimentary tract is twisted from a U shape to a loop; the visceral nerve loop becomes twisted to a figure of eight, with the left cord coming to lie under the alimentary tube and

Order Stylommatophora   land and water snails, land slugs, edible snails; eyes at tip of posterior tentacles.

the right cord over it; and the heart is turned around, the atria coming to lie in front of the ventricle.

The evolution of torsion undoubtedly is an adaptation to the adult way of life. Snails protect themselves by withdrawing into the shell, and inasmuch as the shell has only one opening it is advantageous that both mouth and anus should open there. Also, in a creeping way of life it is probably advantageous that the mouth remains closer to the ground than the anus. Torsion produces these basic characteristics of gastropod structure. Most gastropods pass the trochophore stage as embryos and only the veliger becomes a free larva. In the pulmonate snails even the veliger stage remains embryonic and development in this group is direct.

In the subclass Prosobranchia, the largest group of gastropods, full 180° torsion occurs. In the aspidobranch order the left-right inequality of early growth is not too great, and after torsion snails such as limpets therefore have a caplike, more or less uncoiled shell. In the slipper limpet *Crepidula* the shell is slightly more asymmetric, and in abalones spiral coiling is already fairly pronounced (Fig. 23.8 and see Fig. 23.7B). Limpets are common along stony beaches, attached to or creeping on rocks between tidewater marks. Abalones attain lengths of up to 10 in. The shell

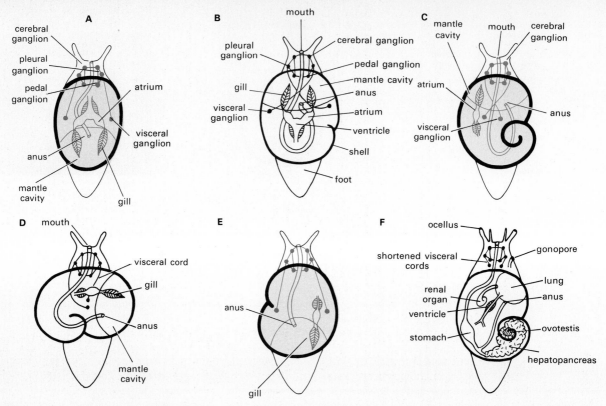

Figure 23.7 Torsion and detorsion. A, original condition, paired organs present, anus posterior. B, after torsion in aspidobranchs. Note figure of eight formed by visceral nerve loop, the left visceral cord passing under the alimentary tract, the right cord over it. Paired organs still of equal size. C, after torsion in mesogastropods. Only right members of paired organs present, lying on left side, anus anterior, visceral cords forming figure of eight. For neogastropod pattern see Fig. 23.9. D, 90° detorsion, as in some opisthobranchs. Mantle cavity, gill, and anus lie on right side, heart lies transversely, visceral cords partially untwisted. E, complete 180° detorsion, as in other opisthobranchs. Pattern resembles A, but only right members of paired organs present. F, condition in pulmonates. Note anteriorly concentrated ganglia with shortened cords, hence absence of figure-of-eight loop.

Figure 23.8 Aspidobranch snails. A, limpets (*Patella*). The shell is hardly coiled. B, abalone (*Haliotis*). Shell coiling slight. Note row of shell perforations for excurrent water.

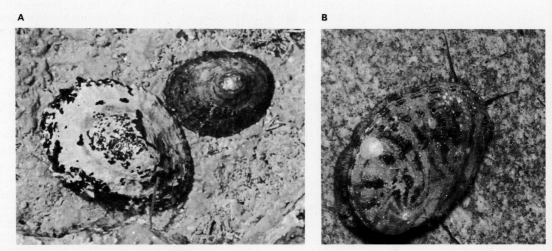

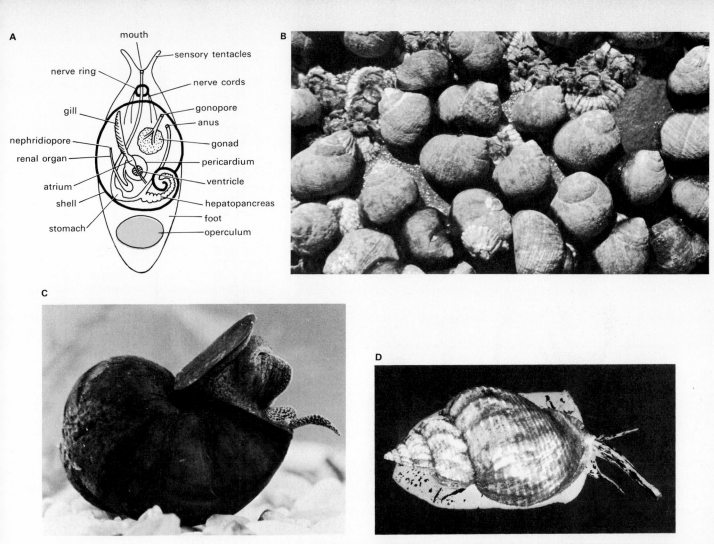

A

mouth
sensory tentacles
nerve ring
nerve cords
gonopore
gill
anus
nephridiopore
gonad
renal organ
pericardium
atrium
ventricle
shell
hepatopancreas
foot
stomach
operculum

B

C

D

Figure 23.9 Mesogastropods and neogastropods. A, the concentrated nerve ring as in neogastropods. Only the right members of originally paired organs are developed and lie on the left side after torsion (as in mesogastropods, see Fig. 23.7C). B, *Littorina,* the common (mesogastropod) periwinkles. C, the freshwater mesogastropod *Paludina,* withdrawing into its shell and closing the operculum. D, the (neogastropod) whelk *Buccinum,* seen from above. The long tube projecting anteriorly is the siphon, which directs water toward the gill. The pair of smaller projections are sensory tentacles on the head.

of an abalone has a row of perforations through which water is expelled from the mantle cavity.

In mesogastropods and neogastropods, by contrast, early growth on the left side is suppressed to·such an extent that the left organs fail to develop. The right organs then come to lie on the left side after torsion and such animals have only one gill, one atrium, and one kidney. Moreover, the visceral hump and the shell are highly coiled (Fig. 23.7C). Gonads here open through separate gonoducts into the mantle cavity. In the nervous system of mesogastropods the paired ganglia are well separated and interconnected by nerve cords that are twisted after torsion. In neogastropods, by contrast, the ganglia lie close together around the anterior nerve ring (Fig. 23.9).

Both mesogastropods and neogastropods have an *operculum,* a horny plate on the dorsal posterior part of the foot. When the tide ebbs such snails withdraw into their shells and secrete a mucus seal around the opercular cover. So protected against drying out the animals then await the return of the tide. Mesogastropods include the periwinkles, exceedingly common along most seashores between

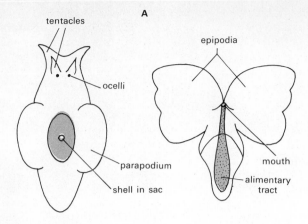

A

tentacles

ocelli

epipodia

parapodium

shell in sac

mouth

alimentary
tract

B

C

Figure 23.10 **Opisthobranchs.** *A,* sea hare (*Aplysia*) at left, sea butterfly (*Cavolinia*) at right. *B, Aplysia,* and *C,* sea butterfly, from life. See also Color Fig. 23.

high- and low-water marks, and the *cowries,* snails with brilliantly pigmented and highly polished shells greatly prized by collectors. Freshwater mesogastropods like *Paludina* are ovoviviparous and without larvae.

Neogastropods comprise the whelks and related types. In the mantle cavity of these animals a tubular extension of the body wall, the *siphon,* directs incoming water toward the gill. Most neogastropods are scavenging and predatory carnivores. They clasp lobsters, crabs, or other mollusks with the foot and bore a hole through the protective armor of the prey by means of the radula. This organ can protrude through the mouth located at the tip of a protrusible *proboscis* (hence the designation "oyster drills" for many of these snails). In some of the neogastropods the salivary glands secrete poison into wounds made by the radula. Male whelks have a penis that can be protruded from the mantle cavity. Female whelks lay eggs in batches of several hundred, each batch enclosed in a capsule. Many capsules are joined to one another, the whole forming a spongy mass.

Advantageous as shells may be, they are distinctly disadvantageous in locomotion. Reduction and complete loss of shells have occurred in the Opisthobranchia, a subclass probably evolved from prosobranchs. But if a shell is reduced or absent, torsion becomes more or less superfluous as well; and opisthobranchs actually exhibit a greater or less reversal of torsion, or *detorsion*. Thus, veligers first undergo torsion and again the organs of the left side fail to develop. But during later development the visceral hump rotates back partly or wholly to its original position (Fig. 23.7*D, E*). Some opisthobranchs undergo a 90° clockwise detorsion, a process that brings the anus and the mantle cavity to the right side. Others undergo a complete 180° detorsion. Here the anus is posterior and the visceral nerve loop no longer has the form of a figure of eight. Although the adults then are perfectly bilateral on the outside, their internal structure and curious development clearly testify to their derivation from asymmetric ancestors.

Full detorsion occurs in, for example, sea hares such as *Aplysia.* These animals have two pairs of tentacles, the anterior pair ear-shaped, the posterior pair each with an eye at the base (Fig. 23.10 and Color Fig. 23). The shell is small and largely covered by upturned portions of the mantle. The foot extends on each side as a *parapodium,* a flap used for swimming. Opisthobranchs also include the pelagic, swimming *pteropods,* or sea butter-flies. The shells of these animals generally are transparent and either coiled or uncoiled and vase-shaped. The foot is expanded as a pair of conspicuous finlike paddles, the *epipodia.* Cilia on these direct planktonic food organisms toward the mouth.

The nudibranchs, or sea slugs, include some of the most beautifully pigmented sea animals (Color Fig. 24). They lack shells and mantle cavities. Detorsion if often complete, as in *Eolis,* and in such cases the anus is posterior. Gills usually are absent, and in many cases breathing occurs through numerous *cerata,* tentaclelike projections from the sides of the animal (Fig. 23.11). In the interior of a ceratum is a pouched extension from the alimentary tract. Some slugs feed on cnidarians, and the cnidoblasts then collect in the cerata. The nudibranch uses these sting cells defensively, just as the original cnidarian would have done.

Prosobranchs are believed to have given rise also to the subclass Pulmonata, the most specialized gastropods (Fig. 23.12). Torsion again takes place here and the animals have a single atrium and kidney. The visceral nerve loop still forms a figure of eight in primitive pulmonates. But in others the longitudinal cords are shortened, and in the most specialized types all ganglia are part of the nerve ring around the esophagus. Such animals have a concentrated, symmetric, untwisted nervous system, the symmetry being achieved by an extreme forward placement of all ganglia, not by a detorsion (Fig. 23.7F). Gills are absent. Instead, the dorsal part of the mantle cavity is highly vascularized and the cavity functions as an air-breathing lung. Some pulmonates have adapted secondarily to a life in fresh water, but such snails are lung breathers nevertheless and must surface periodically for air. Terrestrial pulmonates include shell-less land slugs and the familiar shelled land snails of the genus *Helix.* One species, *H. pomatia,* is edible.

A pulmonate has a single gonad (*ovotestis*) that produces *both* sperms and eggs (see Fig. 23.12). Sperms are formed first and are stored, and eggs developed later are fertilized by sperms from another snail. Sperms are packed together as a *spermatophore* before transfer during copulation. The gonopores lie anteriorly, on the right side behind the head. Mating is triggered when each partner shoots a calcareous *dart* from the female gonopore into the body of the partner. Coming to rest among the internal organs of the snails, the darts appear to act as stimulants for copulation; the penis of each snail is inserted into the vagina of the other and the spermatophores are transferred. Fertilized eggs become enveloped by layers of albumen and calcium deposits secreted by the oviducts. Some pulmonates are ovoviviparous. Others, like *Helix,* lay batches of eggs under leaves or in ground holes, and development then takes place directly. Pulmonates hibernate by creeping into soil burrows and secreting protective calcareous membranes over their shell openings.

Tusk Shells   **Class Scaphopoda**   marine; shell tubular, open at both ends; foot a burrowing organ; gills absent; circulatory system reduced; trochophore and veliger stages.

This small class comprises some 350 species of sand-burrowing mollusks. The animals are elongated in a dorsal direction and are covered by tapered, tubular shells. These give them the appearance of canine teeth or tusks (Fig. 23.13). From the wider, ventral end of a shell projects the muscular, conical foot that serves as a digging organ. Also protruding at the ventral end is a reduced, proboscislike head, to which are attached numerous prehensile tentacles (*captacula*). From one end of the animal to the other passes a channel-like mantle cavity. Water circulates in and out of it through the dorsal shell opening, which projects beyond the burrow of the animal into clear water. The mantle tissues function in breathing. A radula can be protruded through the mouth, and the anus opens posteriorly into the mantle cavity. The nervous system is symmetric, with four pairs of ganglia and connective cords. A heart is absent. The paired kidneys do not open into the coelom, and the right kidney also serves as gonoduct.

Scaphopods appear to represent a separate line of molluscan evolution, on a level of specialization intermediate between that of gastropods and pelecypods.

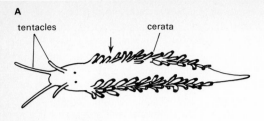

A

tentacles   cerata

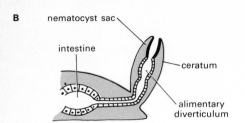

B

nematocyst sac

intestine

ceratum

alimentary
diverticulum

Figure 23.11 Nudibranchs. A,
outline of *Eolis,* showing position
of cerata; arrow points to anus, on
side (90° detorsion). *B,* section
through a ceratum showing terminal opening and alimentary channel in which cnidarian stinging
cells accumulate. See also Color
Fig. 24.

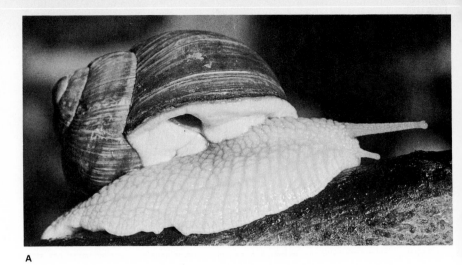

A

B

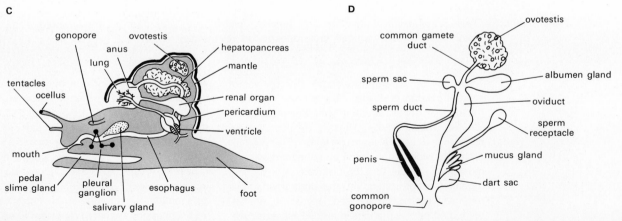

C

gonopore   ovotestis
         anus
    lung   hepatopancreas
tentacles   mantle
   ocellus
                            renal organ
                            pericardium
                            ventricle
mouth
  pedal
  slime gland   pleural   esophagus   foot
               ganglion
        salivary gland

D

                              ovotestis
common gamete
       duct
                              albumen gland
  sperm sac
                              oviduct
  sperm duct
                              sperm
                              receptacle
    penis   mucus gland

                           dart sac
  common
  gonopore

Figure 23.12 Pulmonates. A, Helix, side view showing the opening to the lung chamber under the rim of the shell, *B, Planorbis,* a freshwater pulmonate
with a shell coiled in one plane. *C,* sectional side view of *Helix.* The gonopore opens on left side of anterior region, near head. The ovotestis and the
gonopore are connected by a duct system (as shown in *D*). To preserve clarity of other parts, the alimentary system is not drawn in completely. *D,* the
hermaphroditic reproductive system of *Helix,* simplified. The sperm sac stores sperms produced by the snail itself, whereas the sperm receptacle stores
sperms produced in the mating partner and introduced into the snail during mating.

Clams   ***Class Pelecypoda***   bivalve mollusks; sedentary, mostly marine; laterally compressed, with dorsally hinged valves; head rudimentary, without tentacles; foot usually tongue-shaped and used in burrowing; mouth with labial palps; radula absent, stomach in most cases with *crystalline style;* gills usually expanded to ciliary organs used in filter feeding; sexes nearly always separate; development via trochophore and veliger stages in marine types, via *glochidia* stages in freshwater types.

Subclass Protobranchia   gills for breathing only, feeding by means of labial palps.

Subclass Filibranchia   mussels, scallops; gill filaments usually connected by ciliary junctions; gills primarily for feeding.

Subclass Eulamellibranchia   oysters, soft-shelled mud clams, hard-shelled quahog clams, giant clams, shipworms; gill filaments connected by tissue junctions; gills for feeding primarily.

Subclass Septibranchia   gills organized as muscular septum with pores.

The two shells, or *valves,* of a clam are secreted by mantle lobes, single layers of the body wall (Fig. 23.14). The ventral edges of the lobes either hang free or are fused and enclose the mantle cavity more or less completely. In that case the valves can gape open only to a limited extent. The valves have a dorsal tooth-and-pit hinge and a horny elastic ligament keeps them joined along the hinge and presses them open. Closure is brought about by (usually) two adductor muscles, one anterior and one posterior.

The oldest part of a valve is the *umbo,* a bulge near the hinge, from which growth proceeds in concentric rings. The outermost layer of a valve is a horny *periostracum,* a CO_2-resistant cuticle that protects the shell from dissolution by sea water. Underneath is a calcareous *prismatic* layer, and the innermost part is a pearly *nacreous* layer, composed of calcium carbonate crystals aligned parallel to the mantle surface. Pearls are formed when a foreign object lodges between the valve and the mantle lobe. The object then becomes enveloped by successive layers of nacre secreted by the mantle.

The visceral mass of a clam is suspended from the dorsal midline (Fig. 23.15). Continuous ventrally with the visceral mass is the foot, which in many cases can be protruded between the valves. Two pairs of large, double-layered *gill plates* project into the mantle cavity, one pair on each side of the visceral mass. These gills continue posteriorly as a horizontal partition that divides the posterior region of the mantle cavity into a dorsal compartment and a ventral main compartment. Both compartments open to the outside along the posterior valve edges, the dorsal chamber forming an *excurrent siphon,* the ventral one an *incurrent siphon.* In some clams these siphons are extended as long retractile tubes that project beyond the valves (Fig. 23.16).

A gill plate consists of parallel, double-layered *gill filaments,* each layer of a filament with a blood vessel. Adjacent filaments are fused together, but they leave *gill pores* through which water can pass into *water tubes* between the two layers of a gill. Cilia on the gills draw food-bearing water through the incurrent siphon into the ventral mantle compartment. Water then passes through the gill pores and the water tubes, where it contributes to oxygenation of blood, and leaves by way of the dorsal

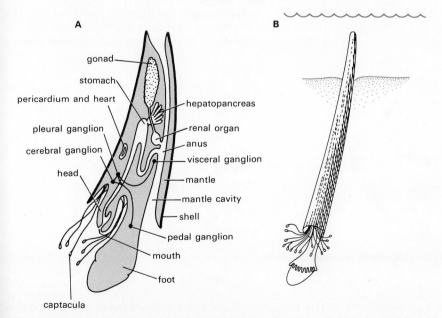

Figure 23.13 Scaphopoda. A, sagittal section through *Dentalium. B,* body proportions and position of animal in its sand burrow. (*A, adapted from Naef.*)

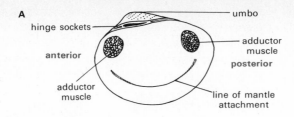

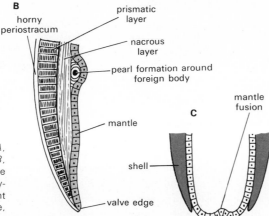

Figure 23.14 Clamshells. *A,* interior view of right valve. *B,* cross section through one valve near ventral edge, to indicate layers. *C,* fusion of left and right mantle layers along ventral midline, as in scallops.

mantle compartment and the excurrent siphon. Food particles are strained out on the gill surface, where they are caught up in secreted mucus. Strands of such mucus are propelled by the cilia anteriorly, toward the *labial palp* on that side. The palps are ciliated tissue flaps that conduct food-containing mucus toward the mouth.

A radula is absent, a superfluous structure in a filter-feeding animal. However, a clam has a well-developed hepatopancreas in which proteins and lipids are digested intracellularly. Also usually present is a *crystalline style,* a gelatinous rod secreted in a stomach pouch (Fig. 23.17). Ciliated cells in this pouch keep the style rotating and pressed against a horny *gastric shield* along the inner surface of the stomach. Tiny particles of style material thus wear off and mix with food. The style substance contains an amylase, the only extracellular digestive enzyme produced by clams. A coiled intestine leads from the stomach through the pericardial coelom to the anus, an opening that empties into the dorsal siphon.

In conjunction with the reduction of the head the nervous system is to some extent reduced, too. A pair of fused cerebropleural ganglia lies anteriorly in a nerve ring around the esophagus. Nerve cords from there connect with a pair of pedal ganglia in the foot and a pair of visceral ganglia near the anus (see Fig. 23.17). Paired renal organs lead from the pericardial coelom to a bladder that opens into the outgoing water current. The circulation is open and the blood is colorless in most cases. A ventricle in the pericardial coelom

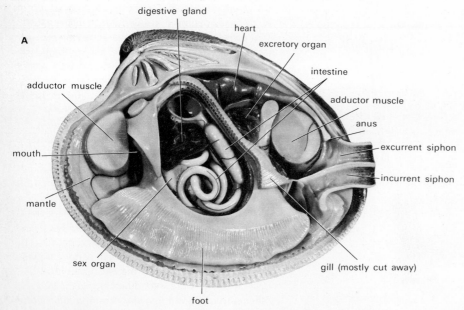

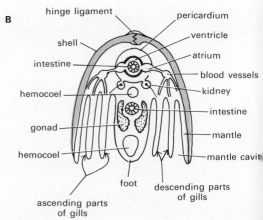

Figure 23.15 Clam structure. *A,* cutaway model indicating general anatomy. *B,* transverse section. A part of the intestine passes through the ventricle of the heart. The hinge ligament is a thickened portion of the periostracum.

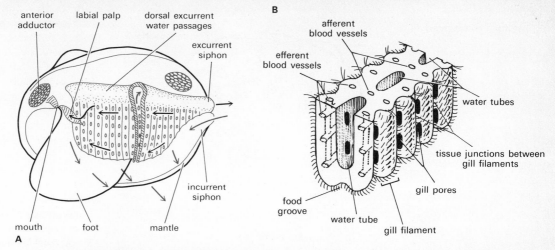

Figure 23.16 Clam structure: food and water paths. *A,* gills of left side, the outer gill shown cut away. Black arrows indicate ciliary paths of food particles toward mouth and of water into excurrent siphon via dorsal gill passage; colored arrows show paths of heavy particles not adhering to gill. *B,* ventral edge of gill, showing ascending and descending gill plates and tissue junctions between individual gill filaments. Afferent vessels carry blood down through tissue partitions between water tubes, and efferent vessels then conduct blood up through ascending and descending portions of a gill. Microscopic food particles collect in mucus strands in the ventral food groove, heavy particles fall off the gill and out of the animal, and water passes through the gill pores into the water tubes within the gill.

surrounds the intestine, and arteries from the ventricle pass to open blood sinuses in all body parts (Fig. 23.18). Blood from there collects in veins that lead to the kidneys. From these organs blood is channeled through the gills and oxygenated blood ultimately returns to the two atria of the heart.

The mantle lobes are important breathing organs in clams, and certain veins from there carry oxygenated blood directly to the heart, bypassing the kidneys and gills. The main vein from the foot has a sphincter muscle that, on contraction, dams up blood and thereby brings about a distension of the foot. In this way a clam can protrude the foot and use it for digging and a slow, push-pull form of locomotion.

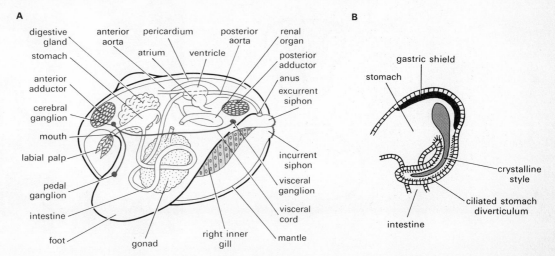

Figure 23.17 Clam structure. A, some of the internal organs in sagittal view. The excretory and reproductive openings exit in the excurrent water passages. *B,* stomach showing crystalline style and gastric shield. (*B, adapted from Barrois.*)

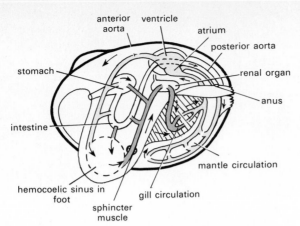

anterior ventricle
aorta

atrium

posterior aorta

stomach

renal organ

anus

intestine

mantle circulation

hemocoelic sinus in
foot

gill circulation

sphincter
muscle

Figure 23.18 Circulation in clams. Arterial blood, light color; venous blood, dark color. Oxygenation of blood takes place in the gills and, most particularly, in the mantle. Arrows indicate direction of flow. The sphincter around the vein in the foot can keep blood in the foot sinuses and thereby can enlarge and stiffen the foot during locomotion.

In most clams the sexes are separate. Gonads are formless masses around the intestinal coils and gonoducts discharge into the outgoing water current. In marine forms fertilization usually occurs externally in open water, and the trochophores and veligers are free-swimming. Freshwater clams typically discharge eggs into the water tubes in the gill plates, where internal fertilization takes place;

sperms enter with the incurrent water. Highly modified veligers, *glochidia* larvae, then develop. These are expelled through the excurrent siphon, and if they are not to perish they must attach themselves within a short time to the gills or fins of fishes. At such sites the glochidia live parasitically until they mature as free-living adults (Fig. 23.19).

Clams lead a semisedentary, burrowing life. They change station occasionally but otherwise remain partly or wholly buried, with the siphon protruding up to clear water. Some clams are permanently attached, and in oysters the larger left valve is cemented to the ground.

In Protobranchia, the most primitive group, the gills are small and function only in breathing (Fig. 23.20). The feeding organs are the labial palps, each of which consists of a proboscislike food-collecting part and of two platelike, ciliated, food-sorting parts. The nervous system of these animals still exhibits the ancestral organization and contains four separate pairs of ganglia. Another primitive trait is the continuity of the gonad with the kidney.

In the Filibranchia (now also classified as two separate subclasses), adjacent gill filaments are held together only by ciliary connections, not by tissue struts as in clams (see Fig. 23.20). The common mussel *Mytilus* has a glandular *byssus pit* at the posterior part of the foot that secretes numerous mucous *byssus threads.* These harden

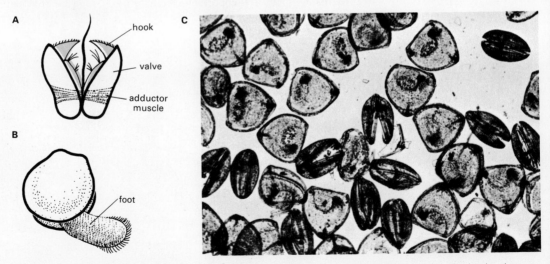

A

hook

valve

adductor
muscle

B

C

foot

Figure 23.19 Development of freshwater clams. A, glochidia larva, seen from anterior end; the hooks clamp on to gill tissue of fish host. B, young free clam, after detaching from fish gills. C, glochidia larvae of a freshwater mussel. (A, B, adapted from Lefevre and Curtis. C, courtesy Carolina Biological Supply Company.)

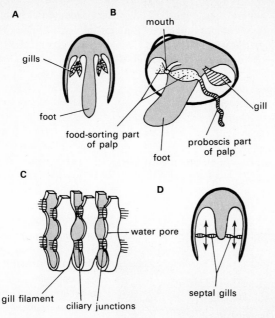

fibers that close the valves, the other with striated fibers that produce a rapid clapping of the valves and enable a scallop to propel itself jetlike for short distances. The highly reduced foot of scallops is used mainly to clean the labial palps and the gills. Scallops are among the few pelecypod hermaphrodites; ovaries are pink, testes are cream-colored (Color Fig. 25 and see Fig. 23.21).

Septibranchia are a small group in which the gills on each side form a horizontal muscular "diaphragm" with pores. As such a septum moves up and down, it circulates water between the dorsal and ventral chambers of the mantle cavity (see Fig. 23.20).

By far the most abundant pelecypods are the eulamellibranchs, or clams proper, identified by fused gill filaments as described (Fig. 23.22 and Color Fig. 26). In this group are the oysters, which in some respects resemble scallops: a foot is lacking, only a single adductor muscle is present, and the animals are hermaphrodites. Indeed the single gonad is an ovotestis that functions alternately as an ovary and a testis, as in the land snail *Helix*. The subclass also includes giant clams such as *Tridacna*, up to 2 yd wide and weighing $\frac{1}{4}$ ton, and rock-boring and wood-boring clams. The latter are noteworthy for their destructiveness. For example, the "shipworm" *Teredo* uses its reduced, roughened valves to bore tunnels in ship bottoms and wharf pilings, reducing them to sawdust. The clams then use the sawdust for food; they secrete cellulase, a unique digestive enzyme among animals (even termites do not digest wood on their own but depend on flagellate protozoa).

Figure 23.20 Pelecypod types. A and *B,* Protobranchia (*Nucula*). *A,* cross section showing small breathing ctenidia. *B,* sectional side view showing food-collecting apparatus and separate breathing gill. *C,* Filibranchia, portion of ventral part of gill, showing ciliary junctions between gill filaments (contrast with eulamellibranch condition, Fig. 23.16.) *D,* Septibranchia, cross section showing perforated partition formed by ctenidia on each side. (*Adapted from Naef, Sedgwick, and other sources.*)

on contact with sea water and attach the animals to rocks and to one another (Fig. 23.21). The scallop *Pecten* has a single (posterior) adductor muscle composed of two parts, one with smooth

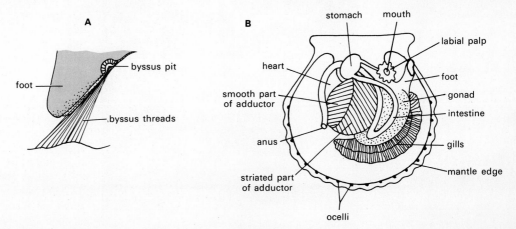

Figure 23.21 Filibranchs. A, the byssus apparatus of the mussel *Mytilus. B,* some of the organs of the scallop *Pecten.* See also Color Fig. 25. (*B, modified from Dakin.*)

Cephalopods   *Class Cephalopoda*   nautiluses, squids, octopuses; marine; elongated in a dorsal direction; head with tentacles; foot partly in form of funnel; shell external and chambered or reduced and internal or absent; nervous system exceedingly well developed; with cartilaginous endoskeleton; circulatory system usually closed; eggs telolecithal, cleavage discoidal; development usually direct, without larvae.

Subclass Tetrabranchiata   extinct nautiloids and ammonites, chambered nautilus; with two pairs of gills; shell external; ink sac absent; circulatory system open.

Order Decapoda   squids, cuttlefish; with 10 tentacles; coelom well developed; shell internal; pelagic.

Order Octopoda   octopuses; with eight tentacles; coelom reduced; shell highly reduced or absent, semisedentary.

Subclass Dibranchiata   with one pair of gills; with ink sac; circulatory system closed.

Cephalopods are the most advanced mollusks and rank among the most highly evolved animals generally. The head of the animals is ventral and the elongated visceral mass dorsal. But since cephalopods swim in a horizontal position, it is convenient to refer to the head end as anterior, to the other as posterior (see Fig. 23.2).

In a squid a thick muscular mantle surrounds the visceral mass (Fig. 23.23). Pigment-containing chromatophores in the mantle can contract and expand and thereby vary the coloration of the animal according to backgrounds or emotional states. Under the mantle on the upper side lies the evolutionary remnant of the shell, a horny, leaf-shaped *pen*. Anteriorly the mantle terminates at a free edge, the *collar*, which fits over a mid-ventral *funnel* (or *siphon*). This muscular tube, representing part of the foot, leads to the mantle cavity. When the mantle musculature is relaxed water passes between the collar and the funnel into and out of the mantle cavity. On contraction of the mantle muscles the collar clamps tightly around the funnel and water is forced out through the funnel tube. The funnel can be bent in various directions and thus permits a squid to change course. In ordinary cruising a squid uses mainly its lateral fins and swims head first; in pursuit of prey or in flight the animal uses its funnel mechanism and jet-propels itself "backward." Squids are among the fastest swimmers and readily match the maneuverability of fishes.

A squid has 10 tentacles, or *arms*, variously regarded as part of the head or part of the foot. The first three and the fifth pairs counted from the upper side are short and studded on the inner surface with stalked, cup-shaped suckers (see Fig. 23.25). The fourth pair is long and has suckers only at the expanded tip. These long arms catch

Figure 23.22 Eulamellibranchs. A, the marine horse clam *Schizothaerus,* with long extensible siphon. *B,* the freshwater clam *Anodonta;* note conspicuous burrowing foot. *C,* the razor clam *Solen. D,* shell of the giant clam *Tridacna.* See also Color Fig. 26.

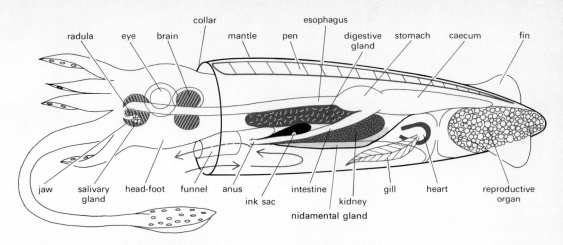

Figure 23.23 Squid structure. Tentacles are shown on left side only. Arrows indicate path of water into and out of mantle cavity, resulting in jet propulsion backward. Note that the head-foot end is anatomically ventral, the high end dorsal; the side where the pen is located is anterior, the opposite side posterior.

prey and bring it to the short arms, which then hold the prey while the horny *jaws* in the mouth bite chunks out of it. A radula is present but probably used little since food is swallowed rapidly. Cephalopods have two pairs of salivary glands, the posterior pair poison-secreting in many cases. The alimentary tract is U-shaped. The stomach connects with a large digestive gland that consists of two parts. A pancreatic part secretes enzymes, and an elongated "liver" absorbs food molecules from blood and stores them. The stomach also communicates with a long *caecum* that extends back to almost the tip of the animal. This pouch separates out solid food particles and absorbs only dissolved nutrients. The anus opens anteriorly in the mantle cavity, on an *anal papilla.* An *ink sac* discharges into the mantle cavity by way of the anus. Expelled through the funnel, a cloud of ink probably distracts an enemy while the squid makes its escape. The ink contains finely dispersed melanin granules.

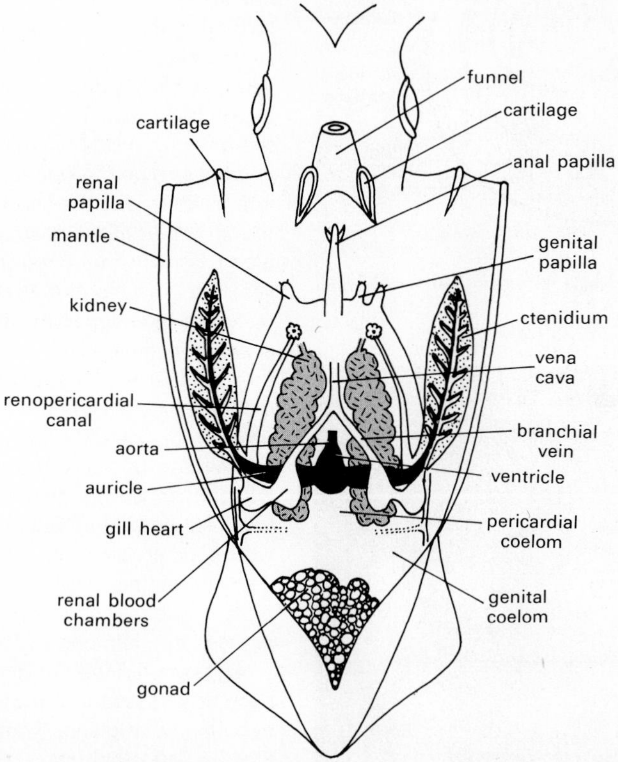

Several fused pairs of ganglia form a complex brain, which is surrounded by a cartilage capsule. Located between the eyes (see Fig. 8.37), the brain represents the dorsal part of a nerve ring that circles the esophagus. From the ring elaborate tracts of nerves branch throughout the body. Dorsally, giant nerve fibers innervate the mantle musculature and control the rapid locomotor activities of the animals.

The coelom is spacious and occupies the hind part of the visceral mass (Fig. 23.24 and see Fig. 23.23). A single gonad leads to a gonoduct that opens in the mantle cavity at a *genital papilla,* to the left of the anal papilla. The genital coelom around the gonad is partially separated by incomplete septa from a pericardial coelom. A pair of kidneys lies just in front of the pericardial cavity. Excretory ducts open into the mantle cavity through *renal papillae,* one on each side of the anal papilla. From the pericardial coelom a pair or *renopericardial canals* leads through kidney tis-

Figure 23.24 Squid structure: ventral dissection.

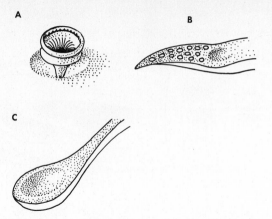

Figure 23.25 *Cephalopod tentacles.* *A*, stalked sucker as on squid tentacles. *B*, squid hectocotylus, showing suckerless depression for spermatophore transfer. In *Loligo* the third right tentacle and in *Sepia* the fifth left tentacle are of this type. *C*, the spoon-shaped hectocotylus of *Octopus.*

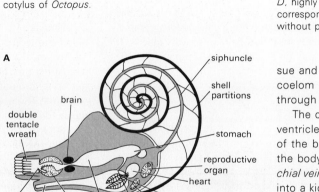

Figure 23.27 *Shell reduction in cephalopods.* *A*, internalized reduced shell in *Spirula;* note presence of siphuncle. *B*, calcified shell of *Sepia*, corresponding to upper part in *A* (part above siphuncle); part below siphuncle is highly reduced, hence siphuncle absent as such. *C* and *D*, highly reduced horny pen of *Loligo* (*D*, surface view), corresponding to uppermost shell part in *A* (the top part without partitions). (*Partly adapted from Naef.*)

Figure 23.26 *The chambered nautilus.* *A*, sectional longitudinal view, showing position of some of the organs. Note double circlet of tentacles and absence of ink sac. Gills shown on one side only. *B*, section through shell. Note compartments partitioned off by septa and the siphuncle, a gas-filled channel. (*A, modified from Naef.*)

sue and opens just behind the renal papillae. The coelom thus communicates with the outside through the excretory pores.

The circulation is closed. Arteries from a heart ventricle carry blood to capillary vessels in all parts of the body, and a *vena cava* returns blood from the body tissues. This vessel splits into two *branchial veins* near the kidneys. Each such vein passes into a kidney and the filtered blood then circulates to a *gill heart* at the base of each gill. Oxygenated blood from the gills returns through atria to the ventricle. The blood contains hemocyanin.

Sexes are separate. In the male gonoduct sperms become enveloped by an elastic tube and form a spermatophore. This packet passes through mantle cavity and funnel to a *hectocotylus*, a particular arm specially constructed to hold a spermatophore. In *Sepia* the hectocotylus is the fifth left arm; in *Loligo,* the third right arm. *Octopus* has a spoon-shaped hectocotylus (Fig. 23.25). In mating the hectocotylus transfers the sperm packet into the mantle cavity of the female. The male of *Argonauta* detaches the whole arm in the process and leaves it in the female.

Eggs are fertilized in the mantle cavity and then become encased in a protective capsule secreted by *nidamental* glands ventral to the kidneys (see Fig. 23.23). Such capsules are shed to the open sea or, more commonly, are attached to rocks or other solid objects. Development is basically schizo-

A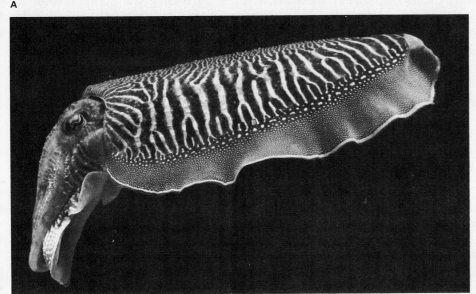

B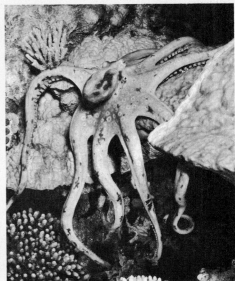

Figure 23.28   Squids and octopuses. A, a cuttlefish. *B,* an octopus. Note funnel below right eye.

coelic, but the pattern is highly modified by the very large amounts of yolk. The telolechithal eggs cleave superficially and a discoblastula results. Larvae generally are not formed and miniature adults eventually emerge from the egg capsules.

The chambered *Nautilus* is the only living survivor of the originally very large subclass of Tetrabranchiata (Fig. 23.26). In nautiloids the numerous and retractile, suckerless tentacles form a double wreath around the head. The animals have two pairs of gills and kidneys, ink sacs are absent, and the eyes are simple ocelli of the cup type (see Fig. 8.37). The circulatory system is open, as in the other molluscan classes, and the external, calcareous shell is a flat, chambered coil. The posterior part of the mantle secretes the curved partitions in the shell. A strand of mantle tissue, the *siphuncle,* perforates all partitions and passes back to the innermost, youngest chambers. Blood vessels in the siphuncle are believed to secrete the airlike gas that fills the unoccupied chambers.

A progressive internalization and reduction of the shell is illustrated in a series of genera of the Dibranchiata (Fig. 23.27). In *Spirula* the shell is still coiled and chambered but is internalized and reduced in size. The cuttlefish *Sepia* has a solid calcareous internal shell, yet the evolutionary derivation from a chambered condition is indicated by oblique, more or less regularly spaced growth planes inside the cuttlebone. In *Loligo* the shell is a horny pen, as noted. *Octopus* has only vestigial remnants of a shell, and in forms like *Idiosepius* all traces of a shell have disappeared.

All squids are more or less streamlined and adapted to pelagic life (Fig. 23.28). Most of them are bioluminescent, many travel in schools, and some are cannibalistic. Giant squids inhabit the deep ocean, where they are preyed on by large fishes and particularly by toothed killer and sperm whales. Octopuses on the contrary tend to be solitary and sedentary, and they have rounded, saclike shapes. Most are quite small, some no more than 2 in. long. The largest have an arm spread of about 30 ft.

Review Questions

1.   Give taxonomic definitions of mollusks and each of the classes. Describe the structural traits of the hypothetical molluscan ancestors. Show in what ways living mollusks still exhibit the ancestral body structure to greater or lesser degree.

2. Describe the structure of chitons. Which features of these animals appear to be specialized and which ancestral? What is the organization of the excretory and breathing systems? Name the orders of Amphineura and their characteristics.

3. Define the subclasses of gastropods. Describe the pattern of embryonic and larval development in gastropods. What is torsion, when and in what groups does it occur, and what are its structural and functional consequences?

4. Distinguish between torsion and coiling. What is the adaptive advantage of torsion? What is detorsion, when and in what groups does it occur, and what are its consequences? What is the adaptive advantage of detorsion?

5. Which gastropod groups have opercula? Which are hermaphroditic? How many atria and kidneys are present in a nudibranch, whelk, and a pulmonate?

6. What are cerata, parapodia, epipodia? Do pulmonates exhibit torsion and detorsion? Describe the structure of scaphopods.

7. Name and define the subclasses of pelecypods. Describe the structure and growth of clam valves. What are the functions of clam gills and how do they execute these functions? What are the siphons of clams and how are they formed?

8. Describe the organization of the alimentary system of a clam and the structure and function of the crystalline style.

9. Describe the nervous, circulatory, and excretory systems of a clam. What are glochidia? How do marine and freshwater clams develop?

10. Define the subclasses and orders of cephalopods. Describe the structure of a squid. What is the organization of the alimentary, nervous, sensory, breathing, circulatory, and excretory systems?

11. What is a renopericardial canal? A spermatophore? A hectocotylus? How does cephalopod mating take place? Contrast the development of cephalopods with that of other mollusks.

12. How do tetrabranch and dibranch cephalopods differ structurally? Show how the skeleton of cephalopods has become reduced during the evolution of these mollusks. What is the adaptive significance of this reduction?

13. What types of mollusks can be found in deep oceans, in the photic zone of the ocean, in the intertidal zone, in the fresh water, and in land environments?

14. Which molluscan groups are permanently sessile, sedentary and semisedentary, and predominantly motile? How many species of mollusks are known and which are the most abundant classes?

15. What is the presumed evolutionary relation of the molluscan classes? How are mollusks as a whole presumably related to lophophorates and to segmented schizocoelomates? Make a diagram indicating these various interrelations and review the evidence on which you base them.

Collateral Readings

Abbott, R. T.: "American Sea Shells," Van Nostrand, Princeton, N.J., 1954. A nice account for both the amateur and professional student of mollusks.

Hyman, L. H.: "The Invertebrates," vol. 6, McGraw-Hill, New York, 1967. The whole sixth volume of this treatise is devoted to amphineurans and gastropods.

Morton, J. E.: "Mollusks," Hutchinson, London, 1958. Strongly recommended for further readings on the phylum.

Mozley, A.: "An Introduction to Molluscan Ecology," Lewin, London, 1954.
A book described sufficiently by its title.

24

ANNELIDS AND ALLIED GROUPS

The evolutionary change from unsegmented to segmented schizocoelomates (Fig. 24.1) appears to have left living evidence of at least one transitional stage, the small phylum *Sipunculida*. Later, after the fully segmented *Annelida* had become established, an offshoot of this phylum apparently gave rise to animals with segmented young stages but secondarily unsegmented adults. This offshoot line is represented today by the small phylum *Echiuroida*. Other annelid offshoots not only retained the segmented state but elaborated it even further, and these ancestors were the source of the arthropods. This transition too appears to have left living evidence, in the form of various annelid- and arthropodlike animals now provisionally grouped as a phylum *Oncopoda*. Sipunculids, annelids, echiuroids, and oncopods are the subject of this chapter.

Sipunculids

Phylum Sipunculida

peanut worms; marine; body with introvert and trunk; alimentary tract recurved and coiled, anus anterodorsal; circulation open and rudimentary or absent; excretion metanephridial; sexes separate; development mosaic, with spiral cleavage, schizocoelic coelom, and trochophore larvae.

The 250 or so species of these worms, $\frac{1}{4}$ in. to 2 ft long, lead a sedentary, burrowing life in sandy and muddy tidal flats. The animals can be identified readily by their long, slender *introvert,* a probocislike anterior tube that can be invaginated and retracted into the plump trunk (Fig. 24.2). At the tip of the introvert is an *oral disk* with the mouth and a wreath of tentacular outgrowths. The worms use the introvert in a lashing type of swimming and in feeding on detritus found in mud.

The body wall consists of a cuticle-covered glandular epidermis, two layers of muscle, some connective tissue, and a peritoneum. Mesenteries in the large, spacious coelom are poorly developed, and the alimentary tract is held in place largely by muscles. This tract passes from the mouth through the introvert to the intestine, which then recurves in the posterior part of the trunk and coils forward to a dorsal anus in the front part of the trunk (Fig. 24.3). The intestinal mucosa is ciliated and folded longitudinally, and a conspicuous ciliated groove runs along the floor of the intestine.

The nervous center is a nerve ring in the anterior portion of the introvert, thickened dorsally as a brain ganglion. A ventral cord with lateral branches passes from the ring posteriorly. In the oral disk lies a sensory *nuchal organ,* a pad of ciliated cells believed to be chemoreceptive. Also in the oral disk are the openings of a pair of

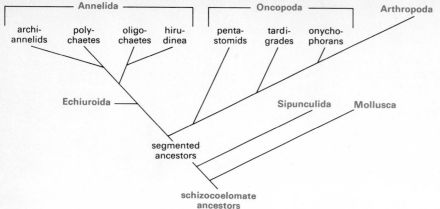

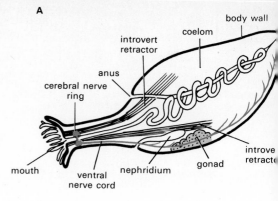

Figure 24.1 *Schizocoelomates*, presumed interrelations.

cephalic tubes that lead backward to a pair of eyecups embedded in the brain ganglion.

Most sipunculids lack a circulatory system; some have a dorsal and ventral open-ended blood vessel and a connecting anterior blood sinus. The coelomic fluid contains red blood cells with hemerythrin, and also interesting microscopic cellular aggregates called *urns*. Some of these are fixed to the peritoneal lining, others swim free by ciliary action. Bacteria, cellular debris, and disintegration products of blood cells can be observed to adhere to the urns; possibly their main role is to collect such debris. Urns are not obviously excretory, though they might serve in this capacity indirectly. Sipunculids have metanephridia anteriorly in the trunk.

Gonads form along the peritoneum and the gametes escape through the coelom and the metanephridia. Development follows the characteristic schizocoelomate pattern, a trochophore larva being formed eventually. This larva later elongates greatly in the region between the mouth and

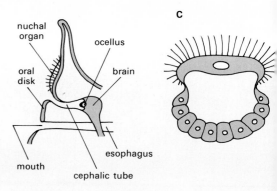

Figure 24.3 *Sipunculid structure. A,* sagittal section through body, showing position of various organs. *B,* sagittal section through dorsal part of oral disk, indicating relation of cephalic tube to ocellus and brain. The nuchal organ is a ciliated pad of tissue. *C,* section through a free coelomic urn, with a large ciliated cell forming a "lid" and a layer of smaller cells forming a cup-shaped vesicle.

the anus, a process that leaves the anus at a progressively more anterodorsal position. The adult thus arises by gradual metamorphosis. Regeneration capacity is extensive in the adult worms.

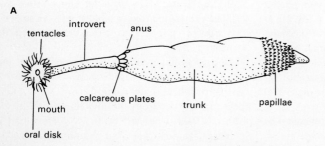

Figure 24.2 *Sipunculids,* external features. The photo depicts *Sipunculus.*

Although sipunculids play a minor ecological role, they are interesting phylogenetically. The worms are clearly related to annelids, as indicated by their entire embryonic and larval development, their annelidlike nervous systems, and their worm-like and coelomate structure in general. However, sipunculids are completely unsegmented as adults, nor do the preadult stages show any signs of segmentation. The worms therefore appear to represent a transition stage on the path of schizo-coelomate evolution that led from unsegmented early groups (with molluscan descendants) to wormlike segmented groups (with annelid descendants).

Annelids   *Phylum Annelida*

metamerically segmented worms, typically with chitinous setae; coelom septate; organ systems arranged on segmental basis; circulatory system usually closed; development typically schizocoelomate; segmentation begins in trochophore stage of embryo or free larva. *Polychaeta, Archiannelida, Oligochaeta, Hirudinea*

Numbering some fifteen thousand species, annelids are important members of all major environments. In these worms metamerism has become evolved in fully elaborated form, presumably for the first time among protostomial coelomates. The first anterior segment typically forms the head, the last contains the anus, and all others are alike developmentally. In given species up to 800 trunk segments can be produced. They often remain more or less alike even in the adult worm, but in numerous instances considerable segmental specialization occurs, particularly in external features. It is this possibility of intersegmental modification that represents one of the adaptive advantages of a metameric body construction (see also discussion in Chap. 7).

Polychaetes   *Class Polychaeta*   largely marine; with segmental parapodia and numerous setae; sexes separate; oviparous, with free-swimming trochophore larvae.

Order Errantia   clamworms, palolo worms, leafworms; pelagic, migratory polychaetes; actively motile, often in temporary tubes or burrows; largely herbivorous; pharynx armed, eversible; body modified in breeding season.

Order Sedentaria   (*a*) parchment worms, feather-duster worms, tube-forming (tubiculous) polychaetes, permanently in secreted tubes; filter-feeding; pharynx unarmed, not eversible. (*b*) lugworms; burrowing polychaetes, permanently in sand or mud burrows; detritus-feeding; pharynx unarmed, eversible.

Polychaetes comprise some two-thirds of all the annelid species, and they undoubtedly represent the primitive class of the phylum. Segmental development begins with a pronounced posterior elongation of the trochophore (Fig. 24.4). This posterior growth is accompanied by a forward proliferation of the mesodermal teloblast bands on each side of the larval body. The bands give rise to anteroposterior pairs of *somites,* cell groups that later hollow out and become schizocoelic coelom sacs (see also Fig. 7.14). As these sacs then enlarge they meet front and back and along the midline, and in this way they form the peritoneal membranes, the mesenteries, and the intersegmental septa.

Segments mature in anteroposterior succession. In each larval segment the ventral ectoderm grows in and produces a segmental portion of the nervous system and a pair of metanephridia. Laterally the ectoderm and mesoderm on each side fold out as a *parapodium,* a flap of body wall. In epidermal pits of a parapodium later develop stiff chitinous bristles, the *setae* characteristic of annelids. The peritoneal mesoderm gives rise to the blood vessels, the muscular system, and the gonads.

An adult polychaete such as the clamworm *Nereis* then has the following segmental organization (Fig. 24.5): The first segment is a preoral head segment, or *prostomium;* it bears one or two pairs of dorsal, lens-containing *eyes,* a pair of

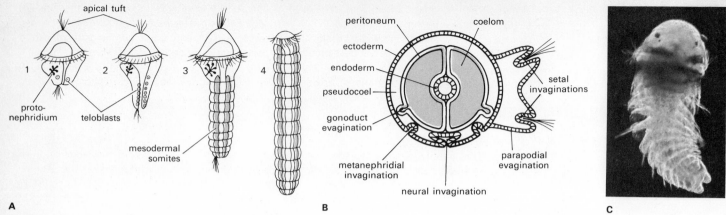

A   **B**   **C**

Figure 24.4 Segmental development of annelids. A, somite formation. 1, trochophore. 2, posterior elongation, proliferation of teloblastic mesoderm bands. 3, mesodermal somites forming. 4, later stage in segmental development, reduction of larval head structures. B, cross section through developing segment, showing formation of ventral, lateral, and ventrolateral organs. C, larva of the marine polychaete *Phyllodoce,* from life, corresponding roughly to stage 3 in A. (A, modified from Woltereck.)

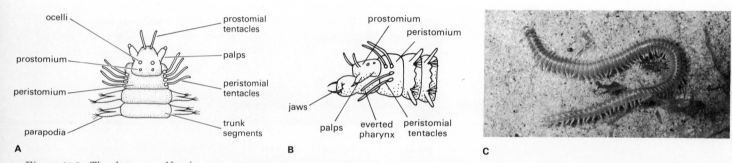

A   **B**   **C**

Figure 24.5 The clamworm Nereis, external features. A, head region, pharynx retracted (and not visible). B, side view of head region, pharynx everted, with left jaw visible. C, the whole animal, from life. See also Color Fig. 27.

anterodorsal sensory *tentacles,* and a pair of antero-ventral sensory *palps.* Behind the head is a *peristomium,* with mouth and four pairs of tentacular palps. The mouth leads to an eversible pharynx, armed with chitinous jaws used in biting off pieces of algae. Behind the pharynx is an esophagus with digestive glands and a straight intestine that terminates at the anus in the last segment. The alimentary tract is the only system not structured metamerically.

The body wall consists of a thin, horny (nonchitinous) cuticle, an epidermis, two segmental muscle layers, and the peritoneum (Fig. 24.6). Laterally, a segmental parapodium is either *uniramous,* composed of a single flap, or *biramous,* composed of a dorsal *notopodium* and a ventral *neuropodium.* Set into epidermal pits of each flap are bundles of chitinous setae. These bristles are

retractile, continuously growing, and used partly for protection but more particularly as locomotor levers and holdfast spikes. A parapodium often is stiffened internally by embedded, nonprojecting setae called *acicula.* Each parapodium typically also bears a dorsal and a ventral projecting *cirrus.* Parapodia do not grow on the prostomium and the peristomium, but the four pairs of peristomial palps of *Nereis* are homologous to the parapodial appendages of the trunk.

On the ventral side each trunk segment contains a pair of ganglia with transverse connectives and lateral branches, as well as longitudinal cords that pass to adjacent segments. Anteriorly the cords form a ring around the pharynx, with a pair of large dorsal brain ganglia in the prostomium. The circulatory system is closed in most cases. It consists of a longitudinal dorsal vessel, a ventral

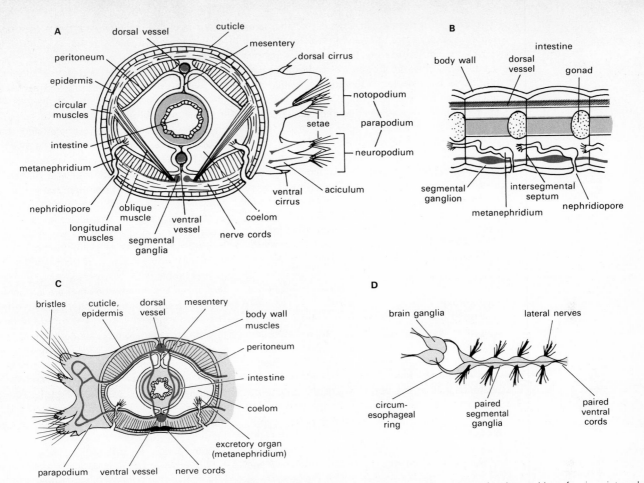

Figure 24.6 *Polychaete structure* (*Nereis*). *A,* cross section through trunk segment. *B,* side view of segments showing position of various internal organs. *C,* cross section of trunk segment, showing pattern of circulation. Blood flows from dorsal vessels to ventral ones in the anterior body segments. *D,* anterior portion of nervous system.

trunk vessel, and transverse interconnections in each segment that include capillary nets in the alimentary tract, the parapodia, and the other segmental organs. Blood flows forward dorsally by wavelike contractions of the dorsal vessel, backward ventrally. Some of the pelagic polychaetes have an open system without capillaries, and some others lack a circulatory system altogether. The blood is colorless or colored; respiratory pigments if present usually are dissolved in plasma. Most polychaetes have hemoglobin, but chlorocruorin is common in the tubiculous types and other pigments occur as well (see Chap. 9).

Breathing occurs mainly through the epidermis, particularly in the parapodia, which wave back and forth and thereby circulate water around the body.

In numerous polychaetes the notopodia of some segments are extended as branched *gills*. The metanephridia open ventrally, near the bases of the parapodia. A few polychaetes have solenocytic protonephridia like their trochophores, and certain others have nephromixia (see Fig. 10.14). Gonads usually are attached to the peritoneal septa and in most polychaetes gonoducts are absent. Gametes leave through the metanephridial channels or simply accumulate in the coelom until a segment bursts.

Pelagic polychaetes swim by sinuous body movements and the parapodia aid as paddles. Neural control of muscles is metameric; segmental reflexes leading to contraction of longitudinal body wall muscles in one segment lead simultaneously to relaxation of the circular muscles of the same segment and to contraction of the longitudinal muscles in adjacent segments. The ventral ganglia exercise the chief local control and the brain ganglia serve as important locomotion *inhibitors;* a decapitated worm moves more violently and continuously but less purposefully and with less control than an intact worm.

Many pelagic polychaetes swim only when necessary and prefer to remain in burrows. *Nereis* is among numerous types that dig simple burrows in sand. Others line their burrows with secreted tubes. In such cases the setal pits in the notopodia produce plastic threads that become matted together as a tubular fabric. Such tube-formers among pelagic polychaetes differ from the true tubiculuous types in that they readily leave their burrows and tubes for feeding and changes of location (hence the name *Errantia*—errant, migratory worms).

During the breeding season, Errantia undergo a number of structural changes: body colors and textures become altered, the parapodia enlarge, and long swimming setae are formed. In *Nereis* such dimorphic changes occur in each sex in the posterior half of the body, where the segmental coeloms concurrently fill with sperms or eggs. On certain fixed nights, determined in part by environmental factors such as the amount of moonlight in a particular month of the year, the posterior sexual parts of the worms become detached and swarm separately to the surface of the sea. These worm sections, called *heteronereis,* then burst and disintegrate and the released sex cells effect fertilization. The nonswarming body parts later regenerate the lost heteronereis regions (Color Fig. 27).

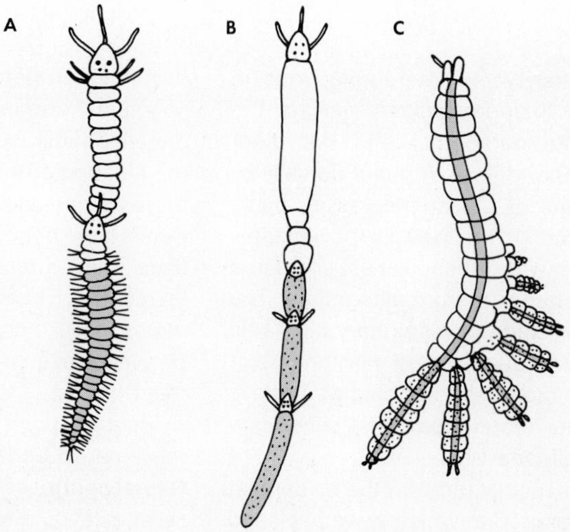

Reef-dwelling palolo worms swarm precisely at one or two predictable nights of the year—during the last quarter of the October-November moon in a South Pacific type (*Eunice*) and the third quarter of the June-July moon in a West Indies type (*Leodice*). These nights are anticipated eagerly by the local islanders, who gather up the swarming palolos and eat them broiled. In certain polychaetes the sexual body parts are not merely posterior sections but whole budded worms. Some types develop single posterior worm buds; others bud off a succession of worms, either in posterior strings or along the side of the body (Fig. 24.7).

The tubiculous *Sedentaria* live in secreted straight or U-shaped tubes that are left only rarely, if at all. Such worms are ciliary plankton feeders. The head segment is greatly reduced but the prostomial and peristomial appendages are highly elaborated and form a wide variety of plankton-catching devices. Among these are the feathery, retractile tentacles of sabellids and serpulids ("feather dusters"; see Color Fig. 1). One of the tentacles often is modified as a stopperlike *operculum* that on retraction closes the tube opening. In many of the tubiculous worms some of the anterior parapodia are modified as gills.

A one-way water flow is maintained through the U-shaped parchmentlike tube of *Chaetopterus* (Fig. 24.8). The worm lies at the bottom of the U with its ventral side downward. The peristomium is enlarged as a tubular cuff around the prostomium, and the notopodia of the anterior segments form food-collecting flaps. More posterior notopodia are fused as large *fans* that keep water flowing through the tube. Plankton in the water is trapped along the ciliated surfaces of the collar and the anterior notopodia. Food particles then are propelled over ciliated grooves, entangled in mucus, and conducted to the mouth. Tubiculous polychaetes do not swarm in the breeding season. Instead, the gametes are shed through the metanephridia and fertilization occurs in free water.

The burrowing polychaetes, exemplified by the lugworm *Arenicola,* dig tunnels in mud or sand along the shore, swallow some of the mud, and live on the organic detritus contained in it. Residual mud is expelled through the anus. The coiled

Figure 24.7 Polychaete budding. Sexual parts in color and stippled. *A, Syllis,* budding of single posterior sexual individual. *B, Autolytus,* budding of posterior chain of sexual individuals. *C, Trypanosyllis,*. formation of posterior lateral sexual buds. The alimentary tract is continuous to the posterior bud. Note that the most posterior individuals are the oldest and thus the most developed.

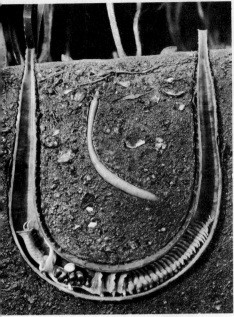

B

Figure 24.8 Polychaetes. A, model of a section through the tube of the parchment worm *Chaetopterus*. The head of the animal is at left. Note the greatly elaborated parapodia. Between the arms of the U tube is a sipunculid. *B,* the lugworm *Arenicola,* a burrowing type.

earth castings of lugworms are familiar sights along beaches. *Arenicola* has a reduced head, an eversible pharynx for digging, and branched gills on the parapodia at midbody. Lugworms virtually lack intersegmental septa, and nephridia and gonads are formed only in the mid-section of the body.

A few polychaete genera are commensalistic (in sponges, for example), and some are parasitic in crabs, echinoderms, or other polychaetes. Feather dusters include species that live in freshwater lakes. Like annelids generally, polychaetes have a highly developed regeneration capacity; a few segments normally suffice to produce a whole new worm. This regenerative potential undoubtedly relates to the metameric construction, for even a single segment virtually represents a complete "subindividual" that contains a full set of internal organs.

Archiannelids   ***Class Archiannelida***   marine, some parasitic; structurally simplified; without external segmentation; parapodia and setae generally absent; with epidermal ciliation and nerve cords in epidermis; sexes separate.

The animals in this very small class once were thought to represent the most primitive annelids (hence the name, "ancient" annelids). It is now recognized, however, that they are probably secondarily simplified offshoots of pelagic polychaete ancestors. The simplification appears to have been achieved by neoteny. Permanently retained trochophoral traits include absence of external segmentation, ciliation on the ventral side of the body, a nervous system that is part of the epidermis, the frequent development of solenocytic protonephridia, and the general absence of parapodia and setae (even though setae do occur in *Saccocirrus*). Internal segmentation is well developed, however, and intersegmental septa are present (Fig. 24.9).

527

mostly terrestrial and freshwater; head reduced, without appendages; parapodia absent, setae few; hermaphroditic, gonads in specific segments only; fertilization internal, development without larvae.

Earthworms probably are descendants of marine polychaete ancestors. The animals dig extensive tunnel systems by swallowing soil as they burrow along. They make use of any food present in the soil and egest the remainder.

Oligochaetes have a reduced prostomium without appendages (Fig. 24.10). In the trunk segments parapodia are absent. But each segment bears four pairs of retractile setae, two pairs on each side, corresponding to the notopodial and neuropodial setae of polychaetes. Internally the alimentary tract consists of pharynx, esophagus, crop, gizzard, and intestine. On the esophagus are paired *lime glands* that secrete calcium ions from blood into the alimentary tract. Blood calcium levels tend to be high because earthworms absorb a great deal of calcium from the soil they swallow. The crop is a storage compartment, and the strongly muscled gizzard grinds swallowed earth. Along the dorsal side of the intestine runs a *typhlosole*, a fold that projects into the intestinal cavity and increases the absorptive area. The nervous system is similar to that of polychaetes.

In addition to a longitudinal dorsal and ventral blood vessel, earthworms also have a ventral subneural vessel between the body wall and the nerve tract. The dorsal vessel is not contractile and contains valves that prevent blood from flowing backward. Circulation is maintained by five pairs of contractile "hearts," connective vessels in segments 7 to 11 that join the dorsal and ventral main channels. The metanephridia are as in polychaetes. Oligochaetes also have excretory *chloragogue* cells in the peritoneum surrounding the gut and the main blood vessels. These cells absorb wastes and then detach and float free in the coelomic fluid. The cells can be eliminated through the metanephridia, or coelomic amoebocytes engulf the chloragogues and migrate to the body wall, where they deposit the excretion products as pigments. In some oligochaetes the nephridia open into the alimentary tract, and in others each segment can contain more than one pair. For example, *Megascolides* has up to two thousand or more "micronephridia" per segment.

The most pronounced differences between polychaetes and oligochaetes are reproductive (Fig. 24.11). Oligochaetes are hermaphrodites (and many develop parthenogenetically). In segments 10 and 11 of *Lumbricus* are paired testes, with pouched extensions that serve as sperm-storing sperm sacs. Sperm ducts lead to ventral gonopores in segment 15. In some oligochaetes (but not *Lumbricus*) the terminal parts of the sperm duct pass through an eversible penis. Segments 9 and 10 of *Lumbricus* contain paired *spermathecae*, receptacles that store sperms from another worm after mating. The female system consists of a pair of ovaries in segment 13 and oviducts (with egg sacs) that open in segment 14. An important reproductive role is played by a conspicuous *clitellum*, a glandular epidermal band in which the segmental divisions are obscured. Formed only during the breeding season in many oligochaetes but present permanently in earthworms, the clitellum of an earthworm extends over segments 31 to 37. A pair of ventral mucus-forming grooves connects the male gonopores in segment 15 with the clitellum.

In mating, two worms come into contact with their heads pointing in opposite directions. In the region between segment 9 and the posterior end of the clitellum each worm secretes a mucus

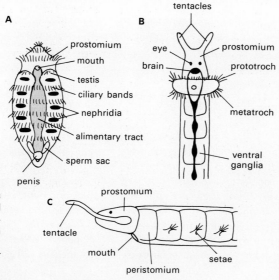

Figure 24.9 Archiannelids. A, *Dinophilus*, male. Note five pairs of nephridia, five ciliary rings, and trochophorelike head. *B, Polygordius*, anterior region. Again note trochophore remnants. *C, Saccocirrus*, side view of anterior region.

A
prostomium
mouth
testis
ciliary bands
nephridia
alimentary tract
sperm sac
penis

B
tentacles
eye
brain
prostomium
prototroch
metatroch
ventral ganglia

C
prostomium
tentacle
mouth
peristomium
setae

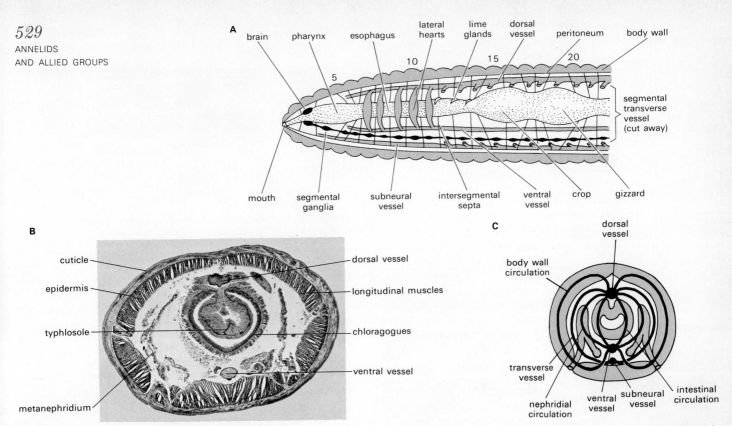

Figure 24.10 Earthworm structure. A, cutaway diagram of anterior body region, segments numbered. Segment 12 and all segments behind it contain on each side a transverse vessel (shown in cutaway view) that interconnects the dorsal and subneural longitudinal blood vessels. *B,* cross section through intestinal region. *C,* dorsoventral circulation pattern in each trunk segment. Blood vessels form extensive capillary beds (not shown) in the body wall, the nephridia, and the gut wall.

sheath around itself. Sperms then released by both worms are propelled in the mucus grooves from segment 15 of one worm to segments 9 and 10 of the other. There the sperms enter the spermathecae and are stored, whereupon the worms separate. The clitellum of each worm now secretes a mucus sheath that peristaltic movements of the body wall propel anteriorly. As this sheath passes segment 14 several eggs are shed into it; and as it continues past segments 9 and 10, sperms from the other worm are deposited in it. Fertilization then takes place inside the mucus sheath, and after the sheath slips over the head of the worm the open ends seal up. An egg *cocoon* forms in this manner. Development follows the typical an-

nelid pattern, but the trochophore stage is greatly abbreviated and a free larval phase is absent.

Lumbricus is a large oligochaete, although the Australian giant earthworms (such as *Megascolides*) attain lengths of 10 ft and diameters of an inch or more. By contrast, aquatic oligochaetes are quite small, some measuring less than 1 mm, and they also appear to be the more primitive types. They live at the bottoms of lakes, ponds, and stagnant waters, heads buried in mud and their posterior ends waving incessantly in the open water. Such movements aid in breathing. Many of these freshwater types can reproduce by spontaneous fragmentation and by posterior budding.

Leeches   ***Class Hirudinea***   mostly freshwater parasites, some carnivorous; with terminal suckers; parapodia and setae absent; number of segments fixed throughout life; coelom packed with mesenchyme; hermaphroditic, with clitellum; copulation and development as in oligochaetes.

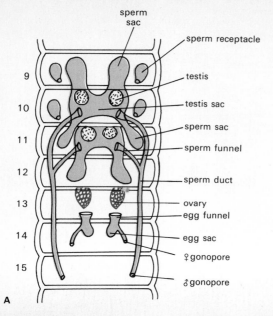

Figure 24.11 *Earthworm reproduction. A*, ventral view of reproductive system, segments numbered. *B*, photo showing position and extent of clitellum. *C*, copulating earthworms. *D*, egg cocoons. (*A, adapted from Hesse; D, courtesy Carolina Biological Supply Company.*)

There is little doubt that leeches represent evolutionary offshoots of oligochaetes that have become ectoparasitic. Transitional types such as salmon leeches (*Acanthobdella*) have two pairs of setae per segment, fewer segments than usual for leeches, and only posterior suckers. In other leeches setae are entirely absent, the number of internal segments is fixed at 34, and suckers typically occur at both ends of the body (Fig. 24.12). The head region, usually the first six segments, bears a pair of eyes and an anterior sucker. The most posterior seven segments form another sucker. A ''segment'' in a leech refers to an *internal* metamere only; each is annulated externally by 2 to 16 superficial cuticular rings, the exact number fixed for each species. Thus there are always more external ''segments'' than true internal ones.

Leeches are classified as three orders. One is exemplified by *Acanthobdella*, in which the pharynx is unarmed and not eversible. In a second order the pharynx is an eversible unarmed proboscis that can be protruded through the mouth and forced into soft tissues of host animals (fish gills, for example). *Placobdella* is a member of this group. In the third, most specialized order, the pharynx is not eversible but is armed with three

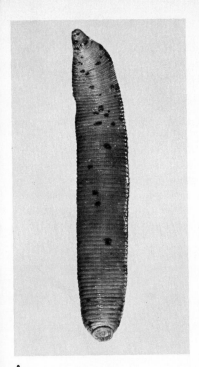

A

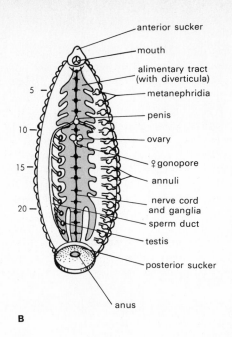

anterior sucker

mouth

alimentary tract
(with diverticula)

metanephridia

5

penis

10

ovary

♀ gonopore

15

annuli

nerve cord
and ganglia

20

sperm duct

testis

posterior sucker

anus

B

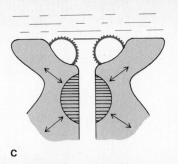

C

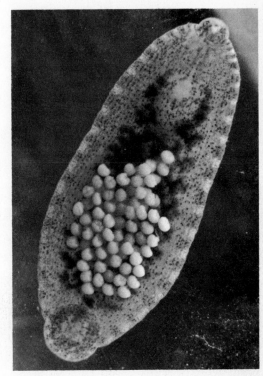

D

Figure 24.12 Leeches. A, note sucker at each end of this leech and the external transverse creases of the body. These are more numerous than the true internal segments. *B,* some of the internal organs. Numbers refer to true internal segments. *C,* the ingestive saw apparatus in anterior sucker. The muscular pharynx behind the saws provides suction for ingestion of blood through an incision made by the saws. *D,* ventral view of a leech of turtles, with developing eggs attached to underside of body. (*A, courtesy Carolina Biological Supply Company.*)

sawlike chitinous jaws. With these a leech can cut a Y-shaped incision through even tough skin or armor such as fish scales (see Fig. 24.12).

Most of these leeches are blood suckers. They have salivary glands that secrete the anticoagulant *hirudin* into a wound cut by the pharyngeal saws. Also present are a crop and numerous intestinal side pockets. These can hold up to ten times as much blood as the weight of the whole leech and enable the animal to survive as long as 9 months from a single feeding. The "medicinal" leech, *Hirudo medicinalis,* was widely used in earlier times for "bloodletting," then thought to be bene-

ficial. Today leeches on occasion still obtain human blood in freshwater environments, but fishes are the more usual hosts. Moreover, many of these leeches are not obligatory blood feeders; they normally survive carnivorously on worms and other invertebrates. Some leeches actually are nonparasitic.

Leeches reproduce like oligochaetes, and a clitellum usually performs a similar cocoon-forming function. To a certain extent leeches are reminiscent of flukes, an instance of parallel evolution under similar conditions of life.

spoon worms; marine; metameres in larva but unsegmented as adults; with proboscislike prostomium; one pair of anteroventral setae; coelom not partitioned; one to three pairs of metanephridia; sexes separate, dimorphic in some cases; development schizocoelic, with trochophore larva.

This small phylum of about sixty species comprises marine plankton feeders that live in coastal sand or mud burrows (Fig. 24.13). Attaining lengths of some 3 in., the plump trunk of such a worm bears an extended, tubular *prostomium* anteriorly. The tip is roughly spoonshaped in *Echiurus* but forms two tissue flaps in *Bonellia*. A highly mobile organ, the prostomium can contract but does not retract into the trunk. At the base of the prostomium is the mouth and a pair of ventral setae. *Echiurus* also has a circlet of bristles around the anus.

Internally a spoon worm has a coiled alimentary tract, a simplified nervous system consisting of anterior nerve ring and longitudinal nerve cords, a circulatory system composed mainly of a dorsal and a ventral vessel, and metanephridia that open anteroventrally on the trunk. A single gonad discharges gametes through the spacious coelom and the metanephridia. Sexes are separate, and in *Bonellia* the dwarfed males are parasitic in females (see discussion of sex determination, Chap. 12). Other genera are not dimorphic in this way.

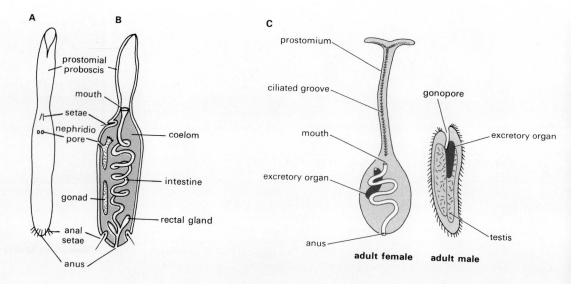

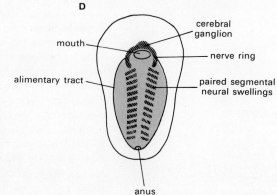

Figure 24.13 Echiuroids. A and *B,* external ventral features and sagittal section of *Urechis. C,* some structural details of *Bonellia.* The reduced and parasitic male is very small and lives permanently within the excretory organ of the female. *D,* ventral view of *Echiurus* larva, showing abortive neural segments. (*A, B, adapted from Fisher; C, D, adapted from Baltzer.*)

The echiuroid larva is a typical trochophore that in later stages exhibits rudimentary segmentation; 15 mesodermal somites arise, and the developing ventral nerve cord acquires a corresponding number of segmental swellings and paired lateral branches. But these beginnings of metamerism never become elaborated further. Indeed they become obscured later and they eventually disappear as the adult stage is reached.

In the past, echiuroids have been grouped loosely with the sipunculids, which they resemble superficially (and even with the priapulids, which they do not resemble). However, in view of the larval metamerism and the sparse setae of the adults, echiuroids now are recognized as a separate and distinct group. Whereas sipunculids probably exemplify an unsegmented *pre*annelid stage of schizocoelomate evolution, echiuroids appear to represent a *post*annelid stage in which segmentation and other annelid traits are reduced or lost.

Oncopods

Phylum Oncopoda

claw-bearing animals, fundamentally segmented; usually with unjointed legs; sexes separate, development various. *Onychophora, Tardigrada, Pentastomida*

These animals represent a ''phylum of convenience,'' as assemblage of three small groups perhaps interrelated very distantly. They have two features in common: claws of a type characteristic of arthropods, and various other traits that are more or less intermediate between those of polychaetes and arthropods. Each of the three groups has been and still is classified in widely different ways, either as a separate phylum or as a subphylum or class with arthropod affinities. The present grouping of all three in one phylum follows a recent proposal and appears to be justified on at least structural grounds. Apart from their classification, the Oncopoda probably represent surviving remnants of three separate and independent lines, all offshoots of a main line that led from annelid ancestors to arthropod descendants (see Fig. 24.1).

Onychophorans

Subphylum Onychophora

terrestrial, wormlike; with many pairs of claw-bearing legs and corresponding internal segments; external segmentation absent; annelid and arthropod traits mixed; sexes separate; mostly ovoviviparous and without larvae.

Of all oncopods the Onychophora are most clearly intermediate between annelids and arthropods. Indeed their very existence provides one of the best proofs that annelids and arthropods are closely related. The group comprises some seventy species, distributed widely in tropical and subtropical regions of the world. About 2 to 3 in. long, forms such as *Peripatus* live in damp, leafy places on the ground and feed mainly on insects (Fig. 24.14).

The caterpillarlike body is covered with a non-chitinous velvet-textured cuticle. Under the epidermis are two muscle layers as in annelids. The head consists of three segments (one in annelids and six in most arthropods). The first bears a pair of annelidlike eyes and a pair of insectlike *preantennae*. A pair of *oral papillae* on the second segment contains the openings of elongated *slime glands*. The secretion of these glands can be shot out explosively to entangle prey or enemy. The third segment contains the mouth, equipped with a pair of horny biting *jaws*.

Along the trunk, the stumpy paired legs each carry two recurved claws (undoubtedly homologous to polychaete parapodia and setae). The legs mark the position of internal segments and segmentally arranged organs. The nervous system is annelidlike, although distinct ganglia are absent. Also annelidlike are the multiple pairs of metanephridia that open ventrally at the bases of the legs. However, the nephridial tubules only drain small coelomic sacs—coeloms are greatly reduced in size and segmental septa are absent, as in arthropods. The hemocoelic blood spaces are corre-

A

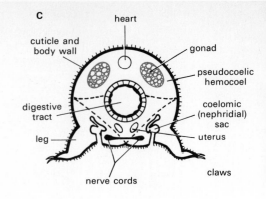

C

heart

cuticle and
body wall

gonad

pseudocoelic
hemocoel

digestive
tract

pseudocoelic
hemocoel

coelomic
(nephridial)
sac

leg

uterus

claws

nerve cords

B

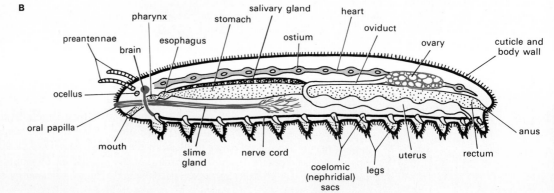

pharynx

salivary gland

stomach

heart

preantennae

brain

esophagus

ostium

oviduct

ovary

cuticle and
body wall

ocellus

oral papilla

mouth

slime
gland

nerve cord

coelomic
(nephridial)
sacs

legs

uterus

rectum

anus

*Figure 24.14 Onychophora. A,
external dorsal view of* Peripatus,
*head at left. B, sagittal section,
and C, cross section, of female*
Peripatus, *showing the position of
the principal organs.*

spondingly large and, again as in arthropods, the blood circulation is open; the only vessel is a single dorsal *heart* with lateral segmental openings, or *ostia*. Another very arthropodlike feature is the breathing system, which consists of *tracheal tubes*. But the openings of these tubes are scattered over the body surface and not arranged segmentally, they lack closing devices, and the tracheal tubes are unbranched.

Gonads are paired, and a common duct opens just in front of the anus. This duct is ciliated internally, as in annelids and in contrast to arthopods, in which cilia are absent. Fertilization occurs internally, and in most cases the zygotes are retained in the female. Some species are truly viviparous—the maternal body contributes to the nutrition of the embryo. As in arthropods, the eggs are centrolecithal and cleave superficially. Later development too follows the arthropod pattern quite closely, with residual annelidlike processes producing the wormlike traits of the adult (Fig. 24.15). The whole animal actually represents as perfect a ''missing link'' between two other animal types as is known in zoology.

Tardigrades   *Subphylum Tardigrada*

water bears; microscopic, mostly terrestrial, with six segments and four pairs of claw-bearing legs; cell numbers constant; without circulatory, excretory, or breathing systems; sexes separate; enterocoelic coelom; development largely oviparous, without larvae.

Tardigrades are a cosmopolitan group of about 350 species. Some are aquatic and largely carnivorous; they pierce nematodes, rotifers, or other tardigrades and suck up their fluid contents. Most species are terrestrial and herbivorous; they live among algae, lichens, mosses, or liverworts and eat the cellular interiors of such plants.

Water bears rarely exceed 1 mm in length. The

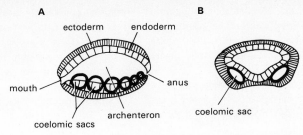

Figure 24.15 *Peripatus development.* A, side view of early embryo with anteroposterior series of coelom sacs. B, cross section of stage as in A. (*Adapted from Sedgwick.*)

stubby body consists of a head and five trunk segments (Fig. 24.16). The first four trunk segments bear short, stumpy legs with claws. On the head segment are the openings of a pair of *oral glands* and also horny, partly chitinous *stylets,* used in piercing the walls of food organisms. The mouth leads to a straight alimentary tract that terminates at the anus in the last segment. The epidermis is covered by a nonchitinous cuticle that

also lines the foregut and hindgut and that is molted several times during the life of the animal. A segmentally arranged body musculature underlies the epidermis. The nervous system is annelid-like and contains four segmental ganglia. A pair of eyecups is embedded on each side of the brain ganglion. The body cavity lacks a peritoneum in the adult and serves as a hemocoelic blood sinus, as in arthropods. Excretion takes place through the oral glands, the alimentary tract, excretory glands that open to the midgut, and the body surface. The epidermis also serves as the main breathing structure.

Like rotifers and other aschelminths, tardigrades exhibit cell constancy, a phenomenon that appears to be common in very small but comparatively complexly structured animals. Again like rotifers, tardigrades have the capacity to dehydrate to cystlike bodies when food, water, or oxygen are in deficient supply. In such a dormant state an animal can remain viable for up to 5 years, and it can be revived some fourteen successive times from a dormant condition before death supervenes. Tardigrades therefore might have a theoretical life span of at least 70 years.

The animals are separately sexed, with a single gonad in a dorsal sac. Fertilization is external and, the eggs also develop externally, directly in most cases. Cleavage follows an irregular pattern, a coeloblastula forms, and gastrulation occurs by delamination. After an embryonic gut has developed, five pairs of coelom sacs form from it by *enterocoelic* lateral evaginations (see Fig. 24.16).

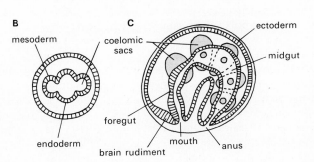

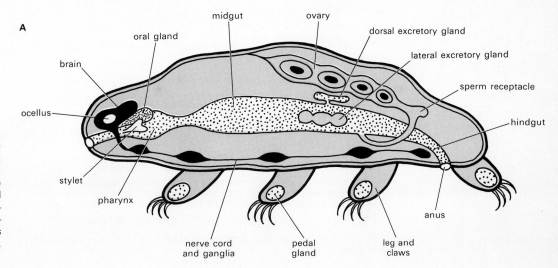

Figure 24.16 *Tardigrades.* A, sagittal section of *Macrobiotus* showing principal organs. The oral gland and stylet each are paired, but only the right members are shown here. B, cross-sectional view, and C, sagittal view of tardigrade embryo, showing enterocoelous formation of coelom sacs from early alimentary tract. (*Adapted from Marcus.*)

The first four pairs of sacs give rise to the musculature, the fifth pair to the reproductive system. By the time the adult stage is attained few remnants of the coelom sacs are left, and the body cavity then remains a pseudocoelic hemocoel.

Tardigrade development and adult structure indicate that the animals combine certain polychaete features, a substantial number of arthropod features, several nematode features, and quite a few features peculiar only to tardigrades. Most unexpected is the enterocoelic method of coelom formation. Conceivably the blastocoel might not be large enough for accumulation of teloblast bands, and direct formation of endodermal coelom sacs could represent an available shortcut.

Pentastomids

Subphylum Pentastomida

blood-sucking endoparasites in vertebrates; larvae with claw-bearing legs; adult wormlike, with two pairs of claws on side of mouth; circulatory, excretory, breathing systems absent.

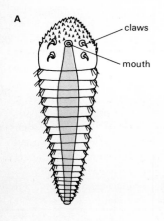

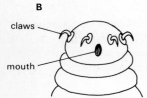

Figure 24.17 Pentastomids. A, Linguatula, from ventral side. B, Porocephalus, head region. (A, adapted from Leuckart; B, adapted from Hill.)

The approximately fifty species of these animals (also known as *Linguatulida*) live primarily in the nasal passages, windpipes, and lungs of amphibia, reptiles, and other vertebrates. The adults are wormlike, up to 5 in. long, with superficial annulations externally (Fig. 24.17). Recurved claws lateral to the mouth enable them to remain attached to host tissues. Air pores are scattered randomly over the body surface. The nervous system is of the ladder type and the reproductive system is elaborate.

Sexes are separate and dimorphic, the female being much larger than the male. Fertilization occurs internally and the zygotes reach aquatic environments through the host mouth. In most species the eggs then must become encysted larvae in intermediate hosts such as fishes or amphibia.

The arthropod affinities of the group are suggested particularly by these larvae. They resemble mites and they have four to six pairs of stumpy legs, each with two terminal claws. Also, the larvae molt their cuticles at intervals. If the intermediate host later is eaten by a final host, the larvae excyst in the host gut and find their way to the breathing system, where they become adults. Some pentastomids complete their life cycles without intermediate hosts.

The evolutionary connection between pentastomids and arthropods is distant and tenuous, and the parasitism of the animals obscures any affinities still further. The claw-bearing legs of the larvae, the larval molts, and also the adult air pores are the main justifications for including these animals among the Oncopoda.

Review Questions

1. Define Sipunculida taxonomically and describe the structure of these animals. Where do sipunculids live and how do they feed and move?

2. What are urns and what is their possible function? On what basis and how can peanut worms be regarded as related to annelids?

3. Give taxonomic definitions of annelids and of each of the classes. Describe the development of a segmented body organization from a trochophore larva. Then describe the segmental structure of an adult polychaete.

4. What is the organization of a parapodium? What are the basic adaptive advantages of metamerism? Describe the nervous, circulatory, and excretory systems of *Nereis*. How is breathing accomplished?

5. Where do polychaete gonads form and how does fertilization occur? What is heteronereis? In what different ways do polychaetes produce swarming body parts?

6. Distinguish between errant, tubiculous, and burrowing polychaetes. Name representative genera of each group. Show how feather-duster and parchment worms are adapted to tubiculous ways of life. Describe the external appearance of a lugworm.

7. In what respects do archiannelids and oligochaetes differ from polychaetes? Describe the external and internal structure of an earthworm and contrast it with *Nereis*. What are typhlosole, crop, lime glands, and what are their functions?

8. Describe the structure of the earthworm reproductive systems. Show how mating and fertilization occur. How does earthworm development take place?

9. Describe the structure of a leech and contrast it with that of an earthworm. In what different ways do leeches feed? What are their adaptations to ectoparasitism? How do leeches mate and develop?

10. Define echiuroids taxonomically and describe the structure of such a worm. In what respects is this structure different from that of an annelid and a sipunculid?

11. How does an echiuroid develop and in what sense do these animals represent a postannelid stage of evolution? Review the development of sexual dimorphism in *Bonellia*.

12. Define Oncopoda and each subphylum taxonomically. Describe the organization of *Peripatus* and list features that are annelidlike, arthropodlike, and unique to onychophorans. How does *Peripatus* develop and where and how does it live?

13. Describe the structure and development of tardigrades. In what respects are these animals rotiferlike?

14. Outline the structural traits of pentastomids. What features are reminiscent of arthropods? Describe the life cycle of a pentastomid.

15. Make a diagram to show the presumed evolutionary relations of all schizocoelomates and all main subgroups in the schizocoelomate phyla. Relate the whole assemblage to lophophorates and to ancestral coelomates.

Collateral Readings

Berrill, N. J.: "Growth, Development, and Pattern," Freeman, San Francisco, 1961. Chapters 12 and 13 in particular contain interesting discussions of segmental growth and development in annelids.

———— and D. Mees: Reorganization and Regeneration in *Sabella, J. Exp. Zool.,* vol. 73, 1936. A representative case study.

Gates, G. E.: On Segment Formation in Normal and Regenerative Growth of Earthworms, *Growth,* vol. 12, 1948. See also the paper by Moment cited below.

Hyman, L.: "The Invertebrates," vol. 5, McGraw-Hill, New York, 1959. Chapter 22 of this volume is a detailed account on sipunculids.

Marcus, E.: Tardigrada, in "McGraw-Hill Encyclopedia of Science and Technology," rev. ed., vol. 13, 1966. A short but very thorough account.

Moment, G. B.: A Study of Growth Limitation in Earthworms, *J. Exp. Zool.,* vol. 103, 1946. The title describes the contents.

25

ARTHROPODS: CHELICERATES

Phylum Arthropoda

jointed-legged animals; metamerically segmented; with chitinous exoskeleton; typically with compound eyes; sexes largely separate; eggs centrolecithal, cleavage superficial and mosaic; larvae various or absent.

† Subphylum Trilobita

trilobites (extinct)

Subphylum Chelicerata

without jaws or antennae.

Class Xiphosurida   horseshoe crabs.   *Class Pantopoda*   sea spiders.

† *Class Eurypterida*   eurypterids (extinct)   *Class Arachnida*   scorpions, spiders, ticks, mi

Subphylum Mandibulata

with jaws and antennae.

Class Crustacea   crustacenas.   *Class Pauropoda*   pauropods.

Class Chilopoda   centipedes.   *Class Symphyla*   symphylans.

Class Diplopoda   millipedes.   *Class Hexapoda*   insects.

General Characteristics

This is the largest phylum not only among animals but all living organisms. Some one million species have been described to date; hence more arthropod types are known than all other living types combined, plants included. About 75 percent of the described species are insects, but according to one estimate some ten million insect species actually may exist. One of the orders of insects, the beetles, alone includes 300,000 species, which makes this order larger than any whole *phylum* of other organisms.

Arthropod sizes range from microscopic mites at one extreme to 5-ft giant crabs with 12-ft leg spans at the other. The animals are more widely and more densely distributed throughout the world than any others, and they occur in all environments and ecological niches, including many not open to other forms. Arthropods also are the main group that other animals must compete with, man not excepted. The chief competitors of arthropods are arthropods themselves, bacteria, and vertebrates. Arthropods in turn depend most on algae in water, on plants on land.

This unrivaled success is a consequence of the arthropod construction, which is based on metameric organization and exoskeletal armor. The original schizocoelomate ancestors from which arthropods evolved is unknown, but there is ample justification to assume that they were primitively polychaetelike, with metameric segments and paired segmental appendages. Such a stock then must have adopted a crawling existence, the parapodial appendages serving as legs. Later, the epidermal potential of forming chitin, restricted at first (and in annelids even today) to the production of setae, evidently became general and led to development of a chitinous cover over the whole body. This exoskeleton not only protected but also strengthened and supported mechanically; it must have permitted the legs to become more elongated and jointed in sections, hence more efficient.

Furthermore, far more so than in the annelid group, the parapodial appendages could become structurally and functionally different in different segments, for the chitinous cover could become molded and elaborated in a variety of permanent shapes. As a result, the segmental appendages of the evolving arthropods became not only walking legs, but also structures adapted to biting, cutting, sucking, piercing, cleaning, grasping, carrying, breathing, swimming, flying, egg laying, sperm transferring, sensing, and even silk spinning. In effect, the possibilities of diversification inherent in a segmental repetition of parts became exploited in an almost infinite variety of ways. Moreover, a chitin cover also served well in the elaboration of complex sensory receptors, including most particularly the unique compound eyes (see Fig. 8.38). Arthropods actually differ from each other far more in their exterior, chitin-covered parts than in their interior construction; and it is just this exterior variability that has made arthropods so hugely diversified.

An exterior chitin envelope puts constraints on body size and growth. Further, the presence of an inert cover in early stages rules out surface ciliation, hence also free-swimming trochophore larvae. Eggs therefore must be large, with enough yolk to form either a nonciliated (nontrochophore) larva or a small adult. Arthropod development actually has deviated drastically from the ancestral pattern of spiral cleavage and trochophores. Although the eggs have remained mosaic, they are centrolecithal and cleave superficially. Coelom sacs later form as in annelids, but they are small and eventually disappear as closed sacs, the body cavity becoming a hemocoel. Segmental appendages develop as outgrowths from ectoderm and mesoderm. As outlined in Chap. 11, larvae metamorphose gradually or abruptly or do not form at all.

The body divisions of an adult arthropod (Fig. 25.1) are *head*, *thorax*, and *abdomen*, and in many cases the first two form a fused *cephalothorax*. Any of the body divisions can be unsegmented externally as a result of fusion of segments in the embryo. Groups of fused or unfused segments often are covered by a *carapace*, an exoskeletal shield. The abdomen typically terminates at a *telson*, a nonsegmental end section.

Any segment can carry paired segmental appendages. All segments and appendages are covered by a continuous chitinous cuticle, secreted by the underlying epidermis (here also called *hypodermis*) and molted periodically during larval growth or throughout life. This cuticle is basically thin and pliable. It forms the joint membranes and

Figure 25.1 Arthropod structure. Lateral view of a wasp showing general segmental structure. Head is externally unsegmented and bears antennae, eyes, and mouth parts. The thorax, consisting of three segments in insects, bears three pairs of legs (one per segment) and two pairs of wings (on the second and third thoracic segments). The abdomen in insects typically consists of 11 segments and lacks appendages. Numbers of segments and types of appendages vary considerably for different arthropod groups, but sets of mutually different segments are present in all.

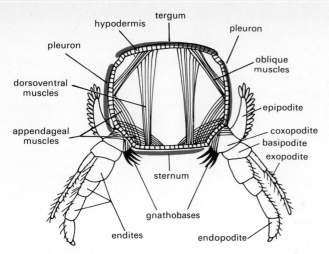

Figure 25.2 *Arthropod skeleton* and segmental appendages. Epipodites typically are gills. Endopodites typically terminate in claws or pincers. Only a few of the trunk and appendicular muscles are shown.

the breathing surfaces and is tucked in at both ends of the alimentary tract as a lining of foregut and hindgut. Between joints the cuticle is thickened as a hard, exoskeletal cover. Hardness results not so much from chitin itself as from secreted impregnating materials. These are horny scleroproteins and, in aquatic forms, calcium salts. Around a segment this hard cover typically forms four plates (*sclerites*): a dorsal *tergum*, a ventral *sternum*, and a *pleuron* on each side. These and the plates of adjacent segments are articulated by segmentally arranged interior muscles. (Fig. 25.2).

An appendage consists of several joints arranged as either a linear set (*uniramous*) or a bifurcated set (*biramous*). From the base forward, a uniramous appendage is composed of *coxopodite, basipodite,* and *endopodite.* The coxopodite often bears a *gnathobase,* a medial protuber-

Figure 25.3 *Internal arthropod structure.* A, lobster. B, grasshopper. See also illustrations in Chap. 26.

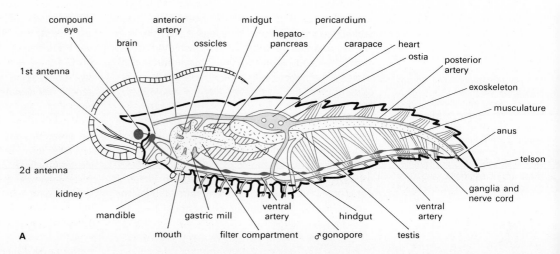

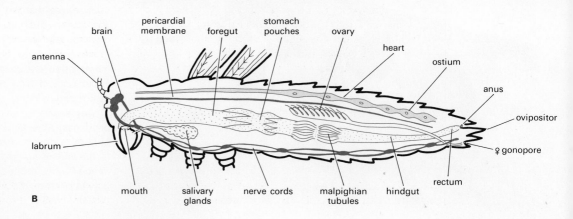

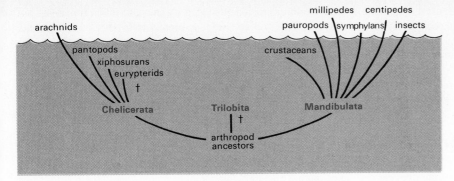

arachnids

pantopods
xiphosurans
eurypterids
†

Chelicerata

millipedes centipedes
pauropods symphylans insects

crustaceans

Trilobita
†

Mandibulata

arthropod
ancestors

Figure 25.4 Arthropod evolution. The presumed interrelations of the main groups are indicated. Primarily aquatic groups are shown below the water surface, primarily terrestrial groups, above.

ance studded with chitinous teeth. In many cases the coxopodite also carries a lateral lobe, or *epipodite*. In a biramous appendage, the basipodite carries both an endopodite and a lateral *exopodite*. All segmental appendages are ventral or ventrolateral outgrowths. Eyes and wings are dorsal outgrowths, not homologous to the segmental appendages.

Internally, the alimentary tract consists of foregut, midgut, and hindgut (Fig. 25.3). The first and last portions, lined with chitin as noted, are ectodermal and only the midgut is endodermal. Salivary glands open into the mouth of the anterior part of the foregut. Digestive glands often are large and conspicuous and connect with the stomach of the midgut. The hindgut usually terminates ventrally in the telson. Breathing is accomplished in most arthropods by gills, book gills, book lungs, or tracheal systems, all chitin-lined metameric outgrowths or ingrowths on body segments or appendages.

The circulation is open. A dorsal tubular heart is the main vessel, but others often are present. Excretion takes place chiefly through two types of structure. One type is represented by organs variously named *coxal glands, green glands,* or *renal glands*. Probably representing metanephridial derivatives, they open at the bases of segmental appendages and occur primarily in aquatic arthropods. Organs of the second type are *Malpighian*

tubules, attached to the hindgut and characteristic of terrestrial arthropods. In such animals the main nitrogenous waste is uric acid and the urine is semisolid. The nervous system basically is of the ladder type, as in polychaetes. In many cases, however, segmental ganglia formed in the embryo become concentrated as one or a few larger adult ganglia. Arthropods typically also have endocrine systems, composed of neurosecretory cells in the brain and a variety of glands in the thorax and at the bases of the compound eyes. The hormone secretions play a major role in the control of molting, in reproduction and development, and also in regulating the activities of epidermal chromatophores.

In the vast majority of arthropods the sexes are separate. Paired dorsal gonads in thorax or abdomen lead to gonopores in various segments of different groups. Arthropods are largely free-living, but symbiotic types of all categories are abundant as well. In their turn arthropods are favorite hosts of many other symbiotic animals. As outlined in Chap. 16, arthropods also exhibit a wide variety of unique behavioral and social traits, and the complexity of their constructed home sites is matched only by the installations of modern man. In effect, there is hardly a single phase or aspect of animal biology in which arthropods are not of foremost significance.

Evolution within the phylum appears to have proceeded along three separate major lines (Fig. 25.4). One gave rise to the subphylum Trilobita, extinct since the end of the Paleozoic. The second line now forms the subphylum Chelicerata, and the third, the subphylum Mandibulata. All three were primitively aquatic but each of the last two later gave rise to terrestrial subgroups. Thus the chelicerates now include the aquatic horseshoe crabs and sea spiders but also the terrestrial scorpions and spiders; and the mandibulates include the aquatic crustaceans but also the terrestrial centipedes, millipedes, and insects.

Trilobites   *Subphylum Trilobita*

extinct; marine; body marked dorsally into three lobes by two longitudinal furrows; paired identical ventral legs on all segments except first two and last; with compound eyes; development with larvae.

Some 2,500 fossil species of these animals are known (Fig. 25.5). They had a *head* with fused

segments and a trunk composed of segmented *thorax*, fused *pygidium*, and terminal *telson*. The first head segment lacked appendages; the second bore a pair of uniramous antennae; the third contained the mouth; and the third and all others carried paired locomotor legs that were biramous and had epipodite gills.

Trilobite development included larvae shaped like circular disks. Almost the whole disk repre-

sented the head. A small posterior pygidium produced thoracic segments in anteroposterior succession, while the pygidium itself grew backward and retained constant length. Segmental growth occurs in precisely this fashion in the embryos or larvae of living arthropods.

Chelicerates *Subphylum Chelicerata*

cephalothorax generally unsegmented, abdomen segmented or not; cephalothorax typically with six pairs of appendages; first pair *chelicerae* (3-jointed), second pair *pedipalps* (6-jointed); last four pairs *walking legs*; jaws or antennae never present; abdominal appendages not primarily locomotor or absent. *Xiphosurida, Eurypterida, Pantopoda, Arachnida*

Horseshoe Crabs *Class Xiphosurida*   marine; cephalothorax with carapace, hinged to fused abdomen; telson a spine; with compound eyes; "trilobite" larvae.

Representing the nearest existing relatives of trilobites, the four surviving species of horseshoe crabs are "living fossils" found along sandy shores, where they lead a borrowing, semi-

sedentary existence. The genus *Limulus* is at least 200 million years old (Fig. 25.6).

In *Limulus* the cephalothorax consists of eight embryonic segments (Table 11). These later fuse and develop a conspicuous, horseshoe-shaped carapace dorsally. On it are two small anterior ocelli and two lateral compound eyes. The first segment is embryonic only and does not develop ventral appendages. Each of the remaining seven cephalothoracic segments has a pair of uniramous appendages. The first pair are the small *chelicerae*. The pair of *pedipalps* behind them is indistinguishable from the next three pairs of *walking legs*. All these are pincer-equipped (*chelate*) in females and young males but terminate in claws in adult males.

All legs and the chelicerae carry prominent *gnathobases*, and the legs are positioned in such a way that the gnathobases surround the mouth. Horseshoe crabs feed on clamworms, soft-shelled clams, and other small animals. Such food is crushed and minced by the gnathobases when the legs are moved, and the food pulp, including sand and pieces of shell, then is pushed into the mouth by the chelicerae. These animals are the only chelicerates that eat solid food; all others subsist on liquids, a diet made necessary by the absence of jaws. The eighth cephalothoracic segment bears a pair of *chilaria*, small appendages without known function. In other chelicerates this eighth segment is embryonic only and lacks appendages.

The cephalothorax as a whole is hinged to the abdomen, composed of a broad anterior region of fused segments and a posterior telson spine. Apart from the telson the abdomen is formed from six

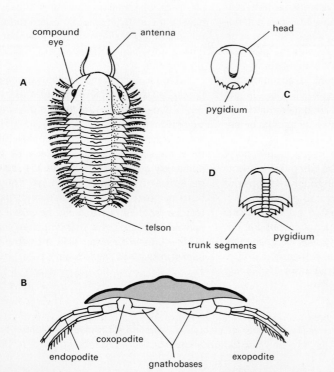

Figure 25.5 Trilobites. A and B, dorsal and cross-sectional views of *Triarthrus* (see also Fig. 15.6). C and D, two larval stages. The pygidium is the segment-generating body part. Segments continue to be formed throughout the larval phase, and thereafter the pygidium itself becomes segmented. (A, B, after Beecher; C, D, after Barrande.)

Figure 25.6 *Limulus,* the horseshoe crab. *A,* living pair on tidal flat. *B,* ventral view. Segments numbered according to sequence in embryo. The operculum is drawn turned forward to show gonopore on underside and to expose the book gills. *C,* dorsal view.

A

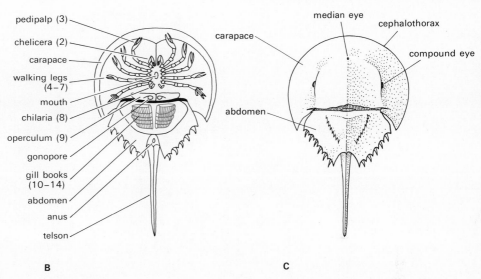

B

C

embryonic segments, each with a pair of biramous appendages. The first pair is fused along the midline and forms a transverse plate, the *operculum,* which covers and protects all posterior appendages. On the back face of the operculum are the gonopores. The next five pairs of appendages are broad, flat *book gills,* each formed from a narrow endopodite and a lateral platelike exopodite. On the back surface are numerous gill plates. By flapping their book gills horseshoe crabs breathe and also swim to some extent.

Internally, the alimentary tract contains a chitin-lined gizzard and a large, highly lobulated digestive gland that occupies much of the space under the carapace (Fig. 25.7). The anus opens at the base of the telson. The ventral ganglia of the cephalothorax are concentrated as an anterior fusion ganglion, and the nervous system as a whole is unusual in that the ganglia and main cords are sheathed by arterial blood vessels. The blood pigment is hemocyanin, dissolved in plasma. Excretion occurs through coxal glands that open at the bases of the last pairs of legs.

Females lay large, yolky eggs in sand burrows and males fertilize the spawn externally. The larvae resemble trilobites and also the adult horseshoe crabs, although a telson is absent. Maturation lasts up to 10 or 11 years, and several (usually annual) larval molts occur during this time. Adults do not molt.

Eurypterids   **Class Eurypterida**   extinct; marine; cephalothorax unsegmented, abdomen segmented; with telson; compound eyes present; segmental structure as in horseshoe crabs, shape of body scorpionlike (Fig. 25.8).

Table 11. *Segments and appendages in trilobites and chelicerate arthropods*

No.	Region	TRILOBITE	HORSESHOE CRAB	EURYPTERID	SCORPION	SPIDER	PANTOPOD
1	Head / Cephalothorax (Append. typic. uniramous)	embryo only	embryo only	embryo only	embryo only	embryo only	embryo only
2		antennae (uniramous)	chelicerae (chelate)	chelicerae (chelate)	chelicerae (chelate, small)	chelicerae (fangs)	chelicerae (chelate)
3		walking legs (biramous)	pedipalps (chelate)	pedipalps	pedipalps (chelate, large)	pedipalps (tactile)	pedipalps (tactile)
4		walking legs (biramous)	walking legs (chelate)	walking legs	walking legs (clawed)	walking legs (clawed)	ovigerous legs
5		walking legs (biramous)	walking legs (chelate)	walking legs	walking legs (clawed)	walking legs (clawed)	walking legs (clawed)
6		walking legs (biramous)	walking legs (chelate)	walking legs	walking legs (clawed)	walking legs (clawed)	walking legs (clawed)
7		walking legs (biramous)	walking legs (chelate)	walking legs	walking legs (clawed)	walking legs (clawed)	walking legs (clawed)
8		walking legs (biramous)	chilaria	embryo only	embryo only	embryo only	walking legs (clawed)
9	Pygidium, thorax / Abdomen (Append. typic. biramous)	walking legs (biramous)	operculum, gonopores	operculum, gonopores	operculum, gonopores	embryo only	reduced abdomen
10		walking legs (biramous)	book gills	book gills	pectines	book lung/tracheae	
11		walking legs (biramous)	book gills	book gills	book lungs	book lung/tracheae	
12		walking legs (biramous)	book gills	book gills	book lungs	spinnerets	
13		walking legs (biramous)	book gills	book gills	book lungs	spinnerets	
14		walking legs (biramous)	book gills	book gills	book lungs	14–18 embryo only	
				15–20 without appendages	15–20 "tail" without appendages		
telson		telson	telson	telson	telson		

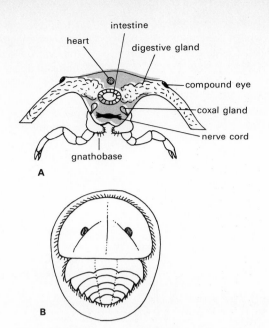

A

B

Figure 25.7 Limulus. A, cross section through cephalothorax showing a few of the organs. *B,* the ciliated "trilobite" larva of *Limulus,* dorsal view, still in its egg membrane; it will later swim free.

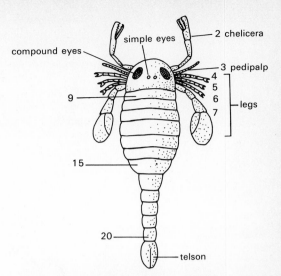

Figure 25.8 Eurypterids. Diagram of dorsal view, only cephalothoracic appendages shown. Segmental numbers take account of embryonic segments as well. See also Figs. 15.3 and 15.6 (*Adapted from Schmidt.*)

Sea Spiders **Class Pantopoda** head and thorax with legs; abdomen vestigial; mouth on proboscis; compound eyes absent; egg-carrying legs on head of both sexes; development via larvae or direct.

The 500 species of these spiderlike animals (also known as *Pycnogonida*) occur at different ocean depths down to the sea bottom. The usually minute bodies of pantopods have exceedingly long and thin legs (Fig. 25.9). On the head are four eyes and a sucking proboscis with the mouth at the tip. Pantopods feed mainly on the tentacles of sea anemones, which they tear open with their chelicerae. The head also bears tactile pedipalps, egg-carrying (*ovigerous*) legs, and a first pair of

walking legs with claws. On the ovigerous legs, which are generally smaller than the others, the *males* carry the developing eggs. The remaining walking legs are on three segments behind the head, and the abdomen is reduced to a tiny posterior protrusion. All legs contain pouches of the alimentary tract and portions of the gonads. The animals breathe through the body surface. Larvae are four- or six-legged.

Arachnids **Class Arachinda** terrestrial, some secondarily aquatic; usually carnivorous, predatory, or ectoparasitic; cephalothorax typically unsegmented, abdomen segmented or not; without compound eyes; breathing by book lungs, tracheae, or both; development largely direct.

Order Scorpionida scorpions; chelicerae small, pedipalps large, both chelate; abdomen segmented, second segment with tactile *pectines;* last six segments elongate, tail-like, terminating in poison sting (telson); ovoviviparous.

Order Pedipalpi whip scorpions; small chelicerae and large pedipalps, both chelate; first pair of legs tactile; four pairs of ocelli; abdomen flattened, segmented; with or without whiplike tail formed by appendages of last segment.

Order Palpigradi   small, scorpionlike; chelicerae strong, pedipalps leglike; without eyes; abdomen segmented, last 15 segments form narrow scorpionlike tail.

Order Araneida   spiders; cephalothorax and abdomen unsegmented, joined by narrow waist; chelicerae fangs, pedipalps leglike; up to four pairs of eyes; abdomen with spinnerets and book lungs, some with tracheae.

Order Solpugida   sunspiders; short pincered chelicerae, pedipalps leglike; two pairs of eyes; first pair of legs sensory; junction of cephalothorax and abdomen not narrow; abdomen segmented, without spinnerets.

Order Pseudoscorpionida   book "scorpions"; scorpionlike; chelicerae spin silk from anterior glands; pedipalps chelate, large; abdomen segmented, rounded, without tail or sting.

Order Ricinulei   spiderlike, small; without eyes; anterior abdominal segments form waist.

Order Phalangida   harvestmen, "daddy longlegs"; body small, cephalothorax and abdomen fused, abdomen segmented; chelicerae short and chelate, pedipalps leglike; one pair of eyes; legs long and thin; without spinnerets; breathing by tracheae.

Order Acarina   mites and ticks; free-living and ectoparasitic; body fused, unsegmented; breathing by trachea; with larvae.

Arachnids comprise well over 30,000 species (Fig. 25.10). Of these, spiders number about 20,000 species, mites and ticks about 6,000, daddy longlegs about 2,500, and scorpions roughly 700. Evolution in the class has resulted

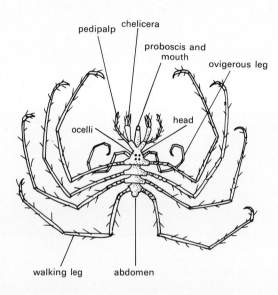

Figure 25.9 Pantopods: the sea spider *Nymphon. (After Mobius.)*

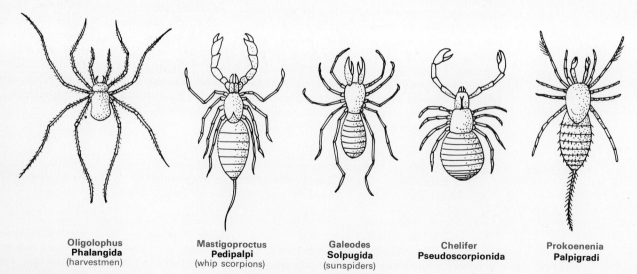

Oligolophus
Phalangida
(harvestmen)

Mastigoproctus
Pedipalpi
(whip scorpions)

Galeodes
Solpugida
(sunspiders)

Chelifer
Pseudoscorpionida

Prokoenenia
Palpigradi

Figure 25.10 Arachnids. Representatives of five of the arachnid orders are illustrated here. The text and the following illustrations describe the three main orders (Scorpionida, Araneida, and Acarina).

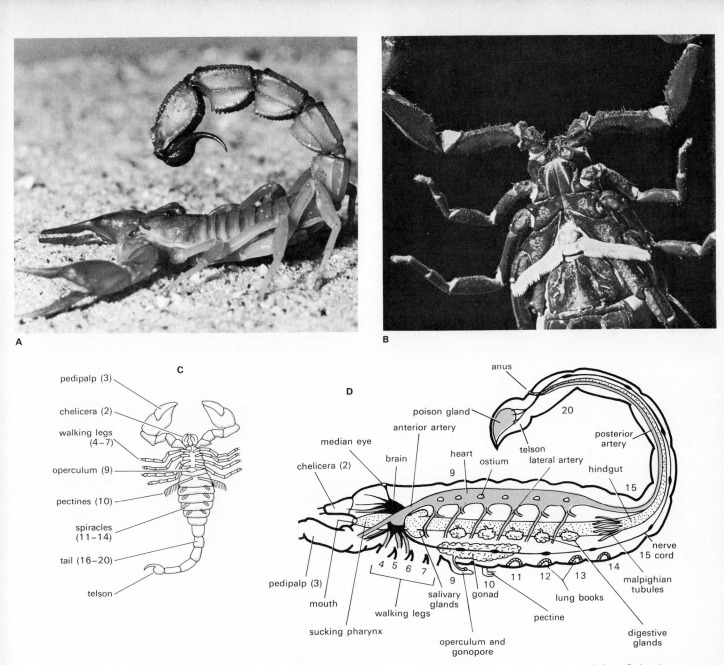

Figure 25.11 *Scorpions*. *A*, *Androctonus*, from life. *B*, ventral view showing pectines. *C*, external segmental structure, ventral view. *D*, interior structure. Segments numbered according to embryonic count. (*C, D, adapted from Leuckart.*)

in a progressively greater fusion of segments and body divisions. Scorpions probably are the most primitive group, spiders are more specialized, and mites and ticks are the most specialized arachnids.

Scorpions

A scorpion has a segmental structure similar to that of a horseshoe crab (Fig. 25.11 and Table 11). The cephalothorax, formed from eight em-

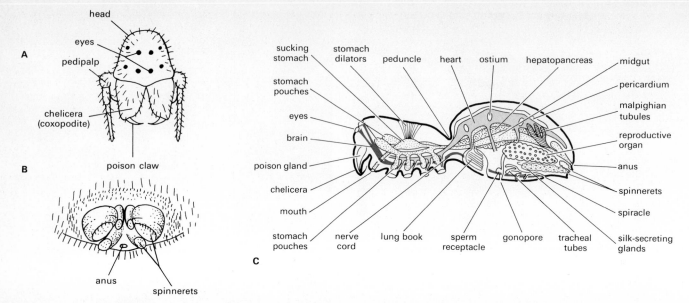

Figure 25.12 Spider structure. A, head from the front, showing chelicerae with fangs. *B,* tip of abdomen with three pairs of spinnerets. *C,* interior structure. The peduncle refers to the "waist" connecting the cephalothorax with the abdomen. *(Adapted from Warburton.)*

bryonic segments, is covered dorsally by a carapace. On the carapace are a pair of median ocelli and from two to five lateral ocelli on each side. Some scorpions are eyeless. The large pedipalps and the first two pairs of legs have gnathobases with a chewing function, as in *Limulus*. Scorpions feed mostly on insects and spiders, which are crushed by the gnathobases. The juices then are sucked into the mouth, located between the second and third segments.

The abdomen consists of 12 segments and a telson. On the first segment is a small operculum with gonopores, and the second bears a pair of comblike *pectines*. These tactile structures brush over the ground when a scorpion walks and serve as touch receptors. On each of the next four abdominal segments is a pair of ventrolateral *spiracles,* slits that lead to the book lung chambers.

Pectines and book lungs are homologous to the book gills of horseshoe crabs. The last five segments form a narrow "tail," and the telson contains a poison gland with a sharp, terminal sting. Scorpions carry their tail arched up and forward, and they use the poison sting to paralyze and kill prey that has been grasped by the pedipalps.

Saliva secreted to the outside predigests and partially liquefies prey. Fluid food then is drawn in by the pharynx, which produces suction when its diameter widens muscularly. A long midgut

connects with several pairs of digestive glands, and the beginning of the hindgut joins with several groups of Malpighian tubules. The nervous system is as in *Limulus*. A large, well-developed heart lies between segments 7 and 13. Each segmental section has a pair of lateral openings, the *ostia,* and also a pair of ventrolateral arteries. The blood contains hemocyanin in plasma. Gonads are diffuse, with gonoducts particularly well developed in females. Fertilization is internal and occurs after elaborate courtship dances. Internal development of the zygotes follows. Some types have highly yolked eggs and such scorpions are ovoviviparous. Others have small, isolecithal eggs that develop viviparously; placental connections are formed in pouches of the uterus.

Spiders

Although a spider and a scorpion are dissimilar in appearance, the basic segmental structure of the two nevertheless is almost identical, at least in the embryos. A spider embryo develops 18 anteroposterior coelomic sacs, each corresponding to an embryonic segment. The first 8 give rise to the cephalothorax, the last 10 to the abdomen (Fig. 25.12 and Table 11). In the adult the segments of both the cephalothorax and the abdomen are fused, and these two body parts are

A

B

D

C

Figure 25.13 Spiders. A, front view of head of wolf spider. Note chelicerae with fangs, large and small eyes. *B,* tarantula; note vertical fangs. *C,* aquatic diving spider, with web and spun air bell. *D,* trap-door spider in sectioned silk-lined burrow.

joined by a narrow *peduncle,* the characteristic ''waist'' of a spider. Up to eight eyes are on the cephalothorax; cave-dwelling spiders are blind.

The chelicerae are poison-gland-containing fangs, with poison ducts opening at the tips. In tarantulas the two fangs point downward, whereas in all other spiders they point toward each other (Fig. 25.13). Pedipalps are tactile, and in males they contain a specialized receptacle for the transfer of sperms during copulation. The gnathobases

of the pedipalps are the main chewing devices of spiders. Food consists chiefly of insects but *Mygale* attacks animals as large as birds and tarantulas readily subdue mice or small lizards. Prey is killed with the poison fangs, chewed and torn with the gnathobases, and predigested for an hour or more by proteinases in secreted saliva. The food juices then are sucked up, suction being produced by the stomach.

Walking legs often end in claws. The second and third abdominal segments develop paired book lung chambers or unbranched bundles of tracheae. Some groups of spiders have two pairs of book lungs only, some have two pairs of tracheal bundles only, and most have an anterior pair of book lungs and a posterior pair of tracheal bundles. Behind the spiracle openings of the breathing structures are the gonopores. Behind them are two or more pairs of *spinnerets,* modified abdominal appendages that provide outlets for the *silk glands* (see Fig. 25.12).

Spider silk is a scleroprotein that hardens on contact with air. All spiders produce silk, but not all construct webs. Those that do form them in different patterns—sheet webs, funnel webs, netted webs, and the beautiful orb webs (Fig. 25.14). Some species construct only horizontal webs, others only vertical ones, still others form their webs only in corners, tree forks, or other specific locations. Silk also is used in numerous

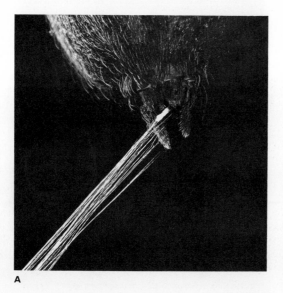

A

B

Figure 25.14 Spider webs. A, abdomen of house spider in process of spinning silk from spinnerets. *B,* an orb web.

other ways—the animals spin egg cocoons, sperm webs, molting sheets, attachment disks, burrow linings, trapdoors with hinges, binding threads for prey, drag lines, and also gossamer for riding air currents at the end of a free-floating thread.

Internally, the midgut connects with a large, dorsal digestive gland and also pouches out as a pair of sacs that pass forward and send branches to the legs. Malpighian tubules open in the hindgut. The heart has three pairs of ostia and gives off an anterior and posterior artery as well as three pairs of lateral vessels. In the nervous system all ventral ganglia are concentrated anteriorly.

Fertilization occurs internally. The male sheds sperms on a leaf or a spun sperm web, then dips a pedipalp into the sperms and fills a special receptacle on that appendage. The tip of the pedipalp then is inserted in the gonopore of the female. Highly elaborate courtship dances usually precede copulation. In some species the females kill and eat the males after sperm transfer. Batches of fertilized eggs usually are laid in spun cocoons, and development is direct.

Among spiders that do not trap prey in webs, the wolf spiders hunt and run after prey and the jumping spiders ambush prey and pounce on it. Trap-door spiders live in burrows that are covered with hinged silk lids (see Fig. 25.13). The animals make brief excursions to the outside for food. Amazonian tarantulas, 3 in. long and with leg spans of 10 in., are the largest spiders. The most dangerous are the black widows, whose poison can be fatal to man.

Mites and Ticks

The acarines mostly are minute animals in which all body divisions are fused together (Fig. 25.15). Free-living types are terrestrial or freshwater predators and scavengers; parasitic types live both on plants and animals, and many are blood suckers. Mites differ from ticks in that, as a group, they are the smaller animals.

In free-living types the chelicerae typically are chelate and the pedipalps tactile and leglike. In parasitic forms, the chelicerae usually are modified as cutting devices and also as elongated, fused, sucking channels. Pedipalps in such cases again can be sensory or form hookline attaching organs. The four pairs of legs are spaced widely apart, a general acarine trait. In blood-sucking forms the salivary gland secrete an anticoagulant, as in

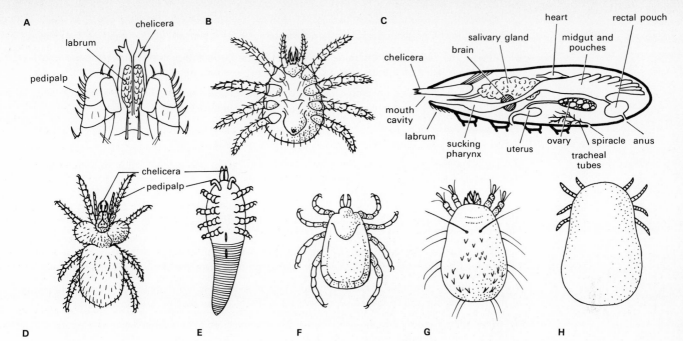

Figure 25.15 Mites and ticks. A, ventral view of acarine head region, showing cutting chelicerae and hooked four-jointed pedipalps. B, generalized ventral view of acarine body, showing relatively wide spacing of leg bases. C, sagittal section through *Dermacentor.* D, the chigger mite *Trombicula,* dorsal view. E, the follicle mite *Demodex,* ventral view. F, *Dermacentor,* dorsal view. G, the itch mite *Sarcoptes,* dorsal view. H, the Texas cattle tick *Boöphilus,* filled with blood.

leeches. Some acarines digest blood intracellularly, in amoeboid lining cells of the stomach. Most mites lack hearts, but ticks generally have a simple dorsal vessel. Breathing occurs through well-developed tracheal tubes.

Some of the blood-sucking parasites complete their entire life cycle on one host, relinquishing their attachment only temporarily at the time of molting. Other forms fall off the host at each molt and must seek a new host thereafter. Larvae can survive for many months on a single blood meal, and adults often live for years from one feeding.

Acarines are notorious carriers of disease organisms that affect vertebrates, mammals and man in particular. For example, mites and ticks are the exclusive carriers of the viruslike *rickettsial* parasites, responsible for Rocky Mountain spotted fever of man (transmitted by a species of *Dermacentor*) and many other rickettsial fevers and "poxes." Similarly, a species of *Boöphilus* carries a sporozoan that causes Texas cattle fever. Acarines also produce diseases on their own. Numerous gall mites damage plants, and various other types produce, for example, itch in man and mange disease in domestic birds and mammals.

Review Questions

1. Characterize the phylum Arthropoda taxonomically and name the subphyla and classes. How are these subgroups presumably interrelated historically?

2. How have the chitinous exoskeleton and the metameric structure probably promoted the broad diversification of arthropods?

3. Describe the basic exterior structure of the arthropod body. What are the skeletal parts of a segment and the joints of a segmental appendage? Describe the organization of each of the 10 animal organ systems in arthropods.

4. Review the structural characteristics of trilobites and show how these animals developed segmentally. Define the chelicerate subphylum taxonomically and similarly define each of the classes.

5. Review the segmental structure of the chelicerate cephalothorax and the structure and function of the appendages of that body division. Describe the structure of a horseshoe crab.

6. How does a horseshoe crab feed, move, breathe, and mate? Where does it live? What is the organization of its excretory system? How does the animal develop?

7. Review the structure of a compound eye and contrast it with that of an arthropod ocellus. Which chelicerate groups have either, neither, or both types of eye?

8. Describe the segmental structure of a eurypterid and a sea spider. Show how their appendages are specifically different. What are ovigerous legs?

9. Name several of the arachnid orders and describe the general characteristics of these animals. Describe the segmental and internal structure of a scorpion. How does a scorpion obtain, ingest, and digest food?

10. Compare the segmental structure of the abdomen of all chelicerate classes. What are chilaria? Pectines? Chelate legs? How many embryonic segments form the body of a spider?

11. Describe the structure of the chelicerae of spiders. What is the meaning of the following sentence: "In Araneida, the gnathobases on the coxopodites of the pedipalps are masticatory."

12. How does a spider breathe? Where are the breathing and the excretory organs located? Describe the segmental and the internal structure of a spider.

13. Describe and contrast the circulatory systems in all chelicerate classes. Do likewise for the nervous systems. What is spider silk chemically and where is it produced? Describe the structure and function of various types of webs.

14. What other uses do spiders make of spun silk? Name various types of spiders and describe their general characteristics. How does a spider mate and develop?

15. Describe the characteristics of mites and ticks. What is their external structure and what are their feeding methods? Name some of the diseases for which acarines are responsible directly or indirectly. When does molting take place in the life histories of the chelicerate arthropods?

Collateral Readings

Baker, E. W., and G. W. Wharton: "An Introduction to Acarology," Macmillan, New York, 1951. A book on mites and ticks.

Comstock, J. H., and W. J. Gertsch: "The Spider Book," 2d ed., Doubleday, New York, 1940.

Gertsch, W. J.: "American Spiders," Van Nostrand, New York, 1949.

Kaston, B. J., and E. Kaston: "How to Know the Spiders," Brown, Dubuque, Iowa, 1953.
    Three basic references on arachnids.

Savory, T. H.: Daddy Longlegs, *Sci. American,* Oct., 1962. An interesting account on a familiar arachnid.

Snodgrass, R. E.: "A Textbook of Arthropod Anatomy," Cornell University Press, Ithaca, N.Y., 1952. One of the standard texts, strongly recommended.

————: Arthropoda, in "McGraw-Hill Encyclopedia of Science and Technology," rev. ed., vol. 1, 1966. A short but thorough synopsis of the biology of the phylum.

Waterman, T. H.: A Light Polarization Analyzer in the Compound Eye of Limulus, *Science,* vol. 111, 1950. Research shows that horseshoe crabs can orient themselves by polarized light from the sky.

26

Phylum Arthropoda

Subphylum Mandibulata

cephalothorax or head unsegmented externally; head segments typically with one or two pairs of antennae, one pair of mandibles, two pairs of maxillae; thoracic appendages two or more pairs; with or without abdominal appendages. *Crustacea, Chilopoda, Diplopoda, Pauropoda, Symphyla, Insecta*

Crustaceans   **Class Crustacea**   marine and freshwater, some terrestrial; head and thorax or cephalothorax, often with carapace; with telson, two pairs of antennae, compound eyes; appendages largely biramous; excretion through antennal or maxillary glands; typically with *nauplius* or other free larval stages.

Subclass Cephalocarida   marine; 20 trunk segments; appendages jointed, not present on abdomen.

Subclass Branchiopoda   brine shrimps, fairy shrimps, water fleas; largely freshwater, with carapace, often bivalved; trunk legs phyllopodial in most.

Subclass Ostracoda   marine and freshwater; carapace bivalved, with adductor; two pairs of trunk appendages.

Subclass Mystacocarida   marine; five pairs of thoracic appendages, last four reduced; six abdominal segments, without appendages; compound eyes absent.

Subclass Copepoda   copepods; marine and freshwater; many parasitic; five or six thoracic segments with appendages, abdomen four segments without appendages; with ocellus, compound eyes absent.

Subclass Branchiura   marine and freshwater; body flat; second maxillae are suckers; ectoparasitic on fish.

Subclass Cirripedia   barnacles; marine; adults sessile, some parasitic; carapace often calcareous; antennae and compound eyes absent in adult; abdomen usually vestigial; hermaphroditic; free-swimming larvae.

Subclass Malacostraca   marine and freshwater; typically eight thoracic segments, six abdominal segments, all with appendages; usually with carapace; female gonopore on sixth thoracic segment, male gonopore on eighth; compound eyes mostly stalked.

Infrasubclass Leptostraca   seven abdominal segments, last without appendages; bivalve carapace with adductor, not fused to thorax, marine.

Infrasubclass Eumalacostraca   six abdominal segments; carapace if present not bivalved, fused to thorax.

Superorder Syncarida   without carapace; in fresh water, subterranean wells.

Superorder Pancarida   brood pouch dorsal, formed by carapace; eyeless; in hot springs, subterranean lakes.

Superorder Peracarida   opossum shrimps, slaters, sow bugs, pill bugs, wood lice, beach fleas, sand hoppers; carapace fused with up to four thoracic segments or absent; first thoracic segment fused with head; brood pouch on thoracic appendages; marine, freshwater, and terrestrial; pelagic, burrowing, on shore, in caves, and some parasitic.

Superorder Hoplocarida   mantis shrimps; carapace fused with three thoracic segments; marine.

Superorder Eucarida   carapace fused with whole thorax; compound eyes stalked; without brood pouch.

Order Euphausiacea   krill; marine, pelagic.

Order Decapoda   marine and freshwater; three pairs of thoracic maxillipeds; five pairs of legs, first pair chelate; with pleural gills in carapace chamber; statocyst in first antennae.
  Macrura   abdomen long; shrimps, prawns, lobsters, crayfishes.
  Brachyura   abdomen short; crabs.

The roughly thirty thousand described species of Crustacea, animals with "crust"-like shells, are among the most ubiquitous living organisms. They range from all depths of the ocean to freshwater lakes over 10,000 ft high, from 50°C waters of hot springs to glacial waters at the freezing point, from briny salt lakes to subterranean cave waters (where the animals are blind); and the sow bugs, pill bugs, and wood lice live on land. Most members of the class are free-living, but commensalistic and parasitic types are quite common. The free-living forms are largely filter feeders on plankton; bristles on the legs strain food particles from water. Some are scavengers, and among the advanced forms are many carnivorous predators. The majority of crustacea are themselves part of plankton, and copepods and krill are so abundant that they form the animal base of most aquatic food pyramids. Most crustacea thus are quite small, but the group also includes the largest arthropods, giant crabs, as well as the heaviest—the lobster *Homarus* can weigh well over 40 lb.

Ancestral crustaceans can be presumed to have had heads with fused segments and numerous similar, trilobitelike trunk segments. These bore similar biramous appendages that served simultaneously in breathing, locomotion, and filter feeding. From such a beginning three major lines appear to have evolved. One is represented by the recently discovered Cephalocarida, a subclass in which the ancestral structure has been largely retained; the only main change has been a loss of abdominal appendages (Fig. 26.1).

The second major line is represented by the subclasses Branchiopoda, Ostracoda, Mystacocarida, Copepoda, Branchiura, and Cirripedia (Fig. 26.2). In this series some of the branchiopods still retain many of the ancestral features, particularly numerous like segments with like appendages serving in locomotion, feeding, and breathing. The other groups exhibit three general kinds of evolutionary changes: (1) reduction of the number of trunk segments and appendages but enlargement of the head appendages, which assume the main feeding and locomotor functions; (2) development of a carapace over progressively more of the body; and (3) adoption of parasitic or sessile ways of life. Water fleas, ostracods, and copepods illustrate the first two of these changes; branchiurans and barnacles illustrate the third.

The third major evolutionary series comprises the subclass *Malacostraca*, the largest crustacean group (Fig. 26.3). Here too three kinds of general deviations from the presumed ancestral organization have occurred: (1) retention of all body segments and appendages, but increasing diversi-

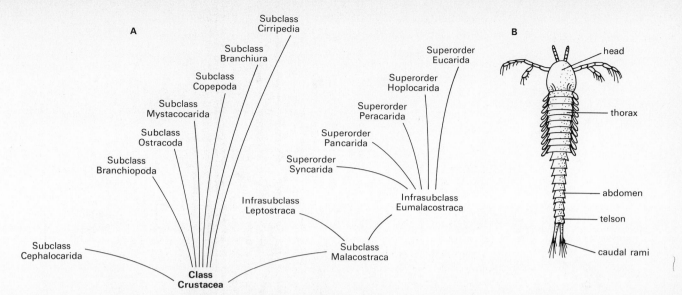

Figure 26.1 Crustacean evolution. A, the main taxonomic groups are arranged according to their presumed interrelations in three evolutionary series. *B,* the cephalocarid line of evolution, as in *Hutchinsoniella,* showing primitive condition of large numbers of similar segments. (*Adapted from Sanders.*)

fication in different body regions; (2) development of a progressively larger carapace that eventually fuses the entire thorax to the head; and (3) substitution of filter feeding by predatory and scavenging ways of life in the advanced groups. The primitive members of this line are represented by the Leptostraca and the Syncarida, the most advanced ones by the Eucarida, particularly the order Decapoda, which includes the shrimps, lobsters, and crabs.

The segmental structure of a decapod is outlined in Table 12 and Fig. 26.4. Head segments typically are alike in all crustacea. A median ocellus is already present in the nauplius larva (and becomes vestigial only in adult Malacostraca). Lateral compound eyes develop later, unstalked (sessile) or stalked. Also present in the nauplius are the first four head segments with their three pairs of appendages; maxillae develop later (see Fig. 26.8). The first antennae basically are uniramous (but secondarily acquire a second branch in lobsters). All other crustacean appendages are biramous in the larva and largely also in the adult. In the mandibles, the chewing jaws are gnathobases. The mouth above the mandibles frequently is guarded by an upper lip (*labrum*) and an underlip (*metastoma*), two flaplike nonsegmental extensions of the body wall. Where present, a carapace always represents an extension of the tergum of the third head segment.

Thorax and abdomen include different numbers of segments in different groups. In primitive types with similar trunk appendages, breathing occurs directly through the body wall, particularly in the appendages. Advanced forms have appendages specialized variously for food-handling, sensory, grasping, swimming, walking, copulatory, sperm-transferring, egg-carrying, and other functions. Breathing then occurs through feathery gills, developed from the epipodites of some or most of the thoracic appendages. In such animals the carapace often extends down on each side as a cover over the leg bases. The gills then lie in lateral chambers, and in certain cases they are attached to the segmental pleural plates on each side. *Pleural gills* of this type occur in lobsters and decapods generally (Fig. 26.5).

The last thoracic segment bears the gonopores, but in some groups (Cirripedia and Malacostraca among them), the female gonopore is on a more anterior segment. In the first or first two abdominal segments of most crustacea the appendages are modified for various reproductive functions. For example, in males the appendages often form copulatory or sperm-transferring organs; in females they form brood pouches or other egg-holding structures. In lobsters eggs are held on *swimmerets,* appendages of the second to fifth abdominal segments. Other groups with posterior abdominal appendages use such structures chiefly

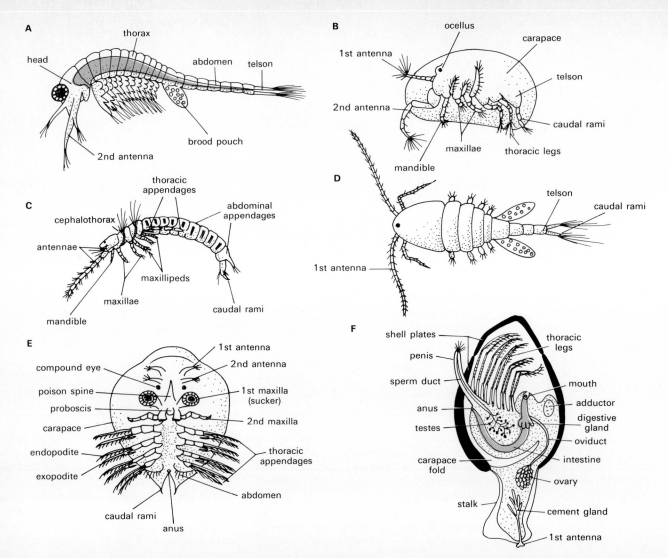

Figure 26.2 Trends in crustacean evolution: progressive reduction of trunk segments and enlargement of head and antennae. *A,* subclass Branchiopoda, the brine shrimp *Artemia;* numerous segments and like appendages, legs locomotor, no carapace. *B,* subclass Ostracoda, genus *Cypris;* antennae locomotor, body enclosed in carapace, thorax and abdomen highly reduced, legs few and not alike. *C,* subclass Mystacocarida, genus *Derocheilocarus. D,* subclass Copepoda, genus *Cyclops;* antennae locomotor, head carapace encloses some of trunk, free trunk segments and legs reduced. *E,* subclass Branchiura, genus *Argulus;* abdomen reduced, limbless. *F,* subclass Cirripedia, the goose barnacle *Lepas;* sessile, carapace forms shell plates, segments reduced, legs ingestive. (*B, after Zenker; C, after Pennak; D, after Hartog; E, after Wilson.*)

in locomotor or breathing functions. The last abdominal appendages in many cases are platelike *uropods* that together with the telson form a tail fan, as in lobsters. In many crustacea the telson typically bears a pair of *caudal rami,* tapered posterior extensions.

The exoskeleton of the larger crustacea is impregnated with calcium salts, and in places it folds into the body as *apodemes,* surfaces where mus-

cles are attached. In decapods the ventral thoracic apodemes are united internally as an *endophragmal skeleton,* a continuous supporting framework (Fig. 26.6). Apodemes in the appendages also provide breakage planes where limbs can snap off by voluntary muscular action of the animal. Decapods are particularly noted for this capacity of *autotomy,* or self-amputation, when a limb becomes injured or trapped. Little blood is lost here,

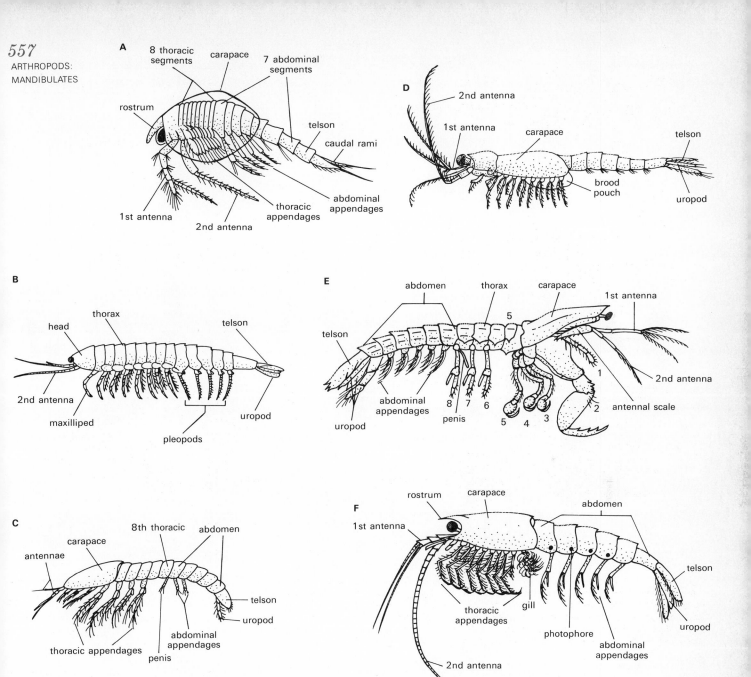

Figure 26.3 Trends in crustacean evolution: retention of all segments and appendages with increasing diversification of them and increase in the size of the carapace. *A*, infrasubclass Leptostraca, genus *Nebalia*. Note seven abdominal segments. In the infrasubclass Eumalacostraca there are six abdominal segments. *B*, superorder Syncarida, the primitive genus *Anaspides*; note absence of carapace, appendages generally alike (pleopod is a general term for any abdominal appendage). *C*, superorder Pancarida, genus *Thermosbaena;* carapace present, appendage diversification in evidence. *D*, superorder Peracarida, genus *Mysis;* carapace large, abdominal appendages reduced. *E*, superorder Hoplocarida, the mantis shrimp *Squilla;* appendage diversification very extensive. *F*, superorder Eucarida, the krill *Nyctiphanes* (order Euphausiacea); photophores on the abdomen are bioluminescent organs. A second large eucarid order, the Decapoda, includes lobsters and crabs, in which appendage diversification and the carapace are most highly developed (see following illustrations). (*A, after Claus; B, after Snodgrass; C, after Brunn; D, after Sars; E, after Calman; F, after Boden, Watase.*)

Table 12. *Segments and appendages in mandibulate arthropods*

	CRUSTACEA			MYRIAPODA				HEXAPODA
	CEPHALOCARIDS	BRINE SHRIMP (ARTEMIA)	DECAPODS	CENTIPEDE (LITHOBIUS)	MILLIPEDE	PAUROPOD	SYMPHYLA	INSECTS
1	—	—	—	—				—
2	antennae	antennae	antennae	antennae	antennae	antennae	antennae	antennae
3	antennae	antennae	antennae	—	—	—	—	—
4	mandibles	mandibles	mandibles	mandibles	mandibles	mandibles	mandibles	mandibles
5	maxillae	maxillae	maxillae	maxillae	—	—	maxillae	maxillae
6	maxillae	maxillae	maxillae	labium (maxillae)	maxillae	maxillae	labium (maxillae)	labium (maxillae)
7	legs	legs	maxillipeds	prehensors	—	—	legs	legs
8	legs	legs	maxillipeds	legs	legs (1 pr)	legs	legs	legs, wings, (spiracles)
9	legs	legs	maxillipeds	legs	legs (1 pr), (gonopores)	legs	legs	legs, wings, (spiracles)
10	legs	legs	chelate legs	legs	legs (1 pr)	legs	legs	(spiracles)
11	legs	legs	legs	legs		legs	legs	(spiracles)
12	legs	legs	legs (♀ gonopore)	legs		legs	legs	(spiracles)
13	legs	legs	legs	legs			legs	
14	legs	legs	legs (♂ gonopore)	legs	variable number of double segments, 2 pr of legs each	legs	legs	(spiracles)
15	legs	legs	—	legs		legs	legs	(spiracles)
16	legs	legs	swimmerets	legs		legs	legs	(spiracles)
17	legs	legs	swimmerets	legs		—	legs	(spiracles), ♀ ovipositor, ♂ copulatory organ (anus)
18	without abdominal appendages	brood pouch	swimmerets	legs		—	legs, 1 pr	
19		penis, (♀ ♂ gonopores)	swimmerets	legs	telson			
20		without abdominal appendages	uropods	legs			spinnerets	(anal cerci)
21			telson	legs				
22				legs				
23				legs (gonopores)				
24								
25				telson				
	telson	telson						

for the apodeme already covers the limb stump almost entirely and clotted blood soon seals it completely. New (but usually smaller) limbs then regenerate readily.

The internal structure is typically arthropod. In the Malacostraca the stomach has two compart-ments. An anterior *gastric mill* equipped with chitinous teeth and other hard outgrowths grinds coarse food. A posterior filter compartment then sorts food: coarse particles are returned to the gastric mill; fine particles pass to the intestine; and liquefied food is taken up in the capacious *hepato-*

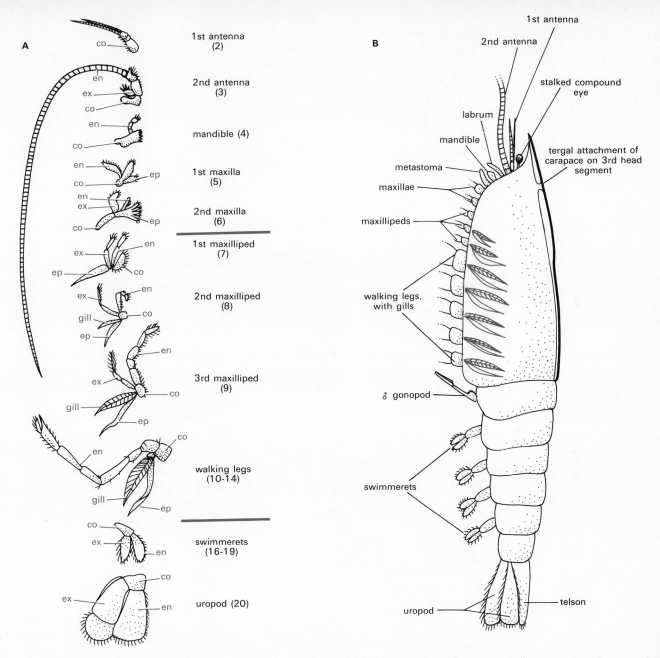

A

1st antenna (2)

2nd antenna (3)

mandible (4)

1st maxilla (5)

2nd maxilla (6)

1st maxilliped (7)

2nd maxilliped (8)

3rd maxilliped (9)

walking legs (10-14)

swimmerets (16-19)

uropod (20)

B

1st antenna

2nd antenna

stalked compound eye

labrum

mandible

metastoma

maxillae

tergal attachment of carapace on 3rd head segment

maxillipeds

walking legs, with gills

♂ gonopod

swimmerets

uropod

telson

Figure 26.4 Segmental structure of a lobster. A, segmental appendages of left side: *co,* coxopodite; *ex,* exopodite; *en,* endopodite; *ep,* epipodite. Segmental numbers counted from anterior end and indicated in parentheses. Note that some appendages lack exopodites. Gills are part of the epipodites. The first abdominal segment (15) bears a copulatory appendage in males (gonopod). *B,* diagram of position of appendages in whole animal. The labrum and metastoma, which guard the mouth, and the eyes are extensions of the body wall, not metameric appendages.

pancreas (''liver'') of the midgut, where the bulk of digestion and absorption occur. The nervous system ranges from a typical ladder-type organization in most groups to a concentrated ventral

ganglionic mass in crabs (Fig. 26.7). Apart from eyes and antennae, sensory structures include tactile and chemoreceptive bristles all over the body, and statocysts at the bases of the first an-

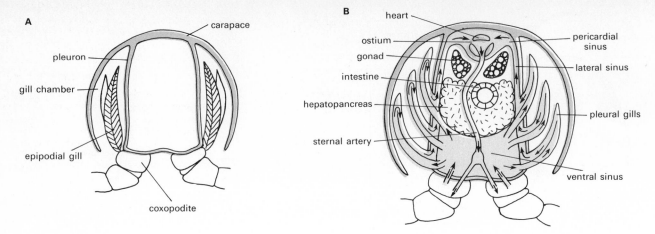

Figure 26.5 *Gills in crustaceans.* *A,* epipodial gills, as in majority of groups. *B,* pleural gills and blood circulation, as in decapods (such as lobsters). In *B,* the sinuses refer to hemocoelic blood spaces. Arrows indicate direction of blood flow. Blood in the ventral sinus and in parts of the gills is venous; blood in the other spaces shown is arterial.

tennae (as in decapods) or in other appendages.

Primitive groups have elongated hearts, with a pair of ostia in each segmental portion. In advanced forms the heart is shortened and has fewer ostia. For example, three pairs of ostia occur in lobster hearts (see Fig. 25.3), one pair in the

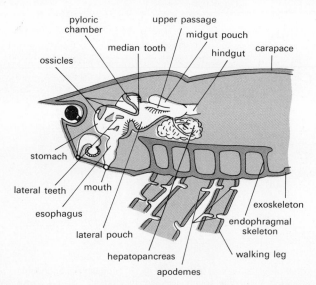

Figure 26.6 *Lobster anatomy.* Sectional side view of cephalothorax, showing endophragmal skeleton, apodemes in appendages, and some structural details of the gastric mill. The ossicles (shaded areas) are calcifications in the stomach wall, and they and the stomach teeth form the gastric mill, which macerates food to fine particles. A lateral pouch on each side of the pyloric chamber then functions as a fine filter; contractions of each pouch press liquefied food to the upper passage and from there to the very short midgut, which communicates with the hepatopancreas through a pair of ducts. Food residues pass through a valve from the midgut into the hindgut.

hearts of water fleas. Many ostracods, copepods, and barnacles lack hearts entirely, and the parasitic copepod *Lernanthropus* has a *closed* vessel system without heart. The blood pigment in that genus is hemoglobin, but the blood of most crustacea either is colorless or contains hemocyanin as in Malacostraca.

Excretion occurs through saclike glands that open at the bases of the second antennae or the maxillae. Nauplii generally develop antennal glands first, then these degenerate, and maxillary glands take over in later larval and adult stages. Decapods and some other groups retain antennal glands throughout life (see Fig. 26.6). The primitive leptostracan *Nebalia* has a renal gland at the base of each thoracic appendage, a metameric array presumably suggestive of the condition in ancestral crustaceans.

Crustacea have a well-developed endocrine system. Important components are a neurosecretory *X-organ* in each eyestalk near the eye, and a *sinus gland* near the base of each eyestalk. This gland appears to be primarily a storage site for hormones produced in the X-organ. Another endocrine component is a nonneural *Y-organ* located in the antennal or maxillary segments of the head. Behind the head is a *postcommissural organ,* near the heart lies a *pericardial organ,* and along the sperm ducts of male crabs are *androgenic organs.* The first two are probably endocrine storage sites like the sinus gland, and the last are nonneural endocrines like the Y-organ.

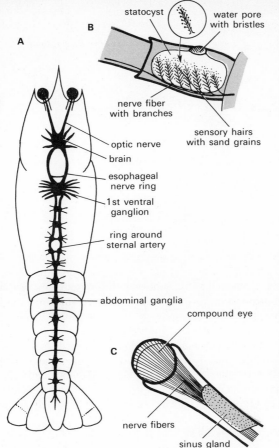

Figure 26.7 *Lobster anatomy.* *A*, general plan of the nervous system. The brain supplies nerves to both pairs of antennae, and the first ventral ganglion innervates the mouth, the maxillae, and the maxillipeds. The second ventral ganglion innervates the large chelate legs, and the remaining thoracic ganglia supply nerves to the last four pairs of legs. *B*, basal piece of first antenna, showing statocyst. *C*, longitudinal section through stalked compound eye, showing location of endocrine sinus gland. See Fig. 25.3 for other anatomic features of lobsters.

One set of hormones produced in the X-organs and stored in the sinus glands controls dispersion of the pigments in epidermal chromatophores. Another set of hormones regulates the secretions of the Y-organs, which in turn promote growth of reproductive systems and also accelerate molting. Molting inhibition or acceleration thus follows the scheme:

$$X\text{-organ} \longrightarrow \text{sinus gland}$$
$$Y\text{-organ} \rightrightarrows \text{molting hormone}$$

Crustacea molt throughout life, but progressively less often as they age. The hormone from the pericardial organ accelerates heartbeat, and still other endocrine effects are likely to be discovered in future. The crustacean endocrine system clearly has a level of complexity comparable to that of insects and vertebrates.

Parthenogenesis is common in water fleas and ostracods, and barnacles and some of the parasitic crustacea are hermaphroditic. In the majority of groups fertilization is internal, often via spermatophores. Eggs develop on or in the females, held in brood pouches or on abdominal appendages. As outlined in Chap. 11, development typically passes through a series of larval stages each separated from the next by a molt (see Fig. 11.18). Some species have a complete series of nauplius, metanauplius, one or more zoaeae, and mysis

Figure 26.8 *Crustacean development.* *A*, nauplius larva, as in *Artemia*, with first three pairs of head appendages; thoracic segments foreshadowed in pygidium as ringlike mesodermal thickenings. *B*, later stage, with first three thoracic segments cut off from pygidium anteriorly, the pygidium having elongated posteriorly by an equivalent amount. *C*, side view of still later stage, showing antero-posterior sequence of appendage buds. After all thoracic segments are laid down, the pygidium itself becomes subdivided into the abdominal segments. See also Fig. 11.18. (*After Weisz.*)

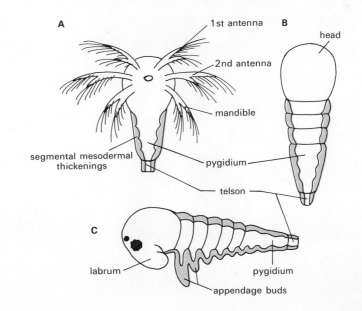

A

2nd antenna

shell gland

heart

ovary

brood pouch

mandible

1st antenna

intestine

thoracic appendages

claw

abdomen

carapace

B

labrum

1st antenna

mandible

maxillae

11th thoracic without appendages

1st thoracic appendage

abdominal segments appendage

abdominal appendages

Figure 26.9 Branchiopoda. A, the water flea *Daphnia;* antennae locomotor, carapace encloses trunk, legs few but alike. The abdomen is reduced. *B,* underside of fairy shrimp *Apus. (A, after Sars; B, after Borradaile.)*

larvae. In other cases some or all of these stages are embryonic, free larval forms are abbreviated or absent, and the animals are ovoviviparous. As in trilobites, segments arise by constricting off a *pygidium* in anteroposterior sequence. The pygidium is located in front of the telson and in an early nauplius it represents the whole trunk (Fig. 26.8). As it then generates segments it retains its length by posterior growth. After a species-specific number of segments has formed the pygidium ceases

to grow, becomes segmented itself, and then constitutes the last segments of the adult abdomen.

The subclass Branchiopoda includes the brine shrimp *Artemia,* found naturally in salt lakes (see Fig. 26.2). This remarkable animal can survive at virtually any salinity from almost distilled water to concentrated brine. Like most of the primitive Branchiopoda, *Artemia* lacks a carapace, legs are on the thorax only, and these legs are *phyllopodial,* leaflike and without joints. More advanced mem-

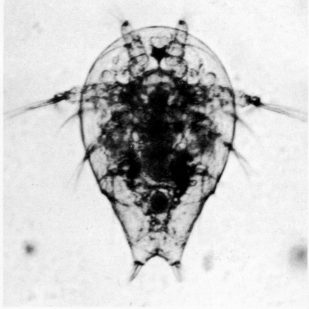

Figure 26.10 Copepods. A, the freshwater type *Cyclops,* with brood pouches. *B,* a nauplius larva hatched from such pouches.

A

B

A

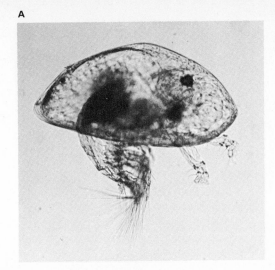

B

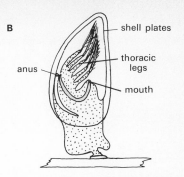

shell plates

thoracic legs

anus

mouth

C

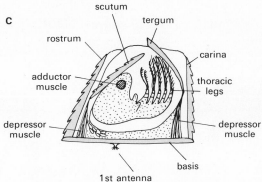

scutum

tergum

rostrum

carina

adductor muscle

thoracic legs

depressor muscle

depressor muscle

1st antenna

basis

D

Figure 26.11 Barnacles. A, the cypris larva of a barnacle (equivalent to a zoaea). *B,* a stage of metamorphosis: attachment by first antenna, shell plates developing. *C,* section through adult showing body position and main shell plates. *D,* mature adult. See also Fig. 26.2 and *Sacculina,* Fig. 17.19.

bers of the subclass include *Daphnia* and other water fleas, which usually have a carapace that covers the whole trunk. Also, appendages are jointed and the second antennae are large, powerful, and serve as the main locomotor organs. Trunk appendages still retain the feeding function (Fig. 26.9).

In the subclass Ostracoda, by contrast, feeding in most cases is the special function of the well-developed head appendages, particularly antennae and maxillae (see Fig. 26.2). The trunk is shortened and the limbs, reduced to two pairs, serve mainly in cleaning the head appendages. These animals in effect are composed almost entirely of head parts. The whole body is enclosed in a bivalved carapace equipped with an adductor muscle. Antennal locomotor and feeding functions are characteristic also of Mystacocarida and Copepoda, groups that lack a carapace. Among copepods, *Calanus* is a main component of marine

plankton; *Cyclops* lives in fresh water. These animals can be identified readily by their large median ocellus and the conspicuous lateral brood pouches in females (Fig. 26.10 and see Fig. 26.2). Numerous copepods have become parasitic, and parasitic types also comprise the small subclass Branchiura.

Cirripedia have free-swimming larvae and modified sessile adults (Fig. 26.11). The zoaealike *cypris* larvae of barnacles settle with their head end to the ground, a permanent attachment being formed by cement organs that open on the first antennae. The antennae and compound eyes then degenerate and the carapace develops a *mantle* around almost the whole animal. Calcareous plates are secreted on the mantle in *Balanus,* the familiar acorn barnacles found on rocks, wharf pilings, and even the backs of large lobsters. The thoracic limbs become highly developed and, protruded from the mantle, are used in food-gathering; a barnacle in

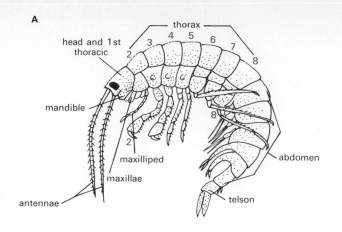

Figure 26.12 Malacostraca. A, Gammarus, a sand hopper (superorder Peracarida, order Amphipoda). Numbers refer to thoracic segments. Of the seven pairs of thoracic appendages, the first two (on segments 2 and 3) are semichelate, the third and fourth help in feeding, and the fifth, sixth, and seventh are turned backward and serve when the animal crawls on its side. Of the five pairs of abdominal appendages, the first three serve in swimming, the last two in jumping. *B, Ligia,* a slater, one of a number of terrestrial crustaceans (superorder Peracarida, order Isopoda). Some of the posterior appendages have thin surface areas where air breathing takes place. *C, D,* superorder Eucarida, order Decapoda. *C, Leander,* a prawn. Eggs are carried in brood pouches. *D,* a land crab. (*A, after Leuckart.*)

effect stands on its head and kicks food to its mouth with its hind legs. The abdomen is usually reduced or absent. Forms such as *Lepas* are stalked, the stalk representing an enlarged anterior region of the head. Quite a number of barnacles are parasitic, and the extreme parasitism of *Sacculina* has already been noted in Chap. 17. Barnacles are hermaphrodites, an adaptation to sessilism and parasitism.

The subclass Malacostraca includes two evolutionary subseries, Leptostraca and Eumalacostraca (Fig. 26.12 and see Fig. 26.3). The main groups are the two superorders Peracarida and Eucarida. Of more than passing interest are the several forms that have adapted to unusual ecological niches, such as subterranean waters, hot springs, and terrestrial environments. The terrestrial types breathe with their abdominal appendages, which have water-retaining recesses or tracheal structures remarkably similar to those of insects. Decapoda are the most familiar crustacea. They are largely scavenging and carnivorous, and the crabs are the most advanced among them.

Myriapods   *Class Chilopoda*   centipedes; carnivorous, predatory; head with 1 pair of antennae, 1 pair of mandibles, 2 pairs of maxillae; trunk with 1 pair of poison-claw-containing prehensors and 15 or more pairs of walking legs; eyes compound or simple or absent; spiracles dorsal or lateral, tracheae branched; gonopores at hind end of trunk.

Class Diplopoda   millipedes; herbivorous, scavenging; head with one pair each of antennae, mandibles, and maxillae; first four trunk segments single, rest double and fused; with ocelli; tracheae unbranched; gonopores on third trunk segment.

Class Pauropoda   head as in millipedes, antennae branched; 12 trunk segments, nine pairs of legs, trunk terga fused in pairs; without eyes, breathing, or circulatory systems; gonopores anterior.

Class Symphyla   centipedelike, in soil, debris, humid environments; head as in centipedes; trunk with 12 pairs of legs, 15 or more tergum plates, and 1 pair of terminal spinnerets; gonopores anterior.

The term "Myriapoda" is often used descriptively for these four classes of superficially similar animals. They probably represent separate descendant lines of quite primitive mandibulate stocks that have become terrestrial. The animals are cosmopolitan but they reach their greatest abundance in humid tropical and subtropical environments.

The head of a centipede is structured segmentally like that of an insect (Fig. 26.13 and Table 12). In the trunk, forms like *Scutigera* and *Lithobius* have 18 embryonic segments. The first develops fanglike prehensors with poison glands and ducts that open at the tips. The next 15 segments each bear a pair of legs. On the seventeenth segment are the gonopores, and the eighteenth contains the anus. Other groups of centipedes are similar except that the number of leg-bearing segments can be greater; types with up to 183 pairs of legs are known. Each trunk segment contains a pair of dorsal (*Scutigera*) or lateral (*Lithobius*) spiracles that, as in insects, lead to branching and intersegmentally connected tracheal tubes.

The nervous system has a primitive ladder-type structure, and the heart extends throughout the trunk. In each segmental portion are a pair of ostia and lateral arteries. Excretion occurs through Malpighian tubules. *Scutigera* has compound eyes, *Lithobius* only simple ocelli, and many other centipedes are blind. Fertilization is internal, often via spermatophores, and the fertilized eggs are laid into ground holes. At hatching the number of segments is complete in some forms, incomplete in others (in which case development is completed during the following molting steps). Centipedes molt throughout life.

The animals typically are rapid runners, an advantage in their predatory way of life. They shun exposure, however, and during the day they tend to hide under stones and leaves or in crevices and crannies. At night they emerge in search of food, which consists of earthworms, insects, and snails.

Figure 26.13 Centipedes. A, Lithobius, from life. B, head structure. Mouth and mandibles are hidden behind the labrum and the maxillae. The prehensors on the first trunk segment are homologous to crustacean maxillipeds. (B, after Snodgrass.)

A

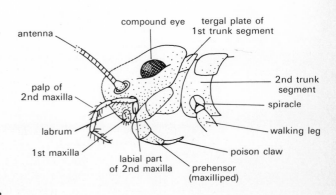

B

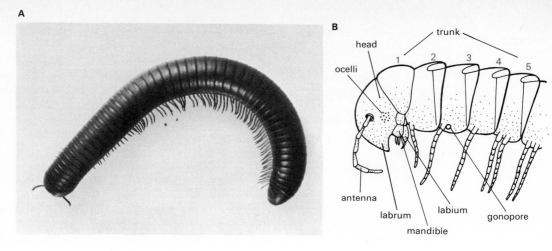

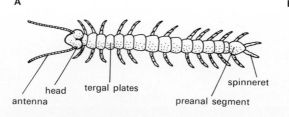

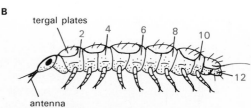

Figure 26.14 Millipedes. A, note two pairs of legs in most trunk segments. *B*, the anterior region. Numbers refer to trunk segments. Note dorsally overlapping plates of segmental exoskeleton. (*A*, courtesy *Carolina Biological Supply Company.*)

Figure 26.15 Myriapods. A, a symphylan, dorsal view. *B*, a pauropod, side view. Numbers refer to trunk segments. (*After Snodgrass.*)

Some centipedes attain lengths of 1 ft, but most members of the class are far smaller. The class includes about 2,500 species.

In millipedes the head contains but a single pair of maxillae (Fig. 26.14 and Table 12). In the first four trunk segments, the second, third, and fourth each bears a single pair of legs, and the gonoducts open on the third. All other trunk segments are fused in pairs. Each such double segment arises from two pairs of coelomic sacs and contains two pairs of ganglia, two pairs of ostia in the heart, two pairs of spiracles, and two pairs of legs. The number of such double segments varies considerably in different groups. Spiracles lead to bundles of unbranched tracheae, another difference from centipedes. Millipedes have simple ocelli, and excretion occurs through Malpighian tubules.

Fertilization is internal, in some cases by means of spermatophores transferred from the mouth of the male to the female. Millipedes are poor runners despite their large number of legs; the herbivorous life of the animals makes rapid locomotion unnecessary. The animals are retiring in habit and when exposed they usually roll up as a ball. More than 8,000 species have been described, but estimates place the number of actually existing species at about 25,000.

Symphyla are a group of only 60 or so species. They are generally centipedelike in appearance and habits, and their head structure is quite similar as well. But they differ from centipedes in the segmental structure of the trunk, in the anterior location of the gonopores, and in several other respects. These animals are believed to be the closest relatives of insects. Pauropods too number only about 60 species; their biology is known only poorly (Fig. 26.15 and Table 12).

Insects   **Class Insecta (*Hexapoda*)**   terrestrial, some secondarily freshwater, exceptional cases marine; head 6 segments, typically with antennae, mandibles, maxillae, labium; thorax 3 segments, typically with three pairs of legs, two pairs of wings, two pairs of spiracles; abdomen typically 11 segments, without locomotor appendages, with eight pairs of spiracles; compound eyes and ocelli present; excretion via Malpighian tubules; fertilization and development external.

Subclass Apterygota (Ametabola)   primitively wingless; without metamorphosis.

Order Protura   piercing mouth parts, concealed in head; without antennae, eyeless; with and without tracheae; abdomen 12 segments, first 3 with small appendages; in moist places; 100 species.

Order Diplura   biting mouth parts, deep in head; with long antennae; eyeless; with tracheae, without Malpighian tubules; abdomen 11 segments, with cerci or forceps; in moist, rotting places; 400 species.

Order Collembola   spring tails; biting mouth parts hidden in head; without compound eyes; six abdominal segments with three pairs of appendages (not legs), modified for jumping and adhesion; without tracheae or Malpigian tubules; in moist, rotting places; 2,000 species.

Order Thysanura   bristle tails; biting mouth parts, exposed; antennae long; 11 abdominal segments and two or three cerci; in buildings, nooks and crannies, book bindings; 750 species, *Lepisma*, silver fish.

Subclass Pterygota (Metabola)   with wings and metamorphosis.

Superorder Exopterygota (Hemimetabola)   wing growth external on larval body; metamorphosis gradual; larvae *nymphs* if terrestrial, *naiads* if aquatic; compound eyes already in larvae.

Order Orthoptera   grasshoppers, locusts, crickets, praying mantises, katydids, walkingstick insects, cockroaches; biting mouth parts; forewings narrow, leathery; hind wings membranous, folding under forewings; many wingless; typically with cerci; herbivorous; 25,000 species.

Order Dermaptera   earwigs; biting mouth parts; forewings short, leathery, hind wings large, membranous; many wingless; cerci form terminal forceps; in dark places, nocturnal; 1,200 species.

Order Plecoptera   stone flies; mouth parts biting or vestigial; wings pleated, hind wings larger and hidden under forewings; weak fliers; with cerci; naiad larvae with tracheal gill tufts behind each leg; 1,500 species.

Order Isoptera   termites; biting mouth parts; wings similar, membranous; social and polymorphic, largely tropical; 2,000 species.

Order Embioptera   embiids, web-spinners; body flattened, elongate; in males wings similar, females wingless; cerci two-jointed, asymmetric in males; colonial, in tunnel networks of secreted silk; mainly tropical; 1,000 species.

Order Odonata   dragonflies, damsel flies; biting mouth parts; wings similar, net-veined; dragonfly wings do not fold back, damsel fly wings fold up at rest; compound eyes conspicuous; abdomen elongate and slender; predatory, legs catch insects in flight; naiads with tracheal gills; 5,000 species.

Order Ephemerida   may flies; biting mouth parts vestigial; wings membranous, folded up at rest, hind wings small; long cerci; naiads with multiple paired tracheal gills, biting mouth parts, long-lived; adults nonfeeding, often live less than 24 hr; 1,500 species.

Order Mallophaga   biting lice; biting mouth parts; body flattened; wingless; eyes reduced or absent; two claws per leg for clinging; ectoparasitic on birds, some mammals, feeding on skin, feather, and hair fragments; 3,000 species.

Order Anoplura   sucking lice, human body lice; piercing-sucking mouth parts, retractable; body flattened; wingless; thorax fused; without eyes; one claw per leg for clinging; ectoparasitic blood suckers on hair of mammals, transmitters of typhus fever, trench fever, and other diseases; 500 species.

Order Psocoptera   book lice, bark lice; biting mouth parts; wings usually without crossveins or absent. 1,300 species.

Order Zoraptera   biting mouth parts; often wingless, eyeless; in warm, dark, rotting places; 21 species.

Order Hemiptera   bugs, water striders; piercing-sucking mouth parts; forewings thickened basally, membranous terminally, crossed flat on body at rest; hind wings membranous, folded under forewings; herbivorous, predatory, and ectoparasitic; some viviparous; 40,000 species.

Order Homoptera   plant lice, scale insects, 17-year locusts, cicadas, leaf hoppers; piercing-sucking mouth parts; wings tentlike at rest, forewings thickened or membranous; sapfeeders; 30,000 species.

Order Thysanoptera   thrips; sucking mouth parts; wings similar, fringed with bristles, veins few or absent; some parthenogenetic; mostly herbivorous on fruit and grasses; 3,500 species.

Superorder Endopterygota (Holometabola)   wing growth internal in larval body; metamorphosis abrupt; immature stages are larvae and pupae; larvae without compound eyes.

Order Mecoptera   scorpion flies; biting mouth parts on turned-down beak; wings similar, with rhomboid venation, roofed over body at rest; 500 species.

Order Neuroptera   dobson flies, alder flies, ant lions, lacewings; biting mouth parts; wings similar, roofed over body at rest; larvae with biting or sucking mouth parts; abdominal gills; carnivorous insect feeders and mite feeders; 5,000 species.

Order Trichoptera   caddis flies; vestigial biting mouth parts; wings roofed over body at rest; wings and body hairy or scaly; larvae aquatic and carnivorous, with biting mouth parts; 7,000 species.

Order Lepidoptera   moths, butterflies, silkworms; sucking mouth parts, with proboscis; wings membranous, scaly; larvae with biting mouth parts; moths, wings horizontal at rest, antennae feathery or filamentous; butterflies, wings folded up at rest, antennae club-shaped or knobbed; 125,000 species.

Order Diptera   flies, gnats, mosquitoes, midges; piercing-sucking-biting mouth parts in proboscis; hind wings reduced to halteres; 100,000 species.

Order Siphonaptera   fleas; piercing-sucking mouth parts; wingless; laterally compressed; ectoparasitic blood suckers; 1,500 species.

Order Coleoptera   beetles, weevils, meal worms, glow worms; biting mouth parts; forewings horny *elytra*, hind wings folded under forewings; 300,000 species.

Order Strepsiptera   aberrant coleopteran derivatives; biting mouth parts reduced to halteres, hind wings membranous; females larvalike, wingless, eyeless, legless; larvae and females parasitic in insects, permanently in body of host; 300 species.

Order Hymenoptera   bees, ants, wasps, sawflies; biting, sucking, and lapping mouth parts; hind wings smaller than forewings; first abdominal segment fused to thorax, waist behind it; ovipositor for piercing, stinging, or sawing, often long and looped forward and downward; social and polymorphic, parthenogenesis common; pupae typically in cocoons; 110,000 species.

An assemblage that constitutes the largest group of animals, with more than ¾ million described species and possibly ten times that many actually in existence, is bound to be replete with more superlative attributes than any other. Within the Coleoptera, the largest order, some 40,000 species comprise the family of the weevils, the largest single family in the world—larger than all but five or six whole phyla of animals. Three other insect orders contain more than 100,000 species each: the Lepidoptera, butterflies and moths; the Hymenoptera, bees and ants; and the Diptera, flies and mosquitoes. Next in abundance are the Hemiptera, or bugs, the Homoptera, or plant lice, and the Orthoptera, or grasshoppers.

Fossil insects are known from the Carboniferous on. Early flying types were as yet unable to fold their wings at rest, and the complicated folding mechanism, basically the same for all groups in which it occurs, developed later. Three levels of

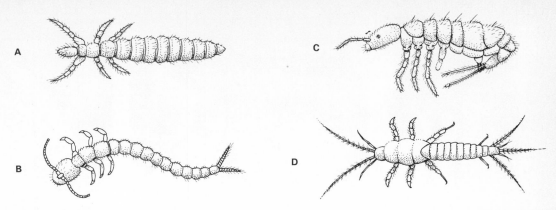

developmental specialization have evolved among insects. The most primitive level is represented by the wingless Apterygota (Fig. 26.16), in which the hatched young are miniature adults; larvae and metamorphosis are essentially absent (hence the alternative name Ametabola; some zoologists actually do not regard the first three apterygote orders listed above as true insects).

The second level is exemplified by the Exopterygota (or Hemimetabola), a group probably evolved from apterygote ancestors such as early Thysanura. Exopterygote larvae (*nymphs* or *naiads*) are wingless initially, but they metamorphose gradually to winged adults in a series of molting steps called *instars.* In the process wing buds on the outside of the body grow progressively larger (Fig. 26.17).

The third and most specialized level is represented by the Endopterygota (or Holometabola), in which the larvae are wormlike, caterpillarlike, and actual caterpillars in Lepidoptera. These undergo a series of molts and the last transforms the larva to a *pupa.* During the pupal phase a drastic metamorphosis takes place, in the course of which

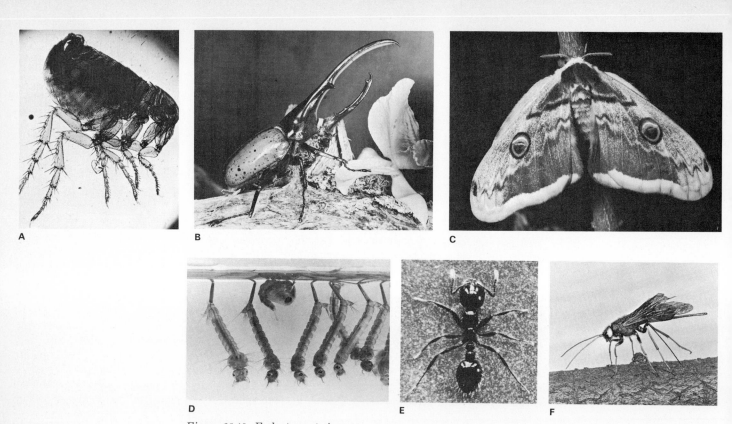

Figure 26.18 Endopterygote insects. A, flea (order Siphonaptera). *B,* Hercules beetle (order Coleoptera). *C,* Saturnia moth (order Lepidoptera). *D,* larvae and one pupa of mosquito suspended from water surface (order Diptera). *E,* tropical fire ant (order Hymenoptera). *F,* wood wasp, with ovipositor (dark rod between hind legs) serving as wood borer (order Hymenoptera). See also Color Fig. 29.

wings and other adult structures arise from internal buds called *imaginal disks.* The adult, or *imago,* then extricates itself from the pupal envelopes. Endopterygotes did not appear in the fossil record until the Permian. Beetles arose only in the Triassic, and the origin of the most advanced endopterygotes, bees, butterflies, and flies, coincided with the Jurassic and Cretaceous expansion of the flowering plants (Fig. 26.18).

The head of an insect consists of six fused segments (Fig. 26.19 and Table 12). The first is embryonic and without appendages; the second bears antennae; the third is embryonic; the fourth develops mandibles (coxopodites only); the fifth bears maxillae; and the sixth similarly grows a pair of maxillae, but these fuse along the midline and form a *labrum,* or underlip. The mandibles and first maxillae are lateral to the mouth, and an upper lip, the *labrum,* protects the mouth anteriorly.

Primitively, labrum, mandibles, maxillae, and labium form *biting-and-chewing* mouth parts. Most exopterygote orders have oral structures of this type.

In more advanced forms these basic oral appendages are modified as *sucking, licking,* or *piercing-and-sucking* structures (Fig. 26.20). In a butterfly, for example, the maxillae are drawn out as an elongated proboscis, each maxilla forming a complete sucking tube. The two tubes are interlocked; they can be extended deep into the nectar-containing region of a flower and rolled up toward the head when not in use. In a bee, the median parts are a pair of elongated protective sheaths, and the maxillae are an additional external jacket. In aphids, the mandibles and maxillae form an elongated retractable tube that pierces plant cells and carries the juices up through the hollow interior. The jointed labium here provides

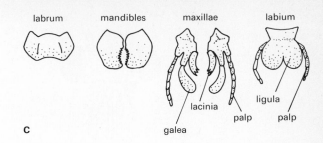

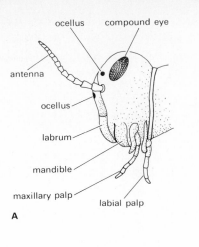

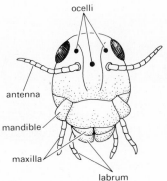

Figure 26.19 Head structure.
A, *B*, grasshopper head. *C*, components of the biting mouth parts. *D*, front view of head of a moth, showing feathery antennae (a diagnostic distinction from butterflies, in which the antennae are filamentous).

a dorsally grooved guiding structure for the piercing-sucking apparatus.

In a female mosquito, the labium forms an elongated protective proboscis with an anterior groove. In this groove lie rapierlike piercing mandibles and maxillae. Two sucking tubes then are inserted into pierced skin: an *epipharynx* formed from the labrum and the mouth roof, and a *hypopharynx* formed from the floor of the mouth. Muscular suction by the pharynx draws up blood through the hypopharynx, and a duct in that tube also carries saliva to the skin wound. In a housefly mandibles are absent and maxillae are reduced, but epipharynx and hypopharynx tubes are present. Also, the labium is a foldable proboscis, expanded terminally as a conspicuous pad with trachealike tubules and fine chitinous rasping teeth. This proboscis shields and guides the sucking tubes. Numerous other modifications of mouth parts occur in different insect groups.

Embryos develop lateral ocelli that become substituted by compound eyes in nymphs and adults of endopterygotes. Adult median ocelli develop, too. The head and the thorax are joined through a narrow neck. Of the three thoracic segments (*prothorax, mesothorax, metathorax*), each carries a pair of legs (Fig. 26.21). These are uniramous and consist of coxopodite (*coxa*), basipodite (*trochanter*), and endopodite (*femur, tibia, tarsus, claws*). Femur and tibia usually are the longest joints. Legs too are variously modified for specialized functions. For example, they are adapted for grasping in a praying mantis, for jumping in a grasshopper, for digging in a mole cricket, for sound production in numerous Orthoptera, for food collection in the pollen baskets of bees, for adhesion in the adhesive pads between the claws of flies, and for swimming in diving beetles.

Each of the second and third thoracic segments typically also bears a pair of wings, flattened folds

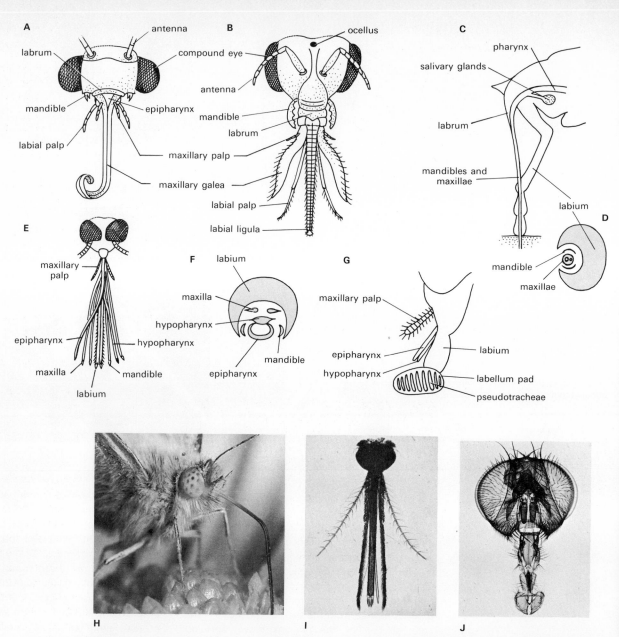

Figure 26.20 Insect mouth parts. A, front view of moth or butterfly head, showing sucking proboscis (galea) formed from maxillary components. *B,* front view of bee head, showing sucking proboscis formed from labial components. *C,* side view of hemipteran head, showing mandibular-maxillary sucking stylet and grooved labial guide. *D,* cross section through hemipteran stylet; in the maxilla, the forward tube conducts food up, the hind tube, saliva down. *E,* front view of mosquito head. The epipharynx is an extension of the labrum, the hypopharynx an extension of the floor of the mouth; together they form a food-conducting channel, as shown in cross section in *F. G,* side view of proboscis apparatus of housefly. *H,* head of cabbage butterfly (*Pieris*), showing sucking proboscis extended. It rolls up toward the head when not in use. *I,* head and sucking structures of a female mosquito. The sucking tube in the center is formed by long extensions of the upper lip (labrum) and the floor of the mouth. The dark bristly structures along each side of the central tube are the mandibles, and the shorter extensions on the outside are maxillae. *J,* front view of proboscis of a housefly. (*A, after Metcalf and Flint; B, after Cheshire; C, after Weber; D, after Imms; E, after Patton and Cragg; G, adapted from Borradaile; I, J, courtesy Carolina Biological Supply Company.*)

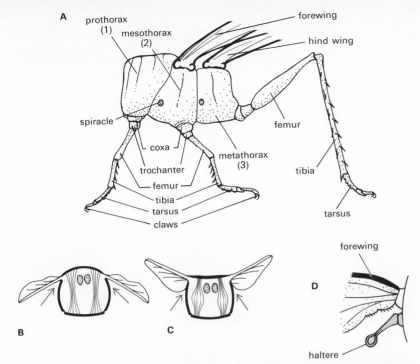

permits the elastic tergum to recoil and brings the wing down.

In butterflies and other groups, the two wings on one side are locked together by bristles or hooks and thus function in unison. In general, wings beat faster the smaller the insect. Either pair of wings can be modified. For example, the hind wings of Diptera are reduced to knobbed stumps, or *halteres* (and their vibrations during flight produce the buzzing sounds made by flies). In Strepsiptera it is the front wings that form halteres. The forewings are hardened as protective covers in Orthoptera, Dermaptera, and Coleoptera, and in the beetles these covers are horny *elytra*. In fleas and many members of other orders both pairs of wings are secondarily absent.

The abdomen typically consists of 11 segments, though the number is often reduced secondarily (Fig. 26.22). In the first seven, appendages begin to form in the embryo but they never mature (except in some of the Apterygota) and are absent in the adult. The appendages of the eighth and ninth segments in females of most insects become *ovipositors*, accessory egg-laying structures, but stings, saws, and piercers in bees and other Hymenoptera. The appendages of the ninth segment in males form copulatory organs. The anus typically is on the eleventh segment, which often also extends as a pair of posterior projections, the *anal cerci* (such as the "forceps" of earwigs).

The last two thoracic and the first eight abdominal segments typically bear a pair of spiracles each, equipped with muscle-operated valves. These air pores lead to a highly developed tracheal system that includes interconnecting tubes, air sacs, and branching tracheae. Contraction of muscles attached to the terga and sterna of the segments bring about exhalation, and relaxation of these muscles leads to recoil of the skeletal plates and thus to inhalation. Such breathing motions of the abdomen can be observed readily in a resting fly or bee. Anterior spiracles tend to be inhalatory; posterior ones, exhalatory. Collembola are skin breathers usually without tracheal systems. The aquatic larvae of the midge *Chironomus* also breathe through the skin, the tracheal system here being nonfunctional in the early larval stages. Most aquatic larvae have *tracheal gills*, expanded, hollow tufts of body wall in various segments. Mosquito larvae breathe through spiracles, however; the larvae hang suspended from the water

Figure 26.21 The insect thorax. A, thorax of grasshopper. The coxa of a leg is homologous to the coxopodite, the trochanter to the basipodite, and all other joints to the endopodite (see also Fig. 7.4). *B* and *C,* the indirect, flight-producing wing musculature. A wing is an extended fold of body wall. Arrows point to the fulcrums of wings, at the dorsal edges of the pleura. In *B,* vertical muscles are relaxed, longitudinal muscles contracted, wings moved down. In *C,* the converse, with wings moved up. *D,* a knobbed haltere, a reduced hind wing as in Diptera. *E,* veins in the wing of a dragonfly.

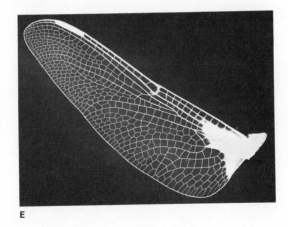

of the body wall that extend out dorsolaterally between tergum and pleuron. Air-filled tubular spaces in the wing form supporting veins; the chitin cuticle is thicker around these tubules than elsewhere. Muscles that attach directly to a wing move the appendage to a flying or a rest position. An indirect musculature between tergum and sternum brings about the actual flying motions. Contraction of these muscles flattens the normally arched tergum and moves the wing up; relaxation

surface. The diving bettle *Dytiscus* periodically carries a fresh air bubble down and breathes in that fashion.

The nervous system of insects is of the ladder type. As in other arthropods, various ganglia often are concentrated as single units. Insects also have a well-developed autonomic nervous system that innervates the alimentary tract, the musculature of the spiracles, and the breathing apparatus. Apart

Figure 26.22 The insect abdomen. A, side view of abdomen as in the grasshopper; numbers refer to segments. *B*, dorsal view of hind end of an earwig (order Dermaptera), showing anal cerci elongated as horny forceps. *C*, the tracheal breathing system as in a grasshopper. Arrows show air movement through spiracles. *D*, a spiracle (center) and the systems of tracheal tubes leading away from it. (*C, after Vinal.*)

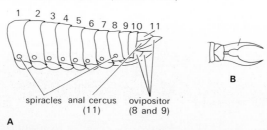

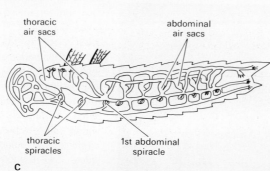

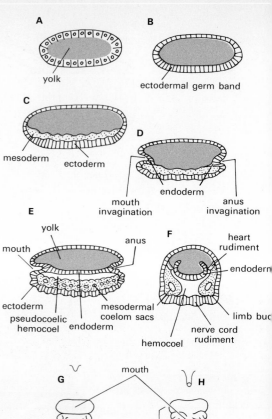

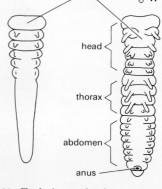

Figure 26.23 Early insect development. A to E, sagittal sections of progressively later embryos, head regions toward left. *A*, superficial cleavage of centrolecithal egg and blastula (see also Fig. 11.8). *B*, late blastula, germ band thickened in ventral ectoderm. *C*, segmental accumulation of mesoderm, formed by various gastrulation methods. *D*, mouth and anus invaginations, and ingrowth from both of these of an endoderm layer between mesoderm and yolk. *E*, appearance of schizocoelic coelom sacs and separation of endoderm from mesoderm, leaving hemocoel. *F*, cross section at stage after *E*. The heart rudiments are mesodermal and later will fuse along the dorsal midline. Yolk mass disappearing. *G* and *H*, ventral views of later embryos (note general annelidlike structure), showing gradual anteroposterior development of segments and segmental appendages. Note movement of mouth backward. (*Adapted from Eastham, Graber.*)

from the eyes, sensory devices include chemo-receptive and tactile "hairs" over most of the body and, in many cases, phonoreceptors at various surface regions (see Chap. 8). Insects also can taste and smell. The antennae are the chief olfactory organs, and in many species they play an important role in detecting members of the opposite sex from considerable distances.

The heart is elongated and contains typically 13 but often fewer pairs of ostia (see Fig. 25.3). It lies dorsal to a usually well-developed pericardial membrane. In aphids and some other forms, accessory hearts are located at the bases of the legs. Blood is generally colorless, but a few types (*Chironomus* among them) have hemoglobin. Excretion occurs through Malpighian tubules that open to the hindgut.

The alimentary system is typically arthropodlike in structure but often highly specialized in function. Salivary glands open on the floor of the mouth cavity. In many groups the pharynx is adapted for sucking fluid foods, and a crop and gizzard frequently form other subdivisions of the chitin-lined foregut. The midgut is short but it usually has numerous lateral surface-increasing pouches. The enzymes secreted by it are geared to the feeding habits of the animal. For example, insects that live largely on proteins (blood suckers, for example) usually lack lipases or carbohydrases or produce such enzymes in limited quantities only. The clothes moth *Tinea*, similarly, eats the protein keratin on sheep hair (wool), and its digestive enzymes are specialized accordingly. Digestive adaptations also occur in most insects that live on unusual foods, such as the cellulase-secreting wood-boring beetles, the paste- and glue-eating book lice, the silk-eating museum beetles, and others. The chitin-lined hindgut of most insects connects with rectal glands that absorb water from the feces and thereby assist in water conservation.

With very few exceptions, insects are separately sexed and fertilization occurs internally, often by means of spermatophore transfer from males. Gonads are paired, and a single gonopore lies ventrally between the ninth and tenth abdominal segments in males, in the eighth, ninth, or tenth in females. Parthenogenesis is common in aphids and the social insects (see Chap. 16). Tsetse flies and a few other types are viviparous, but most insects are oviparous and the ovipositors assist in egg laying.

The eggs generally are large and centrolecithal. Many nuclei are formed by division of the zygote nucleus and they migrate toward the egg surface, where they become the centers of the superficially cleaved blastomeres (Fig. 26.23 and see Fig. 11.8). In a *germ band* on the ventral side gastrulation then takes place, by invagination, delamination, or ingression of a layer of cells. This layer is *mesoderm*, which later forms embryonic (schizo-) coelomic sacs. The endoderm arises from the anterior and posterior invaginations from the ectoderm.

Appendage buds arise in anteroposterior succession on the ventral side. In Apterygota and Exopterygota such buds form adult appendages, all typically in the embryonic phase (Protura excepted). When apterygotes hatch, therefore, they are essentially miniature adults; and when exopterygotes hatch, they resemble adults greatly although wings are still lacking. The wings then grow during the nymphal instars, as exterior folds on the second and third thoracic segments (Color Fig. 28).

In Endopterygota, by contrast, the embryos hatch less fully developed, before all appendages are laid down. The larvae therefore retain an embryonic, annelidlike character. In such larvae pockets invaginate from the body wall and imaginal disks begin to develop at the bottom of the pockets (Fig. 26.24). These disks will give rise to the adult mouth parts, legs, and wings. The larva thus has one set of mouth parts and the adult acquires a new set. Larval mouth parts usually are biting-chewing and adult mouth parts may or may not be of the same type. Transformation to the imago occurs in the pupa. At this stage the pockets with the imaginal disks open out and the developing appendages become exteriorized. The wings, for example, then make their first external appearance (Fig. 26.25 and Color Fig. 29).

Figure 26.24 Later insect development: growth of a wing as an imaginal disk in an interior epidermal pocket, as in endopterygotes. The colored layer is the epidermis.

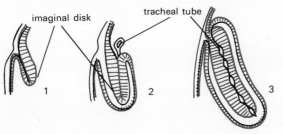

imaginal disk   tracheal tube

1   2   3

A

B

Figure 26.25 Later insect development: stages in the development of the silkworm moth *Bombyx mori*. A, caterpillar. B, spinning of pupation cocoon. Metamorphosis is illustrated in Color Fig. 29.

The nymphal molts of exopterygotes and the larval and pupal molts of endopterygotes are under precise hormonal control. Three types of glands have been shown to participate in these metamorphoses (Fig. 26.26): certain specialized *neurosecretory cells* (NS) in the brain ganglia; a pair of *corpora allata* (CA) in the mandibular segment dorsal to the pharynx; and a pair of *prothoracic glands* (PT) in the first thoracic segment. The NS cells rhythmically secrete an *ecdysiotropic hormone* that stimulates the PT glands, and these in turn secrete a steroid hormone *ecdysone* that acts on the body tissues and induces molting.

The nature of the molt is governed by whether or not *neotenin,* or juvenile hormone, is secreted by the CA glands. These glands are active during larval stages; and under the influence of neotenin, a molt transforms a larva to a later larva. But if neotenin is not secreted, as at the end of larval development, ecdysone acting alone transforms the larva to an adult in the case of exopterygotes. In endopterygotes, a first molt induced by ecdysone alone changes the larva to a pupa, and a second such molt establishes the adult. In the interval between these two molts a pupa frequently passes through a *diapause,* a dormant phase of various durations in different groups. During a diapause the NS cells and the PT glands are inactive and molting does not occur. In some cases emergence from the diapausal condition has been shown to be triggered by low temperatures. The NS cells become reactivated by cold, secretion of PT hormone follows, and the adult-producing imaginal molt then takes place. In temperate-zone insects the low temperatures of the winter season probably prepare the way for the imaginal molt in spring.

Figure 26.26 Hormonal controls in insect development. Neurosecretory cells from the brain (NS) secrete a hormone and stimulate the prothoracic gland (PT) to form ecdysone, a hormone that in conjunction with neotenin from the corpora allata (CA) determines larval molting (step 1). After neotenin ceases to become available (step 2), the next ecdysone-induced molt produces the adult in exopterygotes, the pupa in endopterygotes. The pupa often passes through a dormant phase, the diapause, and the following ecdysone-induced molt then yields the adult. See also Color Figs. 28 and 29.

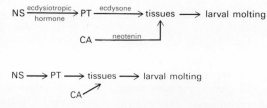

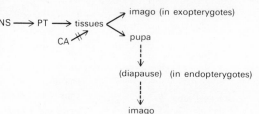

Review Questions

1. Give taxonomic definitions of the mandibulate subphylum and the crustacean class. Name the crustacean subclasses and the superorders of Malacostraca. Name representative animals of all these groups

2. Review the probable evolutionary interrelations of the crustacean groups. What evolutionary trends are in evidence in each broad line of descent?

3. Describe the segmental structure of the crustacean head, thorax, and abdomen. Name and state the functions of all appendages of a lobster.

4. Show how the structures and functions of appendages vary among crustacea. What are apodemes? Where is the sinus gland and what is its function?

5. Describe the internal anatomy of a lobster. In the process review the organization of every organ system. What are pleural and epipodial gills?

6. Show how crustacea develop segmentally through pygidial growth. Review the nature of crustacean larval development.

7. Describe the general characteristics and state the taxonomic position of (a) water fleas, (b) copepods, (c) barnacles. Describe the development and general adult structure of a barnacle.

8. How do crustacea mate? Review the ecology of crustacea and show in what ways certain crustacea (which?) are land-adapted. In what different ways do crustacea feed? What are phyllopodial legs and in which groups do they occur?

9. Give taxonomic definitions of each of the myriapod classes. Describe the segmental head structure of myriapods. Review the trunk structure and contrast the external anatomy of a centipede and a millipede. Which features of which myriapods are similar to those of insects? What are the ways of life of different myriapods?

10. Give taxonomic definitions of the class of insects and of each subclass and superorder. Within each subgroup name as many representative animals (and orders) as you can.

11. Characterize the orders Orthoptera, Homoptera, Hemiptera, Lepidoptera, Diptera, Coleoptera, and Hymenoptera.

12. Describe the segmental structure of an insect and contrast it with that of a myriapod and a crustacean. Review the internal anatomy of an insect and in the process describe the organization of every organ system.

13. Contrast the structure of biting, sucking, piercing, and other types of mouth parts and give specific examples of each. What is the structure of an insect wing and how is flight motion produced? Show how wings and legs are modified in different insect groups.

14. How do aquatic insects breathe? What alimentary specializations occur among insects in adaptation to particular modes of life? How do insects mate? In what environments are insects not found? Compare the species abundance of crustacea, myriapods, and insects.

15. Describe the early development of an insect and show how later development differs according to the subclass or superorder. Describe the hormonal controls of insect molting and development. What are trochanters, elytra, ovipositors, naiads, corpora allata, diapause?

Collateral Readings

Carpenter, R. M.: The Geological History and Evolution of Insects, *Am. Scientist,* vol. 41, 1953. Valuable for basic background on this subject.

Chu, H. F.: ''How to Know the Immature Insects,'' Brown, Dubuque, Iowa, 1949. A useful handbook, particularly in conjunction with the book by Jacques listed below.

Green, J.: ''A Biology of Crustacea,'' Quadrangle, Chicago, 1961. Well worth consulting for more detailed background information.

Hinton, H. E.: Insect Eggshells, *Sci. American,* Aug., 1970. An electron-microscopic and functional study.

Hocking, B.: Insect Flight, *Sci. American,* Dec., 1958. The subject is discussed from a functional standpoint.

Jacques, H. E.: ''How to Know the Insects,'' Brown, Dubuque, Iowa, 1947. Valuable in identifying and characterizing different insect groups.

Johannsen, O. A., and F. H. Butt: ''The Embryology of Insects and Myriapods,'' McGraw-Hill, New York, 1941. A detailed reference.

Johnson, C. G.: The Aerial Migration of Insects, *Sci. American,* Dec., 1963. Seasonal migrations occur in many species, as among birds.

Metcalf, C. L., W. P. Flint, and R. L. Metcalf: ''Destructive and Useful Insects,'' 4th ed., McGraw-Hill, New York, 1962. The ecological and economic biology of insects is examined thoroughly.

Murphy, R. C.: The Oceanic Life of the Antarctic, *Sci. American,* Sept., 1962. The article pays particular attention to krill, a fundamentally important shrimplike crustacean.

Roeder, K. D. (ed.): ''Insect Physiology,'' Wiley, New York, 1953. A collection of authoritative accounts by various authors.

Snodgrass, R. E.: ''Textbook of Arthropod Anatomy,'' Cornell University Press, Ithaca, N.Y., 1952. A basic standard text, strongly recommended.

Van der Kloot, W. G.: Brains and Cocoons, *Sci. American,* Apr., 1956. Neurosurgery on insect pupae reveals some relations between the nervous system and behavior.

Waterman, T. H.: Flight Instruments in Insects, *Am. Scientist,* vol. 38, 1950. The orientation mechanisms of flying insects are examined.

Wigglesworth, V. B.: Metamorphosis and differentiation, *Sci. American,* Feb., 1959. A noted student of insect development discusses the mechanisms of metamorphosis in these animals.

Williams, C. M.: The Metamorphosis of Insects, *Sci. American,* Apr., 1950. A description of some of the original experiments that have led to our present understanding of the subject.

————: Insect Breathing, *Sci. American,* Feb., 1953. The structural and functional aspects of the tracheal system are discussed.

Zahl, P. A.: Mystery of the Monarch Butterfly, *Nat. Geographic,* Apr., 1963. A well-illustrated description of the development of this insect.

Subgrade COELOMATA

ENTEROCOELOMATES   development typically regulative; cleavage radial or bilateral; deuterostomial, with adult anus formed at or near embryonic blastopore; enterocoelic coelom; typically with dipleurulalike stages or larvae; most groups unsegmented, some advanced groups segmented. *Chaetognatha, Pogonophora, Hemichordata, Echinodermata, Chordata*

As among schizocoelomates, the enterocoelomates include unsegmented types that presumably evolved first. They are represented today by the first four phyla listed above and by the most primitive group of the fifth. The remaining groups of the fifth phylum, notably the vertebrates, comprise segmented animals. Nonchordate phyla are the subject of the first chapter of this part, and the three following chapters are devoted to the chordates.

PART 9
COELOMATES: DEUTEROSTOMES

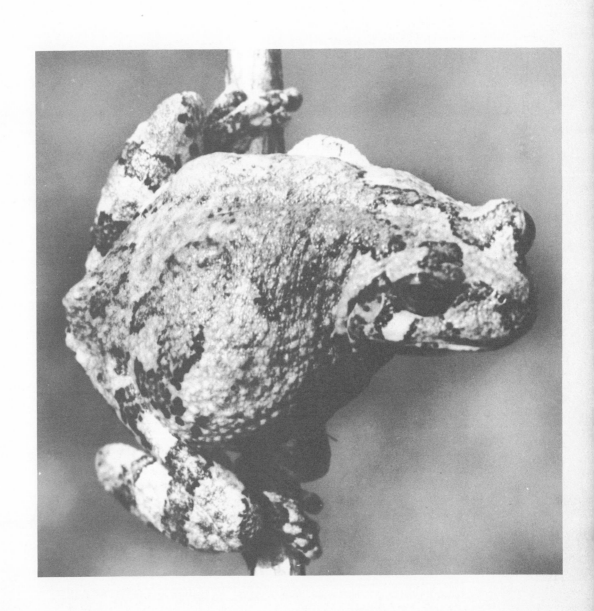

Of the four phyla to be discussed here, the chaetognaths do not appear to be related very closely to the other deuterostomes; they probably exemplify an independent, early branch of deuterostome evolution. The remaining phyla and indeed also the chordates probably do represent a broadly interrelated assemblage. Hemichordates appear to be closest to an ancestral deuterostome line, and beard worms, echinoderms, and chordates well could have arisen from it as independent branch lines (see Figs. 14.17 and 28.1). However, precise phylogenetic interrelations of deuterostome groups are considerably less definitely established than those of protostomes.

Arrowworms

Phylum Chaetognatha

marine, mainly planktonic; head with grasping spines and hood; trunk with lateral and tail fins; coelom subdivided in three compartments; without circulatory, breathing, or excretory systems; hermaphroditic, oviparous, without larvae.

Chaetognaths are small, torpedo-shaped animals, most of them well under 2 in. long (Fig. 27.1). Although they number only about fifty species, some of them, particularly *Sagitta*, are so abundant that they are often among the main components of zooplankton. Arrowworms are carnivorous on planktonic animals such as copepods and are themselves eaten by plankton feeders.

The body of an arrowworm is marked into head, trunk, and tail, and the internal coelomic cavity is subdivided by corresponding partitions. A transverse head-trunk septum lies just behind the head, and a vertical longitudinal septum divides the trunk cavity into two lateral compartments. Another transverse septum partitions the trunk cavity secondarily into an anterior main compartment and a posterior tail compartment. The body wall consists of a nonchitinous cuticle, an epidermis with a basement membrane, and a layer of longitudinal body-wall muscles. A peritoneum is absent (Fig. 27.2). The septa in the body cavity are extensions of the basement membrane. This membrane also extends to the outside as fin rays, supporting strands in the cuticular fins. One or two pairs of trunk fins and a posterior rounded tail fin are present. All are unmuscled and positioned horizontally; they are not used in locomotion but probably serve primarily as stabilizers and buoyancy-promoting devices.

The head bears one or two paired rows of tiny, anterior teeth, a pair of dorsolateral clusters of cup ocelli, and a row of prominent, chitinous *grasping spines* on each side of a ventral mouth. The teeth and the spines are moved by a powerful head

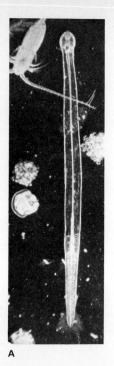

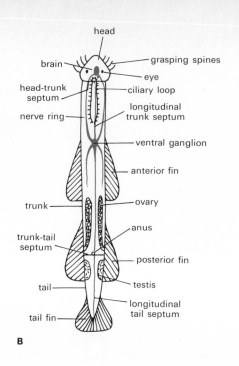

Figure 27.1 Chaetognath structure. A and B, general anatomy of *Sagitta*.

musculature, and the spines are the main food-catching structures. They can be closed in toward the mouth, forming a tiny cage in which food animals are trapped. Head muscles also operate a *hood,* a fold from the body wall at the head-trunk juncture (and containing part of the coelom). This hood can be moved forward over almost the whole head, spines included; it probably streamlines the head when the animal darts about rapidly. All muscles in head and body wall are striated.

The nervous system comprises a dorsal brain ganglion, a ventral trunk ganglion, a connective cord on each side between these (located between epidermis and basement membrane), and posterior cords connected to the trunk ganglion. Partially embedded in the brain dorsally is a pair of glandular *retrocerebral organs,* sacs that open immediately behind the brain at a retrocerebral pore. Just behind this pore the epidermis is modified as a *ciliary loop,* a patch of perhaps sensory or excretory tissue composed of glandular cells and ringed by flagellated cells. In the alimentary system the esophagus is expanded posteriorly as a bulbous enlargement, and the intestine extends as a pair of anteriorly pointing digestive pouches. The anus opens ventrally, just before the trunk-tail septum.

All arrowworms are hermaphrodites. Paired testes are situated just behind the trunk-tail sep-

tum, paired ovaries just in front of it. Gonopores open at corresponding locations. Male openings apparently are formed only during the breeding season. An oviduct has another syncytial tube inside it that leads to the female gonopore, which provides an entry path for sperms (Fig. 27.3). Eggs are shed into the space between the outer and inner tube, and sperms somehow reach this location. Zygotes are shed through temporary openings in oviduct and body wall, not through the female gonopore.

Eggs develop externally, in some cases attached in batches to stones or other objects in water. The eggs are isolecithal, cleavage is radial and regulative, and a coeloblastula develops. Invaginative gastrulation then produces a two-layered embryo. Anteriorly a pair of prospective gamete-forming cells is budded into the archen-

Figure 27.2 Chaetognath structure. A, cross section through trunk of *Sagitta.* B, anterior end of *Sagitta,* seen from ventral side, hood retracted. (*Based on Ritter-Zahony.*)

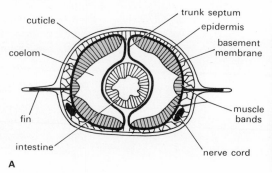

A

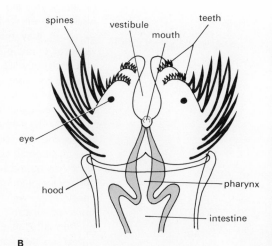

B

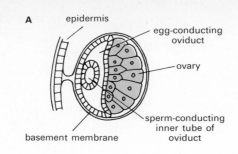

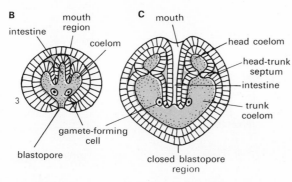

Figure 27.3 Chaetognath development. A, cross section through female reproductive system. Sperms pass from the inner tube to the oviduct channel, where eggs are present and fertilization occurs. *B,* postgastrula, formation of enterocoelic coeloms. *C,* later stage, subdivision of coeloms under way. (*Adapted from Burfield.*)

teron. Also, two invaginated endoderm folds begin to grow from the anterior end backward, until they meet and fuse with the posterior wall of the endoderm. Two coelomic sacs are cut off in this manner, one on each side of the future alimentary tract (see Fig. 27.3). These sacs also partition transversely into head and trunk coeloms, and mouth and anus eventually break through at the ends of the alimentary tube.

All these coelomic and alimentary cavities then fill up temporarily with cells, and the entirely solid embryos hatch at this stage. Later the internal spaces reappear, but it is not quite clear if these adult body cavities correspond exactly to the embryonic ones. The temporarily solid phase explains the absence of an adult peritoneum; the embryonic peritoneum disappears during the solid stage. After hatching the animals resemble miniature adults, and they gradually become mature adults without special larval stages. Chaetognaths regenerate well.

The manner of coelom formation differs from that in other enterocoelomate groups and actually resembles that of some of the brachiopods. Also, the presence of chitin (in the grasping spines) is quite atypical for enterocoelomates, which otherwise lack chitin entirely. It is because of such characteristics that chaetognaths are believed to represent a very early offshoot of enterocoelomate evolution.

Beard Worms ## Phylum Pogonophora (Branchiata)

marine, tube-dwelling, abyssal; with anterior tentacles; mouth and alimentary system absent; circulatory system closed; without breathing system; coelom divided into *protocoel, mesocoel,* and *metacoel;* sexes separate.

This interesting group of 25 known species was first discovered in 1933, in material dredged up from deep sea bottoms. The animals range in length from about 4 in. to 1 ft, but their diameter is not more than about 1 mm. They live in close-fitting secreted tubes that probably stick vertically in the ooze of the ocean floor. Such tubes are composed of polysaccharide-containing tunicin, the same celluloselike material found also in the tunicates among chordates. Beard worms are without any trace of an alimentary system—the only free-living animals so characterized (Fig. 27.4).

The body of a beard worm is divided internally by transverse septa into an anterior *protosome,* a middle *mesosome,* and posterior *metasome.* The coelomic cavities corresponding to these divisions are the *protocoel,* the *mesocoel,* and the *metacoel,* respectively. Such divisions generally are characteristic of all other enterocoelomate phyla. The protocoel probably is homologous to the head coelom of chaetognaths, and the mesocoel and metacoel together correspond to the chaetognath trunk coelom. This internal division of beard worms is not clearly marked externally, but the protosome bears the tentacles, the mesosome forms a short, slightly thickened collarlike section, and the metasome represents the remainder of the body.

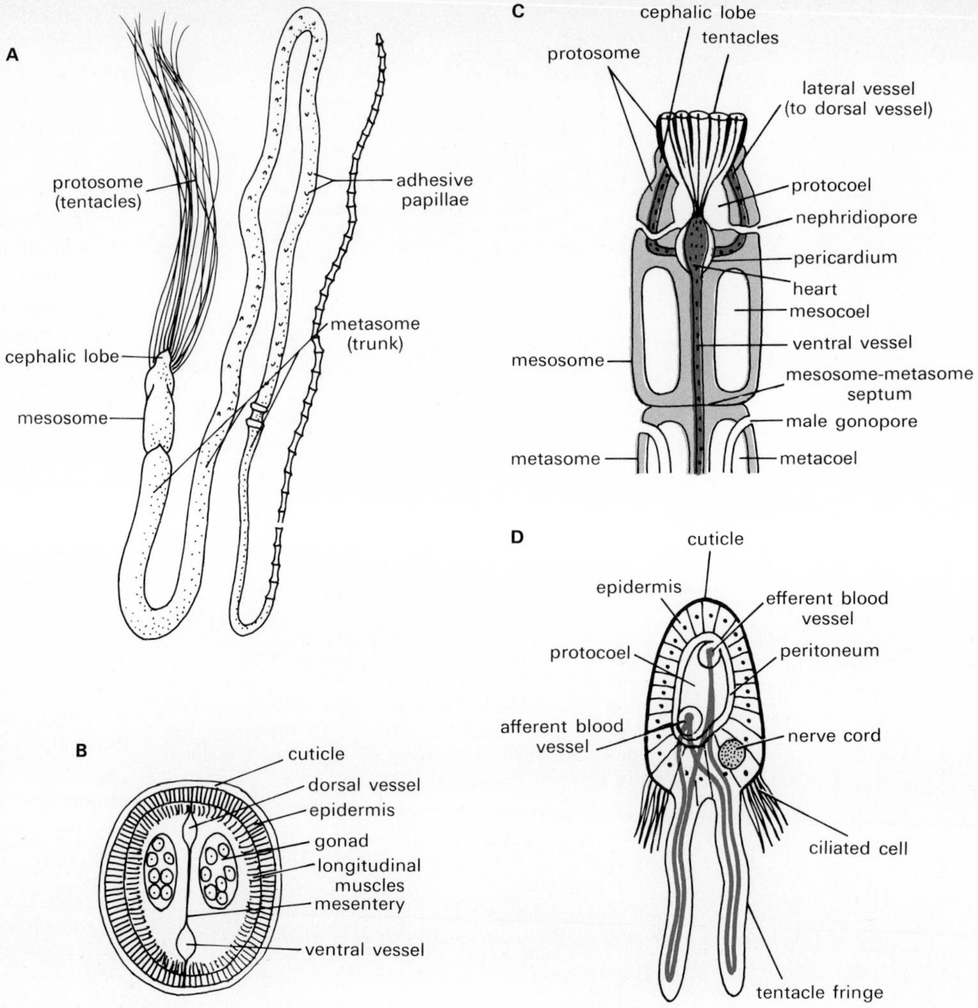

A

protosome
(tentacles)

cephalic lobe

mesosome

adhesive
papillae

metasome
(trunk)

C

cephalic lobe

protosome

tentacles

lateral vessel
(to dorsal vessel)

protocoel

nephridiopore

pericardium

heart

mesocoel

ventral vessel

mesosome-metasome
septum

male gonopore

metacoel

mesosome

metasome

D

cuticle

epidermis

protocoel

afferent blood
vessel

efferent blood
vessel

peritoneum

nerve cord

ciliated cell

tentacle fringe

B

cuticle

dorsal vessel

epidermis

gonad

longitudinal
muscles

mesentery

ventral vessel

Figure 27.4 Pogonophora. A, general appearance. *B,* cross section through trunk. *C,* sectional ventral view of anterior end of body. The lateral blood vessels join dorsal to the heart as the dorsal longitudinal vessel (hidden in this view). *D,* cross section through a tentacle. (*After Ivanov.*)

The body wall consists of a polysaccharide-containing cuticle, a glandular epidermis, and a double layer of muscle. A distinct peritoneum is lacking in the metacoel but is present in the protocoel and the tentacles. The nervous system consists of a primitive intraepidermal plexus and a longitudinal cord that enlarges anteriorly as a ganglionic center. This center lies in a *cephalic lobe* at one side of the tentacle bases.

The protocoel leads to the outside through a pair of presumably nephridial ducts and also extends into the tentacles. These are long, fingerlike outgrowths from the body wall, each with ciliary tracts and numerous fringelike lateral extensions (*pinnules*). Each tentacle also contains a blood vessel loop that connects with a main circulatory loop in the trunk. This closed system is supported in a mesentery that divides the mesocoel and metacoel into lateral compartments. The mesocoel lacks ducts to the outside, but the metacoel opens to the exterior through gonoducts. In males a pair of gonopores opens ventrally just behind the mesocoel-metacoel septum, and in females the gonopores lie near the middle of the trunk.

Beard worms are believed to feed by arranging their tentacles as a tube, the pinnules serving as interlocking devices. The ciliary tracts are assumed to draw food-bearing currents into the tube. Trapped inside, food organisms would be digested extracellularly, and the resulting nutrients would

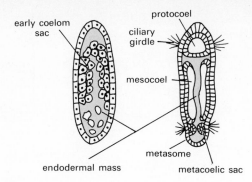

Figure 27.5 Pogonophoran development. The metasome at first is restricted to the posterior portion of the embryo. The three coelomic body divisions arise from the early coelomic sacs. Inasmuch as this mesoderm develops from the endodermal mass, mesoderm and coelom can be considered to form enterocoelically.

be absorbed directly by the tentacular blood vessels. Actual feeding has remained unobserved, however.

In the embryos cleavage is radial or bilateral, a stereoblastula forms, and gastrulation takes place by delamination. A blastopore never arises, but the region where it would be expected becomes the posterior end of the animal (Fig. 27.5). A pair of coelomic sacs cuts off anteriorly from the endoderm, and after they enlarge the sacs constrict into proto-, meso-, and metacoelic portions. The endoderm never acquires a cavity, and an alimentary system therefore does not develop even in rudimentary form. In some genera elongated larvae resembling the adults have been observed. Embryology and adult coelomic structure clearly relate beard worms to the other deuterostomes, hemichordates in particular.

Hemichordates Phylum Hemichordata

marine; wormlike, colonial in secreted housings or solitary in sand burrows; body marked prominently into protosome, mesosome, metasome, and coelom partitioned correspondingly; mostly with gill slits; sexes largely separate; typically with larvae.

Class Pterobranchia colonial, zooids microscopic; with or without gill slits; alimentary tract U-shaped; circulation open; with tentacles on mesosome.

Class Enteropneusta acorn worms; solitary; with many gill slits; alimentary tract straight; circulation closed in some; without tentacles; often with *tornaria* larvae.

The approximately one hundred species of these animals share one important trait with the chordates, gill slits in the pharynx. Of the two hemichordate classes the pterobranchs probably are the more primitive group; but the enteropneusts are the more familiar, and their larvae have great theoretical significance.

Acorn worms live singly in shallow-water burrows and feed on microscopic food in sand (Fig. 27.6). From 1 in. to 2 yd long, the body of such a worm consists of a short or long *proboscis* (protosome), a short *collar* (mesosome), and a long *trunk* (metasome). The proboscis is a muscular burrowing organ, and mucus secreted on its surface entangles sand and food that pass to the mouth. This opening lies ventrally at the base of the proboscis, under a forward flange of the collar. Anteriorly on each side of the trunk are a series of gill slits, usually U-shaped. The body wall here often forms prominent folds that curve up and cover the gill slits partially. The gonads lie within these folds, which are called *genital ridges*. At the end of the trunk is the anus.

Figure 27.6 Hemichordates: model of *Dolichoglossus,* an acorn worm, dorsal view. Note proboscis (protosome), conspicuous collar (mesosome), and row of paired gill glits along anterior portion of trunk (metasome). The lateral edge of this trunk region is folded up on each side as a genital ridge.

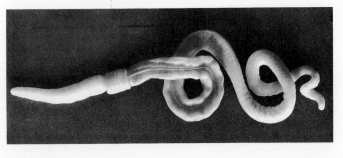

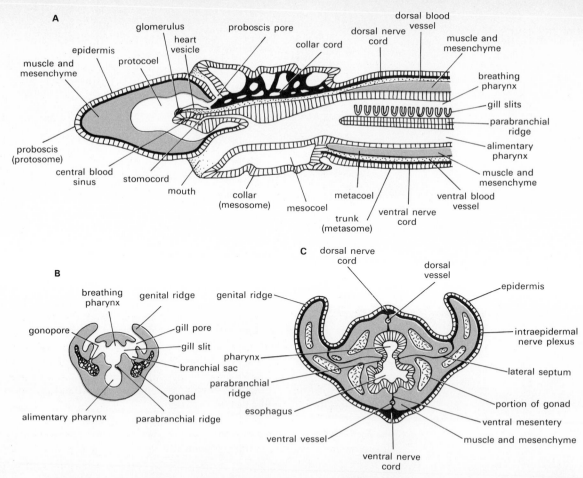

Figure 27.7 *Hemichordate structure. A,* sagittal section of anterior region of *Balanoglossus*. The parabranchial ridge is a fold of endoderm that separates the pharynx into an upper breathing and a lower alimentary chamber. *B,* cross section through gill region. *C,* cross section just behind gill-slit region.

The body wall consists of a glandular ciliated epidermis, a basement membrane, and muscle layers (Fig. 27.7). Much of the coelom is filled with connective tissue and muscle, formed directly from the peritoneum in the embryo and thus accounting for the absence of this membrane in most adults. The nervous system contains a primitive intraepidermal net and a ventral and dorsal cord in the trunk. At the collar-trunk juncture the ventral cord joins the dorsal cord, which then continues forward as a *collar cord.* This section represents the neural center of the animal. In many cases it is hollow, and one or two *neuropores* lead from this neural cavity to the outside of the body.

The proboscis contains a small protocoel that communicates with the outside through a dorsal *proboscis pore.* The mesocoel in the collar is similarly reduced, and it opens posteriorly into the first gill chamber. Both the proboscis and collar coeloms thus have access to sea water (Fig. 27.8). By contrast, the metacoel in the trunk is closed; it contains coelomic fluid with amoeboid cells. The alimentary tract leads from the mouth to a mouth cavity that extends anteriorly as a pouch, the *stomocord.* In the anterior portion of the trunk the alimentary tract contains a long pharynx, perforated dorsally by the paired U-shaped gill slits. Basically the slits are food strainers. Water passes through them to the outside, while food and sand are retained and fall to the ventral gutter of the pharynx. Secondarily such slits also serve in breathing, a specialized function developed

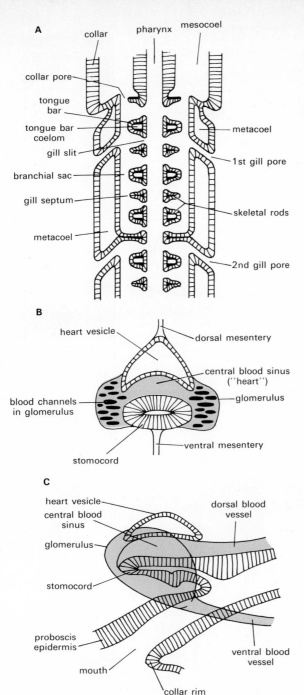

Figure 27.8 Hemichordate structure. A, frontal section through anterior gill region showing relations between mesocoel, branchial sacs, pharynx, and gill pores. B, cross section and C, side view, at level of stomocord, to show circulatory and excretory organs. *(Based on Hill, Van der Horst.)*

most clearly in chordates. Gill slits open into deep, communicating pockets invaginated from the body wall, the *branchial sacs,* or gill chambers. These in turn open to the outside through *gill pores.* The collar coelom on each side of the body communicates with the most anterior gill chamber on that side. Behind the pharynx the alimentary tract continues as esophagus and intestine.

The circulatory system is probably closed, but in parts of it blood passes through sinuses without vessel walls. Blood flows forward in a dorsal longitudinal vessel, backward in a ventral one. Just above the stomocord at the proboscis base the dorsal vessel widens to a noncontractile "heart." Above it is a contractile *heart vesicle,* a muscular coelom sac (probably derived from the right protocoel) that presses rhythmically against the heart and maintains blood circulation (see Fig. 27.8). On each side of the heart is a *glomerulus,* an accumulation of mesodermal tissue believed to have an excretory function. Blood is forced to pass from the heart through the glomeruli and presumably is filtered there. Excretion products probably leave the body through the protocoel, mesocoel, or both. In the pharyngeal region the ventral trunk vessel sends branches to the gill-slit region, and another set of branch vessels leaves that region and enters the dorsal trunk vessel. Oxygenated blood here passes anteriorly. The blood is colorless.

The gonads open into the gill chambers. Fertilization occurs externally during a spawning period. Eggs are isolecithal or telolecithal, cleavage is radial, and the resulting coeloblastula gastrulates by invagination (Fig. 27.9). The anterior part of the archenteron becomes cut off as the embryonic protocoel, and this pouch grows back and constricts off mesocoelic and metacoelic portions on each side. It also develops an opening to the outside, the later proboscis pore.

If development started from a yolky egg, as in *Saccoglossus,* the larva is ellipsoid, with an anterior apical tuft and a posterior telotroch. But if the egg is isolecithal, as in *Balanoglossus,* the resulting larva is a *tornaria,* with apical tuft, telotroch, and a conspicuous ciliary band that winds sinuously over the larval surface. The tornaria is remarkably like the larva of starfishes—a main argument for regarding hemichordates and echinoderms as closely related. Indeed, it is widely believed that both phyla originated from a hypothetical *dipleurula* ancestor, a form postulated to

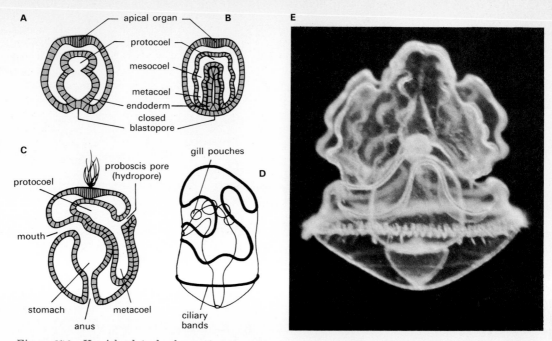

Figure 27.9 Hemichordate development. A, frontal section of postgastrula, showing protocoel being cut off from archenteron. B, later stage, showing posterior enlargement of coelom on each side of archenteron. C, sagittal section through late embryo. Alimentary tract is complete and protocoel communicates through dorsal pore with exterior. D, side view of tornaria larva formed from isolecithal egg, as in *Balanoglossus.* The gill pouches evaginate from the pharynx and become gill slits after acquiring openings through corresponding invaginations from the epidermis. E, tornaria larva of *Glossobalanus,* dorsal view, from life. *(A-D, based on Heider, Davis, Morgan, and other sources.)*

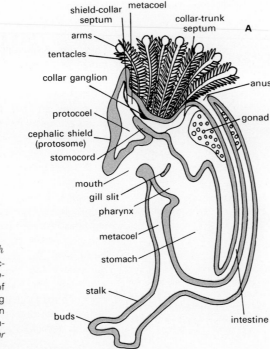

Figure 27.10 Pterobranch hemichordates. A, sagittal section through a zooid of *Cephalodiscus.* B, portion of a colony of *Rhabdopleura,* showing creeping and erect skeletal tubes, position of zooids within, and interconnecting black stolon. *(After Schepotieff.)*

have resembled both the tornaria and the starfish larva (see Fig. 27.13 and below).

Metamorphosis occurs gradually. In the process lateral sacs grow out from the pharynx and corresponding ingrowths arise in the body wall. Gill slits break through in this manner, in anteroposterior sequence. Adult enteropneusts can regenerate a new proboscis readily, and forms such as *Balanoglossus* are known to bud off new individuals at the posterior end of the trunk.

The whole class Pterobranchia consists of just three genera. These animals are colonial and individually of microscopic dimensions. All zooids of a colony arise by budding from a single ancestrula, as in ectoprocts, and a whole colony shares a

common secreted housing. In *Cephalodiscus* the zooids of a colony are separate, but in *Rhabdopleura* they are interconnected by a dark cord, the *black stolon* (Fig. 27.10).

Each zooid consists of a plump metasome with a long slender stalk at the posterior end, a mesosome collar with a set of tentacles anterodorsally, and a shieldlike protosome that tilts down over the mouth area like a lid. Internally only a single pair of gill passages is present and the circulatory system is distinctly open, blood vessels being entirely absent. Sexes are separate in most cases. Eggs are yolky and ciliated larvae with apical tufts are generally formed. At the posterior end of the metasome stalk is a budding zone where numerous new individuals can develop.

Atubaria does not secrete a housing. The housings of *Rhabdopleura* closely resemble those of the fossil *graptolites,* and these fossils show that their once-living inhabitants likewise were joined by interconnecting stolons (see Fig. 15.4). On the basis of such similarities some zoologists regard the graptolites as relatives or even one of the classes of hemichordates.

Hemichordates display echinodermlike traits in some of their larvae and chordatelike traits in their pharynx, and they exhibit embryologic similarities to both these phyla. Presumably therefore the animals represent an evolutionary link between echinoderms and chordates. However, echinoderms and beard worms appear to be related to hemichordates much more closely than the chordates (see Figs. 14.7 and 28.1).

Echinoderms   Phylum Echinodermata

spiny-skinned animals; exclusively marine; larvae bilateral, with protocoel, mesocoel, and metacoel in early stages; adults pentaradial; with calcareous endoskeleton and coelomic water-vascular system; mostly separately sexed; development regulative, typically oviparous, some ovoviviparous and viviparous.

Subphylum Pelmatozoa

oral surface facing up, aboral surface with or without stalk; ambulacral grooves ingestive, tube feet food-catching; both mouth and anus on oral side; main nervous system aboral.

Class Crinoidea　sea lilies, feather stars; body cup-shaped, free or attached; endoskeleton limited to aboral side, oral side membranous; arms branched; *doliolaria* larvae or direct development. *Cenocrinus*, sea lilies, with stalk; *Antedon*, feather stars, without stalk.

Subphylum Eleutherozoa

oral surface facing down or to side; ambulacral grooves not ingestive; tube feet locomotor, with ampullae; mouth oral, anus aboral or absent; main nervous system oral.

Class Holothuroidea　sea cucumber; secondarily bilateral; mouth region with tentacles; ambulacral grooves closed; endoskeleton reduced to ossicles; oral-aboral axis horizontal; *auricularia* larvae or direct development.

Class Asteroidea　sea stars, starfishes; arms with open ambulacral grooves on oral side; with digestive glands; tube feet project between endoskeletal plates; *bipinnaria* larvae or direct development.

Class Ophiuroidea　brittle stars, serpent stars, basket stars; arms highly flexible; ambulacral grooves closed, tube feet reduced; madreporite on oral side; without intestine or anus; *ophiopluteus* larvae or direct development.

Class Echinoidea　sea urchins, sand dollars; without arms; ambulacral grooves closed; endoskeleton fused, nonflexible, with pores for tube feet and with movable spines; *echinopluteus* larvae or direct development.

General Characteristics

The 6,000 species of living echinoderms typically develop via bilateral free-swimming larvae that metamorphose into sessile or sluggish adults. These are organized pentaradially around a comparatively short oral-aboral axis (Fig. 27.11). In crinoids the oral side is directed upward and both mouth and anus are present on that side. In sea cucumbers the oral-aboral axis is horizontal, the mouth marking one end of the axis and the anus the other. In all other groups the oral side is directed downward and the mouth is in the center. Two basic traits are shared by all echinoderms: *an endoskeleton* produced in the dermis, consisting of calcareous plates overlain by epidermis; and a *water-vascular* system, a series of coelomic tubes filled with sea water. The exterior parts of this system are numerous hollow, muscular tube feet, or *podia*, that starfishes, for example, use as little legs.

How could such a unique water-vascular system have evolved? The original function of the system could not have been locomotor, for ancient fossil echinoderms were sessile attached types, and many living crinoids still are. Further, the earliest known fossil echinoderms (the carpoids)

were bilateral animals, and the larvae of all living echinoderms still are bilateral. When crinoid larvae metamorphose they become attached at the anterior, aboral end. The organs near this attached end undergo an internal rotation that brings the mouth to the upper side, near the anus. Concurrently a radial symmetry develops by a suppression of growth on the right side, the left side twisting from a vertical position to a horizontal one and then proliferating radially (Fig. 27.12). Accordingly, echinoderms appear to have evolved from bilateral, free-swimming ancestors whose later descendants became attached sessile types. A secondary radial symmetry developed concurrently, and the animals had radially arranged arms that served as *feeding* tentacles.

However, these tentacles then became clothed with endoskeletal plates, a circumstance that aided in protection but must have severely limited the mobility and thus the food-trapping capacity of the arms. This restriction could be circumvented by the evolution of small mobile branch tentacles on the arms and by development of food passages along the arms to the mouth. Crinoids today actually exhibit such an organization. They have branched skeleton-supported arms, and the mobile tube feet on the oral side of the arms serve entirely in feeding; they trap small organisms that are passed along a ciliated *ambulacral groove* in each arm toward the mouth (see Figs. 27.11 and 27.12). Primitively, therefore the tube feet are food-catching tentacles; and the water-vascular system can be interpreted as a modified tentacular system, evolved in primitive echinoderms in conjunction with sessilism and endoskeleton-induced body rigidity.

But once a tube-foot system had evolved, it could be adapted secondarily to functions other than feeding, most particularly to locomotion. Later echinoderm groups actually relinquished the stalked, attached mode of life and became motile, with the oral side directed downward. Tube feet, and in some cases also endoskeletal spines and the arms themselves, could function in propulsion. The animals then could move to food actively, use the mouth directly, and no longer needed to depend on what the arms could strain out of the water. The ambulacral grooves therefore largely ceased to be of importance, and they actually are closed over by folds of the body wall; in all living echinoderms except the crinoids and asteroids the remnants of the original grooves are *epineural*

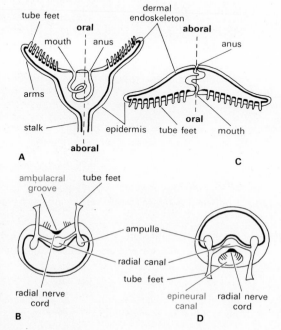

Figure 27.11 Echinoderm body plans. Left, the crinoid pattern, with vertical oral-aboral axis, mouth and tube feet pointing up (*A*), and open ambulacral groove (*B*). *Right,* the ophiuroid-echinoid pattern, with mouth and tube feet pointing down (*C*), and with ambulacral groove closed in as an epineural canal (*D*). Asteroids correspond to pattern *C,* but they have open ambulacral grooves.

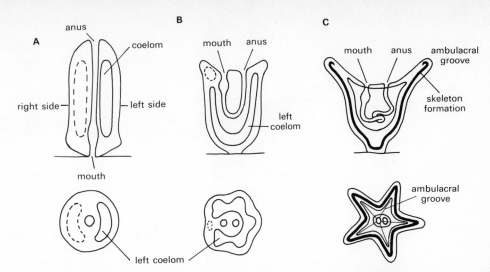

Figure 27.12 Echinoderm radiality: development as in crinoids. Top figures, sagittal sections; bottom figures, cross sections. *A,* oral settling and attachment of bilateral larva, growth suppression on right side. *B,* proliferation of left side, rotation of mouth to new oral position away from attached end, beginning of horizontal radiality. *C,* horizontal pentaradial growth of left coelomic derivatives and body as a whole, formation of horizontal arms, development of enveloping and motion-restricting endoskeleton.

canals just under the body wall (see Fig. 27.11).

Moreover, since the mouth then no longer needed to lie on the same side as the anus, the alimentary tract did not have to be U-shaped. Even so, the organs of echinoderm larvae typically still undergo the internal twisting from a vertical position to a horizontal one, and the development of adult radial symmetry involves a suppression of growth on the right side; the left side of the larva

Figure 27.13 The dipleurula. This hypothetical ancestor of deuterostomes resembles certain actual embryonic and larval stages of hemichordates and echinoderms. *(After Bather.)*

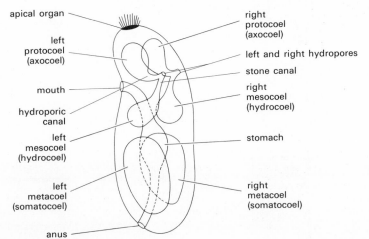

thus becomes the oral side of the adult. Also, in the course of developing the adult water-vascular system, echinoderm larvae still pass through an at least rudimentary tentaclelike stage.

These common features in echinoderm development are widely postulated to represent evolutionary reminders of a hypothetical bilateral ancestor, the *dipleurula* (Fig. 27.13). As noted earlier, this form is assumed to have been a common starting point of both hemichordates and echinoderms: both groups develop protocoels, mesocoels, and metacoels; the tentaclelike stage in echinoderm development is highly reminiscent of pterobranchs; and the tornaria larvae of hemichordates are quite similar to the bipinnaria larvae of asteroids (see Figs. 27.9 and 27.19). A dipleurulalike ancestor thus could have given rise to bilateral hemichordates on the one hand and to secondarily radial echinoderms on the other. Only on such a basis do the complex developmental and adult traits of echinoderms become meaningful.

These traits are exemplified well in the asteroids, which combine both primitive and advanced characteristics. The body of a common starfish such as *Asterias* consists of a central *disk* and, typically, five *arms* (Fig. 27.14). In the center of the disk on the underside, or oral surface, is the mouth. Leading to it along each arm is an ambulacral groove, bordered on each side by a row of

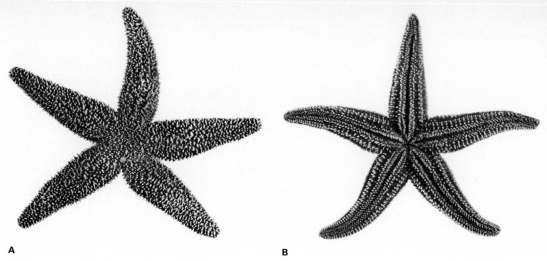

A   **B**

Figure 27.14 Starfishes. A, the aboral side of *Asterias*. The anus lies at the exact center of the disk but is too small to be visible. The buttonlike madreporite is seen excentrically on the disk between the two lower arms. B, the oral side. The mouth is at the center of the disk, and a tube-foot-lined ambulacral groove passes along each of the five arms to the mouth. (*Courtesy Carolina Biological Supply Company.*)

hollow, muscular tube feet, the terminal parts of the water-vascular system. On the upper, aboral side, a tiny anal opening lies in the center of the disk, and near the angle between two of the arms is a reddish *madreporite,* a sieve plate that forms the entrance to the water-vascular system. The body wall consists of cuticle, epidermis, dermis,

muscle layers, and peritoneum. The dermis is a well-developed connective tissue that secretes the endoskeleton. In a starfish this skeleton is made up of separate knobby calcereous plates that are held together by connective tissue and muscles (Fig. 27.15). In places the dermis and overlying epidermis are folded out as microscopic *pedicella-*

Figure 27.15 Starfish structure. A, cross section of an arm. B, the water-vascular system. Polian vesicles are absent in the common starfish *Asterias*.

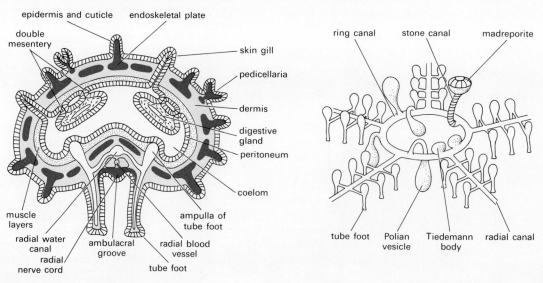

A   **B**

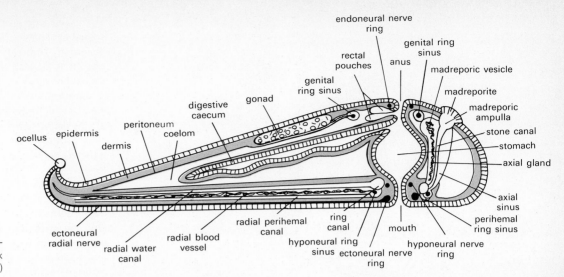

Figure 27.16 Starfish structure. Section through the disk and one arm. (*After Borradaile.*)

riae, muscle-operated pincers that project the *skin gills.* These are microscopic fingerlike projections on the body surface, made up of all layers of the body wall. The gills are hollow; their internal spaces are extensions of the coelom between adjacent skeletal plates.

Sea water enters and leaves the water-vascular system through the madreporite, which leads to a small bulbous cavity, the *madreporic ampulla.* From there a calcified *stone canal* conducts water to a channel that circles the esophagus. Along this *ring canal* usually are one or more sacs, the *polian vesicles,* which probably serve as water reservoirs (not present in *Asterias*), and five pairs of *Tiedemann's bodies,* whose specific function is unknown. From the ring canal also emanate five *radial canals,* one to each arm. There the radial canals give off short lateral branches, each with a saclike *podial ampulla* inside the arm and a tube foot protruding from the arm. Stiffened by water pressed into them from the ampullae, tube feet serve as tiny walking legs. Their ends are used also as suction disks; podia can exert enough steady pull on the shells of a clam to tire the clam and force its shells open.

Clams and oysters are the main diet of starfishes. The mouth leads through a short esophagus to a spacious stomach (Fig. 27.16), which can be everted through the mouth into the soft tissues of a clam. Small food particles and fluid foods pass into a short intestine and from there to five pairs of large digestive glands, the *pyloric caeca.* These occupy most of the free space in the

arms, and each is suspended from the body wall by a double mesentery. Near the anus on the aboral side the intestine connects with two *rectal pouches* that probably have an excretory function.

As in echinoderms generally, the nervous system consists of three subsystems (see Fig. 27.16). The *ectoneural system* contains a *nerve ring* under the body wall on the oral side, near the ring canal of the water-vascular system. From this nerve ring emanate five *radial nerves,* one to each arm. They run along the ambulacral grooves just under the epidermis and are part of a subepidermal nerve net. At each arm tip the radial nerve terminates at an ocellus. The ectoneural system is the main system in all echinoderms except the crinoids. The two other, less well-developed subsystems are the *endoneural* and *hyponeural* systems, each again with nerve ring and radial nerves. The hyponeural system lies orally near the ectoneural one, and the endoneural system (well developed in crinoids) is located under the aboral body wall.

The circulatory system consists of a series of blood channels without walls, situated inside coelomic ducts (Fig. 27.17). The center of this system is an *axial gland* and a *madreporic vesicle,* a meshwork of contractile channels next to the stone canal and traversing a coelomic duct called the *axial sinus.* Pulsations of gland and vesicle presumably maintain a circulation of blood. Aborally the axial gland continues into a circular *genital ring sinus,* which sends paired branch channels to the gonads in each arm. Orally the axial gland connects with another circular channel near the nerve

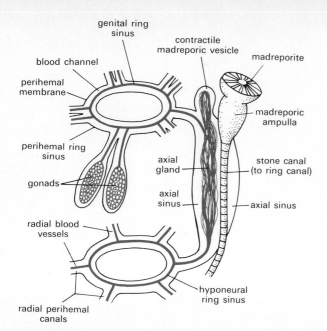

genital ring
sinus

blood channel

perihemal
membrane

perihemal ring
sinus

gonads

radial blood
vessels

radial perihemal
canals

contractile
madreporic vesicle

madreporite

madreporic
ampulla

axial
gland

axial
sinus

stone canal
(to ring canal)

axial sinus

hyponeural
ring sinus

Figure 27.17 Starfish structure. Circulatory and coelomic channels. Only one of the five pairs of gonads is shown. All blood channels are surrounded by a perihemal membrane that encloses a coelomic space (perihemal channel). (*After Hyman.*)

ring, the *hyponeural ring sinus*. From this sinus emanates a *radial sinus* to each arm. The blood is colorless and differs little from coelomic fluid. In this fluid are numerous amoebocytes that probably absorb excretory waste and carry it to the outside via the skin gills.

Starfishes have five pairs of sex organs, one pair per arm. Each gonad opens through a short duct near the arm base. A few asteroids mate by copulation, and some brood their offspring in chambers on the oral side of the disk (Fig. 27.18). But most starfishes spawn, fertilization and development occurring externally. Brooding types have yolky eggs that develop directly. Spawning types shed isolecithal eggs that form free-swimming larvae.

After mesoderm has formed in typical enterocoelic fashion, the mesoderm pouch on each side of the archenteron subdivides into three coelom sacs: protocoel (*axocoel*) anteriorly, mesocoel (*hydrocoel*) behind, and metacoel (*somatocoel*) posteriorly (Fig. 27.19). The left protocoel develops a duct that opens to the outside through a *hydropore*. Events up to this point are essentially identical to corresponding processes in hemichordates (and the hydropore is equivalent to the proboscis pore of acorn worms). The hydropore later forms the madreporite; the left protocoel becomes the madreporic ampulla; the connection between the left protocoel and mesocoel develops as the stone canal; and the left mesocoel gives rise to the ring and radial canals of the water-vascular system. The right protocoel persists as the madreporic vesicle (believed to be homologous to the heart vesicle of hemichordates), and the right mesocoel degenerates. The left and right metacoels will develop as the adult coelom and the coelomic blood channels.

The free-swimming starfish larva is a *bipinnaria*. It has a lobed, buoyancy-increasing ectoderm and ciliary surface bands that serve in locomotion and feeding. Internally, the left and right metacoels enlarge and suspend the archenteron in a vertical mesentery. The anterior part of the archenteron grows toward the ventral side and eventually breaks through the ectoderm as the larval mouth.

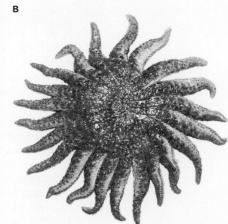

Figure 27.18 Asteroids. A, oral view of starfish with eggs in brood pouch. Note tube feet. *B,* a many-armed sun star, aboral view.

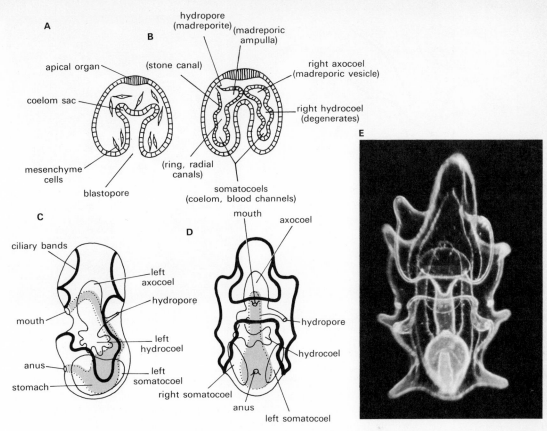

Figure 27.19 Starfish development. Early stages are illustrated in Fig. 11.11. *A*, postgastrula, formation of proto-coelomic sacs. *B*, subdivision of coelom into three sacs on each side; the later fate of each sac is indicated in parentheses. *C*, early bipinnaria larva, seen from left side; the left hydrocoel already has the five lobes that foreshadow the five radial canals of the adult. *D*, later bipinnaria, ventral view. The hydrocoel is in process of growing around the esophagus, foreshadowing the ring canal of the adult. *E*, ventral veiw of bipinnaria larva of *Asterias*, from life. Alimentary tract and epidermal ciliary bands are well visible. (*C, D, based on Horstadius*.)

The left hydrocoel proliferates and forms a closed ring—the later ring canal, and from it evaginate five fingerlike pouches—the later radial canals (see Fig. 27.19). These eventually push out of the left side of the larva as five ectoderm-covered fingers. It is at this general stage that the bipinnaria is reminiscent of a tentacled, pterobranchlike ancestral form. The early radial canals later extend in length and, by lateral branching, give rise to the system of tube feet.

In the course of its free life a bipinnaria acquires additional ectodermal lobes, and this later larva is known as a *brachiolaria* (Fig. 27.20). At its anterior end the brachiolaria develops a special attachment disk by which it settles in preparation for metamorphosis. The whole attachment apparatus and most of the larval ectoderm will be discarded and left behind after metamorphosis; the young adult star forms mainly from the posterior part of the larva. During metamorphosis the larval mouth and anus close over, and most of the larval esophagus and intestine degenerate. From the stomach a new esophagus then grows out through the hydrocoel ring and establishes an adult mouth on the left, oral side. Similarly, another stomach evagination grows toward the right and forms an anus on what becomes the adult aboral side. Evidently, the mouth and anus of the adult do not have any direct relation to the alimentary openings of the larva.

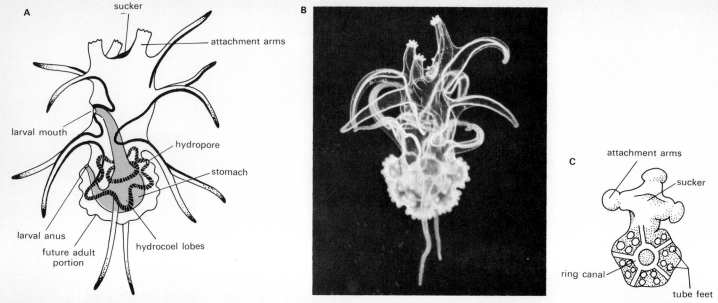

Figure 27.20 *Starfish metamorphosis.* A, brachiolaria larva seen from left side. Only the hydrocoelic portion of the coelom is indicated. The posterior part of this larval stage will become the adult, the anterior part contributes to the formation of a temporary attachment apparatus and then degenerates. B, photo of *Asterias* brachiolaria, from life. At top, note the three fringe-ended arms at which attachment will occur. At bottom, the definitive star is developing. C, young star at metamorphosis. The hydrocoelic lobes form the radial canals, and lateral branches give rise to the first tube feet. (A, after Mead.)

Echinoderm Types

Pelmatozoa

In this subphylum the stalks are hollow aboral extensions of the body wall, covered by calcareous rings (and overlain by epidermis). Rootlike extensions or attachment disks form holdfast devices (Fig. 27.21). At intervals along the stalks are whorls of movable *cirri*, again hollow and protected by skeletal plates. Cirri are grasping organs that aid in maintaining a hold on rocks or seaweeds. In extinct forms the main portion of the body was clothed entirely by endoskeletal plates. In many of the cystoids, for example, the arms projected from a small oral area on the more or less spherical (and often still bilateral) body.

Living crinoids differ from the extinct groups in that the endoskeleton has the form of a cup, the oral side being protected by a leathery body wall without calcareous plates. The arms are usually branched and have small secondary branches, or *pinnules*. Sea lilies are stalked, sessile types, the stalks attaining lengths up to 2 ft (compared with the sometimes 70-ft-long stalks of extinct sea lilies). Of the roughly 700 living species of crinoids, about 100 are sea lilies; the rest are feather stars. The latter are unstalked, but many have long cirri for temporary attachment. Feather stars creep and swim gracefully with their arms (Color Fig. 30).

The crinoid water-vascular system lacks a connection to the outside. Instead, the ring canal connects with numerous stone canals that terminate in the coelom and communicate directly with that cavity. The most developed part of the nervous system is the aboral, endoneural system. The intestine is long and coiled, and pyloric caecae are not present. Gonads lie in the pinnules, which burst open when gametes are ripe. Some crinoids develop from yolky eggs, and larval stages then are abbreviated or omitted. Typically, however, crinoids develop from small, yolk-poor eggs and form a *doliolaria*, a bilateral barrel-shaped larva that has an apical tuft and four or five separate transverse bands of cilia (see Fig. 27.21).

Holothuroids

The horizontal position of the oral-aboral axis in sea cucumbers (Fig. 27.22) introduces a secondary bilateral symmetry, often reinforced by the

598

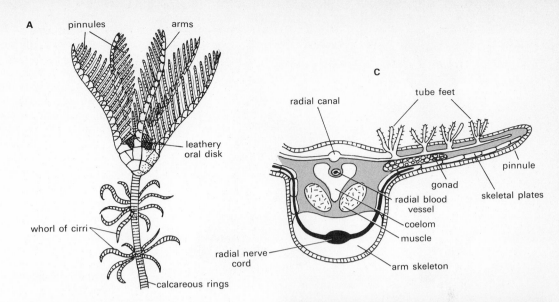

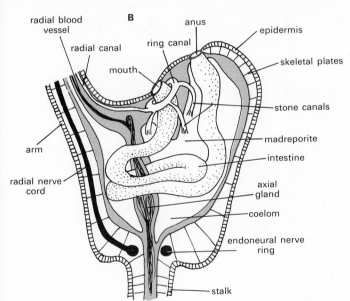

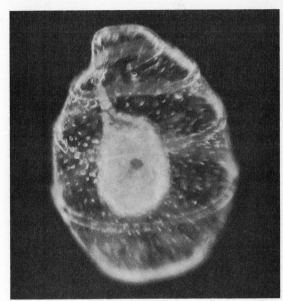

Figure 27.21 Crinoids. A, external features of a living stalked crinoid. B, sagittal section. Only one set of structures leading into an arm is indicated. C, section through an arm and one of its pinnules. One side branch of the water canal in a pinnule leads into a set of three tube feet. D, the doliolaria larva of crinoids, from life. Note ciliary hoops and alimentary tract. See also Color Fig. 30. (A, *after Clark, Carpenter; B, after Reichensperger; C, after Hamann, Chadwick.*)

elaboration of a distinct ventral *sole* on which the animals lie. Tube feet can be in five orderly double rows (as in *Cucumaria*) or along the sole only (as in *Holothuria*) or distributed randomly (as in *Thyone*) or reduced to warty papillae (as in *Synapta*). The endoskeleton is reduced, too, and is represented by microscopic *ossicles* scattered throughout the leathery body wall. The epidermis is glandular and secretes mucus.

Holothurians are sand burrowers and plankton feeders. The feeding organs are the retractile, usually highly branched tentacles around the mouth. They are extensions of the body wall and contain water-vascular canals that branch from the

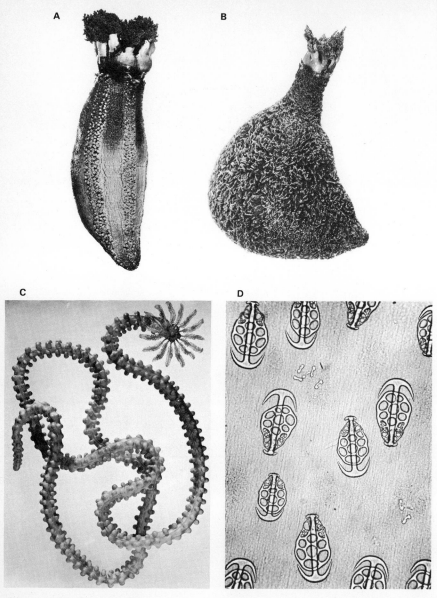

At the base of the tentacles in many sea cucumbers is a dorsal hydropore, the equivalent of a madreporite. However, the stone canal often terminates directly in the coelom, as in crinoids, and a hydropore then is absent in the adult. The main nervous system is the ectoneural system, and an endoneural system actually does not develop in this class. The animals also lack axial glands, but around the pharynx is an oral ring that communicates with two well-developed blood sinuses along the alimentary tract. These sinuses are dorsal and ventral, and the dorsal one is contractile and maintains the blood circulation.

Around the pharynx lies a *calcareous ring* to which the retractor muscles of the tentacles are attached. The alimentary tract is long and looped and connects near the anus with a pair of large, extensively branched outgrowths. These *respiratory trees* serve in breathing and in excretion. Close to the juncture of the trees with the cloaca are bundles of *Cuvierian tubules*. When a sea cucumber is irritated it can expel these tubules through the anus; in some unknown way they apparently distract or scare off potential enemies. Under unfavorable conditions the animals can rupture spontaneously at or near the anus and expel almost all their internal organs. A new set of organs then regenerates from the eviscerated remaining body.

A single gonad discharges through a gonopore near the hydropore. Species with yolky eggs lack larvae, and they brood their young externally on the female or in pockets of the body wall or even directly in the coelom. Most holothurians spawn, however, and external development then includes *auricularia* larvae (Fig. 27.24). These resemble bipinnarias in most respects, and for this reason some zoologists believe sea cucumbers and starfishes to be closely related. In later stages an auricularia transforms to a crinoidlike doliolaria larva, mainly through partial degeneration of the ciliary bands on the surface. Accordingly, holothurians are thought to represent a likely evolutionary link to the crinoids.

Asteroids

This class includes over two thousand living species, most with five arms or multiples of five. In some, however, there are six, seven, or eight arms, as in sun stars such as *Solaster*. Numerous arms, up to 20 or more, identify sun stars such

Figure 27.22   Sea cucumbers. A, Cucumaria, a sea cucumber in which tube feet are arranged in five distinct double rows (two of them visible here). *B, Thyone,* a sea cucumber in which tube feet are not arranged in any orderly pattern. *C, Synapta,* a type in which the body is highly elongate and the podial apparatus is reduced to warty knobs on the body surface. *D,* dermal ossicles in the skin of *Leptosynapta. (A, B, D, courtesy Carolina Biological Supply Company.)*

five radial canals. The latter emanate from the ring canal around the pharynx and continue posteriorly under the rows of actual or reduced tube feet (Fig. 27.23). Between each radial canal and the body surface lies an ectoderm-lined *epineural canal,* the remnant of an ancestral ambulacral groove.

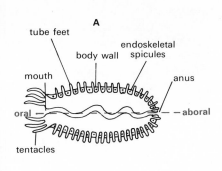

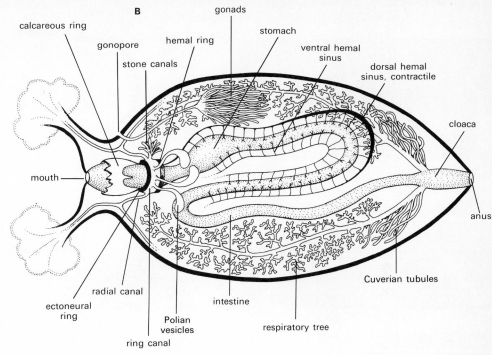

as *Heliaster* (see Fig. 27.18). The open ambulacral grooves and large digestive glands are distinctive traits of the class, not encountered elsewhere in the subphylum Eleutherozoa.

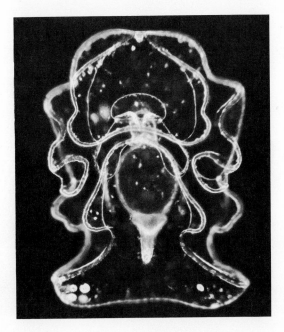

Figure 27.24 The auricularia larva of holothuroids, from life. Such a larva later becomes rather doliolarialike. Compare also with bipinnaria, Fig. 27.19.

Ophiuroids

The two thousand or so species of brittle stars can be regarded as highly specialized variants of starfishes. Ophiuroid arms are marked off sharply from the central disk, and they are long and sinuously mobile (Fig. 27.25). Some species can move the arms only from side to side, but others extend them in all directions and use them as grasping organs. Basket stars such as *Gorgonocephalus* have branched arms. Arm mobility results from a wide spacing of the endoskeletal plates. Since the arms serve adequately in locomotion, tube feet are reduced to warty knobs and podial ampullae are absent. However, ophiuroids do have stone and ring canals, the latter with organs corresponding to Tiedemann's bodies. Radial canals are present, too. Ambulacral grooves are closed over as in holothurians, and the madreporite lies on the oral side of the disk (see Fig. 27.25).

Ophiuroids also differ from asteroids in the absence of pedicellariae and skin gills. Breathing is accomplished instead by five pairs of *bursae,* specialized pouches on the oral side near the arm bases. Further, brittle stars lack intestine or anus, the stomach ending as a blind sac. The mouth is armed with five muscle-operated calcareous teeth.

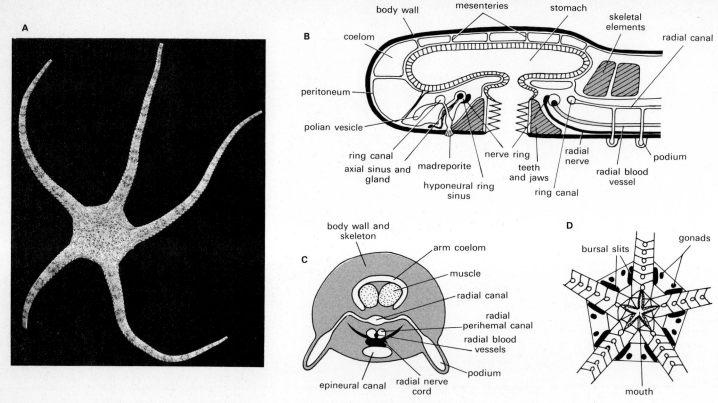

Figure 27.25 Ophiuroids. A, an adult brittle star. See also Color Fig. 30. B, longitudinal section through the disk and the base of one arm. C, cross section through an arm, showing epineural canal. D, plan of the oral disk, to show position of bursal slits and gonads. (*B, adapted from Ludwig; C, based on Hamann, Cuenot; D, based on Reichensperger.*)

Figure 27.26 Ophiuroid development. A, postgastrula, from left side. B, early four-armed ophiopluteus, from left side (note absence of anus). C, eight-armed pluteus in posteroventral view. D, metamorphosing ophiopluteus of *Ophiura*, from life. See also Fig. 11.17.

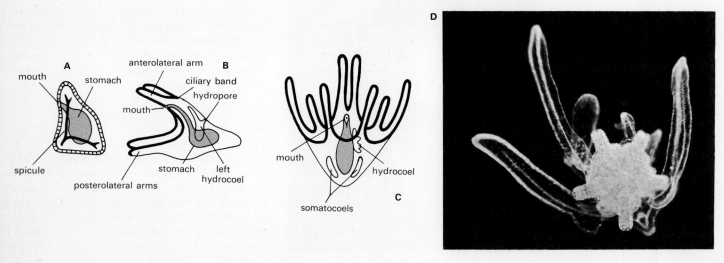

A

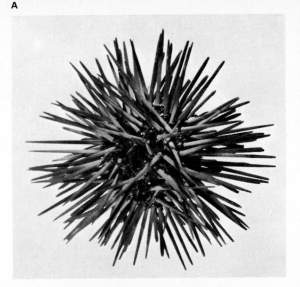

B

C

D

Figure 27.27 Echinoids. A, the sea urchin *Arbacia. B,* fused endoskeletal shell of a sea urchin, oral view, showing the five teeth around the mouth. *C,* sea urchin shell, aboral view. *D,* aboral (left) and oral (right) views of sand dollars. See also Color Fig. 30. *(A, courtesy Carolina Biological Supply Company.)*

Ophiuroids are primarily carnivorous, but they usually eat any kind of food found along the sea bottom. Gonads are attached to the bursae, and in spawning species the gametes escape through the bursal openings. Ovoviviparous and viviparous forms retain their yolky eggs inside the bursae, where they develop directly.

Indirectly developing eggs pass through substantially the same stages as asteroids, but the larvae formed are *plutei.* In such cases the gastrula becomes conical and its left (future oral) side flattens (Fig. 27.26). Anterior and posterior ectodermal outgrowths from this flat side then produce the characteristic arms of a pluteus, each stiffened internally by a calcareous spicule. Such arms increase buoyancy and form in successive pairs; as

many as four pairs are present in older larvae (see Fig. 11.17). During metamorphosis all larval arms and spicules are shed, and the larval anus and intestine degenerate. Adult ophiuroids regenerate well and their arms autotomize readily (hence the name "brittle stars").

Echinoids

Sea urchins are identified by their globular shapes and movable spines (Fig. 27.27). Often up to a foot long, the spines are outgrowths from the endoskeletal plates and are covered by epidermis. In some cases the epidermis at the tip of the spines contains poison glands. Spines have protective and locomotor functions. The rigid skeletal

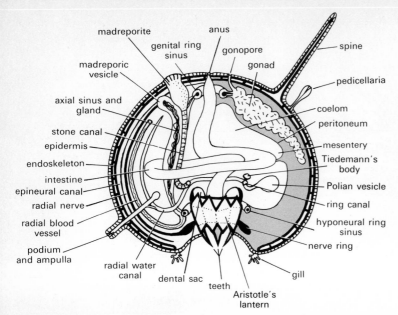

Figure 27.28 Echinoid structure. Section through a sea urchin. For clarity, many parts are here exaggerated or reduced in relative size. Also, only single representatives of organ systems such as tube feet and spines are shown.

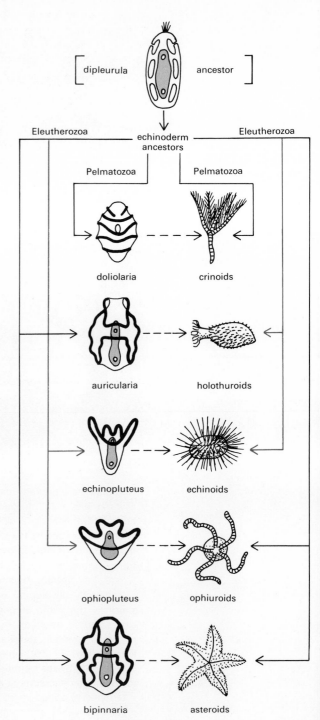

Figure 27.29 Echinoderm phylogeny. Left side, embryology suggests evolution of two eleutherozoan lines, one leading to echinoids and ophiuroids (similar plutei), the other to holothuroids and asteroids (similar bipinnarialike larvae). *Right side,* anatomy and paleontology again suggest evolution of two lines, but one leading to the structurally similar holothuroids and echinoids, the other to the similar ophiuroids and asteroids. In either case, Pelmatozoa probably represent a separate branch of echinoderm evolution. All echinoderms appear to be related to dipleurulalike deuterostome ancestors.

shell consists of 10 pairs of fused meridional rows of plates, and every other pair has pores through which the tube feet protrude. The ambulacral grooves are closed in as epineural canals. Five pairs of skin gills are present along the rim of the small, leathery oral disk (Fig. 27.28). On other parts of the surface breathing occurs through the walls of tube feet. Pedicellariae (which represent modified spines) occur abundantly on the body surface. In many sea urchins the anus lies in the center of the aboral side, but in others this opening has become shifted to an excentric position.

In many respects the internal structure of sea urchins is like that of holothurians, particularly in the region of the pharynx. Thus, corresponding to the calcareous ring of sea cucumbers, sea urchins have an exceptionally well-developed chewing organ, the *lantern of Aristotle*. It consists of 40 separate calcareous ossicles, including five sharp-pointed teeth that converge toward the center of the mouth (see Figs. 27.27 and 27.28). These teeth lie in, and grow continuously from, five *dental sacs* that pouch out from the pharynx. Muscles move the teeth and all other hard parts of the lantern apparatus.

Sea urchins are omnivorous and scavenging, mostly along rocky or stony sea bottoms; they readily can bite through the armor of a dead arthropod. Indeed, sea urchins such as *Paracentrotus* bite through and bore tunnels in rock. Sand dollars (*Clypeaster, Echinarachnius*) are highly flat-tened in an oral-aboral direction, and their movable locomotor spines generally are quite short (see Fig. 27.27). The podial rows occur in an oral and an aboral set, those on the aboral side usually arranged like the outline of flower petals. The tube feet of these animals are not locomotor, breathing being their main function. The anus lies at a point along the edge of the disk. As their name suggests, sand dollars are sand burrowers. Some have broad, flattened lanterns of Aristotle, but in many species such structures are absent.

Echinoids typically have five gonads and five aboral gonopores. Most members of the class spawn and the eggs develop externally. The larvae are *plutei,* hardly distinguishable from those of ophiuroids (see Fig. 11.17). Accordingly, many zoologists have considered echinoids and ophiuroids to be closely related, just as, again on the basis of larval similarities, asteroids and holothuroids have been assumed to be closely related. On the basis of adult structure, however, asteroids clearly resemble ophiuroids most, whereas echinoids have many traits in common with holothuroids. The fossil record likewise suggests such interrelations (Chap. 15). The echinoderm classes thus present a direct conflict between phylogenetic relations based on embryology and those based on adult morphology and paleontology (Fig. 27.29). At present the conflict cannot be resolved, for one view is supported by roughly as many cogent arguments as the other.

Review Questions

1. Give taxonomic definitions of enterocoelomates and of each of the phyla in the group. Discuss the possible evolutionary interrelations of the phyla and the evidence from which such possibilities are deduced. What is the dipleurula hypothesis? Describe the structure of the hypothetical dipleurula.

2. Describe the structure of a chaetognath and review the organization of every organ system. Show how the coelom is subdivided and in what respects these divisions correspond to subdivisions in other deuterostomes.

3. Where do arrowworms live and how and on what do they feed? What are the main sense organs of these animals? How does a chaetognath develop? How does a beard worm feed and where does such an animal live? How does it develop?

4. Describe the structure of a pogonophoran and review the organization of every organ system. What are the external and internal subdivisions of the body? Define and distinguish between the hemichordate classes and name representative genera.

5. Describe the organization of all organ systems in an acorn worm. What structural traits do such worms have in common with chordates? Show how acorn worms develop. What is the possible evolutionary significance of a tornaria?

6. Contrast the structure of an enteropneust with that of a pterobranch. What is the basis for the assumption that pterobranchs and graptolites might be related? On what grounds could pterobranchs be regarded as more primitive than enteropneusts?

7. Define the subphyla and living classes of echinoderms. What are the presumed evolutionary relations of the classes? In what respects is the evidence bearing on such problems conflicting?

8. What evolutionary and adaptive factors have probably led to the development of echinoderms and their water-vascular systems? Describe the symmetries of various echinoderm body organizations.

9. Define podium, ambulacral groove, epineural canal. Describe the structure of a skeletal element and of the whole skeleton of echinoderms. What variant skeletal organizations are encountered in different classes?

10. Describe the external and internal structure of a starfish and review the organization of every organ system. What is the function of the water-vascular system and how are such functions performed? How and on what does a starfish feed?

11. Show how the functional importance of the three neural subdivisions varies for different classes. Distinguish between (a) madreporic ampulla and vesicle, (b) genital and hyponeural ring sinus, (c) radial and axial sinus, (d) axial sinus and gland.

12. Describe the complete development of a starfish, including dipleurulalike, bipinnaria, and brachiolaria stages. Show in detail how the water-vascular system develops.

13. To what hemichordate structures do axocoels, hydrocoels, and somatocoels correspond? What are the developmental fates of these cavities? How does metamorphosis take place? Name the various echinoderm larvae and contrast their structure.

14. Describe the structure of a crinoid and distinguish structurally between a sea lily and a feather star. What are cirri? Pinnules?

15. Can any of the echinoderm groups regenerate? Reproduce vegetatively? How many species are known in each living class? Which echinoderm groups have closed ambulacral grooves?

16. Describe the structure of a sea cucumber and show to what extent this structure corresponds to and differs from that of a starfish. How and where does a holothurian live? What are Cuvierian tubules and what are their functions?

17. Contrast the structure of a starfish and a brittle star. Contrast the development of directly and indirectly developing brittle stars. How do plutei form?

18. In which echinoderm groups is ovoviviparity and viviparity encountered and, in the latter case, where in the adult body does offspring development take place?

19. Describe the structure of a sea urchin and show how it differs from that of other echinoderms. What are the unique features of the skeleton? How and on what does a sea urchin feed?

20. Distinguish sea urchins and sand dollars structurally. How do these animals breathe? Review the paleontological history of echinoderms. If necessary consult Chap. 15.

Collateral Readings

Fell, H. B.: Echinodermata, in "McGraw-Hill Encyclopedia of Science and Technology," rev. ed., vol. 4, McGraw-Hill, New York, 1966. A good, concise discussion.

Hyman, L. H.: "The Invertebrates," vols. 4 and 5, McGraw-Hill, New York, 1955, 1959. These volumes contain advanced accounts on all enterocoelomates except chordates.

Ivanov, A. V.: Pogonophora, Systemat. Zool., vol. 4, 1955.

————: On the Systematic Position of Pogonophora, *Systemat. Zool.,* vol. 5, 1956.

Manton, S. M.: Embryology of Pogonophora and Classification of Animals, *Nature,* vol. 181, 1958.

Moore, R. C.: Echinodermata Fossils, in "McGraw-Hill Encyclopedia of Science and Technology," rev. ed., vol. 4, McGraw-Hill, New York, 1966. A concise review of extinct types.

Morgan, T. H.: The Development of Balanoglossus, *J. Morphol.,* vol. 9, 1894. The classic paper that established the close similarity of tornaria and bipinnaria larvae and thus suggested the evolutionary connection between hemichordates (and chordates) and echinoderms.

Young, J. Z.: "The Life of Vertebrates," 2nd ed., Oxford University Press, Fair Lawn, N.J., 1962. The second and third chapters of this book deal with the hemichordates and their relations to chordates.

28

PROTOCHORDATES

Phylum Chordata

notochord, pharyngeal gill slits, and dorsal hollow nerve cord in preadult stages or throughout life; tailed larvae (*tadpoles*) or direct development.

Acraniates

protochordates; without head. *Urochordata*, tunicates; *Cephalochordata*, amphioxus

Craniates

vertebrates; with head. *Vertebrata*

Because this phylum includes man and the animals most directly important to man, it is unquestionably the most interesting from almost any standpoint. The phylum also has a special evolutionary significance, for it contains by far the most progressive group of animals. Many other phyla have more species and are more diversified, but only chordates include, in one and the same phylum, types as primitively organized as tunicates and as complexly organized as man.

The specific ancestors of chordates are unknown, but a distant and probably indirect affinity to hemichordates, pterobranchs particularly, is suggested by chordate structure. Both hemichordates and chordates exhibit basically similar patterns of embryonic development, and both groups have a pharynx with gill slits used primitively in ciliary plankton feeding. A dorsal hollow nerve cord qualifies as a common trait, too. Two other major

chordate features, notochord and metameric segmentation, have evolved as original innovations within the phylum.

Chordates encompass some fifty thousand species classified as three subphyla: the headless and unsegmented *urochordates*, or *tunicates;* the headless and segmented *cephalochordates*, or *amphioxus;* and the head-possessing and segmented *vertebrates*, or *craniates*. Tunicates undoubtedly are the primitive members of the phylum. Ancestral stocks appear to have given rise to all present tunicates, independently to the amphioxus group, and, again independently, to the vertebrates (Fig. 28.1 and see Fig. 14.17). Much of this evolutionary diversification probably has been achieved through the mechanism of neoteny. Tunicates are marine, amphioxus lives along shores, and vertebrates primitively are freshwater animals. Early chordate evolution thus ap-

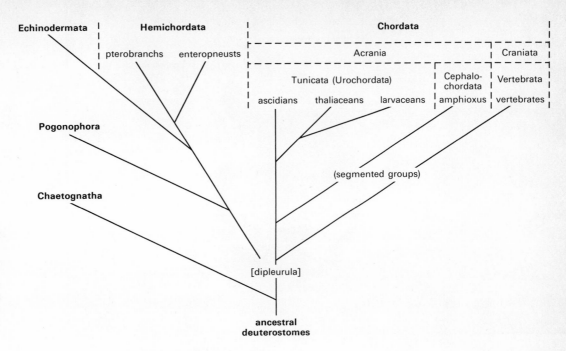

Echinodermata | Hemichordata | Chordata

pterobranchs   enteropneusts | Acrania | Craniata

Tunicata (Urochordata) | Cephalo-chordata | Vertebrata

ascidians   thaliaceans   larvaceans | amphioxus | vertebrates

(segmented groups)

Pogonophora

Chaetognatha

[dipleurula]

ancestral
deuterostomes

Figure 28.1 Evolution of deuterostomial animals. The likely interrelations of the main groups are shown, with emphasis on chordate affinities. Read from bottom up.

pears to have been oriented by shifts in habitat, and such shifts also have played a major role later, during the diversification of vertebrates.

The headless urochordates and cephalochordates often are referred to as *acraniates,* or *protochordates.* These animals are discussed in this chapter, and the *craniates,* or vertebrates, in the next two.

Tunicates
Subphylum Urochordata

marine, sessile or pelagic; with secreted external envelope (*tunic*); unsegmented; coelom not clearly elaborated; pharynx with endostyle and gill slits, used for breathing and ciliary filter feeding; circulatory system open, heart with reversing beat; often colonial through budding; hermaphroditic; development mosaic, indirect or direct. *Ascidiacea, Thaliacea, Larvacea*

Ascidians
Class Ascidiacea   sea squirts; larvae if present free-swimming tadpoles, nonfeeding, with notochord and dorsal nerve cord in tail; adults without tail, sessile, colonial through budding or solitary; gill slits numerous; alimentary tract U-shaped.

Most of the approximately two thousand species in the tunicate subphylum belong to the sessile ascidians. These animals live in sand, mud, or attached to rocks, and many of them form flat, budding colonies.

An individual sea squirt is covered externally by a tunic, or *test,* secreted by the epidermis and composed of *tunicin,* largely cellulose (Fig. 28.2). On the upper, nonattached side the animal has two openings: an incurrent ventral *branchial siphon,* through which water and food enter toward the mouth, and an excurrent dorsal *atrial siphon,*

through which water, elimination products, and reproductive cells leave from a large *branchial cavity,* or *atrium.* Each siphon has ocelli along its rim and can be closed by sphincter muscles.

The mouth leads to a large pharynx, or *branchial basket,* a chamber perforated by dorsoventral rows of numerous gill slits. Blood vessels traverse the pharyngeal wall between the slits. Each slit is elongated and just wide enough to accommodate the cilia around its rim. Cilia also line the inner surface of the pharynx. Along the ventral gutter of the chamber is a band of specialized tissue called

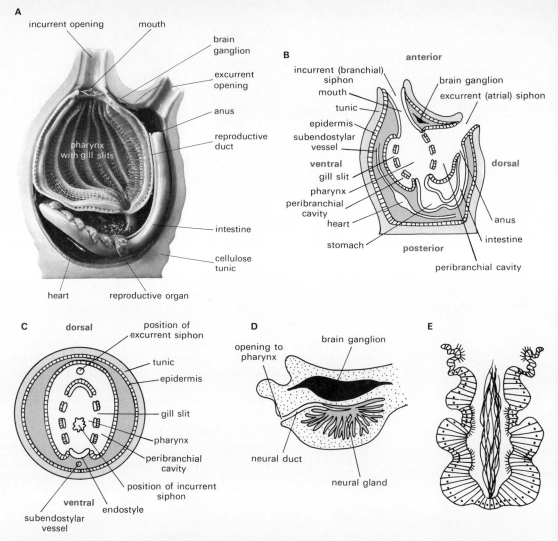

Figure 28.2 Ascidian structure. A, cutaway model of adult. Food-bearing water is drawn into the pharynx through the incurrent siphon. Food passes to the U-shaped alimentary tract and water flows to the outside through the gill slits, atrium, and excurrent siphon. *B,* anatomic orientations and the ectodermal nature of the lining of the branchial sac. The ectoderm layer is drawn with crosslines suggesting cells, the endoderm layer is without such lines. Internal organs such as heart and gonads lie in the tinted regions between the epidermis and the branchial lining. *C,* dorso-ventral (cross) section at a level just below the excurrent siphon, showing how the branchial sac envelops the pharynx; layers and regions drawn as in *B. D,* section through ganglion region, showing neural gland and duct to pharynx. *E,* cross section through the endostyle, showing the glandular and ciliated cells on each side and the flagellated cells in the ventral gutter. *(D, after Roule; E, after Sokolska.)*

the *endostyle.* In it a median strip of cells bears long flagella, and adjacent strips are ciliated and glandular. An endostyle is already foreshadowed in the ventral pharyngeal gutter of hemichordates, and this organ also is the evolutionary source of the vertebrate thyroid gland.

The endostyle secretes mucus continuously, and the flagella and pharyngeal cilia distribute this mucus over the inner surface of the branchial basket. The cilia, particularly those around the gill slits, also draw in a continuous food-bearing water current through the branchial siphon and the

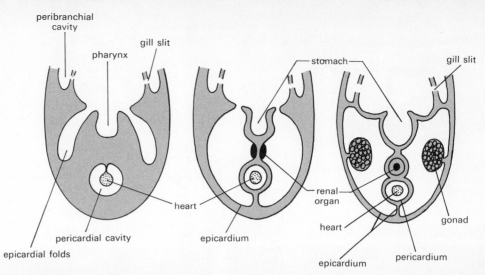

Figure 28.3 Ascidian structure. Left to right, stages in the development of the ventral body region near the posterior end of the pharynx, showing the formation of heart, epicardium, and renal organ. The heart is a tube in the pericardial coelom. The epicardia are a pair of folds evaginated from the posterior end of the pharynx, which eventually envelop all ventral and posterior organs. The renal organ, formed from the epicardia, accumulates a solid excretory concretion in its interior.

mouth. Food particles become entangled in the mucus and water passes through the gill slits to the atrium, oxygenating blood in the process. From the atrium water is expelled through the atrial siphon. Concurrently, food in mucus collects along the dorsal wall of the branchial basket and is

Figure 28.4 Ascidian budding. The sequence outlines bud growth in *Botryllus*, in which buds arise from the atrial body wall. In other cases the inner cells do not always form a neat layer from the start. In all cases, however, the inner cells (shown here with nuclei) eventually do form a vesicle from which all endodermal and mesodermal body parts arise by invagination and evagination. (*Slightly modified from Berrill.*)

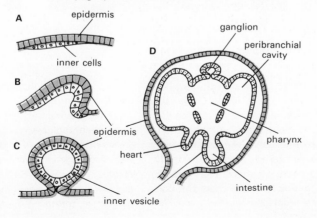

propelled by ciliary action to the esophagus behind the pharynx. The esophagus leads to a stomach at the bottom of the alimentary U, and an intestine terminates at the anus, which opens into the atrial cavity.

The nervous system contains a single neural ganglion between the two siphons. A *neural gland* adjacent to this ganglion leads through a duct to the pharynx, just behind the mouth. This gland appears to function as a smell or taste receptor, and on occasion it has been homologized with the vertebrate pituitary. The circulatory system includes a ventral heart and blood vessels that pass anteriorly to the pharynx and posteriorly to the stomach region. These vessels are open-ended; capillaries are absent and blood flows mostly through hemocoelic spaces. The pericardial space around the heart is the only cavity qualifying as a coelom. The several peculiarities of ascidian blood and the heart have been discussed in Chap. 9. Thus, these presumed ancestors of vertebrates are a most remarkable group. They are virtually unique among animals in having cellulose and they are unique in having reversing hearts, free sulfuric acid in blood cells, and high concentrations of vanadium, an element so dilute in sea water that its presence there can barely be demonstrated.

Many sea squirts have an *epicardium* close to or around the heart. This structure arises as a pair

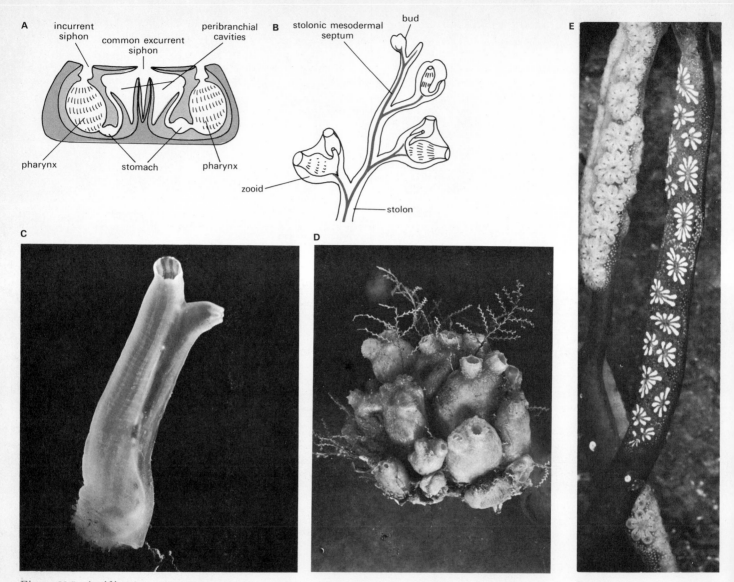

A, incurrent siphon, common excurrent siphon, peribranchial cavities, pharynx, stomach, pharynx

B, stolonic mesodermal septum, bud, zooid, stolon

C

D

E

Figure 28.5 *Ascidian types.* *A,* section through a rosette colony of *Botryllus,* showing common atrial siphon. *B,* stolonic colony of *Perophora,* growing on sea weeds. *C, Ciona,* a solitary type; the incurrent siphon is at top, fully open. *D,* colony of *Ascidia. E,* rosette colonies of *Botryllus,* attached to twigs, rocks, or other objects in water. (*A, based on Delage and Herouard; B, after Listen.*)

of outfoldings from the posterior part of the pharynx (Fig. 28.3). In many cases (as in *Ciona*) the folds enlarge and envelop all the organs in the posterior part of the body, like peritoneum and mesentery. In other cases (as in *Molgula*) the folds fuse and condense as a *renal organ* alongside the heart. This organ stores solid excretory wastes (largely uric acid) that accumulate there throughout the life of the animal. Inasmuch as they are derived from the pharynx and thus endodermal, the epicardial folds might be homologous to true coelomic membranes; but they are formed later than when coeloms normally develop.

In budding, a small epidermal sphere containing pieces of some interior tissue pinches off the parent animal (Fig. 28.4). Such buds can form almost anywhere on the epidermis, and the bud interior can derive from almost any relatively unspecialized

612

adult tissue. In some instances, as in *Botryllus*, buds arise in the body wall of the atrial region. In other cases buds form more posteriorly, and the epicardium then usually contributes the interior bud cells. The surface layer of a bud produces only the epidermal components of the new adult. The interior cells give rise to mesodermal and endodermal components, regardless of the original germ-layer derivation of these cells. If many buds originate simultaneously in a compact group, they can form a colony surrounded by a common tunic and even a common atrial siphon, as in the rosette colonies of *Botryllus* (Fig. 28.5). A single bud also can elongate as a stolon stalk, and from it zooids then can develop at intervals, as in *Perophora*. The capacity to bud is believed to be primitive among ascidians; solitary forms such as *Ciona* are considered to have lost the budding potential.

Male and female gonads lie in the loop of the alimentary tract or in the atrial body wall, and their ducts open to the atrium. Some species are viviparous, and in these the female gonoduct is enlarged as a brood pouch, or uterus. The eggs of such animals are large and yolky. Most ascidians spawn, however, and their smaller, less yolky eggs develop externally. Cleavage is bilateral and mosaic, a unique, exceptional condition among enterocoelomates. After embolic gastrulation, the region around the blastopore proliferates backward and establishes the rudiments of the tail (Fig. 28.6). A group of cells along the roof of the archenteron participates in this posterior proliferation; the cells come to form a single longitudinal row in the tail and represent the future *notochord*. Concurrently a solid band of mesoderm cells buds off from each side of the archenteron. These bands, too, extend back into the tail and later develop as the tail musculature. Coelomic sacs do not form.

A nerve cord arises as in vertebrates by infolding of a neural tube from the dorsal ectoderm. The anterior part of the tube enlarges as a brain vesicle, and in it a neuropore later comes to open to the developing pharynx. This pore, homologous to the neuropore of hemichordates, represents the forerunner of the opening of the adult neural gland. Inside the brain vesicle an ocellus develops as an inward projection from the roof of the brain wall, and a statocyst forms on the floor. Just behind the level of the brain the dorsolateral ectoderm on each side of the larva invaginates and produces a pair of atrial pouches; the two openings to the outside later fuse and form the single atrial siphon of the adult. An endostyle arises in the pharynx, and the first three pairs of gill slits break through and establish communication with the atrial chambers. Additional gill slits form later by subdivision of the first ones.

Anteriorly the tadpole has *adhesive papillae* by which it will eventually become attached. Everywhere else the surface of the larva is enveloped in the rudiments of the tunic, which covers the mouth and thus makes this opening nonfunctional. The tadpole swims about for only a brief period—from minutes to a few hours at most; as noted in Chap. 11, the ascidian tadpole serves primarily in site selection. Metamorphosis involves degeneration of the tail, including notochord, nerve cord, and the tail muscles. Also, a rapid proliferation of the body wall between the attachment area and the mouth shifts the mouth toward the free end of the animal and produces the U shape of the adult alimentary tract. Parts of the brain vesicle persist as the neural ganglion and neural gland. The tunic is resorbed in the region over the mouth and a feeding adult becomes established in this manner (see Fig. 28.7).

It is generally believed today that the sessile ascidian adults probably represent descendants of

Figure 28.6 Early ascidian development. A, sagittal section through gastrula. *B,* cross section through gastrula. *C,* side view of tail bud stage, notochord cells shaded and with black nuclei, lateral mesoderm bands superimposed. *D,* cross section through tail bud stage as in *C. (After Conklin, Van Beneden.)*

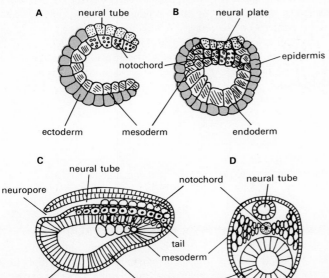

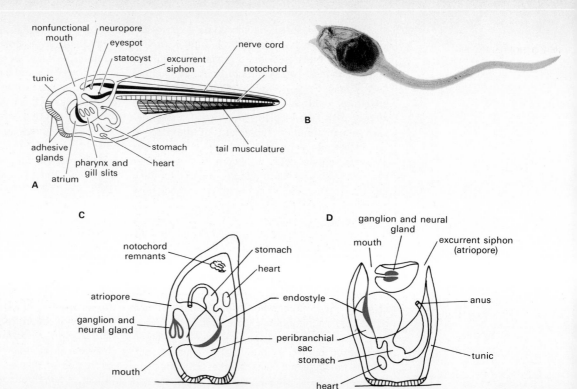

Figure 28.7 *The ascidian tadpole* and metamorphosis. *A*, section through a tadpole. *B*, tadpole of the colonial ascidian *Aplidium* (= *Amaroucium*). *C*, early metamorphosis; tail, notochord, and most of nerve cord resorbed, neuropore becoming opening from neural gland duct into pharynx. *D*, late metamorphosis; rotation of internal organs completed and siphons functional.

hemochordatelike, filter-feeding, chordate ancestors. The main dispersal mechanism of such ancestral chordates would have been budding, and the tadpole larva is envisaged as an original evolutionary invention of this early chordate stock. Being more efficient dispersal agents than buds, tadpoles would have been adaptively highly advantageous to the sessile adults. Moreover, a tailed, well-muscled tadpole could become adapted readily to a prolonged free-swimming existence. Indeed, if the larva developed sex organs precociously, a new, neotenous type of adult actually could be evolved. It is believed that from such a free-swimming neotenous starting point have arisen all the other tunicates, Thaliacea and Larvacea, and all other chordates as well.

Thaliaceans   **Class Thaliacea**   chain tunicates; larvae if present free-swimming, with notochord and dorsal nerve cord in tail; adults without tail, pelagic and free-swimming, locomotion by ''jet'' propulsion; body wall with transverse muscle hoops; gill slits few to numerous; polymorphic, with colonial stages budded in chains from stolons.

The members of this and the following class include some of the most fascinating of all animals. Individual zooids are minute, barrel-shaped or cylindrical, with conspicuous bands of circular body-wall muscles arranged in the form of transverse hoops (Fig. 28.8). The body is organized around a straight alimentary system, with the siphons at opposite ends of the animal. The hoop muscles at these ends serve as sphincters; when one contracts the other relaxes. The result is a

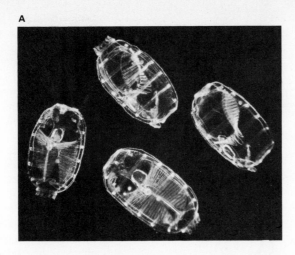

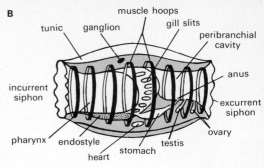

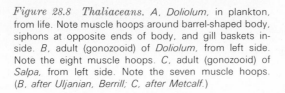

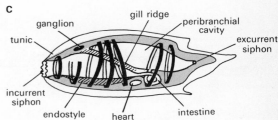

Figure 28.8 Thaliaceans. A, Doliolum, in plankton, from life. Note muscle hoops around barrel-shaped body, siphons at opposite ends of body, and gill baskets inside. *B,* adult (gonozooid) of *Doliolum,* from left side. Note the eight muscle hoops. *C,* adult (gonozooid) of *Salpa,* from left side. Note the seven muscle hoops. *(B, after Uljanian, Berrill; C, after Metcalf.)*

fairly rhythmic expulsion of water from the posterior atrial siphon and a jetlike propulsion of the animal. Pharyngeal cilia maintain the water flow, as in ascidians. Other internal structures too are substantially as in sea squirts.

Pyrosoma, brightly bioluminescent (hence its name), forms gelatinous tubular colonies (Fig. 28.9). Such a tube is closed at one end and the zooids lie in the wall of the tube. Their atrial siphons face inward; hence all excurrent water streams are expelled into the tube and the common jet emerging from the open end propels the colony in the opposite direction.

Fertilization takes place internally. The highly yolky eggs are retained in the parent colony, cleave superficially, and develop directly into *oözooids,* or individuals produced from eggs. From the endostyle of an oözooid then grows a stolon that projects out of the body. On this stolon develops a connected chain of four *blastozooids,* individuals formed by budding. These four become attached around the oözooid in a circlet, and such a young colony then escapes from the parent colony. An offspring colony develops by continued budding from the four blastozooids.

The doliolids, another thaliacean group, undergo one of the most remarkable life cycles (Fig. 28.10). The mature adult is a *gonozooid* that produces externally developing eggs. An egg becomes a torpedo-shaped tailed tadpole enveloped in a thick tunic. The tail later is resorbed and the adult so formed is an *oözooid,* or "nurse," distinguished by a *spur,* a median dorsal projection at the posterior end. Inside the tunic of this oözooid a ventral stolon arises behind the endostyle, and the end of the stolon constricts off a succession of *blastozooid* buds. These migrate to the dorsal spur where they become attached in a median and two lateral rows. The spur then elongates, often up to a $\frac{1}{2}$ yd. The blastozooids that have settled in the lateral rows are stalked and without gonads; they function as a colony of *gastrozooids* that, curiously enough, feed the nurse. Blastozooids in the median row become *phorozooids,* also without gonads but with a *ventral spur.* Some of the wandering blastozooids that continuously bud from the nurse come to settle on these ventral spurs. The phorozooids later break free from the nurse, and each bud carried on their ventral spurs gives rise to a young gonozooid. Such individuals eventually

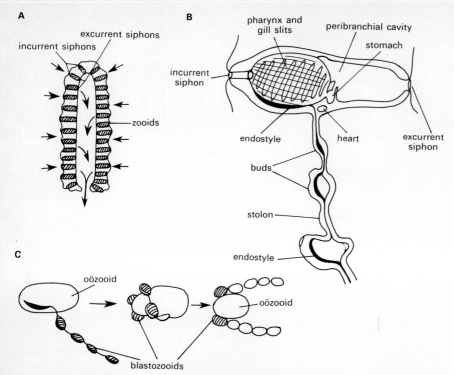

A

incurrent siphons
excurrent siphons
zooids

B

pharynx and gill slits
peribranchial cavity
stomach
incurrent siphon
endostyle
heart
excurrent siphon
buds
stolon
endostyle

C

oözooid
oözooid
blastozooids

Figure 28.9 Pyrosoma. A, section through colony. Note orientation of zooids in tubular colony and flow of water (arrows). *B,* individual zooid, with stolon growing from endostyle and chain of buds formed along stolon. Note that endostylar stolon of one individual forms next individual along chain, including endostylar stolon of that individual. *C,* four blastozooid buds as in *B* become arranged as a circlet around the oözooid and by continued budding give rise to a new colony. *(A, after Metcalf, Berrill; B, after Metcalf, Neumann.)*

detach and develop as new adults, completing the cycle. Evidently, the life history represents a multi-stage succession of budded polymorphic generations, one kind of bud (phorozooid) functioning as transport vehicle for others (wandering buds).

Similarly polymorphic are the *salps,* a third group of thaliaceans. Solitary individuals of these animals are oözooids produced from eggs (Fig. 28.11). They lack gonads but produce a chain-budding stolon, like the oözooids of doliolids. In *Salpa* the buds in a stolonic chain form a linear series, and in *Cyclosalpa* the older, more posterior buds become organized as wheel-like complexes. All such stolonic individuals are blastozooids. They do not develop stolons of their own but acquire gonads. Their eggs develop viviparously in pouches of the oviduct, and the young eventually set free represent new oözooids. Salps thus undergo an alternation of generations: oözooids propagate vegetatively by forming blastozooids, and blastozooids propagate by sexual means and form new oözooids.

Pyrosomas appear to be most closely related to ascidians. Doliolids could have evolved from ancestral pyrosoma stock, and salps from ancestral doliolid stock. Doliolid ancestors also appear to have given rise to the Larvacea.

Figure 28.10 The life cycle of doliolids. An adult gonozooid is illustrated in Fig. 28.8. The oözooid is shown in side view; note the nine muscle hoops. On the dorsal spur, note the paired lateral rows of gastrozooids and the paired median row of phorozooids, each of the latter with a wandering blastozooid migrated to that position from the ventral stolon. In the free phorozooid, only a side view of the posterior end is shown. *(Adapted from Neumann, Berrill.)*

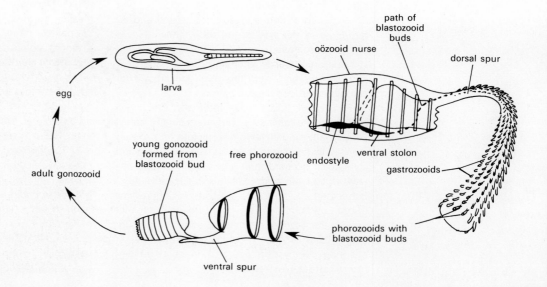

egg
larva
path of blastozooid buds
oözooid nurse
dorsal spur
young gonozooid formed from blastozooid bud
free phorozooid
endostyle
ventral stolon
gastrozooids
adult gonozooid
ventral spur
phorozooids with blastozooid buds

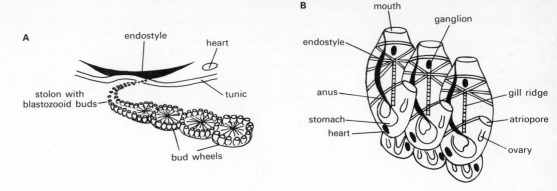

Figure 28.11 *The life cycle of salps. A*, ventral region of parent oözooid, showing endostylar stolon carrying chains and wheel arrays of sexual blastozooids. Terminal bud wheels of this type form in *Cyclosalpa,* straight chains in *Salpa. B*, detail of stolonic chain showing a few paired blastozooids. Eggs from such blastozooids give rise to new adult gonozooids (as in Fig. 28.8). (*A, after Ritter and Johnson, Berrill; B, after Claus.*)

Larvaceans **Class Larvacea** appendicularians; tail with notochord and nerve cord, permanent throughout life; adults larvalike, pelagic; tunic forms complex housing, used with tail in feeding; one pair of gill slits.

These animals are the most specialized tunicates; they are neotenous, permanently tailed forms that use their tail and highly elaborated tunic as a remarkable feeding apparatus.

The body of an appendicularian is minute and the tail is three to seven times longer than the trunk (Fig. 28.12). Also, the tail is twisted 90° on its axis and bends forward horizontally under the trunk. The whole body, tail included, is en-

closed in a roughly tubular tunic that forms a spacious "house," well separated from the surface of the body. Dorsolaterally the house has a pair of windows that are covered with a straining grid of fine fibers. Food-bearing water entering through the windows is directed through a filtering system in the forward part of the house, where food particles separated from water are sucked into the mouth of the animal. Filtered water then

Figure 28.12 *Larvacea. A*, side view of *Oikopleura.* Note complex house, comparatively small body, and long tail of interior animal. *B*, some of the organs of *Oikopleura. C, Oikopleura,* in plankton, from life. (*A, after Lohmann, Berrill.*)

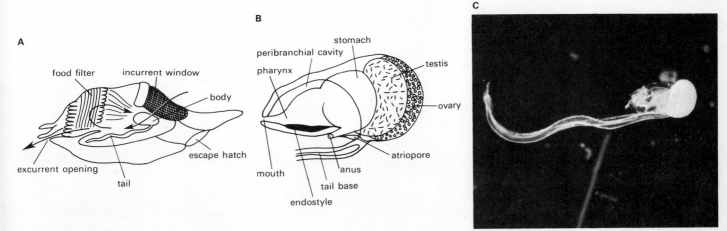

is expelled through a forward door in the house, the outgoing jet propelling the animal in the opposite direction. The entire flow of water is maintained and controlled by undulating movements of the tail along the floor of the house. A posterior ventral trap door serves as an escape hatch; after occupying its house for a period the animal leaves through the hatch and secretes a new house.

Larvacea are hermaphrodites. Eggs escape through temporary ruptures in the body wall and develop directly.

Amphioxus

Subphylum Cephalochordata

lancelets; marine, in sand; notochord and dorsal nerve cord throughout life; head or brain absent; coelom well developed, metamerically segmented; filter-feeding, with atrium, numerous pharyngeal gill slits, and endostyle; circulation open, without heart; excretion through solenocytic protonephridia; sexes separate, fertilization and development external; eggs regulative; with asymmetric larvae.

The whole subphylum consists of about thirty species included in two genera (*Branchiostoma, Asymmetron*). Amphioxus is slender, laterally compressed, 2 to 3 in. long, and pointed at both ends (hence its name; Fig. 28.13). It lives in shallow coastal waters, its anterior end sticking out of a sand burrow. From time to time it swims to a new location by side-to-side undulations of the body.

Externally, a median dorsal fin extends over most of the animal, a median ventral fin occurs in roughly the posterior third, and a tail fin connects the dorsal and ventral fins around the hind end. The dorsal and ventral fins are strengthened internally by boxlike *fin rays* composed of connective tissue. At the junction between the ventral fin and the tail fin lies the anus. The ventral fin terminates anteriorly at the *atriopore,* the exit from the atrium.

The alimentary system begins at an *oral hood,* an epidermal fold fringed with fingerlike *cirri.* These form a coarse screen when the animal feeds. On the inner surface of the oral hood are projections with complex bands of cilia. These structures form a *wheel organ* that produces a spiraling ingoing water current. The longest projection on the dorsal side contains a groove and pit (*Hatschek's groove* and *pit*) believed to be homologous to the protocoelic duct and hydropore of hemichordates, the duct of the ascidian neural gland, and the pituitary of vertebrates. The cavity of the oral hood extends back to the *velum,* a transverse membrane that has the mouth in the center. This opening is fringed by *velar tentacles,* corresponding to the oral tentacles of ascidians (and the mesosomal tentacles of hemichordates).

Behind the mouth, the long pharynx contains 60 to more than 100 pairs of lateral sloping gill slits. The *gill bar* between two adjacent slits is stiffened internally by a skeletal rod. As in ascidians, cilia cover the rim of the gill slits and the inner pharyngeal surface. An endostyle again forms the ventral pharyngeal gutter and a similar ciliated strip lies along the roof of the pharynx. The whole branchial basket operates as in ascidians. Gill slits open to the atrium, a ventral cavity in these animals. Muscles in the floor of the atrium can contract and force water out through the atriopore. The ventrolateral edge of the atrium hangs down on each side as a finlike *metapleural fold.* Behind the pharynx, a straight intestine connects with a digestive gland, a sac that extends forward along the right side of the pharynx. The gland is often called "liver," but it actually appears to function like the vertebrate pancreas.

The notochord extends from one tip of the body to the other, and the hollow nerve cord lies directly over it. Anteriorly the cord contains a brain vesicle only slightly larger than the neural canal. A pigment spot and an olfactory pit are the only sensory structures present. The absence of a distinct brain probably is not primitive but a secondarily simplified trait related with the sedentary habits of the animal. On each side the nerve cord gives off paired segmental nerves in an arrangement that corresponds to the vertebrate pattern. Most of the nerves innervate the lateral segmental musculature. Each muscle segment (*myotome*) is V-shaped, and those on one side of the body alternate with those on the other in a staggered pattern (see Fig. 28.14).

The circulation is open and capillaries are absent. Arteries and veins are arranged substantially

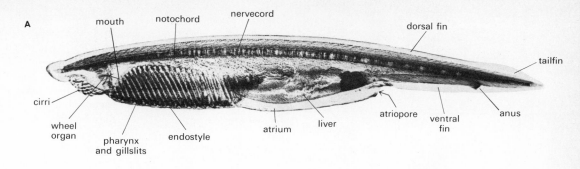

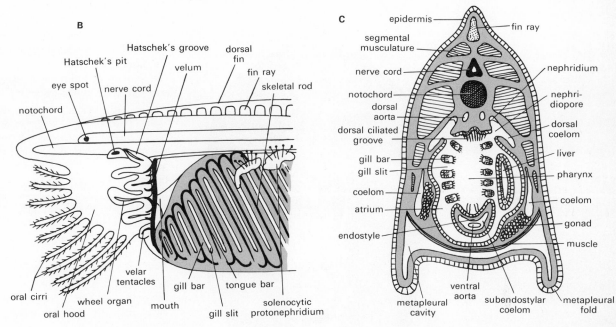

Figure 28.13 Amphioxus. A, general anatomy. *B,* side view of anterior end. *C,* cross section through the pharyngeal region.

as in vertebrates (Fig. 28.14). A contractile ventral aorta below the pharynx pumps blood forward and then up through *aortic arches* between the gill slits. At the ventral end of each aortic arch, a small contractile bulb (*bulbillus*) aids in driving blood through the arch. Blood collects above the pharynx in a pair of dorsal aortae. These two vessels join behind the pharynx as a single systemic aorta that gives off branches to the tissues. From an intestinal network of spaces blood drains as in vertebrates through a hepatic portal vein to the liver. A hepatic vein from there and other veins from blood spaces in various body parts conduct blood to a *sinus venosus.* This chamber continues forward as the ventral aorta. The blood is colorless

and without cells. Oxygenation probably occurs primarily in the atrial walls and only secondarily in the aortic arches.

In the pharyngeal region the coelom is small and consists only of a pair of flattened sacs dorsal to the pharynx. Posteriorly the coelom is represented by the space around the alimentary tract (see Fig. 28.14). The excretory system consists of ectodermal solenocytic protonephridia, encountered elsewhere only in trochophore larvae and certain polychaete annelids. The very puzzling occurrence of such nephridia in amphioxus is regarded by most zoologists as a very remarkable instance of parallel evolution. One such organ lies near the dorsal end of each gill slit. The nephridia

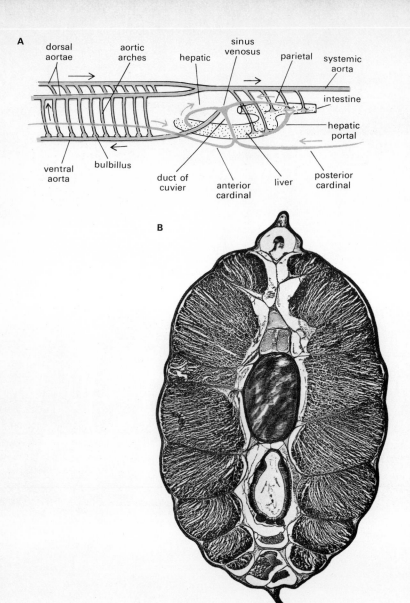

A dorsal aortae — aortic arches — hepatic — sinus venosus — parietal — systemic aorta — intestine — hepatic portal — posterior cardinal — liver — anterior cardinal — duct of cuvier — bulbillus — ventral aorta

B

Figure 28.14 Amphioxus. A, the basic plan of the circulatory system. Veins in gray, arteries in color. Direction of venous blood flow indicated by colored arrows, that of arterial blood flow by black arrows. A bulbillus is a contractile vesicle. The sinus venosus is formed by the confluence of the main veins of the body and the exit of the ventral aorta; it corresponds in position to (and forms part of) the heart of vertebrates. B, cross section through the intestinal region. Note prominent notochord roughly at center, nerve cord above, intestine below. The free space around the intestine is the coelom. Note the left-right staggered placement of the segmental musculature. (*B, courtesy Carolina Biological Supply Company.*)

abut blood vessels in the coelom lining, and the nephridial collecting tubules open to the atrium.

Gonads are paired in *Branchiostoma*, unpaired and on the right side only in *Asymmetron*. They form an anteroposterior succession of saclike organs in the lateral wall of the atrium. Gametes escape through ruptures into the atrium and out through the atriopore. Fertilization takes place externally. Eggs cleave bilaterally, a coeloblastula forms, and gastrulation is embolic. The notochord develops from the roof of the archenteron, the nerve cord by invagination of a dorsal neural tube; an anterior neuropore is present temporarily.

The anterior end of the archenteron then buds off three pairs of coelomic sacs in succession (Fig. 28.15). In the first pair, the right sac becomes the coelom of the most anterior region. The left sac opens to the outside through a hydroporic connection that persists in the adult only as Hatschek's pit and groove. The second pair of sacs becomes the first pair of segmental coeloms; and the third pair later subdivides anteroposteriorly and gives rise to all other segmental coeloms. Each coelomic sac soon becomes subdivided horizontally as a dorsal *myocoel* and a ventral *splanchnocoel*. The myocoels eventually fill up with cells that form the segmental musculature, but the splanchnocoels remain hollow; by fusion they give rise to the continuous coelomic cavity of the adult.

The embryo hatches after the first few muscle segments have formed. In an early larva the mouth lies on the left side. Gill slits begin to form in pairs, all of them on the right side of the body. Later the larva does become symmetric and a gradual metamorphosis follows.

Although amphioxus evidently is quite specialized in many respects, the animal does display numerous vertebrate features in very primitive form and thus suggests what ancient vertebrates might have been like. Most probably, amphioxus exemplifies a transitional stage in the evolution of ancestral ascidians to the first vertebrates. In the ancestors of amphioxus segmentation might have developed originally in the tail musculature and its nerve supply, an advantageous step that would have increased the locomotor power and fine control of the tail. Such added efficiency would have been particularly important if the early chordate shifted its habitat from the open sea to shallow shore waters, especially around river mouths, where influx of minerals promotes rich plankton growth. Waves, surf, undertow, and the currents

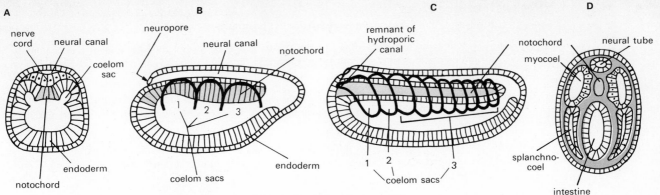

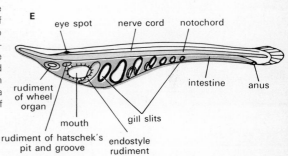

Figure 28.15 Amphioxus development. A and B, transverse and sagittal section through embryo at time of formation of primary coelom sacs. C, later stage, the third primary sac subdivided. The forward outgrowth of the first coelom sac on the left side of the embryo probably corresponds to the hydroporic canal of ancestral deuterostomes. D, cross section through late embryo, at time of subdivision of coelom into dorsal and ventral sacs. In the myocoel, the inner cells (shown with nuclei) give rise to the segmental muscles. E, larva seen from left side, showing asymmetric position of mouth and early gill slits on left. (*After Hatschek.*)

of discharging river water would have made efficient locomotor control exceedingly important. Thus, segmentation could have evolved in adaptation to such conditions.

Through segmented and finely controlled tails, furthermore, the basic means would have become available for swimming upriver. However, actual entry to rivers would have required at least two additional changes, and these did not become perfected in the amphioxus group. Instead, the animals merely adopted a sedentary existence in shallow coastal regions. By contrast, a related line did succeed in evolving the necessary evolutionary adjustments, and these animals became the first vertebrates.

Review Questions

1. Give taxonomic definitions of the phylum Chordata, the acraniate subphyla, and the tunicate classes. Review the presumed evolutionary relations of these various groups. What role has neoteny probably played in chordate evolution?

2. Describe the structure of an adult ascidian and review the organization of every organ system. Describe the feeding process of a sea squirt. What are endostyle, epicardial folds, neural glands? Review the unique attributes of ascidian hearts and blood.

3. How do buds form and develop in ascidians and what types of colonies can arise? Review the embryonic development of ascidians. In what respects are ascidian and annelid eggs and cleavage patterns similar?

4. Show how the ascidian tadpole develops from the embryo, how mesoderm arises, and how adult ascidian structure emerges during larval development and metamorphosis. What is the ecological and probably evolutionary significance of the ascidian tadpole?

5. Describe the structure of *Pyrosoma*. Show how locomotion occurs. By what processes does a colony develop? Distinguish between oözooids and blastozooids.

6. Describe the structure of a doliolid gonozooid and review the life cycle of these animals. What are the differences between oözooids, blastozooids, gastrozooids, and phorozooids? Which thaliacean groups are hermaphroditic?

7. Review the organization of an oözooid salp and outline the life cycle. Contrast this cycle with those of other thaliaceans. Describe the structure of a larvacean and review the feeding and locomotor processes of the animal.

8. Outline the complete structure of amphioxus, including the organization of every organ system. What features appear to be homologous to those of tunicates and hemichordates? How and where does amphioxus feed?

9. Is the persistence of the amphioxus notochord and nerve cord a neotenous trait? Show how cephalochordates develop and how the coelom forms. Are cephalochordates hermaphroditic? Oviparous?

10. What ecological factors might have played a role in the evolution of cephalochordate segmentation? What factors probably contributed to keeping cephalochordates in a marine environment?

Collateral Readings

Berrill. N. J.: Metamorphosis in Ascidians, *J. Morphol.*, vol. 81, 1947.

———: "The Tunicata," Ray Society, London, 1950.

———: Budding in *Pyrosoma, J. Morphol.*, vol. 87, 1950.

———: Budding and Development in *Salpa, J. Morphol.*, vol. 87, 1950.

———: Regeneration and Budding in Tunicates, *Biol. Rev.*, vol. 26, 1951. A series of papers by one of the foremost students of protochordates.

———: "The Origin of the Vertebrates." Oxford University Press, Fair Lawn, N.J., 1955. A masterful documentation of the probable evolution of vertebrates from ascidian ancestors.

———: "Growth, Development, and Pattern," Freeman, San Francisco, 1961. Chapters 13 and 14 contain analyses of growth processes in tunicates, including those taking place during budding.

Conklin, E. G.: The Embryology of Amphioxus, *J. Morphol.*, vol. 54, 1932. An original key paper.

Watkins, M. J.: Regeneration of Buds in *Botryllus, Biol. Bulletin*, vol. 115, 1958. A paper on ascidian budding and regeneration.

Young, J. Z.: "The Life of Vertebrates," 2d ed., Oxford University Press, London, 1962. Chapters 2 and 3 deal with the protochordate groups.

29

VERTEBRATES: FISHES

Subphylum Vertebrata

vertebrate chordates; segmented, with head, trunk, and tail; cranium (skull) enclosing brain; typically with dermal bone, embryo with notochord; adult with notochord and/or vertebral column of cartilage or replacement bone; typically with two pairs of trunk appendages; with pharyngeal gills or lungs; coelom well developed; circulation closed, heart with two, three, or four chambers; excretion pronephric, mesonephric, or metanephric; endocrine system elaborate; sexes usually separate, fertilization various; eggs regulative, development various. *Agnatha,* jawless fishes; *Placodermi,* extinct armored fishes; *Amphibia,* amphibians; *Reptilia,* reptiles; *Aves,* birds; *Mammalia,* mammals

General Characteristics

The overriding orienting factor in vertebrate evolution was the fresh water. Ancestral vertebrates probably invaded freshwater rivers as segmented, tailed, neotenous derivatives of marine ascidian stocks. The necessary evolutionary innovations were strongly muscled larvae that could surmount the force of river currents as soon as they hatched; and excretory systems that could cope with the continuous osmotic inflow of water into the body, unavoidable in a freshwater environment. The first requirement was met through large, very yolky eggs, which could develop without additional food to an advanced, fully segmented stage even before hatching; and the second was met by pronephric and then mesonephric kidneys, specialized particularly for elimination of osmotic water.

An early fundamental consequence of such an excretory pattern was that, although the kidneys could eliminate water well, they were not specially adapted to eliminate salt. Retention of minerals in the body then undoubtedly was counteracted by export through the gills, as still happens today in aquatic vertebrates (see Chap. 9). Yet it is possible that, in early stages of vertebrate evolution, minerals nevertheless could have tended to accumulate internally. Many minerals apparently were disposed of by deposition in the skin, and dermal bone in the form of heavy plates probably evolved in this fashion as a basic vertebrate trait. The earliest known fossil vertebrates were jawless fishes armored in bone (see Chap. 15), and most other vertebrates still have dermal bone in their skulls. Calcium deposits later also came to form replacement bone in the rest of the skeleton, where cartilage developed primitively as a strengthening material around the notochord.

This conversion of an original mineral liability to an adaptive asset could have been facilitated by concurrent evolution of an endocrine system that could control mineral metabolism. The glands of the system would necessarily have to be sensitive to salt entering the body with food and leaving with urine. It is therefore probably not a coincidence that, although other organs contribute as well, mineral-regulating functions in vertebrates are carried out to a large extent by glands situated close to the incurrent alimentary openings and the excurrent excretory organs: the anterior pituitary perhaps evolved from the protocoel canal and the hydropore of ancient deuterostomes; the thyroid and parathyroid glands, evolved from the endostyle in the floor of the chordate pharynx; and the adrenal cortex, along the kidneys.

Like their ancestors, the most primitive living vertebrates still are pharyngeal filter feeders: the larvae and adults of Agnatha are jawless, and in the larvae the ciliation of the pharynx sucks in

Figure 29.1 The vertebrate coelom: cross sections through successive embryonic stages. *A,* neural plate stage, mesoderm formed from endoderm as antero-posterior segmental cell masses. *B,* neural tube formed, mesoderm subdivided into three portions on each side, with coelomic cavities in upper two. The term *myotome* refers to the cellular portion, the term *myocoel* to the cavity, of the coelomic compartment; similarly for *nephrotome* and *nephrocoel. C,* coelomic space formed in lateral plate; nephrocoel communicates with this coelom through ciliated nephrostome. *D,* left and right trunk coeloms envelop intestine and form vertical mesentery; nephrotome differentiated into segmental pronephric and mesonephric components; gonad rudiment formed in dorsal coelomic lining (compare with Fig. 10.15).

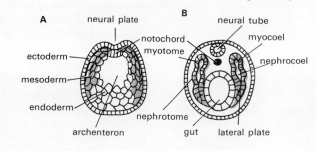

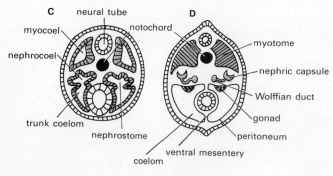

water with small food particles through a round, open mouth (see Fig. 29.3). Such a pattern holds in principle even for adult lampreys, ectoparasites that feed on body fluids and tiny tissue fragments rasped off a host. A considerable evolutionary gap then separates the agnaths from all other vertebrates. In these the main food-collecting organ no longer is the pharynx but the mouth itself, which has jaws and teeth. The nature of the food has changed concurrently, to bulk nutrients, and the ancestral method of filter feeding largely has ceased. This change left the branchial basket primarily as a breathing organ; and since therefore tiny particles no longer needed to be strained out of water, the number of gill slits could become reduced. Such a reduction actually has occurred in parallel with jaw development; the skeletal supports in the most anterior gills became remodeled as jaw supports, a condition encountered even in mammals and man (see Fig. 29.4).

But even though fewer gill slits now remained, these could become more efficient oxygenators; for the circulation had become closed, extensive capillary nets were developed, and an originally two-chambered heart moved blood more rapidly than in any of the chordate ancestors. Such increased circulatory efficiency actually was a necessary corollary to active food hunting and to the still unchanged requirement of excreting osmotic water rapidly. A further corollary to the food-hunting way of life was a new burst of neural evolution that resulted in the elaborate vertebrate brains and sense organs.

The freshwater environment continued to orient also the later stages of vertebrate evolution. Pharyngeal air sacs appear to have been a primitive, original trait of bony fishes, in adaptation to occasional periods of drought. Such sacs then evolved to swim bladders in some of these fishes and to lungs in others (see Fig. 9.23). Other adaptations to at first temporary and later permanent terrestrial life developed as well: fleshy fins, elongated later to legs; yolky eggs and aquatic tadpoles, followed by even yolkier land eggs with shells, amnion and allantois, and direct development; three and then four chambers in the heart, with efficient separation of arterial and venous blood and thus enough oxygen for the increased energy requirements on land; mesonephric and later metanephric kidneys, with reversal from an original water-excreting to a new water-retaining function; a distinct energy-saving neck, with a swivel joint for the head; controls for maintenance of constant body temperature, including insulating surface layers of fat, feathers, and fur; improved breathing and circulatory machinery in mammals, including a diaphragm and a highly efficient form of hemoglobin; and a placental reproductive mechanism, with viviparity and milk for the young.

Because the vertebrate egg typically is large and yolky, the basic pattern of enterocoelic development is to some extent obscured. Gastrulation occurs not by emboly as in amphioxus, but by various combinations of epiboly, involution, and ingression. After gastrulation, a sheet of cells in the roof of the archenteron, called the *chordamesoderm*, gives rise to the notochord along the midline and the mesoderm along each side of it (Fig. 29.1). Each segmental portion of the mesoderm then becomes subdivided into three parts. The dorsal part, closest to the notochord, represents the *myotome* and forms the segmental muscles. The middle part is the *nephrotome,* which contributes to the segmental development of the mesonephric excretory system. And the lowest part is the *lateral plate,* which develops in nonsegmental and essentially schizocoelic fashion; a

split arises inside it and the cavity so formed represents the body coelom. The lateral plate on one side eventually fuses ventrally with the one on the other side and the mesentery forms in this manner.

The eight classes of vertebrates can be grouped as larger units in various ways. For example, *Agnatha* include the jawless fishes and *Gnathostomata* the remaining seven classes in which jaws are present. *Anamniota* are set off from *Amniota,* the group of amnion-possessing reptiles, birds, and mammals. Birds and mammals are *homoiothermic,* or *endothermal,* with blood at constant temperature, and all other classes are *poikilothermic,* or *ectothermal,* or "cold"-blooded. A superclass *Pisces* includes the four classes of fishes and a superclass *Tetrapoda* comprises the four classes of four-legged types. The first of these superclasses is the subject of this chapter.

With at least 25,000 species, the bony fishes are the largest class, roughly half of all vertebrates. Birds form the next largest, with about 10,000 species. Next in order are reptiles (6,000 species), mammals (5,000 species), amphibia (3,000 species), cartilage fishes (600 species), and jawless fishes (50 species).

Jawless Fishes **Class Agnatha**   with notochord throughout life; internal skeleton cartilaginous; without true (dermal) teeth; typically with single nostril.

†*Subclass Ostracodermi*   ostracoderms, extinct; with bony armor or scales.

Subclass Cyclostomata   cyclostomes; without bony armor or scales; paired fins absent; sucking mouth; heart two-chambered; excretion pronephric and mesonephric; gonad single, without duct.

Order Petromyzontiformes   lampreys; seven pairs of gills, each opening separately; without

internal nasal opening; two pairs of semicircular canals; with pineal eye; oviparous, with *amnocoete* larvae! *Petromyzon,* sea lampreys; *Lampetra,* brook lampreys

Order Myxiniformes   hagfishes; 5 to 15 pairs of gills, 2 to 15 pairs of external openings; with internal nasal opening; 1 pair of semicircular canals; without pineal eye; oviparous, development direct. *Myxine*

The extinct agnaths, from 2 in. to 2 ft long, were mostly freshwater bottom inhabitants. Possibly they were weighed down by the often massive dermal armor, which covered the head and part of the trunk (see Fig. 15.8). The rest of the body often was scaled. In some forms the tail was *heterocercal,* the upper lobe larger than the lower one and the notochord extended into it. In other

groups the tail was *hypocercal,* with the lower lobe larger and supported by the notochord. The heterocercal condition is believed to have facilitated downward planing during swimming, and a hypocercal tail probably facilitates upward planing. Paired pectoral fins were present in some groups, and others often had lateral keels or flaps. The living agnaths probably descended from ostraco-

A

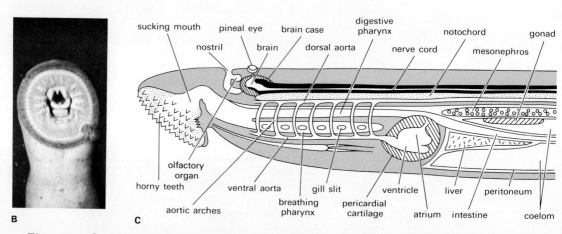

B

C

Figure 29.2 Lampreys. A, side view of *Petromyzon*. Note gill slits, fins. *B*, the sucker mouth, showing the horny epidermal teeth. *C*, sagittal section of anterior part of body. See also Fig. 10.16. (*A*, *B*, courtesy *Carolina Biological Supply Company. C, after Goodrich*.)

derms that had heterocercal tails and single external openings from the nasal organs.

The lampreys of today exhibit many secondarily reduced traits and others that represent special adaptations of the group. Dermal armor, scales, and all bone have been lost; the skin is soft, glandular, and slimy (Fig. 29.2). Paired appendages are absent, and the low median dorsal and caudal fins on the eel-like body are supported by cartilaginous fin rays. The single lobe of the caudal fin corresponds to the upper lobe of the ancestral heterocercal tail.

The round sucking mouth lies in the center of a shallow funnel studded with horny epidermal teeth. This funnel is fringed by short tentacles in lampreys and by long fleshy *barbels* in hagfishes. A pistonlike "tongue," similarly studded with horny teeth, can be protruded through the mouth. In lampreys a single dorsal nostril leads to a closed nasal sac, but in hagfishes the nasal passage opens to the posterior part of the mouth cavity. Just behind the nostril in lampreys lies a functional third eye, the *pineal*. The ears contain only one

or two semicircular canals. Cyclostomes have lateral-line systems and pituitary glands, and the structure of the nervous system conforms to the typical pattern of primitive vertebrates (see Chap. 8). An autonomic system is absent, however.

Along the notochord are segmental cartilage elements representing rudimentary neural arches. Cartilage capsules surround the brain and the heart. The gill chamber is an elongated sac that pouches out from the pharynx ventrally. In lampreys the seven pairs of gill slits open directly to the outside; in hagfishes water from the gills leaves through 2 to 15 pairs of gill pores. A stomach is absent; the esophagus continues directly to the intestine. The circulation pattern is characteristically vertebrate, already foreshadowed in amphioxus. A single atrium and ventricle compose the heart, as in all fishes. The blood too is typically vertebrate, with hemoglobin-containing red corpuscles and leukocytes. A lymph system is also present. The adult excretory system is mesonephric, the coelom is well developed, and a single

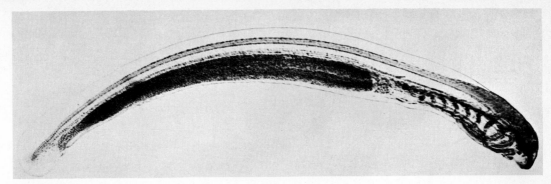

Figure 29.3 Ammocoete larva of jawless fishes. Note gill slits, nerve cord and brain, and position of notochord just underneath nerve cord. This larva is a pharyngeal filter feeder like the ancestors of vertebrates.

gonad discharges to the coelom. Gametes reach the outside through a genital pore along the course of the mesonephric (Wolffian) duct (see Fig. 10.16).

Sea lampreys are blood-sucking ectoparasites that attach themselves to fish by their mouths and rasp through the skin of the hosts. While attached the animals pump water in and out of the gills directly. Lampreys are separately sexed. All spawn in rivers, whether or not they also live in fresh water permanently. An egg develops into an eyeless _ammocoete._ This larva has an endostyle and a pronephric kidney, and it filter-feeds in upstream river bottoms in the ancestral manner (Fig. 29.3). Metamorphosis does not occur for 3 to 7 years, and the young adults then migrate downstream. Sea lampreys head to the open sea where they lead a parasitic mature life for usually not more than a single season. Brook lampreys live in fresh water permanently. The adults are not parasitic; indeed they do not feed at all but spawn within a few days and die.

Hagfishes are marine, oviparous, and develop directly. Some are hermaphroditic.

Placoderms   **† *Class Placodermi***   extinct fishes with jaws and true teeth; typically with dermal armor over head and front of trunk; bone in internal skeleton, notochord persisting and partly ossified; with paired fins, plus one or two dorsal, one caudal, and one anal median fins; tail heterocercal.

The placoderms were the first to exhibit two traits that since have become universal among vertebrates: jaws, and paired appendages on pectoral and pelvic girdles. Jaws probably evolved from the skeletal supports of the most anterior gill arch (Fig. 29.4). In placoderms the upper jaw was jointed with the brain case. Jaw development here affected neither the first pair of gill slits nor the second pair of gill arches, which remained as large as all the others. In later fishes, however, these gill structures did become changed as a result of posterior expansion of the jaws.

The evolution of paired fins is more difficult to interpret. According to a so-called "fin fold" hypothesis, ancestral vertebrates are assumed to have had a continuous ventral pair of fins, just as amphioxus and lampreys still have a continuous dorsal fin. In the descendants of these presumed ancestors, all ventral fin parts except those that were to become the pectorals and pelvics then would have disappeared. This hypothesis probably also accounts for the origin of the median dorsal, caudal, and anal fins.

Placoderms were largely small freshwater bottom dwellers (see Fig. 2.4). The head and anterior part of the trunk were covered with dermal bony armor and the remainder of the body was either scaled or naked. Considerable ossification of the internal skeleton had already occurred, but the gill skeleton remained cartilaginous. The specific placoderm ancestors of later fishes are unknown, but fossil placoderms similar in appearance to the present bony and cartilage fishes have been found.

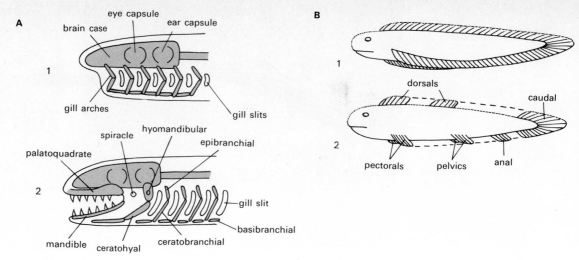

Figure 29.4 Jaw and fin evolution. A, jaw development. 1, scheme of gill slits, skeletal gill arches, and brain case in ancestral jawless vertebrates. 2, scheme in jaw-possessing vertebrates. The main bones of the gill arches are named. The first gill arch has given rise to the main elements of the upper and lower jaw. The upper component of the second gill arch (hyomandibular) can aid in suspending the gill skeleton from the brain case. Also, backgrowth of the jaw has reduced the first gill slit to a small opening, the spiracle. *B,* the fin-fold hypothesis. 1, scheme of continuous fin fold assumed to have characterized the body of ancestral vertebrates. 2, later evolutionary degeneration of parts of the fin fold is postulated to have left only the fins shown. *(A, after Romer; B, after Weidersheim.)*

Cartilaginous Fishes

Class Chondrichthyes   mostly marine; bone entirely absent; notochord reduced but present in adult; scales placoid; tail heterocercal; nostrils paired, internal nares absent; without air sac; with lateral-line system; sexes separate; fertilization internal, development direct; oviparous, ovoviviparous, or viviparous.

Subclass Elasmobranchii (Selachii)   elasmobranchs; with spiracles; five to seven pairs of gills, without operculum; teeth numerous; upper jaw not fused with brain case; dorsal fin nonerectile.

Order Squaliformes   sharks; gill slits lateral; pectorals free anteriorly; swimming by tail; males with pelvic fin claspers. *Squalus,* spiny dogfishes; *Mustelus,* smooth dogfishes; *Rhineodon,* whale sharks; *Cetorhinus,* basking sharks; *Sphyrna,* hammerhead sharks

Order Rajiformes   rays, skates; gill slits ventral; pectorals attached to side of head anteriorly; swimming by pectorals. *Raja,* rays, skates; *Pristis,* sawfishes; *Rhinobatus,* guitarfishes; *Torpedo,* electric ''eels''; *Manta,* devilfishes, manta rays; *Dasyatis,* sting rays

Subclass Holocephali   chimeras; without spiracles; four pairs of gills, with operculum; teeth fused as six pairs of plates; upper jaw fused to brain case; dorsal fin erectile; scales absent in adult; tail whiplike; without ribs or cloaca; males with claspers. *Hydrolagus*

Cartilage fishes once were believed to be more primitive than bony fishes. Such an assumption generally was based on the lack of bone, the presence of spiracles in most cases, and the retention of parts of the notochord. It is now clear, however, that cartilage fishes did not appear any sooner during vertebrate evolution than bony fishes and that they are a highly specialized group.

Although spiracles and notochord are indeed primitive traits (just as they are in early bony fishes), the lack of bone probably is not. Other indications of the specialized nature of these fishes are their almost exclusively marine occurrence, their excretory adaptations to this habitat (see Fig. 9.30), and their exclusively direct development.

In a shark, the skin is studded with tiny *placoid*

A

B

Figure 29.5 Sharks. A, Scylliorhynus, a dogfishlike type, from life. *B,* high-power view of skin, showing the toothlike denticles.

scales that have enamel outside and dentine inside. The scale tapers to a pointed spine, and the whole *denticle* resembles and is homologous to a mammalian tooth (Fig. 29.5 and see Fig. 8.4). The notochord no longer forms a continuous rod but is retained only between adjacent cartilaginous vertebrae. The skull is a cartilaginous box, with lateral capsules for the eyes and ears (Fig. 29.6). It connects with the upper jaws by ligaments and the *hyomandibula,* the upper cartilage of the second gill arch. In chimeras the upper jaws and the skull are fused together, as in the early bony fishes.

Jaw evolution in placoderms has led in elasmobranchs to a backward growth of the most anterior

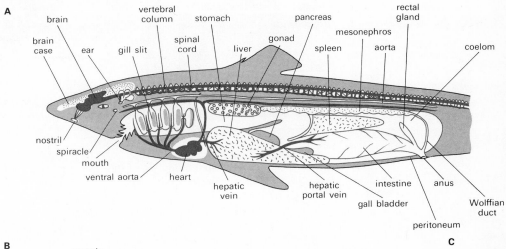

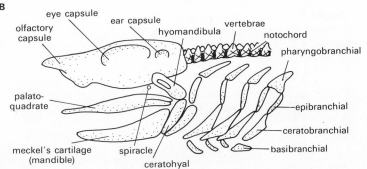

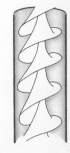

Figure 29.6 Shark structure. A, sagittal section. In a female a uterus (Muellerian duct) is present in addition to the Wolffian duct. The dorsal aorta gives off arteries (not shown) to stomach, intestine, liver, kidneys, and the other internal organs. *B,* the anterior skeleton, all elements composed of cartilage. The main jaw support is the hyomandibular, and the notochord still persists between the vertebrae. *C,* the spiral valve in the intestine. *(B, after Goodrich; C, after James.)*

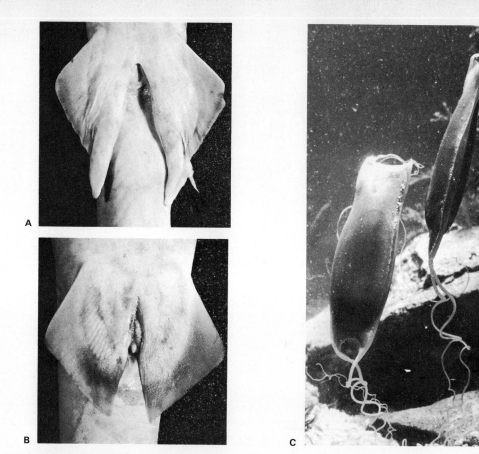

Figure 29.7 Adaptations in sharks. A and B, ventral views of cloacal region of male and female shark, respectively, showing presence of fin claspers in male and their relation to the pelvic fins. C, horny egg case of an oviparous shark. The filamentous extensions from the corners anchor it to objects in water. Note developing embryo faintly visible in case. (A, B, courtesy Carolina Biological Supply Company.)

gill supports and to a concurrent reduction of the first pair of gill slits to small openings, the *spiracles* (see Fig. 29.6). Chimeras lack even these. The gill slits of chimeras are covered by a pair of *opercula,* protective plates attached to the cartilages of the second gill arch. Such covers are not formed in elasmobranchs. In all cartilage fishes the median fins are supported by fin rays, the paired fins by small jointed cartilages. Paired fins are attached to pectoral and pelvic girdles.

The mouth is rimmed by several rows of pointed teeth, homologous to the placoid denticles. Anchored in flesh, the anterior teeth often break loose but are replaced by new ones formed behind. The alimentary tract includes a stomach and, opening to the intestine, a large liver and a separate pancreas. In the intestine is a conspicuous *spiral valve,*

a tissue fold that increases the absorptive surface. Cartilage fishes have a cloaca with openings of the reproductive and excretory ducts. A salt-secreting *rectal gland* is attached to the cloaca. In the many ovoviviparous and viviparous types, the female gonoducts (Muellerian ducts; see Fig. 10.16) are enlarged uteri. Oviparous forms shed eggs protected in rectangular horny cases (commonly seen washed up and dried out along beaches). The males typically have a pair of copulatory organs, the *fin claspers,* between the pelvic fins (Fig. 29.7).

Chimeras, mainly deep-water forms, are represented today by only some two dozen species; all other cartilage fishes are elasmobranchs, sharks and rays (Fig. 29.8). Most sharks are fiercely carnivorous, but the largest, the whale and basking

Figure 29.8 Rays and skates, dorsal and ventral views.

A

B

sharks, are plankton feeders. Basking sharks, with lengths of over 50 ft, are the largest of all fishes. Many rays and skates exhibit interesting offensive and protective adaptations, among them snouts extended as "saws" in the sawfishes, poison spines at the end of the tail in sting rays, and electric organs lateral to the eyes in the torpedo rays. Most rays and skates are bottom inhabitants, as would be expected from their flattened form. Giant mantas become up to 20 ft long.

Bony Fishes   **Class Osteichthyes**   typically with dermal scales, ganoid, cycloid, or ctenoid; spiracles present in primitive groups; with opercula; nostrils paired, with or without internal nares; with lung or swim bladder, the latter with or without duct; with lateral-line system; sexes separate; mostly oviparous with external fertilization; some ovoviviparous or viviparous.

Subclass Sarcopterygii   flesh-finned fishes; paired fins with internal bony skeleton and fleshy exterior; notochord retained; two dorsal fins; often with internal nares; air sac functions as lung.

Order Coelacanthiformes   coelcanths; skull largely cartilaginous; tail diphycercal, with three lobes; without internal nares; extinct types in freshwater, living forms marine. *Latimeria*

Superorder Crossopterygii   lobe-finned fishes; teeth not fused as plates.

Superorder Dipnoi   lungfishes; with three pairs of fused tooth plates.

†*Order Osteolepiformes*   extinct; with pineal eye; tail heterocercal; ancestral to amphibia; freshwater. *Osteolepis*

Order Dipteriformes   tail diphycercal in living forms; ossification reduced; freshwater. *Protopterus, Neoceradotus, Lepidosiren*

Subclass Actinopterygii   ray-finned fishes; paired fins with soft or hard fin rays; single dorsal fin; without internal nares; air sac usually functions as swim bladder.

Superorder Chondrostei   notochord retained; with spiracles; tail heterocercal; paired fins with broad-base attachment; scales ganoid or reduced; lung or swim bladder with duct.

Order Polypteriformes   bichirs; skeleton ossified; with paired ventral lunglike air sac; freshwater. *Polypterus*

Order Acipenseriformes   sturgeons; skeleton largely cartilaginous; scales reduced, skin often with bony plates; swim bladder dorsal, with duct; freshwater and marine, spawning in freshwater. *Acipenser*

Superorder Holostei   notochord retained, replaced in some by bony vertebrae; with or without spiracles; tail shortened, heterocercal; paired fins with narrow-base attachment; scales usually ganoid; swim bladder dorsal, with duct.

Order Semionotiformes   gar pikes; jaws elongated; swim bladder dorsal, with duct. *Lepidosteus*

Order Amiiformes   bowfins; without spiracles; swim bladder dorsal, with duct. *Amia*

Superorder Teleostei   notochord replaced by bony vertebrae; without spiracles; tail homocercal; paired fins with narrow-base attachment; pectorals often high lateral, with pelvics anterior to pectorals; scales cycloid or ctenoid or secondarily reduced; swim bladder with or without duct.

Order Clupeiformes   herrings, sardines, anchovies, tarpons, salmon, trout, whitefishes, pikes; soft-rayed, without fin spines; pelvics abdominal; scales cycloid or reduced; swim bladder with duct; without Weberian ossicles; mostly marine.

Order Myctophiformes   lantern fishes, lizard fishes; pelvics abdominal; swim bladder with duct or absent; without Weberian ossicles; deep sea.

Order Saccopharyngiformes   gulpers; without opercula, swim bladder, pelvics, caudals, scales, or ribs; gill openings minute; mouth tremendous, pharynx highly distensible; eyes tiny; tail slender; deep sea.

Order Cypriniformes   carps, goldfishes, minnows, catfishes, neon fishes, electric "eels"; like clupeiform fishes, but with Weberian ossicles (see Fig. 8.35); most important freshwater order.

Order Anguilliformes   true eels; pectorals often absent, pelvics absent; with or without scales; swim bladder with duct; gill apertures reduced; freshwater and marine, all spawning in sea.

Order Notacanthiformes   spiny eels; pectoral girdle loosely suspended, pelvics abdominal; anal fin long, caudal absent, tail tapered; swim bladder without duct; deep sea.

Order Beloniformes   "flying fishes, needlefishes, pectorals large, high lateral, used in gliding; pelvics abdominal; swim bladder without duct; lateral-line system ventrolateral; mostly marine.

Order Cyprinodontiformes   guppies, swordtails, platyfishes, killifishes; pectorals high lateral, pelvics abdominal or absent; swim bladder without duct; mostly marine.

Order Fasterosteiformes   sea horses, snipes, pipefish, sticklebacks; snout long, mouth at end of tube; with or without fin spines; swim bladder without duct; mostly marine.

Order Gadiformes   cod, haddock, pollock, hake, grenadiers; pelvics jugular; swim bladder without duct; dorsal often subdivided into three parts, anal into two parts; without fin spines; marine.

Order Lampridiformes   ribbonfishes, oarfishes; pelvics thoracic; dorsal with one or two fin spines, rest soft-rayed; swim bladder without duct; body large, serpentine, marine.

Order Percopsiformes   sand rollers; pelvics thoracic; dorsal with one to four fin spines, rest soft-rayed; scales ctenoid; swim bladder without duct; in sluggish fresh water.

Order Beryciformes   squirrel fishes; pelvics thoracic; with fin spines but soft-rayed; scales ctenoid; swim bladder without duct; marine.

Order Zeiformes   John Dorys; pelvics with fin spines but soft-rayed; marine.

Order Perciformes   perches, tunas, mackerels, basses, rockfishes, marlins, sailfishes, blue-fishes, jacks, sunfishes; pectorals high lateral, pelvics thoracic or jugular; fins with spines, hard-rayed; scales ctenoid; swim bladder without duct; most important marine order.

Order Pegasiformes   sea moths, sea dragons; external bony casing, bony snout; without teeth; pectorals horizontal, winglike, pelvics abdominal; swim bladder absent; marine.

Order Pleuronectiformes   halibut, flounders, plaice, sole, flatfishes; on left or right side, with loss of bilaterality; dorsals and anals long; marine, on bottom.

Order Echeneiformes   remoras, shark suckers; dorsal modified as oval adhesive disk; scales cycloid; pectorals high lateral, pelvics thoracic.

Order Tetraodontiformes   puffers, triggerfishes, trunk fishes, porcupine fishes; armored with bony plates and spines, prickly or naked; fin spines present; mostly marine.

Order Gobiesociformes   clingfishes; without fin spines, scales, ribs, or swim bladder; both girdles and pelvics modified as thoracic sucking disk for attachment to bottom; mostly marine.

Order Batrachoidiformes   toadfishes; first vertebra fused with flattened cranium; ribs absent; three pairs of gill arches; pelvics on throat; marine.

Order Lophiiformes   angler fishes; first ray of dorsal modified as angler with luminescent lure; pelvics thoracic or absent; ribs absent; gill aperture small, far back; males sometimes dwarfed, on female; deep sea.

Order Mastacembeliformes   eel-like; median fins continuous, with fin spines; snout long, with tentacles; with or without scales; pelvics absent; swim bladder without duct; fresh water.

Order Synbranchiformes   eel-like; small gill apertures joined across throat; gills reduced, breathing partly by mouth-pharynx pouches; pelvics on throat, small; without pectorals, fin spines, or swim bladder; with or without scales; swamps, caves, sluggish fresh water.

Bony fishes evolved from placoderm ancestors at approximately the same time as (or perhaps even earlier than) the cartilage fishes. As their name suggests, the fishes have a basically bony skeleton, although in many primitive forms the ossification is incomplete. Dermal armor is no longer present, but dermal bones form most of the skull, parts of the girdles, and all the scales.

In primitive groups the scales most often are *ganoid,* diamond-shaped and covered with a layer of hard, glossy enamel. The whole scale is homologous to the denticles of sharks and the teeth of mammals. In other fishes the scales are *cycloid* if their exposed posterior parts are smooth, *ctenoid* if the exposed parts are rough-textured or spiny (Fig. 29.9). The tail of bony fishes is primitively heterocercal but symmetric in most cases. In some symmetric tails the vertebral column extends to the tip (*diphycercal* condition); in others it terminates at the tail base (*homocercal* condition).

The median fins (dorsals, anals, and caudals) are supported by elongated bony fin rays. The support of the paired fins differs sharply in the two great subclasses of bony fishes. In the flesh-finned Sarcopterygii, the pectorals and pelvics contain internal bony skeletons that correspond substantially to the bones of tetrapod limbs. In the ray-finned Actinopterygii, by contrast, paired fins are supported like median fins by bony rays (see Figs. 8.9 and 29.9).

The skull consists of some 60 bones, a number that has become reduced in later vertebrate classes (to about 20 in mammals). In primitive bony fishes the notochord often persists to the adult stage, but in most groups it is replaced partly or (more usually) wholly by bony vertebrae preformed in cartilage. The vertebrae characteristically bear neural and hemal arches as well as ventral ribs (see Fig. 8.10).

The jaws represent the *mandibular* arch, the first of the ancestral skeletal gill arches (Fig. 29.10). Backgrowth of the bones of this arch on each side has resulted as in sharks in a reduction of the first gill slit to a *spiracle.* In most bony

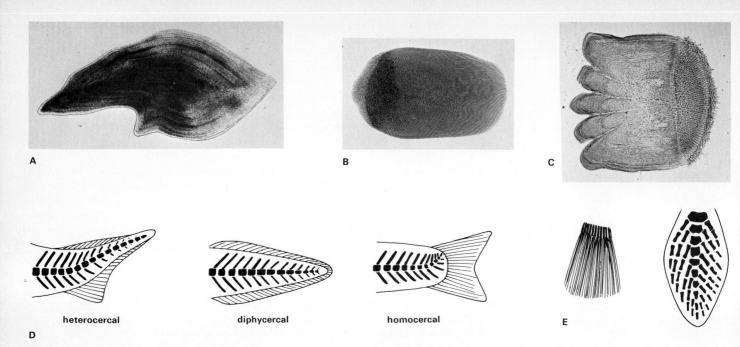

heterocercal diphycercal homocercal

D E

Figure 29.9 · *Scales and fins in bony fishes*. *A*, ganoid scale. *B*, cycloid scale. *C*, ctenoid scale, with smooth portion overlapped by the scale before it and the spiny portion exposed. *D*, the three main tail types. *E*, left, fin of ray-finned bony fish; bony rays support the fin. Right, bony elements in fin of lungfish. See Fig. 8.8 for fin of lobe-finned fish and bony elements of all vertebrate limbs. (*A, B, C, courtesy Carolina Biological Supply Company.*)

Figure 29.10 Jaws in bony fishes. *A*, in most cases the main bony prop between skull and upper jaw is the hyomandibular (though ligaments aid in supporting the palatoquadrate). Position of spiracles is indicated; in most instances this remnant of the first ancestral gill slit is absent. *B*, in primitive sarcopterygians (and amphibia), the upper jaw is fused directly to the brain case; the hyomandibular is free and without supporting function.

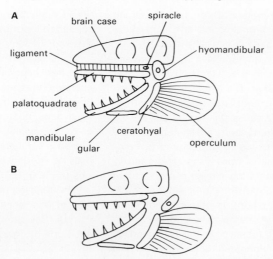

A

brain case spiracle

ligament hyomandibular

palatoquadrate

mandibular ceratohyal

gular operculum

B

fishes, actually, the first gill slits have disappeared completely and not even spiracles are left. In the second, or *hyoid*, arch, the lower bone bears a hinged *operculum*, and the upper, *hyomandibular* bone in most cases serves to hold the upper jaw to the brain case. In some primitive bony fishes the upper jaw is fused directly with the brain case, leaving the hyomandibular without function. The ancestors of amphibia were of this type, and the unused hyomandibular later acquired a new function as the stirrup bone of the middle ear.

The nostrils of bony fishes are paired and lead either to closed nasal sacs or to the mouth cavity through internal nares (see Fig. 9.23). In the pharynx gills usually number four and never more than five pairs. Each gill is a double plate composed of epithelial, highly vascularized *gill filaments* (Fig. 29.11). A skeletal gill arch supports the gill tissue. The inner edges of the gills often extend as comblike *gill rakers* that prevent food from passing through the gill slits. An air sac pouched out from the pharynx serves as a lung primitively and as a swim bladder in more advanced forms.

Both the flesh-finned and ray-finned groups

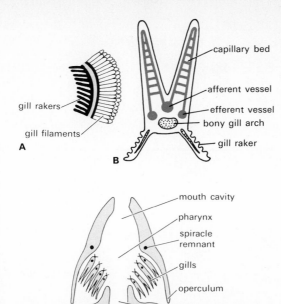

Figure 29.11 Gills in bony fishes. A, left gill in side view showing double row of gill filaments pointing away from pharynx and gill rakers pointing into pharynx. B, horizontal section through a gill filament. The afferent vessel is the part of the aortic arch coming from the ventral aorta, the efferent vessels, the parts leading to the dorsal aorta. C, horizontal section through fish head, showing position of the four pairs of gills. (*Partly after Goodrich.*)

primitively were freshwater forms, and both presumably used their air sacs originally as lungs. The flesh-finned types then continued and perfected the air-breathing habit, but the ray-fins soon came to use the air sac purely as a swim bladder. Flesh-finned fishes also used their skeleton-supported pectorals and pelvics as primitive walking organs. One group among them, the lobe-finned osteolepids, probably gave rise to the amphibia (Fig. 29.12). In the course of this transition the fin skeleton elongated to the typical tetrapod leg.

Amphibia and all later vertebrates also inherited the internal nares from these ancestors, as well as an upper jaw fused directly with the skull.

Osteolepids (Fig. 29.13) are extinct, and the related lobe-finned order of coelacanths is represented by a single surviving genus (*Latimeria,* see Fig. 15.9). The lungfishes, similarly flesh-finned, belong to three surviving genera, *Protopterus* in Africa, *Lepidosiren* in South America, and *Neoceradotus* in southern Australia (see Figs. 15.9 and 29.13). These fishes live in muddy fresh waters. When water dries up seasonally the animals wrap themselves in layers of mud and breathe air until the rains restore an aquatic environment. The whole subclass Sarcopterygii today thus consists of only four genera.

By contrast, the subclass of ray fins now forms the most abundant group of vertebrates. These fishes lack internal nares, they have a single dorsal fin, and, as noted, their paired fins are supported by fin rays. The two primitive superorders, Chondrostei and Holostei (Fig. 29.14) are alike in that the scales are generally ganoid, the tail is basically heterocercal, and the notochord tends to persist. Also, a secondary reduction of ossification is quite common. Both superorders today contain only about forty surviving species. All are either freshwater forms or, like some of the sturgeons, live in the sea but return to rivers for spawning. *Polypterus,* found in Africa, still is a lung-breather (and has formerly been classified erroneously with the lungfishes). In the other genera the air sac is a swim bladder primarily and the duct to the pharynx

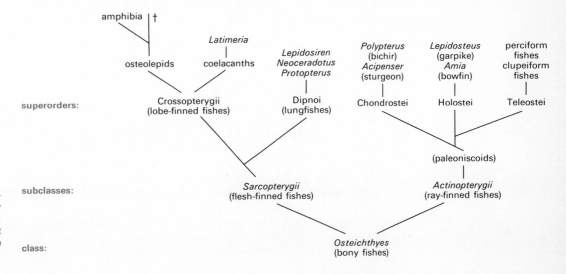

Figure 19.12 Bony fishes: taxonomic and probable evolutionary interrelations of major groups. Presently living genera of all but the teleosts are indicated above each superorder.

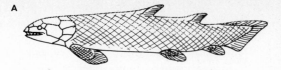

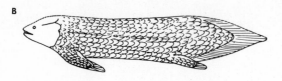

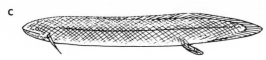

Figure 29.13 Sarcopterygii. *A,* the extinct lobe-fin *Osteolepis.* The living (lobe-finned) coelacanth *Latimeria* is illustrated in Fig. 15.9. *B,* the living lungfish *Neoceradotus.* *C,* the living lungfish *Lepidosiren.* The third of the lungfishes now in existence, *Protopterus,* is illustrated in Fig. 15.9. (*A, after Traquair; B, C, after Norman.*)

is retained. *Amia* is a "living fossil" found exclusively in the larger rivers of North America.

All remaining bony fishes are members of the large and highly diversified superorder Teleostei. In this group the skeleton is fully ossified and the tail is homocercal. Two broad general patterns of teleost structure can be distinguished (Fig. 29.15). In one, the *clupeiform* pattern, fin rays are soft, scales are cycloid, the swim bladder has a persisting duct, and the pelvic fins are in a primitive abdominal, posterior, position. In the other, *perciform* pattern, fin rays are hard and spiny, scales are ctenoid, the swim bladder lacks a duct, the pectoral fins have moved high up on the sides of the body, and the pelvic fins have moved far forward, even as far as the head region (throat or jugular position). Thus the pelvic fins lie *in front* of the pectorals. These two patterns are not always sharply distinct, and many teleosts exhibit mixed characteristics. For example, the fin rays can be soft generally but the first or the first few anterior rays of a fin can be hard and spiny; cycloid scales can occur together with ductless swim bladders, or with pelvics in forward position.

The clupeiform pattern takes its name from the Clupeiformes, the third largest order of teleosts (1,000 species). The pattern also occurs in pure

form in the *cyprinoid* fishes (order Cypriniformes), the second largest teleost order (5,000 species) and the most important freshwater group. Perciform types take their name from the order Perciformes, the largest order not only among fishes but among vertebrates as a whole (8,000 species). These fishes comprise the most important marine groups. In the taxonomic tabulation above, the sequence of orders by and large forms an intergrading series from more or less pure clupeiform to more or less pure perciform types. Thus, all orders down to but not including the Notacanthiformes have swim bladders with ducts; all orders down to but not including the Percopsiformes have cycloid scales; and orders down to but not including Perciformes have basically soft-rayed fins, although hard anterior fin spines are present in many cases.

Figure 29.14 Primitive bony fishes. The bichir *Polypterus* and the sturgeon *Acipenser* are representatives of the superorder Chondrostei. Note the broadbase attachment of the paired fins. The gar pike *Lepidosteus* and the bowfin *Amia* are members of the superorder Holostei. Fins here are attached by narrow bases.

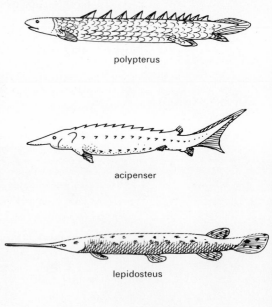

polypterus

acipenser

lepidosteus

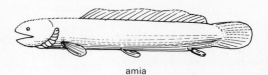

amia

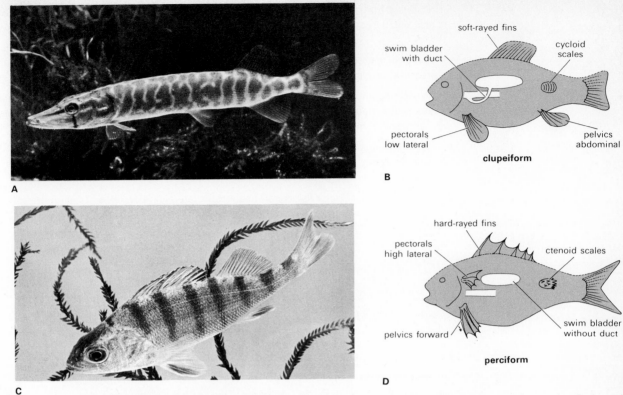

Figure 29.15   Teleosts. A, a pike. Note soft-rayed, smooth-contoured fins and other clupeiform traits. B, the main characteristics of clupeiform teleosts. C, a perch. Note hard-rayed, spiky fins and other perciform traits. D, the main characteristics of perciform teleosts.

Review Questions

1.   Give taxonomic definitions of the vertebrate subphylum and of each of the classes of fishes. Show how most of the basic vertebrate traits are direct adaptations to a life in fresh water. Describe the pattern of mesoderm- and coelom-formation in vertebrates.

2.   Define the class Agnatha and the subclass Cyclostomata. Characterize the extinct ostracoderms and show in what respects modern agnaths are structurally different. What different types of tail fin occur among fishes and which type occurs in which groups? What is a pineal eye? How many species of each class of fishes are known?

3.   Describe the external and internal structure of a lamprey, with attention to all organ systems. Review pertinent data of Chaps. 8–10. In what ways are lampreys adapted to an ectoparasitic existence? What are ammocoetes? Which cyclostomes live permanently in fresh water and which only spawn there?

4.   How did jaws and paired and median fins presumably evolve? What changes in vertebrate anatomy occurred in conjunction with the change from ancestral filter feeding to bulk feeding by mouth? Review the pattern of embryonic development among fishes.

5.   Define the subclasses and orders of chondrychthians. Why is the cartilage skeleton now thought to represent a specialized rather than a primitive trait? In which classes of fishes do the adults have a complete or partial notochord?

6. Contrast the structure of the gill region in sharks and chimeras. What are fin claspers? How do sharks differ from rays? Review the adaptations of sharks and rays to their respective ways of life.

7. Give taxonomic definitions of all subclasses and superorders of bony fishes. Which group was probably ancestral to amphibia? What are the genera of living flesh-finned types? Name and characterize representative types of as many different orders of ray-finned fishes as you can.

8. What different scale types and tail types occur among bony fishes? What changes in vertebrate anatomy occurred in conjunction with jaw elaboration? Which groups have spiracles? How is the upper jaw suspended in placoderms, cartilage fishes, and bony fishes?

9. Which groups of bony fishes have connecting ducts between air sac and pharynx? Internal nares? Air-breathing lungs? Describe the gill structure of a bony fish. Review the organization of every organ system of a bony fish (if necessary consult Chaps. 8–10).

10. Which groups of bony fishes can hear? Have lateral-line systems? Distinguish between clupeiform and perciform fishes and name orders of each type. How do fishes mate? Does viviparity occur among cyclostomes? Bony fishes? Cartilage fishes?

Collateral Readings

Applegate, V. C., and J. W. Moffett: The Sea Lamprey, *Sci. American,* Apr., 1955. The biology and economic harmfulness of this parasitic vertebrate is examined.

Colbert, E. H.: "Evolution of the Vertebrates," Wiley, New York, 1955. Recommended for background on fossil types.

Gilbert, P. W.: The Behavior of Sharks, *Sci. American,* July, 1962. The senses of sharks and their role in feeding behavior are examined.

Millot, J.: The Coelacanth, *Sci. American,* Dec. 1955. An account on the dramatic discovery of this lobe-fin, previously believed to have long been extinct.

Romer, A. S.: "Vertebrate Paleontology," 2d ed., University of Chicago Press, Chicago, 1945. A well-illustrated and highly recommended book on fossil types and vertebrate evolution.

———: "The Vertebrate Body," Saunders, Philadelphia, 1950. Recommended for anatomic background.

———: Major Steps in Vertebrate Evolution, *Science,* vol. 158, p. 1629, 1967.

Young, J. Z.: "The Life of Vertebrates," 2d ed., Oxford University Press, Fair Lawn, N.J., 1962. A large and highly recommended volume on the zoology and evolutionary history of vertebrates.

30

VERTEBRATES: TETRAPODS

Amphibians   ***Class Amphibia***   freshwater and terrestrial; paired appendages are legs; skin without scales in most living species; with internal nares; upper jaws fused to skull; heart three-chambered; breathing through gills, lungs, skin, and mouth cavity; 10 pairs of cranial nerves; sexes separate; fertilization mostly external, with tadpole larvae; some with internal development, ovoviviparous or viviparous.

†*Subclass Stegocephalia*   extinct amphibians; head covered with bony plates, skin often with scales.

Subclass Gymnophiona (Apoda)   cecilians; wormlike; up to 200 vertebrae; without limbs or girdles; skull topped with bone; ribs long; skin smooth, often with embedded dermal scales; eyeless; small tentacles in front of nostrils; tail short; fertilization internal, male with copulatory organ.

Subclass Urodela (Caudata)   salamanders, newts; body with head, trunk, tail; aquatic larvae resemble adults; larvae and some adults with teeth on upper and lower jaws; eyelids typically present in terrestrial adults, absent in aquatic adults.

Order Proteida   mud puppies; permanently aquatic; with permanent gills and lungs; tail with fin.

Order Meantes   mud eels; permanently aquatic; with permanent gills; jaws with horny cover; without pelvic limbs.

Order Mutabilia   conger eels, axolotls, newts, salamanders; adults often aquatic; usually without gills, with lungs, air-breathing.

Subclass Anura (Salientia)   toads, frogs, tailless, without neck; hindlimbs usually long, feet webbed; 10 vertebrae; with eyelids; adults mostly terrestrial.

Order Amphicoela   bell toads; fertilization internal.

Order Opisthocoela   clawed toads, midwife toads; fertilization external; many aquatic, without eyelids.

Order Anomocoela   spade-foot toads; hind feet with horny style.

Order Procoela   true toads; without teeth.

Order Displasiocoela   true frogs; teeth on upper jaw; tongue forked.

The most notable traits of amphibians are those that adapt the animals to an at least partly terres-

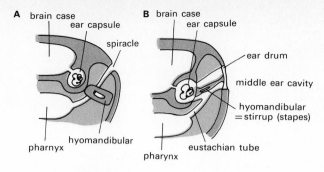

Figure 30.1 The ear in fish and amphibia. Both diagrams illustrate transverse sections through the hind region of the head. *A,* condition in primitive bony fishes, with free hyomandibular (and upper jaw fused to skull), the ear consisting only of the canals and sacs within the bony ear capsule. *B,* amphibian condition, with the hyomandibular (also called *columella* in amphibia) functioning as sound transmitter between tympanum and inner ear. The spiracular duct to the pharynx (remnant of the original first gill slit) has become the middle-ear cavity and the Eustachian tube. (*After Romer.*)

trial life; and most of these traits trace their origin to the ancestral lobe-finned fishes. The paired appendages are elongated to distinct walking legs, each with five (or fewer) toes. Although primitive amphibia still live entirely in water, the majority are air breathers with lungs. Gills occur in the

Figure 30.2 The amphibian tadpole. A, fully formed tadpole, showing characteristics particularly typical also of fishes. *B,* cross section through gill region of later larval stage, at time of transition from external to internal gills. On right side, external gills already have degenerated and the newly developed internal gills are covered by the opercular fold. On left side, external gills still protrude for a time through the spiracle, but ultimately they too degenerate. From then until the lungs become functional, breathing is accomplished through the internal gills. (*Adapted from Goodrich.*)

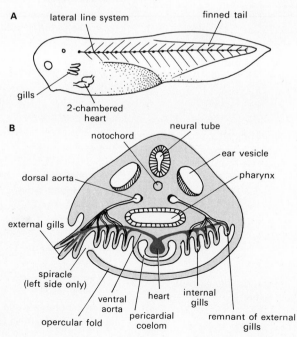

larvae and in exceptional cases also in some of the adults (where lungs are or are not present in addition). In conjunction with air breathing, the internal nares inherited from the lobe-fins provide an important adaptation, for they make breathing possible while the mouth remains closed. As in ancestral lobe-fins, moreover, the upper jaw is fused to the skull. The free hyomandibular bone has become a *columella,* a middle-ear bone. One end of it attaches to an external eardrum flush with the body surface, the other end to the inner ear. This bone transmits aerial sound waves to the ear, clearly adaptive in terrestrial living (Fig. 30.1).

The skin is thin and scaleless, providing a breathing organ auxiliary to gills, lungs, and the lining of the mouth. The skin also is highly glandular and mucus-secreting, a necessary protection against desiccation. Further adaptive in air breathing is the three-chambered heart, capable to some extent of preventing a mixing of oxygenated blood returning from the lungs and venous blood returning from the body. Arterial blood enters one atrium, venous blood the other. The two bloods then do mix in the single ventricle, but not as much as in a two-chambered heart. Where gills are retained in the adults they are reduced to three pairs, and three corresponding pairs of aortic arches are present in such cases; where gills do not persist, the aortic arches have become reduced to a single pair (see Fig. 9.22).

Thus, from a lobe-finned starting point, the amphibian descendant has become able to walk and to breathe air on a permanent basis, even though in some instances these activities still are carried out in an aquatic environment. Actually even the most land-adapted amphibian still requires a moist environment at all times, and although locomotion and breathing are indeed largely land-adapted, most other functions are not. For example, the nervous system still is substan-

A

B

C

Figure 30.3 **Urodeles.** *A,* the conger eel *Amphiuma*. Note reduced appendages. *B,* the Mexican axolotl *Amblystoma mexicanum*. This is a neotenous species, permanently retaining larval traits such as external gills and a larval tail. *C,* a fire salamander. All three animals shown here are members of the urodele order Mutabilia.

tail, a lateral-line system, three pairs of gills and three pairs of aortic arches but in early stages not lungs, a two-chambered heart, a hyomandibular that is not yet a middle-ear bone and in early stages limbs are still absent. In effect, therefore, any terrestrial adaptations exhibited by adult amphibia are acquired anew in each amphibian life cycle (Fig. 30.2).

Of the three living subclasses, the cecilians probably represent a very primitive amphibian stock. However, the roughly seventy species today are secondarily highly specialized. They are mostly tropical, blind, limbless soil burrowers, leading a life reminiscent of earthworms. The males have a penis and fertilization is internal. Eggs are large and they cleave superficially, as in fishes. Some species have aquatic larvae, but most are direct developers and of these many are viviparous. The animals still exhibit one major ancestral trait, dermal scales embedded in the skin.

The roughly 250 species of Urodela include a wide range of types, from fully aquatic ones to more or less terrestrial ones (Fig. 30.3). Mud eels have permanent gills, mud puppies retain both gills and lungs permanently, and both groups are strictly aquatic. Fully aquatic types occur also among newts and salamanders. Thus the conger eel *Amphiuma* has permanent gills and lungs, and the hellbender *Cryptobranchus* has lungs only but also a functional spiracle. The newt *Triturus* and the salamander *Salamandra* similarly have only lungs, yet these animals live both in water and on land. The most nearly terrestrial urodeles are certain internally fertilizing, viviparous species of *Salamandra*. Although these do require a moist environment, they can be independent of natural bodies of water.

Whereas urodeles on the whole are more aquatic than terrestrial, the reverse holds for the Anura, toads and frogs, the largest amphibian group. Most anurans have eyelids, an adaptation to the aerial environment. Also, they often are highly specialized as jumpers, with hind legs far longer and stronger than the forelegs. In conjunction with this locomotor specialization tail and neck are absent, the head being joined directly to the trunk. Indeed the whole body has become squat and foreshortened, and the number of vertebrae is reduced to 10. The tongue of a frog can be hurled out of the mouth at an insect, posterior end first.

Toads are toothless, frogs have teeth on the

tially fishlike, as is the mesonephric excretory system, which remains adapted to eliminate primarily water, not salt. As a result, amphibia are unable to live in a marine or otherwise salty environment and generally have been incapable of colonizing oceanic islands by their own efforts. The skeleton still is basically as in fishes, particularly in the urodeles, even though the appendages have become modified and the number of skull bones has become reduced. Other specialized skeletal alterations have occurred in forms such as frogs.

Above all, reproductive and developmental processes still retain ancestral fishlike characteristics. Eggs typically must be laid in water, and fertilization and development take place externally, as in bony fishes. Some amphibian eggs develop directly and without larvae, inside brood pouches on the backs of certain tree toads and tree frogs (see Fig. 30.4). In most cases, however, development includes aquatic tadpole larvae that are thoroughly fishlike, regardless of whether or not the adults are also aquatic. The tadpole has a finned

A

B

Figure 30.4   Anurans. A, egg-carrying male of the midwife toad *Alytes obstetricans. B,* the tree frog *Hyla.*

upper jaw and the roof of the mouth. None of the anurans has gills as adults and all have lungs. A few toads (*Xenopus,* for example) are permanently aquatic yet air breathing. Most anurans require an outright aquatic environment only for egg laying and tadpole development. A good many are inde-

pendent of water even in these processes, for the eggs develop directly. For example, toads such as *Alytes* and *Pipa* incubate their eggs in brood pouches or in epidermal pits on the back (Fig. 30.4).

Reptiles   **Class Reptilia**   epidermis with cornified scales; limbs five-toed, clawed; breathing by lungs; heart four-chambered, ventricular separation usually incomplete; one pair of aortic arches; 12 pairs of cranial nerves; excretion metanephric; fertilization internal; eggs with extraembryonic membranes and shells; development direct; oviparous, ovoviviparous, or viviparous.

Subclass Anapsida   (cotylosaurian stem reptiles, extinct).

Order Chelonia   turtles, tortoises, terrapins; body encased in bony shell and epidermal cover; teeth absent in living types; single penis in male; oviparous.

†*Subclass Synapsida*   therapsids, mammal-like reptiles, extinct.

†*Subclass Ichthyopterygia*   ichthyosaurs, extinct.

†*Subclass Synaptosauria*   plesiosaurs, extinct.

Subclass Lepidosauria   (lizardlike reptiles, extinct).

Order Rhynchocephalia   tuataras; pineal eye functional; without penis; oviparous. *Sphenodon.*

Order Squamata   skull bones reduced; epidermal scales horny; males with paired penis; oviparous, ovoviviparous, or viviparous.

Suborder Sauria   lizards; eyelids movable; usually with external ear opening; with pectoral girdle, limbs usually present; halves of lower jaw articulated; generally with urinary bladder.

Suborder Serpentes   snakes; eyelids not movable; without external ear openings; without pectoral girdle, limbs absent; halves of lower jaw not articulated, united by ligament; urinary bladder absent.

Subclass Archisauria   (dinosaurs, pterosaurs, extinct).

Order Crocodilia   alligators, crocodiles, caimans, gavials; heavy-bodied, semiaquatic; epidermal scales with dermal reinforcements; webbed feet; teeth in sockets; urinary bladder absent; males with single penis; oviparous.

As the first fully terrestrial vertebrates, reptiles have many traits in common with mammals and birds. The reptilian skin is dry and protected from desiccation by cornified epidermal scales, homologous to fur and feathers. Limbs are walking legs, and they raise the body off the ground more than the amphibian limbs. Reptiles are strictly air breathers, with a breathing system constructed like that of mammals and birds; tracheal and bronchial tubes pipe air to the lungs. The heart is four-chambered, completely so in crocodiles, nearly so in the other groups; it keeps arterial blood separated from venous blood, essentially as in mammals and birds. Aortic arches are reduced to a single pair (see Fig. 9.22). The adult excretory system is metanephric, as in mammals and birds, and is capable of producing a highly hypertonic, water-conserving urine. Secondarily aquatic turtles and alligators excrete urea like mammals, and all terrestrial reptiles excrete a semisolid urine containing uric acid, as in birds.

In the nervous system, 12 pairs of cranial nerves connect with the brain, as in birds and mammals, in contrast to the 10 pairs in amphibia. Reptilian ears still retain an essentially amphibian structure, with a single middle-ear bone and an external eardrum. The eyes of lizards and turtles are capable of recording color. The single most essential adaptation to land life is the shelled,

Figure 30.5 Reptilian skull types. p, parietal; o, postorbital; s, squamosal; j, jugal. In the anapsid type, openings in the skull are not present behind the eye. Synapsid and parapsid skulls each have a different single opening (color), and a diapsid skull has both these two different openings together. (*After Romer.*)

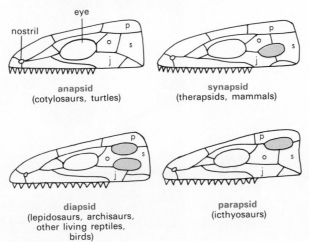

anapsid
(cotylosaurs, turtles)

synapsid
(therapsids, mammals)

diapsid
(lepidosaurs, archisaurs,
other living reptiles,
birds)

parapsid
(icthyosaurs)

amniote egg, as discussed in Chap. 11. The shells are leathery in some cases, calcareous in others. As in birds, the hatching young has a temporary *egg tooth* on its upper jaw with which it cuts itself out of the shell. *All* reptiles lay eggs on land, even secondarily aquatic species that mate in water. Fertilization is always internal, as in mammals and birds; a penis in males can be absent (tuatara), single (turtles, crocodiles), or double (lizards, snakes). These organs are located in or near the cloaca and are protruded during copulation.

Reptiles are distinguished taxonomically mainly by the number and location of lateral windows in the skull bones behind the eyes (Fig. 30.5). A primitive *anapsid* skull, without lateral windows, still occurs in turtles. All other living reptiles have *diapsid* skulls, with two pairs of lateral windows. Reptiles are distinguished also by the nature of the tooth attachment along the jaws, the structure of the pelvis, and other skeletal traits.

With their unique protective body covering, turtles are among the most readily recognizable animals. The shell consists of a dorsal *carapace* and a ventral *plastron*. The carapace is a dome of bone, formed from fused vertebrae and from broadened and fused ribs (Fig. 30.6). Similarly, the bony plate of the plastron is constructed from flattened and fused parts of the pectoral and pelvic girdles and the sternum. The pectoral girdle actually is wholly inside the shell of the rib cage, a condition encountered nowhere else. The bony parts of the shell are overlain by skin, which can be soft but tough, as in the leatherbacks, or hardened to epidermal horn, as in painted turtles and numerous other types.

Since the rigid shell precludes breathing movements of the body wall, the lungs must be ventilated by special internal muscles. Many turtles cannot withdraw their heads into the shells. All living species have horny beaks. In marine turtles the forelimbs are modified as flippers (the term ''turtle'' properly applies only to the aquatic chelonians; terrestrial types are ''tortoises''). The majority of chelonians are aquatic or semiaquatic. The largest are the marine leatherbacks, which are up to 8 ft long and weigh nearly a ton. The giant Galapagos tortoises attain lengths of about 4 ft (and recorded life spans of often well over 60 years). Chelonians are generally omnivorous.

Some 95 percent of all living reptile species comprise the lizards and snakes. Most lizards (Fig. 30.7) subsist on insects. For example, the chame-

A

B

Figure 30.6   Chelonia. A, the turtle skeleton. The ventral part (plastron) is cut away and hinged back, showing the limb girdles underneath the dorsal part (carapace) formed by the rib cage. *B*, the green sea turtle *Chelonia*. (*A, courtesy Carolina Biological Supply Company.*)

Figure 30.7   Reptilia. A, the lizard *Chamaeleon*, catching grasshopper with tongue. *B*, *Sphenodon*, the lizardlike tuatara, sole surviving member of the reptilian order Rhynchocephalia.

B

A

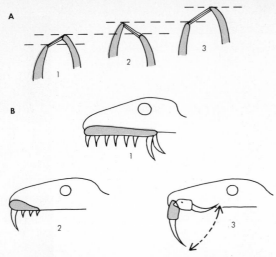

Figure 30.8 Skeletal adaptations in snakes. A, the two halves of the lower jaw, joined by an elastic ligament only, and the alternating forward movement of each half during ingestion. Numbers indicate sequence of movements of jaw halves. B, the three fang positions in venomous snakes. 1, fangs are fixed rigidly at rear of elongated maxillae. 2, fangs are fixed rigidly on front rim of maxillae. 3, a maxilla and its fang are hinged and can be rotated forward and backward.

leons have a long, explosively protrusible tongue (and a well-known ability to change color according to the background foliage of their arboreal environment). Some lizards are herbivorous, and still others, particularly the larger forms, live on mice, frogs, or bulkier vertebrates. The largest lizard is the Komodo dragon *Varanus,* up to 10 ft long. Two species are venomous: the Gila monster *Heloderma suspectum* and the bearded lizard *H. horridum.* Some lizards are limbless soil animals (''glass snakes''), and lizards generally are noted for their capacity of autotomizing the tail. Like snakes, many lizards are ovoviviparous or fully viviparous, but most are oviparous.

Snakes have evolved from lizards, probably several times independently. The two halves of the lower jaws of snakes are not articulated but are held together by ligaments, and the upper and lower jaws often are similarly joined by ligaments only. The mouth therefore can be distended greatly and a snake can ingest an animal several times wider than its own diameter (Fig. 30.8). In venomous snakes, poison fangs can be formed by teeth in the rear or the upper jaw (as in certain tree snakes), by immobile front teeth (as in cobras), or by hinged front teeth that can fold back with the upper jaw bone when not in use (as in vipers and rattlers). In some cases poison from the salivary glands flows along an anterior groove on a fang, in others through a duct inside the fang.

Snakes lack external ear openings but they ''hear'' vibrations transmitted from the ground through the skeleton. Eyes are lidless and covered by a transparent (*nictitating*) membrane. Vertebrae can number up to 400. Sternum, limb girdles, and limbs generally are absent (though vestigial remains of pelvic girdles occur in pythons and related types). Snakes usually employ the free posterior edges of their large, transverse belly scales to obtain locomotor leverage on the ground. Locomotion is achieved in four different ways in different groups, by ''snakelike'' lateral undulations of the body, by straight-line rippling undulations, by accordionlike folding and extension of the body, and by sidewinding (as in the horned rattlesnake, in which the body is thrown at almost right angles to the path of progression; Fig. 30.9).

Anacondas and pythons are the largest snakes; they reach lengths of 30 to 35 ft. The largest poisonous snake is the king cobra, with a length of about 18 ft. Snakes are strictly carnivorous, and as a group they occur in a wide variety of habitats, including trees, the ground, caves, and fresh and salt water.

Crocodilia are the closest living relatives of dinosaurs and birds. These reptiles also exhibit three characteristics that, in different form and for different adaptive reasons, occur in mammals: a four-chambered heart with fully separated ventricles; a palate extended far back, separating the nasal passages from the mouth cavity right to the pharynx (Fig. 30.10); and a peritoneal (nonmuscular) septum that separates the chest cavity from the abdominal cavity. These traits are specific adaptations to a secondarily aquatic existence. They improve breathing efficiency, especially when the mouth is submerged or filled with food or both. Other adaptations to aquatic life include closable nostrils; recessed eardrums that can be covered by a flap of skin under water; webbed feet; and a powerful, laterally flattened tail used in swimming and in offense and defense.

Crocodiles and gavials have narrow jaws, and the fourth pair of teeth on the lower jaw remains exposed when the mouth is closed. Alligator and caiman jaws are broader and rounded anteriorly,

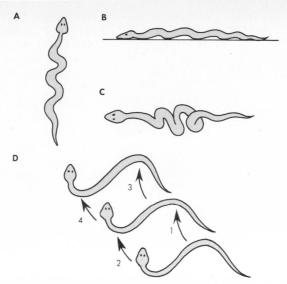

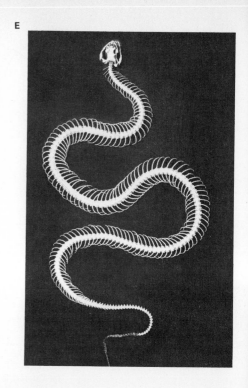

Figure 30.9 Locomotion in snakes. A, lateral undulations. *B,* straight-line rippling. *C,* accordionlike folding. *D,* sidewinding (successive positions are illustrated). *E,* the snake skeleton. Note numerous vertebrae and absence of limbs, limb girdles, and breastbone.

Figure 30.10 Characteristics of Crocodilia: adaptations in head structure to a semiaquatic existence. Note the bones composing the palate.

closable external nares (nostril)
nasal passage
internal nares
premaxilla
maxilla
palatine
pterygoid
recessed eardrums with closable flaps
hard palate

and none of the teeth is exposed after mouth closure. The skin is armored with epidermal horny scales, reinforced along the back by rows of underlying dermal bony plates. Crocodilians are strictly carnivorous and all are oviparous. Gavials live mainly in India, caimans in South America; crocodiles and alligators are distributed more widely in tropical and semitropical regions. The largest members of the order are the American crocodiles and alligators, which attain lengths of about 20 ft, and a marine South Pacific crocodile that can be 25 ft long. The whole order includes only about two dozen species.

Sphenodon punctatus, the tuatara of New Zealand, is the sole surviving species of the order Rhynchocephalia (see Fig. 30.7). It is generally lizardlike, its most distinctive trait being a functional pineal eye, as in lampreys. This reptile is a good example of a living fossil, but in a sense all living reptiles (lizards and snakes perhaps excepted) can be considered as such.

Birds   **Class Aves**   skin with feathers; forelimbs wings, with three fused fingers in hand; hindlimbs legs, each with four or fewer toes; living types with horny beak, teeth absent; heart four-chambered, single aortic arch on right; lungs with extended air pouches; syrinx at base of trachea; 12 pairs of cranial nerves; pelvis fused to sacrum, pubic bones typically not fused ventrally; without external ear lobes, external genitalia, or urinary bladders; urine semisolid; endothermal; fertilization internal; eggs amniote, oviparous.

†*Subclass Archaeornithes* extinct reptilelike birds.

Subclass Neornithes modern birds.

†Superorder Odontognathae extinct; with teeth.

†Superorder Ichthyornithes extinct; without teeth, billed.

Superorder Palaeognathae walking birds, or "ratites"; without teeth; generally flightless, sternum typically without keel.

Order Struthioniformes ostriches; two-toed; pubic bones fused ventrally; feathers without aftershafts; males incubate eggs; omnivorous; largest living birds, 200 lb, 7 ft tall.

Order Casuariiformes emus, cassowaries; wings vestigial; feathers with long aftershafts; three-toed; males incubate eggs; 5 ft tall.

Order Apterygiformes kiwis; long-billed, nostrils at tip; wings vestigial; feathers without aftershafts; four-toed; nocturnal, omnivorous.

†*Order Aepyornithiformes* "rocs," elephant birds; wings rudimentary; extinct for a few centuries only, the largest birds (900 lb; 10 ft tall; eggs 13 in., 20 lb, largest eggs of all animals).

†*Order Diornithiformes* moas; feathers with long aftershafts; four-toed; extinct for a few centuries only; 8 ft tall.

Order Rheiformes rheas; three-toed; feathers without aftershafts; males incubate eggs; 4 ft tall.

Order Tinamiformes tinamous; weak fliers, keel present; good runners; males incubate eggs.

Superorder Neognathae flying birds; without teeth, generally capable of flight, sternum typically with keel.

Order Sphenisciformes penguins; flying in water, not air, forelimbs are water wings; four-toed, feet webbed; feathers scaly; marine.

Order Procellariiformes petrels, fulmars, albatrosses; wings narrow, long; excellent fliers, poor walkers; feet webbed, hind toes absent; nostrils with tubular openings; marine.

Order Gaviiformes loons; diving birds; bill pointed, large; legs short, set far back; toes webbed; mostly marine.

Order Podicipitiformes grebes; diving birds; freshwater; toes lobed; tail vestigial.

Order Pelecaniformes gannets, frigate birds, cormorants, boobies, pelicans; nostrils reduced or absent; with throat pouch; all four toes in foot web; marine and freshwater.

Order Ciconiiformes herons, egrets; storks, ibises, spoonbills, flamingos; wading birds; long-necked, long-legged; feet webbed in some; aquatic or semiaquatic, near marshes.

Order Anseriformes swans, ducks, geese; water fowl; broadbilled, with hard cap at end; feet webbed; tail short.

Order Falconiformes eagles, vultures, buzzards, condors, hawks, falcons, kites, ospreys; bill hooked, sharp-edged; grasping feet with talons; predaceous.

Order Galliformes chickens, turkeys, quail, grouse, ptarmigans, pheasants, partridges, peacocks, hoatzins; game birds; poor fliers; feathers with aftershafts; feet for walking, scratching.

Order Gruiformes cranes, coots, mudhens, rails; long-necked; feathers with aftershafts; marsh birds, some flightless.

Order Charadriiformes gulls, terns, snipes, murres, auks, sandpipers, plovers; wading shore birds, with webbed feet, dense plumage.

Order Columbiformes pigeons, doves, extinct dodoes and passenger pigeons; with large crop secreting "pigeon milk" for young.

Order Cuculiformes   cuckoos, turacos, coucals; two toes in front, two in back, outer back toe reversible; tail long.

Order Psittaciformes   parrots, macaws, parakeets, lovebirds; two toes in front, two in back, outer back toe not reversible; grasping feet; beak hooked, sharp-edged, upper beak hinged on facial bones; plumage brilliantly colored.

Order Strigiformes   owls; head large, eyes directed forward; ear openings often with long feathers; beak short, hooked; grasping feet with sharp claws; predaceous, nocturnal.

Order Caprimulgiformes   nighthawks, whippoorwills; mouth wide, edged with bristly feathers; beak small, legs short; insect-eating, nocturnal.

Order Apodiformes   hummingbirds, swifts; wings pointed, most perfectly aerial birds; some with tubular sucking beak and tongue; includes smallest of all birds.

Order Coliiformes   colies, mousebirds; beaks used in climbing; tail long; all four toes can be directed forward; arboreal, fruit-eating.

Order Trogoniformes   trogons, quetzals; most brilliantly plumed birds; bill short; feet small, first and second toes directed backward; tropical.

Order Coraciiformes   kingfishers, hornbills, motmots, ground rollers, bee-eaters; brilliantly plumed; third and fourth toes fused basally; mostly tropical.

Order Piciformes   woodpeckers, sapsuckers, toucans; beak long, pointed; tongue protrusible; tail feathers pointed; mainly insectivorous.

Order Passeriformes   perching birds; four-toed, three toes in front; over half of all bird species, including all songbirds; crows, larks, jays, magpies, nuthatches, swallows, nightingales, chicadees, titmice, wrens, mockingbirds, robins, bluebirds, thrushes, pipits, shrikes, starlings, sparrows, warblers, orioles, blackbirds, tanagers, finches, grosbeaks.

Because flying requires a structure that cannot deviate too much from fixed aerodynamic specifications, birds are more like each other than the members of most other animal groups. Flying birds are neither overly large nor overly small (and only the flightless birds are very large). Because of the weight limitation birds cannot store much body fat; hence they must eat more or less continuously. Also, a considerable total quantity of food is required to provide the energy, the high (and constant) operating temperature, and the high metabolic level essential for efficient flight. The body temperature of birds is maintained at about 41°C, higher than in mammals.

Feathers are the important heat regulators and also the means of flight. All types of feathers are horny outgrowths from the epidermis (Fig. 30.11). Flight feathers, or *contour* feathers, are located on the wings and usually also on the tail. Each consists of a long *shaft* with primary branches (*barbs*), secondary branches (*barbules*), and tertiary hook-like branches (*barbicels*). Muscles in the arm maintain an overlap between adjacent feathers when the wing moves down; and during the upstroke the feathers are canted like Venetian blinds, letting air pass through with but little drag.

Heat-regulating body feathers and *down* feathers are shorter, with long, flexible, loosely woven barbs. They form a light mat that retains a layer of insulating dead air between the skin and the environment. In many birds an *aftershaft* with barbs is joined to the base of the primary shaft of such feathers. The same smooth muscles in the skin that produce "gooseflesh" in man can erect the down feathers of birds and permit cooling of the body. Birds typically have a dorsal *uropygial gland* at the base of the tail. With their bills the animals spread the oil secreted by this gland over the feathers during preening. The oil keeps the plumage water-repellent and pliable. Like mammals, birds molt their skin cover gradually, but enough feathers usually are left at any given time to maintain flying and thermoregulator capacity.

The skeleton of birds is thoroughly flight-adapted (Fig. 30.12). Bones are light and delicate, with a minimum of spongy bone in the interior and correspondingly larger free spaces. Skull bones are fused and jaw bones are extended as a toothless horn-covered bill or beak. The eyes are rimmed with a circlet of bony *sclerotic plates* that minimize wind pressure on the eyes during flight. As in reptiles, also, the eye can be covered by a

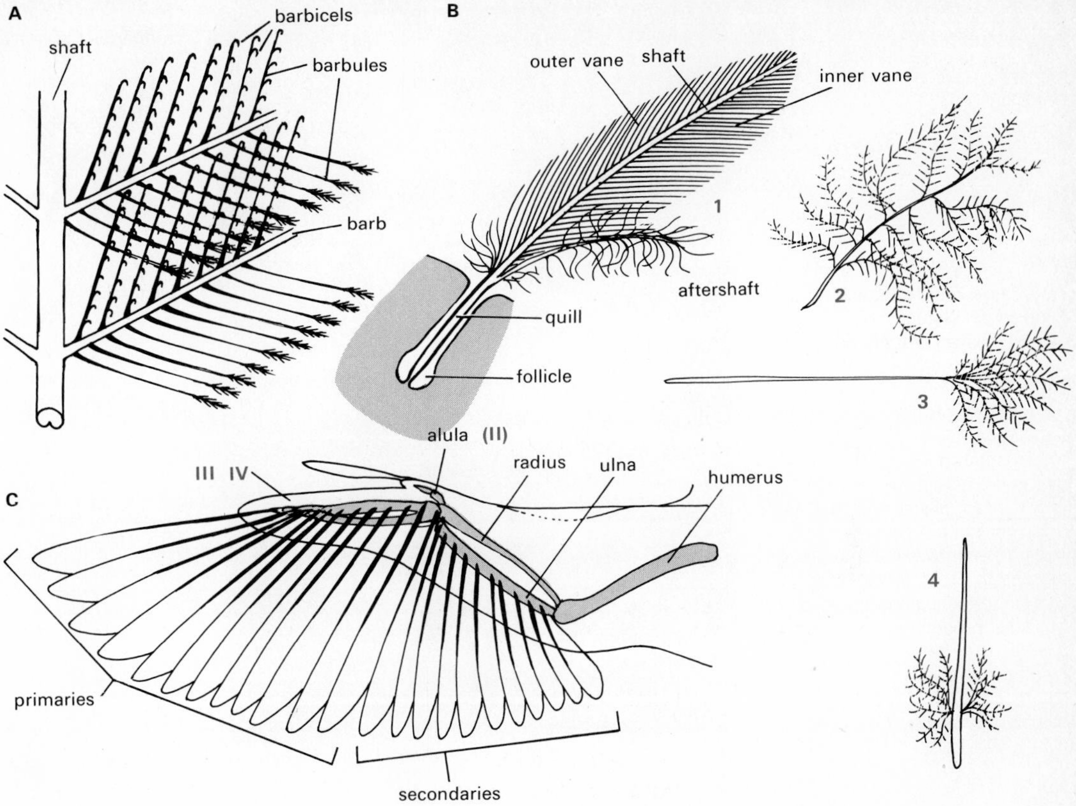

Figure 30.11 Feathers. A, diagram of part of a contour feather showing interlocking of the barbicels of adjacent barbules. *B,* 1, whole view of contour feather, with aftershaft and pattern of insertion in skin. 2, down feather. 3, filoplume. 4, bristle. *C,* arm skeleton with position of flight feathers superimposed. Roman numerals refer to digits of hand. *(A, after Mascha; C, after Pycraft.)*

transparent *nictitating membrane,* drawn over the cornea from the inner corner of the eye. Eardrums are recessed deeply, and the bony canal leading to them usually lacks external lobes or other projections that might disrupt smooth air flow over the head. The neck vertebrae provide extreme head mobility but lock together firmly during flight.

Flying birds have a large sternum with a prominent keel, the attachment surface for the powerful flight muscles. Of these, the *pectoralis major* connects to the underside of the forearm and pulls the wing down. The *pectoralis minor* loops to the upper surface through a canal in the arm socket (see Fig. 30.12). Three digits remain in the hand, a small anterior *alula* and two posterior ones that are elongated and fused. These digits correspond to the index, middle, and ring fingers of man. The alula is lifted up during landing to disrupt the lift-producing air stream over a wing.

The pelvic girdle is fused to the sacral vertebrae and, except in ostriches, the pubic bones are not joined midventrally but leave a wide open passage (for the large eggs). This ''bird-hipped'' condition had already evolved in one group of dinosaurs. Legs typically are four-toed, with one toe directed backward in most cases. In perching birds systems of tendons permit the toes to curl tightly over a tree branch. Body weight alone then ensures maintenance of a tight hold, and the energy expenditure and mental attention required for perching are minimal (see Fig. 30.12). Tail vertebrae are reduced in number and tail feathers function as a rudder.

A remarkable breathing system provides the large quantities of oxygen needed in flying (Fig. 30.13). Lungs actually are fairly small, but they pouch out several large air sacs that occupy much of the space between the internal organs. These sacs not only provide buoyancy but also permit air to flow *past* the breathing surfaces of the lungs proper. As a result, oxygenation of pulmonary blood can occur during both inhalation and exhalation. Some of the sacs extend forward to the neck, where they can play an added role in distending the neck during courtship and other behaviors. The heart is completely four-chambered,

A

B

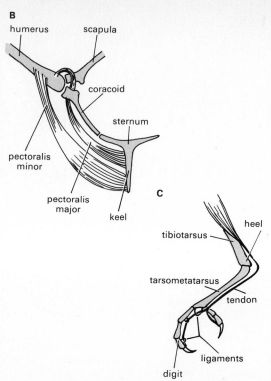

humerus   scapula

coracoid

sternum

pectoralis
minor

pectoralis
major

keel

C

tibiotarsus

heel

tarsometatarsus

tendon

digit

ligaments

Figure 30.12 Flying and perching. A, skeleton of a pigeon. Note large keel for attachment of wing musculature, and ring of sclerotic plates around the eye, for minimizing wind pressure during flight. *B,* arrangement of flight muscles in relation to bones of shoulder girdle and arm. The pectoralis minor pulls the arm up; the pectoralis major pulls it down. *C,* the tendon system in the foot does not stretch. When the leg is straight, as in walking, the tendon permits the toes to move readily. But when the leg and the toes are bent, as in perching on a branch, the tendon becomes taut as a result of the flexed heel and holds the toes tightly curled around the branch. (*Based on Colcott.*)

as in mammals, and the left aortic arch is absent in adult birds (see Fig. 9.22).

The reproductive pattern too is adjusted to the demands of flight. External genitalia are absent (except in the flightless ostriches, in which the males have a penis). This lack, perhaps another adaptation for maintenance of smooth body con-

Figure 30.13 Internal structure of birds. The position of the lungs, air sacs, and coelomic spaces are indicated. Five (paired) air sacs pouch from the lungs. (*After Goodrich.*)

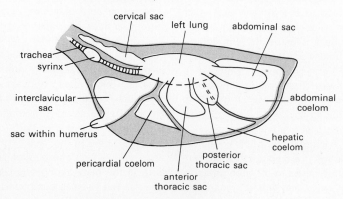

cervical sac

left lung

abdominal sac

trachea
syrinx

interclavicular
sac

sac within humerus

pericardial coelom

anterior
thoracic sac

posterior
thoracic sac

hepatic
coelom

abdominal
coelom

tours, necessitates strong cooperation during mating, for the cloaca of the male must be placed precisely against that of the female. Such cooperation requires behavioral adaptations in the form of elaborate courtship activities, and these in turn are aided by colorful display plumage and color vision. Thus, color vision in birds might be due at least in part to the circumstance that the males lack a penis.

Furthermore, oviparity becomes adaptively most advantageous, for an internally carried offspring would increase the flying weight unduly. But since the young are flightless at first and cannot be made too fat, they must be fed continually and in small doses. The young consequently must be cared for by the parents for extended periods after hatching. Nesting behavior is one adaptive solution; another is the storage crop in the alimentary tract, from which food can be dispensed intermittently. Evidently, the whole structural, functional, and behavioral nature of birds is subordinated to, and explainable in terms of, the basic activity of flying.

The very earliest birds flew and had feathers. Flightless birds living today are descendants of

A   B   C

Figure 30.14   *Birds. A,* emu, one of the flightless ratite types (order Casuariiformes). *B,* Australian lyre bird (order Passeriformes). *C,* shoebill stork (order Ciconiiformes).

flying ancestors, probably evolved by neoteny; the flightless condition of the young is retained to the adult stage. Ratite birds therefore are not primitive (and they do not exemplify an early flightless stage of bird evolution, as had been supposed formerly). Ratites tend to be large (Fig. 30.14). Their wings generally are vestigial, a keel is absent, but their legs are long and strong and the animals are good runners. The eggs of ratites are incubated typically by the males.

Flying birds are among the best known and most familiar of all animals, and the domesticated groups among them are of considerable economic importance. More than five thousand species, over half of the whole class, belong to a single order, the passerine, or perching, birds. This group includes the songbirds, which usually build elaborate nests and in which song and brilliant mating plumage are largely characteristic of the males only.

Mammals   **Class Mammalia**   skin with hair; teeth in sockets; seven neck vertebrae; ears with three middle-ear bones; limbs typically with five digits; heart four-chambered, single aortic arch on left; red corpuscles nonnucleated; coelom divided by muscular diaphragm; larynx at upper end of trachea; pelvis fused to sacrum, pubic bones fused ventrally; with urinary bladder, urine liquid; 12 pairs of cranial nerves, brain elaborate; endothermal; with external genitalia; fertilization internal, eggs amniote; some oviparous, mostly viviparous; young nourished by milk from mammary glands.

Subclass Prototheria   oviparous mammals; adults with horny bill; ears without external lobes; testes in abdomen; nipples absent, numerous ducts of mammary glands open individually; with cloaca; without uterus or vagina; laid eggs with pliable shells.

Order Monotremata   cervical vertebrae with ribs; female urogenital canal leads to cloaca; male urogenital canal forks to separate urinary canal to cloaca, genital duct through penis. *Ornithorhynchus,* duck-billed platypus: aquatic, in burrows, with webbed feet, clawed toes,

651

flattened tail, eggs in nest; *Tachyglossus*, spiny anteaters (*echidnas*): terrestrial, noctural, with tubular beak, protrusible tongue, coarse hair and spines, clawed toes (Figs. 30.15 and 30.16).

Subclass Theria   viviparous mammals; adults with teeth; ears with external lobes; testes in abdomen or scrotum; with nipples; with or without cloaca; with uterus and vagina; young born.

Infraclass Metatheria   pouched mammals; urogenital canal leads to shallow cloaca in both sexes; penis, protruded through cloaca, carries both urine and sperms; young born in immature condition.

Order Marsupialia   opossums, kangaroos, wombats, koala bears; nipples in ventral abdominal pouch (*marsupium*); immature young complete development in pouch, attached to nipples; uterus and vagina double; penis often forked (Fig. 30.17 and see Fig. 30.15).

Infraclass Eutheria   placental mammals; cloaca absent; male urogenital canal leads through penis to single orifice; female urethra and vagina with separate orifices (see Figs. 10.16 and 30.15); uterus double or single, vagina always single; young develop in uterus attached and nourished by chorionic placenta, born in mature condition.

Order Insectivora   shrews, moles, hedgehogs; snout long; clawed toes; arboreal, terrestrial, subterranean.

Order Dermoptera   colugos, "flying lemurs"; aerial gliding; skin web between limbs and tail form parachute.

Order Chiroptera   bats, "flying foxes"; forelimbs and four digits elongated, support flying web with hindlimbs (and with tail in some); first digit of forelimbs and all of hindlimbs clawed; true flight; nocturnal.

Order Edentata   sloths, anteaters, armadillos; teeth absent or only molars, without enamel.

Order Pholidota   scaly anteaters, pangolins; teeth absent; tongue elongate; with overlapping epidermal horny plates.

Order Primates   primates (see Table 5, Chap. 7, and Chap. 15).

Suborder Lemuroidea   lemurs.

Suborder Tarsioidea   tarsiers.

Suborder Anthropoidea   anthropoids.

Superfamily Ceboidea   ceboid monkeys.

Superfamily Cercopithecoidea   cercopithecoid monkeys.

Superfamily Hominoidea   hominoids.

Family Pongidae   apes.

Family Hominidae   men.

Order Lagomorpha   rabbits, hares, cottontails, conies; incisors without roots, growing continuously, two pairs in upper jaw, one pair in lower, enamel on front and back surfaces; canines absent; jaw motion lateral only; elbow nonrotatable.

Order Rodentia   rodents, mice, rats, beavers, guinea pigs, porcupines, chipmunks, squirrels, gophers; incisors one pair per jaw, without roots, growing continuously, enamel on front surface only; canines absent; jaw motion back and forth as well as lateral; elbow rotatable; toes clawed; most numerous mammalian order.

Order Cetacea   whales, dolphins, porpoises; neck absent; forelimbs flippers with internal digits; hindlimbs absent; transverse tail flukes, with median notch; teeth without enamel or absent; ears small, nostrils dorsal; hairs sparse, around mouth.

Order Carnivora   carnivores; canines long; teeth pointed (carnassial); clawed toes.

Suborder Fissipedia   toe-footed carnivores; dogs, wolves, jackals, foxes, raccoons, pandas, bears, wolverines, minks, weasels, ferrets, otters, skunks, badgers, mongoose, hyenas, civets, lynxes, all "cats," including house cats, lions, tigers, jaguars, ocelots, leopards, pumas, cheetahs.

Suborder Pinnipedia fin-footed carnivores; walruses, seals, sea lions.

Order Tubulidentata aardvarks; ears long; snout long, tubular, with protrusible tongue; teeth without enamel; strong claws; insect-eating.

Order Hyracoidea coneys; forelimbs four-toed, hindlimbs three-toed; ears and tail short.

Order Proboscoidea elephants; skin thick, neck short, hairs sparse; nose and upper lip form proboscis; incisors are tusks; molars with transverse enamel ridges, one set of molars present at any time; feet with elastic bottom pads, toes with nail-like small hooves.

Order Sirenia sea cows, manatees, dugongs; forelimbs are paddles, hindlimbs absent; transverse tail flukes without notch; snout blunt, lips fleshy; external ears absent; teeth with enamel; hairs sparse; coastal, estuarine, herbivorous.

Order Perissodactyla odd-toed ungulates; horses, asses, zebras, tapirs, rhinoceroses; one- or three-toed, toes terminating in hoofs.

Order Artiodactyla even-toed ungulates; two- or four-toed, toes terminating in hoofs.

Suborder Bunodonta nonruminant, stomach simple; pigs, boars, warthogs, peccaries, hippopotamuses.

Suborder Pecora ruminant (cud-chewing), stomach four-chambered; bactrian camels (two-humped), dromedaries (one-humped), alpacas, llamas, chevrotains, reindeer, caribous, moose, elk, giraffes, okapis, pronghorns, cattle, yaks, gazelles, elands, impala, buffalo, musk oxen, goats, sheep.

Fossil evidence and the structure of living mammals indicate that the mammalian orders form at least six parallel evolutionary series (Fig. 30.18). The egg-laying monotremes represent one. The pouched marsupials represent another. A third includes the first six placental orders listed above, from insectivores to primates. A fourth series is formed by the rodents and lagomorphs. A fifth, by the cetaceans. And a sixth, by all the remaining orders, from carnivores to hoofed (*ungulate*) mammals. In each of these series independently, the same five general adaptive trends have become elaborated in the course of evolution. All five relate more or less directly to feeding and locomotion.

First, body size has tended to increase, an advantageous change helpful both in searching for food and in avoiding becoming food. Second, the number of teeth has tended to decrease. And, instead of all teeth remaining alike, as in the reptilian ancestors, the fewer teeth have become specialized differently in parallel with a specialization in the types of food eaten. Third, the legs have tended to become longer and stronger, the body and often also the heels being lifted off the ground more and more. The result has been an evident improvement in locomotor efficiency and, indeed, a significant diversification in the types of locomotion. Fourth, in conjunction with newly developed modes of locomotion, a radiation to diverse habitats and ecological niches has occurred. And last, partly as a result of the increase in body size, partly in parallel with the improved motility, the size of the brain has increased. In the course of evolutionary time greater mental capacity has been acquired by mammals of all kinds.

That these adaptive trends have become elaborated independently in the different evolutionary series is exemplified best among the placentals. Thus, as already shown in Chap. 15, the evolutionary progression from insectivore to primate has indeed been accompanied by size increase; by spectacular brain increase; by a locomotor change from four-footed to two-footed methods and to flying; and by ecological diversification from original arboreal to later terrestrial, aerial, and subterranean environments (Fig. 30.19). And with regard to teeth, an insectivore such as a mole has 44, whereas a man has 32, well differentiated as incisors, canines, premolars, and molars (Fig. 30.20).

In the rabbit-rodent series, adaptive change in brain size has not been exceptional, although on a rodent level a rat is an exceedingly brainy animal. The other adaptive trends are more clearly evident. Fossils show that these groups have indeed become larger; rabbits exhibit a very obvious locomotor specialization in their long hind legs adapted to leaping; and dental specialization involves a well-known modification of the incisors (''rabbit teeth''). All these changes are associated with a vast ecological diversification. Rodents comprise

egg-laying mammals

marsupial mammals

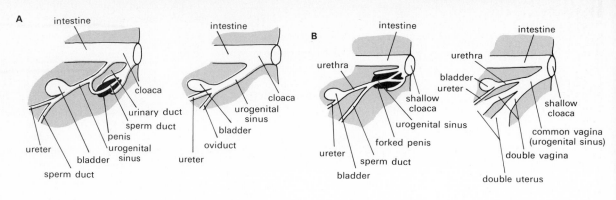

placental mammals

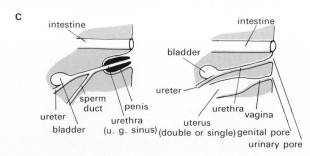

Figure 30.15 The urogenital ducts of mammals. Ducts in males are at left of each pair; ducts in females, at right. During evolution the ducts of the alimentary, excretory, and reproductive systems all started in more or less joined condition (as in *A*) and then progressively acquired separate openings, a condition realized fully only in placental females (as in *C*, above).

Figure 30.16 Egg-laying, oviparous mammals. (monotremes, Prototheria). *A, Ornithorhynchus,* the duck-billed platypus. *B,* nest and eggs of a platypus. *C, Tachyglossus,* the echidna or spiny anteater.

nearly half the number of mammalian species, and each species represents a distinct ecological niche (see Chap. 17).

In the whale series similar trends are in evidence. No group has increased in body size more dramatically or has altered habitat and locomotion more drastically. Tooth reduction has become extreme—in the whalebone whales vertical baleen plates replace teeth altogether. Also, the intelligence of the whale group is well known; porpoises in particular are noted for their mental prowess.

Finally, in the carnivore-ungulate series most groups have exploited the possibilities of a running life on the plains. Leg and foot structure thus show major adaptive modifications. In several subseries within the series, the heel has lifted off the ground, the foot has become elongated, and the animal has come to walk on its toes, as in cats or gazelles, for example. Concurrently, toes have become increasingly hoofed, and the number of toes has

A

B

C

A

B

C

D

Figure 30.17 (above) Pouched, marsupial mammals. (Metatheria) *A*, opossum, the only marsupial surviving in North America. *B*, koala bear native to Australia. *C*, cresttailed marsupial mouse. *D*, Tasmanian devil.

Figure 30.18 The orders of mammals. arranged according to their main evolutionary groupings. Read from bottom up. A representative type is indicated for each order. The eutherian series represent the placental mammals.

Figure 30.19 (below) Mammals of the insectivore-primate group. *A*, a hedgehog (order Insectivora). *B*, *Plecotus* (order Chiroptera), a type with highly elaborated ears for phonoreception. *C*, an Indian pangolin, or scaly anteater (order Pholidota). Other insectivores and primates are illustrated in Chap. 15.

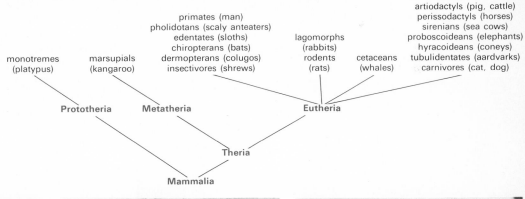

```
                                                          artiodactyls (pig, cattle)
                    primates (man)                        perissodactyls (horses)
               pholidotans (scaly anteaters)                  sirenians (sea cows)
                   edentates (sloths)      lagomorphs     proboscoideans (elephants)
                 chiropterans (bats)       (rabbits)       hyracoideans (coneys)
   monotremes    marsupials   dermopterans (colugos)  rodents            tubulidentates (aardvarks)
   (platypus)    (kangaroo)   insectivores (shrews)   (rats)   cetaceans  carnivores (cat, dog)
                                                               (whales)

         Prototheria     Metatheria                    Eutheria

                              Theria

                         Mammalia
```

A

B

C

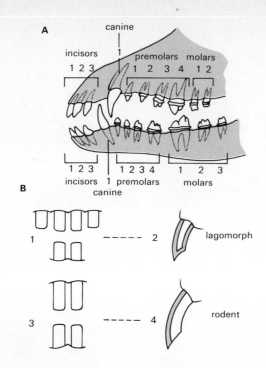

Figure 30.20 Mammalian teeth. A, the dentition of a dog in one half of the head. These teeth are carnassial (pointed), as in carnivores generally. B, the incisors of lagomorphs number four and two in upper and lower jaw respectively (1) and enamel covers both front and back surfaces (2). In rodents, two incisors are present in each jaw (3) and enamel covers only the front surfaces (4).

become reduced. For example, carnivores are five-toed and still nonungulate; coneys have fewer toes; elephants have small nail-like hoofs on each toe; and a fully ungulate condition with variously reduced toe numbers is exhibited by the horse group and the deer-cattle group (Fig. 30.21).

Major changes have occurred also in tooth specializations, in line with the different eating habits and the diversified ecological niches. In carnivores all teeth are *carnassial,* or pointed and

adapted for tearing, and the canines form particularly long fangs. By contrast, the herbivorous ungulates have incisors adapted for cutting, molars adapted for grinding, and canines are reduced or often absent. Body size has increased independently in the several subseries of the group; bears, horses, elephants, and cattle all have evolved from far smaller ancestors. And intelligence likewise has reached higher levels several times—bears, cats, dogs, wolverines, and elephants all are well known as "smart" animals.

Though similar adaptive trends thus characterize all these series, each series has had its own area of special success. The insectivore-primate group has come to excel in aerial and bipedal locomotion and in mental capacities; the rodent group, in ecological diversification; the whale group, in aquatic locomotion and body size; and the carnivore-ungulate group, in ground locomotion and speed. Also, a good deal of convergent evolution has taken place. For example, aquatic types with flippers and finlike tails occur in the cetacean series but also twice in the carnivore-ungulate series (pinnipeds, manatees, Fig. 30.22); ant-eating types with long snouts and protrusible tongues have evolved in the insectivore-primate series (pholidotans, edentates), in the carnivore-ungulate series (aardvarks), and indeed also in the primitive monotreme group (echidnas).

The marsupial mammals duplicate almost all the adaptive trends of the placental mammals in a remarkably parallel manner. Thus, marsupials include arboreal types that are variously insectivorous, slothlike, or aerial gliders ("flying" opos-

Figure 30.21 Feet in the carnivore-ungulate group. A, cat; carnivores have claws but not hoofs. B, elephants, small naillike hoofs present. C, three-toed rhinoceros, and D, one-toed horse; perissodactylans are fully ungulate. E, four-toed pig, and F, two-toed cow; artiodactylans again fully ungulate. Elephants, like men, walk on the whole foot (plantigrade); all other mammals in this illustration walk on the toes (digitigrade).

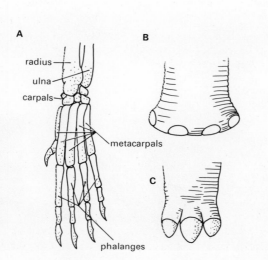

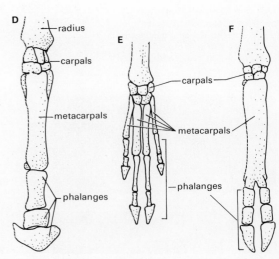

A

B

Figure 30.22 Mammals of the carnivore-ungulate group. A, sea lion, a pinniped carnivore. B, manatees, sirenian mammals. See Chap. 18 for illustrations of other mammalian types.

sums); molelike burrowing types; rodentlike pouched mice and rats; leaping harelike kangaroos; ant-eating types with tubular snouts and long tongues; and carnivorous types variously wolflike, doglike, catlike, and, though now extinct, also lionlike.

Such parallel results offer perhaps the best proof that the same adaptive forces operate universally for all animals, and that, if the environmental opportunities are similar, remarkably similar animals can result several times independently. Furthermore, the primary adaptive forces that have shaped mammalian nature ultimately reduce to just two, feeding and locomotion—the very same that have also shaped the nature of all other animals since the first primeval cell types appeared in the ocean.

Mammals, vertebrates, and chordates as a whole also illustrate exceedingly well that significant evolutionary advance proceeds from the primitive type, not the specialized one. Chordates probably trace back to *primitive* pterobranchlike hemichordates, not to advanced forms such as acorn worms; vertebrates probably originated from *primitive* sessile ascidians, not from the specialized pelagic tunicates; amphibia evolved from early *primitive* lung-breathing fishes, not from the later bony teleosts; reptiles have their ancestry among *primitive* labyrinthodonts, not among later frogs or toads; mammals derive from the *primitive* thecodont reptiles, not from any of the subsequent forms; primates evolved from *primitive* insectivorous stocks, not from modern mammals; and man arose from *primitive* hominoids, not from the later monkeys or apes.

In short, an adaptive radiation usually has evolutionary sources for new types along the primitive branch lines, not along the lines that specialize more rapidly and more spectacularly.

Review Questions

1. Give taxonomic definitions of amphibia and of each amphibian class. Name representative genera of each class. Which amphibian traits trace their origin directly to ancestral fishes? Which group of fishes was ancestral to amphibia?

2. Review the breathing and circulatory changes that take place during amphibian metamorphosis. Review the structural organization of every amphibian organ system.

3. In what respects are amphibia water-adapted and land-adapted? Characterize the nature of cecilians. Name urodeles that have adult gills only, adult gills and lungs, and adult lungs only.

4. Name urodeles that are predominantly aquatic or predominantly terrestrial. In what respects are urodeles neotenous? Describe the skeletal structure of an anuran and show how it differs from that of a urodele. Distinguish between frogs and toads.

5. Define the class Reptilia and each of its living orders. Review the characteristics of the extinct groups. In what ways are reptiles adapted to terrestrial life? Which skull types are represented among reptiles today?

6. Characterize the skeletal structure of Chelonia. In what environments do these animals live and how do they mate? How do lizards differ from snakes? Describe the structural and functional adaptations of snakes to their particular mode of life.

7. How do snakes move? Which reptilian groups have viviparous members? Describe the anatomic traits of crocodiles. Which of these traits are similar to those of mammals? What traits distinguish crocodiles and alligators?

8. How can reptiles be distinguished from amphibia? How many species of each tetrapod class are known?

9. Define Aves taxonomically. Name as many different orders of birds as you can and list representative types of each.

10. Describe the structure of feathers and discuss the role of feathers in heat regulation and flying. Contrast bird and insect flight. What skeletal adaptations to flight occur among birds?

11. Describe the organization of all organ systems of birds and show particularly which traits are adaptations to flying. Which behavioral, reproductive, and ecological traits are consequences of or adaptations to a flying life?

12. What are ratite birds? How do birds manage to sleep on a tree branch without falling off? How do birds produce song and what is the function of song?

13. Define the class of mammals and each subclass and order. Name representative types of each order. Review the different ways in which mammalian embryos and young depend on their maternal parents.

14. Show which mammalian orders form supergroups and describe the similar evolutionary trends in all such supergroups. What are the adaptive advantages of these trends? Indicate variations in dentition, foot structure, locomotion, and food sources among mammals.

15. Describe the structure and function of every organ system of a mammal, drawing on all information accumulated during the study of the whole book. Review the detailed taxonomic classification of man.

Collateral Readings

Colbert, E. H.: "Evolution of the Vertebrates," Wiley, New York, 1955. For data on fossil forms.

Ditmars, R. I.: "Reptiles of the World," Macmillan, New York, 1926.

————: "The Reptiles of North America," Doubleday, Garden City, N.Y., 1936. Key classics on reptiles.

Frieden, L.: The Chemistry of Amphibian Metamorphosis, *Sci. American,* Nov., 1963. The article describes the controlling role of hormones in the metamorphic process.

Gans, C.: How Snakes Move, *Sci. American,* June, 1970. The four modes of locomotion are examined.

Greenewalt, C. H.: How Birds Sing, *Sci. American,* Nov., 1969. An instructive analysis.

Griffin, D. R.: The Navigation of Bats, *Sci. American,* Aug., 1950. A description of the "sonar" mechanism by which bats fly and steer even in the dark.

Klauber, L. M.: "Rattlesnakes: Their Habits, Life History, and Influence on Mankind," University of California Press, Berkeley, 1956. Fully described by the title.

Kortlandt, A.: Chimpanzees in the Wild, *Sci. American*, May, 1962.

Noble, G. K.: "The Biology of the Amphibia," McGraw-Hill, New York, 1931. A key classic.

Pople, C. H.: "Snakes Alive and How They Live," Viking, New York, 1939. The best-known and most authoritative book on snakes.

Romer, A. S.: "Vertebrate Paleontology," 2d ed., University of Chicago Press, Chicago, 1945. For data on fossil types and vertebrate evolution.

———: "The Vertebrate Body," Saunders, Philadelphia, 1950. For anatomic background.

Tucker, V. A.: The Energetics of Bird Flight, *Sci. American*, May, 1969. Wind-tunnel tests show how birds adjust their metabolism during flying.

Wallace, G. J.: "An Introduction to Ornithology," Macmillan, New York, 1955. One of a vast number of books on birds.

Warden, C. J.: Animal Intelligence, *Sci. American*, June, 1951. Psychological tests measure the comparative intelligence of various vertebrates.

Young, J. Z.: "The Life of Vertebrates," 2d ed., Oxford University Press, London, 1962. Highly recommended for background on vertebrate zoology and evolution.

GLOSSARY

A

COMMON PREFIXES, SUFFIXES, AND ANATOMIC TERMS IN ZOOLOGICAL USAGE

NOTE: The meaning of many terms not specifically listed in Section B, below, can be ascertained from Section A. For example, certain fishes are known as the *Actinopterygii*. The parts of this term are the prefix *actino-* and the suffix *-pterygii;* the list below indicates the meanings of these word parts as "ray" and "fin," respectively. Hence the whole term denotes "ray-finned." In general, a very large number of technical designations may be translated into English equivalents by separating the words into parts and consulting the first section of the glossary.

a- [Gr. not]: negates succeeding part of word; for example, *acoel,* without coelom.

ab- [L. away, off]: opposite of *ad;* for example, *aboral,* away from mouth.

acro- (ăk′rō) [Gr. *akros,* outermost]: for example, *acrosome,* body at tip of animal sperm.

actino- (ăk′tĭ·nō) [Gr. *aktis,* ray]: for example, *actinopodial,* ray-footed.

ad- [L. toward, to]: opposite of *ab-;* for example, *adrenal,* at (near) the kidney.

afferent (ăf′ēr·ĕnt) [L. *ad + ferre,* to carry]: to lead or carry toward given position, opposite of efferent; for example, afferent nerve, afferent blood vessel.

amphi- (ăm′fĭ) [Gr. on both sides]: for example, *amphioxus,* pointed at both ends.

an- [Gr. not]: like *a-,* used before vowel or "h"; for example, *anhydride,* compound without hydrogen.

ana- [Gr. up, throughout, again, back]: for example, *analogy,* likeness, resemblance; *analysis,* thorough separation; *anatomy,* cutting apart.

andro- (ăn′drō) [Gr. *aner,* man, male]: for example, *androgen,* male-producing hormone.

anterior, antero- (ăntē′ĭ·ēr, ăn′tĕr·ō) [L. *ante,* before, in front of]: at, near, or toward front end.

antho- (ăn′thō) [Gr. *anthos,* flower]: for example, *Anthozoa,* flowerlike (cnidarian) animals.

anthropo- (ăn′thrō·pō) [Gr. *anthrōpos,* man, human]: for example, *anthropoid,* manlike ape.

apical (ăp′ĭ·kăl) [L. *apex,* tip]: belonging to an apex, being at or near the tip.

apo- (ăp′ō) [Gr. away, off, from]: comp. to L. *ab-;* for example, *apopyle,* far-lying opening.

-apsid (ăp′sĭd) [Gr. *apsis*, loop, arch]: for example, *diapsid*, having two arches or windows in skull.

arch-, archeo- (ärch, är′kē·ō) [Gr. *archos*, chief]: first, main, earliest; for example, *archenteron*, first embryonic gut.

arthro- (är′thrō) [Gr. *arthron*, joint]: for example, *arthropod*, jointed-legged; *arthritis*, joint inflammation.

-aspid, aspido- (ăs′pĭd, ăs′pĭ·dō) [Gr. *aspis*, shield]: for example, *cephalaspid*, head-armored; *anaspid*, without shield.

asc-, asco- (ăs′kō-) [Gr. *askos*, sac]: for example, *aschelminth*, sac worm.

aster-, -aster (ăs′tēr) [Gr. star]: for example, *asteroid*, star-shaped.

auto- (ô′tō) [Gr. same, self]: for example, *autogamy*, self-fertilization; *autotroph*, self-feeding organism.

axo- (ăk′sō) [fr. Gr. *axine*, axis]: pertaining to an axis; for example, *axoneme*, axial filament.

bi- (bī) [L. *bis*, twice, double, two]: for example, *bilateral*; *bicuspid*, having two points.

bio- (bī′ō) [Gr. *bios*, life]: pertaining to life; for example, *biology*; *amphibia*, living in water and on land.

-blast, blast-, blasto- (blăst, blăst′ō) [Gr. *blastos*, embryo]: pertaining to embryo; for example, *blastopore*, embryonic opening.

brachio- (brā′kĭ·ō) [L. *brachium*, arm]: for example, *brachiopod*, arm-footed animal.

-branch, branchio- (brăng′kĭ·ō) [Gr. *branchia*, gills]: for example, *branchial* sac, breathing sac.

cardio- (kär′dĭ·ō) [Gr. *kardia*, heart]: for example, *pericardial*, around the heart.

caudal (kô′dăl) [L. *caudo*, tail]: at, near, or toward the tail.

cephalo- (sĕf′ả·lō) [Gr. *kephalē*, head]: for example, *cephalopod*, head-footed animal.

cerci, cerco- (sûr′sĭ, sûr·kō) [Gr. *kērkos*, tail]: pertaining to tail: for example, *anal cerci*, tail-like appendages, near anus.

cervical (sûr′vĭ·kǎl) [L. *cervix*, neck]: at, near, or toward the neck region.

chaeto- (kē′tō) [Gr. *chaitē*, bristle, hair]: for example, *chaetognath*, bristle-jawed animal.

chloro- (klō′rō) [Gr. *chloros*, green]: for example, *chlorophyte*, green alga.

choano- (kō′ă·nō) [Gr. *choanē*, funnel]: for example, *choanocyte*, funnel-bearing cell.

chondro- (kŏn′drō) [Gr. *chondros*, cartilage]: for example, *Chondrichthyes*, cartilage fishes.

-chord, chorda- (kôrd, kôr′dă) [L. *chorda*, cord, string]: for example, *notochord*, cord along back.

-chrome, chromo-, chroma- (krōm, krō′mō) [Gr. *chroma*, color]: for example, *cytochrome*, cell pigment; *chromosome*, stainable body (in cell nucleus).

-clad, clado- (klăd, klă′dō) [Gr. *klădos*, branch, sprout]: for example, *triclad*, three-branched.

cocco-, cocci- (kŏkō, kŏk′sī- [Gr. *kokkos*, grain]: for example, *coccine*, grainlike; *coccus*, grainlike (spherical) bacterium.

-coel, coela-, coelo- (sēl) [Gr. *koilos*, hollow, cavity]: for example, *pseudocoel*, false coelomic cavity.

cten-, cteno-, (tĕn) [Gr. *kteis*, comb]: for example, *Ctenophora*, comb-bearing animals.

cyano- (sī·ả′nō-) [Gr. *kyanos*, dark-blue]: for example, *cyanophyte*, blue-green alga.

-cyst (sĭst) [Gr. *kystis*, bladder, pouch, sac]: for example, *sporocyst*, spore-containing cyst.

-cyte, cyto- (sīt) [Gr. *kytos*, vessel, container]: pertaining to cell; for example, *cytoplasm*, cell substance; *fibrocyte*, fiber-forming cell.

-dactyl, dactylo- (dăk′tĭl) [Gr. *daktylos*, finger, toe]: for example, *dactylozooid*, fingerlike individual.

de- (dē) [L. away, from, off]: like Gr. *apo-*; for example, *dehydration*, removal of water.

-dent, denti- (dĕnt) [L. *dens*, tooth]: like Gr. *-dont*; for example, *denticle*, little tooth.

dermis, -derm (dûr′mĭs) [Gr. *derma*, skin]: for

example, *ectoderm,* outer skin tissue; *epidermis,* exterior layer.

di- (dī) [Gr. twice, double, two]: like L. *bi-,* for example, *disect,* to cut in two (distinct from *dis-;* see below).

dia- (dī′*å*) [Gr. through, across, thorough]: for example, *diaphragm,* across the midriff.

dino- (dī′nŏ-) [Gr. *dinos,* whirling]: for example, *dinoflagellate; dinosaur,* ''whirling lizard.''

diphy- (dī′fĭ) [Gr. *diphyses,* double]: for example, *diphycercal,* having tail with equal shape above and below vertebral support.

diplo- (dī′plŏ) [Gr. *diploos,* twofold]: for example, *diploid,* with two chromosome sets.

dis- [L. apart, away]: for example, *dissect,* cut apart (distinct from *di-,* see above).

distal (dĭs′tăl): situated away from or far from point of reference (usually the main part of body); opposite of *proximal.*

-dont (dŏnt) [Gr. *odontos,* tooth]: for example, *thecodont,* having encased (socketed) teeth.

dorsal (dôr′săl) [L. *dorsum,* back]: at, near, or toward the back; opposite of *ventral.*

dys- (dĭs) [Gr. hard, bad]: for example, *dysfunction,* malfunction; *dysentery,* ''bad intestine.''

echino- (ĕ·kĭ′nŏ) [Gr. *echinos,* spiny, bristly]: for example, *echinoderm,* spiny-skinned.

eco- (ēkŏ) [Gr. *oikos,* house, home]: for example, *ecology,* study of relationships between organisms and their home territory.

ecto- (ĕk′tŏ) [Gr. *ektos,* outside]: for example, *ectoproct,* outside-anus (that is, exterior to tentacle ring).

-ectomy (ĕk′tŏmĭ) [Gr. *ek,* out of, + *tomein,* to cut]: excision; for example, *thyroidectomy,* excision of thyroid gland.

efferent (ĕf′ĕr·ĕnt) [L. *ex,* out, away + *ferre,* to carry]: to lead or carry away from given position; opposite of *afferent;* for example, efferent nerve, efferent blood vessel.

endo- (ĕn′dŏ) [Gr. *endon,* within]: for example, *endoderm,* inner tissue layer.

entero-, -enteron (ĕn′tērŏ) [Gr. *enteron,* intestine]: for example, *enterocoel,* coelom formed from intestine; *archenteron,* first intestine.

ento- (ĕn′tŏ) [Gr. var. of *endo-,* within]: for example, *entoproct,* inside-anus (that is, inside tentacle ring).

epi- (ĕp′ĭ) [Gr. to, on, over, against]: for example, *epidermis,* outer skin.

erythro- (ĕ·rĭth′rŏ) [Gr. *erythros,* red]: for example, *erythrocyte,* red (blood) cell.

eu- (ū) [Gr. good, well, proper]: for example, *Eumetazoa,* metazoa proper.

ex-, exo-, extero- (ĕks, ĕk′sŏ, ĕks′tēr·ŏ) [L. out, from, exterior]: for example, *exopterygote,* exterior-winged.

-fer, -fera (fĕr, fĕ′rå) [L. *ferre,* to carry]: like Gr. *-phore;* for example, *foraminifer,* hole-carrier; *rotifer,* wheel-carrier; *Porifera,* pore-carrying animals (sponges).

-form, -formes (fôrm, fôr′mēz) [L. *-formis,* having the form of]: for example, *perciform,* perchlike.

frontal [L. *front, frons,* forehead]: in a horizontal plane separating dorsal from ventral half.

gamo-, -gamy (gă′mŏ) [Gr. *gamein,* to marry]: pertaining to gametes or fertilization; for example, *autogamy,* self-fertilization.

gastro- (găs′trŏ) [Gr. *gaster,* stomach]: for example, *gastrozooid,* feeding individual.

-gen, -genic, geno- [Gr. *genēs,* born, created]: for example, *hydrogen,* water-producing; *genotype,* genetic constitution.

-gest, gest- (jĕst) [L. *gestare,* to carry]: for example, *ingest,* to carry in (food).

glosso-, -glossus (glŏs′ŏ) [Gr. *glossa,* tongue]: for example, *glossopharyngeal,* pertaining to tongue and pharynx; *saccoglossus,* sac-tongue-possessing worm (hemichordate).

gluco-, glyco- (gloo′kŏ-, glī′kŏ-) [Gr. *gleukos,* sweet]: pertaining to sugars; for example, *glucogenic,* sugar-producing; *glycolysis,* breakdown of sugar.

gnatho-, -gnath (nă'thō) [Gr. *gnathos*, jaw]: for example, *agnath*, jawless.

gon-, gono-, -gonium (gŏn'ō) [Gr. *gonos*, seed, generation]: pertaining to reproduction; for example, *gonopore*, reproductive opening.

gyn-, -gyne, gyno- (jīn'nō) [Gr. *gynē*, woman, female]: opposite of *andro-*; for example, *gynogenic*, female-producing.

haem-: see *hem-*

haplo- (hăp'lō) [Gr. *haploos*, single]: for example, *haploid*, with one chromosome set.

-helminth (hĕl'mĭnth) [Gr. *helminthos*, worm]: for example, *platyhelminth*, flatworm.

hem-, hemo-, hemato- (hĕm, hē'mō, hĕm'à tō) [Gr. *haima*, blood]: var. of *haem-*; for example, *hemoglobin*, red blood pigment.

hemi- (hĕm'ĭ) [Gr. half]: like *semi-*; for example, *hemichordate*, similar to chordate.

hepato- (hē·păt'ō-) [Gr. *hēpar*, liver]: pertaining to the liver; for example, *hepatopancreas*, liverlike digestive gland.

hetero- (hĕt'ēr·ō) [Gr. *heteros*, other, different]: opposite of *homo-*; for example, *heterotrophic*, feeding on other living things.

hex-, hexa- (hĕks, hĕk'sà) [Gr. six]: for example, *hexapod*, six-legged; *hexose*, six-carbon sugar.

holo- (hŏl'ō) [Gr. *holos*, whole, entire]: for example, *holotrophic*, eating whole (bulk) food.

homo-, homeo-, homoio- (hō'mē, ·ō, hō·moi'ō) [Gr. similar]: for example, *homeostatic*, remaining similar in state; *homoiothermic*, having constant temperature.

hydro- (hī'drō) [Gr. *hydōr*, water]: for example, *hydrolysis*, dissolution by water.

hyper- (hī'pēr) [Gr. above, over]: opposite of *hypo-*; for example, *hypertrophy*, overgrowth.

hypo- (hī'pō) [Gr. under, less]: opposite of *hyper-*; for example, *hypotonic*, less concentrated (than reference system).

ichthyo- (ĭk'thĭ·ō) [Gr. *ichthyos*, fish]: for example, *Osteichthyes*, bony fishes; *ichthyosaur*, fishlike reptile.

inter- (ĭn'tēr) [L. between, among]: for example, *intercellular*, between cells.

intra- (ĭn'trà) [L. within]: for example, *intracellular*, in cells.

iso- (īsō) [Gr. *isos*, equal]: like *homo-*; for example, *isolecithal*, having evenly distributed yolk.

-lecithal (lĕs'ĭ·thăl) [Gr. *lekithos*, egg yolk]: for example, *lecithin*, any of several complex nitrogenous substances found especially in the brain and nerve tissue and in egg yolk.

leuko- (lū'kō) [Gr. *leukos*, white]: for example, *leukocyte*, white (blood) cell.

lip-, lipo- (lĭp-, lī'pō-) [Gr. *lipos*, fat]: pertaining to fats and fatty substances; for example, *lipase*, fat-digesting enzyme.

-logy (lō'jĭ) [Gr. *logos*, discourse, study]: for example, *zoology*, study of animals.

lumbar (lŭm'bēr) [L. *lumbus*, loin]: at, near, or toward loin region.

-lysis, -lytic, -lyte [Gr. *lysis*, a loosening]: pertaining to dissolving; for example, *electrolytic*, dissolution by electricity.

macro- (mă'krō) [Gr. *makros*, long]: opposite of *micro-*; for example, *macromere*, large embryo cell.

mastigo- (măs'tĭ·gō) [Gr. *mastix*, whip]: for example, *Mastigophora*, flagellum-bearing (protozoa).

mega- (mĕg'à-) [Gr. *megas*, large]: opposite of *micro-*; for example, *megascleric*, large-shelled; used like *macro-*.

meri-, mero-, -mere, -mer (mĕr'ĭ, mē'rō, mẹr, mẹr) [Gr. *meros*, part]: for example, *blastomere*, embryo part (cell); *polymer*, chemical of many (similar) parts; *meroblastic*, cleavage into partial blastomeres in embryo; *merozoite*, partly matured individual.

meso- (mĕs'ō) [Gr. *mesos*, middle]: for example, *mesoderm*, middle ''skin''; *mesosome*, middle body region.

meta- (mĕt'à) [Gr. after, behind]: for example, *metacoel*, hind cavity; *Metazoa*, later (advanced) animals.

micro- (mī′krō) [Gr. *mikros,* small]: for example, *micromere,* small embryo cell.

mono- (mŏn′ō) [Gr. *monos,* single]: for example, *monosaccharide,* single sugar (unit).

-morph, morpho- (mŏrf, mŏr′fō) [Gr. *morphē,* form]: for example, *morphology,* study of form (structure); *metamorphosis,* process of acquiring later (that is, adult) structure.

myo- (mī′ō) [Gr. *mys,* muscle]: for example, *myocyte,* muscle cell; *myoneme,* muscle (-like) hair; *myocoel,* muscle-associated coelom.

myx-, myxo-, (mĭks-, mĭk·sō-) [Gr. *myxa,* slime]: for example, *myxophyte,* slime mold.

-neme, nemato- (nē′mĕ, nĕm′à·tō) [Gr. *nema,* thread]: for example, *nematode,* threadlike worm; *axoneme,* axial filament.

nephro- (nĕf′rō-) [Gr. *nephros,* kidney]: for example, *nephric tubule,* excretory tubule.

neuro- (nū′rō) [Gr. *neuron,* nerve]: for example, *neurofibril,* impulse-conducting fibril.

noto- (nō′tō) [Gr. *nōton,* the back]: for example, *notochord,* cord along back.

octo- [Gr. *okto,* eight]: for example, *octopus,* eight-''legged'' animal.

-oid, -oida, -oidea (oid, oid′à, oi′dē·à) [Gr. *eidos,* form]: having the form of; like L. *-form;* for example, *echinoid,* like *Echinus* (sea urchin).

oligo- (ŏl′ĭ·gō) [Gr. *oligos,* few, small]: for example, *oligochaete,* having few bristles.

omni- (ŏm-nĭ-) [L. *omnis,* all]: for example, *omnivore,* animal eating all kinds of food.

onto- (ŏn′tō) [Gr. *on,* being]: for example, *paleontology,* study of ancient (fossil) beings.

oö- (ō′ō) [Gr. *ōion,* egg]: for example, *oöcyte,* egg cell.

oral (ō′răl) [L. *or-, os,* mouth]: at, near, or toward mouth.

osteo- (ŏs′tē·ō) [Gr. *osteon,* bone]: for example, *osteoblast,* bone-forming cell; *periosteum,* tissue layer covering a bone.

ostraco- (ŏs′trà·kō) [Gr. *ostrakon,* shell]: pertaining to a skeletal cover or shield; for example, *ostracoderm,* armor-skinned.

oto-, otic (ō′tō, ō′tĭk) [Gr. *ous,* ear]: for example, *otolith,* ear stone.

ovi-, ovo- (ō′vĭ, ō′vō) [L. *ovum,* egg]: for example, *oviduct; ovary,* egg-producing organ.

paleo- (pã′lē·ō-) [Gr. *palaios,* old]: for example, *paleontology,* study of ancient (fossil) life.

para- (păr′à) [Gr. beside]: for example, *parapodium,* side foot; *Parazoa,* animals on side branch of evolution (sponges).

pectin- (pĕk′tĭn) [L. *pecten,* comb]: for example, *pectine,* comblike organ.

pectoral (pĕk′tō·răl) [L. *pectorale,* breastplate]: at, near, or toward chest or shoulder region.

-ped, -pedia, pedi- (pĕd, pĕd′ĭ·à, pĕd′ĭ) [L. *pes,* foot]: like Gr. *-pod;* for example, *bipedal,* two-footed; *pedipalp,* leglike appendage.

pelvic (pĕl′vĭk) [L. *pelvis,* basin]: at, near, or toward hip region.

pent-, penta- (pĕnt, pĕn′tà) [Gr. *pente,* five]: for example, *pentose,* five-carbon sugar.

peri- (pĕr′ĭ) [Gr. around]: for example, *peristalsis,* wavelike compression around tubular organ (like gut).

phago-, -phage (făg′ō, fāj) [Gr. eating]: for example, *phagocyte,* cell eater; *bacteriophage,* bacterium eater (virus).

phono-, -phone (fōn′ō, fōn) [Gr. *phonē,* sound]: for example, *phonoreceptor,* sound-sensitive sense organ.

phoro-, -phore (fŏr′ō, fōr) [Gr. *phoros,* bearing, carrying]: like L. *-fer;* for example, *trochophore,* ''wheel''-bearing (larva).

photo-, photic (fō′tō, fō′tĭk) [Gr. *photos,* light]: for example, *photosynthesis,* synthesis with aid of light.

-phragm (frăm) [Gr. barrier]: for example, *diaphragm.*

phyllo-, -phyll (fĭl′ō, fĭl) [Gr. *phyllon,* leaf]: for example, *chlorophyll,* green pigment in leaf.

phyto-, -phyte (fī'tō, fīt) [Gr. *phyton,* plant]: for example, *Metaphyta,* later (advanced) plants.

-pithecus (-pĭ·thē'cŭs) [Gr. *pithēkos,* ape]: for example, *Australopithecus,* southern ape.

placo- (plă'kō) [Gr. *plax,* tablet, plate]: for example, *placoderm,* plate-skinned.

-plasm, plasmo-, -plast (plăz'm, plăz'mō, plăst) [Gr. *plasma,* form, mold]: for example, *protoplasm,* first-molded (living matter); *chloroplast,* green-formed (body).

-pleur, pleuro- (ploor, ploor'ō) [Gr. *pleuron,* side, rib]: for example, *pleura,* membrane lining rib cage.

-ploid [Gr. *-ploos,* -fold]: number of chromosome sets per cell; for example, *haploid, diploid.*

poly- (pŏl'ĭ) [Gr. *polys,* many]: for example, *polymorphic,* many-shaped; *polychaete,* many-bristled.

poro- (pō'rō) [Gr. *poros,* pore]: for example, *porous,* full of pores.

post-, postero-, posterior (pōst, pŏs'tēr·ō) [L. behind, after]: opposite of *pre-, antero-;* at, near, or toward hind end or part.

pre- (prē) [L. before, in front of]: opposite of *post-;* for example, *preoral,* in front of mouth.

pro- (prō) [Gr. before, in front of]: like L. *pre-;* for example, *prostomial,* in front of mouth.

-proct, procto- (prŏkt, prŏk'tō) [Gr. *proctos,* anus]: for example, *ectoproct,* having anus outside of ring of tentacles.

proto- (prō'tō) [Gr. *prōtos,* first]: for example, *Protozoa,* first animals.

proximal (prŏk'sĭ·măl) [L. *proximus,* near]: situated near to point of reference (usually the main part of body); opposite of *distal.*

pseudo- (sū'dō) [Gr. *pseudēs,* false]: for example, *pseudocoel,* false coelom; *pseudopodium,* false foot.

ptero-, -ptera, -ptery (tĕr'ō, tĕr'ȧ, tĕr'ĭ) [Gr. *pteron,* wing, fin]: for example, *exopterygote,* exterior-winged.

-pyge, pyg- (pī'jĭ, pīj) [Gr. *pygē,* rump]: for example, *cytopyge,* elimination pore in protozoa; *pygidium,* posterior abdominal region (in arthropods); *uropygial gland,* oil-secreting gland near tail base in birds.

rami-, -ramous (răm'ĭ, ră'mŭs) [L. *ramus,* branch]: for example, *biramous,* two-branched; *ramified,* branched.

renal (rē'năl) [L. *renes,* kidneys]: pertaining to kidney.

rhabdo- (răb'dō) [Gr. *rhabdos,* rod]: for example, *rhabdocoel,* flatworm having straight (rodlike) intestine.

rhizo- (rī'zō) [Gr. *rhiza,* root]: for example, *rhizopod,* having rootlike feet.

-rhynch, rhyncho- (rĭngk, rĭng'kō) [Gr. *rhynchos,* snout]: for example, *kinorhynch,* having movable snout.

sagittal (săj'ĭ·tăl) [L. *sagitta,* arrow]: at, near, or toward plane bisecting left and right halves; in median plane.

sarco-, -sarc (sär'kō, särk) [Gr. *sarx,* flesh]: for example, *coenosarc,* common flesh (living portions).

saur-, -saur (sôr) [Gr. *sauros,* lizard]: for example, *pterosaur,* flying reptile.

schizo- (skĭz'ō) [Gr. *schizein,* to split, part]: for example, *schizocoel,* coelom formed by splitting of tissue layer.

sclero- (sklē̆r'ō) [Gr. *sklēros,* hard]: for example, *scleroprotein,* hard (horny) protein; *sclera,* hard layer.

scypho- (sī'fō) [Gr. *skyphos,* cup]: for example, *Scyphozoa,* cup-shaped animals (jellyfish).

seti-, seta (sĕt'ĭ, sĕt'ȧ) [L. *seta,* bristle]: like Gr. *chaeto-; setiferous,* bristle-bearing.

sipho-, siphono- (sī'fō-) [Gr. *siphōn,* a pipe]: for example, *siphonaceous,* tubular.

-soma, -some, somato- (sō'mȧ, sōm, sō'mȧ·tō) [Gr. *sōma,* body]: for example, *chromosome,* pigmented body; *somatocoel,* body coelom; *somatic mesoderm,* outer mesoderm.

spermo-, sperma-, spermato- (spûr'mō, spûr'mȧ, spûr·mȧ'tō) [Gr. *sperma,* seed]: for example, *spermatophore,* sperm (-bearing) capsule.

splanchno- (splăngk'nō) [Gr. *splanchnon*, entrails]: for example, *splanchnic mesoderm*, inner mesoderm.

sporo- (spō'rō) [Gr. *sporā*, seed]: for example, *Sporozoa*, spore-forming (protozoa); *sporogony*, reproduction by spores.

stato- (stăt'ō) [Gr. *statos*, standing stationary, positioned]: for example, *statolith*, position (-indicating) stone.

stereo- (stĕr'ē·ō) [Gr. *stereos*, solid]: for example, *stereoblastula*, solid blastula.

-stome, -stoma, -stomato-, (stōm, stōm'à, stōm'à·tō) [Gr. *stoma*, mouth]: *peristomial*, around the mouth.

sub-, sus- (sŭb-, sŭs-) [L. under, below]: for example, *subepidermal*, underneath the epidermis; *suspensor*, suspending structure.

sym-, syn-, (sĭm, sĭn) [Gr. *syn*, together, with]: like L. *con-*; for example, *syngamy*, coming together of gametes; *synapse*, looping together (of neurons); *synthesis*, construction, putting together; *symbiosis*, living together.

taxo-, taxi-, -taxis (tăksō, tăksĭ, tăks'ĭs) [Gr. *taxis*, arrangement]: for example, *taxonomy*, ''arrangement'' laws; *taxidermy*, skin arrangement.

tel-, tele-, teleo- (tĕl, tĕl'ē, tĕl'ē·ō) [Gr. *telos*, end]: for example, *telophase*, end phase, *teleost*, (fish with) bony end (adult) state; *teleology*, knowledge of end conditions.

tetra- (tĕr'rà) [Gr. four]: for example, *tetrapod*, four-footed.

theco-, -theca (thē'kō, thē'kà) [Gr. *thēkē*, case, capsule]: for example, *thecodont*, having socketed teeth.

thigmo- (thĭg'mō) [Gr. *thigma*, touch]: for example, *thigmotrophy*, movement due to touch.

thoracic (thō·răs'ĭk) [L. *thorax*, chest]: at, near, or toward chest region, or region between head and abdomen.

-tome, -tomy (tōm, tō'mĭ) [Gr. *tomē*, section, a cutting apart]: for example, *anatomy*, study of structure based on dissection.

trans- [L. across]: for example, *transsection*, crosscut.

transverse (trăns'vûrs) [L. *transversare*, to cross]: at, near, or toward plane separating anterior and posterior; cross-sectional.

tri- (trī) [L. *tria*, three]: for example, *triclad*, three-branched (digestive tract).

-trich, tricho- (trĭk, trĭk'ō) [Gr. *trichos*, hair]: for example, *trichocyst*, hair-containing sac.

-troch, trocho- (trŏk, trŏk'ō) [Gr. *trochos*, wheel]: for example, *trochophore*, (larva) bearing wheel (of cilia).

-troph, tropho- (trōf, trō'fō) [Gr. *trophos*, feeder]: for example, *autotrophic*, self-nourishing.

uro-, ura (ū'rō, ūrà) [Gr. *oura*, tail]: for example, *uropod*, tail foot; *urochordate*, tailed chordate.

ventral (vĕn'trål) [L. *venter*, belly]: opposite to *dorsal*; at, near, or toward the belly or underside.

xantho- (zăn'thō-) [Gr. *xanthos*, yellow]: for example, *xanthophyll*, yellow pigment (of leaf).

zoo-, -zoa, -zoon (zō'ō, zōà, zō'ŏn) [Gr. *zōion*, animal]: for example *protozoon*, first animal; *zooid*, individual animal (in colony).

zygo- (zī'gō) [Gr. *zygon*, yoke, pair]: for example, *zygote*, fertilized egg.

GENERAL LISTING OF TECHNICAL TERMS

NOTE: In most cases in this section where derivations of particular word parts are not given, such parts and their derivations can be found in section A, above.

abdomen (ab′dŏ·mĕn) [L.]: region of animal body posterior to thorax.

Acanthocephala (ă·kăn′thŏ·sĕf′ă·là) [Gr. *akantho,* thorn]: spiny-headed worms, a phylum of pseudocoelomate parasites.

Acarina, acarine (ăkă′rī·nà, ăkă′rĭn) [Gr. *akar,* mite]: (1) the archnid order of mites and ticks; (2) pertaining to mites and ticks.

acicula (à·sĭk′ū·là) [L. dim. of *acus,* needle]: needlelike bristle embedded in polychaete parapodium.

acid (ăs′ĭd) [L. *acidus,* sour]: a substance that releases hydrogen ions in water; having a pH of less than 7.

acoel, acoelomate (ă·sēl′): (1) without coelom; also a group of free-living flatworms without digestive cavity; (2) an animal without coelom; flatworms, nemertine worms.

Acrania (ă·krā′nĭ·ă) [Gr. *kranion,* skull]: headless chordates, including urochordates and cephalochordates.

acromegaly (ăk′rŏ·mĕg′à·lĭ): skeletal overgrowths, particularly in the extremities, produced by excessive growth-hormone secretion from the pituitary.

acrosome (ăk′rŏ·sōm): structure at tip of head (nucleus) of animal sperm, which makes contact with egg during fertilization.

adenine (ăd′ē·nēn): a purine component of nucleotides and nucleic acids.

adenosine (di-, tri-) phosphate (ADP, ATP) (à·dĕn′ŏ·sēn): adenine-ribose-phosphates functioning in energy transfers in cells.

adenylic acid: equivalent to adenosine monophosphate, or AMP.

adipose (ăd′ĭ·pōs) [L. *adipis,* fat]: fat, fatty; fat-storing tissue.

ADP: abbreviation of adenosine diphosphate.

adrenal, adrenalin (ăd·rē′năl, ăd·rĕn′ăll·ĭn) [L. *renalis,* kidney]: (1) endocrine gland; (2) the hormone produced by the adrenal medulla.

adrenergic (ăd′rĕn·ûr′jĭk): applied to nerve fibers that release an adrenalinlike substance from their axon terminals when impulses are transmitted across synapses.

aerobe, aerobic (ā′ēr·ōb, -ō′bĭk) [Gr. *aeros*, air]: (1) oxygen-requiring organism; (2) pertaining to oxygen-dependent form of respiration.

Agnatha (ăg′nȧ·thȧ): jawless fishes, a class of vertebrates including lampreys and hagfishes.

aldehyde (ăl′dē·hīd) [L. abbr. for *alcohol-dehydrogenatum*, dehydrogenated alcohol]: organic compound with —CHO grouping.

aldose (ăl′dōs): one of a series of sugars having a terminal aldehyde grouping.

alga (ăl′gȧ), pl. *algae* (-jē): any member of a largely photosynthetic superphylum of protists.

alkaline (ăl′kȧ·lĭn): pertaining to substances that release hydroxyl ions in water; having a pH greater than 7.

allantois (ȧ·lăn′tō·ĭs) [Gr. *allantoeides*, sausage-shaped]: one of the extraembryonic membranes in reptiles, birds, and mammals; functions as embryonic urinary bladder or as carrier of blood vessels to and from placenta.

allele (ȧ·lēl′) [Gr. *allēlōn*, of one another]: one of a group of alternative genes that can occupy a given locus on a chromosome; a dominant and its associated recessive are allelic genes.

alula (ăl′ū·lȧ) [L. dim. of *ala*, wing]: the first digit (thumb) of a bird wing; reduced in comparative size.

alveolus (ăl·vē′ōl·ŭs), pl. *alveoli* (-lī) [L. dim. of *alveus*, a hollow]: a small cavity or pit; for example, a microscopic air sac of lungs.

ambulacrum, ambulacral (ăm′bū·lā′krŭm, -ăl) [L. walk, avenue]: (1) tube-feet-lined ciliated groove leading over arm to mouth in certain echinoderms; conducts food to mouth; (2) adjective.

amino, amino acid (ȧ·mē′nō): (1) —NH₂ group; (2) acid containing amino group, constituent of protein.

ammocoete (ăm′ō·sēt) [Gr. *ammons*, sand]: lamprey larva.

amnion, amniote, amniotic (ăm′nĭ·ŏn) [Gr. dim. of *amnos*, lamb]: (1) one of the extraembryonic membranes in reptiles, birds, and mammals that forms a sac around the embryo; (2) any reptile, bird, or mammal, that is, any animal with an amnion during the embryonic state; (3) pertaining to the amnion, as in *amniotic fluid*.

amphiblastula (ăm′fĭ·blăst′ū·lȧ): larval stage in certain sponges.

Amphineura (ăm′fĭ·nū′rȧ): a class of mollusks, including the chitons.

ampulla (ăm·pŭl′ȧ) [L. vessel]: enlarged saclike portion of a duct, as in ampullas of semicircular canals in mammalian ear, or in ampullas of echinoderm tube feet.

amylase (ăm′ĭ·lās) [L. *amylum*, starch]: an enzyme that promotes the decomposition of polysaccharides to smaller carbohydrate units.

anaerobe, anaerobic (ăn·ā′ēr·ob, -ō′bĭk): (1) an oxygen-independent organism; (2) pertaining to an oxygen-independent form of respiration.

anamniote (ăn·ăm′nĭ·ōt): any vertebrate other than a reptile, bird, or mammal, that is, one in which an amnion does not form during the embryonic phase.

anaphase (ăn′ȧ·fāz): stage in mitotic division characterized by the migration of chromosome sets toward the spindle poles.

anatomy (ȧ·năt′ō·mĭ): the gross structure of an organism, or the science that deals with gross structure; a branch of the science of morphology.

androgen (ăn′drō·jĕn): one of a group of male sex hormones.

anisogamy (ăn·ĭ′sŏg′ăm·ĭ): sexual fusion in which the gametes of opposite sex types are unequal in size.

Annelida (ăn′ĕ·lĭd·ȧ) [L. *anellus*, a ring]: the phylum of segmented worms.

annulus (ăn′ū·lŭs) [L. *ring*]: a ringlike structure.

antibody (ăn′tĭ·bŏd′ĭ): a protein that combines and renders harmless an antigen, that is, a foreign protein introduced into an animal by infectious processes.

antigen (ăn′tĭ·jĕn): a foreign substance, usually protein in nature, which elicits the formation of specific antibodies in an animal.

Anura (ȧ·nū′rȧ): order of tailless amphibia, including frogs and toads.

apodeme (ăp'ō·dēm) [Gr. *demos*, district]: breakage plane in crustacean appendage where autotomy can occur readily.

apopyle (ăp'ō·pīl) [Gr. *pylē*, gate]: the excurrent opening of a choanocyte chamber in sponges.

Arachnida (à·răk'nĭd·à) [Gr. *arachnē*, spider]: class of chelicerate arthropods including spiders, scorpions, mites, ticks, and other orders.

archenteron (är·kĕn'tĕr·ŏn): the central cavity of a gastrula, lined by endoderm, representing the future digestive cavity of the adult.

arterial (är·tēr'ĭ·ăl): pertaining to arteries; also applied to oxygenated blood.

Arthropoda (är·thrō'pŏd·ä): the phylum of jointed-legged invertebrates.

Artiodactyla (är'tĭ·ō·dăk'tĭl·à) [Gr. *artios*, even]: order of even-toed ungulate mammals; includes cattle, swine, deer, camels.

Aschelminthes (ăs·kĕl·mĭn'thĕs): sac (bladderlike) worms; a pseudocoelomate phylum including rotifers, roundworms, and other groups.

asconoid (ă'skŏn·oid): saclike; refers specifically to a type of sponge architecture.

atom (ăt'ŭm) [Gr. *atomos*, indivisible]: the smallest whole unit of a chemical element; composed of protons, neutrons, and other particles, which form an atomic nucleus, and of electrons, which orbit around the nucleus.

ATP: abbreviation of adenosine triphosphate.

atrium, atrial (ā'trĭ·ŭm, -ăl) [L. yard, court, hall]: entrance or exit cavity; for example, entrance chamber to heart, exit chamber from chordate gill region.

auricle (ô'rĭ·k'l) [L. dim. of *auris*, ear]: ear-shaped structure or lobelike appendage; for example, atrium in mammalian heart, lateral flap near eyes in planarian worms.

auricularia (ô·rĭk'ū·lā'rĭä) [L. *lar*, larva]: larva of holothuroid echinoderms, with earlobelike ciliated bands.

autolysis (ô·tŏl'ĭ·sĭs): enzymatic self-digestion or dissolution of tissue or other part of an animal.

autosome (ô'tō·sōm): a chromosome other than a sex chromosome.

autotroph, autotrophism (ô'tō·trŏf', -ĭz'm): (1) an organism that manufactures organic nutrients from inorganic raw materials; (2) a form of nutrition in which only inorganic substances are required as raw materials.

avicularium (ā·vĭk'ū·lā'rĭ·ŭm) [L. dim. of *avis*, bird]: a specially differentiated polymorphic individual in a colony of ectoprocts, shaped like a bird's head, serving a protective function.

axenic (ā·zĕn'ĭk) [Gr. *xenos*, foreigner]: pertaining to a culture medium in which the available food sources are completely and specifically identified.

axon (ăk'sŏn): an outgrowth of a nerve cell, conducting impulses away from the cell body; a type of nerve fiber.

bacillus (bà·sĭl'ŭs) [L. dim. of *baculum*, rod]: any rod-shaped bacterium.

bacteriophage (băk·tēr'ĭ·ō·fāj) [*bacterium* + Gr. *phagein*, to eat]: one of a group of viruses that infect, parasitize, and eventually kill bacteria.

bacterium (băk·tēr'ĭ·ŭm) [Gr. dim. of *baktron*, a staff]: a small, typically unicellular organism characterized by the absence of a formed nucleus; genetic material is dispersed in clumps through the cytoplasm.

benthos, benthonic (bĕn'thŏs) [Gr. depth of the sea]: (1) collective term for organisms living along the bottoms of oceans and lakes; (2) adjective.

beriberi (bĕr'ĭ·bĕr'ĭ) [Singhalese *beri*, weakness]: disease produced by deficiency of vitamin B_1 (thiamine).

bicuspid (bī·kŭs'pĭd) [L. *cuspis*, point]: ending in two points or flaps, as in bicuspid heart valve; syn. *mitral*.

bioluminescence (bī'ō·lū'mĭ·nĕs'ĕns) [L. *lumen*, light]: emission of light by living organisms.

biome (bī'ōm): habitat zone; for example, desert, grassland, tundra.

biota, biotic (bī·ō'tä, -ŏt'ĭk): (1) the community of organisms of a given region; (2) adjective.

bipinnaria (bī'pĭn·ăr'ĭ·à) [L. *pinna*, feather, fin]: larva of asteroid echinoderms, with ciliated bands suggesting two wings.

blastopore (blăs'tō·pōr): opening connecting archenteron of gastrula with outside; represents future mouth in some animals, future anus in others.

blastula (blăs'tū·là): stage in early animal development, when embryo is a hollow or solid sphere of cells.

blepharoplast (blĕf'à·rō·plăst') [Gr. *blepharon*, eyelid]: the basal granule of a flagellum or cilium; equivalent to *kinetosome*.

brachiopod, Brachiopoda (brā'kĭ·ō·pŏd, brā·kĭ·ŏp'ō·dà): (1) a sessile, enterocoelomate, marine animal with a pair of shells (valves) and a lophophore; (2) phylum name.

bronchus, bronchiole (brŏng'kŭs, brŏng'kĭ·ōl) [Gr. *bronchos*, windpipe]: (1) a main branch of the trachea in air-breathing vertebrates; (2) a smaller branch of a bronchus.

buffer (bŭf'ēr): a substance that prevents appreciable changes of pH in solutions to which small amounts of acids or bases are added.

bursa (bûr'sà) [L. sac, bag]: saclike cavity.

byssus (bĭs'ŭs) [Gr. *byssos*, flax, linen]: silky threads secreted by mussels for attachment to rocks.

caecum (sē'kŭm) [L. *caecus*, blind]: cavity open at one end; for example, the blind pouch at the beginning of the vertebrate large intestine, connecting at one side with the small intestine.

calorie (kăl'ō·rĭ) [L. *calor*, heat]: unit of heat, defined as the amount of heat required to raise the temperature of 1 g of water by 1°C; a *large*, or *dietary*, *calorie* (kilocalorie) is a thousand of the units above.

captaculum (kap·ta'kū·lŭm) [L. *captare*, to capture]: tentaclelike outgrowth from head of tooth-shell mollusks, serving in food capture.

carapace (kăr'à·pās) [fr. Sp. *carapacho*]: a hard case or shield covering the back of certain animals.

carbohydrate, carbohydrase (kär'bō·hī'drāt): (1) an organic compound consisting of a chain of carbon atoms to which hydrogen and oxygen, present in a 2:1 ratio, are attached; (2) an enzyme promoting the synthesis or decomposition of a carbohydrate.

carnassial (kär·năs'ĭ·ăl) [Fr. *carnassier*, flesh-eating]: pertaining to mammalian tooth type adapted for flesh eating.

carnivore, Carnivora (kärnĭv'ō·rà) [L. *carnivorus*, flesh-eating]: (1) any holotrophic animal subsisting on other animals or parts of animals; (2) an order of mammals; includes cats, dogs, seals, walruses.

carotene, carotenoids (kăr'ō·tēn, kà·rŏt'ē·noid) [L. *carota*, carrot]: (1) a pigment producing cream-yellow to carrot-orange colors; precursor of vitamin A; (2) a class of pigments of which carotene is one.

catalysis, catalyst (kà·tăl'ĭ·sĭs) [Gr. *katalysis*, dissolution]: (1) acceleration of a chemical reaction by a substance that does not become part of the endproduct; (2) a substance accelerating a reaction as above.

ceboid (sē'boid): a New World monkey; uses its tail as a fifth limb.

Cenozoic (sē'nō·zō'ĭk) [Gr. *kainos*, recent]: geologic era after the Mesozoic, dating approximately from 75 million years ago to present.

centriole (sĕn'trĭ·ōl): cytoplasmic organelle forming spindle pole during mitosis and meiosis.

centrolecithal (sĕn'trō·lĕs'ĭ·thăl): pertaining to eggs with yolk accumulated in center, of cell, as in arthropods.

centromere (sĕn'trō·mēr): region on chromosome at which spindle fibril is attached during mitosis and meiosis.

Cephalochordata, Cephalopoda, cephalothorax (sĕf'à·lō-): (1) a subphylum of chordates; the lancelets or amphioxus; (2) a class of mollusks; squids, octopuses, nautiluses; (3) the fused head and thorax in certain arthropods.

cercaria (sûr·kā'rĭ·à): a larval stage in the life cycle of flukes; produced by a redia and infects fish, where it encysts.

cercopithecoid (sûr′kō·pĭ·thē′koĭd): Old World monkey; its tail is not used as limb.

cerebellum (sĕr′ē·bĕl′ŭm) [L. dim. of *cerebrum*]: a part of the vertebrate brain; controls muscular coordination.

cerebrum (sĕr′ē·brŭm) [L. brain]: a part of the vertebrate brain; controls many voluntary functions and is seat of higher mental capacities.

Chaetognatha (kē′tŏg·năth·ả): a phylum of wormlike enterocoelomates; animals with curved bristles on each side of mouth.

chelate (kē′lāt) [Gr. *chēlē*, claw]: claw-possessing, esp. a limb or appendage.

chelicera (kē·lĭ′sĕr·ả): a pincerlike appendage in a subphylum of arthropods (chelicerates).

chemolithotroph (kĕm′ō·lĭth′ō·trōf) [Gr. *lithos*, stone]: an organism that manufactures food with the aid of energy obtained from chemicals and with inorganic raw materials.

chemoorganotroph (kĕm′ō·ôr·găn′ō·trōf): an organism that manufactures food with the aid of energy obtained from chemicals and with organic raw materials.

chemosynthesis (kĕm′ō·sĭn′thē·sĭs): a form of autotrophic nutrition in certain bacteria, in which energy for the manufacture of carbohydrates is obtained from inorganic raw materials.

chemotaxis (kĕm′ō·tăk·sĭs): a movement oriented by chemical stimuli.

chitin (kĭ′tĭn): a horny organic substance forming the exoskeleton or epidermal cuticle of many invertebrates (arthropods particularly,) and the cell walls of most fungi.

chloragogue (klō′rả·gŏg) [Gr. *agōgos*, leader]: excretory cell in annelids and some other invertebrates; leads wastes from body fluids to epidermis.

chlorocruorin (klō′rō·kroo′ōr·ĭn) [L. *cruor*, blood, gore]: green blood pigment in plasma of certain annelids.

cholinergic (kō′lĭn·ûrjik): refers to a type of nerve fiber that releases acetyl choline from the axon terminal when impulses are transmitted across synapses.

Chondrichthyes (kŏn·drĭk′thĭ·ēz): fishes with cartilage skeleton, a class of vertebrates comprising sharks, skates, rays, and related types.

Chordata (kôr·kā′tả): animal phylum in which all members have notochord, dorsal nerve cord, and pharyngeal gill slits at some stage of life cycle.

chorion (kō′rĭ·ŏn) [Gr.]: one of the extraembryonic membranes in reptiles, birds, and mammals; forms outer cover around embryo and all other membranes and in mammals contributes to structure of placenta.

choroid (kō′roid): pigmented mid-layer in wall of vertebrate eyeball, between retina and sclera; also blood vessel—carrying membranes in vertebrate brain.

chromatid (krō′mả·tĭd): a newly formed chromosome in mitosis and meiosis.

chromatophore (krō′mả·tō·fōr′): pigment-containing body; refers specifically to chlorophyll-bearing granules in bacteria and to pigment cells in animals.

chromosome (krō′mō·sōm): gene-containing filament in cell nucleus, becoming conspicuous during mitosis and meiosis.

chymotrypsin (kĭ′mō·trĭp′sĭn): enzyme promoting protein digestion; acts in small intestine, produced in pancreas as inactive chymotrypsinogen.

Ciliophora (sĭl′ĭ·ŏf′ôrả) [L. *cilium*, eyelid]: a protozoan subphylum in which member organisms have cilia on body surface.

cilium (sĭl′ĭ·ŭm): microscopic bristlelike variant of a flagellum; functions in cellular locomotion and in creation of currents in water.

cirrus (sĭr′ŭs) [L. tuft, fringe]: a movable tuft or fingerlike projection from a cell or a body surface.

cloaca (klō ā′kả) [L. sewer]: exit chamber from alimentary system; also serves as exit for excretory and/or reproductive system.

Cnidaria (nĭ·dā′rĭ·ả) [Gr. *knidē*, nettle]: coelenterates; the phylum of cnidoblast-possessing animals.

cnidoblast (nī′dō·blăst): stinging cell characteristic of cnidarians; contains nematocyst.

cnidocil (nī′dō·sīl): spike of hair trigger on cnidoblast serving in nematocyst discharge.

coccine (kŏk′sēn) [Gr. *kokkos,* grain]: pertaining to sessile protistan state of existence in which reproduction does not take place during vegetative condition.

cochlea (kŏk′lē·à) [Gr. *kochlias,* snail]: part of the inner ear of mammals, coiled like a snail shell.

coelenterate (sē·lĕn′tēr·āt): an invertebrate animal having a single alimentary opening and tectacles with sting cells; for example, jellyfish, corals, sea anemones, hydroids; a cnidarian.

coelom (sē′lŏm): body cavity lined entirely by mesoderm, especially by peritoneum.

coenosarc (sē′nō·särk) [Gr. *koinos,* common]: the living parts of a cnidarian hydroid colony, as distinguished from external secreted perisarc.

coenzyme (kō·ĕn′zīm): one of a group of organic substances required in conjunction with many enzymes; usually carries and transfers parts of molecules.

colloblast (cŏl′ō·blăst) [Gr. *kolla,* glue]: adhesive cell type in tentacles of ctenophores.

colloid (kŏl′oid): a substance divided into fine particles, where each particle is larger than one of a true solution but smaller than one in a coarse suspension.

colon (kō′lŏn): the large intestine of mammals; portion of alimentary tract between caecum and rectum.

colostrum (kō·lŏs′trŭm): the first, lymphlike secretion of the mammary glands of pregnant mammals.

columella (kŏl′ū·mĕl′à) [L. little column]: an axial shaft; for example, hyomandibular bone in amphibian ear.

commensal, commensalism (kō·mĕn′săl, ĭz′m) [L. *cum,* with, + *mensa,* table]: (1) an organism that lives symbiotically with a host, where the host neither benefits nor suffers from the association; (2) noun.

compound (kŏm′pound) [L. *componere,* to put together]: a combination of atoms or ions in definite ratios, held together by chemical bonds.

conjugation (kŏn·joo·gă′shŭn) [L. *conjugare,* to unite]: a mating process characterized by temporary fusion of the mating partners.

convergence (kŏn·vûr′jĕns) [L. *convergere,* to turn together]: the evolution of similar characteristics in organisms of widely different ancestry.

Copepoda (kō′pē·pŏd·à) [Gr. *kope,* oar]: a subclass of crustaceans.

corona (kō·rō′nà) [L. garland, crown]: any wreath or circlet of cilia, tentacles, or cells.

corpus allatum (kôr′pŭs ă·lā′tŭm) pl. *corpora allata* [L. added body]: endocrine gland in insect head behind brain; secretes hormone inducing larval molt.

corpus callosum (kôr′pŭs kă·lō′sŭm) pl. *corpora callosa* [L. hard body]: broad tract of transverse nerve fibers that unites cerebral hemispheres in mammals.

corpuscle (kôr′pŭs′l) [L. dim. of *corpus,* body]: a small, rounded structure, cell, or body.

corpus luteum (kôr′pŭs lū′tē·ŭm) pl. *corpora lutea* [L. yellow body]: progesterone-secreting bodies in vertebrate ovaries, formed from remnants of follicles after ovulation.

cortex (kôr′tĕks) pl. *cortices* [L. bark]: the outer layers of an organ or body part; for example, adrenal cortex, cerebral cortex.

costa (kŏs′tà) [L. rib, side]: a rib or riblike supporting structure.

cotylosaur (kŏt′ĭ·lō sôr′) [Gr. *kotylē,* something hollow]: a member of a group of Permian fossil reptiles, evolved from labyrinthodont amphibian stock and ancestral to all other reptiles.

covalence (kō′vā′lĕns): chemical bonding by electron-sharing, resulting in a molecule.

coxopodite, coxal (kŏks·ō′pō·dĭt, kŏk′săl) [L. *coxa,* hip]: (1) the first, most basal joint of a segmental appendage of arthropods; (2) adjective.

Craniata (krā′nĭ·ā′tà) [Gr. *kranion,* skull]: head-possessing chordates; vertebrates.

cretinism (krē'tĭn·ĭz'm) [fr. L. *christianus*, a Christian]: an abnormal condition in man resulting from underactivity of the thyroid in the young.

crinoid (krī'noid) [Gr. *krinoeides*, lilylike]: a member of a class of echinoderms; a sea lily or feather star.

Crossopterygii (krŏ'sŏp·tē'rĭ·jē) [Gr. *krossoi*, tassels]: a superorder of bony fishes, within the subclass of Sarcopterygii; the lobe-finned fishes.

Crustacea (krŭs·tā'shē å) [L. *crusta*, shell, rind]: a class of mandibulate arthropods; crustaceans.

crystalloid (krĭs'tăl·oid) [Gr. *kystallos*, ice]: a system of particles in a medium, able to form crystals under appropriate conditions; a true solution.

Ctenophora (tē·nŏf'ŏ·rä): a phylum of radiate animals characterized by comb plates; the comb jellies.

CTP: abbreviation of cytidine triphosphate.

cutaneous (kū·tā'nē·ŭs) [L. *cutis*, skin]: pertaining to the skin; for example, cutaneous sense organ.

cyclosis (sī·klō'sĭs) [Gr. *kyklos*, circle]: circular streaming and eddying of cytoplasm.

cydippid (sī'dĭp·ĭd): a larva of ctenophores.

cyphonautes (sī'fŏ·nôt'ēs) [Gr. *kyphos*, crooked, + *nautēs*, sailor]: a larva of ectoprocts.

cytidine (di-, tri-) phosphates (sī'tĭ·dēn): cytosine-ribose-phosphates (CDP, CTP).

cytidylic acid: equivalent to cytosine monophosphate, of CMP.

cytochrome (sī'tō·krōm): one of a group of iron-containing hydrogen or electron carriers in cell metabolism.

cytolysis (sī·tŏl'ĭ·sĭs): dissolution or disintegration of a cell.

cyton (sī'tŏn): the nucleus-containing main portion (cell body) of a neuron.

cytoplasm (sī'tō·plăz'm): the substance of a cell between cell membrane and nucleus.

cytosine (sī'tō·sēn): a nitrogen base in nucleotides and nucleic acids.

deamination (dē·ămĭ·nā'shŭn): removal of an amino group, especially from an amino acid.

decapod (dĕk'ä·pŏd) [Gr. *deka*, ten]: 10-footed animal, specifically decapod crustacean (for example, lobster), decapod mollusk (for example, squid).

decarboxylation (dē·kär·bŏk'sĭ·lā'shŭn): removal of a carboxyl group (—COOH).

deciduous (dē·sĭd'ū·ŭs) [L. *decidere*, to fall off]: to fall off at maturity, as in trees that shed foliage in autumn.

dedifferentiation (dē'dĭf·ēr·ĕn'shĭ·ā'shŭn): a regressive change toward a more primitive, embryonic, or earlier state; for example, a process changing a highly specialized cell to a less specialized cell.

degrowth (dē'grōth): negative growth; becoming smaller.

dehydrogenase (dē·hī'drŏ·jĕn·ās): an enzyme promoting dehydrogenation.

denaturation (dē·nā'tūr·ā'shŭn): physical disruption of the three-dimensional structure of a protein molecule.

dendrite (dĕn'drīt) [Gr. *dendron*, tree]: filamentous outgrowth of a nerve cell; conducts nerve impulses from free end toward the cell body.

denitrify, denitrification (dē·nī'trĭ·fī): (1) to convert nitrates to ammonia and molecular nitrogen, as by denitrifying bacteria; (2) noun.

denticle (dĕn'tĭ·k'l): [L. *denticulus*, small tooth]: small toothlike scale, as on shark skin.

deoxyribose (dē·ŏk'sĭ·rī'bōs): a 5-carbon sugar having one oxygen atom less than parent—sugar ribose; component of deoxyribose nucleic acid (DNA).

Deuterostomia (dū'tēr·ō·stō'mē·å) [Gr. *deuteros*, second]: animals in which blastopore becomes anus; mouth forms as second embryonic opening opposite blastopore.

diabetes (dī'å·bē'tēz) [Gr. *diabainein*, to pass through]: abnormal condition marked by insufficiency of insulin, sugar excretion in urine, high blood-glucose levels.

diastole (dī·ăs′tō·lē) [Gr. *diastolē,* moved apart]: phase of relaxation of atria or ventricles, during which they fill with blood; preceded and succeeded by contraction, or systole.

diastrophism (dī·ăs′trō·fīz′m) [Gr. *diastrophē,* distortion]: geologic deformation of the earth's crust, leading to rise of land masses.

dichotomy (dī·kŏt′ō mĭ) [Gr. *dicha,* in two + *temnein,* to cut]: a repeatedly bifurcating pattern of branching.

diencephalon (dī′ĕn·sĕf′a·lŏn) [Gr. *enkephalos,* brain]: hind portion of the vertebrate forebrain.

differentiation (dĭf′ĕr·ĕn′shĭ·ā′shŭn): a progressive change toward a permanently more mature, advanced, or specialized state.

diffusion (dĭ·fū′zhŭn) [L. *diffundere,* to pour out]: migration of particles from a more concentrated to a less concentrated region, leading to equalization of concentrations.

dimorphism (dī·môr′fīz′m): difference of form between two members of a species, as between males and females; a special instance of polymorphism.

dipleurula (dī·ploor′ŭ·la): hypothetical ancestral form of most enterocoelomate animals, resembling developmental stage of hemichordates and echinoderms.

diplohaplontic (dĭp′lō·hăp·lŏn′tĭk): designating a life cycle with alternation of diploid and haploid generations.

diploid (dĭp′loid): a chromosome number twice that characteristic of a gamete in a particular species.

diplontic (dĭp·lŏn′tĭk): designating a life cycle with gametogenic meiosis and diploid adults.

disaccharide (dī·săk′a rīd) [Gr. *sakcharon,* sugar]: a sugar composed of two monosaccharides; usually refers to 12-carbon sugars.

dissociation (dĭ·sō′sĭ·ā′shŭn) [L. *dissociare,* to dissociate]: the breakup of a covalent compound in water; results in formation of free ions.

diurnal (dī·ûr′năl) [L. *diurnalis,* daily]: for example, as in daily up and down migration of plankton in response to absence or presence of sunlight.

divergence (dī·vûr′jĕns) [L. *divergere,* to incline apart]: evolutionary development of dissimilar characteristics in two or more lines descended from the same ancestral stock.

diverticulum (dī′vĕr·tĭk′ū lŭm) [L. byway]: branch or sac off a canal or tube; for example, digestive diverticulum.

DNA: abbreviation of deoxyribose nucleic acid.

doliolaria (dŏ·lĭ·ō·lā′rĭ·a) [L. *dolium,* small cask]: yolky barrel-shaped larva of crinoid echinoderms and transient larval stage in holothuroid development.

dominance: a functional attribute of genes; the dominant effect of a gene masks the recessive effect of its allelic partner.

ductus arteriosus (dŭk′tŭs är·tē′rĭ·ō′sŭs): an artery in the embryo and fetus of mammals that conducts blood from the pulmonary artery to the aorta; shrivels at birth, when lungs become functional.

duodenum (dū′ō·dē′nŭm) [L. *duodeni,* twelve each]: most anterior portion of the small intestine of vertebrates.

Echinodermata (ē·kĭ′nō·dûr′ma·ta): the phylum of spiny-skinned animals; includes starfishes, sea urchins.

Echiuroida (ē·kĭ′ūr oi′da): a phylum of wormlike, schizocoelomate animals, characterized by spines at hind end; spoon worms.

ectoderm (ĕk′tō-): outer tissue layer of an animal embryo.

Ectoprocta (ĕk′tō·prŏk·ta): a phylum of sessile coelomate animals, in which the intestine is U-shaped, the mouth is surrounded by a lophophore with ciliated tentacles, and the anus opens outside this lophophore.

egestion (ē·jĕs′chŭn) [L. *egerere,* to discharge]: the elimination of unusable and undigested material from the alimentary system.

elasmobranch (ē·lăs′mō·brăngk) [Gr. *elasmos,* plate]: a member of a subclass of cartilage fishes (sharks and rays); also used as adjective.

electrolyte (ē·lĕk′trō·līt) [Gr. *ēlektron,* amber]: a substance that dissociates as ions in aqueous solution; permits conduction of electric current through the solution.

electron (ĕ·lĕk′trŏn): a subatomic particle that carries a unit of negative electric charge; orbits around atomic nucleus.

electrovalence (ĕ·lĕk′trŏ·vā′lĕns): chemical bonding by electron transfer, resulting in an ionic compound.

element (ĕl′ĕ·mĕnt): one of about 100 distinct natural or man-made types of matter, which, singly or in combination, compose all materials of the universe; an atom is the smallest representative unit of an element.

Eleutherozoa (ĕlū·thĕ·rŏ zō′å) [Gr. *eleutheros*, free]: a subphylum of echinoderms comprising all living classes except the crinoids.

embolus (ĕm′bŏ·lŭs) [Gr. *embolos*, peg, stopper]: blood clot formed within a blood vessel.

emboly (ĕm′bŏ·lĭ): invaginative gastrulation.

embryo (ĕm′brĭ·ō) [Gr. *en* in, + *bryein*, to swell]: an early developmental stage of an organism following fertilization.

emulsion (ĕ·mŭl′shŭn) [L. *emulgere*, to milk out]: a colloidal system in which both the dispersed and the continuous phase are liquid.

endemic (ĕn·dĕm′ĭk) [Gr. belonging to a district]: pertaining to or occurring in a limited locality; opposite of cosmopolitan.

endergonic (ĕn′dĕr·gŏ·nĭk): energy-requiring, as in a chemical reaction.

endocrine (ĕn′·dŏ·krīn) [Gr. *krinein*, to separate]: applied to type of gland that releases secretion not through a duct but directly into blood or lymph; equivalent to hormone-producing.

endoderm (ĕn′dŏ·dûrm): inner tissue layer of an animal embryo.

endoplasm, endoplasmic (ĕn′dŏ·plăz′m): (1) the portion of cellular cytoplasm immediately surrounding the nucleus; contrasts with ectoplasm or cortex, the portion of cytoplasm immediately under the cell surface; (2) adjective.

energy (ĕn′ĕr·jĭ) [Gr. *energos*, active]: capacity to do work; the time rate of doing work is called power.

enterocoel, enterocoelomate (ĕn′tĕr·ŏ·sĕl′): (1) a coelom formed by outpouching of a mesodermal sac from endoderm; (2) an animal having an enterocoel.

enterokinase (ĕn′tĕr·ŏ·kē′nās) [Gr. *kinētos*, moving]: an enzyme in intestinal juice that converts trypsinogen to trypsin.

enteropneust (ĕn′tĕr·ŏ·nūst) [Gr. *pnein*, to breathe]: a member of a class of hemichordates; an acorn worm.

enthalpy (ĕn′thăl·pĭ) [Gr. *enthalpein*, to warm in]: a measure of the amount of energy in a reacting system.

Entoprocta (ĕn′tŏ·prŏk′tå): a phylum of sessile pseudocoelomate animals in which the anus opens inside a ring of ciliated tentacles.

entrainment (ĕn·trān′mĕnt): synchronization of a rhythmic behavior and rhythmic environmental stimuli.

entropy (ĕn′trŏ·pĭ) [Gr. *entropia*, transformation]: a measure of the distribution of energy in a reacting system.

enzyme (ĕn′zīm) [Gr. *en*, in + *zymē*, leaven]: a protein capable of accelerating a particular chemical reaction; a type of catalyst.

ephyra (ĕf′ĭ·rå) [L. a nymph]: free-swimming larval stage in scyphozoan cnidarians; larval jellyfish.

epiboly (ĕ·pĭb′ŏ·lĭ) [Gr. *epibolē*, throwing over]: gastrulation by overgrowth of animal upper region over vegetal lower region of blastula.

epidermis (ĕp′ĭ·dûr′mĭs): the outermost surface tissue of an organism.

epididymis (ĕp′ĭ·dĭd′ĭ mĭs) [Gr. *didymos*, testicle]: the coiled portion of the sperm duct adjacent to the mammalian testis.

epiglottis (ĕp′ĭ·glŏt′ĭs) [Gr. *glōssa*, tongue]: a flap of tissue above the mammalian glottis; covers the glottis in swallowing and thereby closes the air passage to the lungs.

epithelium (ĕp′ĭ·thē′lĭ·ŭm) [Gr. *thēlē*, nipple]: animal tissue type in which cells are packed tightly together, leaving little intercellular space.

esophagus (ĕ·sŏf′å·gŭs) [Gr. *oisō*, I shall carry]: part of alimentary tract that connects pharynx and stomach.

estrogen (ĕs'trō·jĕn) [Gr. *oistros*, frenzy]: one of a group of female sex hormones of vertebrates.

estrus (ĕs'trŭs) [L. *oestrus*, gadfly]: egg production and fertilizability in mammals; for example, estrus cycle, monestrous, polyestrous.

eurypterid (ū·rĭp'tĕr·ĭd) [Gr. *eurys*, wide]: extinct Paleozoic chelicerate arthropod.

eustachian (ū·stā'shŭn): applied to canal connecting middle-ear cavity and pharynx of mammals.

exergonic (ĕk'sĕr·gŏ·nĭk): energy-yielding, as in a chemical reaction.

exocrine (ĕk'sō·krĭn): applied to type of gland that releases secretion through a duct.

exteroceptor (ĕk'stĕr·ō·sĕp'tĕr): a sense organ receptive to stimuli from external environment.

FAD: abbreviation of flavin adenine dinucleotide.

feces (fē'sēz) [L. *faeces*, dregs]: waste matter discharged from the alimentary system.

femur, femoral (fē'mĕr, fĕm'ō·răl) [L. *thigh*]: (1) thighbone of vertebrates, between pelvis and knee; (2) adjective.

fermentation (fûr'mĕn·tā'shŭn): synonym for anaerobic respiration; fuel combustion in the absence of oxygen.

fetus (fē'tŭs) [L. offspring]: prenatal stage of mammalian development following the embryonic stage; in man, roughly from third month of pregnancy to birth.

fiber (fī'bĕr) [L. *fibra*, thread]: a strand or filament produced by cells but located outside cells; also a type of sclerenchyma cell.

fibril (fī'brĭl) [L. dim. of *fibra*]: a strand of filament produced by cells and located inside cells.

fibrin, fibrinogen (fī'brĭn, fī·brĭn'ō·jĕn): (1) coagulated blood protein forming the bulk of a blood clot in vertebrates; (2) a blood protein which on coagulation forms a clot.

fibula (fĭb'ū·lă) [L. buckle]: the usually thinner of the two bones between knee and ankle in vertebrate hindlimbs.

filopodium (fĭ·lō·pō'dĭ·ŭm) [L. *filum*, thread]: a filamentous type of pseudopodium in sarcodine protozoa.

flagellate, flagellum (flăj'ĕ·lāt, -ŭm) [L. whip]: (1) equipped with one or more flagella; an organism or cell with flagella; (2) a microscopic, whiplike filament serving as locomotor structure in flagellate cells.

flavin: multiple-ring compound forming component of riboflavin and hydrogen carriers such as FAD.

follicle (fŏl'ĭ·k'l) [L. *folliculus*, small ball]: ball of cells; as in egg-containing balls in ovaries of many animals or cellular balls at base of hair or feather.

food (fūd): an organic nutrient.

Foraminifera (fō·rămĭ·nĭf'ĕr·ă) [L. *foramen*, hole]: sarcodine protozoa characterized by calcareous shells with holes through which pseudopods are extruded.

fovea centralis (fō'vē·ă sĕn·trā'lĭs) [L. central pit]: small area in optic center of mammalian retina; only cone cells are present here and stimulation leads to most acute vision.

funiculus (fū·nĭk'ū·lŭs) [L. dim of *funis*, rope]: tissue strand, as in attachment of stomach to body wall of ectoprocts.

gamete (găm'ēt): reproductive cell that must fuse with another before it can develop; sex cell.

ganglion (găng'glĭ·ŭn) [Gr. a swelling]: a localized collection of cell bodies of neurons, typically less complex than a brain.

ganoid (găn'oid) [Gr. *ganos*, brightness]: pertaining to shiny, enamel-covered type of fish scale.

gastrin (găs'trĭn): a hormone produced by the stomach wall of mammals when food makes contact with the wall; stimulates other parts of the wall to secrete gastric juice.

Gastropoda (găs·trŏp'ō·dă): a class of mollusks; comprises snails and slugs.

Gastrotricha (găs'trŏt'rĭ·kă): a class of minute, aquatic, pseudocoelomate animals, members of the phylum Aschelminthes.

gastrozooid (găs'trō·zōóid): a feeding individual in a polymorphic colony.

gastrula, gastrulation (găs'troo·lă, -lă'shŭn): (1) a two-layered and later three-layered stage in the embryonic development of animals; (2) the process of gastrula formation.

gel (jěl) [L. *gelare,* to freeze]: quasi-solid state of a colloidal system, where the solid particles form the continuous phase and the liquid is the dispersed phase.

gemmule (jĕm′ūl): vegetative, multicellular bud of (largely freshwater) sponges.

gene (jēn) a segment of a chromosome, definable in operational terms as a unit of biochemical action; repository of genetic information.

genome (jĕn′ōm): the totality of genes in a haploid set of chromosomes, hence the sum of all different genes in a cell.

genotype (jĕn′ō·tīp): the particular set of genes in an organism and its cells; the genetic constitution.

genus (jē′nŭs) [L. race]: a rank category in taxonomic classification between species and family; a group of closely related species.

gestation (jĕs·tā′shŭn): process or period of carrying offspring in uterus.

globulin (glŏb′ū·lĭn): one of a class of proteins in blood plasma of vertebrates; can function as antibody.

glochidia (glō·kĭd′ĭä) [Gr. *glochis,* arrow point]: pincer-equipped bivalve larvae of freshwater clams, parasitic on fish.

glomerulus (glō·měr′ū·lŭs) [L. dim. of *glomus,* ball]: small meshwork of blood capillaries or channels, as in a vertebrate nephron.

glottis (glŏt′ĭs) [Gr. *glōssa,* tongue]: slitlike opening in mammalian larynx formed by vocal cords.

glucogenic (glōo′kō·jĕn′ĭk): glucose-producing, esp. amino acids which, after deamination, metabolize like carbohydrates.

glucose (gloo′kōs): a 6-carbon sugar; main form in which carbohydrates are transported from cell to cell.

glycerin (glĭs′ēr·ĭn): an organic compound with a 3-carbon skeleton; can unite with fatty acids and form a fat; syn. *glycerol.*

glycogen (glī′kō·jĕn): a polysaccharide composed of glucose units; a main storage form of carbohydrates.

glycolysis (glī·kŏl′ĭ·sĭs): respiratory breakdown of carbohydrates to pyruvic acid.

goiter (goi′tēr) [L. *guttur,* throat]: an enlargement of the thyroid gland.

Golgi body (gôl′jē): a cytoplasmic organelle playing a role in the manufacture of cell secretions.

gonad (gōn′ăd) [Gr. *gonē,* generator]: animal reproductive organ; collective term for testes and ovaries.

gonozooid (gōn′ō·zō′oid): a reproductive individual in a polymorphic colony.

gradation (grā·dā′shŭn) [L. *gradus,* step]: leveling of land by geologic effects of erosion.

graptolite (grăp′tō·līt) [Gr. *graptos,* written]: one of a phylum group of fossil exoskeletons of uncertain affinities; may be related to pterobranch hemichordates.

guanine (gŭ′ă·nēn): purine component of nucleotides and nucleic acids.

guanosine (di-, tri-) phosphates (gū·ă′nō·sēn): guanine-ribose-phosphates (GDP, GTP).

guanylic acid: equivalent to guanosine monophosphate, or GMP.

haem-: see *hem-.*

haploid (hăp′loid): a chromosome number characteristic of a mature gamete of a given species.

haplontic (hăp·lŏn′tĭk): designating a life cycle with zygotic meiosis and haploid adults.

hectocotylus (hĕk′tō kŏt′ĭ·lŭs) [Gr. *hekto-,* hundred, + *kōtylē,* cup]: modified arm of male cephalopod mollusks serving in sperm transfer to female.

helix (hē′liks) [L. a spiral]: spiral shape; for example, polypeptide chain, snail shell.

heme (hēm): an iron-containing red blood pigment.

Hemichordata (hĕm′ĭ·kŏr dă′tá): a phylum of enterocoelomate animals.

hemoglobin (hē′mō·glō′bĭn) [L. *globus,* globe]: oxygen-carrying constituent of blood; consists of red pigment heme and protein globin.

hemophilia (hē′mō fĭl′ĭá) [Gr. *philos,* loving]: hereditary disease in man characterized by excessive bleeding from even minor wounds; clotting mechanism is impaired by failure of blood platelets to rupture.

hepatic (hē·păt′ik) [Gr. *hēpar*, liver]: pertaining to the liver; as in hepatic vein, hepatic portal vein.

herbivore (hûr′bĭ·vōr) [L. *herba*, herb + *vorare*, to devour]: a plant-eating animal.

hermaphrodite (hûr·măf′rō·dīt) [fr. Gr. *Hermes* + *Aphrodite*]: an organism that contains both male and female reproductive structures.

heterotroph, heterotrophism (hĕt′ĕr·ō·trŏf): (1) an organism dependent on both inorganic and organic raw materials from the environment; (2) form of nutrition characteristic of heterotrophs.

heterozygote (hĕt′ĕr·ō·zī′gōt): an organism in which a pair of alleles for a given trait consists of different (for example, dominant and recessive) kinds of genes.

holothuroid (hŏl·ō·thū′·roid) [L. *holothuria*, water polyp]: a member of a class of echinoderms; a sea cucumber.

holotroph, holotrophism (hō′lō·trŏf): (1) a bulk-feeding organism; nutrition usually includes alimentation; (2) form of nutrition characteristic of holotrophs.

hominid (hŏm′ĭ·nĭd) [L. *homo*, man]: a living or extinct man or manlike type; the family of man or pertaining to this family.

hominoid (hŏm′ĭ·noid): superfamily including hominids, the family of man, and pongids, the family of apes.

homology (hō·mŏl′ō·jĭ) [Gr. *homologia*, agreement]: similarity in embryonic development and adult structure, indicative of common evolutionary ancestry.

homozygote (hō′mō·zī′gōt): an organism in which a pair of alleles for a given trait consists of the same (for example, either dominant or recessive, but not both) kinds of genes.

hormone (hôr′mōn) [Gr. *hormaein*, to excite]: a secretion produced in an organism and affecting another part of that organism.

humerus (hū′mĕr·ŭs) [L. shoulder]: the bone of the vertebrate upper forelimb, between shoulder and elbow.

humoral (hū′mĕr·ăl) [L. *humor*, moisture, liquid]: pertaining to body fluids, esp. biologically active chemical agents carried in body fluids; for example, hormones or similar substances.

humus (hū′mŭs) [L. soil]: the organic portion of soil.

hybrid (hī′brĭd) [L. *hibrida*, offspring of tame sow and wild boar]: an organism heterozygous for one or more (usually many) gene pairs.

hydranth (hī′drănth): flowerlike terminal part of coelenterate polyp, containing mouth and tentacles; a feeding polyp.

hydrolysis (hī·drŏl′ĭ·sĭs): dissolution through the agency of water; esp. decomposition of a chemical by addition of water.

hyperparasitism (hī′pĕr-): infection of a parasite by one or more other parasites.

hypertonic (hī′pĕr·tŏn′ĭk): exerting greater osmotic pull than the medium on the other side of a semipermeable membrane; hence having a greater concentration of particles and acquiring water during osmosis.

hypothalamus: forebrain region containing various centers of the autonomic nervous system.

hypothesis (hī·pŏth′ē·sĭs) [Gr. *tithenai*, to put]: a guessed solution of a scientific problem; must be tested by experimentation.

hypotonic (hī′pō·tŏn′ĭk): exerting lesser osmotic pull than the medium on the other side of a semipermeable membrane; hence having a lesser concentration of particles and losing water during osmosis.

ichthyosaur (ĭk′thĭ·ō·sôr): extinct marine Mesozoic reptile, with fish-shaped body and porpoise-like snout.

imago, imaginal (ĭ·mā′gō, ĭ·măj′ĭ·năl) [L. image]: (1) an adult insect; (2) adjective.

induction (ĭn·dŭk′shŭn) [L. *inducere*, to induce]: process in animal embryo in which one tissue or body part causes the differentiation of another.

ingestion (ĭn·jĕs′chŭn) [L. *ingerere*, to put in]: intake of food from the environment into the alimentary system.

instar (ĭn′stär) [L. likeness, form]: period between consecutive molts in insect development.

insulin (ĭn′sū·lĭn) [L. *insula,* island]: hormone produced in the pancreas; promotes conversion of blood glucose to tissue glycogen.

integument (ĭn·tĕg′ū·mĕnt) [L. *integere,* to cover]: covering; external coat; skin.

intermedin (ĭn·tĕr·mē′dĭn): hormone produced by the mid-portion of the pituitary gland.

interoceptor (ĭn′tĕr·ō·sĕp′tĕr): a sense organ receptive to stimuli generated in the interior of an organism.

invagination (ĭn·vaj′ĭ·nā′shŭn) [L. *in,* + vagina, sheath]: local infolding of a layer of tissue, leading to formation of pouch or sac; as in invagination during a type of embolic gastrulation.

invertase (ĭn·vûr′tās) [L. *invertere,* to invert]: enzyme promoting a splitting of sucrose into glucose and fructose.

involution (ĭn′vō·lū′shŭn) [L. *involutio,* a rolling or folding in]: inrolling of a tissue layer underneath another one, as in a type of embolic gastrulation.

ion, ionization (ī′ŏn, -ĭ·zā′shŭn) [Gr. *ienai,* to go]: (1) electrically charged atom or group of atoms; (2) addition or removal of electrons from atoms.

isogamy (ī·sŏg′à·mĭ): sexual fusion in which the gametes of opposite sex types are structurally alike.

isolecithal (ī′sō·lĕs′ĭ·thăl): pertaining to animal eggs with yolk evenly distributed throughout egg cytoplasm.

isomer (ī′sō·mēr): one of a group of compounds identical in atomic composition but differing in structural arrangement.

isotonic (ī′sō·tŏn′ĭk): exerting same osmotic pull as medium or other side of a semipermeable membrane, hence having the same concentration of particles; net gain or loss of water during osmosis is zero.

isotope (ī′sō·tōp) [Gr. *topos,* place]: one of several possible forms of a chemical element differing from other forms in atomic weight but not in chemical properties.

karyogamy (kăr′ĭ·ŏg′·à·mĭ) [Gr. *karyon,* nut]: fusion of nuclei during fertilization.

kenozooid (kē′nō·zō′oid) [Gr. *koinos,* common, communal]: stalk- and stolon-forming polymorphic individuals of an ectoproct colony.

keratin (kĕr′à·tĭn) [Gr. *keratos,* horn]: a protective protein formed by the epidermis of vertebrate skin.

ketogenic, ketone (kē′tō·jĕn′ĭk, -tōn): (1) keto-acid-producing, esp. amino acids which after deamination metabolize like fatty acids; (2) organic compound with a —CO— group.

kilocalorie: see *calorie*

kinesis (kĭ·nē′sĭs): locomotor movement that changes in intensity in direct proportion with the intensity of a stimulus.

kinetosome (kĭ·nĕt′ō·sōm) [Gr. *kinētos,* moving]: granule at base of flagellum, presumably motion-controlling.

Kinorhyncha (kĭn′ō·rĭng′kà): a class of pseudocoelomate animals in the phylum Aschelminthes.

labium, labial (lā′bĭ·ŭm, -ăl) [L. lip]: (1) any liplike structure, esp. underlip in insect head; (2) adjective.

labrum (lā′brŭm) [L. lip]: a liplike structure; esp. upper lip in arthropod head.

labyrinthodont (lăb′ĭ rĭn′thō·dŏnt) [Gr. *labyrinthos,* labyrinth]: extinct, late-Paleozoic fossil amphibian.

lacteal (lăk′tē·ăl) [L. *lactis,* milk]: lymph vessel in a villus of intestinal wall of mammals.

lactogenic (lăk′tō jĕn′ĭk): milk-producing; as in lactogenic hormone, secreted by vertebrate pituitary.

lagena (là·jēn′à) [L. large flask]: portion of the primitive vertebrate (fish) ear in which sound is translated into nerve impulses; evolutionary forerunner of cochlea.

larva (lär′vä) pl. *larvae* (-vē) [L. mask]: period in developmental history of animals between hatching and metamorphosis.

larynx (lăr′ĭngks) [Gr.]: voice box; soundproducing organ in mammals.

lemniscus (lĕm·nĭs′·kŭs) [L. a ribbon hanging down]: an elongated, paired, interior extension of the body wall in the anterior region of acanthocephalan worms.

leukocyte (lū′kō·sīt): a type of white blood cell in vertebrates characterized by a beaded, elongated nucleus; formed in bone marrow.

leukemia (ū·kē′mĭ·ȧ): a cancerous condition of blood, characterized by overproduction of leukocytes.

lipase (lī′pās): enzyme promoting conversion of fat to fatty acids and glycerin or reverse.

lipid (lĭp′ĭd): fat, fatty, pertaining to fat; syn. *lipoid*.

lithosphere (lĭth′ō·sfēr) [Gr. *lithos*, stone]: collective term for the solid, rocky components of the earth′s surface layers.

littoral (lĭt′ō·rȧl) [L. *litus*, seashore]: the sea floor from the shore to the edge of the continental shelf.

lophophore (lō′fō·fōr) [Gr. *lophos*, crest]: tentacle-bearing food-trapping arm in anterior region of certain coelomates (lophophorate animals).

lorica (lō·rī′kȧ) [L. sheath, cover]: secreted protective covering, as in some ciliate protozoa.

luciferse, luciferin (lū·sĭf′ēr·ās, -ĭn) [L. *lux*, light]: (1) enzyme contributing to production of light in organisms; (2) a group of various substances essential in the production of bioluminescence.

lymph (lĭmf) [L. *lympha*, goddess of moisture]: the body fluid outside the blood circulation.

lymphocyte (lĭm′fō·sīt): a type of white blood cell of vertebrates characterized by a rounded or kidney-shaped nucleus; formed in lymphatic tissues.

macromolecule (măk′rō-): a molecule of very high molecular weight; refers specifically to proteins, nucleic acids, polysaccharides, and complexes of these.

macronucleus (măk′rō·nū′klē·ŭs): a large type of nucleus found in ciliate protozoa; controls all but sexual functions.

madreporite (măd′rē·pō′rīt) [It. *madre*, mother, + *poro*, passage]: a sievelike opening on the surface of echinoderms; connects the water-vascular system with the outside.

Malacostraca (măl′ä·kŏs′trȧ·kȧ) [Gr. *malakostraka*, soft-shelled]: a subclass of crustaceans.

maltose (môl′tōs): a 12-carbon sugar formed by the union of two glucose units.

mandible (măn′dĭ·b′l) [L. *mandibula*, jaw]: in arthropods, one of a pair of mouth appendages, basically biting jaws; in vertebrates, the main support of the lower jaw.

marsupial (mär·sū′pĭ·ȧl) [Gr. *marsypion*, little bag]: a pouched mammal, member of the mammalian subclass Metatheria.

mastax (măs′tăks) [L. *masticare*, to chew]: horny, toothed chewing apparatus in pharynx of rotifers.

Mastigophora (măs′tĭ gō′fōrȧ): the subphylum of flagellate protozoa; zooflagellates.

maxilla (măk·sĭl′ȧ) [L.]: in arthropods, one of the head appendages; in vertebrates, one of the upper jawbones.

maxilliped (măk·sĭl′ĭ·pĕd): one of three pairs of segmental appendages in lobsters, located posterior to the maxillae.

medulla (mē·dŭl′ȧ) [L.]: the inner layers of an organ or body part; for example, adrenal medulla; the *medulla oblongata* is a region of the vertebrate hindbrain that connects with the spinal cord.

medusa (mē·dū′sȧ): free-swimming stage in the life cycle of cnidarians; a jellyfish.

meiosis (mī·ō′sĭs) [Gr. *meioun*, to make smaller]: nuclear division in which the chromosome number is reduced by half; compensates for the chromosome-doubling effect of fertilization.

melanin (mĕl′ȧ·nĭn) [Gr. *melas*, black]: black pigment in organisms.

menopause (mĕn′ō·pôz) [Gr. *menos*, month + *pauein*, to cause to cease]: the time at the end of the reproductive period of (human) females when menstrual cycles cease to occur.

menstruation (měn'stroo·à'shŭn) [L. *mensis*, month]: discharge of uterine tissue and blood from the vagina in man and apes at the end of a menstrual cycle in which fertilization has not occurred.

mesencephalon (měs'ĕn·sĕf'à·lŏn) [Gr. *enkephalos*, brain]: the vertebrate midbrain.

mesenchyme (měs'ĕng·kĭm) [Gr. *enchyma*, infusion]: mesodermal connective tissue cells; often jelly-secreting.

mesogloea (měs'ō·glē'à) [Gr. *gloisos*, glutinous substance]: the often jelly-containing layer between ectoderm and endoderm of cnidarians and comb jellies.

mesonephros (měs'ō·něf'rōs): the adult kidney of fishes and amphibia.

metabolism (mě·tăb'ō·lĭz'm) [Gr. *metabolē*, change]: a group of life-sustaining processes including mainly nutrition, respiration, and synthesis of more living substance.

metabolite (mě·tăb'ō·lĭt): any chemical participating in metabolism; a nutrient.

metamorphosis (mět'à môr'fō·sĭs) [Gr. *metamorphoun*, to transform]: transformation of a larva to an adult.

metaphase (mět'à·fāz): a stage during mitotic division in which the chromosomes line up in a plane at right angles to the spindle axis.

Metaphyta (mě·tăf'ĭ·tà): a major category of living organisms comprising the phyla Bryophyta and Tracheophyta; plants.

Metazoa (mět'à·zō'à): a major category of living organisms comprising all animals.

metencephalon (mět'ĕn·sĕf'à·lŏn) [Gr. *enkephalos*, brain]: anterior portion of vertebrate hindbrain.

micron (mī'krŏn) pl. **microns, micra**: one-thousandth part of a millimeter, a unit of microscopic length.

micronucleus (mī'krō·nū'klē·ŭs): a small type of nucleus found in ciliate protozoa; controls reproductive functions and macronucleus.

mictic (mĭk'tĭk) [Gr. *mixis*, act of mixing]: pertaining to fall and winter eggs of rotifers, which if fertilized produce males and if not fertilized produce females.

mimicry (mĭm'ĭk·rĭ) [Gr. *mimos*, mime]: the superficial resemblance of certain animals, particularly insects, to other more powerful or more protected ones, or to leaves and other plant parts.

mineral (mĭn'ĕr·ăl) [L. *minera*, ore]: an inorganic material.

miracidium (mī'rà sĭd'ĭ·ŭm): larval stage in the life cycle of flukes; develops from an egg and gives rise to a sporocyst larva.

mitochondrion (mī'tō·kŏn'drĭ·ŏn) [Gr. *mitos*, thread, + *chondros*, grain]: a cytoplasmic organelle serving as site of respiration.

mitosis (mī·tō'sĭs): nuclear division characterized by complex chromosome movements and exact chromosome duplication.

mitral (mī'trăl) [fr. *miter*]: applied to valve between left atrium and ventricle of mammalian heart; syn. *bicuspid*.

mole (mōl) [L. *moles*, mass]: the gram-molecular weight of a substance; its weight in grams equal to its molecular weight.

molecule (mŏl'ē·kūl) [L. *moles*, mass]: a compound in which the atoms are held together by covalent bonds.

Mollusca, mollusk (mŏ·lŭs'kà, mŏl'ŭsk) [L. *molluscus*, soft]: (1) a phylum of nonsegmented schizocoelomate animals; (2) a member of the phylum Mollusca.

Monera (mŏn·ē'rà) [Gr. *monos*, alone]: a major category of living organism comprising bacteria and blue-green algae.

monestrous (mŏn·ĕs'trŭs) [Gr. *oistros*, frenzy]: having a single estrus (egg-producing) cycle during a given breeding season.

monophyletic (mŏn'ō·fī·lĕt'ĭk) [Gr. *phylon*, tribe]: developed from a single ancestral type; contrasts with polyphletic.

monosaccharide (mŏn'ō·săk'à·rĭd) [Gr. *sakcharon*, sugar]: a simple sugar, such as 5- and 6-carbon sugars.

morphogenesis (môr'fō·jĕn'ē·sĭs): development of size, form, and other architectural features of organisms.

morphology (môr·fŏl'ŏ·jĭ): the study or science of structure, at any level of organization; for example, cytology, study of cell structure; histology, study of tissue structure; anatomy, study of gross structure.

morula (môr'ū·là) [L. little mulberry]: solid ball of cells resulting from cleavage of egg; a solid blastula.

mucosa (mū·kō'sà) [L. *mucosus*, mucus]: a mucus-secreting membrane; for example, the inner lining of the intestine.

Muellerian duct: the gonoduct of female vertebrates; forms oviduct, uterus, in mammals.

mutation (mū·tā'shŭn) [L. *mutare*, to change]: a stable change of a gene or gene part (muton); the changed condition is inherited by offspring cells.

myelencephalon (mī'ĕ·lĕn·sĕf'àlŏn) [Gr. *myelos*, marrow]: the most posterior part of the vertebrate hindbrain, confluent with the spinal cord; the medulla oblongata.

myelin (mī'ĕ·lĭn): a fatty material surrounding the axons of nerve cells in the central nervous system of vertebrates.

myofibril (mīŏ·fī'brĭl): a contractile filament inside a cell.

myosin (mī'ŏ·sĭn): a muscle protein.

myxedema (mĭk'sē·dē'mà) [Gr. *oidēma*, a swelling]: a disease resulting from thyroid deficiency in the adult characterized by local swellings in and under the skin.

nacre, nacreous (nă'kēr, -krē·ŭs) [Pers. *nakdra*, pearl oyster]: (1) mother-of-pearl; (2) adjective.

NAD: abbreviation of nicotinamide-adenine-dinucleotide.

NADP: abbreviation of nicotinamide-adenine-dinucleotide-phosphate.

nauplius (nô'plĭ·ŭs) [L. shellfish]: first in a series of larval phases in crustacea.

nekton (nĕk'tŏn) [Gr. *nēktos*, swimming]: collective term for the actively swimming animals in the ocean.

Nematoda (nĕm'ä tō'dà): the class of roundworms in the phylum Aschelminthes.

Nematomorpha (nĕm'ä tō·môr'fà): the class of hairworms in the phylum Aschelminthes.

Nemertina (nĕm·ēr tīn'à): ribbon or proboscis worms, an acoelomate phylum (also called Rhynchocoela).

neoteny (nē·ŏt'ē·nĭ) [Gr. *neo*, new + *teinein*, extend]: retention of larval or youthful traits as permanent adult features.

nephric, nephron (nĕf'rĭk, -rŏn): (1) pertaining to a nephron or excretory system generally; (2) a functional unit of the vertebrate kidney.

nephromixium (nĕf'rō·mĭks'ĭ·ŭm): joint excretory and reproductive unit in certain polychaete annelids, composed of solenocytic nephridia and a gonoduct from the coelom which opens into the nephridial tubule.

neritic (nē·rĭt'ĭk) [fr. Gr. *Nereus*, a sea god]: oceanic habitat zone, subdivision of the pelagic zone, comprising the open water above the continental shelf, i.e., above the littoral.

neuron (nū'rŏn) [Gr. nerve]: nerve cell.

neutron (nū'trŏn): a subatomic particle with a unit of mass; it is uncharged and occurs in an atomic nucleus.

nicotinamide: a derivative of nicotinic acid (niacin, one of the B vitamins), a component of the hydrogen carriers NAD and NADP.

nictitating (nĭk'tĭ·tāt'ĭng) [L. *nictare*, wink]: pertaining to thin transparent eyelidlike membrane in many vertebrates, which opens and closes laterally across cornea.

nidamental (nĭd'à·mĕnt·ăl) [L. *nidus*, nest]: pertaining to gland in female cephalopod mollusks that secretes protective capsule around eggs.

nitrify, nitrification (nī'trĭ·fī, -fĭ·kā'shŭn): (1) to convert ammonia and nitrite to nitrate; (2) noun.

notochord (nō'tō·kôrd): longitudinal elastic rod serving as internal skeleton in the embryos of all chordates and in the adults of some.

nuchal (nū'kăl) [L. *nucha*, nape of neck]: pertaining to nape of neck; for example, nuchal ligament in tetrapods, aiding in holding head in horizontal position.

nucleic acid (nū·klē'ĭk): one of a class of molecules composed of joined nucleotides; DNA or RNA.

nucleolus (nū·klē'ō·lŭs): an RNA-containing body in the nucleus of a cell; a derivative of chromosomes.

nucleoprotein (nū'klē·ō-): a molecular complex composed of nucleic acid and protein.

nucleotide (nū'klē·ō·tĭd): a molecule consisting of a phosphate, a 5-carbon sugar (ribose or deoxyribose), and a nitrogen base (adenine, guanine, uracil, thymine, or cytosine).

nucleus (nū'klē·ŭs) [L. a kernel]: an organelle in all cell types except those of the Monera; consists of external nuclear membrane, interior nuclear sap, and chromosomes and nucleoli suspended in the sap; also the central body of an atom.

nutrient (nū'trĭ·ĕnt) [L. *nutrire,* to nourish]: a substance usable in metabolism; a metabolite; includes inorganic materials and foods.

ocellus (ō·sĕl'ŭs) [L. dim. of *oculus,* eye]: eye or eyespot, of various degrees of structural and functional complexity; in arthropods, a simple eye, as distinct from a compound eye.

olfaction, olfactory (ŏl·făk'shŭn), -tō·rĭ) [L. *olfacere,* to smell]: (1) the process of smelling; (2) pertaining to smell.

ommatidium (ŏm'å·tĭd'ĭ·ŭm) [Gr. *omma,* eye]: single visual unit in compound eye of arthropods.

omnivore (ŏm'nĭ·vōr) [L. *omnis,* all]: an animal living on plant foods, animal foods, or both.

Oncopoda (ŏn·kō'pō·då) [Gr. *onkos,* bulk]: a small phylum comprising schizocoelomate animals related to arthropods.

Onychophora (ŏnĭ·kō'fŏr·å) [Gr. *onych,* claw]: a subphylum of Oncopoda, comprising *Peripatus* and related types.

oögamy (ō·ŏg'å·mĭ): sexual fusion in which the gametes of opposite sex type are unequal, the female gamete being a nonmotile egg; the male gamete, a motile sperm.

operculum (ō·pûr'kŭ·lŭm) [L. a lid]: a lidlike structure.

ophiuroid (ŏf'ĭ·ū·roid) [Gr. *ophis,* snake]: a member of a class of echinoderms; a brittle star.

organ (ôr'găn) [Gr. *organon,* tool, instrument]: a group of different tissues joined structurally and cooperating functionally to perform a composite task.

organelle (ôr·găn·el'): a structure or body in a cell.

organic (ôr·găn'ĭk): pertaining to compounds of carbon of nonmineral origin.

organism (ôr'găn·ĭz'm): an individual living creature, either unicellular or multicellular.

ornithine (ôr'nĭ·thēn) [Gr. *ornithos,* bird]: an amino acid which, in the liver of vertebrates, contributes to the conversion of ammonia and carbon dioxide to urea.

osmosis (ŏs·mō'sĭs) [Gr. *ōsmos,* impulse]: the process in which water migrates through a semipermeable membrane, from the side containing a lesser to the side containing a greater concentration of particles; migration continues until concentrations are equal on both sides.

ossicle (ŏs'ĭ·k'l) [L. dim. of *ossis,* bone]: a small bone or hard bonelike supporting structure.

Osteichthyes (ŏs·tē·ĭk'thĭ·ēz): the vertebrate class of bony fishes.

ostium (ŏs'tĭ·ŭm) [L. door]: orifice or small opening; for example, one of several pairs of lateral pores in arthropod heart, pore for entry of water in certain sponges.

ovary (ō'vå·rĭ): the egg-producing organ of animals.

oviparity, oviparous (ō'vĭ·păr'ĭ·tĭ, ō·vĭp'årŭs) [L. *parere,* to bring forth]: (1) animal reproductive pattern in which eggs are released by the female and offspring development occurs outside the maternal body; (2) adjective.

ovoviviparity, ovoviviparous (ō'vō·vĭv'ĭ·păr'ĭ·tĭ, ō'vō·vĭ vĭp'å·rŭs): (1) animal reproductive pattern in which eggs develop inside the maternal body, but without nutritive or other metabolic aid by the female parent; offspring are born as miniature adults; (2) adjective.

ovulation (ō'vū·lā'shŭn): expulsion of an animal egg from ovary and deposition of egg in oviduct.

oxidation (ŏk'sĭ·dā'shŭn): one half of an oxidation-reduction (redox) process; the process is exergonic and the endproducts are more stable than the starting materials; often takes the form of removal of hydrogen (or electrons) from a compound.

paleoniscoid (pā′lē·ō·nĭs′koid): extinct Devonian bony fish, ancestral to modern bony fishes, lungfishes, and lobe-fin fishes.

paleontology (pā′lē·ŏn·tŏl′ō·jĭ): study of past geologic times by means of fossils.

Paleozoic (pā′lē·ō·zō′ĭk): geologic era between the Precambrian and the Mesozoic, dating approximately from 500 to 200 million years ago.

palp (pălp) [L. *palpus*, feeler]: a feelerlike appendage.

Pantopoda (păn·tŏ′pōdá) [Gr. *pantos*, all]: a subphylum of Oncopoda.

papilla (pá·pĭl′á) [L. nipple]: any small nipplelike projection.

parapodia (păr′á·pō′dĭ·á): fleshy segmental appendages in polychaete worms; serve in breathing, locomotion, and creation of water currents.

parasite (păr′á·sīt) [Gr. *sitos*, food]: an organism living symbiotically on or in a host organism, more or less detrimental to the host.

parasympathetic (păr′ä·sĭm′pá·thēt′ĭk): applied to a subdivision of the autonomic nervous system of vertebrates; centers are located in brain and most anterior part of spinal cord.

parathyroid (păr′á·thī′roid): an endocrine gland of vertebrates, usually paired, located near or in the thyroid.

parenchyma (pä·rĕng′kĭ·má) [Gr. *para* + *en*, in + *chein*, to pour]: name applied to mesenchyme tissues of acoelomate animals.

parthenogenesis (pär′thē·nō·jĕn′ē·sĭs) [Gr. *parthenos*, virgin]: development of an egg without fertilization; occurs naturally in some animals (for example, rotifers) and can be induced artifically in others (for example, frogs).

pathogenic (păth′ō·jĕn′ĭk) [Gr. *pathos*, suffering]: disease-producing.

Pauropoda (pô·rŏ′pō dá) [Gr. *pauros*, small]: a small class of myriapod mandibulate arthropods, probably related to millipedes.

pectine (pĕk′tīn) [L. *pecten*, comb]: one of a pair of comblike segmental appendages on scorpion abdomen, tactile in function.

pedicellaria (pĕd′ĭ·sĕl á′rĭ·á) [L. *pedicellus*, little stalk]: a pincerlike structure on the surface of echinoderms; protects skin gills.

pedipalp (pĕd′ĭ·pălp): one of a pair of head appendages in chelicerate arthropods.

peduncle (pē·dŭng′k′l) [L. *pedunculus*, little foot]: a stalk or stemlike part.

pelagic (pē·lăj′ĭk) [Gr. *pelagos*, ocean]: oceanic habitat zone comprising the open water of an ocean basin.

Pelecypoda (pĕ′lē·sĭp′ō·dá): a class of the phylum Mollusca, comprising clams, mussels, oysters.

pellicle (pĕl′ĭ·k′l) [L. dim. of *pellis*, skin]: a thin, membranous surface coat, as on many protozoa.

Pelmatozoa (pĕl·mă′tō·zō′á) [Gr. *pelma*, sole of foot]: a subphylum of echinoderms, comprising only the crinoids among living groups.

penis (pēnĭs) [L.]: a copulatory organ containing the terminal portions of the sperm duct.

pepsin (pĕp′sĭn) [Gr. *peptein*, to digest]: a protein-digesting enzyme in gastric juice of vertebrates.

peptidase (pĕp′tĭ·dās): an enzyme promoting the liberation of individual amino acids from a whole or partially digested protein.

peptide (pĕp′tīd): the type of bond formed when two amino acid units are joined.

Perissodactyla (pē·rĭs′ō·dăk′tĭ·lá) [Gr. *perissos*, odd]: the order of odd-toed ungulate mammals; includes horses, tapirs, rhinoceroses.

peristalsis (pĕr′ĭ·stăl′sĭs) [Gr. *peristaltikos*, compressing]: successive contraction and relaxation of tubular organs such as the alimentary tract, resulting in a wavelike propagation of a transverse constriction.

peritoneum (pē·rĭ′tō·nē′ŭm) [Gr. *peritonos*, stretched over]: a mesodermal membrane lining the coelom.

permeability (pûr′mē·á·bĭl′ĭ·tĭ) [L. *permeare*, to pass through]: penetrability, as in membranes that let substances pass through.

pH: a symbol denoting the relative concentration of hydrogen ions in a solution; pH values normally run from 0 to 14, and the lower the value, the more acid is a solution.

pharynx (făr'ĭngks) [Gr.]: the part of the alimentary tract between mouth cavity and esophagus.

phenotype (fē'nŏ·tīp) [Gr. *phainein,* to show]: the physical appearance of an organism resulting from its genetic constitution (genotype).

Phoronida (fŏ·rŏn'ĭ·dȧ): a phylum of wormlike lophophore-possessing animals.

phosphagen (fŏs'fȧ·jĕn): collective term for compounds such as creatine-phosphate, which store and may be sources of high-energy phosphates.

phosphorylation (fŏs'fŏ·rĭ·lă'shŭn): the addition of a phosphate group (—O—H$_2$PO$_3$) to a compound.

photolithotroph (fŏ'tŏ·lĭth'ŏ·trŏf) [Gr. *lithos,* stone]: an organism that manufactures food with the aid of light energy and with inorganic raw materials.

photoorganotroph (fŏ'tŏ·ôr·găn'ŏ·trŏf): an organism that manufactures food with the aid of light energy and with organic raw materials.

photoperiod, photoperiodism (fŏ'tŏ·pęr'ĭ·ŭd, -ĭz'm): (1) day length, (2) the responses of organisms to different day lengths.

photosynthesis (fŏ'tŏ·sĭn'thē·sĭs) [Gr. *tithenai,* to place]: process in which light energy and chlorophyll are used to manufacture carbohydrates out of carbon dioxide and water.

phrenic (frĕn'ĭk) [Gr. *phrenos,* diaphragm]: pertaining to the diaphragm; for example, phrenic nerve, which innervates the diaphragm.

phylogeny (fī·lŏj'ē·nĭ) [Gr. *phylon,* race, tribe]: the study of evolutionary descent and interrelations of groups of organisms.

phylum (fī'lŭm), pl. *phyla:* a category of taxonomic classification, ranked above class.

physiology (fĭz'ĭ·ŏl'ŏ·jĭ) [Gr. *physis,* nature]: study of living processes, activities, and functions in general; contrasts with morphology, the study of structure.

phytoplankton (fī'tŏ·plăngk'tŏn) [Gr. *planktos,* wandering]: collective term for the plants and plantlike organisms in plankton; contrasts with zooplankton.

pilidium (pī·lĭd'ĭ·ŭm) [L. *pilus,* hair]: a larval type characteristic of many nemertine worms.

pinacocyte (pĭn'ä·kŏ·sīt') [Gr. *pinax,* tablet]: an epithelial cell type on the body surface of sponges.

pineal (pĭn'ē·ȧl) [L. *pinea,* pine cone]: a structure in the brain of vertebrates; functions as a median dorsal eye in lampreys and tuataras.

pinna, pinnule (pĭn'ȧ, pĭn'ūl) [L. feather, fin]: a featherlike structure (for example, the pinnules on the arms of crinoids).

pinocytosis (pī'nŏ·sī·tō'sĭs) [Gr. *pinein,* to drink]: intake of fluid droplets through cell surface.

pituitary (pī·tū'ĭ·tērĭ) [L. *pituita,* phlegm]: a composite vertebrate endocrine gland ventral to the brain; composed of anterior, intermediate, and posterior lobes, each a functionally separate gland.

placenta (plȧ·sĕn'tȧ) [L. cake]: a mammalian tissue complex formed from the inner lining of the uterus and the chorion of the embryo; serves as mechanical, metabolic, and endocrine connection between adult female and embryo during pregnancy.

placoderm (plăk'ŏ·dûrm) [Gr. *plakos,* flat plate]: a member of an extinct class of Devonian fishes.

planarian (plȧ·nâr'ĭ·ăn) [L. *planarius,* level]: any member of the class of free-living flatworms.

plankton (plăngk'tŏn) [Gr. *planktos,* wandering]: collective term for the largely microscopic, passively floating or drifting flora and fauna of a body of water.

planula (plăn'ü·lȧ) [L. dim. of *planus,* flat]: basic larval form characteristic of cnidarians.

plastron (plăs'trŏn) [It. *piastrone,* breastplate]: a ventral shell part, as in turtles.

Platyhelminthes (plăt'ĭ·hĕl·mĭn'thēz) [Gr. *platys,* flat]: flatworms, a phylum of acoelomate animals; comprises planarians, flukes, and tapeworms.

plesiosaur (plē'sĭ·ŏ·sôr) [Gr. *plesios,* near]: a long-necked, marine, extinct Mesozoic reptile.

plexus (plĕk'sŭs) [L. braid]: a network, esp. nerves or blood vessels.

pluteus (ploot'ē·ŭs) [Gr. *plein*, to sail, float, flow]: the larva of echinoids and ophiuroids; also called echinopluteus and ophiopluteus, respectively.

pneumatophore (nū·mă'tō·fōr) [Gr. *pneumatos*, air, wind]: the air-filled float of siphonophoran hydroid cnidarians.

Pogonophora (pō·gŏ'nŏ·fōr'ȧ) [Gr. *pōgōn*, beard]: beard worms, a phylum enterocoelomate deep-sea animals.

poikilothermic (poi'kĭ·lŏ·thûr'mĭk) [Gr. *poikilos*, multicolored]: pertaining to animals without internal temperature controls; "cold-blooded."

polyclad (pŏl'ĭ·klăd): a member of an order of free-living flatworms, characterized by a digestive cavity with many branch pouches.

polyestrous (pŏl'ĭ·ĕs'trŭs): having several estrus (egg-producing) cycles during a given breeding season.

polymer (pŏl'ĭ·mēr): a large molecule composed of many like molecular subunits.

polymorphism (pŏl'ĭ·môr'fĭz'm): differences of form among the members of a species; individual variations affecting form and structure.

polyp (pŏl'ĭp) [L. *polypus*, many-footed]: the usually sessile stage in the life cycle of cnidarians.

polyphyletic (pŏl'ĭ·fī·lĕt'ĭk) [Gr. *phylon*, tribe]: derived from more than one ancestral type; contrasts with monophyletic.

polyploid (pŏl'ĭ ploid): having many complete chromosome sets per cell.

polysaccharide (pŏl'ĭ·săk'a·rīd): a carbohydrate composed of many joined monosaccharide units; for example, glycogen, starch, cellulose, all formed out of glucose units.

Porifera (pō·rĭf'ēr·ȧ): the phylum of sponges.

porocyte (pō'rŏ·sīt): a cell type in certain sponges characterized by a pore or canal passing through it; serves in water intake.

Priapulida (prĭ'ă·pū'·lĭ·dȧ): a phylum of probably coelomate animals, with uncertain evolutionary status.

primordium (prī·môr'dĭ·ŭm) [L. beginning]: the earliest developmental stage in the formation of an organ or body part.

proboscis (prŏ·bŏs'ĭs) [L.]: any tubular process or prolongation of the head or snout.

progesterone (prŏ·jĕs'tēr·ōn): hormone secreted by the vertebrate corpus luteum and the mammalian placenta; functions as pregnancy hormone in mammals.

proglottid (prŏ·glŏ'ĭd): a segment of a tapeworm.

prophase (prŏ'fāz'): a stage during mitotic division in which the chromosomes become distinct and a spindle forms.

proprioceptor (prŏ·'prĭ·ŏ·sĕp'·tēr) [L. *proprius*, one's own]: sensory receptor of stimuli originating in internal organs; a stretch receptor.

prosimian (prŏ·sĭm'ĭ·ăn) [L. *simia*, ape]: an ancestral primate.

prosopyle (prŏ'·sŏ·pīl) [Gr. *proso*, forward]: one of several incurrent openings of a choanocyte chamber in sponges.

protein (prŏ'tē·ĭn) [Gr. *prōteios*, primary]: one of a class of organic compounds composed of many joined amino acids.

proteinase (prŏ'tē·ĭn·ās): an enzyme promoting the conversion of protein to amino acids or the reverse; also called protease.

prothrombin (prŏ·thrŏm'bĭn) [Gr. *thrombos*, clot]: a clotting factor in vertebrate blood plasma; converted to thrombin by thrombokinase.

Protista (prŏ·tĭs'tȧ) [Gr. *prōtistos*, first]: a major category of living organisms, including algae (except blue-greens), slime molds, protozoa, and fungi.

proton (prŏ'tŏn): a subatomic particle with a unit of positive electric charge and a mass of 1; a component of an atomic nucleus.

protoplasm (prŏ'tŏ·plăz'm): synonym for living matter, living material, or living substance.

Protostomia (prŏ'tŏ·stŏ'mē·ȧ): animals in which blastopore becomes mouth; anus forms as second embryonic opening opposite blastopore.

protozoon (prŏ'tŏ·zŏ'ŏn): a member of either of four subphyla (Mastigophora, Sarcodina, Ciliophora, Sporozoa) of a protistan phylum.

pseudocoel, pseudocoelomate (sū'dō·sēl, -ō·māt): (1) an internal body cavity lined by ectoderm and endoderm; (2) an animal having a pseudocoel.

pseudopodium (sū'dō·pō'dĭ·ŭm): a cytoplasmic protrusion from an amoeboid cell; functions in locomotion and feeding.

pterosaur (tĕr'ō·sôr): extinct Mesozoic flying reptile.

pulmonary (pŭl'mō·nĕr'ĭ) [L. *pulmonis,* lung]: pertaining to the lungs.

pupa (pū'på) [L. doll]: a developmental stage, usually encapsulated or in cocoon, between larva and adult in holometabolous insects.

purine (pū'rēn): a nitrogen base such as adenine or guanine; a component of nucleotides and nucleic acids.

pylorus (pī·lō'rŭs) [Gr. *pylōros,* gatekeeper]: the opening from stomach to intestine.

pyrimidine (pī·rī'mĭ·dēn): a nitrogen base such as cytosine, thymine, or uracil; a component of nucleotides and nucleic acids.

Radiata (rā·dĭ·ā'tå) [L. *radius,* ray]: a taxonomic grade within the Eumetazoa, comprising cnidarians and ctenophores.

Radiolaria (rā'dĭ·ō·lār'ĭ·å): sarcodine protozoa characterized by silicon-containing shells.

radula (răd'ū·lå) [L. *radere,* to scrape]: a horny rasping organ in the mouth of many mollusks.

recessive (rē·sĕs'ĭv) [L. *recedere,* to recede]: a functional attribute of genes; the recessive effect of a gene is masked if the allelic gene has a dominant effect.

rectum (rĕk'tŭm) [L. *rectus,* straight]: a terminal nonabsorptive portion of the alimentary tract in many animals; opens via the anus.

redia (rē'dĭ·å): a larval stage in the life cycle of flukes; produced by a sporocyst larva and gives rise to many cercarias.

reduction (rē·dŭk'shŭn) [L. *reducere,* to lead back]: one half of an oxidation-reduction (redox) process; the phase that yields the net energy gain; often takes the form of addition of hydrogen (or electrons) to a compound.

reflex (rē'flĕks) [L. *reflectere,* to bend back]: the unit action of the nervous system; consists of stimulation of a receptor, interpretation and emission of nerve impulses by a neural center, and execution of a response by an effector.

renal (rē'nål) [L. *renes,* kidneys]: pertaining to the kidney.

rennin (rĕn'ĭn) [Middle Engl. *rennen,* to run]: an enzyme in mammalian gastric juice, promotes coagulation of milk.

respiration (rĕs'pĭ·rā'shŭn) [L. *respirare,* to breathe]: liberation of metabolically useful energy from fuel molecules in cells; can occur anaerobically or aerobically.

reticulum (rē·tĭk'ū·lŭm) [L. little net]: a network or mesh of fibrils, fibers, filaments, or membranes, as in *endoplasmic reticulum.*

retina (rĕt'ĭ·nå) [L. *rete,* a net]: the innermost tissue layer of the eyeball; contains the photoreceptor cells.

rhabdocoel (răb'dō·sēl) [Gr. *rhabdos,* rod]: member of a group of free-living flatworms having a straight, unbranched digestive cavity.

Rhodophyta (rō'dŏf'ĭ·tå) [Gr. *rhodon,* red]: the phylum of red algae.

ribosome (rī'bō·sōm): an RNA-containing cytoplasmic organelle; the site of protein synthesis.

ribotide (rī'bō·tīd): a nucleotide in which the sugar component is ribose.

rickettsia (rĭk·ĕt'sĭ·å) [after H. T. Ricketts, American pathologist]: a type of microorganism intermediate in nature between a virus and a bacterium, parasitic in cells of insects and ticks.

RNA: abbreviation of ribonucleic acid.

Rotifera (rō·tĭf'ĕrå) [L. *rota,* wheel]: a class of microscopic animals in the phylum Aschelminthes.

rudimentary (roo'dĭ·mĕn'tå·rĭ) [L. *rudis,* unformed]: pertaining to an incompletely developed body part.

saccule (săk'ūl) [L. *sacculus,* little sac]: portion of the inner ear of vertebrates containing the receptors for the sense of static balance.

saprotroph (săp′rō·trōf) [Gr. *sapros*, rotten]: an organism subsisting on dead or decaying matter.

Sarcodina (sär′kō·dī′nȧ): a subphylum of protozoa; amoeboid protozoa.

Sarcopterygii (sär′kō·tĕr′ĭ·jē): flesh-finned fishes, a subclass of bony fishes comprising the lobe-finned fishes and lungfishes.

Scaphopoda (skä·fŏp′ō·dȧ) [Gr. *skaphē*, boat]: tooth shells, a class of the phylum Mollusca.

schizocoel, schizocoelomate (skĭz′ō·sēl): (1) coelom formed by splitting of embryonic mesoderm; (2) an animal having a schizocoel.

sclera (sklē′rȧ): the outermost coat of the eyeball, continuous with the cornea.

scolex (skō′lĕks) [L. worm, grub]: the head of a tapeworm.

scrotum (skrō′tŭm) [L.]: external skin pouch containing the testes in most mammals.

sebaceous (sē·bā′shŭs) [L. *sebum*, tallow, grease]: pertaining to sebum, an oil secreted from skin glands near the hair bases of mammals.

seminal (sĕm′ĭ·nȧl) [L. *semen*, seed]: pertaining to semen or sperm-carrying fluid.

septum, septate (sĕp′tŭm, -tāt) [L. enclosure]: (1) a complete or incomplete partition; (2) adjective.

sere (sēr) [fr. L. *series*, series]: stage in an ecological succession of communities, from the virginal condition to a stable climax community.

serum (sē′rŭm) [L.]: the fluid remaining after removal of fibrinogen from vertebrate blood plasma.

simian (sĭm′ĭ·ȧn) [L. *simia*, an ape]: pertaining to monkeys; also used as noun.

sinus (sī′nŭs) [L. a curve]: a cavity, recess, space or depression; for example, blood sinus, bone sinus.

siphon (sī′fŏn) [Gr. *siphōn*, a pipe]: tubular structure for drawing in or ejecting fluids, as in mollusks, tunicates.

siphonoglyph (sī·fŏn′ō·glĭf) [Gr. *glyphein*, to carve]: flagellated groove in pharynx of sea anemones; creates water current to gastrovascular cavity.

siphuncle (sī′fŭngk′l): gas-filled canal passing through coiled shell of chambered nautilus.

Sipunculida (sī′pŭng·kū′lĭ·dȧ): a phylum of wormlike schizocoelomate animals; peanut worms.

sol (sŏl): quasi-liquid state of a colloidal system, where water forms the continuous phase and solid particles the dispersed phase.

solenocyte (sō·lēn′ō·sīt) [Gr. *sōlēn*, channel, pipe]: excretory cell in a type of protonephridium.

somatic (sō·măt′ĭk): pertaining to the animal body generally.

somite (sō′mīt): one of the longitudinal series of segments in segmented animals; especially an incompletely developed embryonic segment or a part thereof.

species (spē′shĭz) pl. *species* (spē′shēz) [L. kind, sort]: a category of taxonomic classification, below genus rank, defined by breeding potential or gene flow; interbreeding and gene flow occur among the members of a species but not between members of different species.

specificity (spĕs′ĭ·fĭs′ĭ·tĭ): uniqueness, esp. of proteins and genes in a given organism and of enzymes in given reactions.

spectrum (spĕk′trŭm) [L. image]: a series of radiations arranged in the order of wavelengths; for example, solar spectrum, visible spectrum.

sphincter (sfĭngk′tēr) [Gr. *sphingein*, to bind tight]: a ring-shaped muscle capable of closing a tubular opening by constriction; for example, pyloric sphincter, which closes the opening between stomach and intestine.

spicule (spĭk′ūl) [L. *spiculum*, little dart]: a slender, often needle-shaped secretion of sponge cells; serves as skeletal support.

spiracle (spī′rȧ·k′l) [L. *spirare*, to breathe]: reduced evolutionary remnant of first gill slit in fishes; also surface opening of breathing system in terrestrial arthropods.

spore (spōr): a reproductive cell capable of developing into an adult directly.

sporine (spō′rĕn): pertaining to a sessile state of protistan existence in which cell division can occur during the vegetative condition.

sporocyst (spō′rō·sĭst): a larval stage in the life cycle of flukes; produced by a miracidium larva and gives rise to many redias.

Sporozoa (spō′rō·zō′a̯): a subphylum of parasitic protozoa.

sternum (stûr′nŭm) [Gr. *sternon,* chest]: vertebrate breastbone, articulating with ventral ends of ribs on each side.

sterol, steroid (stĕr′ōl, stĕr′oid): one of a class of organic compounds containing a molecular skeleton of four fused carbon rings; includes cholesterol, sex hormones, adrenocortical hormones, and vitamin D.

stimulus (stĭm′ū·lŭs) [L. goad, incentive]: any internal or external environmental change that activates a receptor structure.

stolon (stō′lŏn) [L. *stolo,* shoot, branch]: in colonial animals, a (usually) horizontal branch or runner from which upright individuals can bud.

strobilation (strō·bī·lā′shŭn) [L. *strobilus,* pine cone]: process of segmentlike budding in sessile scyphistoma larvae of scyphozoan cnidarians, resulting in cutting off of successive free-swimming ephyra larvae.

stroma (strō′ma̯) [Gr. couch, bed]: the connective tissue network supporting the epithelial portions of animal organ.

style, stylet (stīl, stī′lĕt) [Gr. *stylos,* pillar]: a stalklike or elongated body part, often pointed at one end.

substrate (sŭb′strāt) [L. *substratus,* strewn under]: a substance that is acted on by an enzyme.

symbiont, symbiosis (sĭm′bī·ŏnt, sĭm′bī·ō′sĭs): (1) an organism living in symbiotic association with another; (2) the intimate living together of two organisms of different species, for mutual or one-sided benefit; the main variants are mutualism, commensalism, and parasitism.

sympathetic (sĭm′pa̯·thĕt′ĭk): applied to a subdivision of the autonomic nervous system; centers are located in the mid-portion of the spinal cord.

synapse (sĭ·năps′): the microscopic space between the axon terminal of one neuron and the dendrite terminal of an adjacent one.

syncytium (sĭn·sī′shĭ·ŭm): a multinucleate animal tissue without internal cell boundaries.

syngen (sĭn′jĕn): a mating group (or variety) within a protozoan species; mating can occur in a syngen but not usually between syngens; a functional (as distinct from taxonomic) ''species.''

synthesis (sĭn′thē·sĭs) [Gr. *tithenai,* to place]: the joining of two or more molecules resulting in a single larger molecule.

syrinx (sĭr′ĭngks) [Gr. a pipe]: the vocal organ of birds, located where the trachea branches into the bronchi.

systole (sĭs′tō·lē) [Gr. *stellein,* to place]: the contraction of atria or ventricles of a heart.

taiga (tī′ga̯) [Russ.]: terrestrial habitat zone characterized by large tracts of coniferous forests, long, cold winters, and short summers; found particularly in Canada, northern Europe, and Siberia.

tardigrade (tär′dĭ·grād) [L. *tardigradus,* a slow stepper]: a member of a subphylum of Oncopoda; water bears.

tarsus (tär′sŭs) [Gr. *tarsos,* sole]: in insects, the terminal parts of a leg; in vertebrates, the ankle.

taxon (tăks′ŏn) pl. *taxa:* the actual organisms in a taxonomic rank.

taxonomy (tăks·ŏn′ō·mĭ) [Gr. *nomos,* law]: classification of organisms, based as far as possible on natural relationships.

tectorial membrane (tĕk·tō′rĭ·al) [L. cover, covering]: component of the organ of Corti in cochlea of mammalian ear.

telencephalon (tĕl′ĕn·sĕf′a̯·lŏn) [Gr. *enkephalos* brain]: the vertebrate forebrain.

telolecithal (tĕl′ō·lĕs′ĭ·thăl): pertaining to eggs with large amounts of yolk accumulated in the vegetal (lower) half; for example, as in frog eggs.

telophase (tĕl′ō·fāz): a stage in mitotic division during which two nuclei form; usually accompanied by partitioning of cytoplasm.

telson (tĕl′sŭn) [Gr. boundary, limit]: terminal body part of an arthropod (not counted as a segment).

template (tĕm′plĭt): a pattern or mold guiding the formation of a duplicate.

temporal lobe (těm′pō·rál) [L. *tempora*, the temples]: a part of the vertebrate cerebrum; contains centers for speech and hearing.

tergum (tûr′gŭm) [L. the back]: the dorsal exoskeletal plate of a body segment in arthropods.

testis, pl. **testes** (těs′tĭs, -tēs) [L.]: sperm-producing organ in animals.

tetrad (tět′răd): a pair of chromosome pairs during the first metaphase of meiosis.

tetrapyrrol (tět′rå·pī′rŏl): a molecule consisting of four joined rings of carbon and nitrogen; heme and chlorophyll pigments are of this type.

thalamus (thǎl′å·mŭs) [Gr. *thalamos*, chamber]: a lateral region of the diencephalic portion of the vertebrate forebrain.

theory (thē′ō·rĭ) [Gr. *theōrein*, to look at]: a scientific statement based on experiments that verify a hypothesis; the usual last step in scientific procedure.

therapsid (thē·răp′sĭd) [Gr. *thērion*, beast]: extinct Mesozoic mammal-like reptile.

Theria (thēr′ĭ·å) [Gr. *thērion*, beast]: a subclass of mammals comprising marsupials and placentals; the viviparous mammals.

thorax (thō′răks) [L.]: part of animal body between neck or head and abdomen; chest.

thrombin (thrŏm′bĭn) [Gr. *thrombos*, clot]: a clotting factor in vertebrate blood; formed from prothrombin and in turn converts fibrinogen to fibrin.

thrombokinase (thrŏm′bō·kĭn′ās): enzyme released from vertebrate blood platelets during clotting; transforms prothrombin to thrombin; also called thromboplastin.

thrombus (thrŏm′bŭs): a blood clot within the circulatory system.

thymidine (*di-*, *tri-*) **phosphates** (thī′mĭ·dēn): thymine-deoxyribose-phosphate (TDP, TTP).

thymidylic acid (thī·mĭ·dĭ′lĭk): equivalent to thymine monophosphate (TMP).

thymine (thī′mēn): a pyrimidine component of nucleotides and nucleic acids.

thymus (thī′mŭs) [Gr.]: a lymphatic gland in most young and many adult vertebrates; disappears in man at puberty; located in lower part of throat and upper part of thorax.

thyroxin (thī·rŏk′sĭn): the hormone secreted by the thyroid gland.

tibia (tĭb′ĭ·å) [L.]: in vertebrates, usually the larger shinbone of the hindlimb between knee and ankle; in insects, the leg portion between femur and tarsus.

tissue (tĭsh′ū) [L. *texere*, to weave]: a group of cells of similar structure performing similar functions.

tornaria (tôr·nā′rĭ·å) [L. *tornus*, lathe, chisel]: the larva of certain enteropneust hemichordates.

trachea, tracheal (trā′kē·å) [Gr. *trachys*, rough]: (1) air-conducting tube, as in windpipe of mammals and breathing system of terrestrial arthropods; (2) adjective.

transduction (trăns·dŭk′shŭn): transfer of genetic material from one bacterium to another through the agency of a virus.

triclad (trī′klăd): a member of a group of free-living flatworms, characterized by a digestive cavity with three branch pouches; a planarian.

tricuspid (trī·kŭs′pĭd) [L. *cuspis*, a point]: ending in three points or flaps, as in tricuspid valve of mammalian heart.

trilobite (trī′lō·bīt): an extinct marine Paleozoic arthropod, marked by two dorsal longitudinal furrows into three parts or lobes.

triploid (trĭp′loid) [Gr. *triploos*, triple]: having three complete chromosome sets per cell.

trochanter (trō·kăn′tēr) [Gr. *trechein*, to run]: in insects, the part of a leg adjoining the coxa, equivalent to the basipodite of arthropods.

trochophore (trŏk′ō fōr): a free-swimming ciliated marine larva, characteristic of schizocoelomate animals.

tropic, tropism (trŏp′ĭk) [Gr. *tropē*, a turning]: (1) pertaining to behavior or action brought about by specific stimuli; for example, phototropic (light-oriented) motion, gonadotropic (stimulating the gonads); (2) noun.

trypsin (trĭp′sĭn) [Gr. *tryein*, to wear down]: enzyme promoting protein digestion; acts in small intestine, produced in pancreas as inactive trypsinogen.

tundra (to͞on'drȧ) [Russ.]: terrestrial habitat zone between taiga and polar region, characterized by absence of trees, short growing season, and frozen ground during much of the year.

Turbellaria (tûr'bĕ·lär'ĭ·ȧ) [L. *turba*, disturbance]: the class of free-living flatworms; planarians.

turgor (tûr'gŏr) [L. *turgere*, to swell]: the distension of a cell by its fluid content.

typhlosole (tĭf'lŏ·sōl) [Gr. *typhlos*, blind]: dorsal fold of intestinal wall projecting into gut cavity in oligochaete annelids (earthworms).

umbilicus (ŭm·bĭl'ĭ·kŭs) [L.]: the navel of mammals; during pregnancy, an umbilical cord connects the placenta with the offspring, and the point of connection with the offspring later becomes the navel.

umbo (ŭm'bō) [L. boss of shield]: rounded prominence near hinge of clam valve; oldest part of shell.

ungulate (ŭng'gū·lāt) [L. *ungula*, hoof]: hoofed, as in certain orders of mammals.

uracil (ū'rȧ·sĭl): a pyrimidine component of nucleotides and nucleic acids.

urea (ū·rē'ȧ) [Gr. *ouron*, urine]: compound formed in the mammalian liver out of ammonia and carbon dioxide and excreted by the kidneys.

ureter (ū·rē'tēr) [Gr.]: duct carrying urine from a mammalian kidney to the urinary bladder.

urethra (ū·rē'thrȧ) [Gr.]: duct carrying urine from the urinary bladder to the outside of the body; in the males of most mammals the urethra also leads sperms to the outside during copulation.

uridine (di-, tri-) phosphates (ū'rĭ·dēn-): uracil-ribose-phosphates (UDP, UTP).

uridylic acid (ū·rĭ·dĭl'ĭk): equivalent to uridine monophosphate (UMP).

Urochordata (ū'rŏ·kôr·dā'tȧ): a subphylum of chordates; comprises the tunicates.

Urodela (ū'·rŏ·dē'lȧ) [Gr. *dēlos*, visible]: a subclass of tailed amphibia, comprising newts and salamanders.

uropod (ū'rŏ·pŏd): an abdominal appendage in lobsters and other crustaceans.

uterus (ū'tēr·ŭs) [L. womb]: enlarged region of a female reproductive duct in which animal embryo undergoes all or part of its development.

utricle (ū'trĭ·k'l) [L. *utriculus*, little bag]: portion of the vertebrate inner ear containing the receptors for dynamic body balance; the semicircular canals lead from and to the utricle.

vacuole (văk'ū ōl) [L. *vacuus*, empty]: a small, usually spherical space in a cell, bounded by a membrane and containing fluid, solid matter, or both.

vagina (vȧ·jī'nȧ) [L. sheath]: the terminal, penis-receiving portion of a female reproductive system.

vagus (vā'gŭs) [L. wandering]: the 10th cranial nerve in vertebrates.

valence (vā'lĕns) [L. *valere*, to have power]: a measure of the bonding capacity of an atom; bonds can be electrovalent, formed through electron transfer, or covalent, formed through electron-sharing.

vasomotion (văs'ŏ·mō'shŭn) [L. *vasum*, vessel]: collective term for the constriction (vasoconstriction) and dilation (vasodilation) of blood vessels.

veliger (vēl'ĭ·jēr) [L. *velum*, veil]: posttrochophoral larval stage in many mollusks.

velum (vē'lŭm): a membranous curtainlike band of tissue; for example, on underside of many hydroid medusae.

venous (vē'nŭs) [L. *vena*, vein]: pertaining to veins; also applied to oxygen-poor, carbon dioxide–rich blood.

ventricle (vĕn'trĭ·k'l) [L. *ventriculus*, the stomach]: a heart chamber that receives blood from an atrium and pumps out blood from the heart.

vestigial (vĕs·tĭj'ĭ·ȧl) [L. *vestigium*, footprint]: degenerate or incompletely developed, but more fully developed at an earlier stage or during the evolutionary past.

villus (vĭl'ŭs) pl. *villi* [L. a tuft of hair]: a microscopic fingerlike projection from the intestinal lining (mucosa) into the cavity of the mammalian gut.

virus (vī'rŭs) [L. slimy liquid, poison]: a submicroscopic noncellular particle, composed of a nucleic acid core and a protein shell; parasitic inside host cell.

viscera (vĭs′ēr·á), sing, *viscus* [L.]: collective term for the internal organs of an animal.

vitamin (vī tá·mĭn) [L. *vita,* life]: one of a class of growth factors contributing to the formation of coenzymes.

vitreous (vĭt′rē·ŭs) [L. *vitrum,* glass]: glassy; as in vitreous humor, the clear transparent jelly filling the posterior part of the vertebrate eyeball.

viviparity, viviparous (vĭv′ĭ·păr′ĭ·tĭ, vī·vĭp′á·rŭs) [L. *vivus,* living + *parere,* to bring forth]: (1) animal reproductive pattern in which eggs develop inside female body with nutritional and other metabolic aid of maternal parent; offspring are born as miniature adults; (2) adjective.

volvant (vŏl′vănt) [L. *volvere,* to roll, twist]: closed-ended adhesive thread in a type of nematocyst of cnidarians.

Wolffian duct: the collecting duct of a mesonephric kidney; carries both sperms and urine in male fishes and amphibia.

Xiphosura (zĭf′ō·sū′rá) [Gr. *xiphos, sword*]: a class of chelicerate arthropods; the horseshoe crabs (*Limulus*)

zoaea (zō·ē′ä) pl. *zoaeae:* a larval form of crustaceans.

zooid (zō′oid): an individual animal in a colony; often physically joined with fellow zooids and can be a polymorphic variant.

zooplankton (zō′ō·plăngk′tŏn): collective term for the nonphotosynthetic organisms in plankton; contrasts with phytoplankton.

zygote (zī′gōt) [Gr. *zygōtos,* paired together]: the cell resulting from sexual fusion of two gametes; a fertilized egg.

INDEX